W9-BHS-790

HAESTAD METHODS

STORMWATER CONVEYANCE MODELING AND DESIGN

First Edition

HAESTAD METHODS

STORMWATER CONVEYANCE MODELING AND DESIGN

First Edition

Authors
Haestad Methods
S. Rocky Durrans

Managing Editor
Kristen Dietrich

Contributing Authors
Muneef Ahmad, Thomas E. Barnard,
Peder Hjorth, and Robert Pitt

Peer Review Board
Roger T. Kilgore (Kilgore Consulting)
G. V. Loganathan (Virginia Tech)
Michael Meadows (University of South Carolina)
Shane Parson (Anderson & Associates)
David Wall (University of New Haven)

Editors
David Klotz, Adam Strafaci, and Colleen Totz

HAESTAD PRESS
Waterbury, CT USA

STORMWATER CONVEYANCE MODELING AND DESIGN
First Edition

Index and proofreading: Beaver Wood Associates

Special thanks to *The New Yorker* magazine for the cartoons throughout the book.

Page 5 – (2000) Liza Donnelly
Page 12 – (1986) Al Ross
Page 58 – (1990) J. B. Handelsman
Page 102 – (1963) Robert Day
Page 131 – (1988) Peter Steiner
Page 140 – (2002) Robert Mankoff
Page 145 – (1979) William Steig
Page 198 – (1988) Jack Ziegler
Page 209 – (1975 Dean Vietor
Page 214 – (2003) Sidney Harris
Page 237 – (1993) Mick Stevens
Page 254 – (1990) Danny Shanahan
Page 280 – (2003) Mary Lawton
Page 288 – (1995) Peter Steiner
Page 306 – (2003) Ted Goff
Page 324 – (1988) Warren Miller
Page 365 – (1991) Mort Gerberg
Page 376 – (2002) Lee Lorenz
Page 400 – (2003) P. C. Vey
Page 434 – (1987) Lee Lorenz
Page 481 – (1992) Henry Martin
Page 501 – (1988) J. B. Handelsman
Page 525 – (1988) Arnie Levin
Page 567 – (1996) Robert Day
Page 581 – (2000) Jack Ziegler
Page 596 – (1970) Vahan Shirvanian
Page 599 – (1970) James Stevenson
Page 613 – (1971) Lee Lorenz
Page 632 – (2000) David Sipress

10 9 8 7 6 5 4 3 2

Library of Congress Control Number: 2003106704
ISBN: 0-9657580-8-7

Haestad Methods, Inc.
37 Brookside Rd.
Waterbury, CT 06708-1499
USA

Phone: +1-203-755-1666
Fax: +1-203-597-1488
e-mail: info@haestad.com
Internet: www.haestad.com

To all of my parents: Lola de, Dennis, Richard, Margaret, Harold, and Charlotte; for your sacrifices, tolerance, support, and encouragement.

-S. Rocky Durrans

"Theory can leave questions unanswered, but practice has to come up with something."

-Mason Cooley

"This book is a unique blending of at least six books that I currently use. . . . I wish it had been available when I took the Professional Engineer's Exam!"

Shane Parson, Ph.D., P.E.
Anderson & Associates, Inc.

"Haestad Press has once again published a book that is one of the best I've seen. It will surely become a 'first to reach for' when performing stormwater conveyance design."

Robert M. Steele
LPA Group

"*Stormwater Conveyance and Design* was written by persons who have 'been there and done that' and who work closely with design professionals. As such, there is much knowledge and practical information in the book. While reviewing the book, I've learned something new from every chapter—information that will help me do my job better.."

Michael Meadows, Ph.D., P.E.
University of South Carolina

Acknowledgments

Stormwater Conveyance Modeling and Design is the culmination of the efforts of many individuals, the most important of which are the thousands of stormwater engineering professionals who have dedicated their careers to managing stormwater and protecting the environment. Haestad Methods has been serving stormwater engineers around the world for over 20 years, and our discussions and meetings with the community have provided the inspiration to write this book.

Haestad Press is grateful to all of the talented people who contributed to the creation of this book. In particular we are grateful for the work of S. Rocky Durrans, the primary author. Several others contributed content in specific areas, including Robert Pitt (stormwater quality management and U.S. stormwater regulations), Muneef Ahmad (Canadian stormwater regulations), and Peder Hjorth (international stormwater policies and practices). Tom Barnard made significant contributions in organizing the content from these multiple sources and adding additional material. Mr. Barnard also authored many of the end-of-chapter exercises. Shane Parson contributed material for a number of the sidebar topics found throughout the book, and Bob Steele provided practical modeling and design examples.

Many engineers at Haestad Methods assisted with editing, reviewing, adding content, working problems, and checking for accuracy, making this project a company-wide effort. Their input played an enormous role in shaping the text into its current form. Peter Martin led the effort to create and assemble the graphics used throughout the book. Chris Dahlgren, Caleb Brownell, Cal Hurd, John Slate, and several others contributed to the illustration efforts. Special thanks go to the organizations that provided us with reprint permissions for their graphics. Also, thanks to Jm Brozena and Tom Walski, who provided a number of photographs, especially for Chapter 13.

Others involved in the production process for the book include Lissa Jennings (publishing logistics and proofreading assistance), Rick Brainard and Jim O'Brien (cover design), Ben Ewing (web page design), Jeanne Moody (index and proofreading), John Shirley (formatting and page layout), Wes Cogswell (accompanying CD-ROM), and Amanda Irvin and David Sutley (example problem unit conversions).

Several engineering professionals reviewed the book and provided us with insightful feedback and suggestions for improvement. Huge thanks go to our peer reviewers: Roger Kilgore, G.V. Loganathan, Mike Meadows, Shane Parson, and David Wall.

Other core members of the Haestad Press team who put in many hours on this project and provided a huge amount of support for this effort are David Klotz, Adam Strafaci, Colleen Totz, and Tom Walski.

Special acknowledgments go to Haestad Methods' Niclas Ingemarsson, executive vice president, for his management guidance, and Keith Hodsden, general manager of sales and technical support, for his assistance in organizing staff resources. Finally, thank you to the company president, John Haestad, whose vision it was to create this book, for his unwavering support of this project.

Kristen Dietrich
Managing Editor

Authors and Contributing Authors

Stormwater Conveyance Modeling and Design represents a collaborative effort that combines the experiences of ten contributors and peer reviewers and the engineers and software developers at Haestad Methods. The authors and contributing authors are:

Authors

Haestad Methods
Waterbury, Connecticut

S. Rocky Durrans
University of Alabama

Contributing Authors

Muneef Ahmad
GM Sernas & Associates

Thomas E. Barnard
Haestad Methods

Peder Hjorth
Lund University

Robert Pitt
University of Alabama

Haestad Methods

The Haestad Methods Engineering Staff is an extremely diverse group of professionals from six continents with experience ranging from software development and engineering consulting, to public works and academia. This broad cross section of expertise contributes to the development of the most comprehensive software and educational materials in the civil engineering industry. In addition to the specific authors credited in this section, many at Haestad Methods contributed to the success of this book.

S. Rocky Durrans, Ph.D., P.E.

S. Rocky Durrans, Ph.D., P.E., is a professor of Civil and Environmental Engineering at the University of Alabama, where he is responsible for the hydrologic and hydraulic engineering programs. Prior to his academic career, Dr. Durrans spent many years in the consulting arena where he gained extensive experience in the design of stormwater and major urban drainage systems. He is regarded as an authority on flood and rainfall analysis and has many publications to his credit. Dr. Durrans received a B.S. in Civil Engineering and an M.S. in Civil Engineering (Water Resources) from the University of Colorado at Denver, and a Ph.D. in Civil Engineering (Water Resources) from the University of Colorado at Boulder.

Muneef Ahmad, P. Eng.

Muneef Ahmad, P. Eng., is a water resources engineer with the Sernas Group, Inc., a civil engineering consulting firm based in Southern Ontario, Canada. The firm provides a broad range of services throughout Southern Ontario, including municipal, water resources, land use planning, transportation planning, class environmental assessments, and power and utility design. Project-related experience for Mr. Ahmad includes work for public and private sector clients in hydrologic modeling, hydraulic design, erosion remediation works, and stormwater management facility design.

Thomas E. Barnard, Ph.D., P.E.

Thomas E. Barnard, Ph.D., P.E., is a senior engineer with Haestad Methods. He has more than 20 years of experience in environmental engineering working as a consultant, researcher, educator, and author. He holds a B.S. in civil engineering from the University of Vermont, an M.S. in environmental engineering from Utah State University, and a Ph.D. in environmental engineering from Cornell University. His expertise includes water and wastewater treatment, hazardous waste management, surface water hydrology, and water quality monitoring systems. Dr. Barnard's experience includes hydrologic and water quality modeling of a delta lake in Egypt. He is a registered professional engineer in Pennsylvania.

Peder Hjorth, Ph.D.

Dr. Peder Hjorth is a faculty member in the Department of Water Resources Engineering at Lund University, where he is responsible for teaching and conducting research in hydrology, hydraulics, and water resources engineering. He has also served some 15 years as Course Director of an international training program on water resources management in developing countries. He received his M.S. degree in civil engineering from Chalmers University of Technology and his Ph.D. in water resources engineering from Lund Institute of Technology/Lund University. He has served as chairman of the International Association for Hydraulic Research (IAHR) Technical Section on Water Resources Management.

Robert Pitt, Ph.D., P.E.

Robert Pitt, Ph.D., P.E., is the Cudworth Professor of Urban Water Systems and Director of Environmental Engineering Programs at the University of Alabama. He currently teaches classes in urban water resources, stressing the integration of hydrology and water quality issues. He previosly taught at the University of Alabama at Birmingham for 14 years and previously was a senior engineer with industry and government for 16 years. Dr. Pitt obtained his Ph.D. from the University of Wisconsin, Madison, his M.S. in civil engineering from California State University, San Jose, and his B.S. from Humboldt State University. He has been active in stormwater research for most of his professional career, having conducted research for many national, state, and local agencies, and some industries. He has published numerous reports and papers, including eight books. He is a Diplomate of the American Academy of Environmental Engineers, and a member of several professional societies. In 2002, he received a Distinguished Service Award from the University of Wisconsin recognizing his service to the engineering profession.

Foreword

Every day, stormwater management decisions are made concerning topics such as land use, site planning, impervious cover, and infrastructure. These decisions, made by developers, engineers, property owners, and government officials, are often made on a micro scale, even though cumulatively, they have macro-scale effects—not the least of which is the cost of constructing and maintaining stormwater facilities. Stormwater management hydraulic models are essential tools for designing and analyzing these facilities, and through their appropriate application, one can ensure adequate flood protection without wasting financial resources through over-design or poorly conceived designs.

The common paradigm for stormwater management has always been the efficient removal of stormwater. Design has focused on curb and gutter, ditches, and storm drain systems. Efficient, in this context, has meant the rapid removal of stormwater from the point where it occurs. This way of thinking assumes that stormwater is something to be disposed of rather than used as a resource. Over the years, however, we have realized that such efficiency frequently causes downstream flooding. This realization inaugurated a period during which regional detention ponds were considered the cost-effective means of attenuating flood peaks from upstream development. Because regional ponds tended to be large facilities, the role of municipal government increased through study of regional stormwater management needs, identification of sites, and coordination of financing through public and private sources.

We then realized that such large facilities, frequently located on live drainageways, impaired the natural stream function. Furthermore, it became apparent that these regional ponds did little to enhance water quality and may have exacerbated channel degradation problems. As a result, design practices have evolved to consider infiltration, storage, sediment transport, and water quality issues. Adapting regional facilities to meet water quality objectives, through techniques such as extended detention and wetland retention systems, were among the responses to the newly identified problems. Regional systems also began to give way to dispersed on-site infiltration and retention systems that are designed to not only reduce peaks of flood events but also to mimic the runoff characteristics for the full range hydrological events. The goal became to capture and treat water on-site before releasing it downstream.

These changes in thinking have required those involved in stormwater management to adapt existing modeling tools and to develop new ones to meet the new demands. The suite of stormwater management facilities that we now have to choose from has grown. We can expect that, as our experience grows with our current stormwater management philosophy, we will continue to adapt as we recognize new limitations of standard practices. It may be expected that this will primarily come in the form of land use and development management strategies rather than new forms of constructed facilities.

As the undesired effects of past stormwater management practices are being more fully realized, stormwater management regulations are also evolving. Rather than focusing on the infrequent events for flood protection, emphasis is increasingly placed on facility design that mimics pre-development channel-forming discharges as well as environmental goals such as fish passage. Designing for a channel forming discharge requires an understanding of the amount of work that a stream can do in transforming its vertical profile and horizontal pattern. This, in turn, requires mastery of sediment transport characteristics as well as flow frequency relationships. With respect to fish passage, we are concerned with defining discharges, depths, and velocities of lower flow events. Some of these new demands on designers require an understanding of continuous simulation modeling in addition to event-based design. Modeling tools are also evolving to meet today's regulatory environment.

In spite of the changes in stormwater management practices and regulations and the evolution of modeling tools, stormwater modeling, analysis, and design are still based on the same fundamentals. They still require tools that allow for the representation of precipitation, conveyance, storage, and treatment of stormwater. And, they still require tools that allow for the geographic variation in rainfall and runoff patterns as well as jurisdictional variation in policies and regulations.

In 1889, Emil Kuichling introduced the rational method to the United States in his paper "The Relation Between the Rainfall and the Discharge of Sewers in Populous Districts." In the intervening century, we have kept what is good and discarded what did not meet our demands, and we have continued to adapt. Whatever the specific stormwater requirements and techniques, we still require analysts and designers who understand the fundamental theory behind the tools, and who can appropriately apply stormwater management models. This book from Haestad Press is a valuable contribution to keeping the practice of stormwater modeling alive and vigorous.

Roger T. Kilgore, P.E.
Kilgore Consulting

Table of Contents

Preface

Design and analysis of an urban stormwater management system is a fascinating and complex blend of heuristics and practicality, and of complex hydrologic and hydraulic analyses associated with runoff-producing storm events and the resulting mix of overland, open-channel, and confined flows. Designers and analysts of stormwater conveyance features must keep in mind not only the engineering goals of efficient and cost-effective solutions, but a host of other goals and constraints pertaining to legal and institutional requirements, public safety, aesthetics, and maintainability.

Designers and managers of urban stormwater conveyance systems are increasingly required to consider innovative approaches that minimize runoff production through practices such as on-site infiltration, and that improve the water quality of the runoff before discharging it to downstream properties or receiving waters. The greatest benefits of these new practices tend to be associated with low to moderate intensity storms that occur quite frequently, as they contribute the bulk of the runoff and pollutant load from an urban drainage basin on an annual or long-term basis. Although these new practices and mandates will undoubtedly improve water quality in urban areas, stormwater management systems will continue to be required to safely and efficiently convey the large flows resulting from high-intensity storm events. The general public may demand and expect good water quality in an area, but it is highly unlikely that they will be willing to sacrifice their safety and well-being to accomplish those goals. Thus, the need for efficient and cost-effective stormwater conveyance facilities is not being supplanted by new environmental regulations—it is instead being augmented by them and making the challenge of design increasingly exciting.

This text includes elements of design pertaining to stormwater quality, mainly through its discussions of environmental regulations, best management practices, and introduction to detention pond design, but its main focus is on the more traditional aspects of hydrologic prediction and hydraulic design and analysis of conveyance facilities. In writing the text, an attempt has been made to produce a volume that integrates materials and methods not usually found in a single reference source. Thus, it is hoped that not only will the experienced design professional and/or manager find it useful as a desk reference, but that students and academics will find it useful for courses dealing with stormwater management system design.

Chapter Overview

Chapter 1 of this book discusses the need for stormwater management and the general design process. It also presents a history of the development of wet-weather flow management technologies.

Chapter 2 discusses the various components of stormwater management systems and introduces the various types of models and tools, including GIS and CAD, that are used in their design and analysis. It also summarizes the main elements of the design process, from master planning through final design and construction contract development. The experienced designer will find this chapter informative, although it may be most valuable to the inexperienced individual who wants to learn about stormwater management systems and their design and analysis methodologies.

Nearly all hydrologic and hydraulic methods used for design and analysis of stormwater management systems are applications of the fundamental physical laws of conservation of mass, energy, and momentum. Chapter 3 reviews these laws as a basis for further developments and applications of them in later chapters.

Design of any stormwater system component, such as a culvert or pipe, must be preceded by estimating the design discharge (flow rate) that the component is to convey. Hydrology is the science by which such estimates are developed, and, in that context, involves estimating the runoff rate from a drainage basin for a chosen duration and frequency of rainfall. Chapter 4 presents sources and methods for modeling design rainfall events, and is followed by methods for conversion of rainfall into runoff in Chapter 5.

The flow of stormwater through various conveyance system components occurs generally as confined (pipe) flows and as open-channel flows. Hydraulic methods of analysis of each of these fundamental types of flow are presented in Chapters 6 and 7, respectively, which build on the fundamental laws presented in Chapter 3.

With the fundamental tools of hydrologic and hydraulic analysis established in earlier chapters, Chapters 8 through 12 delve into detailed design procedures for various components of stormwater conveyance systems. Chapter 8 addresses design of open channels, and Chapter 9 addresses culvert design. Chapters 10 and 11 cover design of the most common conveyance system elements, namely roadway gutters, stormwater inlets, and pipe systems and their outfalls. Chapter 12 addresses design of stormwater detention facilities, which are often mandated for attenuation of peak runoff rates from developed lands or for water quality management.

It is occasionally true that stormwater requires pumping from low-lying areas such as highway underpasses and from areas protected by levees. Chapter 13 provides an introduction to the design of stormwater pumping facilities.

Chapter 14 provides a broad overview of the laws and regulations under which stormwater management systems must be developed, and includes both U.S. and international perspectives. The presentation is necessarily broad, as each political jurisdiction has its own specific requirements. The reader must acknowledge this, and must take it upon him or herself to become familiar with the requirements in their locale.

Stormwater management professionals are increasingly being required to address issues of runoff quality and receiving water impacts, and to design system components that can control the pollutants in small, frequently occurring rainfall events. Chapter 15 presents examples of best management practices (BMPs) that are commonly applied in the Denver, Colorado (USA), metropolitan region for stormwater management. Similar methods are often used in many other parts of the United States and the world.

Conventions

Because of the diversity of sources from which material has been gathered, and because of the differences in mathematical notation from one source to another, choices have been made regarding notation. Remaining faithful to the notation used in original references would have given rise to the same notation being used to represent different things in various places throughout the book. An example of this is the variable Q, used in most hydraulic literature to represent the volumetric discharge of a flow, but used in NRCS (SCS) hydrologic methods to represent the effective precipitation depth during a storm. Because of the confusion that could be caused, we opted instead to use a single set of notation throughout, even though this sometimes results in notation that differs from that found in the original reference sources.

Continuing Education and Problem Sets

Also included in this book are a number of hydraulics and modeling problems to give students and professionals the opportunity to apply the material covered in each chapter. Some of these problems have short answers, and others require more thought and may have more than one solution. The accompanying CD-ROM in the back of the book contains academic versions of Haestad Methods' StormCAD, PondPack, CulvertMaster, and FlowMaster software (see "About the Software" on page xvii), which can be used to solve many of the problems. However, we have endeavored to make this book a valuable resource to all modelers, including those who may be using other software packages, so these data files are merely a convenience, not a necessity.

If you would like to work the problems and receive continuing education credit in the form of Continuing Education Units (CEUs), you may do so by filling out the examination booklet available on the CD-ROM and submitting your work to Haestad Methods for grading. For more information, see "Continuing Education Units" on page xv.

Haestad Methods also publishes a solutions guide that is available for a nominal fee to instructors and professionals who are not submitting work for continuing education credit.

Feedback

The authors and staff of Haestad Methods have strived to make the content of *Stormwater Conveyance Modeling and Design* as useful, complete, and accurate as possible. However, we recognize that there is always room for improvement, and we invite you to help us make subsequent editions even better.

If you have comments or suggestions regarding improvements to this textbook, or are interested in being one of our peer reviewers for future publications, we want to hear from you. We have established a forum for providing feedback at the following URL:

www.haestad.com/peer-review/

We hope that you find this culmination of our efforts and experience to be a core resource in your engineering library, and wish you the best with your modeling endeavors.

Continuing Education Units

With the rapid technological advances taking place in the engineering profession today, continuing education is more important than ever for civil engineers. In fact, it is now mandatory for many, as an increasing number of engineering licensing boards are requiring Continuing Education Units (CEUs) or Professional Development Hours (PDHs) for annual license renewal.

Most of the chapters in this book contain exercises designed to reinforce the hydraulic and hydrologic principles previously discussed in the text. Many of these problems provide an excellent opportunity to become further acquainted with software used in stormwater modeling. Further, these exercises can be completed and submitted to Haestad Methods for grading and award of CEUs.

For the purpose of awarding CEUs, the chapters in this book have been grouped into several units. Complete the following steps to be eligible to receive credits as shown in the table. Note that you do not need to complete the units in order; you may skip units or complete only a single unit.

Unit	Topics Covered	Chapters Covered	CEUs Available (1 CEU = 10 PDHs)	Grading Fee* (US $)
1	Introduction, Basic Components, and Fundamental Laws and Units	Chapters 1, 2, & 3	1.0	$50
2	Modeling Rainfall and Runoff	Chapters 4 & 5	1.5	$75
3	Flow in Closed Conduits and Open Channels	Chapters 6 & 7	1.5	$75
4	Design of Open Channels and Culverts	Chapters 8 & 9	1.0	$50
5	Gutter Flow, Inlet Design, and Storm Sewer Design	Chapters 10 & 11	1.0	$50
6	Stormwater Detention and Pumping	Chapters 12 & 13	1.0	$50
7	Regulatory Issues and Stormwater Quality Management	Chapters 14 & 15	0.5	$25
All Units	All	All	7.5	$375

*Prices subject to change without notice.

1. Print the exam booklet from the file *exam_booklet.pdf* located on the CD-ROM in the back of this book,

   - or -

   contact Haestad Methods by phone, fax, mail, or e-mail to have an exam booklet sent to you.

   Haestad Methods
   37 Brookside Road
   Waterbury, CT 06708
   USA
   ATTN: Continuing Education

   Phone: +1 203 755 1666
   Fax: +1 203 597 1488
   e-mail: ceu@haestad.com

2. Read and study the material contained in the chapters covered by the Unit(s) you select.
3. Work the related questions at the end of the relevant chapters and complete the exam booklet.
4. Return your exam booklet and payment to Haestad Methods for grading.
5. A Haestad Methods engineer will review your work and return your graded exam booklet to you. If you pass (70 percent is passing), you will receive a certificate documenting the CEUs (PDHs) earned for successfully completed units.
6. If you do not pass, you will be allowed to correct your work and resubmit it for credit within 30 days at no additional charge.

Notes on Completing the Exercises

- Some of the problems have both an English units version and an SI version. You need only complete one of these versions.
- Show your work where applicable to be eligible for partial credit.
- Many of the problems can be done manually with a calculator, while others are of a more realistic size and will be much easier if analyzed with stormwater software.
- To aid in completing the exercises, a CD-ROM is included inside the back cover of this book. It contains academic versions of Haestad Methods' StormCAD, PondPack, CulvertMaster, and FlowMaster software, and software documentation. For detailed information on the CD-ROM contents and the software license agreement, see the information pages in the back of the book.
- You are not required to use the included Haestad Methods software to work the problems.

About the Software

The CD-ROM in the back of this book contains academic versions of Haestad Methods' StormCAD Stand-Alone, PondPack, CulvertMaster, and FlowMaster software. The following provides brief summaries of the software functionality. For detailed information on how to apply it to solve stormwater problems, see the help systems and tutorial files included on the CD-ROM. The software packages included with this textbook are fully functional but are intended for academic purposes only. Professional or commercial application is prohibited (see license agreement in the back of this book).

STORMCAD STAND-ALONE

StormCAD is a program to assist civil engineers in the design and analysis of storm sewer systems. StormCAD Stand-Alone has a CAD-like interface but does not require the use of third-party software in order to run.

StormCAD can automatically compute inlet and pipe flows with the rational method, or the user can input flows developed using another technique. For inlets with multiple subareas, StormCAD can compute the total area and weighted runoff coefficient. Pipe travel time is computed automatically and added to the upstream time of concentration to determine controlling time of concentration.

StormCAD provides intuitive access to the tools needed to model both surface collection system elements and subsurface storm sewer hydraulics. Surface flow calculations include gutter spread and depth and inlet capture and bypass flows. Subsurface hydrualic analysis capabilities include gradually varied flow profile calculations for mild, steep, horizontal, and adverse pipes; surcharged pipe flow calculations; and several methods for computing minor losses through inlet and manhole structures.

StormCAD's automated design features can assist engineers by setting pipe inverts and sizes based on constraints such as cover, slope, and percent full requirements, and sizing inlets based on allowable spread and depth for inlets in sag or minimum efficiency for inlets on grade. Multiple design and analysis scenarios can be created and compared within the same project file.

Other features of StormCAD include the ability to handle a variety of pipe cross-section shapes (circular, elliptical, arch, and box); tools to import and export data from databases, GIS, and CAD; a cost analyzer; and a variety of options for reporting results (profile plots, color coding, annotation, and tabular reports).

PONDPACK

PondPack is a program for performing hydrologic analysis and detention facility design. It is capable of analyzing a tremendous range of situations, from simple sites to complex networked watersheds with multiple ponds, outlet structures, diversions, and outfalls. PondPack's graphical interface and intuitive dialogs simplify data entry.

Either synthetic or gauged rainfall data can be used to create design storms in PondPack, or IDF curve data can be used to create a balanced storm or compute the rational method discharge. The NRCS (SCS) 24-hour synthetic distributions and Bulletin 70/71 synthetic distributions for several midwestern U.S. states have been pre-entered and are ready to use.

A number of methods for computing time of concentration are available, including Carter, Eagleson, Espey/Winslow, FAA, Kerby/Hathaway, Kirpich, Length/Velocity, SCS Lag, and TR-55. Alternatively, the user can manually enter a time of concentration computed using another method. The Horton, Green-Ampt, and curve number equations, as well as a constant loss rate, are available to account for infiltration. Hydrographs can be developed using the SCS unit hydrograph or other unit hydrograph, a rational method template, the modified rational method, or Santa Barbara Urban Hydrograph procedure.

The program can analyze pre- and post-developed watershed conditions and automatically size detention facilities. It can also compute rating curves simple and compsite outlet strucutures with or without tailwater effects, including culverts, weirs, orifices, inlet boxes, and standpipes. Hydrograph attenuation in channels can be accounted for using the Muskingum or Modified Puls method. The program can also model hydraulically interconnected ponds with downstream tidal outfalls or rating tables.

CULVERTMASTER

CulvertMaster is used to analyze and design simple and complex culvert crossings, including those with roadway overtopping. The program can handle cirular, elliptical, arch, and box culvert cross sections with inlet and outlet control calculations for the the full range of end treatments presented in FHWA's HDS-5 publication. The "Quick Culvert Calculator" component can solve for unknown variables such as headwater depth, discharge, or culvert size. "Culvert Designer" can be used to develop and compare multiple design alternatives. The "Culvert Analyzer" component can evaluate more complex systems, such as those with multiple non-matching culverts in parallel and roadway overtopping.

FLOWMASTER

FlowMaster is a helpful, easy-to-learn program for perfoming quick analyses and designs of individual hydraulic elements, inlcluding pressure and gravity pipes; triangular, trapezoidal, rectangular, and irregular open channels; rectangular, triangular, Cipoletti and broad-crested weirs; orifices; simple and composite gutter sections; and curb, grate, combination, and slotted-drain inlets.

FlowMaster's "Hydraulic Toolbox" can solve or rate an unknown variable using fromulas such as the Manning, Hazen-Williams, Kutter, Darcy-Weisbach, and Colebrook-White equations. FlowMaster's inlet computations strictly comply with the latest FHWA Hydraulic Circular Number 22 inlet computation guidelines. The program can store multiple design trials and compare the results in customizable tabular reports, performance curves, and rating tables.

Erosive forces due to high stormwater flows in this masonry-constructed channel resulted in the undercutting and partial collapse of the adjacent roadway.

CHAPTER

1

Introduction to Stormwater Conveyance Modeling and Design

0Stormwater is water from rain, snowmelt, or melting ice that flows across the land surface. Conveyance systems for dealing with stormwater are everywhere—in backyards, streets, parking lots, and parks. They frequently affect people but typically go unnoticed unless they fail. The consequences of failure range from nuisance flooding of yards, basements, and roadway travel lanes, to temporary road or bridge closure and minor property damage, to widespread destruction and even loss of life.

Another consideration when dealing with stormwater that has received increased attention in recent years is stormwater quality. In urban areas, stormwater often comes into contact with pollutants such as oil from roadways and parking lots, which it then carries into surface waters such as lakes and streams. Sediment loads from construction sites can also be problematic, as can runoff from agricultural lands exposed to fertilizers, pesticides, and animal waste.

This book presents information necessary to appropriately analyze and design stormwater conveyance systems, including the following topics:

- Basic components of stormwater systems and conveyance modeling
- Creation of project design storms
- Determination of the runoff rates to the system that will result from these storms
- Conveyance structure design alternatives
- Hydraulic analyses to determine which alternatives meet design criteria
- Practices for improving stormwater quality

Other factors to be considered, such as cost and available budget and the upstream and downstream impacts of a project, are also discussed.

This chapter begins the book by citing statistics and providing background on the factors driving the need for stormwater conveyance systems. Section 1.2 gives a brief history of stormwater management, and Section 1.3 provides an overview of issues that must be addressed in system design.

1.1 NEED FOR STORMWATER CONVEYANCE SYSTEMS

The need to manage stormwater runoff to protect vulnerable areas against destructive flooding is an age-old problem. Flooding poses a threat to developed areas, agricultural crops and livestock, human life, and highways and other transportation systems. Flooding happens across all spatial scales—from individual properties and neighborhoods along small creeks, to entire regions bordering major rivers.

Of course, the largest part of the losses is due to major flooding along large streams and rivers. Therefore, proper designs of storm sewers and culverts, which are the focus of this book, will only partly cure the growing cost of flooding in terms of monetary damage and loss of life. (Chapter 2 provides an overview of the types of stormwater facilities presented in this text.) Nevertheless, careful stormwater conveyance designs that consider not only the service of an immediate local area, but also how this area merges into and interacts with regional-scale stormwater and flood protection plans and facilities (sometimes called a "major system" of drainage conveyance facilities), can alleviate many of the damages associated with frequent storm events.

The magnitude of the flooding problem can be appreciated by citing a few statistics. Figure 1.1 illustrates the trend in annual flood losses in the United States over the period from 1903 to 1999, where all losses are expressed in 1997 dollars. The general upward trend is significant in view of the fact that the vertical axis is logarithmic. Over the period from 1992 to 1996, it has been estimated that losses due to natural hazards averaged $54 billion per year, or over $1 billion per week (National Science and Technology Council, 1997). Flooding alone accounted for about 26 percent of that total, or roughly $270 million per week.

It is staggering to realize that these huge losses have occurred despite the fact that for most of the twentieth century, the United States has been largely spared the expense of a catastrophic natural disaster in a highly populated and developed area. For example, a great earthquake of magnitude 8 or more on the Richter scale has not struck a major metropolitan area since the one in San Francisco, California, in 1906. A Class 4 or 5 hurricane has not struck a major urban area directly since 1926 in Miami, Florida (van der Vink et al., 1998). However, even though hurricanes have largely spared major population centers in recent years, they do pose threats every year and have caused substantial property damage, mainly to the Gulf and Atlantic coasts.

Sobering also is the cost of natural disasters in terms of loss of life. Figure 1.2 shows estimates published by the U.S. National Climatic Data Center and cited by the National Research Council (1991). The estimates indicate that during the period from 1941 to 1980, U.S. population-adjusted death rates due to flooding have generally increased, but those due to lightning and tornadoes have dropped dramatically. Popu-

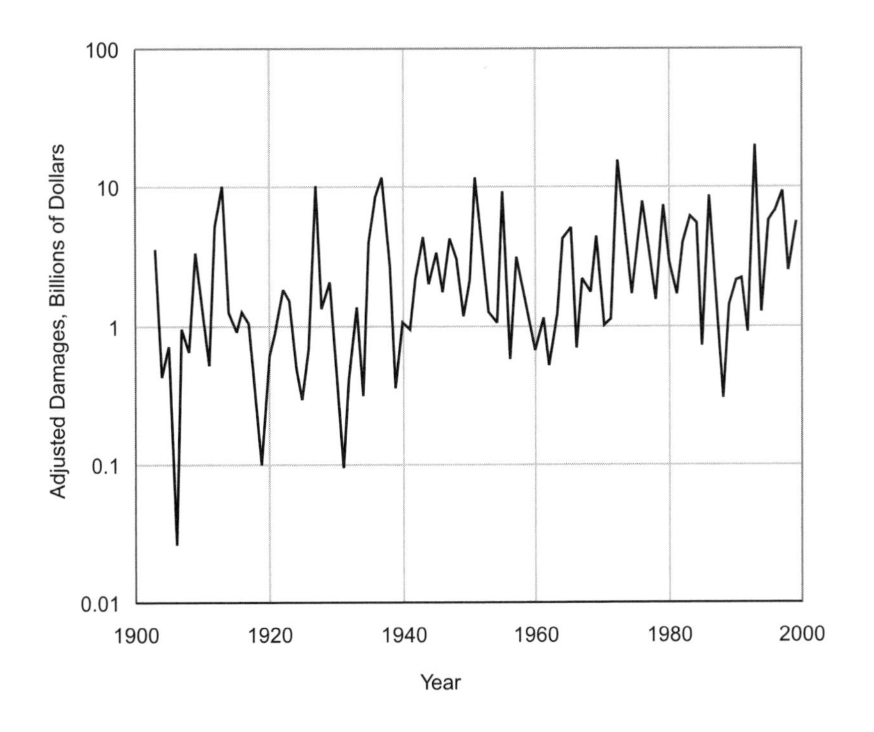

Figure 1.1
Annual flood losses in the United States from 1903–1999 (as reported by the U.S. National Weather Service, 2000)

lation-adjusted death rates due to tropical cyclones (hurricanes) have remained fairly constant over that time period. As of 1980, the population-adjusted death rate due to flooding alone was about as great as that due to lightning, tornadoes, and tropical cyclones combined. Although the greater share of the loss of life from floods indicated in Figure 1.2 is due to major flooding, a substantial portion is due to flooding of drainageways in urbanized areas. Since such drainageways are often engineered as part of an urban stormwater management and conveyance system, they fall within the scope of this text.

Many factors contribute to the upward trends in flood damages. However, the escalation is due largely to the increased occupancy and use of floodprone lands and coastal areas. Despite good intentions, public financing and implementation of so-called "flood control" projects are partly to blame for this state of affairs. The public, including politicians and decision-makers, are accustomed to hearing the term "flood control" without fully understanding the limitations of such projects. Often, one consequence of these projects is a false sense of security, so that when the inevitable large flood does occur, damages are greater than they would have been otherwise.

History has shown that floods cannot be controlled completely. Technical experts have never intended the term "flood control" to be interpreted in an absolute sense, but the phrase is often misleading to the public. To avoid public misconceptions about the level of protection provided by flood control projects, terminology such as "excess water management," which is less suggestive of absolute control, should be considered.

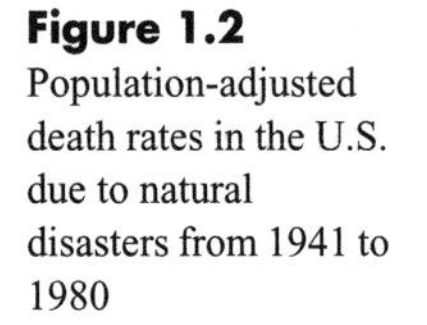

Figure 1.2
Population-adjusted death rates in the U.S. due to natural disasters from 1941 to 1980

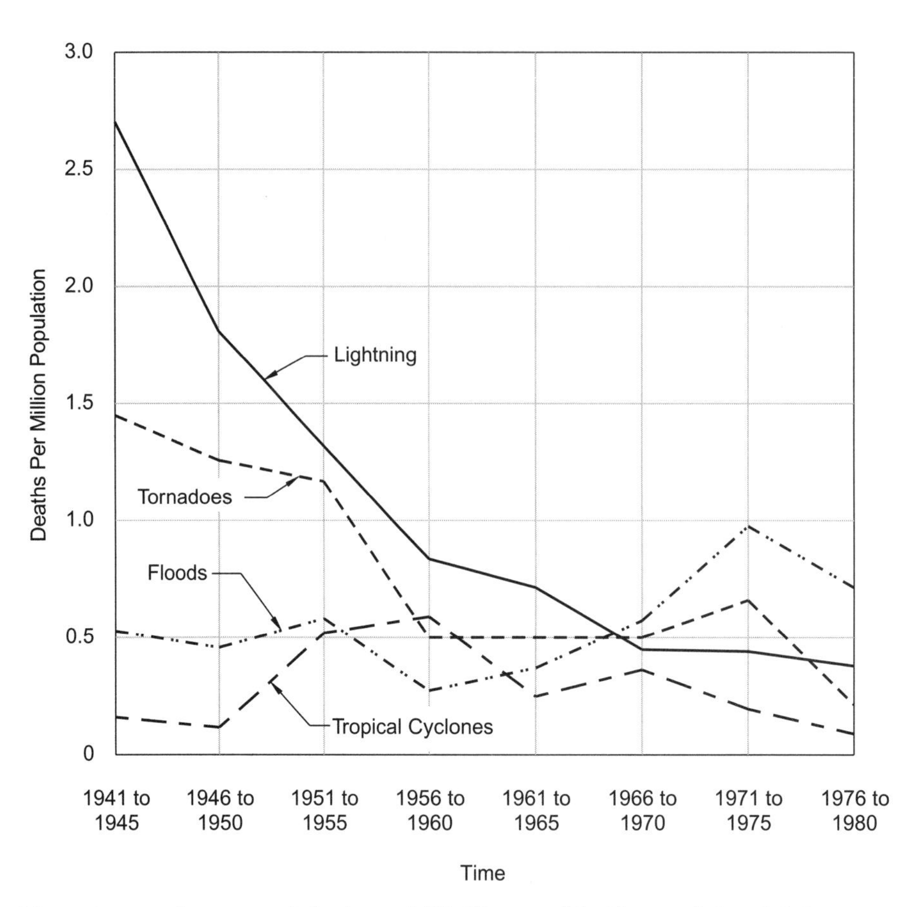

The monetary damage and the loss of life illustrated in Figures 1.1 and 1.2, respectively, are aggregates representing totals for the entire United States. Similar information may be obtained for other parts of the world. The Natural Hazards Center at the University of Colorado, Boulder is a clearinghouse for data and other information on natural hazards worldwide, and provides Internet links to many useful sources. The Web site for the Hazard Reduction and Recovery Center at Texas A&M University also provides useful research tools.

Damage and loss of life will continue to be among the most important factors driving stormwater management. These factors are also the most tangible indicators of the success (or failure) of the stormwater management process. Other considerations that drive the need for stormwater management are environmental and water quality concerns, aesthetics, and legal and regulatory considerations. The complexity of the driving factors and the consequent need for evaluation of the trade-offs among competing alternatives make stormwater management an exciting and challenging process.

*"At this point, it's still not classified as a hurricane—
it's still being called a raindrop."*

1.2 HISTORY OF STORMWATER MANAGEMENT SYSTEMS

As noted in Section 1.1, the need for dealing with stormwater runoff is an age-old problem. A number of accounts of historical developments in wet-weather flow management have been published over the years; the summary by Burian et al. (1999) is one of the most recent and serves as a model for the briefer presentation here. Additional historic information relevant to regulatory requirements is presented in Chapter 14.

Ancient Practices

Several ancient civilizations constructed successful surface-water drainage systems. Some civilizations incorporated the removal of sanitary wastes into the surface runoff system to form a combined system of sewage. For example, around 3000 B.C., dwellers of the Indus civilization in the city of Mohenjo-Daro (now in Pakistan) constructed combined sewers consisting of a simple sanitary sewer system with drains to remove stormwater from streets. The masonry work and clever design of this system demonstrate that, in some instances, more care was taken with the sewage system than with construction of some of the buildings. Other ancient sewer systems were constructed by the Mesopotamian Empire in Assyria and Babylonia (circa 2500 B.C.), and by the Minoans on the island of Crete (3000–1000 B.C.). Yet others were constructed in parts of the city of Jerusalem (circa 1000 B.C.); and in central Italy by the Etruscans around 600 B.C. Ruins of some major cities in China indicate that partially underground systems were constructed around 200 A.D. Other examples of ancient sewage-system builders include the Greeks in Athens, the Macedonians and Greeks under the rule of Alexander the Great, and the Persians.

Of all the societies of western Asia and Europe, from antiquity until the nineteenth century, only the Romans built a carefully planned road system with properly drained surfaces. Specific drainage facilities used by the Romans included occasional curbs and gutters to direct surface runoff into open channels alongside roadways. They also graded roadway surfaces to enhance drainage to the channels. This channel-based sewer system collected not only stormwater runoff from roadway surfaces but also sanitary and household wastes. During periods of dry weather when flows were insufficient to flush the channels, the wastes accumulated and caused unsanitary conditions to develop. Therefore, the channels were covered, evolving into combined sewer systems. These covered drains ultimately developed into the Roman *cloacae*, or underground sewers (see Figure 1.3). These sewers were a source of civic pride and symbolized the advanced stature of the Roman civilization.

Figure 1.3
Outlet of the Cloaca Maxima on the Tiber River, Rome (Blake, 1947)

Post-Roman Era to the 1700s

From the time of the Roman Empire through the 1700s, European sewage management strategies experienced little advancement and even regressed in terms of sanitation. Along with the decreasing populations of cities after the fall of the Roman Empire came the abandonment of municipal services such as street lighting, running water, and sewer systems.

During the Dark Ages, Western society was indifferent to such issues. Most citizens ignored matters of hygiene and cleanliness, for example. As the Middle Ages progressed, drainage and sanitary services were developed in response to aggravating conditions and disease outbreaks. Unfortunately, these systems were disjointed and poorly planned. Paris and London exemplify European cities with piecemeal systems at that time.

Resurgence in the development of planned sewage systems from the haphazard systems of the Middle Ages occurred between the fourteenth and nineteenth centuries. The sewers constructed in Europe during this time, such as those in London and Paris, were simply open ditches. Besides being conveyances for stormwater, these channels became receptacles for trash and other household and sanitary wastes, the accumulation of which caused overflows. As a remedy, the channels were covered to create combined sewers, essentially replicating the practice of the Romans some 1500 years earlier.

In Paris, the first planned covered sewer dates to 1370, when Hugues Aubriot constructed the Fosse de St. Opportune. This "beltway sewer" discharged into the Seine River and acted as a collector for the sewers on one side of the Seine. However, for the most part, sewer systems before the 1700s lacked proper design and were conducted in a piecemeal fashion. Maintenance and proper operation were virtually neglected. The systems were poorly functioning and subject to repeated blockages, resulting in nuisance flooding.

The Nineteenth Century

At the end of the 1700s, the outlook was improving for wet-weather flow management in Europe. Society held a belief in progress that became increasingly linked to technology throughout the 1800s. Until the 1820s, European sewers were constructed of cut stone or brick with rectangular or roughly rounded bases that contributed to deposition problems. Substitution of millstone and cement mortar stone made construction of curved and smooth sewer floors easier, thus improving both the self-cleansing properties and hydraulic efficiency. A variety of new pipe shapes were developed, including egg-shaped, oval, and v-notch cross sections. Additional improvements in the early 1800s consisted of increased attention to sewer maintenance and the adoption of minimum slope and velocity criteria to provide adequate flushing during dry-weather periods.

Comprehensive planning was the next major step in sewage system design. In 1843, William Lindley, an Englishman living in Hamburg, Germany, was commissioned to plan and design a system following a major fire (see Figure 1.4). London followed Hamburg's successful example by commissioning Joseph Bazalgette to plan and design its system. Construction of the Main Drainage of London was completed in 1865. In 1858, E. S. Chesbrough designed the first comprehensive wet-weather flow management system in the United States for the City of Chicago, Illinois. About the same time, J. W. Adams designed a comprehensive sewer system for Brooklyn, New York. The designs implemented in the United States often made use of empirical data obtained from European practice. Because rainfall intensities and storm durations are highly location dependent, as are factors such as time of concentration and soil type, the topographic and climatological differences between the United States and Europe ultimately contributed to design deficiencies.

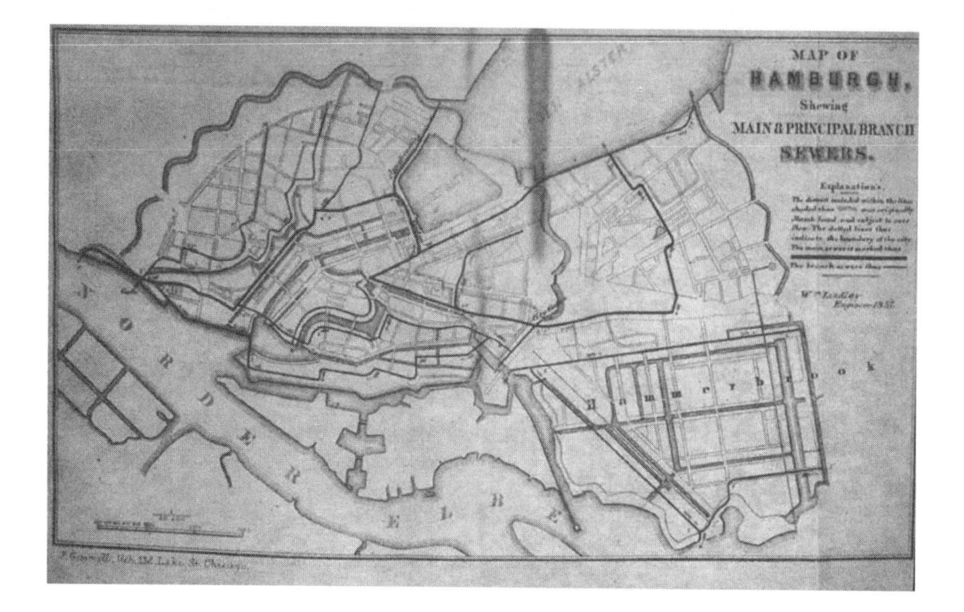

Figure 1.4
Map of Hamburg, Germany sewer system from 1857 (Chesbrough, 1858)

In the middle to late 1800s, rapid urbanization in the United States led to increases in water consumption and consequent overwhelming of privy-vault and cesspool systems for wastewater disposal. After much debate, city councils, engineers, and health groups generally agreed that centralized sewage systems were more cost-effective and provided greater benefits than did other options.

The question arose, however, as to which type of sewage system should be constructed—separate or combined. Initially, the combined sewer scheme was widely implemented for several reasons, including the fact that there was no European precedent for successful separate systems. Further, a belief existed that combined systems were cheaper to construct.

There were opponents to this practice, however. For example, Colonel George E. Waring designed relatively successful separate sewer systems in Lenox, Massachusetts, and Memphis, Tennessee. Around the same time (about 1880), at the request of the U.S. National Board of Health, Rudolph Hering visited Europe to investigate practices there. Hering's report recommended combined systems when serving extensive and densely developed areas, and separate systems for areas in which rainwater did not require removal underground. By the end of the 1800s, engineers had embraced Hering's ideas, and combined sewers were recommended for most urbanized areas. This philosophy did not change until the 1930s and 1940s when more wastewater treatment was required, because the contribution of stormwater volume translated into higher treatment costs. Consequently, separate sewers then became the more commonly recommended option.

Regardless of the type of sewage (separate or combined), control and treatment of wastewater discharges were limited. It was commonly assumed that cities could safely dispose of their wastes into adjacent waterways, with "dilution" being the "solution to pollution." By the middle to late 1800s, epidemiological research by John

Snow and William Budd, which showed water's ability to transmit infectious disease, and germ and bacterial research by Louis Pasteur and Robert Koch, were beginning to challenge the effectiveness of dilution strategies. Despite the fact that this fundamental scientific research demonstrated a connection between polluted waters and disease, wastewater treatment was not widely practiced at this time. Debate centered on whether it was better to treat wastewater before discharging it to a stream or for downstream users of the water to treat it before distribution as potable water. Most engineers subscribed to the latter approach, effectively ignoring the effects on the habitats and recreational uses afforded by receiving waters.

In the mid-1800s, estimation of surface stormwater runoff, which was needed to design appropriately sized conveyance structures, was based mainly on empirical results. Among the tools used was Roe's table, a simple look-up table indicating the pipe size required to convey runoff for a given drainage area and slope based on Roe's observations of London sewers (Metcalf and Eddy, 1928). Other methods were based on empirical and, frequently, ill-founded formulas developed by Adams, McMath, Talbot, and others. The empirical nature of these methods and their development for site-specific conditions often resulted in very poor estimates and designs when they were applied under conditions different from those for which they had been developed. For example, Talbot's formula predicts the required cross-sectional area of a drainage pipe as follows (Merritt, 1976):

$$A = C^4 M^{3/2} \tag{1.1}$$

where A = required cross-sectional area of drainage pipe (ft^2)
C = coefficient ranging from 0.2 for "flat country not affected by accumulated snow or severe floods" to 1.0 for "steep, rocky ground with abrupt slopes"
M = the drainage basin area (ac)

Talbot's formula was based on a rainfall intensity of 4 in/hr, but could be adjusted for other intensities by proportion. The formula does not consider pipe slope or roughness, factors that are now known to greatly influence required conveyance pipe diameters.

In the second half of the nineteenth century, hydrologic and hydraulic methods were enhanced considerably through the research of such notable figures as Antoine Chézy and Robert Manning, who developed uniform flow formulas, and T. J. Mulvaney and Emil Kuichling, who developed what is known today as the rational method for stormwater runoff rate estimation. In 1899, Arthur Talbot performed some of the initial work on rainfall estimation, producing curves that eventually evolved into present-day intensity-duration-frequency curves.

The Twentieth Century

By the 1920s, the growing accumulation of rainfall records enabled the use of *design storms* in which the time variation of rainfall intensity during a storm event was explicitly acknowledged and accounted for. Additional advancements during the twentieth century consisted of the development of unit hydrograph theory by

L. K. Sherman in 1932 and studies of rainfall abstractions and infiltration by W. H. Green and G. A. Ampt in 1911 and by Robert Horton in 1933. By the middle to latter part of the century, the U.S. Soil Conservation Service (now the Natural Resources Conservation Service) had developed simple but effective methods for runoff estimation for both rural and urban areas. Methods for statistical and probabilistic analysis of floods, droughts, and other hydrologic, hydraulic, and water quality variables also advanced rapidly during the twentieth century.

With the development of the digital computer came rapid advancements in technical tools and methods for wet-weather flow management. Included among these advances were developments of specialized hydrologic and hydraulic computer programs (see Figure 1.5), as well as more general computational and data processing technologies such as spreadsheets, databases, pre- and postprocessors, and geographic information systems. The computational power of computers enabled engineers to efficiently optimize their designs through the use of advanced mathematical optimization and statistical techniques. Whereas just a few decades ago it was uncommon to find even a single computer in most design engineering offices, a computer is now an indispensable tool on every engineer's desk.

Figure 1.5 Computer programs for designing storm sewers became available in the latter part of the twentieth century

Perhaps some of the most significant developments that have occurred in the past few decades have been related to the recognition of the impacts of stormwater runoff on receiving waters. By the early 1970s, the U.S. Congress had passed the Water Pollution Control Act, and increasingly strict water quality regulations have since been developed. To mitigate wet-weather flow impacts on receiving waters and habitats, methods of control and treatment of urban stormwater runoff are required. Among these methods are physical, chemical, and biological treatment processes and storage and treatment combinations. It should be noted, however, that water quality is not the only wet-weather factor contributing to habitat degradation downstream of urban areas. Other factors, including channelization, debris and cover removal, and channel straightening, must be addressed more adequately in the future.

Summary and Outlook

Most of the tools and methods currently used for prediction of runoff discharges and volumes and for appropriately sizing required conveyance facilities have been developed within the last 200 years. That ancient civilizations were able to effectively deal with stormwater runoff indicates that advanced knowledge and sophisticated tools are not necessary for relatively simple conveyance systems. This is not the case in modern society, however, where competing infrastructure needs vie for resources and multiple goals require evaluation of several designs. Computers and automated control technologies enable modern designers and system operators to optimize and maintain complex systems and make the design and operation much more efficient than they would be otherwise.

What the future will bring is always uncertain. A report by the U.S. EPA's Urban Watershed Management Branch (Field et al., 1996) describes five areas in which research is needed. They are

- Characterization and problem assessment
- Watershed management
- Impacts and control of toxic substances
- Control technologies
- Infrastructure improvement

The third of these areas, which deals with pollutants, has received the least attention in the past. However, it is beginning to receive considerable attention in the research community and can be expected to become more of an issue in practice in the coming years. A growing research area that is absent from this list is sustainable development, which is concerned with preserving water quality for human use and the environmental preservation into the future.

1.3 OVERVIEW OF STORMWATER CONVEYANCE SYSTEM DESIGN

Design of a stormwater conveyance system is a trial-and-error process. The "art" of design involves defining the objectives and determining, from among a number of feasible options, the one that best meets those objectives. The design engineer optimizes the use of resources (materials, construction labor, operation and maintenance costs, and so on) to meet the objectives while satisfying certain constraints (such as project budgets, right-of-way limitations, and design storm criteria).

Design is different from *analysis*. Design is more complex than analysis. In a design, the engineer is not limited to the evaluation of any single combination of inputs in seeking to meet the desired objectives. An analysis, on the other hand, is simply a technical evaluation of a possible choice to determine whether it meets the objectives while satisfying the constraints. The design process may be viewed as consisting of many analyses, each of which determines whether the design is feasible (that is,

"Wow! What a day to be a workaholic!"

whether it satisfies the design constraints) and whether it meets the project objectives. Finally, from among all the feasible designs, the "best" one is selected.

With this view of the design process in mind, several questions need to be addressed:

- What is (are) the objective(s) of the design?
- What resources need to be allocated?
- What are the constraints that need to be satisfied?
- How does one define the "best" feasible design? What makes one choice better than another?

Only when these questions are explicitly acknowledged and addressed can the design be considered acceptable. The following subsections are intended to clarify these questions. However, the issues surrounding these questions are specific to each design.

Design Objectives

The *design objectives* are the project goals that the engineer attempts to meet as much as possible without violating the design constraints. There may be more than one objective associated with the design of a stormwater conveyance system. The most obvious objective is probably adequate conveyance of stormwater runoff. However, in the face of specific regulations regarding return periods for which design flows must be computed, stormwater conveyance capacity is more likely to be viewed as a constraint than as an objective.

It is usually not possible to develop a design that is optimal for all objectives. In other words, if a particular design alternative is optimal for one objective, then it will usually be less than optimal for the other objectives. Therefore, some weighting or tradeoff analysis to quantify the relative importance of the objectives is necessary. For further reading on multiobjective planning, the reader is referred to Loucks, Stedinger, and Haith (1981); Goodman (1984); and Grigg (1996).

Many possible objectives may be adopted for a particular project. Historically, minimization of the total life-cycle cost of the project has been an attractive one, probably due in part to the relative ease with which it can be quantified. For cases in which the engineer is faced with developing a design for which the capital cost may not exceed an available budget, cost may not be a design objective, but instead a design constraint.

Other possible objectives are

- Minimizing flood damages or traffic hazards during a storm event of a magnitude greater than that for which the primary, or minor, storm sewer system is designed
- Minimizing aesthetic, physical, chemical, and biological impacts to existing receiving waters
- Maximizing potential recreational and aesthetic benefits provided by wet detention facilities

The engineer should seek to define the design objectives with input from the entity funding the project, other relevant regulating and/or operation and maintenance agencies, the general public, and special-interest groups.

Resources

In conveyance system design, resources allocated to a project (or project costs) consist of raw materials and labor for construction. Operation and maintenance resources (equipment, manpower, and so forth) required over the project lifetime should also be included. For instances in which natural channels, wetlands, or other areas where ecosystems are disturbed, the loss of habitat and the associated biotic communities should also be considered as "resources" allocated to a project.

Other types of resources that should be considered are those related to acquisitions of rights-of-way and/or temporary or permanent easements. In some cases, it is necessary to purchase or condemn residential or other properties so that they can be demolished to make way for the planned project improvements. In these instances, resources allocated to the project should include purchase and relocation costs, but these costs may be offset in part by less frequent flooding in the project locale.

Constraints

As already noted, a limited budget within which a project must be implemented is one type of constraint. Additional constraints may be geometric in nature, such as limited pipe burial depths, existing underground utilities that cannot be relocated, and limited rights-of-way. Other constraints consist of federal, state, and local regulations within which a project must be developed, as well as laws regarding water rights and flood protection.

As a general rule, constraints can be of two types. One type of constraint specifies a requirement that cannot be violated at any cost. The second type of constraint can be violated, but the cost associated with its violation may be high. For example, if a design constraint were to say that an existing underground gas transmission line cannot be moved, the reality is that the line could be moved, but the cost could be prohibitive.

What Is Best?

After the engineer identifies a number of feasible designs, the task turns to the issue of selection of the best design. But how should one define what is *best?* What makes one design better than another? The answer depends on many factors, but the degree to which each design meets the stated project objective(s) is certainly one basis for comparison. For example, if cost minimization were the only objective, then the lowest-cost alternative would be chosen. The design engineer does not necessarily make the final selection of the design. Often, the engineer's client or a regulatory body is the decision-maker. In such instances, the role of the designer is to provide the decision-maker with options and quantifications of costs, benefits, and impacts. The designer essentially acts as an advisor on complex technical issues.

Factors that might be considered in judging the suitability of the various choices are issues such as safety and aesthetics. Other measures might relate to equity in the distribution of project benefits. For example, if a citywide stormwater drainage improvement project were to make improvements only in new or politically powerful districts, while overlooking needs in poor and disadvantaged areas, matters of fairness and equitability would not be satisfied.

The design process as outlined by this section may seemingly imply that the storm sewer and culvert design process can be viewed as a mathematical optimization problem (Loucks, Stedinger, and Haith, 1981). That approach, however, is usually not formally applied in design practice because the degree of quantification necessary to do so is usually not possible. The practitioner must instead rely heavily on experience and knowledge of the problem at hand to identify suitable alternatives for detailed evaluation. Nevertheless, the general idea of optimizing the use of resources to achieve adopted objectives within certain constraints does succinctly characterize the overall design process. After the preferred design is identified through this heuristic process, one can prepare detailed construction plans and specifications.

1.4 CHAPTER SUMMARY

The need to protect against flooding is driven primarily by the loss of life and property that flooding causes. Most of the losses can be associated with major flooding of streams and rivers, and the stormwater conveyance systems described in this text have a limited effect on events of this magnitude. However, these primary systems significantly reduce the damages caused by smaller but more frequent events.

In modern society, storm drainage design is complicated by competing infrastructure needs and goals that require careful balancing of resources and evaluation of multiple possible designs. Computers have greatly simplified this process. Considerations such as water quality and the need to deal with pollutants are receiving more attention in recent years.

The design of stormwater conveyance systems is an iterative process that involves defining project objectives and determining the design alternative that best meets them, but does not exceed available resources or violate design constraints. The design process can be viewed as a series of analyses, or technical evaluations of individual design alternatives. The "best" design is then selected from among these alternatives.

REFERENCES

Blake, M. E. 1947. *Ancient Roman Construction in Italy from the Prehistoric Period to Augustus.* Washington, D.C.: Carnegie Institution of Washington.

Burian, S. J., S. J. Nix, S. R. Durrans, R. E. Pitt, C-Y. Fan, and R. Field. 1999. "Historical Development of Wet-Weather Flow Management," *Journal of Water Resources Planning and Management* 125, no. 1: 3–13.

Chesbrough, C. S. 1858. "Chicago sewerage report of the results of examinations made in relation to sewerage in several European cities in the winter of 1856–57." Chicago, Illinois.

Field, R., M. Borst, M. Stinson, C-Y. Fan, J. Perdek, D. Sullivan, and T. O'Connor. 1996. *Risk Management Research Plan for Wet Weather Flows.* Edison, N. J.: U.S. Environmental Protection Agency.

Goodman, A. S. 1984. *Principles of Water Resources Planning.* Englewood Cliffs, N. J.: Prentice Hall.

Grigg, N. S. 1996. *Water Resources Management: Principles, Regulations, and Cases.* New York: McGraw-Hill.

Loucks, D. P., J. R. Stedinger, and D. A. Haith. 1981. *Water Resource Systems Planning and Analysis*, Englewood Cliffs, N. J.: Prentice Hall.

Merritt, F. S., ed. 1976. *Standard Handbook for Civil Engineers*, 2d ed. New York: McGraw-Hill.

Metcalf, L., and H. P. Eddy. 1928. *Design of Sewers.* Vol. 1 of *American Sewerage Practice.* New York: McGraw-Hill.

National Science and Technology Council. 1997. Committee on Environment and Natural Resources: Fact Sheet, Natural Disaster Reduction Research Initiative.

National Research Council. 1991. *Opportunities in the Hydrologic Sciences.* Washington, D.C.: Committee on Opportunities in the Hydrologic Sciences, Water Science and Technology Board, National Academy Press.

National Weather Service. 2000. "Flood Losses: Compilation of Flood Loss Statistics." National Weather Service, Office of Hydrologic Development. www.nws.noaa.gov/oh/hic/flood_stats/Flood_loss_time_series.htm.

van der Vink, G., R. M. Allen, J. Chapin, M. Crooks, W. Fraley, J. Krantz, A. M. Lavigne, A. LeCuyer, E. K. MacColl, W. J. Morgan, B. Ries, E. Robinson, K. Rodriquez, M. Smith, and K. Sponberg. 1998. "Why the United States is Becoming More Vulnerable to Natural Disasters," *EOS, Transactions of the American Geophysical Union* 79, no. 44: 533–537.

Workers construct a large, triple-barrel box culvert (top). Low flows are temporarily diverted around the disturbed area through bypass piping (bottom).

CHAPTER

2

System Components, Models, and the Design Process

Stormwater is managed with a network of structures designed to collect, convey, detain, treat, and discharge runoff. Examples of these structures are storm sewer conduits, culverts, drainage ditches, and detention ponds. Generally, these structures act in a passive mode, relying on the force of gravity to move water. Occasionally, however, stormwater is actively managed with devices such as pumps, gates, or valves.

In this book, *models* are defined as representations of the physical behavior of hydrologic and/or hydraulic systems. They may be physical, electric-analog, or mathematical in nature. Mathematical models are based on the laws of physics and typically apply the principles of conservation of mass, energy, or momentum. *Computer models* (such as those on the CD-ROM accompanying this book) are collections of automated mathematical models.

Models are used to predict how stormwater structures will perform. The basic elements that must be considered when modeling stormwater management systems are hydrology (specifically, watershed response to storms), hydraulic capacities of the system structures, and upstream and downstream impacts (such as the effect of tailwater elevation on storm sewer system hydraulics or the effect of a storm sewer system on downstream discharges).

Engineers use models in the design process to simulate the performance of a stormwater collection and conveyance system under a range of flows and conditions. The tradeoffs between size and performance can then be evaluated. Hydraulic performance in terms of resulting flows, headwater elevations, exit velocities, and so forth is the primary focus of this text. Other factors that must be considered in selecting system components are construction cost, structural loads, maintenance requirements, aesthetics, seepage flow (from or into the groundwater), and biological habitat. The final design is based on compliance with regulations, standards, and, ultimately, the professional judgment of the engineer.

Sections 2.1 and 2.2 of this chapter describe stormwater conveyance system components and modeling approaches for the purpose of assisting those new or inexperienced in the field of stormwater management in getting a brief overview of basic concepts, common terminology, and analysis methods. They also serve as a "road map" for the book by directing the reader where to find further information on various topics. The stages of the design process, including master planning, preliminary design, and detailed design, are presented in Section 2.3. Section 2.4 provides a brief discussion of the integration of computerized stormwater models with CAD and GIS. The chapter concludes with an overview of model input data sources in Section 2.5.

2.1 Hydrologic Components

Hydrology is the study of the properties, distribution, and effects of water as it cycles through the earth's surface, subsurface, and atmosphere. This book is concerned with hydrology in terms of loading stormwater management systems. Coverage of the topic is therefore limited to quantifying *precipitation* and *runoff* for that purpose.

This section briefly describes the modeling of rainfall and runoff. These topics are covered in detail in Chapters 4 and 5.

Precipitation

To design or evaluate a storm sewer system, the engineer must determine the recurrence frequency of the storm event(s) that the system should be designed to handle. These events, also called *design storms*, will typically be dictated by local practice or regulations. In other instances, it may be up to the engineer to determine what storm events should be used given the type of project, its impact on the surrounding area, and the consequences of system failure.

Rainfall intensity varies with time over the course of a storm event. Design storms are frequently computed by applying a total storm rainfall depth to a synthetic rainfall distribution (see Section 4.5, page 91). Rainfall distributions may also be developed from actual precipitation records for an area. Evaluation of multiple storm events for a single design is often desirable, especially if stormwater will be detained or pumped.

Once the design storms are selected and developed, rainfall data can be combined with drainage basin characteristics to determine runoff volumes and discharges.

Runoff

Runoff or *rainfall excess* is the portion of the rainfall that is not lost to interception, evapotranspiration, or infiltration (see Section 5.1, page 108). It "runs off" over the ground surface and into streams, ditches, and other stormwater management structures. Thus, runoff is the hydrologic component of primary interest in stormwater conveyance modeling.

The methods and models available for computing runoff are numerous. They range from simple peak discharge calculations for sizing pipes to more complex *hydrograph* methods that account for total runoff volume and the change in discharge over time. The *rational method* is an example of the former; the *unit hydrograph method* is an example of the latter. Methods for computing runoff from snowmelt, which may be a consideration in some cases, are also available.

Spreadsheets can assist in runoff hydrograph calculation, and many computer programs for automating the calculations are also available. Computer programs such as HEC-HMS and PondPack (Haestad, 2003a) offer several models from which the user can choose.

2.2 Hydraulic Components

Many types of structures are used in stormwater management systems. A conveyance structure may be as simple as an earthen ditch or as complex as an extensive system of inlets and subsurface piping that delivers water to a storage facility from which it is pumped to a treatment plant or receiving stream.

The subsections that follow describe some of the basic properties of open channels, culverts, storm sewer systems, and detention/retention facilities. Brief descriptions of commonly used materials and applicable mathematical models are included, as well as examples of computer software available for model calculations. Table 2.1, which is located at the end of this section, summarizes the various stormwater system components, methods used in modeling these components, and typical model input requirements. It also provides text references for obtaining more detailed information.

Open Channels

Open channels include natural stream channels, as well as natural *swales* (depressions) along which water runs following rainfall events. In the constructed or urbanized environment, open channels include street gutters, drainage ditches, pipes, lined or unlined stormwater collection channels, and natural or modified stream channels. Figure 2.1 shows an example of flow in an open channel with a trapezoidal cross section.

Flow in a pipe is classified as open-channel flow if there is a water surface at atmospheric pressure (that is, a *free water surface*). For the purposes of this book, the term *open-channel flow* refers to any flow having a free water surface, regardless of whether it occurs in a closed conduit such as a culvert or storm sewer, aboveground channel, or other structure. Open-channel flow theory is presented in detail in Chapter 7. However, when the text refers simply to an *open channel* in the context of physical characteristics or design approaches, as in Chapter 8, the term is meant to describe nonpipe conveyances only.

Figure 2.1
Open-channel flow

Prismatic versus Nonprismatic Channels. Open channels exist in a wide variety of cross-sectional shapes. For analytical purposes, channels may be classified as prismatic or nonprismatic. In a *prismatic channel* the variables that describe the channel geometry—such as its width, side slopes, and longitudinal slope—remain constant along the length of the channel. In a *nonprismatic channel* one or more of these geometric variables changes along the channel length. Natural stream channels, whose widths and other channel properties are variable, are examples of nonprismatic open channels. Treatment of open-channel flow in this text deals primarily with prismatic, man-made channels. Nevertheless, certain aspects associated with the hydraulics of flow in nonprismatic channels will be addressed where appropriate.

Materials. Open channels can also be classified on the basis of whether they are lined or unlined. The bottom and sides of an *unlined channel* consist of natural geologic materials such as earth, gravel, or rock. In a *lined channel*, the channel bottom and sides are covered with an erosion-resistant material such as concrete, asphalt, riprap, or vegetation.

As shown in Chapter 7, the channel's lining material (or lack thereof) affects hydraulic performance characteristics such as flow depth and velocity. In turn, these same characteristics affect the erosive forces of the flow, thereby influencing the lining selection. The selection of the channel lining material is therefore a significant part of the hydraulic design process.

Model Representation. To model open-channel flow, the engineer must know the flow condition. With *steady flow,* flow characteristics such as discharge and velocity are constant over time at a given point in the channel. When these characteristics do vary over time as a point, the condition is *unsteady.*

Steady flow can be further classified as either uniform flow or varied flow. *Uniform flow* is a condition in which the depth, velocity, cross-sectional area, and discharge are constant along the channel length. As a practical matter, this can happen only in a prismatic channel and only if the flow is steady. In uniform flow, the depth of flow is called *normal depth*.

Two models that can be used to describe uniform flow are the Manning equation and the Chézy equation. In their simplest forms, either of these equations may be used to calculate velocity or discharge as a function of channel slope, roughness and geometry. In design, these models are used in a variety of ways. For example, for a given channel geometry, slope, roughness, and depth, the discharge may be calculated. Alternatively, if the channel geometry, material roughness, discharge and maximum allowable depth are specified, the required slope can be calculated. An example of a computer model used for uniform flow calculations is FlowMaster. Modeling of uniform flow is covered in more detail in Section 7.3.

Varied flow is a condition in which the depth and velocity of flow change along the length of the channel. If the depth and velocity change only gradually, the flow is said to be *gradually varied flow (GVF)*. If the depth and velocity change quickly over a short distance, as in a hydraulic jump, the flow is said to be a *rapidly varied flow (RVF)*. Gradually varied flow occurs in all nonprismatic channels and in prismatic channels under the influence of a flow control other than normal depth. Because the curvatures of flow streamlines are small in a gradually varied flow, the pressure forces can be assumed to be hydrostatic. Constitutive relationships developed for uniform flow, such as the Manning equation, can be assumed to be valid for computing friction losses in gradually varied flow.

Because of the variable channel geometry and the presence of flow controls, the data required for gradually varied flow analysis are considerably more complex than for uniform flow analysis. The input includes channel cross-section geometry, reach lengths, channel roughness, and the water surface elevation at the control section. HEC-RAS is an example of a computer program frequently used in gradually varied flow profile calculations. For more information on gradually varied flow calculations, see Section 7.8.

Channel routing is a procedure by which the outflow at the downstream end of a channel reach is computed from inflow data and channel characteristics. It accounts for the effects of channel storage and travel time on discharge rates; thus, it is a type of unsteady flow modeling. Channel routing is applicable to both open channels and sewer systems when channel storage and/or travel time affect flow rate to a significant degree.

Channel routing may be accomplished by using either *hydrologic* or *hydraulic routing* methods. Hydrologic routing is computationally simpler and less data-intensive, but usually is also less accurate. Hydrologic routing algorithms are based on the physical principle of conservation of mass and on an assumed storage relationship. Hydraulic methods are more physically based and employ conservation of mass and conservation of momentum. Their use requires considerable data on channel geometry and roughness variables and on initial and boundary conditions of the flow itself. The Muskingum, Modified Puls, and Convex routing methods are examples of hydrologic routing models. (See Section 5.8 and Appendix C.) Hydraulic routing involves a solution of the Saint-Venant equations, and may consist of kinematic, diffusion, or dynamic wave routing. Such methods are not well-suited to manual calculations but are coded into software packages such as HEC-RAS.

Culverts

A culvert is a relatively short underground water conveyance conduit. The primary purpose of a culvert, like that of a bridge, is to provide a means whereby the water in a stream or other open channel can pass through an obstruction such as a highway or railway embankment. Figure 2.2 shows a small culvert passing under a driveway.

Materials. Culverts come in a variety of shapes, sizes, and materials. As shown in Figure 2.3, cross-sectional shapes associated with culverts include circular, box, elliptical (horizontal or vertical orientation), and arched.

Common materials used for culvert design are reinforced precast or cast-in-place concrete and corrugated steel. Other materials include corrugated aluminum, polyethylene, and polyvinyl chloride (PVC). Discussions of these materials, including available shapes and sizes and sources of further information, are presented in Section 9.2.

The inlet and/or outlet ends of a culvert often receive special treatment. A culvert end may simply project from the embankment, or it may be mitered to conform to the embankment slope. It may also be fitted with a special flared end section or with a headwall and wingwalls. Section 9.3 discusses various types of end treatments and their hydraulic performance characteristics, as well as special inlet configurations that can be used when allowable headwater depths are limited.

Model Representation. A complete theoretical analysis of the hydraulics of a particular culvert installation is complex. Flow conditions vary from one culvert to the next, and they also vary over time. The barrel of the culvert may flow full or partially full, depending on upstream or downstream conditions, barrel characteristics, and inlet geometry. A commonly used method of analysis of culvert hydraulics was devel-

Figure 2.2
A corrugated metal pipe culvert under a driveway

oped by the Federal Highway Administration and published as "Hydraulic Design Series No. 5: Hydraulic Design of Highway Culverts," often referred to as HDS-5 (Norman, Houghtalen, and Johnston, 2001). Modeling software such as CulvertMaster follows the calculation methods set forth in HDS-5.

The following list describes the principal terms and concepts in culvert hydraulics More detailed information on culvert hydraulic analysis and HDS-5 methodologies is provided in Section 9.5 and Section 9.6.

- *Headwater depth* is the depth (relative to the culvert's upstream invert elevation) of flow just upstream of the culvert entrance. Any increase in energy required to push an increased discharge through a culvert translates into a greater headwater depth.

- *Tailwater depth* is the depth of water just downstream of the culvert outlet. The tailwater surface elevation may be dictated by downstream channel characteristics, by obstructions, or by a receiving-water elevation. An exit condition in which the tailwater depth is significantly less than the depth of flow in the culvert, and thus does not affect the upstream hydraulics, is called a *free outfall*.

Figure 2.3
Common cross-sectional shapes for culverts

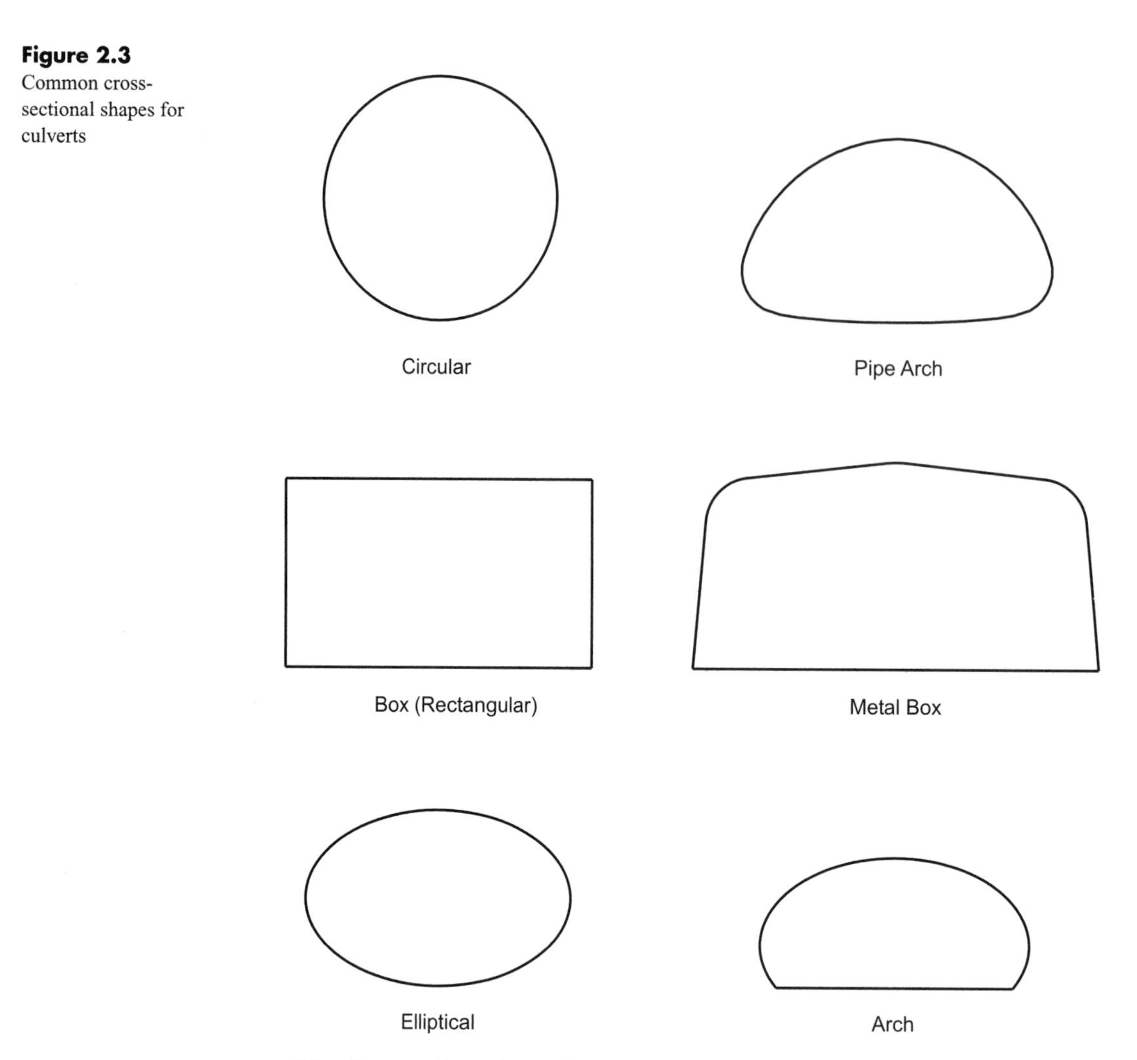

- The *flow condition* in a culvert can be characterized as either *full (pressure) flow* or *partially full flow* with a free water surface. Both free-surface and pressure flow can occur simultaneously at different locations within the same culvert. Uniform flow and gradually varied flow, as discussed earlier for open channels, apply to free-surface flow in culverts as well.
- The flow in a culvert may be controlled by conditions at either the inlet or the outlet of the culvert, so the type of control is classified as either *inlet control* or *outlet control*. In the case of inlet control, the hydraulic capacity of the culvert entrance limits the amount of water conveyed for a given headwater depth; therefore, hydraulic characteristics downstream of the inlet control section do not affect the culvert capacity. In an outlet control condition, the culvert barrel capacity or downstream tailwater elevation limits the amount of flow that can be conveyed by the culvert for a particular headwater elevation. The control section for outlet control is located at the barrel exit or further downstream.

- The flow velocity at the exit of a culvert is typically higher than that of the stream channel into which the culvert discharges. High outlet velocity can cause streambed scour and bank erosion in the vicinity of the culvert outlet. Minor problems occasionally can be avoided by increasing the barrel roughness, but structures such energy dissipators and/or outlet protection such as riprap are often necessary.

Storm Sewer Systems

A *storm sewer system* is a network of primarily subsurface structures for the collection and conveyance of stormwater runoff. A typical storm sewer system consists of *inlets* through which stormwater enters from the land surface, *pipes* that convey this water, *junction and manhole structures* that serve as connection points for pipes and provide access to the subsurface system, and *outlets* or *outfalls* where stormwater is discharged to treatment facilities, to detention/retention structures, or into a larger conveyance such as an open channel.

Components and Materials. Storm sewer systems almost always include gutters, inlets, manholes, pipes, and outlet structures. Some systems may include storage and/or pumping facilities and other structures such as weirs. The following subsections briefly describe each of these components.

Although roadway gutters are not always considered a part of a stormwater management system, they are in fact an integral part of the system and are responsible for conveying runoff that collects in streets to inlets, where it enters the subsurface conveyance system. A *gutter* may be formed by a curb along the edge of a street, or, for roads without curbs, it may consist of a shallow swale along the road edge. The various gutter types are presented in Section 10.1.

Figure 2.4 is a photo of a concrete curb-and-gutter section. The width of flow in the gutter is called the *spread*. Depending on the roadway classification and posted speed limits, allowable spread widths may extend beyond the roadway shoulder width and into the travel lane.

Curbs are typically composed of either precast, poured-in-place, extruded, or asphalt concrete. In some cases, a depressed section of gutter may be integrated with the concrete curb, and the remainder of the gutter section is composed of the roadway pavement material.

A stormwater inlet is a structure for intercepting stormwater on the ground surface or in a roadway gutter and conveying it to the subsurface storm sewer piping system. Common inlet types are grate inlets (Figure 2.5), curb inlets (Figure 2.6), and combination inlets (Figure 2.7).

For each of these three types, inlets can be further described as either continuous-grade inlets or sag inlets. A *sag inlet* is constructed in a low point where water drains to the inlet from all directions. With a *continuous-grade* inlet (also referred to as an inlet on grade), surface water may flow away from the inlet opening in one or more directions.

Figure 2.4
Concrete curb and gutter

Below the ground surface, inlets consist of box-like structures that support the inlet opening (for example, the grate) and connect to the system piping. The inlet box is typically constructed of precast reinforced concrete. Other materials include poured-in-place concrete, corrugated metal, brick, and block. A *catch basin* is a special type of inlet box designed to trap sediment and debris.

Although not considered inlets *per se*, runoff from streams and channels can enter storm sewer systems through culvert-type entrances such as headwalls. Additional specialized inlet types are discussed in Section 10.2.

Figure 2.5
A grate inlet in a triangular gutter section

Figure 2.6
A large curb inlet

Figure 2.7
A combination inlet

Manholes, sometimes called access holes, are usually installed in sewers at locations where pipes join, change direction, or change size. They provide access to the system for inspection and maintenance. Manholes for relatively small-diameter sewers generally consist of a concrete or masonry base, which ideally has been formed to aid in reducing the energy loss associated with water traveling from the inlet pipe(s) to the outlet pipe. Figure 2.8 shows a precast concrete manhole structure prior to installation. *Junction structures,* like manholes, are used where two or more pipes join one another. However, junction structures tend to be associated with larger or more complex installations and are often custom-made.

For storm sewer *pipes,* circular pipe is the most typical cross-sectional shape, but, as with culverts, a variety of cross-sectional shapes are available (see Figure 2.3). Elliptical and arch-shaped cross sections may be used for instances in which vertical or horizontal clearances are limited. Rectangular concrete box sections are also available. Concrete pipe (shown in Figure 2.9) is available in both nonreinforced and reinforced varieties and in a range of wall thicknesses and strength classes and is probably the most commonly used pipe material for storm sewer construction.

Figure 2.8
A precast manhole structure

Corrugated metal pipe (or *CMP*) is also used frequently in storm sewers (see Figure 2.10). Like concrete pipe, CMP is manufactured in a number of cross-sectional shapes and sizes and in strict accordance with standard specifications. Corrugated steel products include circular and arch pipes, structural plate pipe-arches, and structural plate arches, with the last requiring bolted assembly in the field. Various coatings and steel thicknesses are available.

Although concrete and corrugated steel pipes are by far the most common in storm sewer and culvert applications, other materials, such as corrugated aluminum, polyethylene, and polyvinyl chloride (PVC), are used as well. PVC pipes are shown in Figure 2.11.

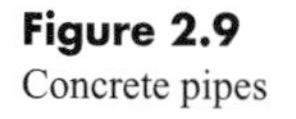

Figure 2.9
Concrete pipes

Figure 2.10
Corrugated metal pipe (CMP)

Storm sewers may *outfall* to the ground surface, into a natural or man-made body of water, or into detention or treatment facilities. Exit velocities at storm sewer outfalls are often large enough to cause erosion problems and potential undermining of the outfall pipe. *Outlet structures* dissipate the excess energy of the flow and prevent undermining. An outlet structure may be as simple as a concrete headwall with a riprap and filter blanket downstream of the outfall point, or it may consist of a more complex energy dissipator or stilling basin.

Figure 2.11
PVC pipes

In most cases, stormwater management systems are designed as gravity systems. Pumping of stormwater is generally undesirable because it significantly increases the cost of stormwater conveyance but cannot be avoided in some instances. For stormwater pumping purposes, the most common types of pumps are axial-flow pumps, radial-flow pumps, and mixed-flow pumps. Both wet-well and dry-well installations are found in stormwater pumping applications. Chapter 13 provides information on the design of stormwater pumping systems.

Model Representation. Many components are involved in modeling a storm sewer system, and various levels of model complexity exist for their analysis. The hydraulic design or analysis of a storm sewer can typically be broken down into two major parts: surface hydraulics (that is, gutter flow and inlet capacity) and subsurface hydraulics (pipe capacity and energy losses). Other possible components are pump stations and storage facilities.

Levels of analysis that may be performed on storm sewers include:

- Use of a steady-state model that ensures the system can handle peak flows.
- Use of an extended-period simulation (EPS) analysis that applies hydrologic routing and that may also simulate inter-storm time periods.
- Use of a dynamic model that solves the full Saint-Venant equations.

A steady-state analysis is typically sufficient for design applications that do not involve system flooding or storage facilities and for which the peak flows from various subbasins arrive at the outlet at roughly the same time. This type of analysis is the primary focus of this text. The paragraphs that follow describe basic steady-state model requirements for surface systems, subsurface systems, and pumping facilities.

The surface portion of a storm sewer system consists of gutters and other open channels, and inlet structures. Gutter hydraulics are typically modeled as uniform flow in an open channel. Specialized forms of the Manning equation have been developed to facilitate this calculation and are given in Section 10.1. The data required are the discharge, the geometry of the street (longitudinal and transverse slopes) the height of the curb, and the roughness of the pavement.

Inlets in sag locations operate as weirs for shallow flows and as orifices for greater flow depths. The forms of the weir and orifice equations used for computing the capacities of curb-opening and grate inlets are presented in Section 10.2 of this text. For grated inlets, information on the grate perimeter and open area is required. For curb inlet calculations, the dimensions of the curb opening and the local depression are necessary.

Calculations for inlets on grade are somewhat more complex, as all of the flow that comes toward the inlet may not be intercepted. A portion of the flow in the gutter, termed *bypass flow*, may flow over or around the inlet and continue downgradient. The calculation of *intercepted flow* and bypass flow is performed using empirical methods that take into account the velocity of the flow in the gutter relative to the inlet length, the amount of inlet depression, and the inlet's *splash-over velocity* (for grate inlets, the critical velocity at which a portion of the flow begins to splash over the inlet). Section 10.2 provides the equations describing the capacity of inlets on grade.

Inlet bypass flows form a surface network for stormwater runoff similar to the way sewer piping forms a subsurface network. Manual calculations for inlet capture and bypass can be quite tedious if a project has a large number of inlets on grade. Some computer programs used for computing inlet and gutter hydraulic characteristics are simple equation-solvers that look at each structure individually. An example of this type of program is FlowMaster, which is useful for performing quick calculations on small systems.

Other computer programs allow the user to develop a network of on-grade inlets, inlets in sag, and gutters. These programs analyze the surface system using the same methodologies as the simpler programs, but alterations in the loading of gutters and inlets due to upstream bypass flows are automatically taken into account. Typically, this type of program will also perform the subsurface hydraulic calculations. Many software packages are available for this application, but one popular package is StormCAD (Haestad, 2003b).

Flow in the subsurface system is modeled as one of two flow conditions (see Table 2.1). When a pipe is not full and the water surface is open to the atmosphere, the flow is modeled as open-channel flow. A friction loss equation such as the Manning or Chézy equation may be used to determine flow depths and velocities. When a pipe is flowing full (that is, it is *surcharged*), closed conduit pressure flow calculation methods are used. The data requirements for these calculations are discharge, pipe size, roughness, length, and slope.

In a storm sewer network, the minor energy losses that occur at inlets, at locations where pipe size or alignment changes, and at outlets can be significant. Data requirements for computing these losses include the geometry of the structure and the entrance and exit velocities. Several methods for computing minor losses are available (see Sections 6.3 and 11.4), and computer models such as StormCAD can compute these losses as part of a network analysis.

The operating characteristics of a pump, when installed in a piping system, can be ascertained from graphs of pump and system curves. These characteristics include the energy head produced by the pump, the corresponding pump discharge, the brake horsepower requirements for the pump motor(s), and the net positive section head required to avoid cavitation damage to the impeller and other internal parts. Methods for determining these operating characteristics, for both single- and multiple-pump installations, are presented in Section 13.3.

Stormwater Detention and Retention

A common objective in stormwater management is to maintain the peak runoff rate from a developing area to a value no greater than the predevelopment rate. This practice will reduce local flooding, soil erosion, sedimentation, and pollution. Local governing bodies typically establish specific design criteria for peak flow attenuation. *Detention basins* reduce the magnitude of peak flows by temporarily storing the runoff and releasing it over an extended period of time. Additionally, ponds often function as sedimentation basins, reducing the concentration of suspended materials in the discharge.

Detention Basin Types and Components. *Wet ponds* are designed to maintain a minimum water depth between storms. *Dry ponds* are designed to completely discharge stored runoff between storms. The configuration of a detention pond is generally dictated by site conditions. A dry detention pond is shown in Figure 2.12.

The most common detention facility designs are earthen basins or ponds located downstream of the collection area and upstream of the discharge point. The detention pond's outlet structure controls the rate at which runoff may discharge from the facility. Common outlet structure devices include weirs, orifices, culverts, inlet boxes, and standpipes. Most ponds are also equipped with an emergency overflow spillway to handle flows if the primary outlet becomes clogged, or if a storm has a magnitude greater than the pond's design event.

Model Representation. Detention ponds are modeled using the routing procedures described in Section 12.7. The basic data requirements are an inflow hydrograph, the depth to volume relationship of the pond, and the hydraulic characteristics of the outlet structure.

Figure 2.12
Dry detention pond

Capabilities of computer programs for stormwater detention modeling vary widely. The most basic programs allow for hydrograph input into a single pond with a simple outlet such as an orifice or weir, and the pond is routed using a technique such as storage-indication routing or numerical integration. More powerful models, such as PondPack (Haestad Methods, 2003a), can analyze networked watersheds and multiple ponds, interconnected ponds, and complex outlet structures. Fully dynamic stormwater modeling programs allow storage facilities to be modeled together with other elements of the stormwater conveyance system.

Summary of Hydraulic Models for Various Components of Stormwater Management Systems

A number of computer modeling software programs are available for virtually all of the methods presented in this text, and some examples of these programs have been mentioned in this chapter. However, because of the wide variety of structure types and available calculation techniques used in the field of stormwater management, there is great variation and specialization among the available programs. The primary thrust of this text as it relates to computer modeling is to aid the reader in understanding the methodologies underlying the programs. Such understanding is necessary if the engineer is to apply the computer programs correctly and confidently.

Table 2.1 lists the numerical models frequently used to describe the behavior of various components of stormwater conveyance systems, the input data required by these models, and references to the sections of this book where each model is discussed. It should be noted that most software programs can implement a collection of models from this list. For instance, modeling software for storm sewer systems would likely include solvers for uniform, gradually varied, and pressure flow; gutter flow; inlet performance; and minor losses in manholes.

Table 2.1 References for common stormwater conveyance system component models

Component	Modeling Methods	Typical Input data	Section of Text
Channel with uniform flow	Manning equation, Chézy equation, or Darcy-Weisbach equation	Channel roughness Channel geometry Channel slope Discharge or velocity	7.3
Channel with gradually varied flow	Direct-step method or standard-step method	Geometry of each cross section Starting water surface elevation	7.8
Channel with flow routing	Muskingum	Historical flood data or routing coefficients Inflow hydrograph	5.8
	Modified Puls	Geometry of outlet Inflow hydrograph	5.8
	Saint-Venant equations	Channel geometry Inflow hydrograph	Appendix C
Culvert with inlet control	HDS-5 methodology	Inlet geometry Discharge	9.5
Culvert with outlet control	HDS-5 methodology	Pipe geometry Pipe material Inlet geometry Discharge Tailwater elevation	9.5
Roadway overtopping	HDS-5 methodology	Roadway profile coordinates Type of road Roadway width Discharge	9.6
Gutter flow	Manning equation or Chézy equation	Roadway slopes Curb height Pavement roughness Discharge	10.1
Storm sewer pipes with open channel flow	Manning equation or Chézy equation; Saint-Venant equations	Pipe roughness Pipe geometry Pipe slope Discharge or velocity	7.3; 7.4; 7.5; 7.8 11.5; Appendix C
Storm sewer pipes with pressure flow	Darcy-Weisbach, Manning, Hazen-Williams, or Chézy equation; Saint-Venant equations	Pipe roughness Pipe geometry Pipe slope Discharge or velocity Tailwater conditions	6.2; 6.3; 11.5; Appendix C
Inlets	Weir and orifice equations	Discharge Inlet type and geometry	10.2
Manholes	Minor loss equations	Entrance and exit velocities Structure geometry	6.3; 11.4
Diversions/overflows	Weir and/or orifice equations	Geometry Water surface elevations	7.6; 7.7
Pumps	Energy equation System equilibrium	Pump head-discharge curves System curve(s)	13.3; 13.7
System outlets	Weir equations Orifice equations Minor loss equations Energy dissipation model	Boundary conditions such as tailwater data or downstream channel characteristics Geometric characteristics of outlet	6.3; 9.5; 11.7
Detention ponds	Hydrologic routing and Hydraulic routing	Inflow hydrograph Depth vs. volume for pond Geometry of outlet or stage-discharge curve	12.6; 12.7

Note: This table is repeated in Appendix D. That version lists the software program on the accompanying CD-ROM that can be used with each type of model component.

2.3 THE DESIGN PROCESS

The urban stormwater infrastructure can be viewed as consisting of a minor system and a major system. The minor system is designed to handle frequent events with return periods typically on the order of 2 to 10 years. The minor system consists of, for instance, roadway gutters, inlets, and pipes.

The major system consists of the pathways taken by flows in excess of the capacity of the minor system. It can be thought of as inundated roadways, swales and depressions, and natural and man-made open channels. A major system always exists, even when a minor system does not. Unfortunately, major systems are often neglected and do not receive the attention that they warrant, especially in light of the high probability of operational failure of the minor system.

The overall process of designing a storm drainage system can be broken down into a number of phases. These phases are

1 Master planning

2 Concept and preliminary design development

3 Detailed design

4 Preparation of construction drawings, specifications, and contract documents

Each of these phases is described more fully in the sections that follow.

Master Planning

Master planning, often conducted for an entire urban area by a drainage-system authority or review agency, provides a holistic view of the urban drainage system and how its various components interact. The need for master planning is particularly acute where an urbanized area spans a number of different political jurisdictions, each having its own objectives and design criteria. In such instances, master planning can help ensure that systems that cross jurisdictional boundaries are consistent with one another. Even if only one political jurisdiction is involved, master planning provides useful guidance for constructing new drainage systems that are consistent not only with one another, but also with existing facilities.

The issues that should be addressed by a master plan are, at a minimum, delineations of major urban drainageways (whose floodplains might be mapped for flood insurance purposes) and approximate limits and locations of storm sewers in the contributing drainage basin(s). The master plan should make a clear distinction between the minor and major drainage systems, and it should also take whatever steps are politically and legally possible to ensure that the major system is functional and provides a reasonable degree of protection from severe storm events.

Technically, a master plan should contain enough detail to be an effective guide to the ultimate construction and/or rehabilitation of storm drainage facilities in the area covered by the plan. Hydrologic analyses should have sufficient detail to provide reasonable estimates of required system conveyance capacities and should address estimates of expected future flows in developing drainage basins. Zoning master plans, where available, are invaluable for providing estimates of ultimate land uses in a drainage basin.

The master plan report and other documents should clearly state the design storm-recurrence intervals for which planned drainage facilities are to be designed and should be referred to frequently as individual components of the overall system are designed and constructed. A listing of typical design storm return periods, as recommended by the American Society of Civil Engineers, is presented in Table 2.2.

Table 2.2 Typical return periods for design of stormwater drainage systems (ASCE, 1992)

Land Use	Return Period (years)
Minor drainage systems	
Residential	2 to 5
High-value generation commercial	2 to 10
Airports (terminals, roads, aprons)	2 to 10
High-value downtown business	5 to 10
Major drainage system elements	Up to 100

In areas where existing storm-drainage facilities require rehabilitation and/or retrofitting, and especially in cases where additional right-of-way acquisitions may be required, the master plan should identify such needs and should provide estimates of the costs associated with the upgrades.

Preliminary Design Development

As new subdivision developments are built or as drainage-improvement projects are begun, development of preliminary designs should precede detailed final design activities. Preliminary designs should be consistent with the broad system outlines established during the master planning phase, but they should provide additional detail on locations of individual drainage structures and features. In the case of a new subdivision, for example, the preliminary design can be indicated on a preliminary plat showing the proposed street and parcel layout of the development. This preliminary design provides an opportunity for jurisdictional authorities to review the consistency of the proposed facilities with the broader-scale master plan and ensure that both minor and major drainage-system considerations have been dealt with.

Activities associated with preliminary design entail more detailed hydrologic analyses than those conducted for master planning purposes. They should also explicitly account for how the proposed facilities fit into larger-scale regional plans. Design alternatives should receive detailed evaluation in this phase, as should methods for dealing with upstream flows and discharges to downstream areas.

At a minimum, a preliminary design drawing and report should consist of topographic mapping on which streets, land parcels, and proposed storm-drainage facilities have been superimposed. The storm-drainage layout should indicate pipe, manhole, and inlet locations, as well as locations of other features such as detention and/or retention ponds. Drainage basins contributing to each inlet should be shown and referenced to accompanying hydrologic estimates of design flows in the report. Finally, the preliminary design report should explicitly address the major drainage system, including the general pathways and directions of major system flow on the design drawing.

Detailed Design

Following the approval of the preliminary design by the appropriate jurisdictional authorities, the engineer can proceed with the detailed design. As the name implies, this phase involves detailed engineering analyses that will serve as the basis for the design described by the construction plans and specifications.

Detailed hydrologic and hydraulic analyses must be performed at this stage to finalize design discharges and determine the dimensions of hydraulic structures, such as pipes, inlets, ponds, and energy dissipators. Structural analyses are necessary to determine required pipe strength classes or wall thicknesses and may also be required for other project-specific hydraulic structures. Geotechnical analyses are necessary to locate shallow bedrock that could affect construction costs; to determine groundwater elevations that could affect trenching, buoyancy forces on submerged pipes and structures, and pipe infiltration; and to determine pipe-bedding requirements.

It is essential that complete and well-organized files be maintained to document each activity and decision in the detailed design phase. Should unforeseen circumstances occur during construction, or should litigation arise after construction of the system has been completed, these files will provide a record of the adequacy of the design and its adherence to accepted engineering practices.

Construction Drawings, Specifications, and Contract Documents

Construction drawings should be prepared as the final design is completed and should go hand in hand with that activity. Construction plans consist of plan and profile sheets showing horizontal and vertical alignments of new and existing facilities, including all dimensions required for a survey crew to lay out the system in the field. Plan views should indicate the locations, elevations, and dimensions of all proposed pipes, inlets, manholes, and other system features, and should show at least approximate locations of existing utilities that could pose potential conflicts. Existing land-surface features, such as curbs, walks, and other structures requiring demolition and subsequent replacement, should also be shown.

Profile views should include required pipe slopes and invert elevations, existing and proposed ground-surface elevations, approximate limits of bedrock, and approximate elevations of underground utilities. If the construction contractor is to perform utility relocations, then construction requirements for the relocations must also be included in the plans. It is sometimes desirable to show energy and hydraulic grade-line profiles on construction plans; they are certainly not needed for construction, but may be required for regulatory submittals.

In addition to plan and profile drawings, construction plans should clearly identify required materials and material classes and should provide construction details, such as reinforcing steel layouts, for any special structures. Some construction details may already be available in sets of standard plans published by, for example, transportation departments. As a courtesy to project bidders, construction plans may also include one or more sheets containing tabulations of estimated material quantities.

Technical specifications are the heart of any set of construction documents, spelling out in detail the required construction methods and material types, classes, and testing requirements. It is common to reference standard construction specifications published by jurisdictional agencies, but care should be taken to avoid conflicting requirements in the plans and technical specifications. There may be conflicts between the engineer's specifications and standard specifications, between standard specifications and construction plans, or both. There is no such thing as a "standard project." Standardized technical specifications must be viewed with a wary eye.

The contract documents for a construction project consist of the plans, technical specifications, and other "boiler-plate" provisions. These additional provisions generally consist of bidding documents, bonds, official notices to proceed and notices of project acceptance, and general and special conditions of the contract. The general conditions typically specify such things as owner-engineer-contractor responsibilities and job-site requirements relating to issues such as materials storage, sanitation, site cleanup, and conformance with applicable laws and regulations. The general conditions are typically a standard document used for all construction contracts within a jurisdictional area. The "Standard General Conditions of the Construction Contract" document from the Engineers Joint Contract Documents Committee (EJCDC, 2002) is a widely used set of general conditions in the United States Special conditions, which are project-specific and assembled by the engineer, identify requirements other than those specified in the general conditions, and can also be used to modify provisions of standardized technical specifications.

2.4 INTEGRATION OF STORMWATER CONVEYANCE MODELING WITH CAD SOFTWARE AND GIS

The integration of conveyance modeling with computer-aided design (CAD) programs and geographic information systems (GIS) has a number of benefits. Hydraulic and hydrologic design and analysis software often includes both CAD and GIS-type features. In some cases, analysis applications can run directly within an organization's preferred CAD or GIS platform. This section describes some of the benefits of leveraging CAD and GIS capabilities in hydraulic design and analysis.

CAD Integration

CAD software can be useful in both the initial and final stages of a hydraulic design project. A CAD drawing typically provides the base plan for the layout of stormwater management facilities such as sewer networks, channels, and detention ponds. After the facilities have been designed, this plan is updated to reflect proposed structures for inclusion in the construction documents.

Software applications interact with CAD systems in a variety of ways. A graphics-based application such as StormCAD may be capable of displaying a background CAD file for use in system layout, or the hydraulic analysis application may run directly inside of a CAD program such as AutoCAD (see Figure 2.13) or Microstation. CAD features can be used to fine-tune system layout and supply data on the topography and horizontal and vertical structure coordinates. Many CAD programs used by civil engineers are equipped with special tools for assisting engineers in creating roadway grading plans and drawing sewer profiles.

Figure 2.13 Storm sewer plan and profile created using hydraulic design software running inside of AutoCAD

Engineers can choose from various approaches when integrating hydraulic design with a CAD application. Several possible options, ordered from least to greatest degree of integration, are as follows:

- No automated data exchange occurs between the CAD program and the hydraulic modeling software. The engineer manually enters schematic data on the hydraulic system into the analysis software and designs the system. The system characteristics resulting from the design are then manually input into the CAD drawing for inclusion in construction documents.
- A CAD drawing containing the background information necessary to lay out the stormwater facilities is used as a background map in the hydraulic design software. After the system's layout is completed and its elements designed, the layout is imported into the CAD drawing. Other element characteristics are added to the drawing manually.
- The initial hydraulic system layout is created in the CAD drawing and imported into the analysis program. Depending on the features of the CAD

software, the data imported may include, in addition to the horizontal layout (that is, coordinate and/or connectivity data), elevation information, structure types, preliminary structure sizes, and other characteristics. Once the design is final, system data can be transferred back into the CAD application through automated import/export methods.

- An analysis application that runs directly within the CAD platform is used. In the design process, hydraulic structures are inserted directly into the drawing file, and data are readily available from both the CAD program and hydraulic design tool. This technique is the most efficient of the various integration approaches.

Typically, increased integration of CAD and hydraulic design and analysis applications translates to less data duplication, decreased opportunities for the introduction of errors and omissions, and increased time savings.

GIS Integration

A *geographic information system (GIS)* is a powerful configuration of computer hardware and software used for compiling, storing, managing, manipulating, analyzing, and mapping (displaying) spatially referenced information. It combines the functionalities of a computer graphics program, an electronic map, and a database.

Like CAD, a GIS can be used in many stages of stormwater conveyance modeling. Although CAD programs provide an excellent format for developing technical drawings, they generally do not provide a means of storing data associated with those drawings. For example, a CAD drawing may show a stormwater pipe network for a neighborhood. However, if the engineer is to model the network using information from the drawing, the drawing must contain detailed annotations for each structure describing characteristics such as pipe material, size, and inlet and outlet elevations. A GIS, by comparison, can store model information in an internal database associated with each GIS data layer. The database functions of a GIS also allow data to be stored and maintained externally and accessed when needed through a dynamic database link.

GIS software typically has tools for drawing (digitizing), map display, data storage, and data analysis. The *digitizing tools* allow for conversion of hard-copy drawings to an electronic format and are almost identical to those of CAD programs. *Map display tools* allow users to change the appearance of different types of features (points, lines, polygons, and text) according to values in the database. [For instance, within a single layer, all 6-in. (150-mm) pipes may be shown in red, with concrete pipes displayed as solid lines and corrugated steel pipes displayed as dashed lines.] The *data storage tools* are similar to those of many database programs, with data lookup, query, and sort functions.

The *data analysis tools* of GIS make it well-suited to supporting stormwater conveyance modeling. These tools allow users to consider both spatial and data relationships to develop new data or to provide model inputs. For example, one may be trying to evaluate the runoff response of a watershed by looking at information about a storm sewer trunk line, which may be made up of multiple pipes of different sizes and materials. Data analysis tools in GIS could aid in determining the hydrograph's time to peak by spatially defining a connected flow path, and then using stored data on pipe characteristics to solve the hydraulic equations necessary to calculate the runoff response times.

As may be inferred from the previous example, two important ways in which GIS software supports stormwater conveyance modeling are

- By assisting hydrologic modeling
- By storing data from stormwater system inventories.

GIS and Hydrologic Modeling. GIS can be integrated with hydrologic modeling in a number of ways, depending on the type of work being performed. Many local government agencies maintain extensive GIS systems, and these can be excellent sources of hydrologic and hydraulic model input data, such as

- Data for physical features like rivers and streams, land cover, geology, and soils
- Specialized topographic data including spot elevations, surveyed cross-sections, contours, and digital elevation models (DEMs)
- Data for man-made features like roadways and other impervious areas, bridges, storm sewers, and other utilities
- Tax parcel or property boundaries, addressed buildings, and ownership information
- Governmental maps including municipal boundaries, census data, and zoning

GIS software packages offer various specialized tools for managing the topographic data used in delineating watersheds. However, actual watershed boundaries can differ from those delineated using only surface topography (for instance, storm sewers may cross surface watershed boundaries). For this reason, the engineer must take care to address both natural and man-made features when delineating watersheds and ensure that the boundaries are correctly placed in the GIS. Figure 2.14 shows a watershed delineated using a GIS.

After the watershed boundary is delineated, the next task is often to model the runoff response of the basin. GIS data analysis tools can assist in processing rainfall data and determining area-weighted averages for parameters such as runoff coefficient (see Section 5.2, page 118). For more complex coefficients based on multiple data layers, the GIS can also use an *overlay* tool to develop new mapping for areas with unique combinations of characteristics. For example, the curve number method predicts runoff response based on both land use and hydrologic soil group (see Section 5.2, page 119). Areas with unique curve numbers can be delineated by overlaying a land use map on a soils map and then using a lookup table.

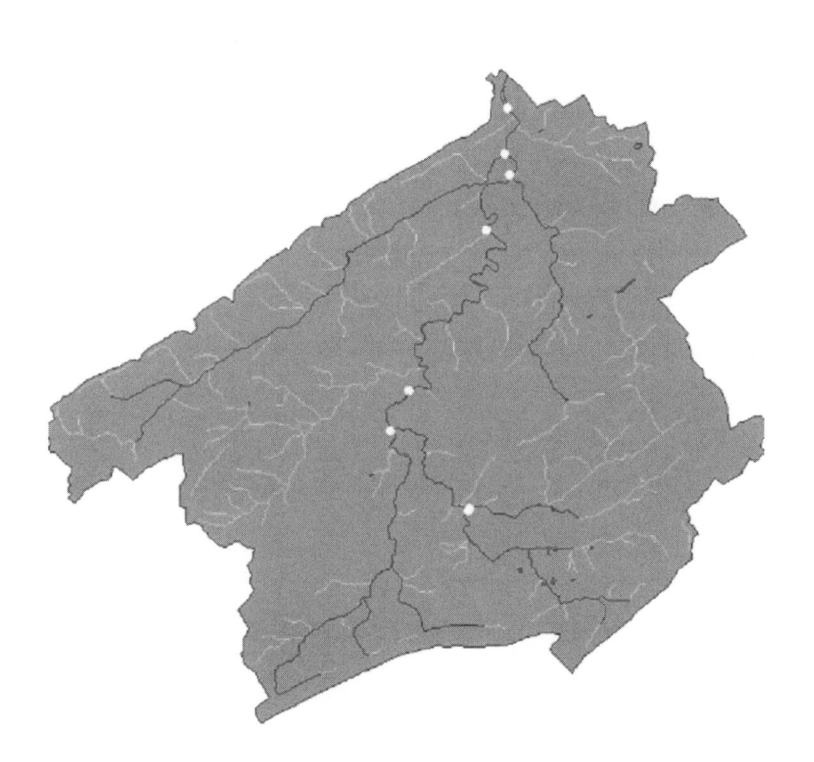

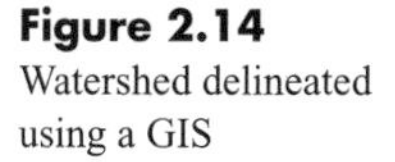

Figure 2.14
Watershed delineated using a GIS

GIS data can also play a critical role in finding suitable sites for new stormwater structures and systems. As an area develops, land for stormwater management structures such as large detention ponds may become more scarce and difficult to find. A GIS is ideally suited to performing the spatial and data analyses needed to locate these sites. For example, hydrologic analysis may show that a 10-acre (4-ha) pond needs to be located along a 3-mi (5-km) stream reach. Any site for the pond must be within 300 ft (90 m) of the stream, have a minimum of 15 ac (6 ha) of available space, and possess suitable soils and proper zoning. Further, the pond cannot impact environmental or historical sites. Through a series of data analysis steps, a GIS could identify sites that meet these criteria and rank these sites according to anticipated land cost based on the current tax assessment.

GIS for Stormwater System Inventories. GIS can also be used to store the inventory information for stormwater management systems. Even simple stormwater systems may have a variety of gutters, inlets, outlet, manholes, culverts, pipes, channels, and other facilities. The database capabilities of a GIS allow each component to have a unique database structure and setup for use in storing individual structure information.

GIS inventories can play an important role in both the hydraulic modeling and maintenance of a stormwater system. Data about structures that could be stored and used in modeling would likely include

- Types
- Locations
- Materials

Data useful in maintenance efforts may include

- Inspection reports
- Maintenance logs
- Public works complaint records

The database linking capability of the GIS is important, because many of the maintenance records may be stored in databases of several different local government departments.

GIS integration with stormwater modeling, like CAD integration, can range from no automated data exchange to complete integration within a GIS platform. However, the data analysis and database capabilities that can be performed using a GIS are more complex and, in many cases, can substantially reduce the number of data-entry steps. The data exchange between a model and a GIS is two-way: a GIS can provide existing information to a model, and data from a model of a new system can be imported into a GIS to update a stormwater system inventory. Some hydraulic analysis applications (for example, StormCAD) provide a means of automatically synchronizing data with a GIS. Figure 2.15 shows storm sewer system data stored in a GIS.

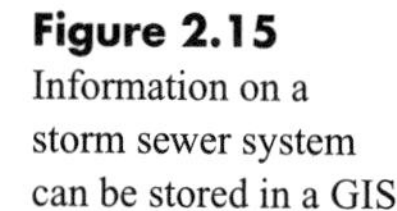

Figure 2.15
Information on a storm sewer system can be stored in a GIS

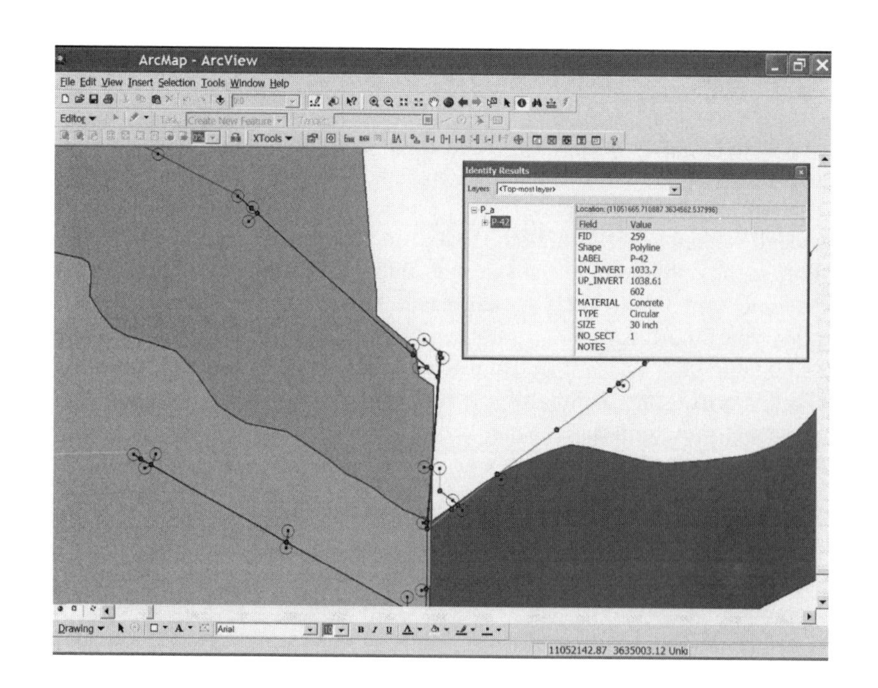

2.5 Sources of Model Data

The typical stormwater conveyance project, whether it is a new design or a study of an existing system, uses data from a variety of sources. At the beginning of a project, the engineer should consider what the data requirements for the project will be, including data types, quality, and quantity. Once these requirements have been defined, it is usually a straightforward effort to identify the data sources. When multiple sources are available for a specific data type, the choice of which source to use is based on balancing cost with data quality.

Proposed Systems

For proposed systems, most of the model input is developed by the engineer as part of the design process and will be subject to constraints dictated by project goals, site conditions, and local design standards. Common sources of data utilized in the design of new systems are described in the paragraphs that follow.

Surveys. For proposed systems, survey data are used as the base upon which the proposed layout and grading plans, and therefore stormwater conveyance designs, are developed. A survey also provides information on existing watercourses, drainage systems, utilities, and various obstacles in the project area, and aids the engineer in avoiding conflicts with these features. Of particular interest to stormwater conveyance designers is survey information on the site's outfall(s). The outfall location provides the engineer with a controlling elevation that the proposed system must tie into. For cases in which a proposed sewer system will tie directly to an existing sewer system, information on existing structure characteristics is required.

Layout and Grading Plans for the Proposed Site. The design of the stormwater conveyance system for a project is typically performed concurrently with or following the development of the layout and grading plans. These plans serve as the primary source of information for development of the preliminary drainage system layout. The proposed grading reveals the site's major catchment areas and provides the ground elevations required by the model.

Hydrologic Data Sources. Hydrologic data are available from a variety of sources. For most designs, local rainfall data and design storm information can be obtained from the local review agency. Transportation departments are another possible source of hydrologic data, as are governmental organizations such as the National Weather Service and Natural Resource Conservation Service in the United States. Often, data are readily available through an organization's web site. Further information on hydrologic data sources is presented in Chapters 4 and 5.

Local Design Standards. Local design standards provide a number of constraints in drainage design and may originate from more than one source (for instance, in addition to city or county requirements, state environmental regulations and transportation department standards may also be applicable to a single design project). Requirements frequently include

- Minimum pipe or culvert size
- Acceptable pipe or culvert materials
- Approved inlet types
- Design storm event return periods
- Maximum gutter spread and depth
- Minimum inlet capture efficiency

- Maximum distance between manholes
- Minimum freeboard requirement
- Stormwater detention requirements
- Accepted analysis techniques and applicable procedures and coefficients
- Submittal requirements for permitting

When beginning a drainage design in an unfamiliar jurisdiction, it is best to contact the local reviewing agency or agencies early in the design process to discuss the applicable requirements and obtain a copy of the local stormwater regulations/design guidelines, if available. This preparation can save considerable time later in the process.

Existing Systems

The sources of information used in evaluating the performance of an existing stormwater facility such as a sewer system, detention basin, or roadway culvert are often the same as those used in new designs, but the approach to the project will differ somewhat. On the surface, the analysis of an existing system may appear to be an easier task than designing new facilities: the geometry of the system is known, and topographic information can be taken directly from a survey rather than from a proposed design. However, the analysis of an existing system often presents other challenges, as such studies almost always stem from problems such as undersized piping and flooding in the area. The modeler is faced with the forensic problem of trying to re-create what happens in a real system, as opposed to simply sizing system components for a synthetic design event.

Surveys. Survey data are relied on much more extensively in the analysis of an existing system than in the design of a new system. For instance, a number of storm sewer manholes and inlet structures must often be field-located and their covers removed to determine pipe sizes, orientations, and invert elevations. Ponds and depressions that serve as detention basins must be carefully surveyed, and these areas are often wet or overgrown, making access difficult. Detention outlet structures must also be surveyed carefully to model the basin accurately, including planned or unplanned overflow spillways. If a culvert roadway crossing is being examined, the culvert entrance type must be noted and the roadway profile surveyed for overtopping analysis. A detailed topographic survey may also be needed to determine drainage subbasin areas.

Site Reconnaissance. When conducting a study of an existing drainage system, it is advisable for the engineer to make at least one site visit. A visit to the site provides an opportunity to

- Resolve ambiguities in the survey
- Examine and delineate contributing drainage areas that may lie outside of the survey area
- Observe the condition of the existing infrastructure

- Look for project-related problems such as soil erosion, debris in drainageways, and water quality issues
- Observe watershed and drainageway characteristics to determine model parameters such as channel roughnesses, runoff coefficients, and culvert entrance types
- Photograph the site to assist other engineers involved in the project, provide answers to questions that may arise after the site visit, and provide a record of site conditions

Original Plans and Design Calculations. A great deal of information useful in the study of an existing system can be obtained from original construction documents and design calculations. The compliance of the constructed system with the original plans must still be verified through a survey, but these plans do help the engineer to understand the original intent of the designer. They are an excellent starting point for the engineer in deciding what information the surveyor should gather. The engineer can also use the plans to identify problems in the system resulting from differences in the intent of the designer and the constructed system—problems that may be alleviated by altering the site to bring it into closer agreement with the original plan—or to identify deficiencies in the original design.

The original hydraulic and hydrologic design calculations can also be of help to the engineer. The model input data and calculations can be verified for accuracy, which can aid the discovery of why the system is not providing the anticipated level of protection. For instance, a contributing drainage area upstream of the project's conveyance system may have been developed since the project's construction and be contributing more runoff than it was previously.

As-Built Surveys/Drawings. When available, as-built drawings can be an excellent source of information when conducting a study of an existing system. These drawings may be obtained from the property owner, the engineer who did the original design, the surveyor who performed the as-built survey, or the local review agency.

If the accuracy of the as-built drawings can be verified through site reconnaissance, and depending on the level of detail required by the study, as-built drawings can greatly reduce or even eliminate the need for additional survey data. The engineer must make allowances for the level of detail of the as-built drawings and any changes made to the site since its construction when deciding how much to rely on this source of information.

Structure Inventories and Atlases. Local public works departments or stormwater utilities, and state transportation departments, often maintain mapping and other data on stormwater conveyances, especially those within the public right-of-way. Frequently, this information is stored within a GIS. These data can be useful when performing a hydraulic analysis, especially to help the engineer understand what is happening off-site and learn about the potential downstream effects of a project.

Existing Flood Studies. Existing hydraulic studies are available for many streams, especially in developed areas. In the United States, these studies may have been performed by the U.S. Army Corps of Engineers or, increasingly, by private consulting engineers under contract with government agencies or developers. In the United States, these studies are usually submitted to the Federal Emergency Management Agency (FEMA) and, if approved, become part of regulatory flood maps.

For projects involving large, previously studied conveyances, detailed reports and hydrologic and hydraulic calculations can often be obtained through the contracting or reviewing agency. The calculations can be used to understand the performance of conveyances in and around the project area, and they can be modified to reflect changes to these conveyances resulting from the proposed project.

When using existing flood study data, it is important to keep in mind the study's level of accuracy. Many of these studies are out-of-date and were subject to tight budgetary constraints. Thus, the level of detail that often characterizes these large-scale studies is typically insufficient when examined on the smaller scale of many stormwater conveyance projects. For instance, flows may have been computed using a regression formula having a high degree of uncertainty associated with it, or stream cross-sections may be very far apart.

Also important to be aware of when referring to an older study are changes in the state of engineering practice since the study was performed. More accurate modeling techniques may have become available since that time. Further, the study area may have undergone significant changes since the study was performed, such as alterations to land use, stream cross-sections, and roadway crossings.

Floodplain Modeling Using HEC-RAS (Dyhouse et al., 2003) provides detailed information on flood studies.

2.6 CHAPTER SUMMARY

Stormwater drainage systems involve many types of structures; thus, a number of mathematical models (both hydraulic and hydrologic) are necessary to describe these systems adequately. Facility types used in stormwater management include open channels, culverts, storm sewers, pump stations, and detention basins, and various software packages are available for their analysis and design.

Hydraulic software that allows for integration with CAD, GIS, and databases already used by an organization can greatly facilitate the model-building process, the creation of construction documents, and the maintenance of records on stormwater facilities.

Data that can be used in modeling stormwater systems come from a variety of sources. The sources chosen depend somewhat on whether the project consists of the study of an existing system or the design of a new system. For proposed systems, data come primarily from surveys, site plans, and local design standards. Surveys take on increased importance when modeling existing systems, as do site reconnaissance and record drawings. Original plans and design calculations and structure inventories may also be used when available.

REFERENCES

American Society of Civil Engineers (ASCE). 1992. *Design and Construction of Urban Storm Water Management Systems,* New York: ASCE.

Dyhouse, Gary, J. A.Benn, David Ford Consulting, J. Hatchett, and H. Rhee. 2003. *Floodplain Modeling Using HEC-RAS.* Waterbury, Connecticut: Haestad Methods.

Engineers Joint Contract Documents Committee (EJCDC). 2002. *Standard General Conditions of the Construction Contract,* Reston, Virginia: ASCE.

Haestad Methods. 2003a. *PondPack: Detention Pond and Watershed Modeling Software.* Waterbury, Connecticut: Haestad Methods.

Haestad Methods. 2003b. *StormCAD: Storm Sewer Design and Analysis Software.* Waterbury, Connecticut: Haestad Methods.

Normann, J.M., R.J. Houghtalen, and W.J. Johnston. 2001. *Hydraulic Design of Highway Culverts.* 2d ed. Washington, D.C.: Federal Highway Administration (FHWA).

A hydraulic jump is formed in a laboratory tank. For a short jump, momentum is the same upstream (left) and downstream (right). Energy is lower downsteam because of dissipation by turbulence.

CHAPTER

3

Fundamental Laws and Units

Scientists and engineers have developed equations to describe hydrologic processes and the hydraulic properties of stormwater conveyances. The engineering calculations used in the design and evaluation of stormwater systems are, to a large extent, applications of the fundamental physical laws of conservation of mass, energy, and momentum. The first three sections of this chapter provide reviews of these fundamental laws. The fourth section addresses some of the common units of measurement employed in hydrologic and hydraulic engineering.

3.1 CONSERVATION OF MASS

The familiar principle of conservation of mass simply states that matter is neither created nor destroyed. The mass entering a system is equal to the mass leaving that system, plus or minus the accumulation of mass (that is, storage) within the system. Without doubt, this principle is the single most important concept that must be applied in hydrologic and hydraulic engineering. In the engineering analyses that must be performed, the properties of flow change both through space and over time. For example, the amount of fluid present in a given channel reach may change over time; thus, accounting for the accumulation (or loss) of fluid during any time interval is essential.

As a practical matter, the application of the principle of conservation of mass is simply an accounting procedure. Consider, for example, a bank account and assume that the account contains $100 at the beginning of the month. Suppose also that deposits made during the month amount to $50, and withdrawals amount to $70. Clearly, the account balance at the end of the month is $80 (= $100 + $50 – $70). If the ending balance had been specified instead of the monthly withdrawals, then a similar calculation could be applied easily to determine the withdrawals. In any case, the change in the account balance is equal to the difference between deposits and withdrawals.

Now, consider an *incompressible fluid* (that is, a fluid having a constant density) that flows through a fixed region (a *control volume*) in space and denote the volume of fluid, or *storage*, in that region at time t by the quantity $S(t)$. The fixed region might correspond to a lake or reservoir, to a water tank, or to a length of river or stream. If $I(t)$ represents a volumetric flow rate into the region, then a volume $I(t)\Delta t$ of additional fluid flows into the fixed region during the time interval of duration Δt. Representing the volumetric flow rate out of the system as $Q(t)$, a volume of fluid equal to $Q(t)\Delta t$ flows out of the region during the time interval. [Note that $I(t)$ and $Q(t)$ may be functions of time.]

According to the principle of conservation of mass, at the end of the time interval (that is, at time $t + \Delta t$), the storage $S(t + \Delta t)$ of fluid in the fixed region will be equal to what was there initially, plus the additional fluid that entered during that time, less the amount of fluid that left during that time. Expressed mathematically (with example units),

$$S(t+\Delta t) = S(t) + I(t)\Delta t - Q(t)\Delta t \tag{3.1}$$

where t = time (s)
$S(t + \Delta t)$ = storage at the end of the time interval (ft^3, m^3)
$S(t)$ = storage at the beginning of the time interval (ft^3, m^3)
$I(t)$ = volumetric inflow rate at time t (cfs, m^3/s)
$Q(t)$ = volumetric outflow rate at time t (cfs, m^3/s)

A volumetric flow rate, typically denoted by Q, is referred to as a *discharge* in hydrologic and hydraulic engineering.

A rearrangement of Equation 3.1 yields

$$\frac{S(t+\Delta t) - S(t)}{\Delta t} = I(t) - Q(t) \tag{3.2}$$

As Δt approaches zero, this equation becomes

$$\frac{dS}{dt} = I(t) - Q(t) \tag{3.3}$$

Equation 3.3 is the differential form of the mathematical expression for conservation of mass. An integral form may be obtained by multiplying both sides of the expression by dt and integrating:

$$\int_{S1}^{S2} dS = S_2 - S_1 = \int_{t1}^{t2} I(t)dt - \int_{t1}^{t2} Q(t)dt \tag{3.4}$$

If a flow is *steady*—that is, if the flow characteristics do not change with time—then the time derivative in Equation 3.3 is zero, and conservation of mass may be expressed by simply stating that the inflow and outflow discharges to and from a control volume are equal. In other words, $I(t) = Q(t)$.

The *volumetric flux* of a fluid is defined as the volume of fluid that passes a unit cross-sectional area per unit time. It can be expressed as

$$Volumetric\ Flux = \frac{S}{At} \tag{3.5}$$

where S = fluid volume (ft^3, m^3)
A = cross-sectional area (ft^2, m^2)
t = time (s)

Because Q is a fluid volume per unit time, the volumetric flux can be restated as

$$Volumetric\ Flux = \frac{Q}{A} \tag{3.6}$$

where Q = flow (cfs, m^3/s)

It can be seen from this definition that volumetric flux has units of length per unit time. Noting that velocity has the same units, one can express the *average velocity* of flow as

$$V = \frac{Q}{A} \tag{3.7}$$

where V = average velocity (ft/s, m/s)

Comparing Equations 3.6 and 3.7, average flow velocity is the same as volumetric flux.

In steady flow, the discharges into and out of a control volume must be equal, and Equation 3.7 can be used to write

$$(AV)_{inlow} = (AV)_{outflow} \tag{3.8}$$

The reader should recognize that the equations presented in this section have been cast in terms of fluid volumes and volumetric flow rates even though the discussion concerns conservation of mass. The use of volumes and volumetric flow rates is valid as long as the fluid in question is *incompressible* (has a constant density). This assumption is valid for nearly all civil and environmental engineering applications dealing with water. For instances in which this assumption is not valid, the volumes and volumetric flow rates in the above equations should be replaced with *fluid masses* and *mass flow rates*. A fluid mass is found by multiplying the fluid's volume by its density, and a mass flow rate is found similarly by multiplying the volumetric flow rate by the density.

Example 3.1 – Determining Discharge and Velocity Based on Conservation of Mass. A storm sewer is being designed for a location where the slope of the sewer must decrease because of local topography. A manhole will be positioned at the location where the pipe slope changes. Upstream of the manhole, the topography is relatively steep, and the pipe diameter is 600 mm. Downstream of the manhole, where the topography is flatter, a 750-mm diameter pipe is required to convey the flow. No additional flow enters the sewer at the location of the manhole. The pipes in the sewer are flowing full both upstream and downstream of the manhole, and the discharge in the upstream pipe is 0.5 m^3/s. The flow is steady. Determine the discharge and velocity in the downstream pipe.

Solution: Consider the manhole as a control volume into which flow enters from the upstream pipe and from which flow departs through the downstream pipe. Because the flow is steady, the inflow and outflow discharges must be equal, and thus the discharge in the downstream pipe is 0.5 m^3/s. The velocity in the downstream pipe can be found using Equation 3.7. The cross-sectional area of the 750-mm diameter downstream pipe is $A = 0.442$ m^2, and thus the velocity is

$$V = Q/A = 0.5/0.442 = 1.13 \text{ m/s}$$

Example 3.2 – Applying Conservation of Mass to a Detention Pond. A discharge hydrograph is a graph showing how the discharge in a pipe or channel changes over time. The discharge is usually shown on the vertical axis of the graph, and time is shown on the horizontal axis. For any time interval on the horizontal axis, the area under the curve shown on the graph represents the volume of the flow over that time interval.

Figure E3.2.1 shows the inflow and outflow hydrographs for a detention pond. The inflow hydrograph is represented by the solid line, and the outflow hydrograph is shown as a dashed line. Assuming that the pond is initially empty, the outflow hydrograph is triangular in shape, and the pond is again empty at time t_{end}, determine the following:

a) the total volume of water passing through the pond
b) the time t_{end}
c) the time at which the storage in the pond is at a maximum
d) the maximum storage in the pond

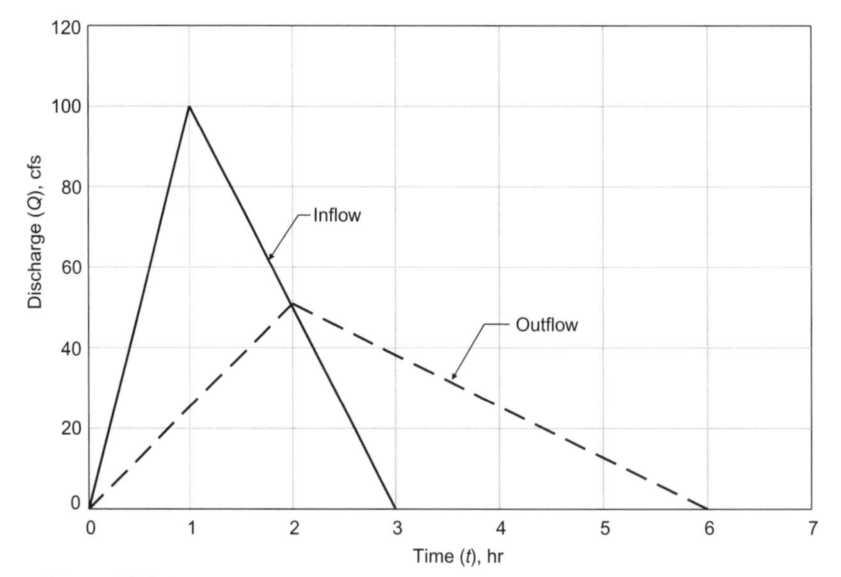

Figure E3.2.1 Inflow and outflow hydrographs for a detention pond

Solution to (a): The total volume of water passing through the pond is equal to the total volume of water entering the pond because it is empty at time zero and again at time t_{end}. This volume of water is equal to the area under the inflow or outflow hydrograph. In this case, the volume can be computed using the formula for the area of a triangle as

$$\forall_{inflow} = 0.5(3 \text{ hr})(100 \text{ cfs})(3{,}600 \text{ s/hr}) = 540{,}000 \text{ ft}^3$$

Solution to (b): The time t_{end} at which the pond is again empty may be found using Equation 3.4 and noting that $S_1 = S_2 = 0$ when one takes $t_1 = 0$ and $t_2 = t_{end}$. Hence, the flow volumes represented by the inflow and outflow hydrographs must be equal. The inflow volume was found in the solution to part (a) to be 540,000 ft^3; thus,

$$\forall_{outflow} = \forall_{inflow} = 540{,}000 \text{ ft}^3 = 0.5(50 \text{ cfs})(t_{end})$$

Solving,

$$t_{end} = 21{,}600 \text{ s} = 6 \text{ hr}$$

Solution to (c): Inspection of Figure E3.2.1 shows that during the time interval from 0 to 2 hr, the discharge into the pond is greater than the discharge out of it. Thus, the storage in the pond must be increasing during that time interval. After $t = 2$ hr, the outflow discharge is greater than the inflow discharge, so the storage in the pond must be decreasing. Therefore, the time at which storage in the pond is a maximum is $t = 2$ hr.

Solution to (d): Apply Equation 3.4 with $t_1 = 0$ and $t_2 = 2$ hr. Because the pond is empty at time zero, $S_1 = 0$. The maximum storage volume S_2 at time t_2 is equal to the difference between the volumes represented by the inflow and outflow hydrographs over the time interval from 0 to 2 hr:

$$S_2 = [0.5(1 \text{ hr})(100 \text{ cfs}) + 0.5(1 \text{ hr})(100 \text{ ftcfs} + 50 \text{ cfs}) - 0.5(2 \text{ hr})(50 \text{ cfs})](3600 \text{ s/hr})$$

$$S_2 = 270{,}000 \text{ ft}^3$$

3.2 Conservation of Energy

In the most formal sense, a discussion of conservation of energy for a fluid would begin with the First Law of Thermodynamics. That law states that the rate of change of stored energy in a fluid system is equal to the rate at which heat energy is added to the system, minus the rate at which the fluid system does work on its surroundings. The stored energy is composed of *kinetic energy* due to the motion of the fluid, *potential energy* due to its position relative to an arbitrary datum plane, and *internal energy.*

In the vast majority of civil and environmental engineering applications, flow can be considered to be steady and incompressible, and it is generally sufficient to apply the principle of conservation of energy in a much simpler way using what is commonly known as the *energy equation*. The expression most commonly applied expresses energy on a unit weight basis (that is, as energy per unit weight of fluid) and may be stated as

$$\frac{p_1}{\gamma} + Z_1 + \frac{\alpha_1 V_1^2}{2g} = \frac{p_2}{\gamma} + Z_2 + \frac{\alpha_2 V_2^2}{2g} + h_L - h_P + h_T \tag{3.9}$$

where p = fluid pressure (lb/ft^2, Pa)
γ = specific weight of fluid (lb/ft^3, N/m^3)
Z = elevation above an arbitrary datum plane (ft, m)
α = velocity distribution coefficient
V = fluid velocity, averaged over a cross section (ft/s, m/s)
g = acceleration of gravity (ft/s^2, m/s^2)
h_L = energy loss between cross sections 1 and 2 (ft, m)
h_P = fluid energy supplied by a pump between cross sections 1 and 2 (ft, m)
h_T = energy lost to a turbine between cross sections 1 and 2 (ft, m)

The first three terms on each side of this equation represent, respectively, the internal energy due to fluid pressure, potential energy, and kinetic energy. When the dimension of each term is given in units of length as in Equation 3.9 (resulting from division of energy units by fluid weight units), the three terms are generally called the *pressure head*, the *elevation head*, and the *velocity head*. The terms on the left side of the equation with the subscript 1 refer to an upstream cross section of the fluid, and those on the right side with the subscript 2 refer to a downstream cross section.

Appendix A contains tabulations of the physical properties of water at standard atmospheric pressure. The specific weight and other properties of water may be determined from those tables.

Strictly speaking, the energy equation applies only to a particular streamline within a flow. Because the velocities along individual streamlines are generally different due to the effects of pipe walls or channel sides, the velocity head terms in Equation 3.9 should be corrected using the velocity distribution coefficient (α) if one chooses to use an average cross-sectional velocity. Chapter 6 discusses this coefficient in more detail. Chapter 6 also provides more details related to the energy loss, pump, and turbine terms of Equation 3.9.

Because the dimension of each term in Equation 3.9 has units of length, qualitative aspects of the energy equation can be shown graphically. Figure 3.1 shows the profile of a pipe in which a venturi meter has been installed to measure flow. The pipe diameter downstream of the venturi is smaller than the pipe diameter upstream of the venturi. An arbitrary datum is shown below the pipe profile.

In Figure 3.1, the elevation head (Z) represents the vertical distance from the datum to the pipe centerline. The *hydraulic grade line (HGL)* represents the height to which a column of water would rise in a standpipe placed anywhere along the length of the pipe. The height of the HGL is sometimes called the *piezometric head*. The vertical distance from the pipe centerline to the HGL is the pressure head, p/γ, and the distance from the HGL to the *energy grade line (EGL)* is the velocity head, $V^2/2g$. The vertical distance from the datum to the EGL is representative of the sum of the first three terms on each side of Equation 3.9 and is called the *total head*. Flow always occurs in a direction of decreasing total head (not necessarily decreasing pressure), and hence is from left to right in the figure.

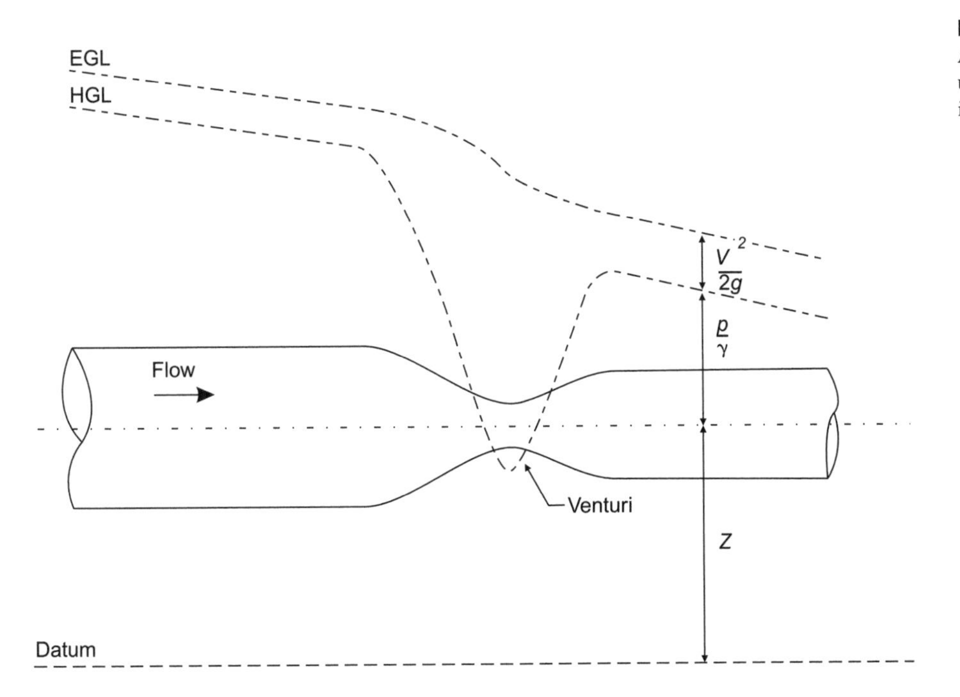

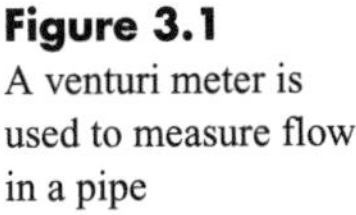

Figure 3.1
A venturi meter is used to measure flow in a pipe

According to conservation of mass, the velocity must be greater in the pipe downstream of the venturi than in the upstream pipe; thus, the distance between the HGL and EGL is greater for the downstream pipe. At the throat of the venturi itself, where the flow velocity is the highest, the distance between the EGL and HGL is the greatest. In the regions upstream and downstream of the venturi where the HGL is above the pipe centerline, the fluid pressure (gauge pressure) in the pipe is positive. In this example, the HGL falls below the pipe centerline at the venturi, and the fluid gage pressure is negative. Note that although the fluid pressure just downstream of the venturi is greater than that in the throat of the venturi, flow is still from left to right.

The EGL in the figure has a nonzero slope because energy is lost due to friction and turbulent eddies as fluid moves along the length of the pipe. In other words, the total head changes along the pipe and decreases in a downstream direction. In Figure 3.1, because neither a pump nor turbine is present, the difference in the elevation of the EGL between any two locations (cross sections) along the pipe is representative of the head loss h_L between those two cross sections. The loss rate of total energy in a fluid flow increases with the velocity of the flow. Thus, in the figure, the magnitude of the slope of the EGL is larger for the downstream pipe than for the upstream pipe.

Example 3.3 – Determining Flow Direction from the Energy Equation. Two cross sections along a constant-diameter pipe are denoted by A and B. The pipe centerline elevation at cross section A is 30 m, and the fluid pressure in the pipe at that location is 210 kPa. At cross section B, the pipe centerline elevation is 21 m, and the fluid pressure is 240 kPa. Determine the direction of flow in the pipe. The specific weight of water is 9.81 kN/m^3.

Solution: The total energy head at each of the two cross sections is the sum of the elevation head (that is, the pipe centerline elevation), the pressure head, and the velocity head. Flow occurs in the direction of decreasing total energy head.

From Equation 3.9, the total energy head H_A at cross section A is

$$H_A = \frac{p_A}{\gamma} + Z_A + \frac{\alpha V_A^2}{2g} = \frac{210}{9.81} + 30 + \frac{\alpha V_A^2}{2g} = 51.41 + \frac{\alpha V_A^2}{2g}$$

At cross section B, the total energy head is

$$H_B = \frac{240}{9.81} + 21 + \frac{\alpha V_B^2}{2g} = 45.46 + \frac{\alpha V_B^2}{2g}$$

Because by continuity, the discharge and velocity must be the same at each of the two cross sections, the velocity head terms on either side of Equation 3.9 are equal and cancel out. Thus, in this case, flow is in a direction of decreasing piezometric head. Because the piezometric head of 51.41 m at cross section A is higher than the piezometric head of 45.46 m at cross section B, flow is from A to B. Note that this is true even though the fluid pressure is higher at cross section B.

3.3 CONSERVATION OF MOMENTUM

The principle of conservation of momentum states that the summation of the external forces acting on a system is equal to the time rate of change of momentum for the system. The physical principle of conservation of momentum is generally more difficult to apply in practice than either conservation of mass or conservation of energy. The added complexity is the consequence of momentum being a vector-valued quantity (that is, it has both magnitude and direction), whereas mass and energy are scalars (represented by magnitude only). Momentum equations must therefore be written for each coordinate direction separately.

Friends often dropped by and subjected Sir Isaac to a little good-natured ridicule.

In its most general form, conservation of momentum is expressed as

$$\sum F_x = \frac{\partial}{\partial t} \int_{cv} \rho V_x dS + \sum (M_{out})_x - \sum (M_{in})_x \qquad (3.10)$$

where F_x = force acting on the water in a control volume in the x-direction (lb, N)
t = time (s)
cv = control volume (ft^3, m^3)
ρ = fluid density ($slugs/ft^3$, kg/m^3)
V_x = x-component of the velocity of the fluid in the control volume (ft/s, m/s)
S = fluid volume (ft^3, m^3)
$(M_{out})_x$ = x-component of momentum outflow rate from the control volume (lb, N)
$(M_{in})_x$ = x-component of momentum inflow rate into the control volume (lb, N)

This expression, which is written for the coordinate direction x, can be written for other coordinate directions as needed. It states that the sum of the external forces acting on the water in a control volume is equal to the time rate of change of momentum within the control volume, plus the net momentum flow rate (in the x-direction) from the control volume. In this expression, the sums of $\mathbf{M}_{out}$ and $\mathbf{M}_{in}$ account for the possibility of more than one inflow and/or outflow pathway to/from the control volume. A *momentum flow rate*, in the x-direction, can be written as

$$M_x = \beta \rho Q V_x \qquad (3.11)$$

where M_x = momentum flow rate in the x-direction (lb, N)
β = velocity distribution coefficient
Q = discharge (ft^3/s, m^3/s)

The numerical value of β is close to 1 for turbulent flow. Sections 6.1 and 6.4 provide an explanation of turbulent flow and more detail on β.

If a fluid flow is steady, such that its characteristics do not change with time, Equation 3.10 reduces to

$$\sum F_x = \sum (M_{out})_x - \sum (M_{in})_x \qquad (3.12)$$

Further, if there is only a single inflow stream (that is, a single inflow pipe) and a single outflow stream from the control volume, Equation 3.12 can be rewritten as

$$\sum F_x = \rho Q \Delta (\beta V)_x \qquad (3.13)$$

Equation 3.13 states that the vector sum of the x-components of the external forces acting on the fluid within a fixed control volume (including the pressure forces) is equal to the fluid density times the discharge times the difference between the x-components of the outgoing and incoming velocity vectors from and to the control volume. As in Equation 3.9, the velocity terms should be modified using a velocity distribution coefficient if one chooses to use the average velocity at a cross section.

Example 3.4 – Determining the Force Exerted on a Bend from the Momentum Equation. A pipe with flow in a northerly direction has a bend in which flow is turned to 45 degrees east of north (see Figure E3.4.1). The discharge is 25 cfs, the pipe diameter is 24 in., and the fluid pressures upstream (cross section 1) and downstream (cross section 2) of the bend are 25 psi and 22 psi, respectively. Assuming that the pipe is flowing full and $\beta = 1$, determine the force that the water exerts on the pipe bend.

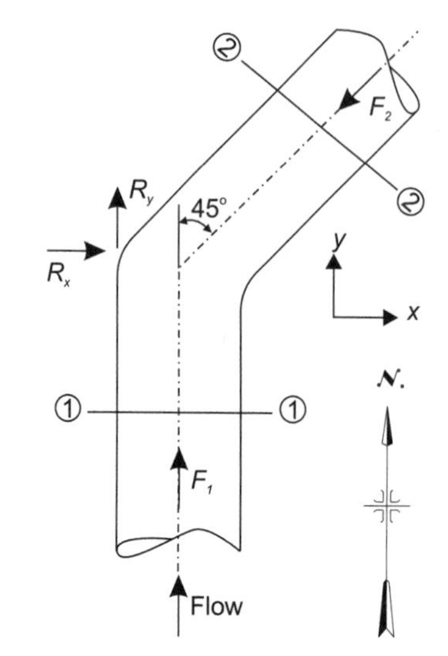

Figure E3.4.1 Forces on a pipe bend

Solution: Take a control volume as the region of the pipe between cross sections 1 and 2 shown in the figure. Let the x-direction be taken as due east, and the y-direction as due north. External forces exerted on the water in the control volume in the x- and y-directions consist of the forces F_1 and F_2 due to the fluid pressures upstream and downstream of the bend, and the forces R_x and R_y exerted on the flow by the pipe. The fluid pressure forces can be found by multiplying the fluid pressures by the pipe cross-sectional area:

$$F_1 = p_1 A = (25 \text{ psi})(3.14)(12 \text{ in})^2 = 11{,}300 \text{ lb}$$

$$F_2 = p_2 A = 9{,}950 \text{ lb}$$

The velocity in the pipe at both cross-sections is

$$V = Q/A = (25 \text{ cfs})/(3.14 \text{ x } 1 \text{ ft}^2) = 7.96 \text{ ft/s}$$

An application of Equation 3.13 in the x-direction leads to

$$\Sigma F_x = R_x - 9{,}950 \sin 45° = 1.94(25)(7.96 \sin 45° - 0)$$

A similar application in the y-direction leads to

$$\Sigma F_y = R_y + 11{,}300 - 9{,}950 \cos 45° = 1.94(25)(7.96 \cos 45° - 7.96)$$

Solving these expressions yields $R_x = 7{,}310$ lb and $R_y = -4{,}380$ lb. The negative sign on R_y means that its direction is actually to the south rather than the assumed northerly direction (see figure). The net force acting on the water has a magnitude of

$$R = \sqrt{7{,}310^2 + 4{,}380^2} = 8{,}520 \text{ lb}$$

The direction in which this net force acts is

$$\theta = \arctan(4{,}380/7{,}310) = 30.9° \text{ to the south of due east}$$

The forces determined in this example are those acting on the water. The forces exerted on the pipe by the water are equal to these, but in the opposite directions. Thrust blocks or special pipe jointing methods may be required in some instances to resist these forces.

3.4 COMMON UNITS OF MEASUREMENT

Hydrologic and hydraulic engineering is a field marked by many different units of measurement; therefore, one must frequently convert from one set of units to another. All of the fundamental dimensions of mass (or force), length, and time are routinely used. For example, rainfall and rainfall losses are often expressed in units of inches or millimeters of rainfall per hour (in/hr, mm/hr), and discharges in pipes and open channels are often expressed in units of cubic feet or meters per second (cfs, m^3/s). Drainage basin areas might be expressed in units of acres (ac), hectares (ha), square miles (mi^2), or square kilometers (km^2); and volumes of water might be expressed in cubic feet or meters (ft^3, m^3), or in units such as acre-feet (ac-ft). Table 3.1, "Common units in stormwater applications," on page 61 summarizes some of the more frequently used units in stormwater applications, and Appendix B presents conversion factors for changing from one set of units to another.

Table 3.1 Common units in stormwater applications

Measurement	U.S. Customary Units	SI Units
Length	foot (ft), inch (in.), mile (mi)	meter (m), millimeter (mm), kilometer (km)
Mass	slug, pound-mass (lbm)	kilogram (kg), gram (g)
Time	hour (hr), minute (min), second (s)	hour (hr), minute (min), second (s)
Temperature	degree Fahrenheit (°F)	degree Celcius (°C)
Force	slug, pound (lb)	Newton (N), kilonewton (kN)
Area	sq. foot (ft^2), acre (ac), square mile (mi^2)	sq. meter (m^2), hectare (ha), sq. kilometer (km^2)
Volume	cubic foot (ft^3), gallon (gal), acre-foot (ac-ft)	cubic meter (m^3), liter (l or L)
Pressure/Stress	pounds per sq. ft (lb/ft^2), pounds per sq. inch (psi)	kilonewtons per sq. m (kN/m^2), grams per sq. meter (g/m^2)
Flow	cubic feet per second (cfs or ft^3/s), gallons per minute (gpm)	cubic meters per second (m^3/s), liters per second (L/s)
Concentration	milligrams per liter (mg/l)	

A common situation encountered in hydrologic and hydraulic engineering is the need to determine the units associated with the area under a graphed curve (that is, with the integral of some function). This problem was encountered in Example 3.2. The solution is accomplished by simply multiplying the units of the two quantities represented on each axis of the graph. If, for example, a hydrograph shows discharge on the vertical axis in ft^3/s and time on the horizontal axis in seconds, an area under the plotted hydrograph represents cubic feet of water.

Example 3.5 – Determining the Equivalent Depth of Runoff. The detention pond of Example 3.2 is located at the outlet of a 140-ac drainage basin. Given the information in Example 3.2, determine the equivalent depth of runoff from the drainage basin.

Solution: From Example 3.2, the volume of inflow to the detention pond is 540,000 ft^3, or 12.4 ac-ft. Dividing this volume of water by the area of the drainage basin yields the equivalent depth of runoff as

$$d = 12.4 \text{ ac-ft}/140 \text{ ac} = 0.886 \text{ ft} = 1.06 \text{ in.}$$

This 1.06 in. of runoff may be viewed as the depth of effective precipitation causing the runoff to occur. As will be described in Chapter 4, effective precipitation is what remains of total precipitation after surface infiltration and other rainfall abstractions have been accounted for.

Example 3.6 – Determining the Volume of Water from a Surface Area versus Elevation Function. The area of the water surface of a detention pond, graphed as a function of the water surface elevation, can be used to determine the volume of water in the pond. Figure E3.6.1 shows water surface area as a function of water surface elevation for a pond with a bottom elevation of 196.0 ft. The surface areas have been determined by digitizing a contour map of the pond at 2-ft contour intervals (for example, the area shown for a water surface elevation of 200.0 ft has been found by determining the area represented within the 200-ft contour on the map. Using the values shown in the figure, determine the volume of water in the pond when the water surface elevation is 200.0 ft.

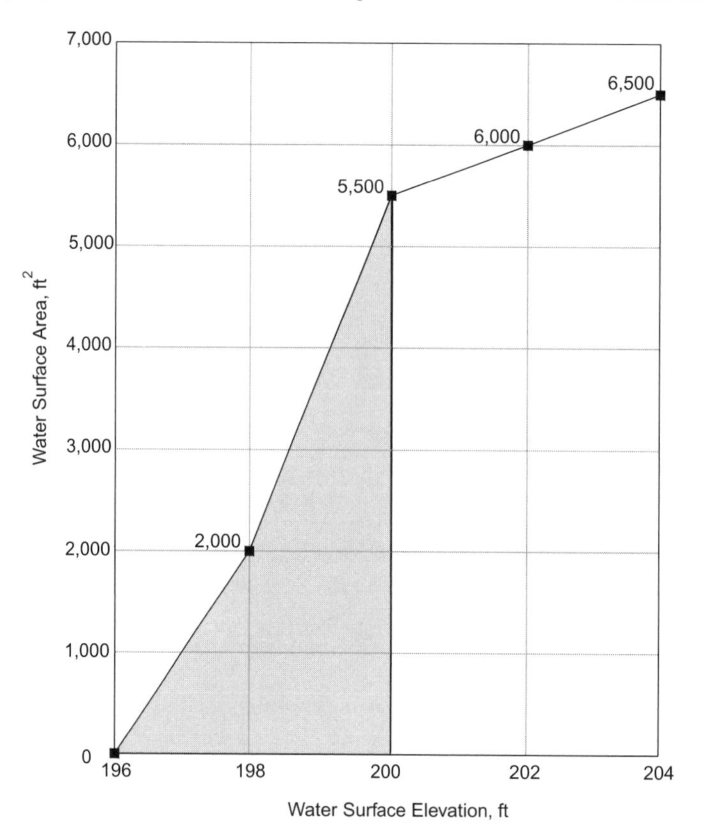

Figure E3.6.1 Water surface area versus water surface elevation

Solution: An incremental volume dS of water in the pond can be expressed as

$$dS = A(h)dh$$

where $A(h)$ = water surface area at elevation h
dh = incremental change in water surface elevation

Integration of both sides of this expression shows that the volume of water is simply the area under a curve of surface area expressed in terms of elevation. Using the trapezoidal rule to approximate the area under the curve shown in Figure E3.6.1, the volume of water in the pond when the water surface elevation is 200.0 ft (represented by the shaded area in Figure E3.6.1) is found as

$$S = 0.5(2{,}000\ \text{ft}^2)(2\ \text{ft}) + 0.5(2{,}000\ \text{ft}^2 + 5{,}500\ \text{ft}^2)(2\ \text{ft}) = 9{,}500\ \text{ft}^3 = 0.22\ \text{ac-ft}$$

3.5 CHAPTER SUMMARY

The calculations used in analyzing and designing stormwater conveyance systems are typically derived from one or more of the fundamental physical laws of conservation of mass, momentum, and energy.

As it applies to hydraulics, conservation of mass basically states that, for an incompressible fluid, the amount of fluid in a control volume at the end of a time interval is equal to the amount of fluid in the control volume at the beginning of the time interval, plus the amount of fluid entering the control volume during the time interval, and minus the amount leaving the control volume during the time interval.

If flow can be considered to be steady and incompressible and changes in thermal energy negligible, the energy at a point is taken as the sum of the pressure head (internal fluid energy), elevation head (potential energy), and velocity head (kinetic energy). Conservation of energy between two points can be applied in the following form:

$$\frac{p_1}{\gamma} + Z_1 + \frac{\alpha_1 V_1^2}{2g} = \frac{p_2}{\gamma} + Z_2 + \frac{\alpha_2 V_2^2}{2g} + h_L - h_P + h_T \qquad (3.14)$$

According to the principle of conservation of momentum, the summation of the external forces acting on a system is equal to the time rate of change of momentum for the system. It is a vector-valued quantity that must be evaluated separately for each direction. For the x-direction, momentum flow rate is given by

$$M_x = \beta \rho Q V_x \qquad (3.15)$$

A variety of units of measurement are common in the field of hydrologic and hydraulic engineering. Rainfall and rainfall losses are often expressed in units of inches or centimeters of rainfall per hour (in/hr, cm/hr), and discharges are often expressed in units of cubic feet or meters per second (cfs, m^3/s). Drainage basin areas are frequently in acres (ac), hectares (ha), square miles (mi^2), or square kilometers (km^2), and volumes of water are often given in cubic feet (ft^3), cubic meters (m^3), or acre-ft (ac-ft).

REFERENCES

Abbott, M. B. 1979. *Computational Hydraulics: Elements of the Theory of Free Surface Flows.* London : Pitman.

Chow, V. T., D. L. Maidment, and L.W. Mays. 1988. *Applied Hydrology.* New York: McGraw-Hill

Prasuhn, A. L. 1980. *Fundamentals of Fluid Mechanics.* Englewood Cliffs, N. J.: Prentice-Hall.

Shames, I. H. 1992. *Mechanics of Fluids.* 3d ed. New York: McGraw-Hill.

Simon, A. L. 1981. *Practical Hydraulics.* 2d ed. New York: John Wiley & Sons.

White, F. M. 1999. *Fluid Mechanics.* 4th ed. New York: McGraw-Hill.

PROBLEMS

3.1 A pipe discharges from a pond at a rate of 40,000 gpm. What is the flow in cfs?

3.2 During an emergency release, the water level in a 12,000-ac reservoir dropped by 1 inch in 6 hours. What was the discharge rate in cfs?

3.3 The last flow gaging station on the Susquehanna River before it enters the Chesapeake Bay is at the Conowingo Dam. The average flow recorded at this station is 2,500 m^3/s. What is the total amount of water (in m^3) that flows into the Chesapeake Bay from the Susquehanna River in an average year?

3.4 A square detention pond has a 100 × 100-ft, level bottom and 3H:1V side slopes. What volume of water can the pond hold if the depth is 5 ft?

3.5 The discharge in a stream is 125 cfs. Downstream of the monitoring point, a 30-in. storm sewer is discharging into the stream. The pipe is flowing full and the average velocity is 5 ft/s. What is the total discharge in the stream downstream of the storm sewer discharge?

3.6 The shape of a runoff hydrograph from a subdivision for a particular storm event can be approximated as a triangle. The duration of the event was 18 hours. The peak flow was 3.4 m^3/s and it occurred 4 hours after the start of the runoff hydrograph. What was the total volume of runoff produced?

3.7 A rainfall event is quantified as follows:

- 30 minutes at 0.25 in/hr
- 60 minutes at 0.50 in/hr
- 45 minutes at 0.15 in/hr

The precipitation occurs over a 500-ac watershed. What are the total volume of rainfall (in ft^3) and the equivalent rainfall depth (in in.)?

3.8 Water is flowing in an open channel at a depth of 4 ft and a velocity of 7.5 ft/s. It then flows down a chute into another open channel where the depth is 2 ft and the velocity is 30 ft/s. Assuming frictionless flow and $\alpha = 1$, determine the difference in elevation between the channels.

3.9 During a flood, runoff flows into a large reservoir where the velocity is 0. At the reservoir's dam, the water overtops a rectangular spillway at a depth of 0.6 m and a velocity of 2.4 m/s. The water then flows down a long spillway and makes a smooth transition to a discharge channel. The bottom of the discharge channel is 13.7 m below the top of the spillway and the depth of flow is 3.7 m. The flow is assumed to be frictionless.

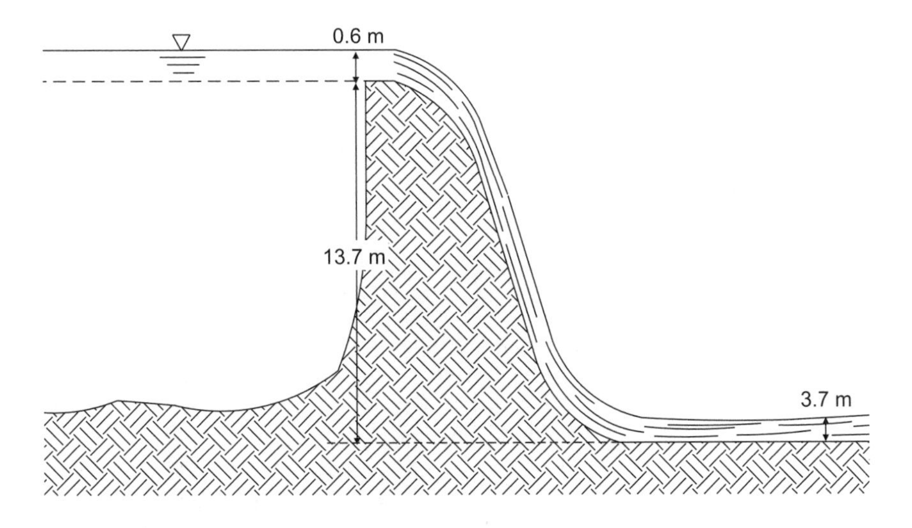

a) Calculate the velocity head and total head at locations (1) upstream of the spillway (in the reservoir), (2) at the crest of the spillway, and (3) in the discharge channel.

b) Plot the hydraulic and energy grade lines over the path of the water.

3.10 A 36-in. diameter pipe has a 30° horizontal bend and a flow of 100 cfs. The inlet pressure is 35 psi and the outlet pressure is 33 psi. Calculate the direction and magnitude of the forces required to stabilize the bend.

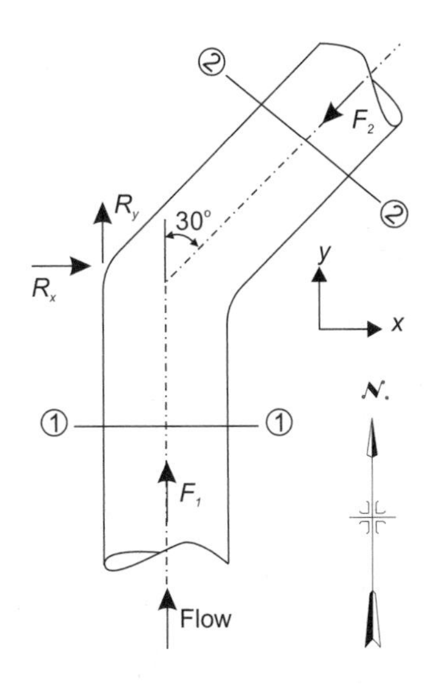

Some rainfall is intercepted by plants, thereby reducing the fraction of rainfall that becomes runoff.

CHAPTER

4

Modeling Rainfall

Rainfall data are a fundamental building block for determining the amount of stormwater runoff generated during a particular event. This chapter provides an introduction to surface water hydrology concepts needed for the design and analysis of stormwater conveyance systems, with emphasis on rainfall processes and their representations. Chapter 5 presents methods for modeling runoff from rainfall data.

A simple but practical way of viewing the subject of hydrology is as a means of estimating the *loads* (that is, the discharges) for which stormwater conveyance systems must be designed. For example, when a structural engineer designs a beam, he or she must know or estimate the loads that the beam will be required to support. In a similar vein, the design of a storm sewer system or highway culvert requires an estimation of the stormwater loads that the system will be required to convey. Hydrology is the science by which these estimates can be made.

In general, hydrologic prediction methods can be classified as being either *event-based* or *continuous*. Event-based methods are concerned with the prediction of discharge and/or volume of runoff resulting from a single rainfall event. Continuous methods simulate runoff on an uninterrupted basis and include both wet- and dry-weather periods. Event-based methods are the traditional methods by which stormwater conveyance systems have been designed. However, with increasing attention being paid to water quality issues, application of continuous simulation methods is becoming more common. This book covers event-based runoff prediction methods.

Section 4.1 of this chapter reviews the various processes that make up the hydrologic cycle, and Sections 4.2 and 4.3 describe the various type and characteristics of precipitation. Methods of obtaining rainfall data, including rainfall measurement and sources of historic data, are presented in Section 4.4. Several formats for presenting rainfall data are described in Section 4.5. Finally, Section 4.6 briefly summarizes the rainfall data requirements of several runoff prediction methods.

4.1 THE HYDROLOGIC CYCLE

The amount of water on Earth is constant and finite and is in constant circulation throughout a space called the *hydrosphere*. Although its dimensions vary from one part of the globe to another, the hydrosphere extends approximately 9 mi (14 km) into the atmosphere and about 0.6 mi (1 km) into the Earth's crust. Circulation of water through the hydrosphere, known as the *hydrologic cycle*, takes place through a number of pathways. Movement of water through the hydrologic cycle is driven by solar energy, air currents, Coriolis forces (caused by the rotation of the Earth on its axis), and gravitational and capillary forces. As water moves through the hydrologic cycle, heat energy stored within the water also moves, and thus the hydrologic cycle is a major factor in the Earth's global heat balance.

Many different representations of the hydrologic cycle can be found in textbooks that cover the topic. Some are pictorial (as shown in Figure 4.1), and others are more system-oriented and take the form of link/node diagrams similar to flowcharts (see Figure 4.2). Both representations show in extremely simplified terms the various pathways that water takes as it passes through the hydrologic cycle. However, not all of these pathways need to be considered in every engineering problem. For example, in the design of a stormwater drainage system for a paved parking lot, infiltration and evaporation are usually ignored because of the imperviousness of the pavement material and the negligible amount of evaporation expected to occur over the relatively short time interval associated with a storm.

Figure 4.1
Pictorial representation of the hydrologic cycle

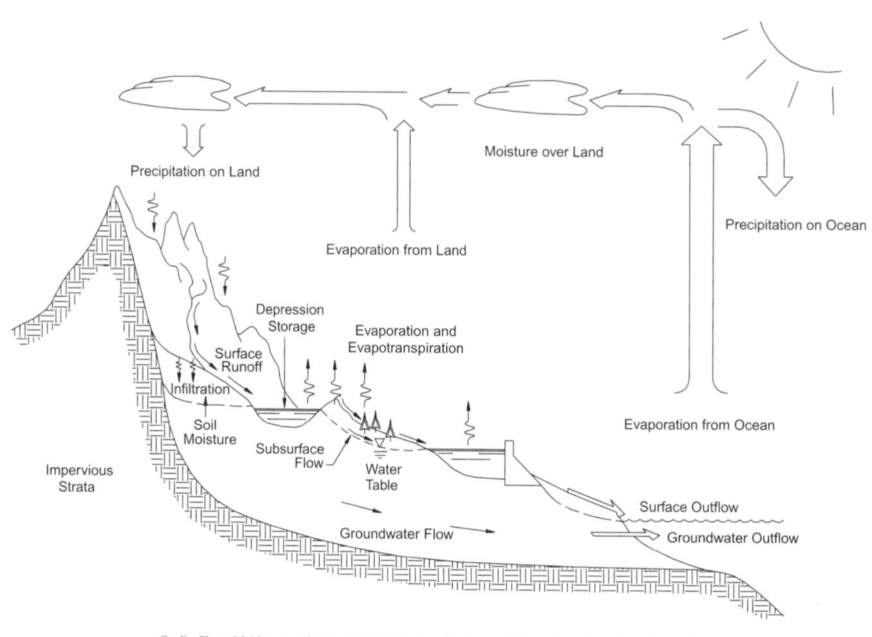

Credit: Chow, Maidment, and Mays (1988). Reproduced with permission of the McGraw-Hill Companies.

There is no beginning or end to the hydrologic cycle; water circulates through the cycle continuously. Thus, a description of its various components could begin at any point. It is common, however, to begin with *precipitation*. Precipitation can occur in

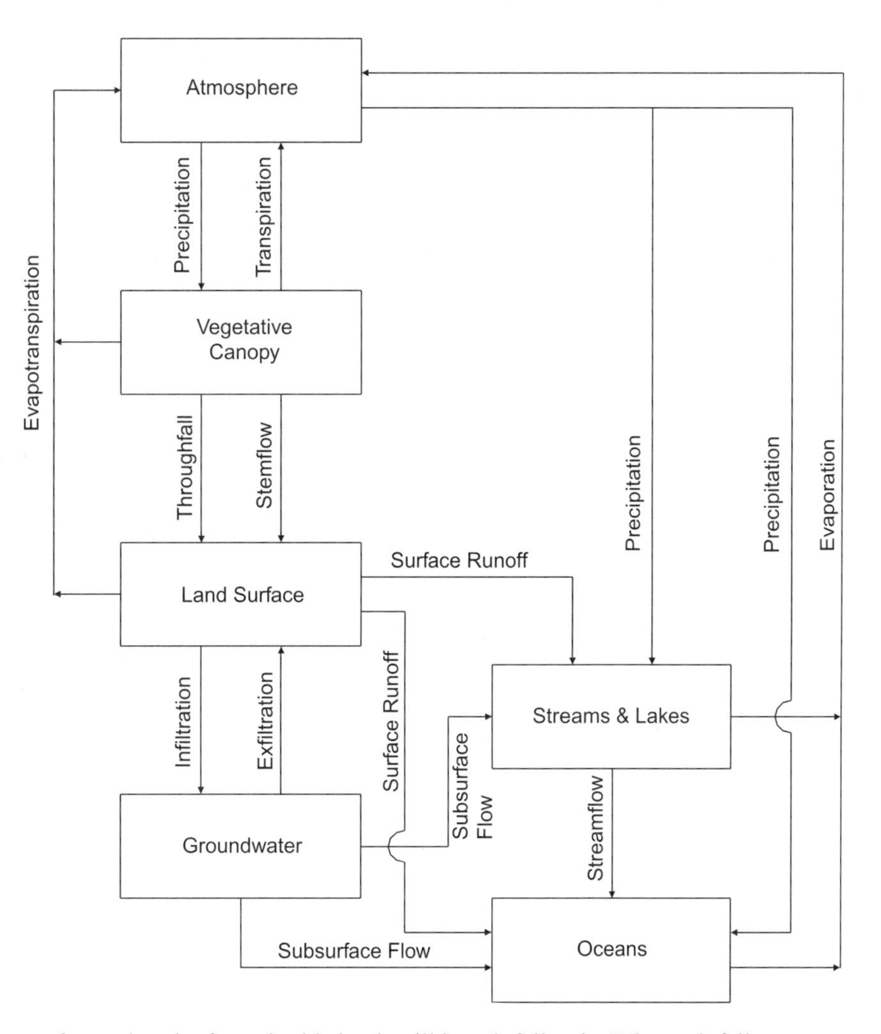

Figure 4.2 Flowchart representation of the hydrologic cycle

many forms, but the focus in this book will be rainfall only. When rainfall occurs over a drainage basin, a certain amount of it is caught by vegetation before it reaches the ground surface; this portion is referred to as *interception*. The rainfall that does make its way to the ground surface can take any of a number of pathways, depending on the nature of the ground surface itself. Some water may *infiltrate* into the ground, and some may be trapped in low areas and depressions. The water that neither infiltrates nor gets trapped as *depression storage* contributes to *direct surface runoff*. This surface runoff may combine with *subsurface flow* and/or groundwater *outflow (base flow)* to become *streamflow* supplying rivers, lakes, and oceans. Water is returned to the atmosphere by *evaporation* from vegetation, soils, and bodies of water, and by *transpiration* of vegetation. (Evaporation and transpiration from vegetation may together be referred to as *evapotranspiration*.)

4.2 PRECIPITATION TYPES

Forms of precipitation include rainfall, hail, and snowfall. Precipitation occurs when water vapor in the atmosphere is converted into a liquid or solid state. It takes place only when water vapor is present in the atmosphere (which is virtually always true) and a cooling mechanism is present to facilitate condensation of the vapor. The cooling and resulting precipitation can take place in any of several ways, as described in the subsections that follow. As stated previously, the focus in this text will be on rainfall only. Readers needing to address other forms of precipitation, especially snow, are referred to Singh (1992) for useful introductory coverage.

Frontal Precipitation

Frontal precipitation occurs when a warm air mass, moved by wind currents and atmospheric pressure gradients, overtakes and rises above a cooler air mass. The rising of the warm, moisture-laden air to a higher altitude causes it to cool (unless there is a temperature inversion), and the water vapor condenses. The precipitation arising from this process often extends over large areas, and the interface between the warm and cool air masses is called a *warm front*, as illustrated in Figure 4.3.

A *cold front* occurs when a cold air mass overtakes a warmer one and displaces the warm air upwards. Again, the rising and cooling of the warm air is what causes condensation to occur. Unlike warm fronts, the precipitation arising from a cold front is frequently spotty and often covers relatively small areas.

Figure 4.3
Frontal precipitation

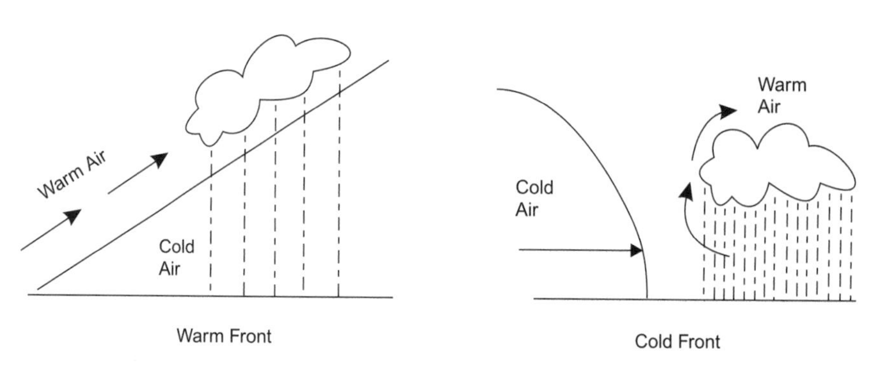

Cyclonic Precipitation

A *cyclone* is a region of low pressure into which air flows from surrounding areas where the pressure is higher. Flow around the low pressure center moves in a counterclockwise direction in the northern hemisphere and in a clockwise direction in the southern hemisphere. The direction of the rotation is determined by *Coriolis forces*, which are forces caused by the tangential velocity of the Earth at the equator being faster than its tangential velocity at other latitudes.

As air masses converge on the low pressure area, the incoming mass of air must be balanced by an outgoing one. Because air is entering from all directions horizontally, the outgoing air has no choice but to move vertically upward. The upward motion of the outgoing air mass leads to cooling and condensation of water vapor. Precipitation of this type is called *cyclonic precipitation* (see Figure 4.4).

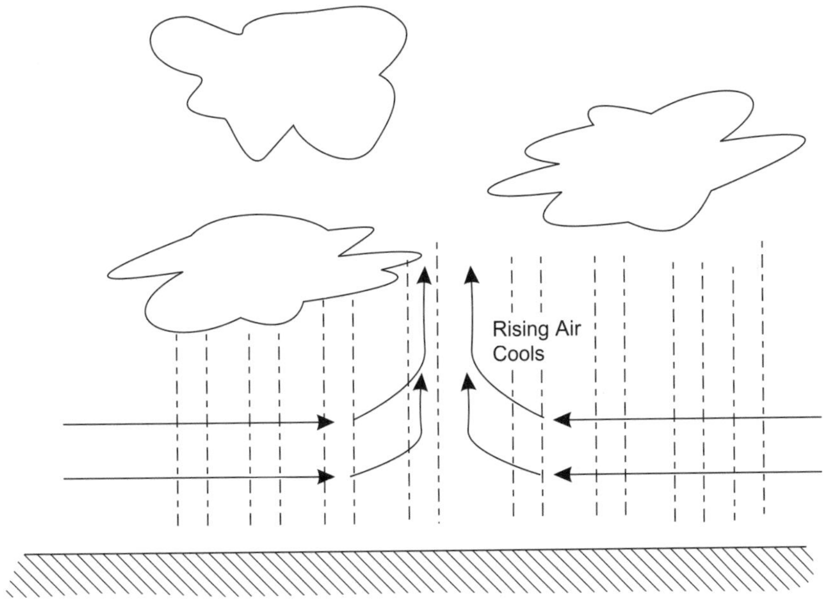

Figure 4.4
Cyclonic precipitation

Convective Precipitation

Convective precipitation tends to take place in small, localized areas (often no larger than a square mile or a few square kilometers) and is caused by differential heating of an air mass, as illustrated in Figure 4.5. This differential may occur in urban areas or during summer months when air near the ground surface becomes heated and rises with respect to the cooler surrounding air. This rise can be quite rapid and often results in thunderstorms. These events are very difficult to predict because they occur in short time frames. They can yield very intense rainfall.

Orographic Precipitation

Orographic precipitation occurs when an air mass is forced by topographic barriers to a higher altitude where the temperature is cooler, as illustrated in Figure 4.6. This type of precipitation is common in mountainous regions where air currents are forced up and over the tops of the mountains by wind movement. When the air rises to a cooler altitude, condensation occurs. Orographic precipitation can be quite pronounced on the windward side of a mountain range (the side facing the wind), while there is often a rain shadow where there is relatively little precipitation on the leeward side.

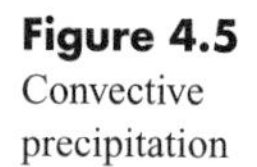

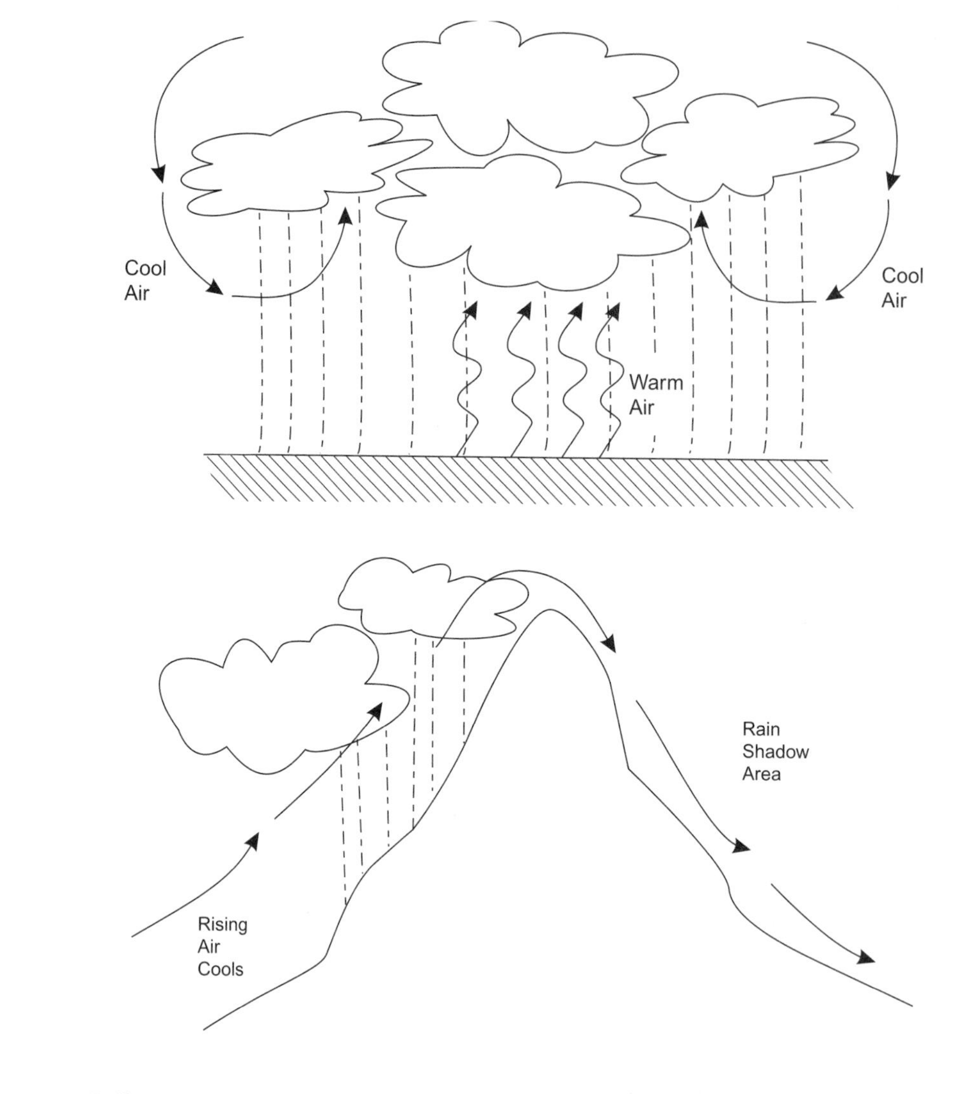

Figure 4.5
Convective precipitation

Figure 4.6
Orographic precipitation

4.3 BASIC RAINFALL CHARACTERISTICS

For the design engineer, the most important characteristics of rainfall are

- The *depth*, or volume, of rainfall during a specified time interval (or equivalently, its average *intensity* over that time interval)
- The duration of the rainfall
- The area over which the rainfall occurs
- The temporal and spatial distributions of rainfall within the storm
- The average recurrence interval of a rainfall amount

Depth and Intensity

The *depth* of rainfall is the amount of rainfall, in inches or millimeters, that falls within a given duration or time period. The rainfall *intensity* represents the rate at which rainfall occurs. The average intensity for a period is simply the rainfall depth divided by the time over which the rainfall occurs.

Temporal Distributions

The *temporal rainfall distribution*, shown in Figure 4.7, is a good visualization tool for demonstrating how rainfall intensity can vary over time within a single event. The y-axis is represented by a simple rain gauge that fills over a certain period of time, which is shown on the x-axis. *Total depth* is simply the final depth in the gauge. Average *intensity* during any segment of the storm is represented by the slope of the rainfall curve. A steeper slope for a given curve segment corresponds to a greater average intensity during that segment.

The temporal distribution shown in Figure 4.7 can also be represented using a bar graph that shows how much of the total rainfall occurs within each time interval during the course of an event. A graph of this nature is called a *rainfall hyetograph*. Hyetographs can be displayed in terms of incremental rainfall depth measured within each time interval (Figure 4.8), or as the average intensity calculated for each interval by dividing incremental depth by the time interval (Figure 4.9). The latter approach, shown in Figure 4.9, is recommended.

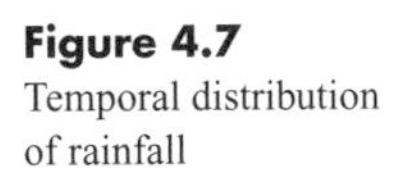

Figure 4.7
Temporal distribution of rainfall

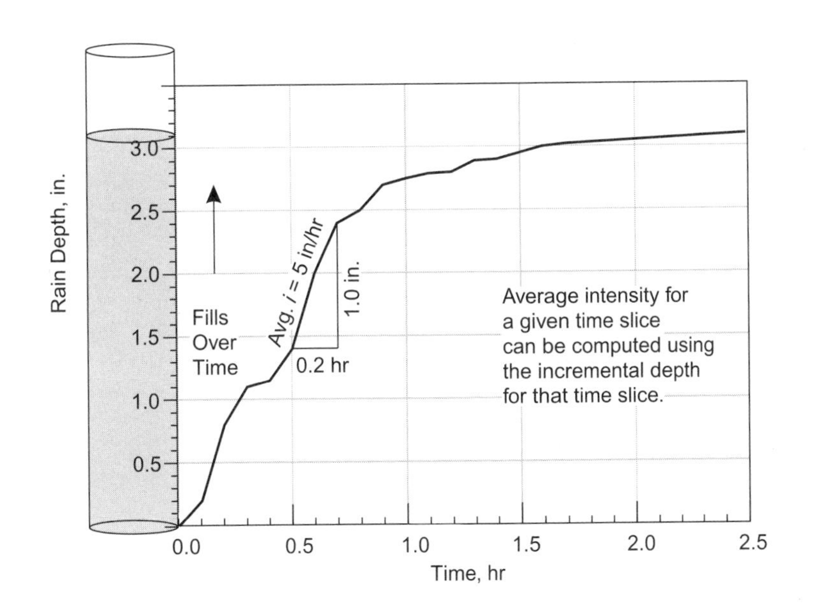

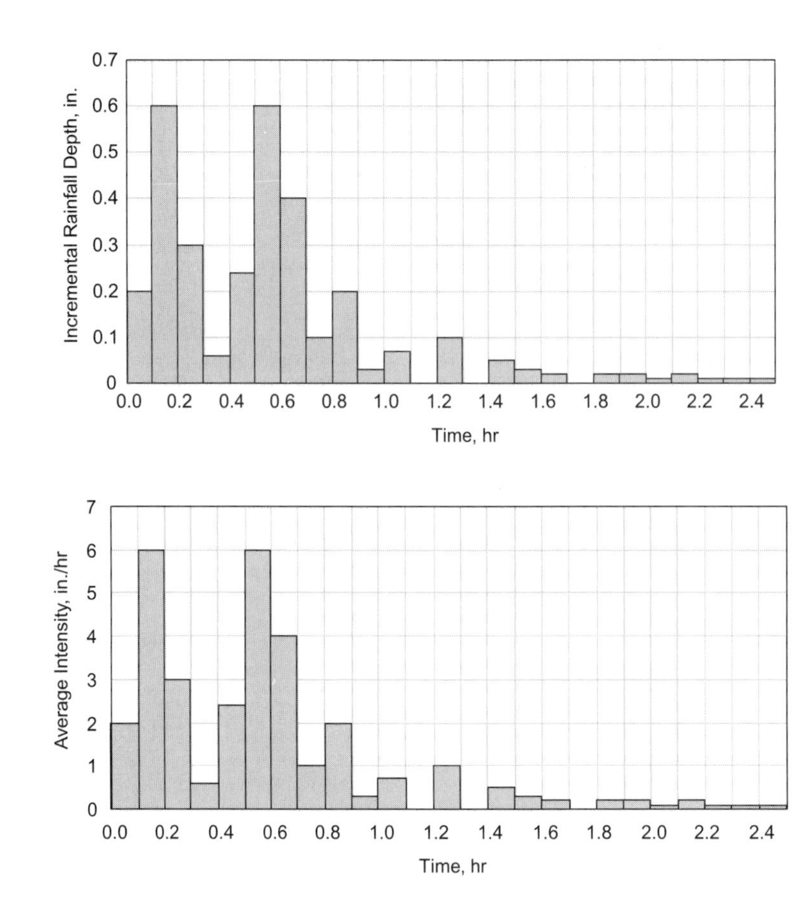

Figure 4.8
A hyetograph of incremental rainfall depth versus time

Figure 4.9
A hyetograph of incremental rainfall intensity versus time

Spatial Distributions

The spatial distribution of rainfall relates to the issue of whether the rainfall depths or intensities at various locations in a drainage basin are equal for the same event. The spatial aspect of rainfall is not covered in detail in this text because, in practice, spatial variations for relatively small drainage basins can be neglected. However, areal rainfall reduction factors are presented later in this chapter to demonstrate the adjustment of point rainfall values for application with large drainage areas.

Recurrence Interval

The probability that a rainfall event of a certain magnitude will occur in any given year is expressed in terms of *recurrence interval* (also called *return period* or *event frequency*). The recurrence interval is the *average* length of time expected to elapse between rainfall events of equal or greater magnitude. It is a function of geographic location, rainfall duration, and rainfall depth. Although recurrence interval is expressed in years, it is actually based on a storm event's *exceedance probability*, which is the probability that a storm magnitude will be equaled or exceeded in any

Time-Series Data for Rainfall

Statistical analyses of raw rainfall data are used to assign recurrence intervals to various rainfall depths or intensities for selected storm durations. To accomplish the statistical analyses, one must first extract either an annual series of data or a partial-duration series of data from the raw rainfall data records.

An *annual series* is constructed by extraction of the single largest depth or intensity within each year of record for a chosen duration. If one has n years of data, the resulting annual series contains n data points—one from each year. To construct a *partial-duration series,* one begins by choosing a precipitation threshold value. Then, any depth or intensity exceeding the chosen threshold for the chosen duration is extracted from the records. Thus, if n years of record are available, one may end up with more or fewer than n data points depending on the magnitude of the chosen threshold. (The U.S. National Weather Service implicitly chooses the threshold so that the partial-duration series has n data points if there are n years of record.) In any case, the partial-duration series approach explicitly recognizes that more than one large event may occur in any year, and that the second or even third largest event in a year may be larger than the largest event in another year.

In practice, the results of annual and partial-duration series analyses are found to lead to essentially the same results for recurrence intervals of 10 years or more. That is, the precipitation depth or intensity for a selected storm duration and recurrence interval is essentially the same regardless of the type of data series used. However, for recurrence intervals smaller than 10 years, partial-duration series analyses generally give rise to higher rainfall depths than do annual series analyses. Thus, in the interest of obtaining conservatively high estimates of rainfall amounts, an engineer might choose to use results based on partial-duration series analyses.

given year. The relationship between recurrence interval and exceedance probability is given by

$$T = 1/P \tag{4.1}$$

where T = return period (years)
P = exceedance probability

For example, a 25-year return period represents a storm event that is expected to occur once every 25 years, on average. This does not mean that two storm events of that size will not occur in the same year, nor does it mean that the next storm event of that size will not occur for another 25 years. Rather, a 4-percent chance of occurrence exists in any given year.

Example 4.1 – Computing Recurrence Intervals. What is the recurrence interval for a storm event that has a 20-percent probability of being equaled or exceeded within any given year? A 2-percent probability? A 200-percent probability?

Solution: For a storm event with a 20-percent probability of being equaled or exceeded in a given year, the recurrence interval is computed as

$$1/0.20 = 5 \text{ yr}$$

For an event with a 2-percent probability, the recurrence interval is

$$1/0.02 = 50 \text{ yr}$$

For an event with a 200-percent probability, the recurrence interval is

$$1/2.00 = 0.5 \text{ yr} = 6 \text{ months}$$

4.4 OBTAINING RAINFALL DATA

A number of types of gauges are available for measuring rainfall, and these measurements have varying degrees of accuracy. In most locations, rainfall data from gauging stations are compiled by a government agency and are readily available to the engineer for use in modeling. Very often, the raw rainfall data have already been processed and used to develop simplified techniques for engineers to use in generating design storms. This section presents information on methods of rainfall collection and where to obtain rainfall data.

Rainfall Data Collection Methods

In the United States, as in most countries, a national network of gauging sites exists through which rainfall data are measured and recorded on a regular basis. Historically, these data were collected using non-recording gauges at a frequency of once per day. Many gauging installations are now set up to collect data on an hourly basis using automatic recording gauges. In some instances, gauging sites are set up to collect data at even smaller time intervals (usually 15 minutes), but the number of these sites is small compared to the numbers of hourly and daily recording sites.

The *standard nonrecording gauge* used by the U.S. National Weather Service is an 8-in. (20.3-cm) diameter cylinder with a funnel mounted inside its top. The funnel directs rainwater into a cylindrical measuring tube where its depth is measured with a dipstick. The cross-sectional area of the measuring tube is one-tenth that of the 8-in. diameter gauge opening, so that 0.01 in. (0.254 mm) of precipitation over the gauge appears as 0.10 in. (2.54 mm) of water in the measuring tube. Measurements of the depth of accumulated water in the standard gauge are made once daily, usually by volunteer observers. An illustration of the standard gauge is shown in Figure 4.10.

Three types of commonly used recording rain gauges are weighing gauges, float gauges, and tipping-bucket gauges. *Weighing gauges*, shown in Figure 4.11, employ a scale or spring balance to record the cumulative weight of the collection can and precipitation over time. The weight of the precipitation is then converted into an equivalent volume and depth of rainfall using the specific weight of water and the cross-sectional area of the gauge opening.

Float gauges record the rise of a float as the water level rises in a collection chamber. A tipping-bucket gauge, shown in Figure 4.12, operates by means of a two-compartment bucket. When a certain quantity (depth) of rainfall fills one compartment, the weight of the water causes it to tip and empty, and the second compartment to move into position for filling. Similarly, the second bucket tips when full and moves the first compartment back into position. This process continues indefinitely. Records of the bucket tip times, combined with knowledge of the compartment volumes, are used to estimate rainfall depths and rates.

Another emerging technology for measuring rainfall is the *optical gauge* or *disdrometer*. The basic principle behind the optical gauge is the use of light sources and line-scan cameras to measure the size/shape and velocity of raindrops. As shown in Figure 4.13, the intersection of orthogonal light sources defines the measuring area.

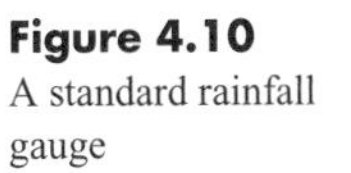
Figure 4.10
A standard rainfall gauge

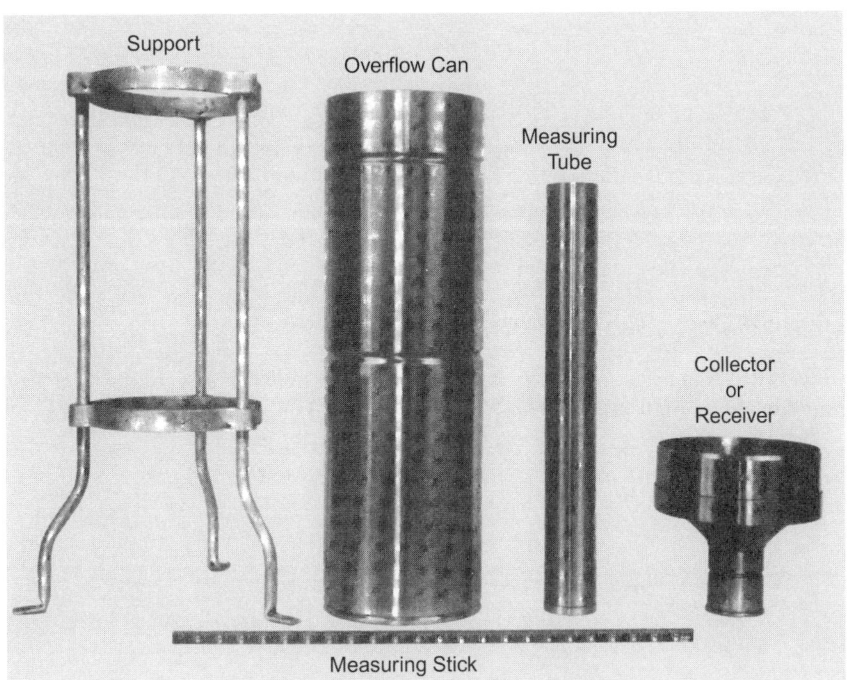

Credit: Linsley, Kohler, and Paulus, 1982. Reproduced with permission of the McGraw-Hill Companies.

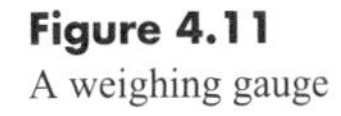
Figure 4.11
A weighing gauge

Credit: Chow, 1964. Reproduced with permission of the McGraw-Hill Companies.

The raindrops that fall through these beams cast a shadow that is recorded by the cameras. The image is processed to determine the size and shape of the drops. Because the two light sources are separated by a small distance, the vertical velocity of the raindrop can also be determined. The velocity and size and shape of the raindrop can then be used to estimate the rainfall intensity.

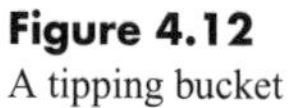

Figure 4.12
A tipping bucket gauge

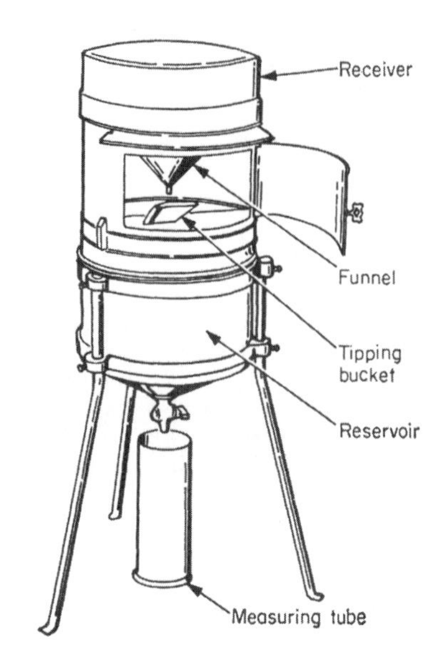

Credit: Chow, 1964. Reproduced with permission of the McGraw-Hill Companies.

Figure 4.13
Camera and light source configuration for a two-dimensional video disdrometer

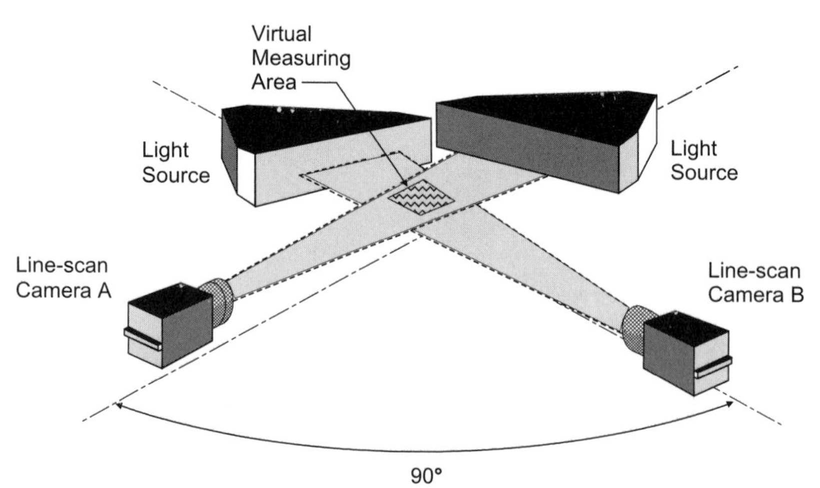

Credit: A. Kruger, W.F. Krajewski, and M.J. Kundert, IIHR–Hydroscience and Engineering, The University of Iowa, Iowa City, Iowa.

Regardless of the type of rainfall gauge used, errors inevitably occur in rainfall measurements. As a general rule, these errors lead to underestimation of the actual rainfall. Errors arise from many sources, including reading of depths for small amounts of rain, dipstick displacement, recording of tipping times in tipping bucket gauges, evaporation, and frictional effects in weighing and float gauges. However, the greatest source of error in rainfall gauging is the effect of wind, which may cause gauge catch deficiencies of more than 20 percent (Larson and Peck, 1974). Windshields of various designs have been developed to reduce the effects of wind on gauge catch, but errors persist nevertheless.

In recent decades, much attention has been given to measurement of precipitation using remote sensing technologies, the most prominent of which is *radar* (see Figure

4.14). The U.S. National Weather Service, as a part of its recent modernization program, has installed a national network of radar sites. This network often provides much better spatial resolution of the extent of precipitation events than gauges do, but greater rainfall depth uncertainty exists with radar information. Radar is particularly useful for tracking the movement of storms and tornadoes. Attention is also being given to the use of radar data for flash flood prediction and flood warning systems (Sweeney, 1992; Johnson et al., 1998).

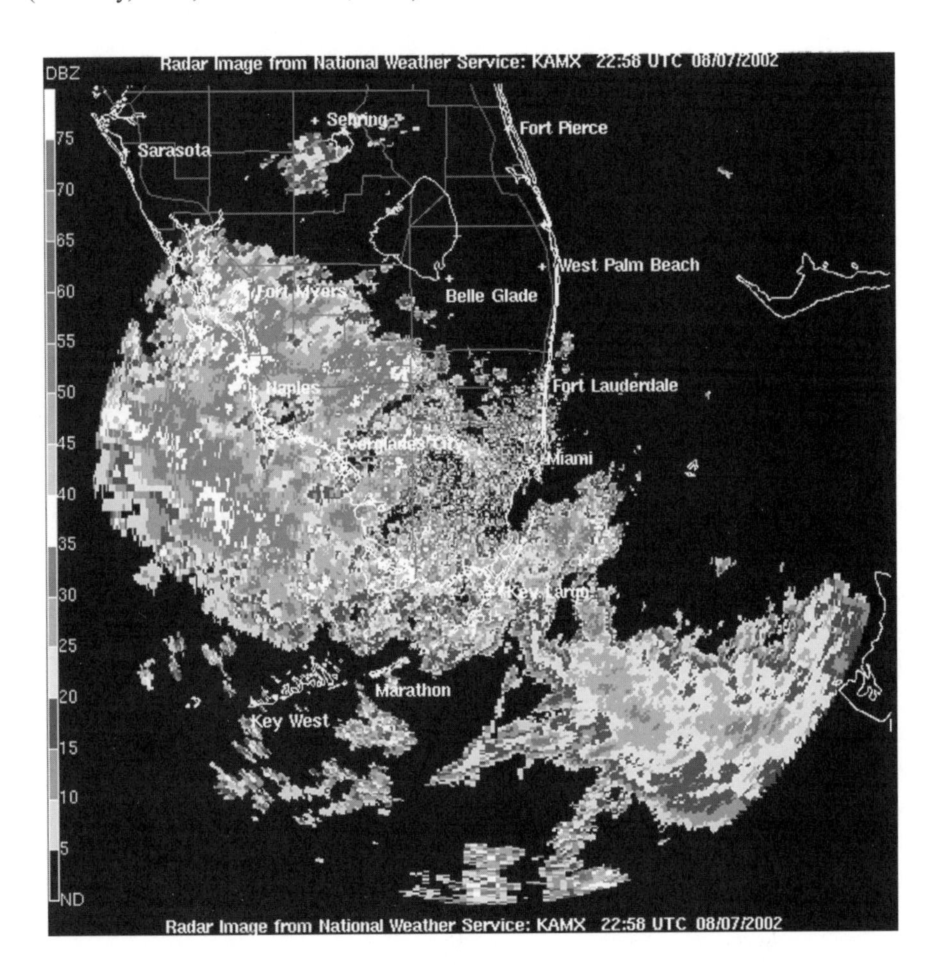

Figure 4.14
U.S. National Weather Service radar image showing rainfall over southern Florida

Rainfall Data Sources

Rainfall data can be obtained from a number of sources. Some sources provide raw rainfall data that has not been subjected to analyses other than those related to quality control. Others provide products that have been derived from various methods of processing of raw data. Processed data are used in most engineering analyses, but it should be recognized that many of the sources of processed data are now somewhat dated. Raw data sources can be used to obtain detailed information about a particular rainfall gauging location, and of course are the most basic information used in all types of rainfall analyses.

Raw rainfall data typically can be obtained from the meteorological bureau of the country of interest. In the United States, for example, raw rainfall data for any gauging site within the national network can be obtained from the National Climatic Data Center (NCDC), which is operated by the National Oceanic and Atmospheric Administration (NOAA). The NCDC has made much of the raw data available on the Internet. Global data are also available through the NCDC website.

In addition to national meteorological bureaus, other government organizations, as well as private organizations, may be useful rainfall data sources. The U.S. Geological Survey (USGS), for instance, as a part of its research programs, occasionally collects rainfall data at a number of sites in a given geographical area. These data may exist in "unit values" files and can be obtained by contacting the nearest office of the USGS. Over the last decade or so, raw data also have been made available on CD-ROM by private organizations such as EarthInfo, Inc. of Boulder, Colorado, a company that supplies climate and precipitation data for stations throughout the world.

For most practical engineering design and analysis purposes, the main sources of precipitation information consist of processed rainfall data. This processed information may be available from government organizations involved in hydrologic analysis, or from local institutions that conduct this type of research. In the United States, such information may be found in several National Weather Service publications. For example, the NWS publication TP 40 (Hershfield, 1961) presents maps showing precipitation depths in the United States for storm durations from 30 minutes to 24 hours and for recurrence intervals from 1 to 100 years. TP 40 was partially superseded by a later publication known as HYDRO-35 for the central and eastern United States (Frederick, Myers, and Auciello, 1977), and by the NOAA Atlas 2 for the 11 coterminous western states (Miller, Frederick, and Tracey, 1973). Again, these documents contain maps showing precipitation depths for given durations and recurrence intervals. However, unlike TP 40, they can be used to obtain information for durations as short as 5 minutes. Figures 4.15 through 4.20 illustrate the six maps contained in HYDRO-35.

Note that TP 40, HYDRO-35, and the NOAA Atlas 2 are now rather dated. A number of efforts to update the information presented in these publications have been initiated in recent years. Durrans and Brown (2001) summarize a number of recent studies that have been performed in the United States. Practitioners should be on the watch for new developments in the locale(s) where they practice.

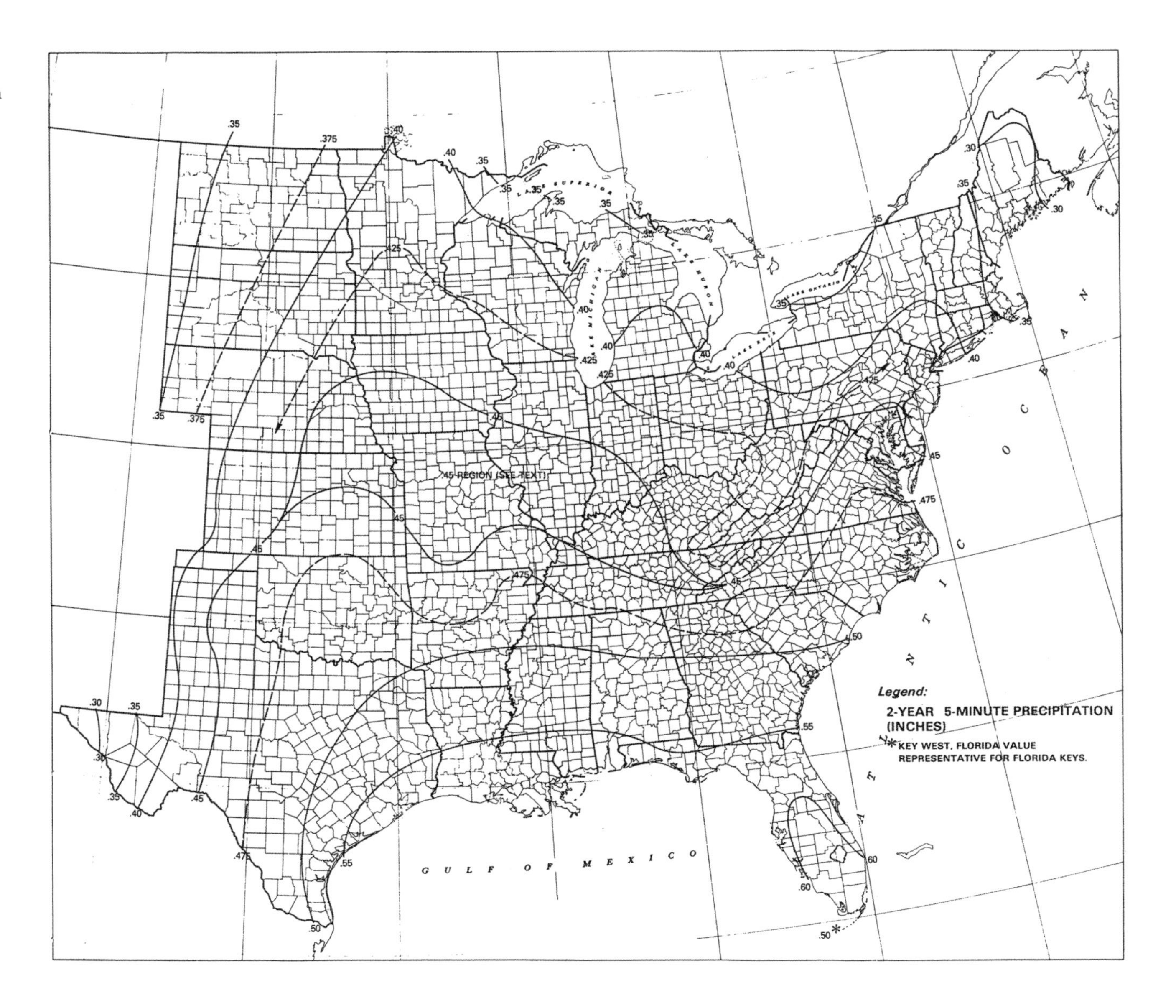

Figure 4.15 2-year, 5-minute precipitation depths in inches (Frederick, Myers, and Auciello, 1977)

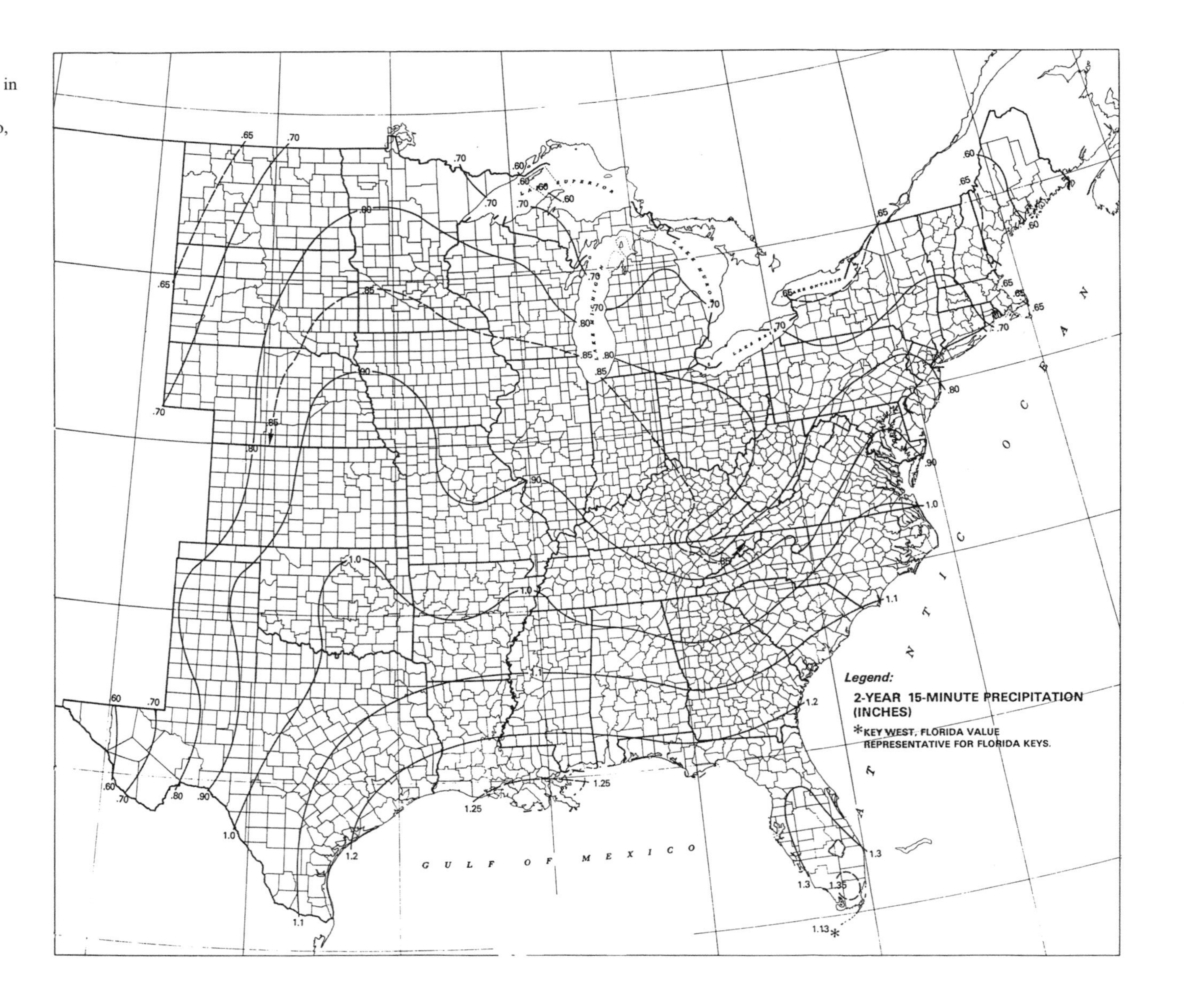

Figure 4.16 2-year, 15-minute precipitation depths in inches (Frederick, Myers, and Auciello, 1977)

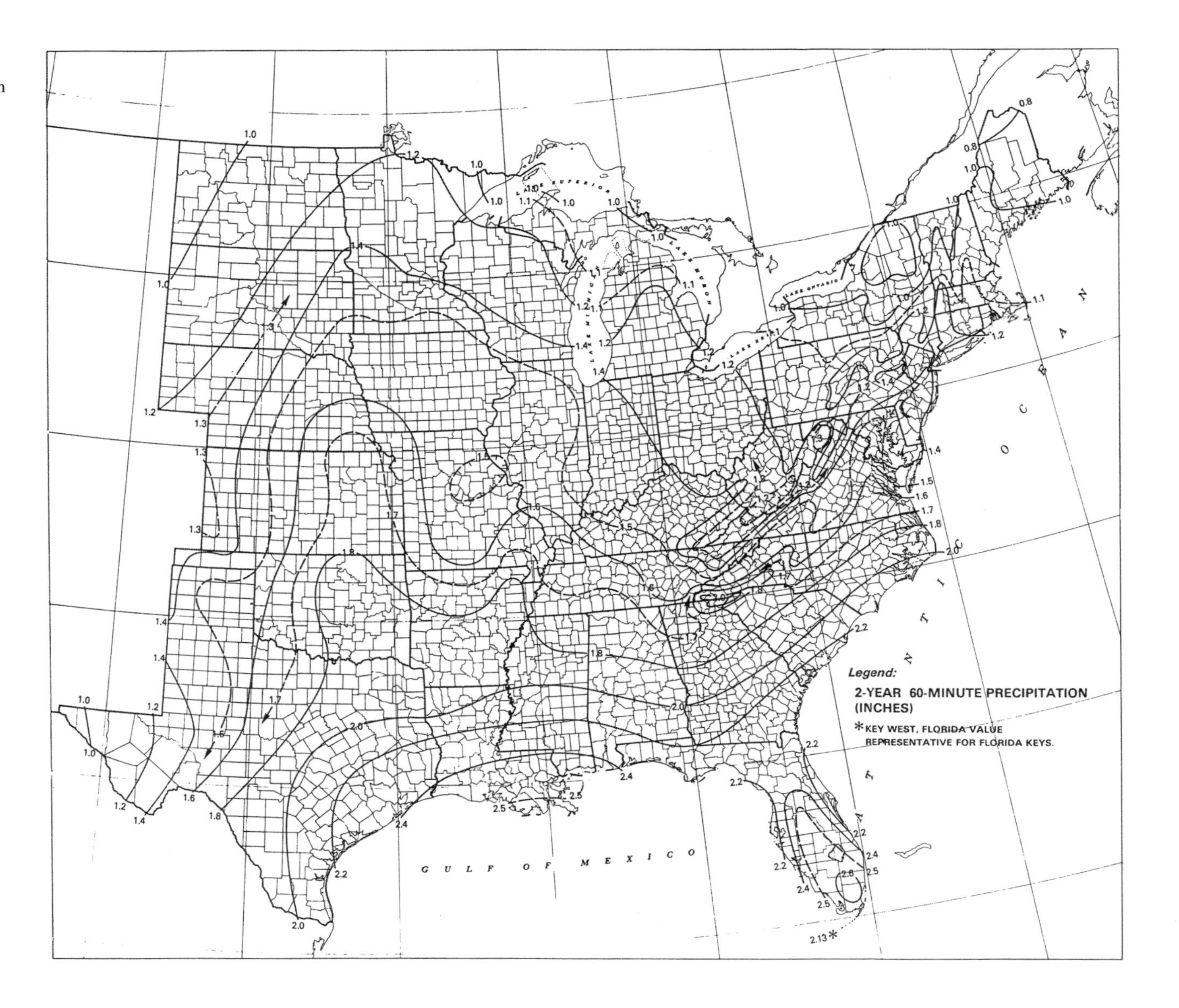

Figure 4.17
2-year, 60-minute precipitation depths in inches (Frederick, Myers, and Auciello, 1977)

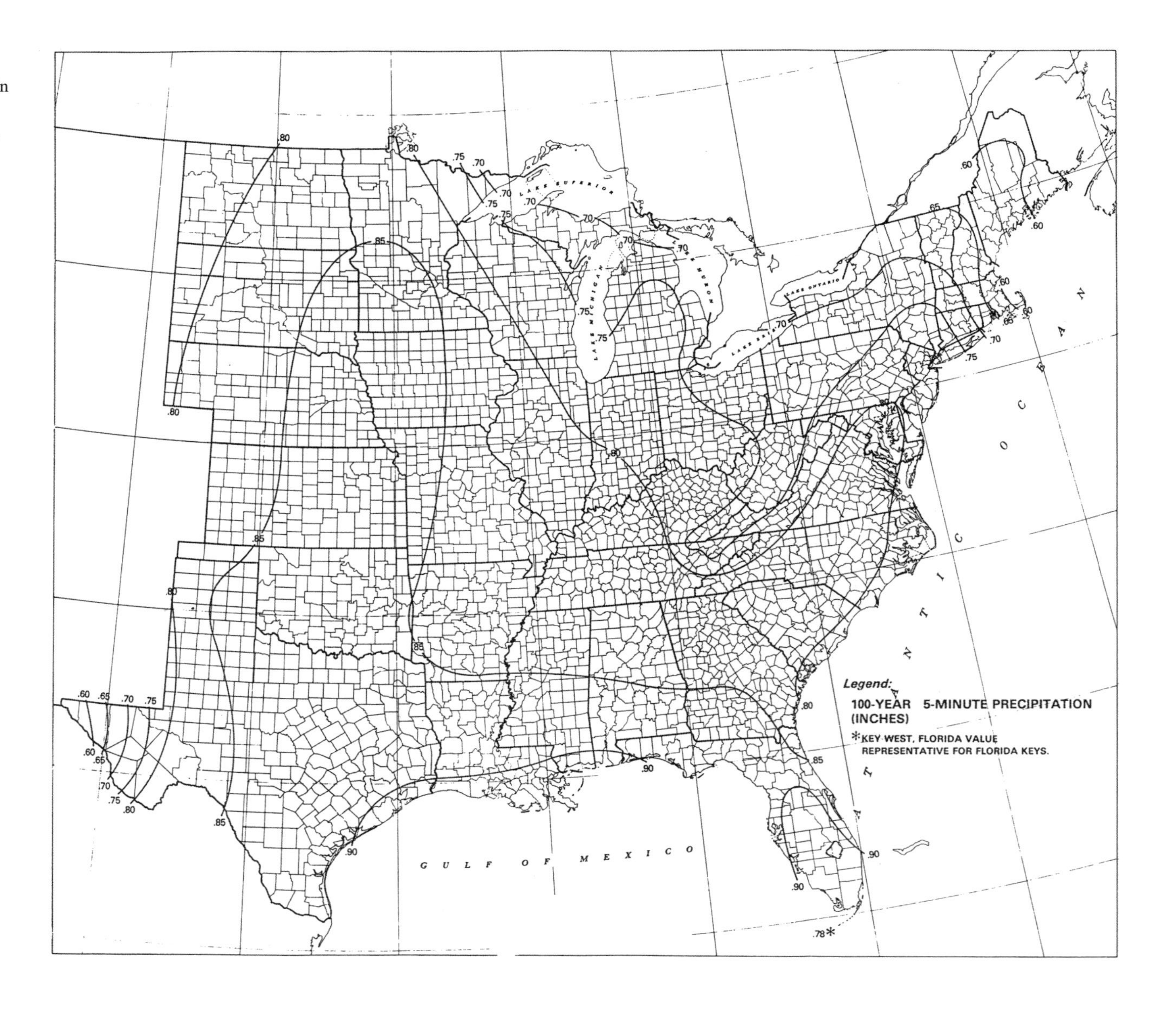

Figure 4.18 100-year, 5-minute precipitation depths in inches (Frederick, Myers, and Auciello, 1977)

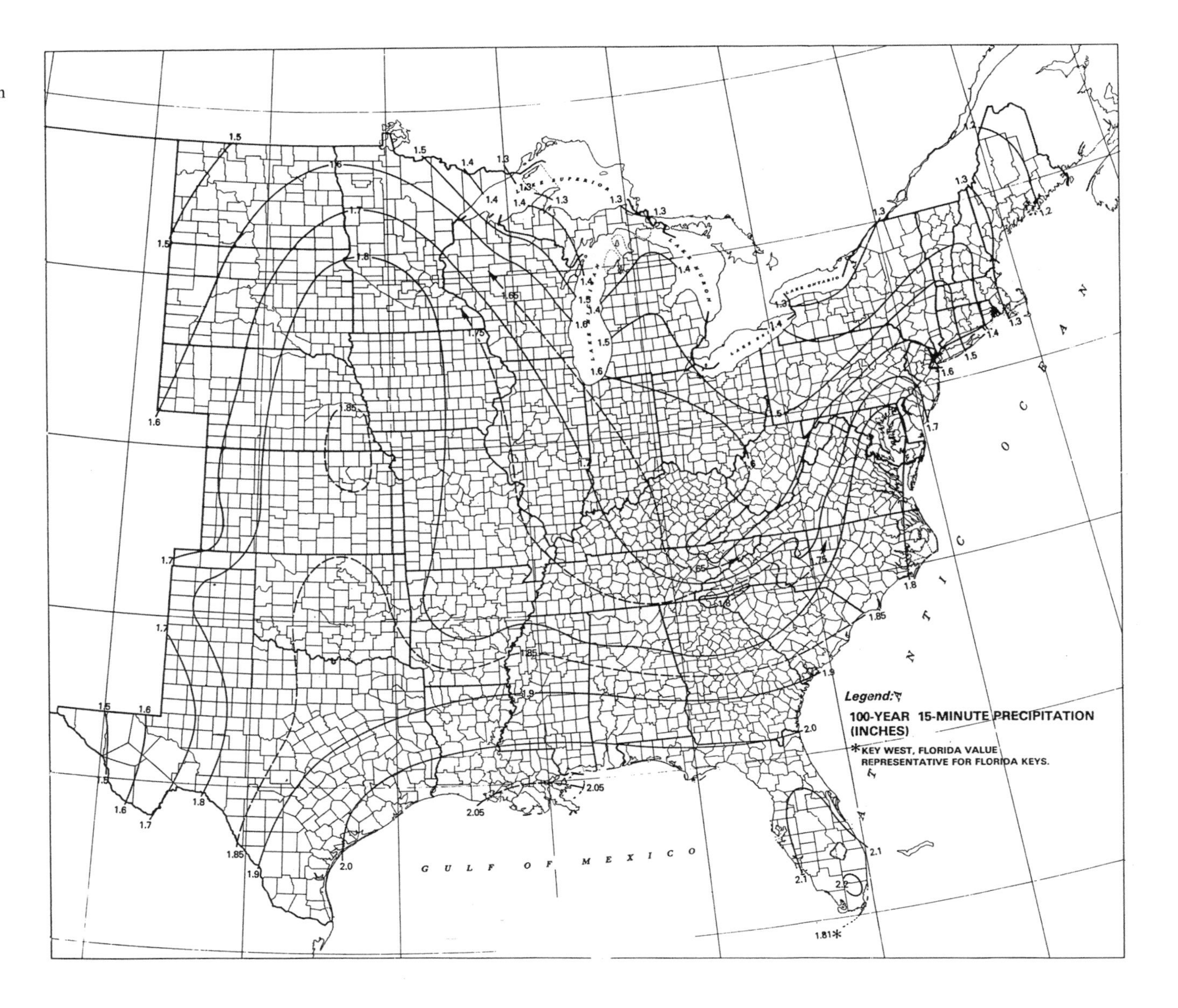

Figure 4.19
100-year, 15-minute precipitation depths in inches (Frederick, Myers, and Auciello, 1977)

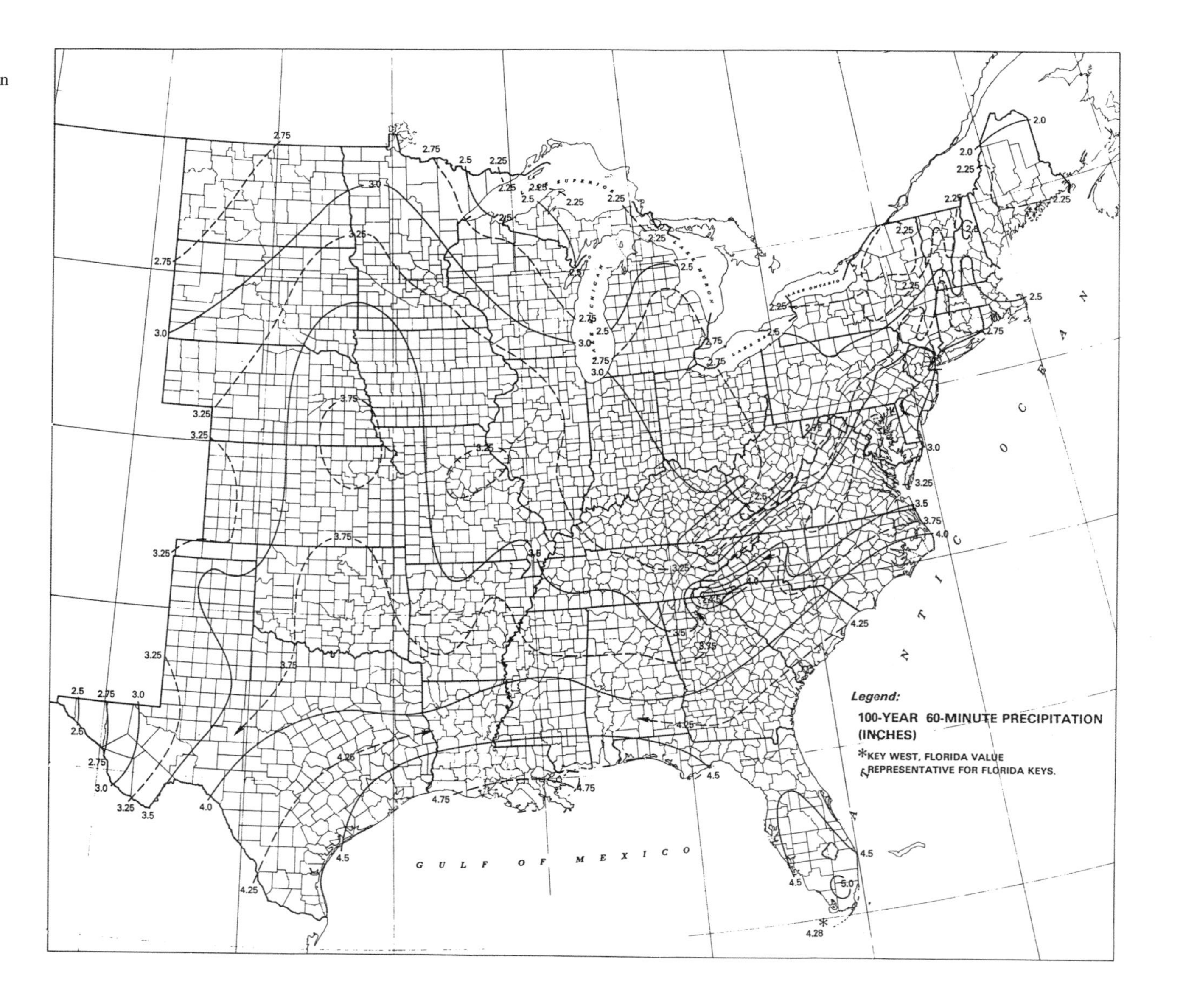

Figure 4.20 100-year, 60-minute precipitation depths in inches (Frederick, Myers, and Auciello, 1977)

4.5 TYPES OF RAINFALL DATA

Information on rainfall is available in a variety of forms. IDF curves provide average rainfall intensity information for particular storm events. Hyetographs provide rainfall depths or intensities at successive time intervals over the course of a storm. They can describe recorded rainfall events, or they may be generated using synthetic temporal rainfall distributions. The subsections that follow describe rainfall data types in more detail.

Intensity-Duration-Frequency Curves

For a selected storm duration, a rainfall intensity exists that corresponds to a given exceedance probability or recurrence interval. A rainfall *intensity-duration-frequency curve* (also referred to as an *IDF curve*) illustrates the average rainfall intensities corresponding to a particular storm recurrence interval for various durations.

Figure 4.21 shows sample IDF curves for 2-, 5-, 10-, 25-, 50-, and 100-year storms. These curves are the result of the statistical analysis of rainfall data for a particular area. Given the information on this graph, one can determine that the average one-hour rainfall intensity expected to be equaled or exceeded, on average, once every 25 years is 3.0 in/hr (76 mm/hr).

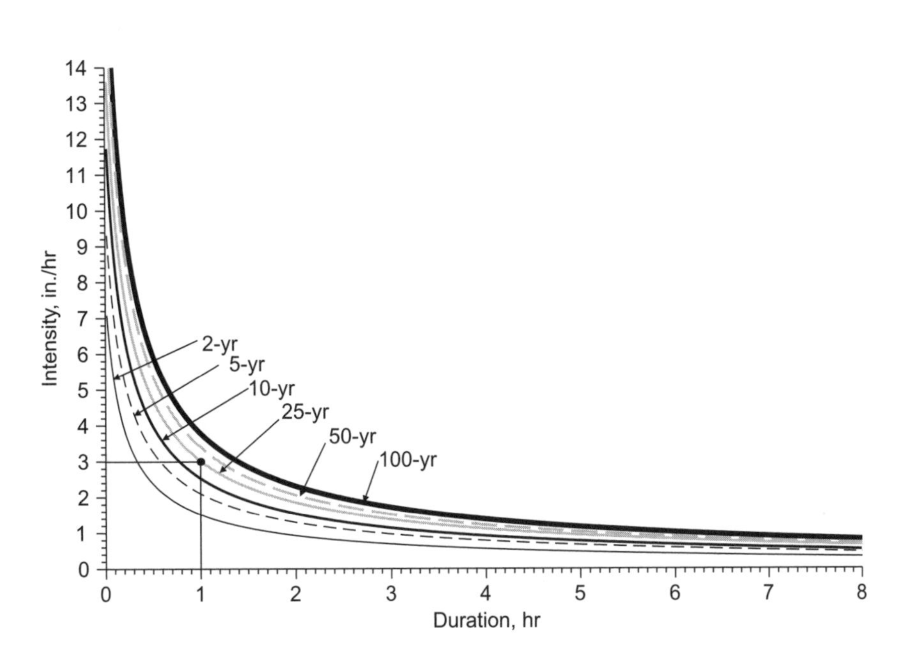

Figure 4.21
IDF curves for 2- through 100-year return period events

IDF curves can also be used to determine the recurrence interval associated with an observed storm rainfall intensity. That is, if one were to observe the intense gauged rainfall event shown in Figure 4.22, a set of IDF curves for that locale (shown in Figure 4.21) could be used to estimate the recurrence interval associated with that storm event. The storm in Figure 4.22 has an overall average intensity equal to 5 in./4 hr = 1.25 in/hr (32 mm/hr), but there would be periods during the 4-hour duration with both higher and lower intensities than the average. From Figure 4.21, an average intensity of 1.25 in/hr (32 mm/hr) over a duration of 4 hours is approximately a 50-year event.

Figure 4.22
A sample gauged rainfall event

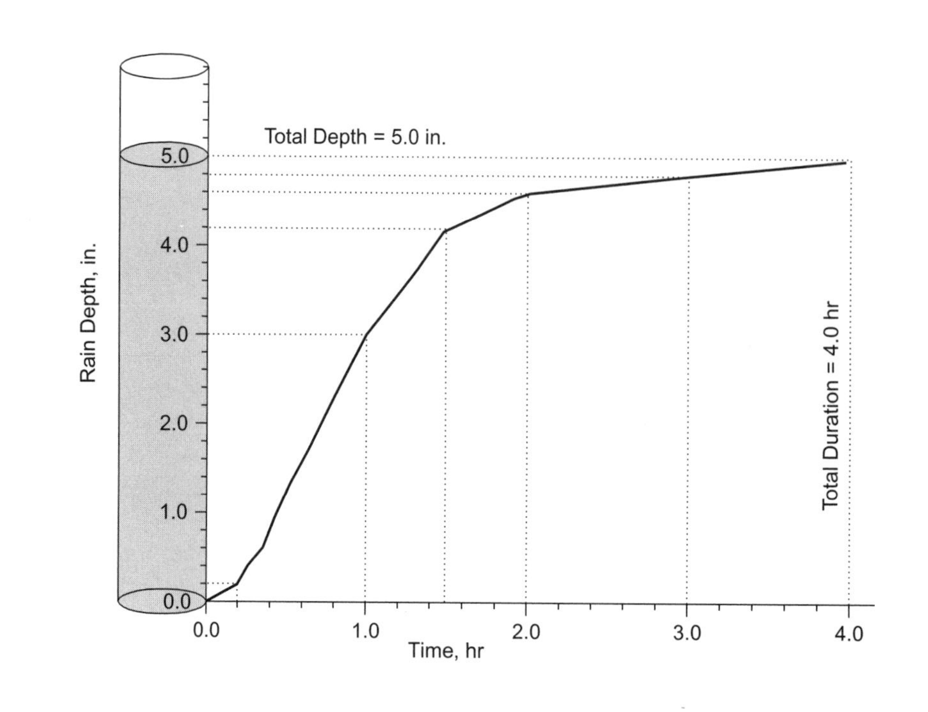

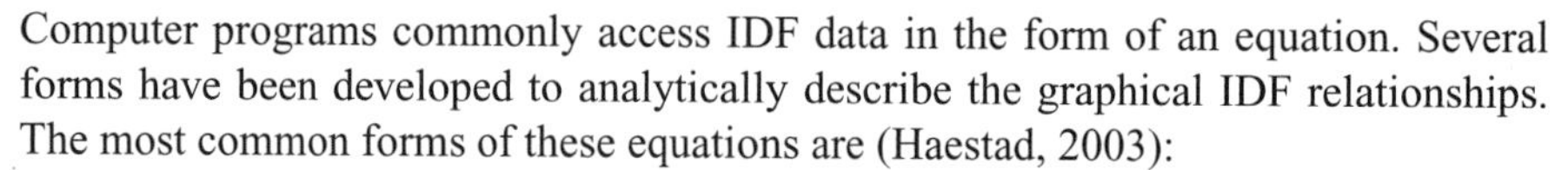

Computer programs commonly access IDF data in the form of an equation. Several forms have been developed to analytically describe the graphical IDF relationships. The most common forms of these equations are (Haestad, 2003):

$$i = \frac{a}{(b+D)^n} \tag{4.2}$$

$$i = \frac{a(R_P)^m}{(b+D)^n} \tag{4.3}$$

$$i = a + b(\ln D) + c(\ln D)^2 + d(\ln D)^3 \tag{4.4}$$

where i = intensity of rainfall (in/hr, mm/hr)
D = rainfall duration (min)
R_P = return period (yr)
a, b, c, d, m, and n are coefficients used to describe the IDF relationship

IDF Data Sources. As stated earlier in this chapter, hydrology can be thought of as providing the loading for the design of stormwater conveyance systems. One of the most widely used methods for estimating peak stormwater runoff rates for small drainage basins is the *rational method* (see Section 5.4, page 140 for more information). To use the rational method, one must have available a set of IDF curves developed for the specific geographic location where the conveyance system is to be built. A single set of IDF curves may be applicable over a fairly large area (for instance, an entire city or county), given that there are no significant topographic changes (such as large ridges) that cause rainfall characteristics across the area to vary.

Many drainage jurisdictions and agencies can provide the engineer with IDF data recommended for their particular geographic location. Engineers should understand when and by whom the IDF curves were created because more recently updated resources may be available.

Creating I-D-F Curves from HYDRO-35. A widely used procedure for developing IDF curves for the central and eastern United States is presented in HYDRO-35. The procedure uses the six maps shown in Figures 4.15 through 4.20. For a chosen location, one determines the depths of precipitation for the various combinations of storm durations and recurrence intervals represented by the six maps. Storm depths for other durations and return periods are then found by interpolation between the map values. For a fixed recurrence interval, rainfall depths for the 10-minute duration are computed from those for 5- and 15-minute durations as follows:

$$P_{10} = 0.41P_5 + 0.59P_{15} \tag{4.5}$$

Similarly, for a given recurrence interval, depths for the 30-minute duration are computed as

$$P_{30} = 0.51P_{15} + 0.49P_{60} \tag{4.6}$$

When the storm duration is fixed, depths for recurrence intervals other than 2 or 100 years are computed as

$$P_{T\text{-yr}} = aP_{2\text{-yr}} + bP_{100\text{-yr}} \tag{4.7}$$

where *a* and *b* are coefficients given in Table 4.1.

The coefficients *a* and *b* in this last expression depend on the recurrence interval, *T*, as shown in Table 4.1. The coefficients *a* and *b* in the table and the coefficients in the P_{10} and P_{30} estimation equations apply to all locations in the central and eastern United States covered by HYDRO-35.

After the depths of rainfall are determined for various storm durations and recurrence intervals by using the procedures described previously in this section, they can be converted into average intensities by dividing each depth by its corresponding duration. The IDF curves are then constructed by graphing the average intensities as functions of duration and recurrence interval.

Table 4.1 Coefficients for computation of IDF curves (Frederick, Myers, and Auciello, 1977)

T (yr)	*a*	*b*
5	0.674	0.278
10	0.496	0.449
25	0.293	0.669
50	0.146	0.835

Example 4.2 – Developing IDF Curves from HYDRO-35 Isohyetal Maps. Using the procedures outlined in HYDRO-35, develop a set of IDF curves for the U.S. city of Memphis, Tennessee, for durations from 5 minutes to 1 hour and for recurrence intervals from 2 to 100 years.

Solution: From the six isohyetal maps contained in HYDRO-35 and shown in Figures 4.15 through 4.20, the following rainfall depths are obtained for Memphis (located at the southwest corner of the state of Tennessee):

2-yr, 5-min depth:	0.49 in.	100-yr, 5-min depth:0.83 in.
2-yr, 15-min depth:	1.01 in.	100-yr, 15-min depth:1.79 in.
2-yr, 60-min depth:	1.79 in.	100-yr, 60-min depth:3.65 in.

Table 4.2 contains these depths (underlined) as well as other depths that have been interpolated between them using Equations 4.5 through 4.7. Within any of the last five columns of the table, precipitation depths for recurrence intervals other than 2 or 100 years are determined from the 2- and 100-year values using Equation 4.3 and the coefficients in Table 4.1. Depths in the third column are then determined from the adjacent depths in the second and fourth columns by using Equation 4.1, and depths in the fifth column are determined from the adjacent depths in the fourth and sixth columns by using Equation 4.2.

Table 4.2 Total Rainfall Depths (in.)

	Duration				
T (yr)	**5 min**	**10 min**	**15 min**	**30 min**	**60 min**
2	<u>0.49</u>	0.80	<u>1.01</u>	1.39	<u>1.79</u>
5	0.56	0.93	1.18	1.69	2.22
10	0.62	1.02	1.30	1.90	2.53
25	0.70	1.17	1.49	2.22	2.97
50	0.76	1.28	1.64	2.46	3.31
100	<u>0.83</u>	1.40	<u>1.79</u>	2.70	<u>3.65</u>

Intensities are computed from the table of rainfall depths by dividing the rainfall depths in each column by their corresponding duration and expressing the result in units of inches per hour. These computations lead to the results in Table 4.3.

Table 4.3 Intensity (in/hr)

	Duration				
T (yr)	**5 min**	**10 min**	**15 min**	**30 min**	**60 min**
2	5.88	4.78	4.04	2.78	1.79
5	6.73	5.55	4.71	3.38	2.22
10	7.39	6.13	5.22	3.81	2.53
25	8.39	7.01	5.97	4.43	2.97
50	9.18	7.69	6.57	4.92	3.31
100	9.96	8.38	7.16	5.40	3.65

A graph of the values in Table 4.3 is the desired set of IDF curves, shown in Figure E4.2.1.

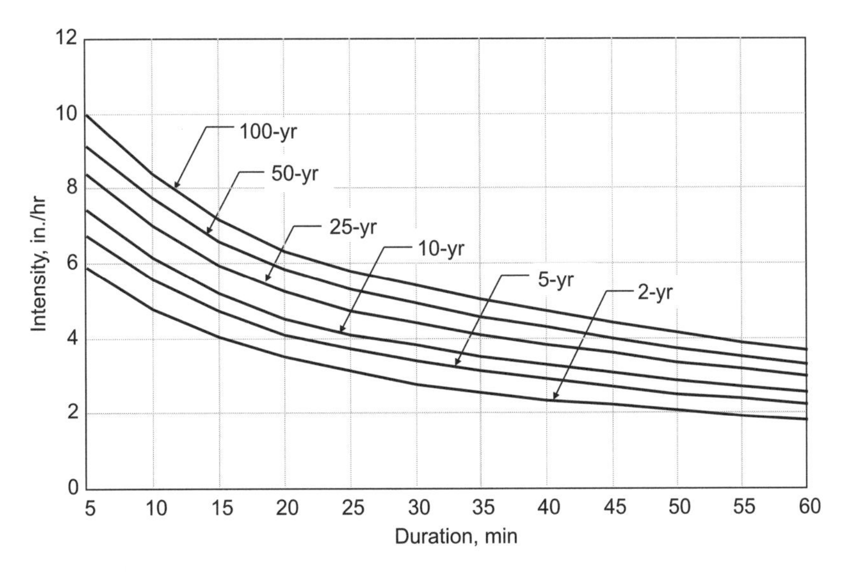

Figure E4.2.1 Derived IDF Curves

Note that rainfall amounts obtained from TP 40, HYDRO-35, and the NOAA Atlas 2, and hence from a set of IDF curves developed on the basis of those publications, represent rainfall depths and intensities at a single point in space. If one requires an average intensity over a large area (such as a large drainage basin), the point depths and intensities should be reduced somewhat to reflect the reduced probability of a storm occurring with the same intensity over this large area. Reduction percentages are indicated in Figure 4.23 (U.S. Weather Bureau, 1958) and depend on the area and the storm duration of interest. For example, the 10-year, 1-hour point rainfall intensity at a particular location should be multiplied by a factor of approximately 0.72 if it is to be applied to a drainage basin with an area of 100 mi^2 (259 ha). It is generally accepted that the curves shown in Figure 4.23 can be applied for any recurrence interval and for any location in the United States, although some differences have been noted between different regions.

Temporal Distributions and Hyetographs for Design Storms

Some types of hydrologic analysis require the distribution of precipitation over the duration of the storm. For example, the depth of a 100-year, 1-hour rainfall event in Tuscaloosa, Alabama, is estimated from HYDRO-35 to be about 3.8 in. (96.5 mm). But how much of this rainfall should be considered to occur in, say, the first 10 minutes of the storm? What about the second 10 minutes, or the last 10 minutes?

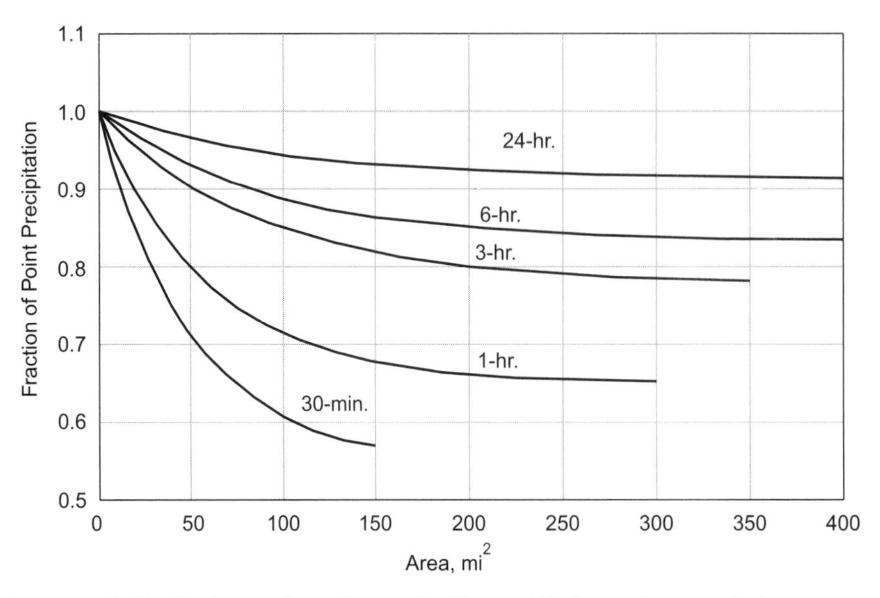

Figure 4.23
Reduction of point precipitation based on the area of the drainage basin (U.S. Weather Bureau, 1958)

A temporal distribution such as the one in Figure 4.7 shows the cumulative progression of rainfall depth throughout a storm. A *rainfall hyetograph* (Figures 4.8 and 4.9) shows how the total depth (or intensity) of rainfall in a storm is distributed among various time increments within the storm. Figures such as these can be used to represent the pattern of an actual recorded storm event.

When the design of a stormwater management facility requires a complete rainfall hyetograph, the engineer is faced with selecting the appropriate storm distribution to use in designing and testing for unknown, future events. In design situations, engineers commonly use *synthetic* temporal distributions of rainfall. Synthetic distributions are essentially systematic, reproducible methods for varying the rainfall intensity throughout a design event.

The selected length of the time increment, Δt, between the data points used to construct a temporal rainfall distribution, depends on the size (area) and other characteristics of the drainage basin. As a rule of thumb, the time increment should be no larger than about one-fourth to one-fifth of the basin lag time, or about one-sixth of the time of concentration of the basin. The Natural Resources Conservation Service (NRCS, formerly known as the Soil Conservation Service, or SCS) recommends using $\Delta t = 0.133t_c = 0.222t_L$ [see Section 5.3 for explanations of basin lag time (t_L) and time of concentration (t_c)], but notes that a small variation from this is permissible (SCS, 1969). The smallest time increment for which rainfall data are generally available is about 5 minutes. In small urban drainage basins where it is often necessary to use time increments as small as 1 or 2 minutes, this data must be interpolated.

In addition to selecting an appropriate Δt, the engineer must select the total duration to be used when developing a design storm hyetograph. In many cases, the storm duration will be specified by the review agency having jurisdiction over the area in which a stormwater conveyance facility will be built; this approach promotes consistency from one design to another.

When the design storm duration is not specified, the design engineer must select it. The recommended way to accomplish this is to review rainfall records for the locality of interest to determine typical storm durations for the area. Seasonal characteristics of storm durations may exist, with longer storms tending to occur in one season and shorter storms in another. An alternative approach is to perform runoff calculations using a number of assumed design storm durations. With this approach, the engineer can select the storm duration giving rise to the largest calculated flood peak and/or runoff volume.

U.S. NRCS (SCS) Synthetic Temporal Distributions. Many methods have been proposed for distributing a total rainfall depth throughout a storm to develop a design storm hyetograph. The NRCS developed a set of synthetic distributions commonly used in the United States to create 24-hour synthetic design storms (SCS, 1986). The dimensionless distribution data from the NRCS given in Table 4.4 provides fractions of the total accumulated rainfall depth over time for storms with 24-hour durations. (Figure 4.24 depicts this table graphically.) The storms are classified into various types, with each type being recommended for use in a certain U.S. geographic region, as shown in Figure 4.25. If necessary, interpolation may be employed to obtain values not shown in Table 4.4. Nonlinear interpolation methods are recommended for this purpose, although many practicing engineers use simpler linear interpolation instead. The NRCS (SCS, 1992) and many computer programs provide data in increments of 0.1 hours.

Table 4.4 NRCS dimensionless storm distributions (adapted from SCS, 1992)

t (hr)	Type I	Type IA	Type II	Type III
0	0.000	0.000	0.000	0.000
1	0.017	0.020	0.011	0.010
2	0.035	0.050	0.022	0.020
3	0.054	0.082	0.035	0.031
4	0.076	0.116	0.048	0.043
5	0.100	0.156	0.063	0.057
6	0.125	0.206	0.080	0.072
7	0.156	0.268	0.099	0.091
8	0.194	0.425	0.120	0.114
9	0.254	0.520	0.147	0.146
10	0.515	0.577	0.181	0.189
11	0.623	0.624	0.235	0.250
12	0.684	0.664	0.663	0.500
13	0.732	0.701	0.772	0.750
14	0.770	0.736	0.820	0.811
15	0.802	0.769	0.854	0.854
16	0.832	0.801	0.880	0.886
17	0.860	0.831	0.902	0.910
18	0.886	0.859	0.921	0.928

Table 4.4 (cont.) NRCS dimensionless storm distributions (adapted from SCS, 1992)

t (hr)	Type I	Type IA	Type II	Type III
19	0.910	0.887	0.938	0.943
20	0.932	0.913	0.952	0.957
21	0.952	0.937	0.965	0.969
22	0.970	0.959	0.977	0.981
23	0.986	0.980	0.989	0.991
24	1.000	1.000	1.000	1.000

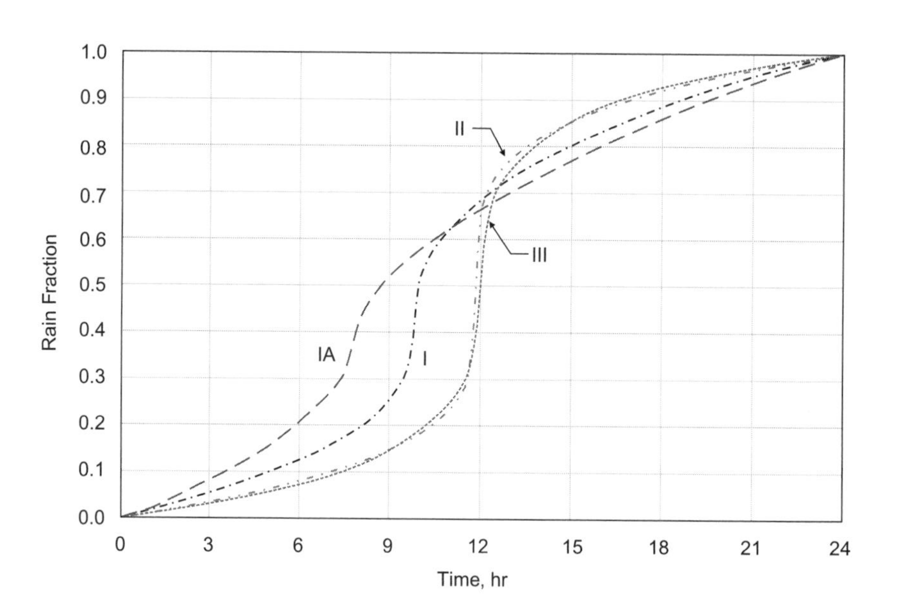

Figure 4.24 Graphical representation of NRCS (SCS) rainfall distributions

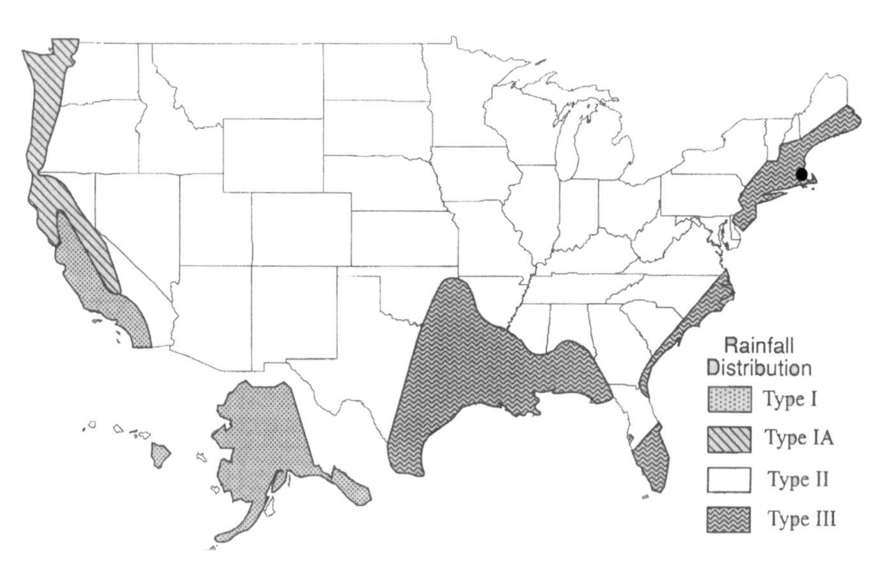

Figure 4.25 Coverage of NRCS (SCS) rainfall distributions (SCS, 1986)

Example 4.3 – Developing a Design Storm Hyetograph from SCS Distributions. Develop a design storm for a 50-year, 24-hour storm in Boston, Massachusetts. Assume that $\Delta t = 0.1$ hr is a reasonable choice for the drainage basin to which the design storm will be applied, and that the total rainfall depth is 6.0 in.

Solution: Figure 4.25 illustrates that a Type III storm distribution is a reasonable choice for Boston. Table 4.5 illustrates the calculation of the storm hyetograph.

The first column of the table is the time, in hours, since the beginning of the storm, and is tabulated in 1-hr increments for the total storm duration of 24 hours. In actuality, the Δt used in the calculations would be 0.1 hr; the 1-hr increment is used here for brevity.

The second column is the fraction of the total storm depth that has accumulated at each time during the storm. These values are obtained from Table 4.4 for the Type III storm distribution. (The values would be obtained by interpolation in the case of a 0.1-hr increment.) The third column contains the cumulative rainfall depths for each time during the storm and is obtained by multiplying each fraction in the second column by the total storm depth of 6.0 in. The fourth column contains the incremental depths of rainfall for each time interval during the storm; these values are computed as the difference between the concurrent and preceding values in the third column.

Table 4.5 50-year, 24-hour storm hyetograph for Boston, Massachusetts

t (hr)	Fraction	Cum. P (in.)	Incr. P (in.)
0	0.000	0.000	—
1	0.010	0.060	0.060
2	0.020	0.120	0.060
3	0.031	0.186	0.066
4	0.043	0.258	0.072
5	0.057	0.342	0.084
6	0.072	0.432	0.090
7	0.091	0.546	0.114
8	0.114	0.684	0.138
9	0.146	0.876	0.192
10	0.189	1.134	0.258
11	0.250	1.500	0.366
12	0.500	3.000	1.500
13	0.750	4.500	1.500
14	0.811	4.866	0.366
15	0.854	5.124	0.258
16	0.886	5.316	0.192
17	0.910	5.460	0.144
18	0.928	5.568	0.108
19	0.943	5.658	0.090
20	0.957	5.742	0.084
21	0.969	5.814	0.072
22	0.981	5.886	0.072
23	0.991	5.946	0.060
24	1.000	6.000	0.054

The resulting graph of cumulative precipitation is shown in Figure E4.3.1, and the hyetograph is shown in Figure E4.3.2. The height of each bar on the hyetograph is the average rainfall intensity during that time interval, and the area of each bar is the incremental rainfall depth during that time interval. Because the time increment on the graph is 1 hr (a 1-hr increment is shown in the graphic for simplicity, though the actual Δt is 0.1 hr) the value for the height of the bar (in units of in./hr) is equal to the incremental depth for that time increment (in in.).

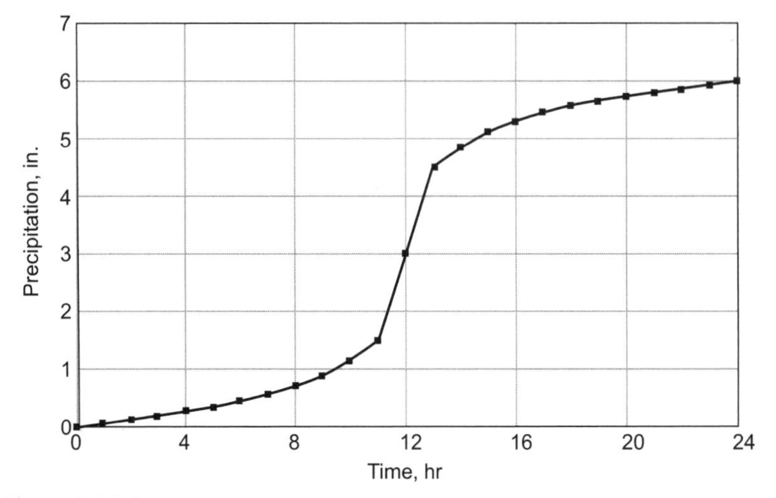

Figure E4.3.1 Graph of derived design storm cumulative precipitation

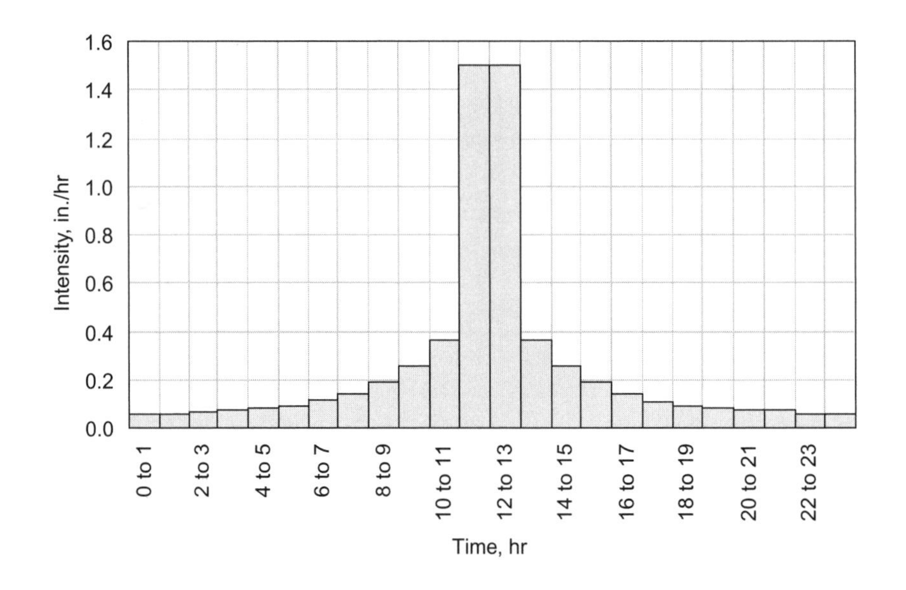

Figure E4.3.2 Derived design storm hyetograph

Alternating Block Method. Another method for developing design storm hyetographs is called the *alternating block method* and is illustrated in Example 4.4. With this method, the resulting hyetograph is consistent with the IDF curves for a particular location, and is in fact developed from those IDF curves. This method can be used for cases where the storm duration is not 24 hours, as is required for the SCS method.

Example 4.4 – Developing a Design Storm Hyetograph Using the Alternating Block Method. Using the IDF curves developed in Example 4.2, develop a design storm hyetograph for a 25-yr, 1-hr storm in Memphis, Tennessee using the alternating block method. Use $\Delta t = 10$ min for development of the hyetograph.

The first two columns in Table 4.6 are duration and intensity values from the Memphis IDF curves for a recurrence interval of 25 years. The durations and intensities are tabulated every $\Delta t = 10$ min for the total storm duration of 60 min. Values in the third column are cumulative depths associated with each duration and are equal to the duration multiplied by the intensity. Incremental depths are shown in the fourth column and are obtained by subtracting sequential values in the third column.

Table 4.6 Calculation of incremental precipitation depths for 25-yr, 1-hr storm

Duration (min)	Intensity (in./hr)	Cum. P (in.)	Incr. P (in)
10	7.02	1.17	1.17
20	5.26	1.75	0.58
30	4.42	2.21	0.46
40	3.84	2.56	0.35
50	3.36	2.8	0.24
60	2.97	2.97	0.17

Table 4.7 demonstrates the essence (and the source of the name) of the alternating block method. The first column is a tabulation of each time interval during the 1-hr design storm, and the second is the corresponding rainfall depth in each time interval. The second column is formed by taking the largest incremental depth from the last column of Table 4.6 (1.17 in.) and placing it about one-third to one-half of the way down the column (in this example, in the 20-to-30-min. time interval). The next largest incremental depth from Table 4.6 (0.58 in.) is then placed immediately after the largest incremental depth (in this case, in the 30 to 40-min. time interval). The third largest depth (0.46 in.) is placed before the largest depth (in the 10 to 20-min. interval). The remainder of the second column is filled by proceeding in this fashion, taking the largest remaining value from the last column of Table 4.6 and placing it in an alternating way either after or before the other entries in Table 4.7. Finally, the last column in Table 4.7 is computed by dividing each of the entries in the second column by Δt and expressing the result in units of in./hr. Figure E4.4.1 illustrates the resulting hyetograph.

Table 4.7 Formulation of the design storm hyetograph (alternating block method)

Time (min)	Incr. *P* (in.)	Intensity (in./hr)
0–10	0.24	1.44
10–20	0.46	2.76
20–30	1.17	7.02
30–40	0.58	3.48
40–50	0.35	2.1
50–60	0.17	1.02

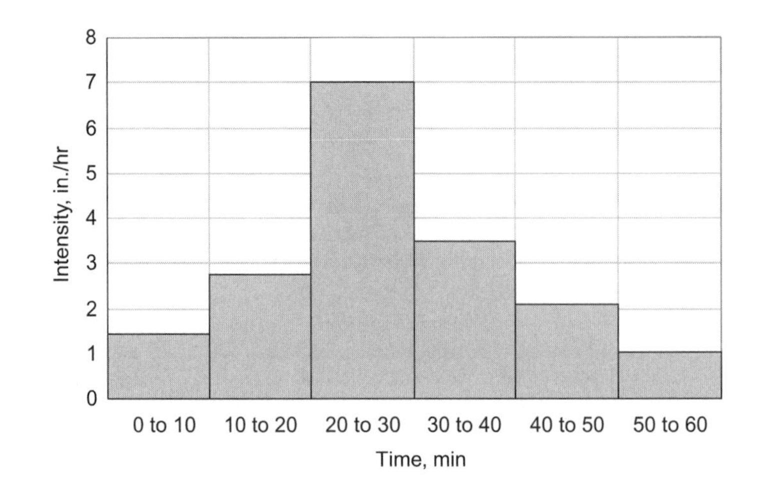

Figure E4.4.1 Design hyetograph derived from the alternating block method

---

Other Synthetic Temporal Distributions. Other methods of developing design storm hyetographs also exist. Notable among them are the quartile-based storm patterns presented by Huff (1967) and the instantaneous intensity method proposed by Keifer and Chu (1957). The engineer should use care when applying more recently developed synthetic design storms, because they will typically result in significantly different runoff hydrographs and may not be acceptable to the review agency having jurisdiction.

The original studies by Huff were updated and expanded for the Illinois State Water Survey Bulletin 71 (Huff and Angel, 1992), which provides rainfall data for nine states in the U.S. Midwest: Illinois, Indiana, Iowa, Kentucky, Michigan, Minnesota, Missouri, Ohio, and Wisconsin. Figure 4.26 shows some of the temporal distributions from Bulletin 71. Because both axes are dimensionless, the engineer can vary both the total depth of rainfall and the storm duration. The four temporal distributions correspond to different duration ranges (t_d) and the quartile of the storm during which the peak intensity occurs. As a general rule of thumb, these curves indicate that longer-duration events typically yield a lower overall intensity (which corresponds to a flatter slope on the distribution curve).

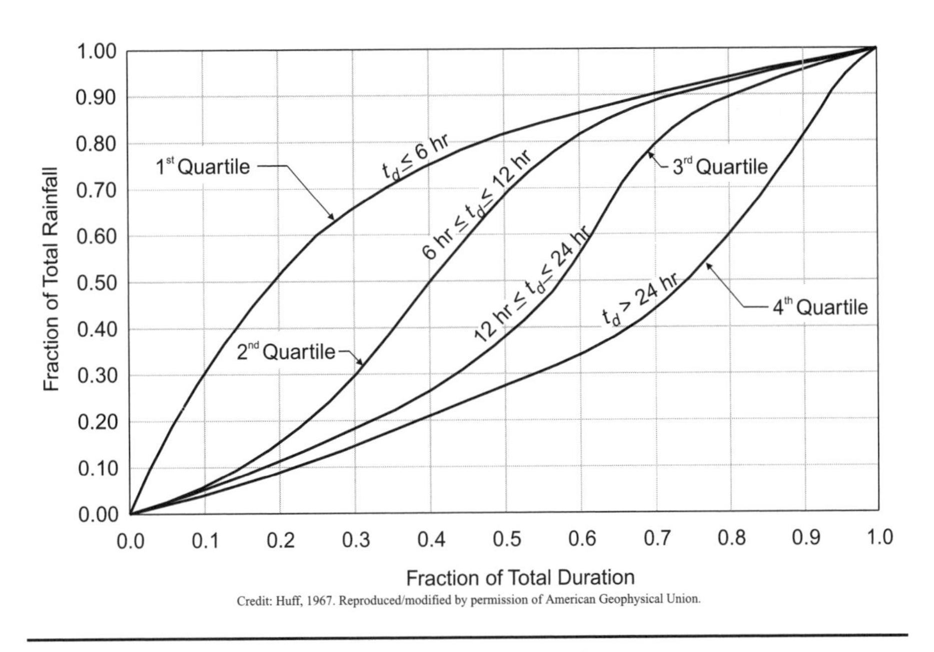

Figure 4.26 Synthetic temporal distributions for the U.S. Midwest (Huff and Angel, 1992)

Example 4.5 – Developing a Design Storm Hyetograph from a Bulletin 71 Distribution. Develop a design storm hyetograph for an event with a duration t_d = 3 hr and a total depth of 50 mm using the Bulletin 71 temporal distributions. Assume that Δt = 0.15 hr is a reasonable choice for the drainage basin to which the design storm will be applied. The dimensionless first-quartile rainfall distribution is given in Table 4.8.

Table 4.8 Bulletin 71 1st-quartile distribution

t/t_d	Fraction
0.00	0.00
0.05	0.16
0.10	0.33
0.15	0.43
0.20	0.52
0.25	0.60
0.30	0.66
0.35	0.71
0.40	0.75
0.45	0.79
0.50	0.82
0.55	0.84
0.60	0.86
0.65	0.88
0.70	0.90
0.75	0.92
0.80	0.94
0.85	0.96
0.90	0.97
0.95	0.98
1.00	1.00

Solution: First, the t/t_d column from Table 4.8 is multiplied by the duration (3 hr), and the rainfall fractions are multiplied by the total rainfall depth of 50 mm to obtain cumulative rainfall depths. The results are shown in Table 4.9.

Table 4.9 Cumulative depth for Example 4.5.

t (hr)	Cum. P (mm)
0.00	0.0
0.15	8.0
0.30	16.5
0.45	21.5
0.60	26.0
0.75	30.0
0.90	33.0
1.05	35.5
1.20	37.5
1.35	39.5
1.50	41.0
1.65	42.0
1.80	43.0
1.95	44.0
2.10	45.0
2.25	46.0
2.40	47.0
2.55	48.0
2.70	48.5
2.85	49.0
3.00	50.0

Next, the incremental precipitation depths are obtained by subtracting successive values for cumulative precipitation. Table 4.10 and Figure E4.5.1 show the resulting hyetograph.

Table 4.10 Incremental depth for Example 4.5

t (hr)	Incr. P (mm)
0.00–0.15	8.0
0.15–0.30	8.5
0.30–0.45	5.0
0.45–0.60	4.5
0.60–0.75	4.0
0.75–0.90	3.0
0.90–1.05	2.5
1.05–1.20	2.0
1.20–1.35	2.0
1.35–1.50	1.5
1.50–1.65	1.0
1.65–1.80	1.0
1.80–1.95	1.0
1.95–2.10	1.0
2.10–2.25	1.0
2.25–2.40	1.0

Table 4.10 Incremental depth for Example 4.5

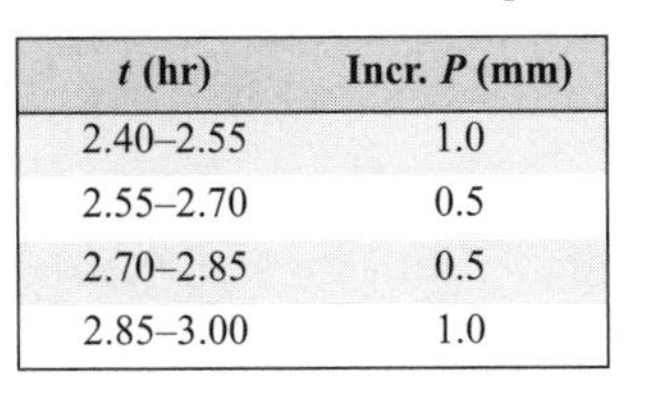

t (hr)	Incr. P (mm)
2.40–2.55	1.0
2.55–2.70	0.5
2.70–2.85	0.5
2.85–3.00	1.0

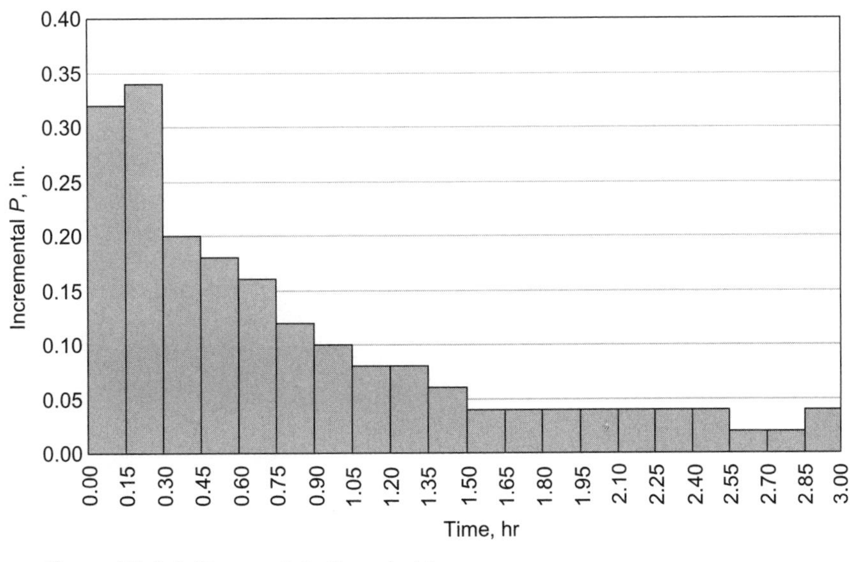

Figure E4.5.1 Hyetograph for Example 4.5

4.6 RAINFALL REQUIREMENTS FOR MODELING RUNOFF

This chapter has presented several ways in which rainfall information may be represented for stormwater runoff modeling and management. Which method to use in any system design or analysis depends on the approach used for transformation of the rainfall input into a prediction of the resulting stormwater discharge and on local regulations. Chapter 5 reviews a number of different runoff prediction methods and their input requirements.

Examples of various runoff prediction methods and their corresponding rainfall requirements are as follows:

- **Rational Method** – This peak discharge method is commonly used for smaller projects such as inlet design and simple storm sewers. Application of the rational method requires IDF curves, either provided by a review agency or created from other sources.

- **SCS Peak Discharge Method** – This method produces only a peak discharge (no hydrograph). Its use requires the 24-hour total rainfall depths for the selected recurrence interval.
- **SCS Unit Hydrograph** – Hydrographs are required for modeling detention ponds and complex watershed systems. Any rainfall distribution can be used with the SCS unit hydrograph method; however, the 24-hour total rainfall depths and the 24-hour rainfall temporal distribution (Type I, IA, II, or III) applicable to the geographic location of the project is most typical.
- **General Unit Hydrograph Methods** – A number of methods may be applied to synthesize a unit hydrograph for a given watershed. Because unit hydrograph analyses are time-based, applications of these methods require use of a rainfall hyetograph or a temporal distribution of some type. As a general rule, no limitations exist on the methods applied to estimate the hyetograph or temporal distribution. A number of "standardized" methods exist for design storm development, such as the SCS method, but hydrograph analyses are by no means limited to use with such methods. As a good practice, it is recommended that engineers compute stormwater runoff hydrographs using a number of plausible rainfall distributions, as this will enlighten them as to the sensitivities of their results and the corresponding uncertainties in their designs and analyses.

"All we can do now is sit tight and pray."

4.7 CHAPTER SUMMARY

Before designing or evaluating any stormwater drainage system, the engineer must first determine the flows, and therefore the rainfall event or events, that the system must be capable of handling.

In evaluating a rainfall event, the rainfall characteristics that must be considered are the depth/volume, duration, area, and average recurrence interval of the precipitation, as well as its temporal and spatial distributions. Various types of rainfall gauges are used to collect rainfall data and develop precipitation records. Rainfall data records are available from agencies such as the NWS in the United States, and much information is available on the Internet. Often, rainfall data has been processed into a form more readily accessible to the engineer and is presented in publications such as HYDRO-35 or TP-40, or as IDF curves for the locale of interest. IDF curves can also be used to approximate the recurrence interval for an actual event, and equations found in HYDRO-35 can be used to develop IDF curves for areas where little data is available.

Temporal distributions describe the variation in rainfall intensity over the course of a storm event. Temporal distributions may be actual gauged rainfall amounts, or they may be synthetic in nature. Examples of synthetic storm distributions are the NRCS (SCS) 24-hour distributions for the United States. Temporal distributions are used to develop rainfall hyetographs, which show incremental rainfall depths or intensities for storm events.

Once the appropriate hyetograph for a design or study has been developed, this information can be used to compute the runoff flow rates that will be used in drainage system analysis and design.

REFERENCES

Chow, V. T. 1964. *Handbook of Applied Hydrology.* New York: McGraw-Hill.

Chow, V. T., D. R. Maidment, and L. W. Mays. 1988. *Applied Hydrology.* New York: McGraw-Hill.

Durans, S. R. and P. A. Brown. 2001. "Estimation and Internet-Based Dissemination of Extreme Rainfall Information." *Transporation Research Record* no. 1743: 41–48.

Frederick, R. H., V. A. Myers, and E. P. Auciello. 1977. *Five- to 60-Minute Precipitation Frequency for the Eastern and Central United States.*NOAA Technical Memorandum NWS HYDRO-35. Silver Spring, Maryland: National Weather Service, Office of Hydrology.

Haestad Methods. 2003. *StormCAD User Manual.* Waterbury, Connecticut: Haestad Methods.

Hershfield, D. M. 1961. *Rainfall Frequency Atlas of the United States for Durations from 30 Minutes to 24 Hours and Return Periods from 1 to 100 Years.* Technical Paper No. 40. Washington, D.C.: Weather Bureau, U.S. Dept. of Commerce.

Huff, F. A. 1967. "Time Distribution of Rainfall in Heavy Storms." *Water Resources Research* 3, no. 4: 1007–1019.

Huff, F. A. and J. R. Angel. 1992. "Rainfall Frequency Atlas of the Midwest." *Illinois State Water Survey Bulletin* 71: 141.

Johnson, L. E., P. Kucera, C. Lusk, and W. F. Roberts. 1998. "Usability Assessments for Hydrologic Forecasting DSS." *Journal of the American Water Resources Association* 34, no. 1: 43–56.

Keifer, C. J., and H. H. Chu. 1957. "Synthetic Storm Pattern for Drainage Design." *Journal of the Hydraulics Division, ASCE* 84, no. HY4: 1–25.

Larson, L. W., and E. L. Peck. 1974. "Accuracy of Precipitation Measurements for Hydrologic Modeling." *Water Resources Research* 10, no. 4: 857–863.

Miller, J. F., R. H. Frederick, and R. J. Tracey. 1973. *Precipitation-Frequency Atlas of the Coterminous Western United States.* NOAA Atlas 2, 11 vols. Silver Spring, Maryland: National Weather Service.

Singh, V. P. 1992. *Elementary Hydrology.* Englewood Cliffs, NJ: Prentice Hall.

Soil Conservation Service (SCS). 1969. Section 4: Hydrology. In *National Engineering Handbook.* Washington, D.C.: U.S. Dept. of Agriculture.

Soil Conservation Service (SCS). 1986. *Urban Hydrology for Small Watersheds.* Technical Release 55. Washington, D.C.: U.S. Department of Agriculture.

Soil Conservation Service (SCS). 1992. *TR-20 Computer Program for Project Formulation Hydrology* (revised users' manual draft). Washington, D.C.: U.S. Department of Agriculture.

Sweeney, T. L. 1992. *Modernized Areal Flash Flood Guidance.* NOAA Technical Memorandum NWS HYDRO 44. Silver Spring, Maryland: National Weather Service, Office of Hydrology.

U.S. Weather Bureau. 1958. *Rainfall Intensity-Frequency Regime, Part 2 – Southeastern United States.* Technical Paper 29. Washington, D.C.: U.S. Department of Commerce.

PROBLEMS

4.1 The total volume of water stored in the atmosphere is estimated to be 12,900 km^3. The average rate of global evaporation and transpiration is 505,000 km^3/yr. What is the average amount of time (in days per year) that moisture spends in the atmosphere?

4.2 Using the methods given by HYDRO-35 and the data given in the table below, construct a set of IDF curves for return intervals of 2, 5, 10, 25, 50, and 100 years and durations of 5, 10, 15, 30, and 60 minutes.

	Rainfall Intensity (in/hr) for duration (min)		
T (yr)	**5**	**15**	**60**
2	5.1	3.3	2.4
100	8.4	6.6	3.4

4.3 Using Figures 4.15 through 4.20, complete the following table for Boston, Massachusetts.

	Rainfall Intensity (in/hr) for duration (min)		
T (yr)	**5**	**15**	**60**
2			
100			

4.4 Given the following data from a storm, plot the rainfall hyetograph (intensity versus time), calculate the total rainfall depth for the storm, and calculate the average rainfall intensity.

Interval (min)	Intensity (in/hr)
0–15	0.25
15–30	0.30
30–45	0.40
45–60	0.25
60–75	0.10
75–90	0.05

4.5 Use the appropriate 24-hour NRCS (SCS) synthetic rainfall distribution to develop a design storm hyetograph for a 10 year, 24-hour storm in Atlanta, Georgia. Use a total rainfall depth of 8.5 in. (from TP 40).

Depression storage, such as the shallow ponding in this grassed area, reduces the amount of runoff discharged from a watershed.

CHAPTER

5

Modeling Runoff

A number of procedures for developing estimates of stormwater runoff rates and other quantities of interest in hydrology are available to the engineer, but a set of universally accepted, "cookbook" procedures does not exist. Some government agencies have developed their own hydrologic methods and computer modeling tools, and some professional societies and organizations have developed manuals of practice, but ultimately the design professional needs to judge the suitability of any particular method for use in a given practical application. The professional must apply his or her knowledge and understanding of the problem at hand in piecing together a set of analytical tools to solve that problem. Experience clearly plays a large part in selecting and applying the proper rainfall and runoff methods—activities that can prove difficult for practitioners who are new to hydrology. Although the lack of strict procedures can be unsettling, the resourcefulness required to perform hydrologic engineering makes it an exciting field of practice.

The hydrology coverage in this text began in Chapter 4, which provided information on developing and obtaining precipitation data. Precipitation information in formats such as IDF curves or rainfall hyetographs is required input for the surface water hydrology models presented in this chapter. The primary focus of this chapter is the runoff generation processes and flow routing methods needed when determining the loads used in stormwater conveyance system design and analysis. A number of examples detailing the hydrologic calculations for many of these methods are presented to help the reader understand the theoretical concepts and assumptions inherent in the various models. However, engineers usually perform these calculations with the aid of computer programs.

Section 5.1 defines the various types of rainfall abstractions and presents some of the equations used in quantifying them. Section 5.2 presents factors affecting the amount of effective precipitation (runoff) and methods for quantifying it, as well as effective rainfall hyetograph development. Methods for estimating the response time of a drainage basin to a storm event are described in Section 5.3.

Section 5.4 extends the concepts presented in earlier sections to estimation of the peak discharge from a watershed. Often, the peak runoff rate for a given storm is sufficient to load and analyze the performance of a conveyance such as a storm sewer or culvert. Section 5.5 presents the application of the unit hydrograph concept to compute the complete runoff hydrograph resulting from a storm. Runoff hydrographs are necessary for situations in which both runoff volume and flow rate must be considered, as in the design of detention ponds.

Section 5.6 introduces the concept of base flow in a stream, and Section 5.7 provides a basic explanation of the concepts related to modeling the contribution of snowmelt to runoff, which is an important consideration in many regions. The Muskingum method, which is a technique for computing the effect of storage in a channel on the shape of the downstream hydrograph, is presented in Section 5.8.

5.1 RAINFALL ABSTRACTIONS

Recalling the discussion of the hydrologic cycle in Section 4.1, only a portion of the total rainfall occurring over a drainage basin contributes to surface runoff and stream flow. For example, a simple comparison of rainfall and runoff records for most locations in the United States shows that the equivalent depth of runoff (that is, stream flow) is typically about 30 to 50 percent of the precipitation depth on an annual basis. During intense storm events, the equivalent depth of runoff is often a much larger fraction of the total precipitation depth.

Rainfall that does not contribute to direct surface runoff may be intercepted by vegetation, infiltrated into the ground surface, retained as depression storage in puddles and small irregularities in the land surface, or returned to the atmosphere through transpiration and evaporation. Collectively, these losses of rainfall are called *abstractions*. The rainfall that remains after abstractions have occurred comprises the surface runoff and is called *effective precipitation* or *effective rainfall*. This section describes the various types of abstractions that occur in natural and urbanized drainage basins. Later sections present several commonly used methods for quantifying abstractions and effective precipitation.

Four basic types of abstractions are generally acknowledged, but only three of these—*interception*, *depression storage*, and *infiltration*—typically need to be addressed in the design and analysis of stormwater conveyance systems. The fourth abstraction type is *evaporation and transpiration* (usually referred to as *evapotranspiration*). Although they are not commonly considered in modeling stormwater conveyances, losses due to evapotranspiration are taken into account in safe yield calculations for facilities such as water supply reservoirs. Abstraction types are illustrated in Figure 5.1.

Interception

Interception refers to the capture of rainfall on the leaves and stems of vegetation before it reaches the ground surface. Water intercepted by vegetation is returned to the atmosphere by evaporation during dry-weather periods.

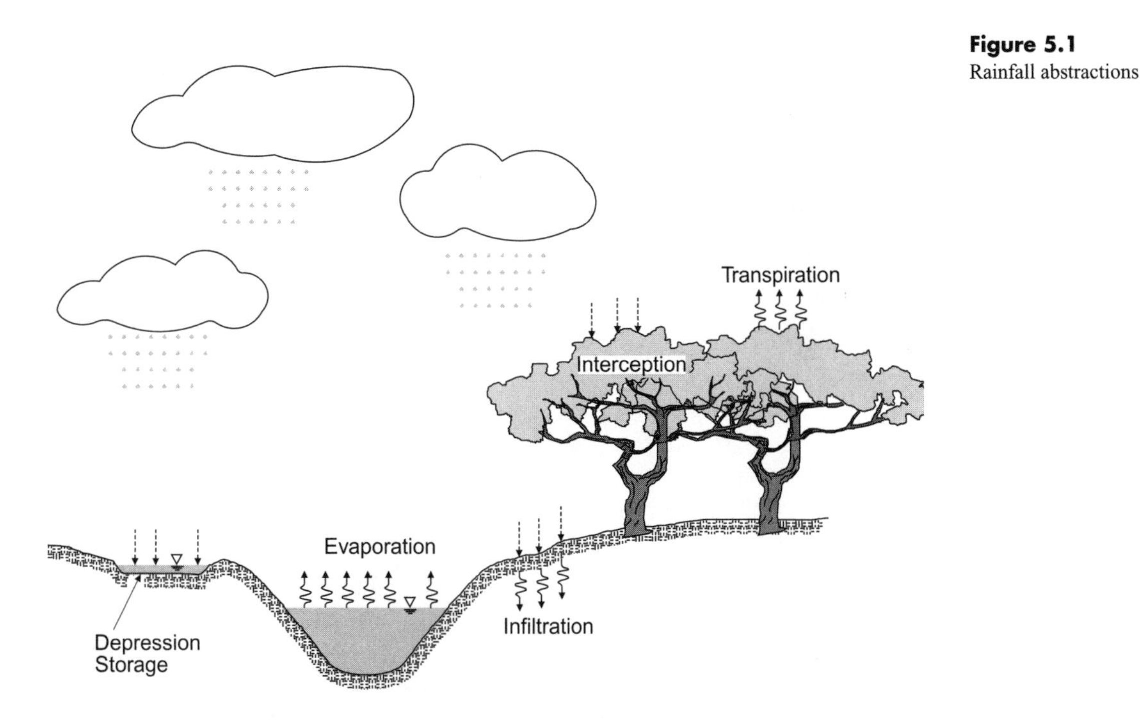

Figure 5.1
Rainfall abstractions

On an annual basis, interception can be quite significant and in some areas may approach 20 to 30 percent of the total rainfall (Helvey and Patric, 1965; Helvey, 1967; Zinke, 1967). During the relatively short and intense storm events of interest for rainfall and runoff studies, the percentage is often much smaller. Although some attempts to develop predictive formulas for computing interception have been made (Horton, 1919; Merriam, 1960), it is frequently assumed that interception is no more than an average equivalent rainfall depth of about 0.1 to 0.3 in. (2.5 to 7.6 mm) over the drainage basin.

Interception tends to be greater for coniferous trees than for deciduous trees (Patric, 1966). At their maximum growth, grasses may intercept as much rainfall as trees during individual storms (Merriam, 1961). In temperate regions, interception has a distinctly seasonal aspect because of the loss of leaves from vegetation during the cold season.

Depression Storage

Excess water begins to pond on the land surface when the rainfall intensity exceeds the infiltration capacity of the soil during a storm event. The ponded water fills small depressions and irregularities in the ground surface, and additional water is held on the surface through the phenomenon of surface tension. The water held in depressions and on the surface because of surface tension is called *depression storage,* and it either evaporates during dry-weather periods or infiltrates into the soil. Noninfiltrated rainfall that remains after surface depressions and irregularities have been filled contributes to surface runoff.

The depression storage capacity of a drainage basin is usually expressed in terms of an equivalent average depth of rainfall over the basin. Values for depression storage capacity range from about 0.01 in. (0.25 mm) for paved surfaces (Viessman, Knapp, and Lewis, 1977) to about 0.3 in. (7.6 mm) for forest litter (ASCE, 1992). These values are for moderately sloped surfaces. Values tend to be higher for flatter surfaces and lower for steeper surfaces.

Infiltration

When rainfall occurs on a pervious surface, some of the rainwater infiltrates into the ground in response to gravitational and capillary forces. The infiltrated water may contribute to groundwater recharge, or it may be taken up by the roots of vegetation and subsequently transpired through stomata (openings) in leaves. Infiltrated water may also be evaporated from the soil during dry-weather periods between storm events, or it may move laterally through the near-surface soils and reappear as surface water in a stream.

Infiltration capacity depends to a great extent on soil type. If the rate at which water can infiltrate a soil is greater than the rate at which rainfall is supplied to the soil surface, all rainfall is lost to infiltration. However, if the rainfall rate is greater than the infiltration capacity of the soil, surface ponding and/or surface runoff occurs. Sandy and gravelly soils generally have higher infiltration capacities than do silts and clays. For all soils, the rate at which infiltration can occur decreases with time and approaches a constant rate as the soil becomes wetter.

Two widely used methods for modeling infiltration, the Horton and Green-Ampt methods, are presented in the following subsections.

Horton Equation. A widely used method of representing the infiltration capacity of a soil is the Horton equation (Horton, 1939). The Horton method was empirically developed to describe field observations reflecting an exponential decay of infiltration rate over time as the soil becomes more saturated. For conditions in which the rainfall intensity is always greater than the infiltration capacity (that is, when rainwater supply for infiltration is not limiting), this method expresses the infiltration rate $f(t)$ as a function of time, as follows:

$$f(t) = f_c + (f_0 - f_c)e^{-k(t-t_0)} \quad (5.1)$$

where $f(t)$ = the infiltration rate (in/hr, mm/hr) at time t (min or hr)
f_c = a steady-state infiltration rate (in/hr, mm/hr) that occurs for sufficiently large t
f_0 = the initial infiltration rate at the time that infiltration begins (in/hr, mm/hr)
k = a decay coefficient (min^{-1} or hr^{-1})
t_0 = time at which infiltration begins (min or hr)

It can be shown theoretically that the steady-state infiltration rate f_c is equal to the saturated vertical hydraulic conductivity of the soil.

Note that the exponential term in Equation 5.1 is dimensionless. The units associated with t, t_0, and k may be taken in either minutes or hours, provided they are consistent. The units on the remaining terms in the equation are completely independent of the units used in the exponent term.

Estimation of the parameters f_c, f_0, and k in Equation 5.1 can be difficult because of the natural variabilities in *antecedent moisture conditions* (that is, the amount of moisture present in the soil prior to the rainfall event of interest) and soil properties. Region-specific data may be available, such as the values recommended by Rawls, Yates, and Asmusse (1976) in Table 5.1, but such tabulations should be used with caution. Singh (1992) recommends that f_0 be taken as roughly 5 times the value of f_c.

Table 5.1 Typical values of Horton infiltration parameters (Rawls, Yates, and Asmusse, 1976)

	f_0		f_c		
Soil Type	**(in/hr)**	**(mm/hr)**	**(in/hr)**	**(mm/hr)**	**k (min^{-1})**
Alphalpha loamy sand	19.0	483	1.4	36	0.64
Carnegie sandy loam	14.8	376	1.8	46	0.33
Dothan loamy sand	3.5	89	2.6	66	0.02
Fuquay pebbly loamy sand	6.2	157	2.4	61	0.08
Leefield loamy sand	11.3	287	1.7	43	0.13
Tooup sand	23.0	584	1.8	46	0.55

Often, the rainfall intensity during the early part of a storm is lower than the potential infiltration capacity (rate) of the soil; thus, the supply of rainwater is a limiting factor on the infiltration rate. During the time period when the water supply is limiting, the actual infiltration rate is equal to the rate at which rainwater is supplied to the ground surface. This effect is illustrated in Figure 5.2. Later in the storm, when the rainfall rate is greater than the infiltration rate, the actual infiltration rate will be greater than that predicted by Equation 5.1 because infiltration was limited early in the storm.

An integrated version of the Horton method can account for the underestimation of the infiltration rate due to limiting rainfall intensity early in a storm (Viessman, Knapp, and Lewis, 1977; Bedient and Huber, 1992; Chin, 2000), as can more complicated infiltration models such as the Green-Ampt (Green and Ampt, 1911) model presented in the next subsection. Nevertheless, the simple Horton model represented by Equation 5.1 is often used in practice because it yields a larger amount of effective precipitation than does the integrated version of the Horton model and is thus conservative for stormwater conveyance design. Depending on selected parameter values, Equation 5.1 may or may not yield more effective rainfall than do other models (for example, the Green-Ampt model).

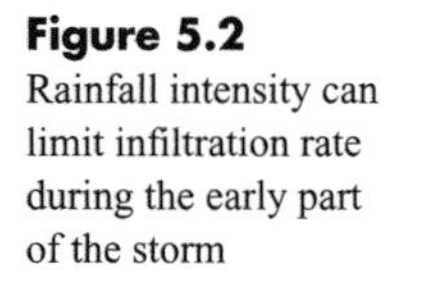

Figure 5.2
Rainfall intensity can limit infiltration rate during the early part of the storm

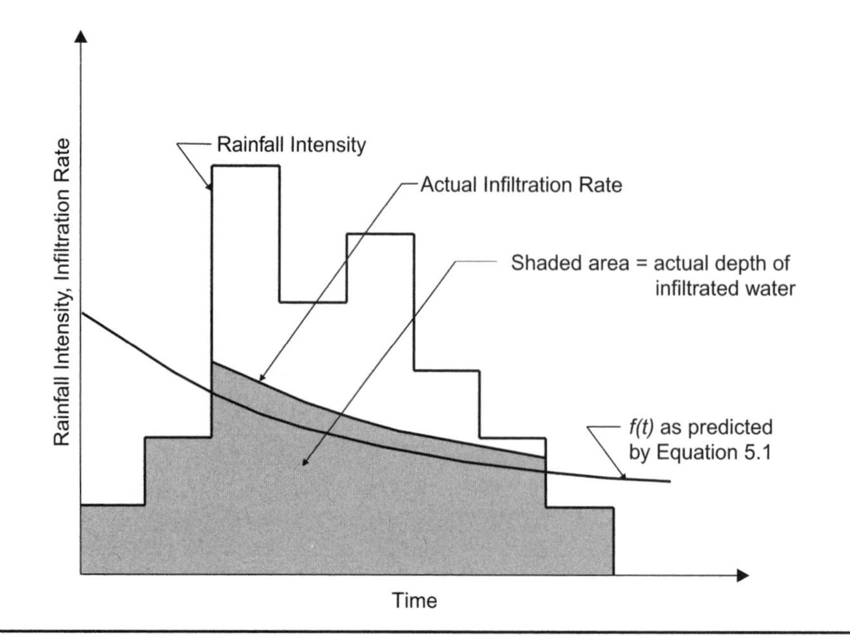

Example 5.1 – Using the Horton Equation to Determine Infiltration Rate. Using Horton's equation, find the infiltration rate at time = 2.0 hours for a Dothan loamy sand. The rainfall duration is 4 hours, and for this event, it takes 0.05 hours to reach the interception capacity, at which time infiltration begins.

From Table 5.1,

f_0 = 3.5 in/hr; f_c = 2.6 in/hr; and k = 0.02 min^{-1} = 1.2 hr^{-1}

Applying Equation 5.1,

$f(2\ \text{hr}) = 2.6\ \text{in/hr} + (3.5\ \text{in/hr} - 2.6\ \text{in/hr})e^{-1.2(2\ \ 0.05)} = 2.69\ \text{in/hr}$

The computed infiltration rate is between the values of f_0 and f_c, as it must be.

Green-Ampt Equation. In contrast to the empirically developed Horton equation, the Green-Ampt method is based on a theoretical application of Darcy's law (which relates flow velocity to the permeability of the soil) and conservation of mass. The resulting equation inversely relates the infiltration rate f to the total accumulated infiltration F as (Chow, Maidment, and Mays, 1988)

$$f = K_s\left(\frac{\psi(\theta_s - \theta_i)}{F} + 1\right) \tag{5.2}$$

where f = infiltration rate (in/hr, cm/hr)
K_s = saturated hydraulic conductivity (permeability) (in/hr, cm/hr)
ψ = capillary suction (in., cm)
θ_s = volumetric moisture content (water volume per unit soil volume) under saturated conditions
θ_i = volumetric moisture content under initial conditions

F = total accumulated infiltration (in., cm)

The benefit of the Green-Ampt method is that the infiltration rate can be calculated based on physical, measurable soil parameters, as opposed to the more empirical coefficients of Horton. For more information on these parameters, see Chow, Maidment, and Mays (1988).

To calculate the infiltration rate at a given time, the total infiltration up to that time must be calculated. This value can be determined by integrating Equation 5.2 with respect to time (starting at $t = 0$) and solving for F.

$$F = K_s t + \psi(\theta_s - \theta_i) \cdot \ln\left(1 + \frac{F}{\psi(\theta_s - \theta_i)}\right) \tag{5.3}$$

where t = time (hr)

Equation 5.3 cannot be explicitly solved and thus requires the application of a numerical method such as the Newton-Raphson or bisection method to solve for F. Also, the equation assumes that the rainfall intensity is always greater than the infiltration rate at a given time step. If the intensity is less than the associated infiltration rate, then the infiltration is equal to the rainfall amount for that time step.

Table 5.2 provides average values for ψ, K_S, and θ_s for the eleven U.S. Department of Agriculture (USDA) soil textures, which can serve as initial estimates for these parameters.

Table 5.2 Green-Ampt parameters (Rawls et al., 1993)

Soil Texture Class	Volumetric Moisture Content, θ_s	Capillary Suction, ψ (in.)	Capillary Suction, ψ (cm)	Saturated hydraulic conductivity[a] K_s (in/hr)	Saturated hydraulic conductivity[a] K_s (cm/hr)
Sand	0.437 (0.374–0.500)	1.95 (0.38–9.98)	4.95 (0.97–25.36)	9.28	23.56
Loamy sand	0.437 (0.363–0.506)	2.41 (0.53–11.00)	6.13 (1.35–27.94)	2.35	5.98
Sandy loam	0.453 (0.351–0.555)	4.33 (1.05–17.90)	11.01 (2.67–45.47)	0.86	2.18
Loam	0.463 (0.375–0.551)	3.50 (0.52–23.38)	8.89 (1.33–59.38)	0.52	1.32
Silt loam	0.501 (0.420–0.582)	6.57 (1.15–37.56)	16.68 (2.92–95.39)	0.27	0.68
Sandy clay loam	0.398 (0.332–0.464)	8.60 (1.74–42.52)	21.85 (4.42–108.0)	0.12	0.30
Clay loam	0.464 (0.409–0.519)	8.22 (1.89–35.87)	20.88 (4.79–91.10)	0.08	0.20
Silty clay loam	0.471 (0.418–0.524)	10.75 (2.23–51.77)	27.30 (5.67–131.50)	0.08	0.20
Sandy clay	0.430 (0.370–0.490)	9.41 (1.61–55.20)	23.90 (4.08–140.2)	0.05	0.12

Table 5.2 (cont.) Green-Ampt parameters (Rawls et al., 1993)

Soil Texture Class	Volumetric Moisture Content, θ_s	Capillary Suction, ψ (in.)	Capillary Suction, ψ (cm)	Saturated hydraulic conductivity[a] K_s (in/hr)	Saturated hydraulic conductivity[a] K_s (cm/hr)
Silty clay	0.479 (0.425–0.533)	11.50 (2.41–54.88)	29.22 (6.13–139.4)	0.04	0.10
Clay	0.475 (0.427–0.523)	12.45 (2.52–61.61)	31.63 (6.39–156.5)	0.02	0.06

a. K_s can be modified to obtain the Green-Ampt K. For bare ground conditions, K can be taken as $K_s/2$.

The following example illustrates common usage of the Green-Ampt method in calculating the infiltration volume (expressed as depth) over the course of an unsteady storm. For another example of the application of the Green-Ampt method, see Chow, Maidment, and Mays (1988). Computer programs for performing these calculations are available as well.

Example 5.2 – Computing Infiltration Volumes with Green-Ampt. Use the Green-Ampt method to solve for the amount of rainfall infiltrated for each time step of the rainfall hyetograph in Table 5.3.

Table 5.3 Rainfall Data for Example 5.2

Time (hr)	Intensity (in./hr)
0.0	0.3
0.1	1.9
0.2	3.2

Assume the following soil parameters:

$\psi = 3.5$ in.

$K_s = 0.13$ in./hr

$\theta_s - \theta_i = 0.434$

Solution: The key to finding the infiltration volume is keeping track of the position on two timelines. The first timeline is the rainfall hyetograph. The second timeline is the curve relating total infiltration to time as generated by Equation 5.2. At $t = 0$, both of these timelines correspond.

If the rainfall intensity is less than or equal to the infiltration rate for a particular time step, then all rainfall during that time step will infiltrate into the ground. If the intensity is greater than the infiltration rate during part or all of the time step, then ponding or runoff occurs during that time step. The total accumulated infiltration volume can be calculated by using Equation 5.3. As the total infiltration increases, the time on the second timeline increases, as well. If the intensity is greater than the infiltration rate, then the storm timeline and accumulated infiltration timeline proceed at the same pace. If intensity is less than the infiltration rate, then the timelines proceed at different paces over the course of the time step.

The steps taken in solving this example follow.

Time Step 1

a) Assume that all the rainfall infiltrates during the first time step:

   0.3 in./hr × 0.1 hr = 0.03 in.

b) Calculate the infiltration rate at the end of the time step as if all the rainfall infiltrates using Equation 5.2.

$$f = (0.13)\left[\frac{3.5(0.434)}{0.03} + 1\right] = 6.71 \text{ in./hr}$$

c) Because 6.71 in./hr > 0.3 in./hr up to an accumulated infiltration of 0.03 in., all rainfall during the time step is infiltrated.

Time Step 2

a) Again, assume all the rainfall for the time step infiltrates:

   1.9 in/hr ×0.1 hr = 0.19 in.

   The total accumulated infiltration is then computed as

   0.19 in. + 0.03 in. = 0.22 in.

b) The infiltration rate that would occur at the end of time step 2 assuming 0.22 in. of accumulated infiltration is computed to be

$$f = (0.13)\left[\frac{3.5(0.434)}{0.22} + 1\right] = 1.03 \text{ in/hr}$$

c) Because this infiltration rate is less than the intensity for time step 2 of 1.9 in/hr, ponding or runoff must begin sometime between 0.1 hr and 0.2 hr. The accumulated infiltration for which the rainfall intensity matches the infiltration rate must be computed. Substituting f = 1.9 in./hr and solving Equation 5.2 for F:

$$1.9 = 0.13\left[\frac{3.5(0.434)}{F} + 1\right]$$

   F = 0.112 in.

d) The time between 0.1 hr and 0.2 hr when the intensity equals the infiltration rate must be determined. Intensity equals infiltration when the total accumulated infiltration is 0.112 in. Keeping in mind that 0.03 in. infiltrated during the first time step, runoff begins at

   (0.112 in. – 0.03 in.) /1.9 in./hr = 0.043 hr after the beginning of the second time step

e) Next, the position on the second teemingly when 0.112 in. of accumulated infiltration would occur must be determined. Equation 5.3 is rearranged and solved for t.

$$t = \frac{1}{0.13}\left\{0.112 - (3.5)(0.434)\cdot \ln\left(1 + \frac{0.112}{3.5(0.434)}\right)\right\} = 0.03 \text{ hr}$$

f) To find the amount of infiltration during the time step, the time on the second timeline corresponding to the end of the second time step of the storm must be determined. Runoff begins after

   0.1 hr + 0.043 hr = 0.143 hr

when the total accumulated infiltration equals 0.112 in. This point corresponds to 0.03 hr on the second timeline. There is

   0.2 hr – 0.143 hr = 0.057 hr

remaining in the second time step for the storm timeline (first timeline). Thus, the end of time step 2 corresponds to a time of

0.03 hr + 0.057 hr = 0.087 hr

on the second timeline.

g) Now, solve for the total accumulated infiltration at 0.087 hr using Equation 5.3.

$$F = (0.13)(0.087) + (3.5)(0.434) \cdot \ln\left[\frac{F}{3.5(0.434)} + 1\right]$$

As mentioned above, F cannot be solved for explicitly and must be determined using a numerical method. Using a root finder, F is computed to be 0.193 in. at the end of time step 2.

h) The infiltration that occurred during time step 2 is the difference between the total accumulated infiltration at the end of step 1 and the total accumulated infiltration at the end of step 2:

0.193 in. – 0.03 in. = 0.163 in.

Time Step 3

a) Unlike the previous two time steps, the infiltration rate at the beginning of time step 3 is calculated for $F = 0.193$ in.

$$f = (0.13)\left[\frac{3.5(0.434)}{0.193} + 1\right] = 1.153 \text{ in./hr}$$

Because 3.2 in/hr > 1.153 in/hr, runoff must occur over the entire time step.

b) The next step is to determine the point on the second timeline that equates to the end of time step 3. Because runoff occurs over the entire time step (unlike step 2), 0.1 hr can simply be added to the point on the second timeline at the end of time step 2:

0.1 hr + 0.087 hr = 0.187 hr

Therefore, 0.187 hr on the second timeline corresponds to 0.3 hr on the storm hyetograph.

c) The total accumulated infiltration at 0.187 hr is computed as using Equation 5.3.

$$F = (0.13)(0.187) + (3.5)(0.434) \cdot \ln\left[\frac{F}{3.5(0.434)} + 1\right] = 0.288 \text{ in.}$$

d) The infiltration during the time step equals 0.288 in. – 0.193 in. = 0.095 in.

5.2 DETERMINATION OF EFFECTIVE PRECIPITATION (RUNOFF)

During and shortly after a rainfall event, the total discharge flowing in a stream channel generally consists of *direct runoff* from contributing land surfaces, *interflow* or *subsurface storm flow* resulting from lateral movement of water through shallow soil layers, and *base flow* or *groundwater outflow.* During dry-weather periods between storm events, a stream may not have a discharge at all. If a discharge does exist during those periods, that discharge consists solely of base flow.

As noted earlier, *effective precipitation* is that portion of the total precipitation during a storm event that is not lost to abstractions—it is the precipitation that becomes direct runoff. The *volume* of direct runoff caused by a storm event is equal to the product of the effective precipitation depth and the land surface area on which the precipitation occurred (that is, the drainage basin area).

Design of storm sewers and other elements of stormwater conveyance systems usually involves the estimation of direct runoff only, as stormwater conveyance elements often do not have significant base flow discharges. Section 5.6 discusses base flow estimation for cases in which it cannot be neglected.

The depth of effective precipitation may be determined in a number of ways; however, in its most general form, the relationship of direct runoff (effective precipitation) to total precipitation and abstractions is

$$D_r = D_p - D_{li} - D_i - D_s - D_e \quad (\text{if } D_r > 0) \tag{5.4}$$

Otherwise,

$$D_r = 0 \tag{5.5}$$

where D_r = total depth of direct runoff (effective precipitation)
D_p = total depth of precipitation (rainfall)
D_{li} = total initial loss, sometimes called initial abstractions
D_i = total depth infiltrated after initial losses
D_s = total depression storage depth
D_e = transpiration and evaporation losses (often ignored for short duration stormwater events)

These depths are presented graphically in Figure 5.3.

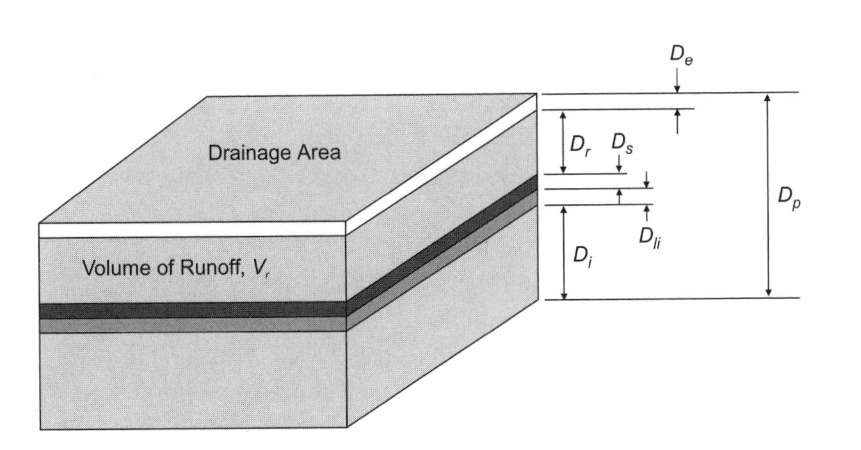

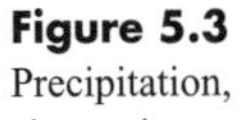

Figure 5.3
Precipitation, abstraction, and runoff volumes

As pointed out previously, the volume of direct runoff is expressed by

$$V_r = D_r \times A \tag{5.6}$$

where V_r = direct runoff volume
A = contributing drainage area

The area term, A, in Equation 5.6 is the contributing drainage area, or *watershed* area, for which the direct runoff volume is being evaluated. A watershed is a land area that drains to a single point of discharge. Typically, contour maps are used to delineate watershed boundaries and determine the area. Because water flows downhill, delineating a watershed is simply a matter of identifying an outfall point of interest and locating the watershed boundary such that any rain that falls within the boundary will be directed toward that point of discharge. The delineated area is then measured using a planimeter or by graphical or computer-aided methods.

Because river and stream systems collect water, a watershed may have any number of subwatersheds within it. The focus of the analysis and the determination of whether subwatersheds must be analyzed separately depend on the scope and purpose of the project at hand. Figure 5.4 shows the collection channels for two typical natural watersheds.

Figure 5.4
Typical natural watersheds with collection channels

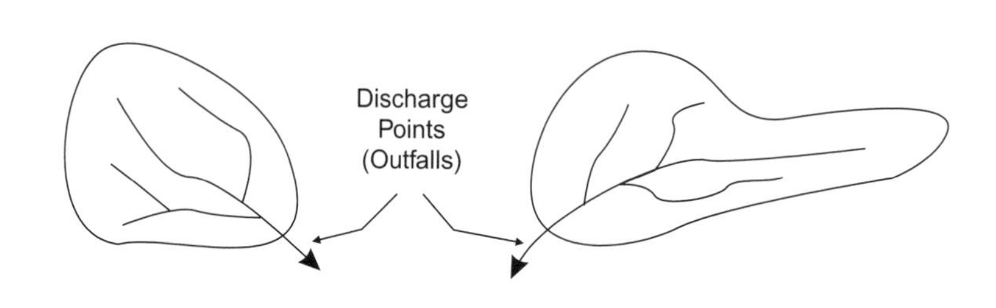

Example 5.3 – Computing Direct Runoff Volume. A 2.5-hr storm deposits 50 mm of rainfall over a drainage basin with an area of 1.5 ha. If the interception capacity is 8 mm, the depression storage depth is 5 mm, the calculated total infiltration is 18 mm, and evaporation is negligible, find the total depth of effective precipitation and the total direct runoff volume for the drainage area.

Solution: Applying Equation 5.4,

$$D_r = 50 - 8 - 18 - 5 - 0 = 19 \text{ mm}$$

Applying Equation 5.6,

$$V_r = (19 \text{ mm}/1000 \text{ mm/m})(1.5 \text{ ha}) = 0.0285 \text{ ha-m} = 285 \text{ m}^3$$

The concept of computing the direct runoff volume by subtracting abstractions from a precipitation depth is straightforward. However, the determination of the quantities for the rainfall abstractions, and thus the amount of *rainfall excess* (that is, direct runoff), is more complex. The following sections describe several possible methods for accounting for rainfall abstractions. The method that the engineer chooses can depend on a variety of factors, including the size of the site, land use and characteristics, data availability and needs, and acceptable local practice.

Runoff Coefficient

A simple approach to estimating runoff is to apply a coefficient that represents the ratio of rainfall that produces runoff. The use of the runoff coefficient C assumes that the effective rainfall intensity (i_e) is a fraction of the gross rainfall intensity (i), or

$$i_e = C \times i \tag{5.7}$$

where C = runoff coefficient ($0 \le C \le 1$)
i = rainfall intensity (in./hr, mm/hr)

This expression also can be written in terms of rainfall depths:

$$P_e = C \times P \tag{5.8}$$

where P_e = effective precipitation depth (direct runoff depth) (in., mm)
P = total precipitation depth (in., mm)

A good way to visualize the runoff coefficient is to think of it as a percentage of rainfall. For example, a C of 0.85 would yield a direct runoff (effective precipitation) depth that is 85 percent of the gross rainfall depth. The runoff coefficient is used with the rational method for peak runoff rate estimation, which is described in Section 5.4 (page 140).

Example 5.4 - Using the Runoff Coefficient to Compute Effective Rainfall Intensity. For a rainfall event with an average intensity of 50 mm/hr falling on a drainage area having a runoff coefficient = 0.70, determine the effective rainfall intensity.

$i_e = Ci = 0.70 \times 50$ mm/hr = 35 mm/hr

The runoff coefficient is also used with the rational method to compute peak runoff flow rate. A table containing ranges of values of C for various land uses, Table 5.11, is included with the rational method discussion in Section 5.4.

NRCS (SCS) Curve Number Method

In the 1950s, the U.S. Department of Agriculture Soil Conservation Service (now the NRCS) developed a procedure to partition the total depth of rainfall represented by a design storm hyetograph into initial abstractions I_a, retention F, and effective rainfall (runoff) P_e (SCS, 1969). These components are illustrated in Figure 5.5.

Initial abstractions consist of all rainfall losses occurring before the beginning of surface runoff, including interception, infiltration, and depression storage. *Retention* refers to the continuing rainfall losses following the initiation of surface runoff, which are predominantly due to continuing infiltration. Conservation of mass requires that

$$F = P - I_a - P_e \tag{5.9}$$

where F = equivalent depth of retention (in., mm)
P = total rainfall depth in storm (in., mm)
I_a = equivalent depth of initial abstractions (in., mm)
P_e = depth of effective precipitation (runoff) (in., mm)

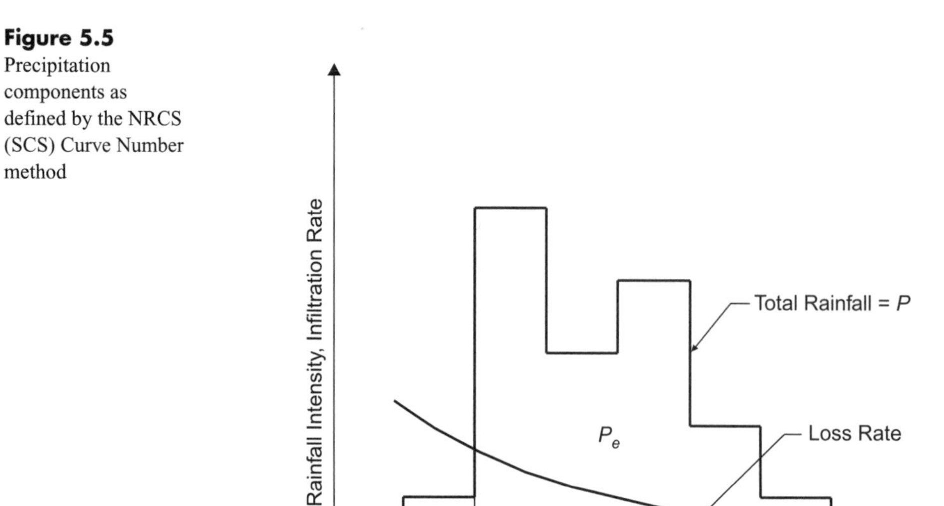

Figure 5.5 Precipitation components as defined by the NRCS (SCS) Curve Number method

By rearranging the terms, Equation 5.9 yields

$$P_e = (P - I_a) - F \tag{5.10}$$

This equation is very similar to the generalized infiltration and runoff Equation 5.4 (for that equation, $I_a = D_{li}$, $F = D_i$, $D_s = 0$, and $D_e = 0$).

The *curve number (CN)* referred to in the NRCS curve number method is a parameter used to estimate the *maximum possible retention (S)* of the soil in the area of interest. Its value depends on factors such as soil type, land use, vegetative cover, and moisture content prior to the onset of the storm event. S does not include initial abstractions (I_a). Using the equations that follow and the curve number (typical values for which are presented later in this section), the effective precipitation (runoff) resulting from a storm event can be computed.

An assumption made in the development of the curve number method is

$$\frac{F}{S} = \frac{P_e}{P - I_a} \tag{5.11}$$

where S = maximum possible retention (in., mm)

In essence, the assumption represented by Equation 5.11 is that the ratio of actual retention to maximum possible retention of water during a storm is equal to the ratio of effective rainfall to maximum possible effective rainfall (total rainfall less initial abstractions). Substitution of Equation 5.9 into Equation 5.11 yields

$$P_e = \frac{(P - I_a)^2}{(P - I_a) + S} \tag{5.12}$$

which is valid for values of $P > I_a$. Data analyzed by the NRCS indicated that I_a is related to S, and on average supported the use of the relationship $I_a = 0.2S$. Thus, Equation 5.12 becomes

$$P_e = \frac{(P - 0.2S)^2}{P + 0.8S} \tag{5.13}$$

when $P > 0.2S$ ($P_e = 0$ when $P \leq 0.2S$). Because the initial abstraction I_a consists of interception, depression storage, and infiltration prior to the onset of direct runoff, it may be appropriate in some applications to assume that $I_a = 0.1S$ or $I_a = 0.3S$ instead of $I_a = 0.2S$. For example, the relationship $I_a = 0.1S$ might be appropriate in a heavily urbanized area where there is little opportunity for initial abstractions to occur. Equation 5.13 must be modified when the relationship between I_a and S is assumed to be different from $I_a = 0.2S$.

The use of Equation 5.12 or 5.13 in estimating the depth of effective rainfall during a storm requires an estimate of the maximum possible retention S. NRCS conducted research to approximate S for various soil and cover conditions. To provide engineers with tables having a manageable range of coefficients from 1 to 100, the original values for S were modified using the following simple relationship:

$$CN = \frac{1000}{S + 10} \tag{5.14}$$

where CN = runoff curve number
S = maximum possible retention (in.)

Practical values of the curve number CN range from about 30 to 98, with large values being associated with impervious land surfaces. The NRCS has tabulated curve numbers as a function of soil type, land use, hydrologic condition of the drainage basin, and antecedent moisture condition.

Rearranging Equation 5.14, S is related to the runoff curve number, CN, as

$$S = \frac{1000}{CN} - 10 \tag{5.15}$$

Figure 5.6 provides a graphical solution to Equation 5.13 for various rainfall depths and curve numbers.

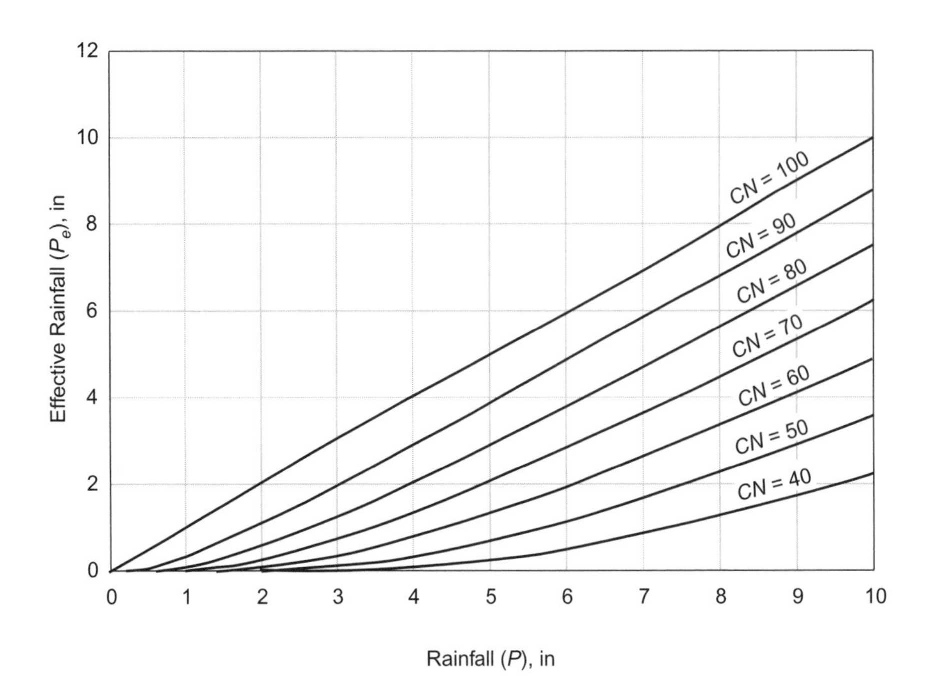

Figure 5.6
Graphical solution for effective precipitation (runoff) using the curve number method and assuming $I_a = 0.2S$

Soil Groups. Soils in the United States have been classified by the NRCS into four hydrologic groups: A, B, C, and D. *Group A* soils have high infiltration rates (low runoff potential), even when they are thoroughly wetted. Typical Group A soils are well-drained sands and gravels. *Group D* soils are at the opposite end of the spectrum, having low infiltration rates (high runoff potential). Typical Group D soils are clays, shallow soils over nearly impervious material, and soils with a high water table. *Group B* and *Group C* soils are in the midrange of the spectrum.

Information on the hydrologic soil group or groups present within a drainage basin in the United States may be found by contacting the nearest NRCS office and obtaining the soil survey of the county in which the project is located. If the hydrologic soil group is not provided in the soil survey, it can be found based on the soil name using Appendix A of TR-55 (SCS, 1986). When a drainage basin undergoes urbanization, the hydrologic soil group may change due to compaction of the soil by heavy construction equipment or mixing of soils as a consequence of grading operations.

Cover Type/Land Use. The surface conditions of a drainage area have a significant impact on direct runoff. For example, in the case of a sandy (Group A) soil completely paved with asphalt, the soil itself will have no impact on the amount of runoff. Even for pervious conditions, cover type plays a significant role in the amount of direct runoff from a site. For example, a heavily forested area will yield runoff volumes that differ from those of a lawn or plowed field.

Hydrologic Condition. The *hydrologic condition* of rangeland, meadow, or pasture is defined to be *good* if it is lightly grazed and has vegetative cover on more than 75 percent of the area. Conversely, a *poor* hydrologic condition corresponds to a heavily grazed area with vegetation covering less than 50 percent of the surface.

Sources of Soils Data for Stormwater Modeling

Soils information plays an important role in stormwater modeling and design. The ability of water to flow through the soil, typically indicated by the saturated hydraulic conductivity, K_s, has an important impact on how precipitation becomes runoff. K_s is used with methods such as Green-Ampt to compute infiltration. Dry soils with high K_s values may produce little or no runoff even from heavy storms. Conversely, wet soils that are highly compacted will result in almost all precipitation becoming runoff.

Another important soil property descriptor used in stormwater modeling is the hydrologic soil group (HSG), which has four soil classifications: A, B, C, and D. The NRCS Curve Number Method uses the HSG and land use to determine the runoff curve number. The table below shows the infiltration rate ranges and textures associated with each hydrologic soil group (HSG).

Type	Infiltration Rate	Texture
A	0.30–0.45 in/hr	Sands and gravels
B	0.15–0.30 in/hr	Course to moderately fine
C	0.05–0.15 in/hr	Moderately fine to fine
D	< 0.05 in/hr	Clays with high swelling, high water tables

Values for K_s, HSG, and many other soil properties for locations in the United States can be obtained from the NRCS. NRCS's SOILS web site (soils.usda.gov) contains information and links to soils data at varying levels of detail. Information on obtaining local soils data is provided (including links to PDF-format U.S. county soil survey manuscripts when available), as well as online access to the *State Soil Geographic Database (STATSGO)*. STATSGO contains generalized soil information on entire states at a level of detail appropriate for broad planning and management uses covering state, regional, and multi-state areas (map scale is 1:250,000).

The most detailed electronic information available from the NRCS is the *Soil Survey Geographic Data Base (SSURGO)* soils data. SSURGO data are the digital equivalent of information contained in the printed county soil surveys developed by the NRCS. The NRCS is currently converting the tabular and mapping information from the soil surveys into the digital SSURGO data. The SSURGO web site (www.ftw.nrcs.usda.gov/ssur_data.html) lists counties that have had their soils data converted to a GIS format and SSURGO database format.

Information for soils at this detail level (scales ranging from 1:12,000 to 1:63,360) is organized into Soil Map Units denoted by symbols like 10A or OcB. The first part of the symbol (10 or Oc) represents the soils map unit name, usually based on the abbreviated name of one or several official soil series that make up that map unit. The second part of the symbol (A or B) represents a slope class category *for that county*. One important thing to remember when working with soil maps is that each county's soil survey and resulting SSURGO data is unique to that county. Slope classes often differ between counties, even neighboring counties, because of the range of slopes in that county. Also, because soils surveys have been performed over a number of years and soil series standards have changed over time, similarly named map units from neighboring counties can have different characteristics.

Antecedent Moisture Condition. When rainfall events occur in quick succession, the time period between storms may be too short for the soils to dry to their average or normal moisture conditions. When rainfall occurs on soils that are already wet, the net result is that direct runoff volumes and peaks will be higher than normal. The NRCS (SCS) method accounts for this possibility by allowing the curve number to depend on an *antecedent moisture condition (AMC)*. Three AMC classifications exist. Normal conditions correspond to AMC-II. AMC-I corresponds to a drier condition, and AMC-III to a wetter condition. The *National Engineering Handbook (NEH-4)* (SCS, 1969) provided guidance on AMC selection in its Table 4.2; however, AMC is dependent on the size and location of the watershed, and this table was eliminated in the 1993 edition of Chapter 4 (NRCS, 1993). Often, typical local practice will influence AMC selection.

Table 5.4 is from Chapter 10 of NEH-4 (Mockus, 1972) and relates *CN* values for AMC-II to corresponding *CN* values for AMC-I and AMC-III. Column 5 represents the initial abstraction (I_a) for AMC-II conditions. The phrase "Curve starts where $P =$" indicates the cumulative depth of precipitation that must occur for direct runoff to begin (this occurs when $P = I_a$).

Alternatively, the curve numbers corresponding to AMC-I and AMC-III conditions can be computed from AMC-II curve numbers using Equations 5.16 and 5.17 (Chow, Maidment, and Mays, 1988). The computed *CN* may be rounded to the nearest whole number.

$$CN_{\text{I}} = \frac{4.2CN_{\text{II}}}{10 - 0.058CN_{\text{II}}} \tag{5.16}$$

$$CN_{\text{III}} = \frac{23CN_{\text{II}}}{10 + 0.13CN_{\text{II}}} \tag{5.17}$$

where CN_{I}, CN_{II}, and CN_{III} = curve numbers for AMC-I, -II, and -III, respectively

Curve Number Tables. Tables 5.5 through 5.8 provide listings of curve numbers that account for both cover conditions and soil type for normal antecedent moisture conditions (AMC-II) and $I_a = 0.2S$ (Mockus, 1969). The curve numbers shown for the urban and suburban land use conditions in Table 5.5 are based on the percentages of directly connected impervious areas in the drainage basin as shown in the table, and should be used with caution when the actual percentage of imperviousness in a drainage basin differs from this assumed value. When necessary, a composite curve number can be developed as an area-weighted average of individual curve numbers. The composite *CN* may be rounded to the nearest whole number.

Additional information on hydrologic condition and curve numbers for land uses other than those contained in Tables 5.5 through 5.8 may be found in Chapter 9 of NEH-4 (Mockus, 1969) and in TR-55 (SCS, 1986).

Table 5.4 Curve numbers (*CN*) and constants for the case $I_a = 0.2S$ (from Mockus, 1972)

(1) *CN* for Condition II	(2) *CN* for Conditions I	(3) *CN* for Conditions III	(4) S Values[a] (in.)	(5) Curve[a] Starts Where P = (in.)
100	100	100	0	0.00
99	97	100	0.101	0.02
98	94	99	0.204	0.04
97	91	99	0.309	0.06
96	89	99	0.417	0.08
95	87	98	0.526	0.11
94	85	98	0.638	0.13
93	83	98	0.753	0.15
92	81	97	0.870	0.17
91	80	97	0.989	0.20
90	78	96	1.11	0.22
89	76	96	1.24	0.25
88	75	95	1.36	0.27
87	73	95	1.49	0.30
86	72	94	1.63	0.33
85	70	94	1.76	0.35
84	68	93	1.90	0.38
83	67	93	2.05	0.41
82	66	92	2.20	0.44
81	64	92	2.35	0.47
80	63	91	2.50	0.50
79	62	91	2.66	0.53
78	60	90	2.82	0.56
77	59	89	2.99	0.60
76	58	89	3.16	0.63
75	57	88	3.33	0.67
74	55	88	3.51	0.70
73	54	87	3.70	0.74
72	53	86	3.89	0.78
71	52	86	4.08	0.82
70	51	85	4.29	0.86
69	50	84	4.49	0.90
68	48	84	4.71	0.94
67	47	83	4.93	0.99
66	46	82	5.15	1.03
65	45	82	5.38	1.08
64	44	81	5.63	1.13
63	43	80	5.87	1.17
62	42	79	6.13	1.23
61	41	78	6.39	1.28
60	40	78	6.67	1.33
59	39	77	6.95	1.39
58	38	76	7.24	1.45
57	37	75	7.54	1.51
56	36	75	7.86	1.57
55	35	74	8.18	1.64
54	34	73	8.52	1.70
53	33	72	8.87	1.77
52	32	71	9.23	1.85
51	31	70	9.61	1.92
50	31	70	10.0	2.00
49	30	69	10.4	2.08
48	29	68	10.8	2.17
47	28	67	11.3	2.26
46	27	66	11.7	2.35
45	26	65	12.2	2.44
44	25	64	12.7	2.55
43	25	63	13.3	2.65
42	24	62	13.8	2.76
41	23	61	14.4	2.88
40	22	60	15.0	3.00
39	21	59	15.6	3.13
38	21	58	16.3	3.26
37	20	57	17.0	3.41
36	19	56	17.8	3.56
35	18	55	18.6	3.71
34	18	54	19.4	3.88
33	17	53	20.3	4.06
32	16	52	21.3	4.25
31	16	51	22.3	4.45
30	15	50	23.2	4.67
25	12	43	30.0	6.00
20	9	37	40.0	8.00
15	6	30	56.7	11.33
10	4	22	90.0	18.00
5	2	13	190.0	38.00
0	0	0	∞	∞

a. For *CN* in column 1

Table 5.5 Runoff curve numbers for urban areas (Mockus, 1969)[a]

Cover Description		Curve Numbers for Hydrologic Soil Group			
Cover Type and Hydrologic Condition	Average Percent Impervious Area[b]	A	B	C	D
Fully developed urban areas (vegetation established)					
Open space (lawns, parks, golf courses, cemeteries, etc.)[c]:					
Poor condition (grass cover < 50%)		68	79	86	89
Fair condition (grass cover 50% to 75%)		49	69	79	84
Good condition (grass cover > 75%)		39	61	74	80
Impervious areas:					
Paved parking lots, roofs, driveways, etc. (excluding right-of-way)		98	98	98	98
Streets and roads:					
Paved; curbs and storm sewers (excluding right-of-way)		98	98	98	98
Paved; open ditches (including right-of-way)		83	89	92	93
Gravel (including right-of-way)		76	85	89	91
Dirt (including right-of-way)		72	82	87	89
Western desert urban areas:					
Natural desert landscaping (pervious area only)[d]		63	77	85	88
Artificial desert landscaping (impervious weed barrier, desert shrub with 1 to 2 in. sand or gravel mulch and basin borders)		96	96	96	96
Urban districts:					
Commercial and business	85	89	92	94	95
Industrial	72	81	88	91	93
Residential districts by average lot size:					
1/8 acre (506 m^2) or less (town houses)	65	77	85	90	92
1/4 acre (1,012 m^2)	38	61	75	83	87
1/3 acre (1,349 m^2)	30	57	72	81	86
1/2 acre (2,023 m^2)	25	54	70	80	85
1 acre (4,047 m^2)	20	51	68	79	84
2 acres (8,094 m^2)	12	46	65	77	82
Developing urban areas					
Newly graded area (pervious areas only, no vegetation)[e]		77	86	91	94
Idle lands (CNs are determined using cover types similar to those in Table 5.6)					

a. Average runoff condition, and $I_a = 0.2S$.

b. The average percent impervious area shown was used to develop the composite *CN*s. Other assumptions are as follows: impervious areas are directly connected to the drainage system, impervious areas have a *CN* of 98, and pervious areas are considered equivalent to open space in good hydrologic condition.

c. *CN*s shown are equivalent to those of pasture. Composite *CN*s may be computed for other combinations of open space cover type.

d. Composite *CN*s for natural desert landscaping should be computed using Figure 2.3 or 2.4 (in TR-55) based on the impervious area percentage ($CN = 98$) and the pervious area *CN*. The pervious area *CN*s are assumed equivalent to desert shrub in poor hydrologic condition.

e. Composite *CN*s to use for the design of temporary measures during grading and construction should be computed using Figure 2.3 or 2.4 (in TR-55) based on the degree of development (impervious area percentage) and the *CN*s for the newly graded pervious areas.

Table 5.6 Runoff curve numbers for cultivated agricultural lands[a] (Mockus, 1969)

Cover Description			Curve Numbers for Hydrologic Soil Group			
Cover Type	Treatment[b]	Hydrologic Condition[c]	A	B	C	D
Fallow	Bare soil	--	77	86	91	94
	Crop residue cover (CR)	Poor	76	85	90	93
		Good	74	83	88	90
Row crops	Straight row (SR)	Poor	72	81	88	91
		Good	67	78	85	89
	SR + CR	Poor	71	80	87	90
		Good	64	75	82	85
	Contoured (C)	Poor	70	79	84	88
		Good	65	75	82	86
	C + CR	Poor	69	78	83	87
		Good	64	74	81	85
	Contoured & terraced (C&T)	Poor	66	74	80	82
		Good	62	71	78	81
	C&T + CR	Poor	65	73	79	81
		Good	61	70	77	80
Small grain	SR	Poor	65	76	84	88
		Good	63	75	83	87
	SR + CR	Poor	64	75	83	86
		Good	60	72	80	84
	C	Poor	63	74	82	85
		Good	61	73	81	84
	C + CR	Poor	62	73	81	84
		Good	60	72	80	83
	C&T	Poor	61	72	79	82
		Good	59	70	78	81
	C&T + CR	Poor	60	71	78	81
		Good	58	69	77	80
Close-seeded or broadcast legumes or rotation meadow	SR	Poor	66	77	85	89
		Good	58	72	81	85
	C	Poor	64	75	83	85
		Good	55	69	78	83
	C&T	Poor	63	73	80	83
		Good	51	67	76	80

a. Average runoff condition, and $I_a = 0.2S$.

b. Crop residue cover applies only if residue is on at least 5% of the surface throughout the year.

c. Hydrologic condition is based on a combination of factors that affect infiltration and runoff, including density and canopy of vegetative areas, amount of year-round cover, amount of grass or close-seeded legumes in rotations, and degree of surface roughness. "Poor" indicates that there are factors that impair infiltration and tend to increase runoff. "Good" indicates that there are factors that encourage average and better than average infiltration and tend to decrease runoff.

Table 5.7 Runoff curve numbers for other agricultural lands[a] (Mockus, 1969)

Cover Description		Curve Numbers for Hydrologic Soil Group			
Cover Type	Hydrologic Condition	A	B	C	D
Pasture, grassland, or range-continuous forage for grazing[b]	Poor	68	79	86	89
	Fair	49	69	79	84
	Good	39	61	74	80
Meadow-continuous grass, protected from grazing and generally mowed for hay		30	58	71	78
Brush—brush-weed grass mixture with brush the major element[c]	Poor	48	67	77	83
	Fair	35	56	70	77
	Good	30[d]	48	65	73
Woods-grass combination (orchard or tree farm)[e]	Poor	57	73	82	86
	Fair	43	65	76	82
	Good	32	58	72	79
Woods[f]	Poor	45	66	77	83
	Fair	36	60	73	79
	Good	30[d]	55	70	77
Farmsteads—buildings, lanes, driveways, and surrounding lots		59	74	82	86

a. Average runoff condition, and $I_a = 0.2S$
b. *Poor:* less than 50% ground cover or heavily grazed with no mulch. *Fair:* 50 to 75% ground cover and not heavily grazed. *Good:* more than 75% ground cover and lightly or only occasionally grazed
c. *Poor:* less than 50% ground cover. *Fair:* 50 to 75% ground cover. *Good:* more than 75% ground cover
d. Actual curve number is less than 30; use $CN = 30$ for runoff computations
e. *CN*s shown were computed for areas with 50% woods and 50% grass (pasture) cover. Other combinations of conditions may be computed from the *CN*s for woods and pasture.
f. *Poor:* Forest litter, small trees, and brush are destroyed by heavy grazing or regular burning. *Fair:* Woods are grazed but not burned, and some forest litter covers the soil. *Good:* Woods are protected from grazing, and litter and brush adequately cover the soil

Table 5.8 Runoff curve numbers for arid and semiarid rangelands[a] (Mockus, 1969)

Cover Description		Curve Numbers for Hydrologic Soil Group:			
Cover Type	Hydrologic Condition[b]	A[c]	B	C	D
Herbaceous-mixture of grass, weeds, and low-growing brush, with brush the minor element	Poor		80	87	93
	Fair		71	81	89
	Good		62	74	85
Oak-aspen-mountain brush mixture of oak brush, aspen, mountain mahogany, bitter brush, maple, and other brush	Poor		66	74	79
	Fair		48	57	63
	Good		30	41	48
Pinyon-juniper-pinyon, juniper, or both; grass understory	Poor		75	85	89
	Fair		58	73	80
	Good		41	61	71
Sagebrush with grass understory	Poor		67	80	85
	Fair		51	63	70
	Good		35	47	55
Desert shrub-major plants include saltbush, greasewood, creosote-bush, blackbrush, bursage, palo verde, mesquite, and cactus	Poor	63	77	85	88
	Fair	55	72	81	86
	Good	49	68	79	84

a. Average antecedent moisture condition, and $I_a = 0.2S$. For range in humid regions, use Table 5.7.
b. *Poor:* less than 30% ground cover (litter, grass, and brush overstory). *Fair:* 30 to 70% ground cover. *Good:* more than 70% ground cover.
c. Curve numbers for group A have been developed for desert shrub.

Example 5.5 – Estimating Direct Runoff Depth and Volume Using the NRCS (SCS) Curve Number Method (Modified from SCS, 1986). Estimate the curve number, depth of direct runoff (effective precipitation), and direct runoff volume for a 400-ha drainage basin if the total depth of precipitation is 127 mm. All soils in the basin are in hydrologic soil group C. The proposed land use is 50 percent detached houses with 0.1-ha lots; 10 percent townhouses with 0.05-ha lots; 25 percent schools, parking lots, plazas, and streets with curbs and gutters; and 15 percent open space, parks, and schoolyards with good grass cover. Use an antecedent soil moisture condition of AMC-III. The detached housing and townhouse areas have directly-connected impervious area percentages corresponding to the ranges assumed in Table 5.5.

Solution: The composite curve number corresponding to AMC-II conditions is computed as a weighted average of the curve numbers presented in Table 5.5.

Land Use	Area (ha)	*CN*	Area × *CN*
Detached houses (0.1-ha lots)	200	83	16,600
Townhouses (0.05-ha lots)	40	90	3,600
Streets, plazas, etc.	100	98	9,800
Open space, parks, etc.	60	74	4,400
Sums	**400**		**34,400**

Thus, the composite CN is

$$CN = 34{,}400 / 400 = 86$$

From Equation 5.17, the curve number corresponding to AMC-III moisture conditions is found to be

$$CN_{\text{III}} = \frac{23(86)}{10 + 0.13(86)} = 93$$

Using Equation 5.15 and converting inches to millimeters, the maximum possible retention for this basin at AMC-III is

$$S = (1/0.0394)(1000/93 - 10) = 19 \text{ mm}$$

Initial abstractions are estimated to be

$$I_a = 0.2S = 3.8 \text{ mm}$$

Because $P > I_a$, Equation 5.13 is used to estimate the depth of direct runoff (effective precipitation) as

$$P_e = [127 - 0.2(19)]^2/(127 + 0.8(19)] = 107 \text{ mm}$$

From Equation 5.6, the total volume of direct runoff is

$$V_r = (107 \text{ mm} /1000 \text{ mm/m})\ (400 \text{ ha}) = 42.68 \text{ ha-m}$$

Directly Connected and Unconnected Impervious Area Adjustment Factors. Some of the curve numbers given in Table 5.5 are actually composite values based on the assumption that a certain percentage of the drainage area is impervious area with a CN of 98, and the remainder of the area has a CN corresponding to open space in good hydrologic condition [see footnote (b) of Table 5.5]. Further, it is assumed that the impervious portions of the drainage area are *directly connected,* meaning that runoff from these areas flows directly into the drainage system without first crossing a pervious area as unconcentrated flow. A directly connected impervious area can be contrasted with an *unconnected impervious area,* which is an impervious area whose runoff has more opportunity to infiltrate because it must flow over a pervious area before it becomes shallow concentrated flow or enters the drainage system. An example of an unconnected impervious area could be a tennis court surrounded by a grassy park area.

An engineer faced with a drainage area for which all impervious areas are directly connected, but the percentage of impervious area differs from the values in Table 5.5, can compute an adjusted curve number, CN_c, similar to the way a weighted runoff coefficient is computed:

$$CN_c = \frac{(CN_P)(A - A_i) + 98A_i}{A} \quad (5.18)$$

where CN_c = adjusted curve number
CN_p = curve number for pervious area
A = total area (ac, ha)
A_i = impervious area (ac, ha)

If the drainage area contains unconnected impervious areas, or both unconnected and directly connected impervious areas, the adjusted curve number is computed as

$$CN_c = CN_p + \left(\frac{A_i}{A}\right)\left(98 - CN_p\right)\left[1 - 0.5\left(\frac{A_u}{A_i}\right)\right] \quad (5.19)$$

where A_i = impervious area, connected and unconnected (ac, ha)
A = total area (ac, ha)
A_u = unconnected impervious area (ac, ha)

Method Limitations. The curve number method is widely used in the United States, mainly because of its simplicity and ease of use. It should be noted, however, that the curve number method has some significant limitations. Perhaps the most serious of these is that the dimension of time is not explicitly considered. In Example 5.5, for instance, it should matter whether the 5 in. (127 Mm) of rain fell in a time span of 24 hours or 2 hours, but the method has no way of accounting for this difference. Kibler (1982) observed that the method should be used only with 24-hour rainfall durations (for which it was originally developed) and advises caution when analyzing incremental amounts of rainfall occurring during a storm.

The CN method has several recognized limitations:

- The method describes average conditions, which makes it useful for design purposes, but the method's accuracy decreases for historical events.
- The curve number equation is not time-dependent, and it thus ignores differences resulting from varying rainfall duration and intensity.
- The common assumption that $I_a = 0.2S$ is generalized from data for agricultural watersheds; it may overestimate losses for impervious areas and underestimate losses for surface depressions.

- The method is not applicable when computing runoff due to snowmelt or rain on frozen ground.
- The method is less accurate for runoff depths of less than 0.5 in. (13 mm).
- The CN method only computes direct runoff; it does not consider subsurface flow or groundwater effects.
- If the weighted *CN* is less than 40, a different procedure must be used.

Another criticism of the curve number method is that its functional form implies that the infiltration rate $f = dF/dt$ can be expressed as (Morel-Seytoux and Verdin, 1981)

$$f = \frac{dF}{dt} = \frac{iS^2}{(P + 0.8S)^2} \qquad (5.20)$$

where $i = dP/dt$, the rainfall intensity (in/hr, mm/hr)

Because P and S are constants for a particular storm and drainage basin, this equation implies that the infiltration rate must rise and fall with the rainfall intensity. This implication is clearly unrealistic from a physical perspective.

A final criticism of the curve number method is that $dP_e/dP = 1$ for any curve number and sufficiently large P. The implication is that the infiltration rate approaches zero during sufficiently long storms, which again is unrealistic physically unless the water table rises to the ground surface or the soil is very poorly drained. As noted earlier, the infiltration rate should approach the saturated vertical hydraulic conductivity of a soil during long storms.

Creating Direct Runoff (Effective Rainfall) Hyetographs

Provided the data are available, the most accurate method of accounting for rainfall abstractions and developing a direct runoff (effective rainfall) hyetograph is to model abstractions and infiltration within each time step for a rainfall event. The process consists of subtracting the interception from the beginning of a gross rainfall hyetograph, subtracting infiltration from what remains after interception has been accounted for, and subtracting depression storage from what remains after both interception and infiltration have been accounted for. This approach is physically based and is not limited to use with any particular storm duration or rainfall hyetograph shape. The procedure is illustrated in the following example.

Example 5.6 – Computing an Effective Rainfall Hyetograph. Develop a direct runoff (effective rainfall) hyetograph for a watershed with an initial loss (interception capacity) of 0.3 in., a depression storage capacity of 0.2 in., and Horton infiltration parameters of f_0 = 1.5 in/hr, f_c = 0.3 in/hr, and k = 0.04 min^{-1}. The rainfall hyetograph, tabulated for both incremental depth and average intensity, is as follows:

t (min)	*P* (in.)	Avg. *i* (in./hr)
0–10	0.24	1.44
10–20	0.46	2.76
20–30	1.17	7.02
30–40	0.58	3.48
40–50	0.35	2.10
50–60	0.17	1.02

Solution: The interception capacity of 0.3 in. is subtracted first. Because 0.24 in. of rainfall occurs during the first 10 minutes of the storm, all of that rainfall plus an additional 0.06 in. of the rainfall occurring in the second 10 minutes of the storm is lost to interception. The rainfall hyetograph after accounting for interception is given in column 3 in the following table.

(1)	(2)	(3)
t (min)	*P* (in.)	*P* – 0.3 (in.)
0–10	0.24	0
10–20	0.46	0.4
20–30	1.17	1.17
30–40	0.58	0.58
40–50	0.35	0.35
50–60	0.17	0.17

The infiltration rate $f(t)$ can be calculated and tabulated as a function of t using Equation 5.1, where t_0 is the time at which rainwater first begins to infiltrate (t_0 = 10 min in this example, because rainfall prior to that time is lost to interception and hence is not available for infiltration).

$$f(t) = 0.3 + 1.2e^{-0.04(t-10)}$$

Column 1 in the table below is the time since the beginning of rainfall, and column 2 is the time since the beginning of infiltration. Column 3 is the infiltration rate computed using Equation 5.1. Column 4 contains incremental infiltration depths for each 10-minute period during the storm. For example, the first value of F is computed as the average of the current and preceding infiltration rates multiplied by the time interval Δt = 10 min = $^1/_6$ hr or, 0.22 = [(1.50 + 1.10)/2]/6.

(1)	(2)	(3)	(4)
t (min)	$t - t_0$ (min)	$f(t)$ (in/hr)	Incr. F (in.)
0			
10	0	1.5	
20	10	1.1	0.22
30	20	0.84	0.16
40	30	0.66	0.13
50	40	0.54	0.10
60	50	0.46	0.08

Subtraction of the infiltration depth in each time interval (column 3) from the corresponding rainfall depth remaining after interception (column 2) leads to the hyetograph in column 4 (any negative values produced should be set equal to zero) below. Finally, subtraction of the depression storage capacity of 0.2 in. leads to the effective rainfall hyetograph in column 5. Column 6 shows the effective rainfall hyetograph converted to intensities.

(1)	(2)	(3)	(4)	(5)	(6)
t (min)	$P - 0.3$ (in.)	Incr. F (in.)	P (in.)	P_e (in.)	i_e (in./hr)
0–10	0		0	0	0
10–20	0.4	0.22	0.18	0	0
20–30	1.17	0.16	1.01	0.99	5.94
30–40	0.58	0.13	0.45	0.45	2.70
40–50	0.35	0.10	0.25	0.25	1.50
50–60	0.17	0.08	0.09	0.09	0.54

The rainfall hyetograph and the effective rainfall hyetograph are illustrated in Figure E5.6.1 and Figure E5.6.2. Note that the effective rainfall intensities are less than the actual rainfall intensities. Note also that effective rainfall, and hence direct runoff, does not begin (in this example) until 20 minutes after the beginning of the storm.

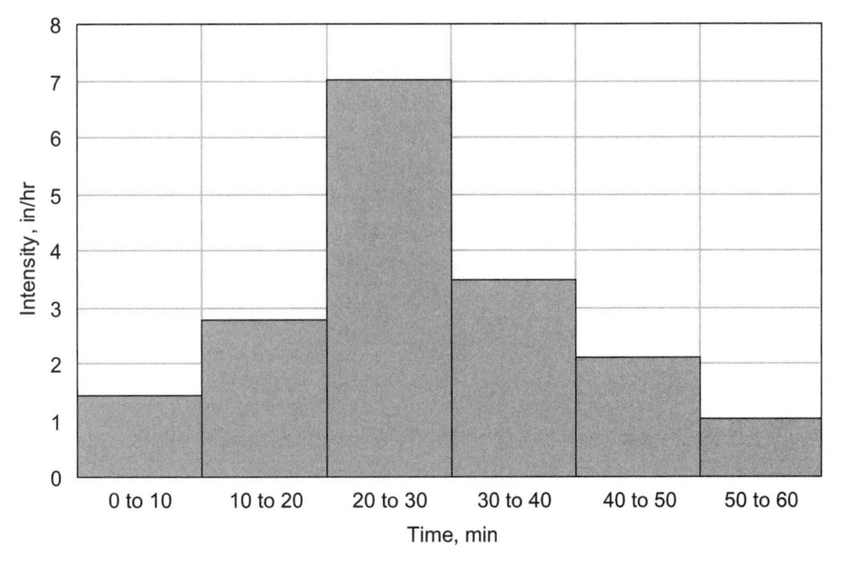

Figure E5.6.1 Rainfall hyetograph for Example 5.6

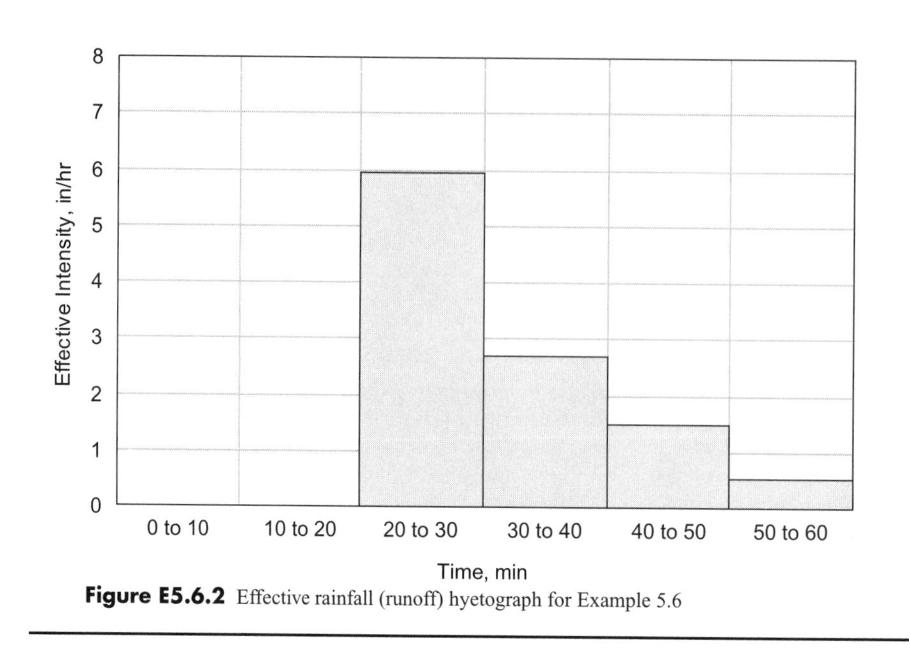

Figure E5.6.2 Effective rainfall (runoff) hyetograph for Example 5.6

5.3 MEASURES OF BASIN RESPONSE TIME

The maximum amount of flow discharged from a watershed at its outlet is related to the amount of time required for the entire watershed to be contributing to the flow. In modeling stormwater conveyance systems, the basin outlet may be taken as the location of an inlet or some other point of interest in the system. It can take minutes, hours, or even days from the onset of a rainfall event for the water falling in some parts of a watershed to be contributing to flow at a point of interest.

Because some points in a watershed are hydraulically closer to the outlet point than others, flow originating from different locations in the watershed will have differing *travel times* to the outlet. The *response time* of a drainage basin is usually considered to be the largest of all possible travel times, although it is sometimes taken as an average of all possible travel times. Estimates of peak runoff rates resulting from a rainfall event are quite sensitive to estimates of basin response time and vary inversely with them. That is, all else being equal, long response times are associated with small peak discharges and vice versa. Estimates of basin response time are also relevant to selection of the computational time step size Δt used for runoff prediction, as discussed in Section 4.5 (page 91).

The first of two common measures of basin response time is the *time of concentration*, denoted by t_c. The most widely adopted definition of time of concentration is the time required for a drop of effective rainfall falling at the most hydraulically remote point in a drainage basin to reach the basin outlet. The most hydraulically remote point from the outlet is usually, but not always, the most geographically remote point in the drainage basin. The significance of the time of concentration is easily recognized when one realizes that it is the minimum amount of time that must elapse before all parts of the drainage basin contribute to the flow at the basin outlet.

The second measure of response time commonly used in runoff estimation is the *basin lag time*, denoted by t_L. Often called simply the *basin lag* or *lag time*, this response time can be thought of as an approximate average of the possible travel times for runoff in a drainage basin. In practice, the basin lag is usually assumed to be the amount of time between the center of mass of a pulse of effective rainfall and the peak of the resultant direct runoff hydrograph (see Figure 5.7; t_p denotes time of peak discharge). The basin lag time is often used when estimating a complete runoff hydrograph as opposed to merely the peak runoff rate.

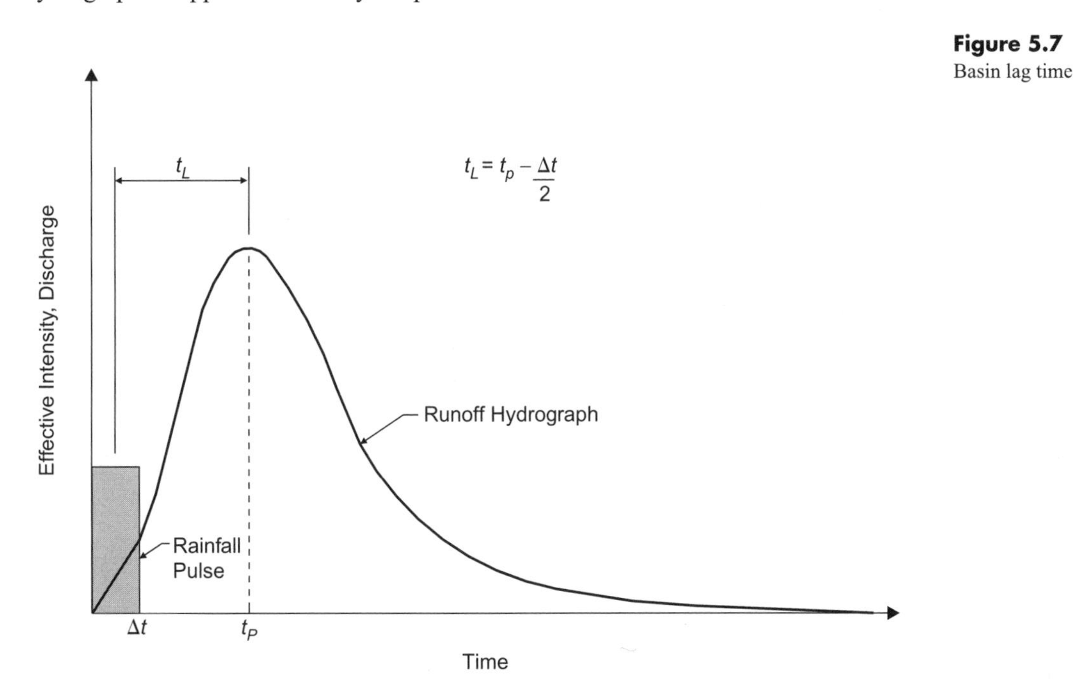

Figure 5.7
Basin lag time

Basin Response Time Estimation Methods

Many methods for estimating the time of concentration and basin lag are given in various private, federal, and local publications. Although each of these methods is different (in some cases only slightly), all are based on the type of ground cover, the slope of the land, and the distance along the flow path. In most localities, there is also a minimum t_c (typically 5 to 10 minutes) recommended for small watersheds such as a section of a parking lot draining to a storm sewer. Some methods predict the response time directly, and others predict the velocity of flow. The predicted velocity coupled with estimates of the flow path length can then be used to estimate the response time.

With few exceptions, methods for the prediction of basin response time are empirical in nature. Consequently, large errors in response time estimates can be expected to occur if these methods are not carefully selected and applied. These errors can significantly affect peak runoff estimates. The method selected for estimation of t_c or t_L should be one that was developed for basin conditions similar to those existing in the drainage basin for which an estimate is desired. McCuen, Wong, and Rawls (1984) compared a number of methods for estimating the time of concentration and developed measures of their reliability. As a general rule, methods that compute individual travel times for various types of flow segments (for example, overland flows and channelized flows), and then sum the individual travel times to estimate the total travel time, are thought to be the most reliable.

Physically, the response time of a drainage basin depends on, at a minimum, the length of the flow path, the slope of the basin, and the surface roughness. Additional factors included in some prediction methods are rainfall intensity and a measure of the basin shape. Because urbanization of a watershed tends to reduce surface roughness and often changes flow path lengths and slopes, a change in basin response time (nearly always a decrease) and corresponding increases in peak runoff rates should be expected as a consequence of urbanization. One way to help reduce the increase in peak runoff rate caused by urbanization is to increase the time of concentration and/or basin lag through practices such as terracing of land surfaces.

Table 5.9 lists several commonly used methods for estimating basin lag time. Table 5.10 lists several commonly used methods for estimating the time of concentration of a drainage basin. Figure 5.8 illustrates average overland flow velocities as a function of land use characteristics and surface slope. When flows are channelized in gutters, open channels, or storm sewers, Manning's equation may be used to estimate the velocity of flow. For more information, see Section 6.2 (page 200), Section 7.2, and Section 10.1.

Table 5.9 Commonly used methods for estimation of basin lag time, in hours

Equation	Source	Remarks
$t_L = C_t(LL_{ca})^{0.3}$	Snyder (1938), Linsley (1943)	C_t = empirical coefficient [typical range between 1.8 (steeper basins) and 2.2 (flatter basins)], L = basin length (mi), and L_{ca} = length along main channel to a point adjacent to the basin centroid (mi).
$t_L = 0.6\exp\{0.212D(LL_{ca})^{-0.36}\} / S^{0.5}$	Taylor and Schwarz (1952)	D = drainage density, L = basin length (mi), L_{ca} = length along main channel to a point adjacent to the basin centroid (mi), and S = average channel slope (ft/ft).
$t_L = L_w^{0.8}(1000 / CN - 9)^{0.7} / 1900S^{0.5}$	Soil Conservation Service (1986)	L_w = length of drainage basin (ft), CN = curve number of drainage basin, and S = average basin slope (percent).
$t_L = 0.6t_c$	Kent (1972)	Lag is approximated as $0.6t_c$ for use in SCS Unit Hydrograph computations.

Table 5.10 Commonly used methods for estimation of the time of concentration, in minutes

Equation	Source	Remarks
$t_c = 60LA^{0.4} / DS^{0.2}$	Williams (1922)	L = basin length (mi), A = basin area (mi^2), D = diameter (mi) of a circular basin of area A, and S = basin slope (percent). The basin area should be smaller than 50 mi^2.
$t_c = KL^{0.77} / S^n$	Kirpich (1940)	Developed for small drainage basins in Tennessee and Pennsylvania, with basin areas from 1 to 112 ac. L = basin length (ft), S = basin slope (ft/ft), K = 0.0078 and n = 0.385 for Tennessee; K = 0.0013 and n = 0.5 for Pennsylvania. The estimated t_c should be multiplied by 0.4 if the overland flow path is concrete or asphalt, or by 0.2 if the channel is concrete-lined.
$t_c = (2LN / 3S^{0.5})^{0.47}$	Hathaway (1945), Kerby (1959)	Drainage basins with areas of less than 10 ac and slopes of less than 0.01. This is an overland flow method. L = overland flow length from basin divide to a defined channel (ft), S = overland flow path slope (ft/ft), and N is a flow retardance factor (N = 0.02 for smooth impervious surfaces; 0.10 for smooth, bare packed soil; 0.20 for poor grass, row crops, or moderately rough bare surfaces; 0.40 for pasture or average grass; 0.60 for deciduous timberland; and 0.80 for coniferous timberland, deciduous timberland with deep ground litter, or dense grass).
$t_c = 300(L / S)^{0.5}$	Johnstone and Cross (1949)	Developed for basins in the Scotie and Sandusky River watersheds (Ohio) with areas between 25 and 1,624 mi^2. L = basin length (mi), and S = basin slope (ft/mi).
$t_c = \{(0.007I + c) / S^{0.33}\} \times (IL / 43{,}200)^{-0.67} L / 60$	Izzard (1946)	Hydraulically derived formula. I = effective rainfall intensity (in/hr), S = slope of overland flow path (ft/ft), L = length of overland flow path (ft), and c is a roughness coefficient (c = 0.007 for smooth asphalt, 0.012 for concrete pavement, 0.017 for tar and gravel pavement, and 0.060 for dense bluegrass turf).
$t_c = 0.94I^{-0.4} (Ln / S^{0.5})^{0.6}$	Henderson and Wooding (1964)	Based on kinematic wave theory for flow on an overland flow plane. I = rainfall intensity (in/hr), L= length of overland flow (ft), n = Manning's roughness coefficient, S = overland flow plane slope (ft/ft).
$t_c = 1.8(1.1 - C)L^{0.5} / S^{0.333}$	Federal Aviation Agency (1970)	Developed based on airfield drainage data. C = rational method runoff coefficient, L = overland flow length (ft), and S = slope (percent).
$t_c = \frac{1}{60} \sum (L / V)_i$	Soil Conservation Service (1986)	Time of concentration is developed as a sum of individual travel times. L = length of an individual flow path (ft) and V = velocity of flow over an individual flow path (ft/s). V may be estimated by using Figure 5.8 or by using Manning's equation.

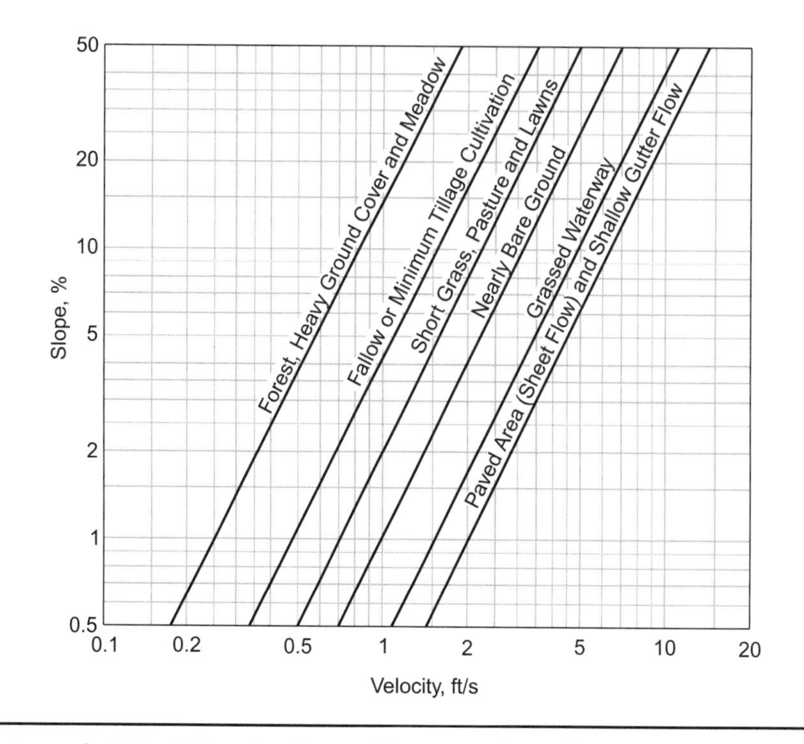

Figure 5.8 Average overland flow velocities as a function of land use characteristics and surface slope (Kent, 1972)

Example 5.7 – Estimating Time of Concentration Using NRCS (SCS) Methods.

An urbanized drainage basin is shown in Figure E5.7.1. Three types of flow conditions exist from the furthest point of the drainage basin to its outlet. Estimate the time of concentration based on the following data:

Reach	Flow Description	Slope (%)	Length (ft)
A to B	Overland (forest)	7	500
B to C	Overland (shallow gutter)	2	900
C to D	Storm sewer with manholes, inlets, etc. (n = 0.015, diam. = 3 ft)	1.5	2,000
D to E	Open channel, gunite-lined, trapezoidal (B = 5 ft, y = 3 ft, z = 1:1, n = 0.019)	0.5	3,000

Solution: For the reach from A to B, the average flow velocity is $V = 0.7$ ft/s (from Figure 5.8). The travel time for that reach is therefore

$$t_{AB} = L/V = 500/0.7 = 700 \text{ s}$$

Similarly, for the reach from B to C, the average flow velocity is $V = 2.8$ ft/s. The travel time for that reach is therefore

$$t_{BC} = L/V = 900/2.8 = 320 \text{ s}$$

To compute the travel time in the storm sewer from C to D, Manning's equation is employed to compute the pipe-full velocity:

$$V = \frac{1.49}{n}\left(\frac{D}{4}\right)^{2/3} S^{1/2} = \frac{1.49}{0.015}\left(\frac{3}{4}\right)^{2/3} (0.015)^{1/2} = 10 \text{ ft/s}$$

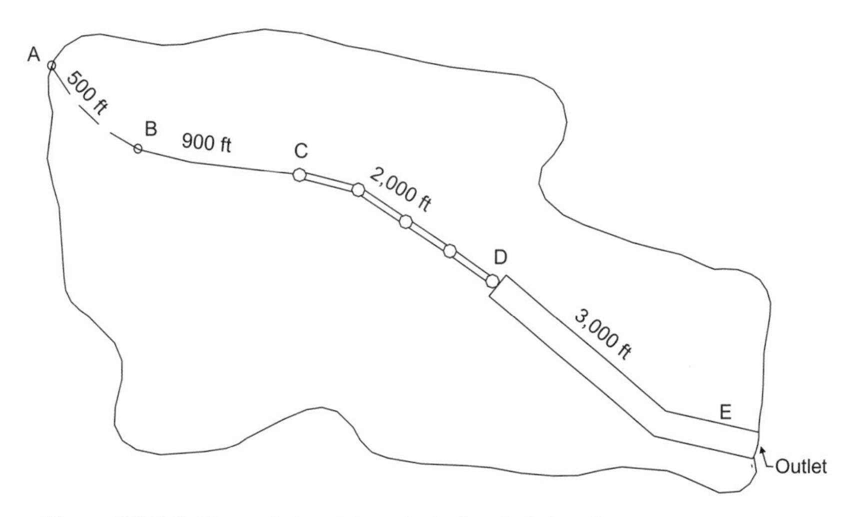

Figure E5.7.1 Flow paths in a drainage basin for calculation of t_c

The travel time for that reach is therefore

$$t_{CD} = L/V = 2000/10 = 200 \text{ s}$$

Travel time in the open channel from D to E is computed using the bank-full velocity, again found via the Manning equation:

$$V = \frac{1.49}{n} R^{2/3} S^{1/2} = \frac{1.49}{0.019} (1.78)^{2/3} (0.005)^{1/2} = 8.2 \text{ ft/s}$$

The travel time for that reach is therefore

$$t_{DE} = L/V = 3{,}000/8.2 = 370 \text{ s}$$

The time of concentration is the sum of the four individual travel times and is

$$t_c = 1{,}590\ s = 0.44 \text{ hr}$$

Time of Concentration and Basin Lag

Because the time of concentration is the longest of all the travel times in a drainage basin, and because the basin lag time can be thought of as an average of all the possible travel times, the time of concentration is longer than the basin lag time. The NRCS has suggested that $t_c = 1.67t_L$, or that $t_L = 0.6t_c$. Overton and Meadows (1976) indicated that $t_c = 1.6t_L$.

The NRCS, in Technical Release 55 (TR-55) (SCS, 1986) describes a procedure for estimating the time of concentration of a drainage basin as the sum of up to three types of individual travel times consisting of sheet flow, shallow concentrated flow, and channelized flow. The method presented therein is essentially the same as that represented by the last entry in Table 5.10 in that it involves estimation of flow velocities and then deduces travel times by considering the length of the flow paths. Conceptually, Example 5.7 illustrates the same type of procedure.

"Well, we needed the rain."

5.4 PEAK FLOW ESTIMATION

To design stormwater conveyance or detention systems, the engineer must first estimate runoff rates, as these are the discharges for which the conveyance facilities must be designed. Two basic levels of analysis exist. The first level is a peak flow calculation to determine the maximum runoff rate at a given point resulting from a storm event. This level of analysis is often sufficient for designing storm sewers and culverts whose only function is to convey runoff away from areas where it is unwanted. The second level, which is more complex, consists of the generation of a runoff *hydrograph* to provide information on flow rate versus time and runoff volume. This type of information is necessary when drainage basins are too large or too complex to be treated by peak flow estimation methods, or when the analysis of natural or artificial detention or retention facilities is required. Section 5.5 describes hydrograph estimation methods.

A number of methods are commonly used for the estimation of peak runoff rates. Three popular methods are (1) the rational method, (2) the NRCS (SCS) peak flow estimation method, and (3) regression-based methods developed by the U.S. Geological Survey. Each of these three methods is described in the subsections that follow.

Rational Method

The rational method, also called the Lloyd-Davies method in the United Kingdom, was developed in 1851 by Mulvaney. It is an equilibrium-based approach to peak flow estimation that uses rainfall intensity data and watershed characteristics to predict peak flows for a rainfall event. This method was originally presented in American hydrologic literature by Kuichling (1889) and has been a staple of American hydrologic practice since that time. The rational method is especially popular in storm sewer design because of its simplicity, and because storm sewer design typically requires only peak discharge data.

At the most fundamental level, the rational method assumes that an equilibrium (that is, a steady state) is attained such that the effective rainfall inflow rate of water onto a drainage basin is equal to the outflow rate of water from the basin. If one expresses the volumetric effective inflow rate as the product of the basin area A and the effective rainfall intensity i_e, then the outflow rate Q is obtained as $Q = i_e A$. Further, if one accounts for abstractions using a runoff coefficient, then the effective intensity is a product of the actual rainfall intensity and the runoff coefficient, resulting in

$$Q = CiA \tag{5.21}$$

where Q = runoff rate (ac-in/hr, ha-mm/hr)
C = runoff coefficient (see Table 5.2)
i = rainfall intensity (in/hr, mm/hr)
A = drainage area (ac, ha)

Because 1 ac-in/hr = 1.008 cfs $\approx$ 1 cfs, engineers performing calculations by hand in U.S. customary units typically ignore the conversion factor and simply assume that the discharge Q is in units of cfs. This conversion factor is not ignored in computer applications.

Table 5.11 lists recommended runoff coefficients corresponding to various land uses. It should be noted, however, that some locales have developed runoff coefficient tables that also consider soil type and/or storm return period. Coefficients should be selected carefully for proper application to a particular locale.

When a drainage basin consists of a mixture of land uses, a *composite runoff coefficient* may be computed for the basin by weighting individual runoff coefficients for each land use by their respective areas, as demonstrated in Example 5.8.

The time of concentration used to find i_e is the smallest time for which the entire basin is contributing runoff to the basin outlet; therefore, the storm duration must be at least as long as the time of concentration if a steady-state condition is to be achieved. Also, steady-state conditions dictate that the storm intensity be spatially and temporally uniform. It is not reasonable to expect that rainfall will be spatially uniform over a large drainage basin, or that it will be temporally uniform over a duration at least as long as the time of concentration when t_c (and hence A) is large. Therefore, these conditions limit the applicability of the rational method to small drainage basins. An upper limit of 200 acres (80.9 ha) has been suggested by some, but the limit should really depend on the storm characteristics of the particular locale. These local characteristics may limit the applicability of the rational method to basins smaller than 10 acres (4 ha) in some cases.

Table 5.11 Runoff coefficients for use in the rational method (Schaake, Geyer, and Knapp, 1967)

Type of Area or Development	C
Types of Development	
Urban business	0.70–0.95
Commercial office	0.50–0.70
Residential development	
Single-family homes	0.30–0.50
Condominiums	0.40–0.60
Apartments	0.60–0.80
Suburban residential	0.25–0.40
Industrial development	
Light industry	0.50–0.80
Heavy industry	0.60–0.90
Parks, greenbelts, cemeteries	0.10–0.30
Railroad yards, playgrounds	0.20–0.40
Unimproved grassland or pasture	0.10–0.30
Types of Surface Areas	
Asphalt or concrete pavement	0.70–0.95
Brick paving	0.70–0.80
Roofs of buildings	0.80–0.95
Grass-covered sandy soils	
Slopes 2% or less	0.05–0.10
Slopes 2% to 8%	0.10–0.16
Slopes over 8%	0.16–0.20
Grass-covered clay soils	
Slopes 2% or less	0.10–0.16
Slopes 2% to 8%	0.17–0.25
Slopes over 8%	0.26–0.36

Example 5.8 – Determining the Weighted Runoff Coefficient. Estimate the runoff coefficient for a drainage basin that is made up of 6 ha of park and 12 ha of medium-density, single-family housing.

Solution: From Table 5.11, the runoff coefficients for the park and residential areas are estimated to be 0.20 and 0.40, respectively. The composite runoff coefficient for the entire drainage basin of 18 ha is therefore

$$C = [6(0.20) + 12(0.40)]/18 = 0.33$$

Additional assumptions associated with the rational method are that the runoff coefficient is a constant and does not change during the duration of the storm, and that the recurrence intervals of the rainfall and corresponding runoff are equal (see, for example, Schaake, Geyer, and Knapp, 1967).

When several drainage basins (or subbasins) discharge to a common facility (such as a storm sewer or culvert), the time of concentration should be taken as the longest of all the individual times of concentration. Further, the total drainage area served (the sum of the individual basin areas) should be no larger than the 200-acre limit (or smaller where applicable) of the rational method.

The basic steps for applying the rational method are as follows:

Step 1: Estimate t_c and apply I-D-F data. Develop or obtain a set of intensity-duration-frequency (IDF) curves for the locale in which the drainage basin resides (see Chapter 4). Estimate the time of concentration of the basin using one of the techniques presented in Section 5.3 or other applicable method. Assume that the storm duration is equal to the time of concentration and use the IDF curves to determine the precipitation intensity for the recurrence interval of interest. Note that the assumption that the storm duration and time of concentration are equal is conservative in that it represents the highest intensity for which the entire drainage area can contribute.

The rational method is frequently used in the design of storm sewer systems, and an important part of the hydraulic analysis is determining the time of concentration for each component in the system so that design flows can be computed correctly. The time of concentration for an inlet's contributing drainage area is used to compute the inlet design flow. However, finding the appropriate time of concentration for use in computing the design flow for a pipe additionally involves the consideration of travel time through any upstream piping. The longest total travel time to the pipe is the controlling time of concentration used in design flow calculation. An illustration of this approach is provided later in the text in Example 11.1 (page 424).

Step 2: Compute watershed area. The basin area A can be estimated using topographic maps, computer tools such as CAD or GIS software, or by field reconnaissance.

Step 3: Choose C coefficient. The runoff coefficient C can be estimated using Table 5.11 if the land use is homogeneous in the basin, or a composite C value can be estimated if the land use is heterogeneous (see Example 5.5).

Step 4: Solve for peak flow. Finally, the peak runoff rate from the basin can be computed using Equation 5.21.

The following example illustrates the use of the rational method for several subbasins draining into a common storm sewer system.

Example 5.9 – Computing Peak Flow with the Rational Method. A proposed roadway culvert near Memphis, Tennessee is to drain a 5-ha area with a runoff coefficient of 0.65. The time of concentration for the drainage area is 15 min. Use the rational method to compute the peak discharge (in m^3/s) for a 25-year design storm. From Example 4.2 (page 90), the 15-min intensity for a 25-year storm in Memphis is 5.96 in/hr (151.4 mm/hr).

Solution: From Equation 5.21, the peak flow is computed as

$$Q = CiA = (0.65)(151.4 \text{ mm/hr})(5 \text{ ha}) = 492.05 \text{ ha-mm/hr}$$

Converting the units gives $Q = 1.37\ m^3/s$.

Additional Duration and Area Considerations. Differences in land surface characteristics within a drainage basin can give rise to peak discharges larger than those computed assuming basin homogeneity. For example, the impervious surfaces within a basin can be subdivided into two parts, one consisting of *directly connected impervious areas* and the other consisting of *indirectly connected* or *unconnected impervious areas*. In a residential subdivision, the directly connected impervious areas consist of roadway and driveway surfaces from which runoff flows directly to stormwater inlets. Runoff from indirectly connected impervious areas passes over pervious areas prior to reaching channels, gutters, or inlets; thus, an infiltration opportunity exists for runoff from these areas. Rooftop areas are a common example of unconnected impervious areas in a residential subdivision.

One problem with using the rational method as previously described is that it may under-predict peak flow for heterogeneous drainage basins. This under-prediction occurs because the time of concentration for the basin as a whole is longer than that for the directly connected impervious areas alone. Thus, the intensity for the directly connected impervious area considered alone will be higher than that of the basin as a whole. In fact, it may be so much higher that the peak flow computed for the directly connected impervious area only is greater than that computed for the whole basin, despite the fact that its area is smaller. The higher peak flow estimate should govern in such cases. Example 11.2 (page 427) illustrates this procedure.

NRCS (SCS) Peak Flow Estimation

The NRCS (SCS) has developed a simple procedure for estimating peak runoff rates for small drainage basins. Graphical and tabular solutions to this procedure are presented in Technical Release 55 (TR-55) (SCS, 1986), and tabular and regression-based solutions are presented in the Federal Highway Administration's Hydrologic Engineering Circular No. 22 (HEC-22) (Brown, Stein, and Warner, 2001). The HEC-22 regression procedure is described in this section.

For use of this method, the drainage basin should have a fairly homogeneous distribution of CN, and the composite CN should be at least 40. The basin should also have one main channel, though branches are acceptable if they have approximately equal times of concentration. Further, this method was developed for use with 24-hour storm rainfall depths. Its use with other storm durations is not advisable.

This method is implemented by first applying Equation 5.12 or 5.13 to estimate the depth of effective precipitation during a storm. The peak discharge is then estimated as

$$Q_p = q_u A P_e \tag{5.22}$$

where Q_p = peak runoff rate (cfs, m^3/s)
q_u = unit peak runoff rate (cfs/mi^2/in, m^3/s/km^2/cm)
A = drainage basin area (mi^2, km^2)
P_e = depth of effective precipitation (in., cm)

The unit peak runoff rate is estimated as

$$q_u = C_f \cdot 10^K \tag{5.23}$$

where C_f = conversion factor (1.00 for U.S. units; 4.30×10^{-3} for SI units)
K = exponent given by Equation 5.24

The exponent K is computed as

$$K = C_0 + C_1 \log_{10} t_c + C_2 (\log_{10} t_c)^2 \tag{5.24}$$

where C_0, C_1, and C_2 are coefficients listed in Table 5.12
t_c = time of concentration (hr)

The values of C_0, C_1, and C_2 are given in Table 5.12 as a function of the SCS 24-hour design storm distribution type and the ratio I_a/P, where P is the 24-hour rainfall depth and I_a is the depth of initial abstractions, as covered in Section 5.2 (page 119). For convenience, the ratio I_a/P is tabulated in Table 5.13 as a function of the curve number. The ratio I_a/P should not be less than 0.1, nor larger than 0.5. Further, the time of concentration should be in the range of 0.1 to 10 hours.

When ponding or swampy areas exist in a drainage basin, it is recommended that the peak discharge produced by Equation 5.22 be reduced to account for the temporary storage of runoff. In this case, an adjusted peak flow Q_{pa} should be computed as

$$Q_{pa} = F_p Q_p \tag{5.25}$$

where Q_{pa} = adjusted peak flow (cfs, m^3/s)
F_p = adjustment factor obtained from Table 5.14

Ponds and swampy areas lying along the flow path used for computation of the time of concentration should not be included in the area percentage in Table 5.14. Neither channel nor reservoir routing can be accommodated with this method.

Storm Developing

Table 5.12 Coefficients for SCS (NRCS) peak discharge method by storm type I_a/P (Brown, Stein, and Warner, 2001)

Storm Type	I_a/P	C_0	C_1	C_2
I	0.10	2.30550	–0.51429	–0.11750
	0.15	2.27044	–0.50908	–0.10339
	0.20	2.23537	–0.50387	–0.08929
	0.25	2.18219	–0.48488	–0.06589
	0.30	2.10624	–0.45696	–0.02835
	0.35	2.00303	–0.40769	–0.01983
	0.40	1.87733	–0.32274	0.05754
	0.45	1.76312	–0.15644	0.00453
	0.50	1.67889	–0.06930	0
IA	0.10	2.03250	–0.31583	–0.13748
	0.15	1.97614	–0.29899	–0.10384
	0.20	1.91978	–0.28215	–0.07020
	0.25	1.83842	–0.25543	–0.02597
	0.30	1.72657	–0.19826	0.02633
	0.35	1.70347	–0.17145	0.01975
	0.40	1.68037	–0.14463	0.01317
	0.45	1.65727	–0.11782	0.00658
	0.50	1.63417	–0.09100	0
II	0.10	2.55323	–0.61512	–0.16403
	0.15	2.53125	–0.61698	–0.15217
	0.20	2.50928	–0.61885	–0.14030
	0.25	2.48730	–0.62071	–0.12844
	0.30	2.46532	–0.62257	–0.11657
	0.35	2.41896	–0.61594	–0.08820
	0.40	2.36409	–0.59857	–0.05621
	0.45	2.29238	–0.57005	–0.02281
	0.50	2.20282	–0.51599	–0.01259
III	0.10	2.47317	–0.51848	–0.17083
	0.15	2.45395	–0.51687	–0.16124
	0.20	2.43473	–0.51525	–0.15164
	0.25	2.41550	–0.51364	–0.14205
	0.30	2.39628	–0.51202	–0.13245
	0.35	2.35477	–0.49735	–0.11985
	0.40	2.30726	–0.46541	–0.11094
	0.45	2.24876	–0.41314	–0.11508
	0.50	2.17772	–0.36803	–0.09525

Table 5.13 I_a/P for selected rainfall depths and curve numbers (adapted from Brown, Stein, and Warner, 2001)

	Curve Number, *CN*											
P (in.)	**40**	**45**	**50**	**55**	**60**	**65**	**70**	**75**	**80**	**85**	**90**	**95**
0.50	0.50	0.50	0.50	0.50	0.50	0.50	0.50	0.50	0.50	0.50	0.44	0.21
1.00	0.50	0.50	0.50	0.50	0.50	0.50	0.50	0.50	0.50	0.35	0.22	0.11
1.50	0.50	0.50	0.50	0.50	0.50	0.50	0.50	0.44	0.33	0.24	0.15	0.10
2.00	0.50	0.50	0.50	0.50	0.50	0.50	0.43	0.33	0.25	0.18	0.11	0.10
2.50	0.50	0.50	0.50	0.50	0.50	0.43	0.34	0.27	0.20	0.14	0.10	0.10
3.00	0.50	0.50	0.50	0.50	0.44	0.36	0.29	0.22	0.17	0.12	0.10	0.10
3.50	0.50	0.50	0.50	0.47	0.38	0.31	0.24	0.19	0.14	0.10	0.10	0.10
4.00	0.50	0.50	0.50	0.41	0.33	0.27	0.21	0.17	0.13	0.10	0.10	0.10
4.50	0.50	0.50	0.44	0.36	0.30	0.24	0.19	0.15	0.11	0.10	0.10	0.10
5.00	0.50	0.49	0.40	0.33	0.27	0.22	0.17	0.13	0.10	0.10	0.10	0.10
5.50	0.50	0.44	0.36	0.30	0.24	0.20	0.16	0.12	0.10	0.10	0.10	0.10
6.00	0.50	0.41	0.33	0.27	0.22	0.18	0.14	0.11	0.10	0.10	0.10	0.10
6.50	0.46	0.38	0.31	0.25	0.21	0.17	0.13	0.10	0.10	0.10	0.10	0.10
7.00	0.43	0.35	0.29	0.23	0.19	0.15	0.12	0.10	0.10	0.10	0.10	0.10
7.50	0.40	0.33	0.27	0.22	0.18	0.14	0.11	0.10	0.10	0.10	0.10	0.10
8.00	0.38	0.31	0.25	0.20	0.17	0.13	0.11	0.10	0.10	0.10	0.10	0.10
8.50	0.35	0.29	0.24	0.19	0.16	0.13	0.10	0.10	0.10	0.10	0.10	0.10
9.00	0.33	0.27	0.22	0.18	0.15	0.12	0.10	0.10	0.10	0.10	0.10	0.10
9.50	0.32	0.26	0.21	0.17	0.14	0.11	0.10	0.10	0.10	0.10	0.10	0.10
10.00	0.30	0.24	0.20	0.16	0.13	0.11	0.10	0.10	0.10	0.10	0.10	0.10
10.50	0.29	0.23	0.19	0.16	0.13	0.10	0.10	0.10	0.10	0.10	0.10	0.10
11.00	0.27	0.22	0.18	0.15	0.12	0.10	0.10	0.10	0.10	0.10	0.10	0.10
11.50	0.26	0.21	0.17	0.14	0.12	0.10	0.10	0.10	0.10	0.10	0.10	0.10
12.00	0.25	0.20	0.17	0.14	0.11	0.10	0.10	0.10	0.10	0.10	0.10	0.10
12.50	0.24	0.20	0.16	0.13	0.11	0.10	0.10	0.10	0.10	0.10	0.10	0.10
13.00	0.23	0.19	0.15	0.13	0.10	0.10	0.10	0.10	0.10	0.10	0.10	0.10
13.50	0.22	0.18	0.15	0.12	0.10	0.10	0.10	0.10	0.10	0.10	0.10	0.10
14.00	0.21	0.17	0.14	0.12	0.10	0.10	0.10	0.10	0.10	0.10	0.10	0.10
14.50	0.21	0.17	0.14	0.11	0.10	0.10	0.10	0.10	0.10	0.10	0.10	0.10
15.00	0.20	0.16	0.13	0.11	0.10	0.10	0.10	0.10	0.10	0.10	0.10	0.10

Table 5.14 Adjustment factor, F_p, for ponds and swampy areas (Brown, Stein, and Warner, 2001)

Pond and Swamp Area (percent)	F_p
0	1.00
0.2	0.97
1.0	0.87
3.0	0.75
5.0	0.72

Example 5.10 – Using the NRCS (SCS) Method to Compute Peak Discharge. Compute the peak discharge from the 1,000-acre (1.56 mi^2) drainage basin described in Example 5.9, but assume AMC-II applies. The rainfall distribution is an SCS Type II, and the time of concentration of the drainage basin is 0.9 hr. No ponds or swampy areas exist in the basin.

Solution: From Example 5.5, P = 5.0 in. and CN = 86 for the AMC-II condition. For this AMC-II condition, the maximum possible retention S is 1.63 in., and the effective precipitation P_e is 3.46 in.

From Table 5.13, the ratio I_a/P is 0.10; and from Table 5.12, the applicable coefficients are C_0 = 2.55323, C_1 = –0.61512, and C_2 = –0.16403. Substitution of these values into Equation 5.24 yields

$$K = 2.55323 - 0.61512\log_{10}(0.9) - 0.16403[\log_{10}(0.9)]^2 = 2.58104$$

From Equation 5.23, the unit peak runoff rate is

$$q_u = 10^{2.58104} = 381 \text{ cfs/mi}^2\text{/in.}$$

The peak discharge from the drainage basin is computed from Equation 5.22 as

$$Q_p = 381(1.56)(3.46) = 2{,}057 \text{ cfs}$$

USGS Regression-Based Formulas

The U.S. Geological Survey (USGS), in its efforts related to estimation of flood frequencies for ungauged locations, developed formulas for the prediction of peak discharges corresponding to various recurrence intervals. These formulas were developed on a state-by-state basis but may vary from one region of a state to another. Separate formulas are usually presented for rural and urbanized areas, and for different recurrence intervals. The nearest office of the USGS may be contacted to obtain information for a desired location. Jennings, Thomas, and Riggs (1994) present a nationwide summary of the rural equations, including a computer program for their implementation. Sauer et al. (1983) present equations permitting rural peak flows to be converted to urbanized peak flows.

Failure Probability and Hydraulic Structure Design

Although statistical methods, probability theory, and risk analysis are beyond the scope of this text, it is worthwhile to examine an application of such methods to stormwater management system design. As an illustration, a culvert having a design return period of 50 years and an expected life of 75 years can be shown using the binomial probability distribution to have a failure probability of 0.78, or 78 percent. This probability does not reflect the likelihood of a structural failure such as collapse; rather, it is the probability that the design discharge capacity of the culvert will be exceeded *one or more times* during its lifetime.

The probability of failure determined for the culvert in this example seems exceedingly high. It is certainly unacceptable for a structural engineer to design facilities with such high structural failure rates. However, hydrologic and hydraulic engineering differs in that requiring a small failure probability can be prohibitively expensive. If the culvert described previously were required to have a failure probability of 5 percent or less, then the design storm return frequency for an expected culvert life of 75 years would need to be at least 1,500 years!

It should be clear from these examples that a trade-off must occur in the selection of the recurrence interval for which a stormwater conveyance system is to be designed. If the recurrence interval is small and the system capital costs are correspondingly low, there is a very high probability of hydraulic failure of the system during its lifetime. Conversely, if the probability of failure must be low, the recurrence interval becomes large, and the capital costs of system construction are also large.

In practice, engineers often adopt design recurrence intervals specified by the regulatory and/or review agency having jurisdiction over the area in which a conveyance system will be built. In the interest of minimizing the cost of any one system (so as to permit resources to be allocated more or less equally among the many systems in the jurisdictional area), recurrence intervals are typically set fairly low. The recurrence interval used in the first example above is typical of real-world practice.

The high probability of operational failure, and thus overflow, typically associated with stormwater conveyance systems means that significant attention needs to be paid to the issue of what happens when a flood does occur. The operation of the major drainage system (the often unplanned, naturally occurring system that takes over when the minor flow system is overtaxed), which historically has not received much attention, needs some careful investigation and planning. It is not enough to simply call an exceedance of a system design capacity an Act of God, especially when such Acts have such a large probability of occurring.

To develop the predictive formulas, the USGS begins by estimating flood frequencies at stream flow gauging locations throughout a state. These estimates are accomplished by fitting the log Pearson Type III probability distribution to data series of annual flood peaks. Using the fitted distribution, flood discharges corresponding to various recurrence intervals are predicted. It is then assumed that flood discharges for a particular recurrence interval, and for any location in an assumed homogeneous region, can be predicted on the basis of easily measured drainage basin characteristics.

Stepwise multiple regression is used to develop the predictive relationship for each recurrence interval and region. Predictor variables typically examined include basin area, main channel slope and length, basin elevation, percentage of basin covered by swamps and lakes, land use and soil conditions, and precipitation. The drainage basin area is usually the most significant predictor, and is often the only one retained in the relationship. For example, the predictive formulas developed for rural basins in northern Alabama for the 2-, 5-, 10-, 25-, 50-, and 100-year storms are (Olin, 1984):

$$Q_2 = 182A^{0.706} \quad Q_5 = 291A^{0.711} \quad Q_{10} = 372A^{0.714}$$

$$Q_{25} = 483A^{0.717} \quad Q_{50} = 571A^{0.720} \quad Q_{100} = 664A^{0.722}$$

In these equations, Q_T is the peak discharge, in cfs, corresponding to the T-year recurrence interval, and A is the drainage basin area in mi^2. These northern Alabama formulas are limited to use for drainage basins with areas between 1 and 1,500 mi^2 (259 and 388,500 ha) and have standard errors that range from 29 to 36 percent.

When applying regression-based formulas to compute peak runoff rates from a drainage basin, the engineer must check the range of areas used in formula development [1 mi^2 to 1,500 mi^2 (259 and 388,500 ha) in the Alabama case] to ensure that the equation is applicable to the basin being evaluated. If the basin area is outside of this range, the predicted peak discharge for a relatively high recurrence interval can be smaller than the predicted peak discharge for a lower recurrence interval. Similarly, the predicted peak discharge for an urban area can be less than the predicted peak discharge for a rural area. Finally, it should be recognized that the standard errors associated with regression-based formulas for rural basins are typically on the order of 30 percent (35 to 50 percent for urban basins). Therefore, significant errors may arise when using this method.

Example 5.11 – Using a Regression Equation to Compute the Peak Discharge Rate. Use the rural equations for northern Alabama to estimate the 50-year peak discharge from a drainage basin with an area of 25 mi^2. The standard error for this formula is 33 percent (Olin, 1984). For this basin, what is the standard error of the estimate in cfs?

Solution: The peak discharge is

$$Q_{50} = 571(25)^{0.720} = 5{,}800 \text{ cfs}$$

The standard error S_e associated with the estimate is

$$S_e = 0.33(5{,}800) = 1{,}900 \text{ cfs}$$

5.5 HYDROGRAPH ESTIMATION

A *hydrograph* represents the runoff rate (discharge) as it varies over time at a particular location within a watershed. The integrated area under a hydrograph represents the *volume* of runoff. *Base flow*, if present, represents subsurface flow from groundwater that discharges into the conveyance channel. Many storm conveyance channels are dry at the beginning of a rainfall event, which equates to a base flow of zero as long as the water table does not rise into the channel during the storm. Figure 5.9 displays the various components of a surface runoff hydrograph for a case in which the base flow is zero and demonstrates that the area under the hydrograph is equal to the runoff volume as initially presented in Section 5.2.

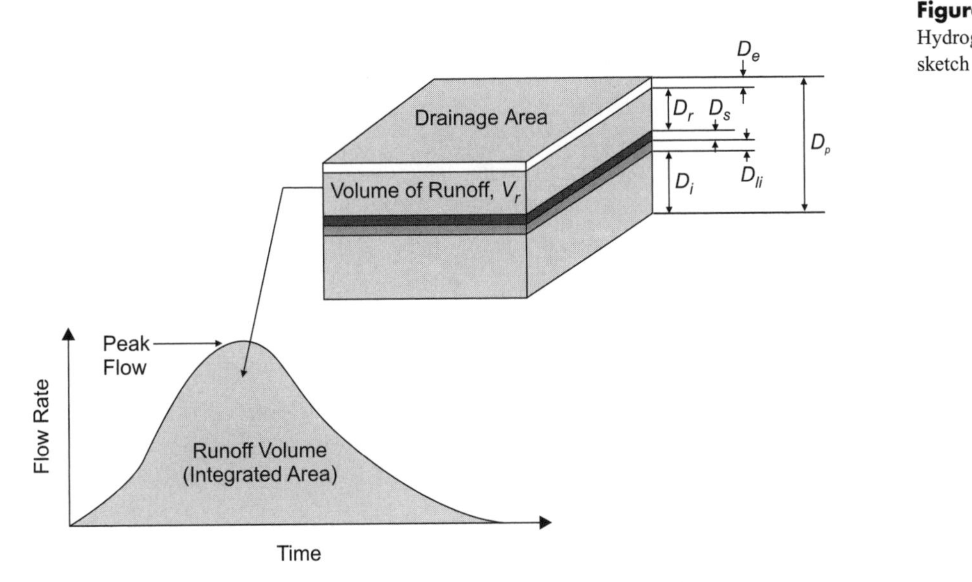

Figure 5.9
Hydrograph definition sketch

Estimation of a runoff hydrograph, as opposed to merely the peak rate of runoff, is necessary to account for the effects of storage in a drainage basin. Because a hydrograph accounts for volume and flow variations over an entire rainfall event, it is useful for analyzing complex watersheds and designing detention ponds. Of particular importance to the design engineer is assessment of the effects of storage associated with natural ponds and lakes, and with constructed stormwater detention and/or retention facilities. Hydrograph estimation is also necessary in the assessment of the impacts of storm duration and/or hyetograph shape on runoff production, and in cases of two or more adjacent drainage basins (or subbasins) discharging to a common stream.

Most approaches to hydrograph estimation are based on the concept of a *unit hydrograph*, which is a hydrograph produced by a unit depth of runoff (usually 1 in. or 1 cm) distributed uniformly over a basin for defined period of time. The unit hydrograph must be *convolved* (combined) with an effective precipitation (runoff) hyetograph to obtain the direct runoff hydrograph from a basin. Direct runoff hydrograph estimation using unit hydrograph approaches is described in detail in this section.

Several alternatives to the unit hydrograph method are used for runoff hydrograph estimation, including hydraulic methods such as the kinematic wave model. This method is a more physically-based, hydraulic approach to hydrograph estimation but requires that the surface of a drainage basin be idealized as a set of overland flow planes and channels. A brief introduction to the kinematic wave model is presented later in this section (page 162).

Overview of Direct Runoff Hydrograph Estimation Using a Unit Hydrograph

Development of a unit hydrograph is an important intermediate step in computing a direct runoff hydrograph for design or analysis purposes. After a unit hydrograph has been established for a particular basin, it can be used to compute the direct runoff hydrographs resulting from different rainfall events.

Estimation of a direct runoff hydrograph using a unit hydrograph approach involves several steps. The basic procedure is given below, with references to the specific sections of this text and examples detailing each step.

1 Establish a design storm rainfall hyetograph (see Section 4.5).

2 Convert the design storm rainfall hyetograph into a direct runoff (effective rainfall) hyetograph by subtracting rainfall abstractions (see Section 5.2, page 132) and Example 5.6 (page 132).

3 Estimate a unit hydrograph for the drainage basin of interest (see the next section and Example 5.12, page 157).

4 Merge the direct runoff (effective rainfall) hyetograph and the unit hydrograph using discrete convolution to produce a direct runoff hydrograph, which represents the surface runoff hydrograph for the complex storm pattern represented by the rainfall hyetograph [see Section 5.5 and Example 5.13 (page 161)].

5 If appropriate, add base flow to the direct runoff hydrograph to obtain the total hydrograph representing both surface runoff and base flow (see Section 5.6).

Figure 5.10 illustrates the basic process of developing a hydrograph without baseflow.

Unit Hydrographs

The concept of the unit hydrograph was introduced to the hydrologic community by Sherman (1932). The basic theory rests on the assumption that the runoff response of a drainage basin to an effective rainfall input is linear; that is, it may be described by a linear differential equation. Practically speaking, this means that the concepts of proportionality and superposition can be applied. For example, the direct runoff volume and discharge rates resulting from 2 in. of effective rainfall in a given time interval are four times as great as those caused by 0.5 in. of effective rainfall in the same amount of time. It also means that the total amount of time for the basin to respond to each of these rainfall depths is the same. Consequently, the base length of each of the direct runoff hydrographs is the same.

The unit hydrograph approach to runoff estimation is a *spatially lumped* approach, meaning that it assumes that no spatial variability exists in the effective rainfall input into a drainage basin. In cases where the effective rainfall input varies from one location to another within a basin, the unit hydrograph approach requires the subdivision of the basin into smaller subbasins, with routing of the runoff from each subbasin to obtain the integrated basin response.

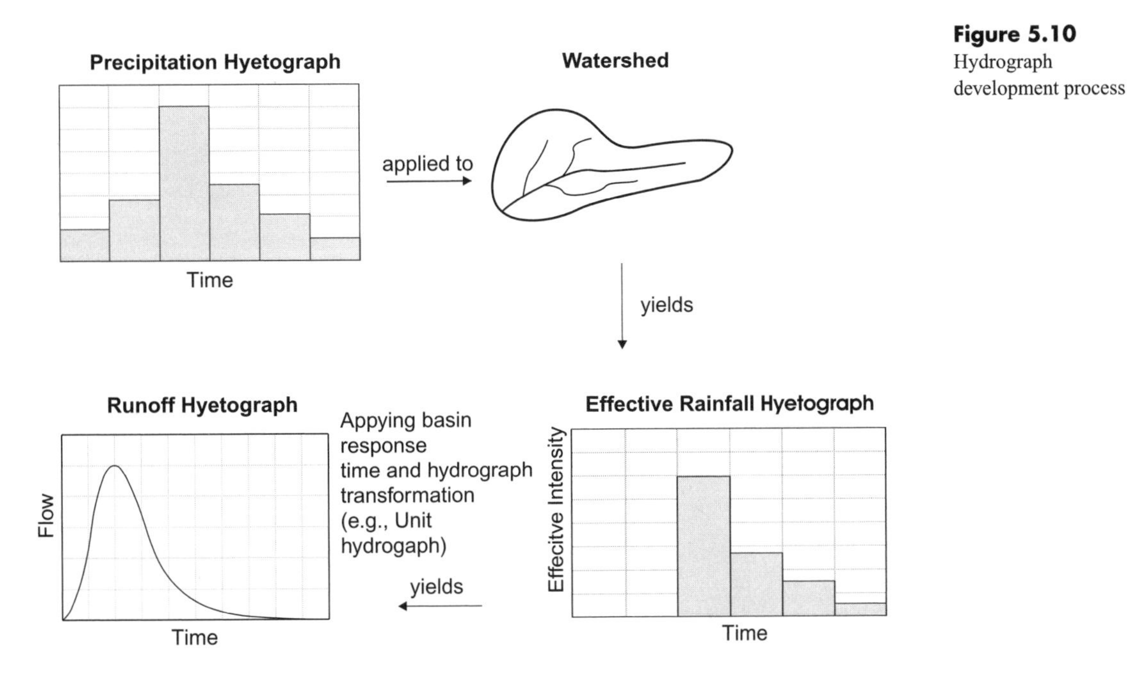

Figure 5.10 Hydrograph development process

Visualizing and Developing Unit Hydrographs. By definition, the Δt-hour (or Δt-minute) unit hydrograph of a drainage basin is the direct runoff hydrograph produced by 1 in. (or 1 cm in the case of SI units) of effective rainfall (runoff) falling uniformly in time and space over the drainage basin during a duration Δt time units. By conservation of mass, the volume of direct runoff produced (represented by the area under a graph of the unit hydrograph) must be equal to 1 in. (or 1 cm) of runoff (effective rainfall) times the drainage basin area (see Figure 5.11). It may be observed from this definition that the unit hydrographs for two drainage basins must be different from one another, because the basin areas (as well as other factors) are different.

Figure 5.12 helps to demonstrate the principles of a unit hydrograph. A sprinkler system distributes water uniformly over an entire drainage area. Constant rainfall is modeled by using the sprinklers to apply a constant rate of water until exactly one inch of direct runoff (effective precipitation) is generated, at which time the sprinklers are turned off. The duration of time required for the sprinklers to generate the required depth of effective precipitation is shown as Δt. A stream gauge is placed on the downstream end of the contributing drainage area to record the runoff hydrograph resulting from the 1 in. (or 1 cm) of runoff. The recorded hydrograph represents the *unit hydrograph* generated for the drainage area.

For a given drainage basin, there are an infinite number of durations over which 1 in. of direct runoff (effective rainfall) could be generated depending on the intensity of the rainfall being applied to the area. For instance, 1 in. of runoff could occur over Δt = 30 minutes, over Δt = 1 hour, or over some other time interval. Thus, for any given drainage basin, one could speak of its 30-minute unit hydrograph, its 1-hour unit hydrograph, or its unit hydrograph of any other duration. Note that the duration

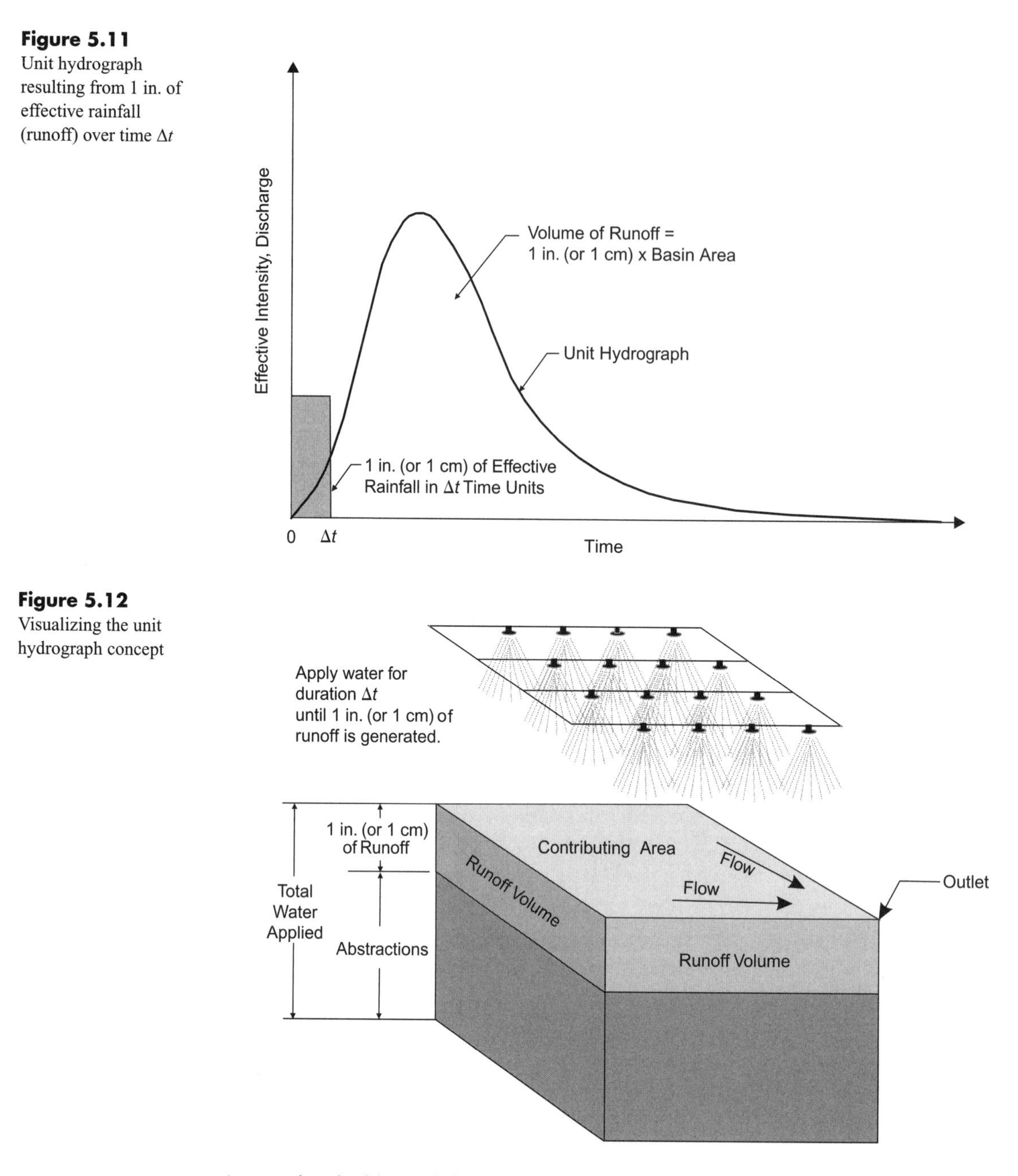

Figure 5.11
Unit hydrograph resulting from 1 in. of effective rainfall (runoff) over time Δt

Figure 5.12
Visualizing the unit hydrograph concept

Δt associated with a unit hydrograph is the duration over which the 1 in. (or 1 cm) of effective rainfall occurs and not the duration over which the corresponding runoff drains from the area.

Two basic categories of unit hydrographs exist: (1) unit hydrographs for gauged watersheds and (2) synthetic unit hydrographs (described in the next section). When concurrent rainfall and runoff records are available for a drainage basin, methods of *deconvolution* (separation) may be applied to estimate a unit hydrograph for the basin on the basis of those records (Singh, 1988). More commonly, rainfall and runoff records for a drainage basin do not exist, and one must resort to synthesis of a unit hydrograph based on information that can be gathered about the basin. Extensive literature exists on various ways to calculate synthetic unit hydrographs, including procedures proposed by Clark, Snyder, and Singh (Snyder, 1938; Singh, 1988). This text provides an introduction to the development of synthetic unit hydrographs by applying the NRCS (SCS) method.

NRCS (SCS) Synthetic Unit Hydrographs.. The NRCS (SCS) analyzed a large number of unit hydrographs derived from rainfall and runoff records for a wide range of basins and basin locations and developed the average *dimensionless unit hydrograph* shown in Figure 5.13 (Snider, 1972). The times on the horizontal axis are expressed in terms of the ratio of time to time of peak discharge (t/t_p), and the discharges on the vertical axis are expressed in terms of the ratio of discharge to peak discharge (Q/Q_p). Table 5.15 lists the dimensionless unit hydrograph ordinates.

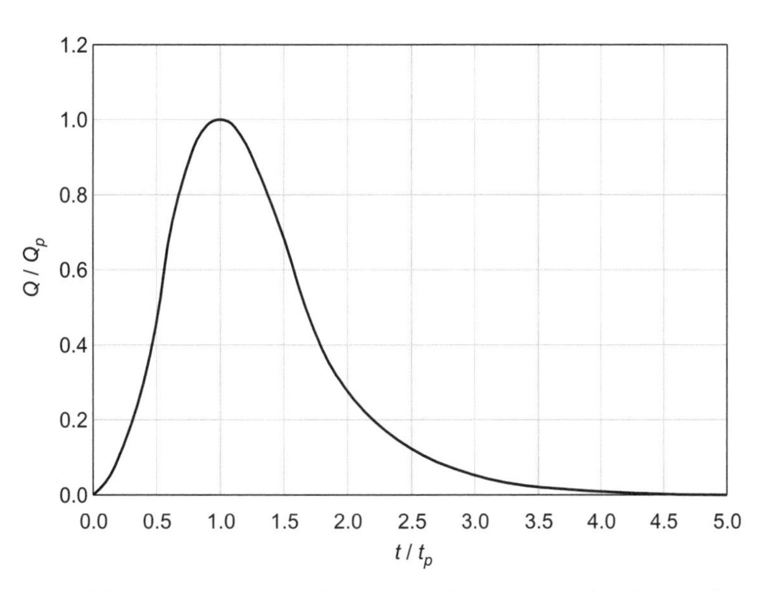

Figure 5.13 NRCS (SCS) dimensionless unit hydrograph (from Snider, 1972)

Application of the dimensionless unit hydrograph involves estimating the lag time t_L of the drainage basin. The lag time can be estimated by relating it to an estimate of the time of concentration, or it can be estimated directly (see Section 5.3). The time to peak of the synthetic unit hydrograph of duration Δt is then computed as

$$t_p = \frac{\Delta t}{2} + t_L \qquad (5.26)$$

The NRCS recommends that Δt be equal to $0.133t_c$, or equal to $0.222t_L$ (Snider, 1972). A small variation from this value is acceptable. The Δt chosen for the development of the synthetic unit hydrograph must be consistent with the Δt chosen for the design storm and effective rainfall hyetographs (see Section 4.5, page 91).

Table 5.15 Ordinates of the NRCS (SCS) dimensionless unit hydrograph (from Snider, 1972)

t/t_p	Q/Q_p	t/t_p	Q/Q_p
0	0.000	1.7	0.460
0.1	0.030	1.8	0.390
0.2	0.100	1.9	0.330
0.3	0.190	2.0	0.280
0.4	0.310	2.2	0.207
0.5	0.470	2.4	0.147
0.6	0.660	2.6	0.107
0.7	0.820	2.8	0.077
0.8	0.930	3.0	0.055
0.9	0.990	3.2	0.040
1.0	1.000	3.4	0.029
1.1	0.990	3.6	0.021
1.2	0.930	3.8	0.015
1.3	0.860	4.0	0.011
1.4	0.780	4.5	0.005
1.5	0.680	5.0	0.000
1.6	0.560		

The peak discharge Q_p for the synthetic unit hydrograph is calculated as

$$Q_p = \frac{C_f KAQ}{t_p} \tag{5.27}$$

where Q_p = peak discharge (cfs, m^3/s)
C_f = conversion factor (645.33 U.S., 2.778 SI)
K = 0.75 [a constant based on geometric shape of dimensionless unit hydrograph (Snider, 1972)]
Q = runoff depth for unit hydrograph calculation (1 in. U.S., 1 cm SI)
A = the drainage basin area (mi^2, km^2)
t_p = the time to peak (hr)

Simplifying Equation 5.27 yields

$$Q_p = \frac{484A}{t_p} \quad \text{(for U.S. units)} \tag{5.28}$$

or

$$Q_p = \frac{2.08A}{t_p} \quad \textit{(for SI units)} \tag{5.29}$$

The coefficients 484 and 2.08 appearing in the numerators of Equation 5.28 and Equation 5.29 include a unit conversion factor and are average values for many drainage basins. These values may be reduced to about 300 and 1.29, respectively, for flat or swampy basins, or increased to about 600 and 2.58, respectively, for steep or mountainous basins. Care should be taken when changing this coefficient, as the base length and/or shape of the synthetic unit hydrograph must also be changed to ensure that it represents a volume of water equivalent to 1 in. or 1 cm of effective rainfall over the drainage basin area.

After t_p and Q_p are estimated using Equation 5.26 and Equation 5.27, the desired synthetic unit hydrograph may be graphed or tabulated using the dimensionless unit hydrograph shown in Figure 5.13 and Table 5.15.

Example 5.12 – Developing an NRCS (SCS) Synthetic Unit Hydrograph. Develop a synthetic unit hydrograph for a 1.5-mi^2 drainage basin in Memphis, Tennessee, having a time of concentration of 90 min. Assume that the basin slopes are moderate so that the factor 484 can be applied in computing Q_p.

Employing the guidelines given above for estimation of Δt, its value should be about

$$0.133t_c = 0.133(90) = 12 \text{ min}$$

A duration of Δt = 10 min is selected, and the synthetic unit hydrograph will be a 10-min unit hydrograph. The basin lag is estimated as

$$t_L = 0.6t_c = 54 \text{ min}$$

Using Equation 5.26, the time to peak of the synthetic unit hydrograph is

$$t_p = (10/2) + 54 = 59 \text{ min} = 0.98 \text{ hr}$$

The peak discharge is estimated using Equation 5.28 as

$$Q_p = 484(1.5)/0.98 = 740 \text{ cfs}$$

Ordinates of the synthetic unit hydrograph are tabulated in Table 5.16 and plotted in Figure E5.12.1. The first column of the table is the time t, in minutes, and is tabulated in Δt = 10 min intervals. The second column is the dimensionless time ratio t/t_p, where t_p = 59 min. The third column is the dimensionless discharge ratio, and is determined using the dimensionless time ratio and interpolation from Table 5.15. The fourth and last column contains the ordinates of the 10-min unit hydrograph, which are computed as the products of the dimensionless discharge ratios and Q_p = 740 cfs.

Table 5.16 SCS (NRCS) synthetic unit hydrograph computations for Example 5.12

(1)	(2)	(3)	(4)
t (min)	t/t_p	Q/Q_p	Q (cfs)
0	0.00	0.00	0
10	0.17	0.08	59
20	0.34	0.24	178
30	0.51	0.49	363
40	0.68	0.79	585
50	0.85	0.96	710
60	1.02	1.00	740
70	1.19	0.93	688

Table 5.16 (cont.) SCS (NRCS) synthetic unit hydrograph computations for Example 5.12

(1)	(2)	(3)	(4)
t (min)	t/t_p	Q/Q_p	Q (cfs)
80	1.36	0.81	599
90	1.53	0.64	474
100	1.69	0.47	348
110	1.86	0.35	259
120	2.03	0.26	192
130	2.20	0.21	155
140	2.37	0.16	118
150	2.54	0.12	89
160	2.71	0.09	67
170	2.88	0.07	52
180	3.05	0.05	37
190	3.22	0.04	30
200	3.39	0.03	22
210	3.56	0.02	15
220	3.73	0.02	15
230	3.90	0.01	7
240	4.07	0.01	7
250	4.24	0.01	7
260	4.41	0.01	7
270	4.58	0.00	0

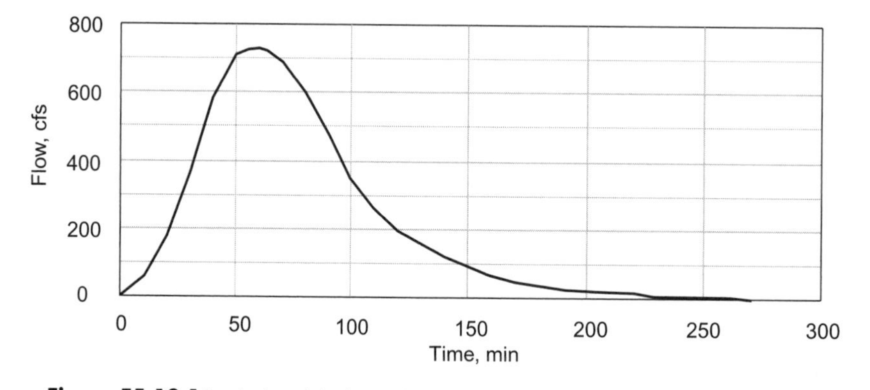

Figure E5.12.1 Synthetic unit hydrograph for Example 5.12

Discrete Convolution

A unit hydrograph represents the runoff hydrograph from a drainage basin subjected to 1 in. or 1 cm of direct runoff (effective precipitation) applied over duration Δt. However, a runoff (effective rainfall) hyetograph typically contains many Δt time periods, each of which has its own associated runoff depth (see Example 5.6 on page 132). *Discrete convolution* is the process by which the direct runoff hydrograph resulting from a complete rainfall hyetograph is computed by applying a unit hydrograph to each discrete time step within the hyetograph.

To conceptualize the process of discrete convolution, refer to Figure 5.14 and Figure 5.15. Figure 5.14 illustrates a runoff (effective rainfall) hyetograph consisting of n_p = 3 rainfall pulses, each of duration Δt = 10 min, with the depth of each pulse denoted by P_i (i = 1, 2...n_p). Figure 5.15 shows a unit hydrograph with n_u = 11 non-zero ordinates shown at Δt time intervals, with the non-zero unit hydrograph ordinates denoted by U_j (j = 1, 2 ... n_u). [Note also that the ordinates of the unit hydrograph are in units of cfs per inch of direct runoff (effective precipitation), and thus the notation of this subsection is different from the previous one in which unit hydrograph ordinates were denoted by Q in cfs.] The time increment Δt used for development of the runoff (effective rainfall) hyetograph must be the same as the Δt duration of excess rainfall used to create the unit hydrograph.

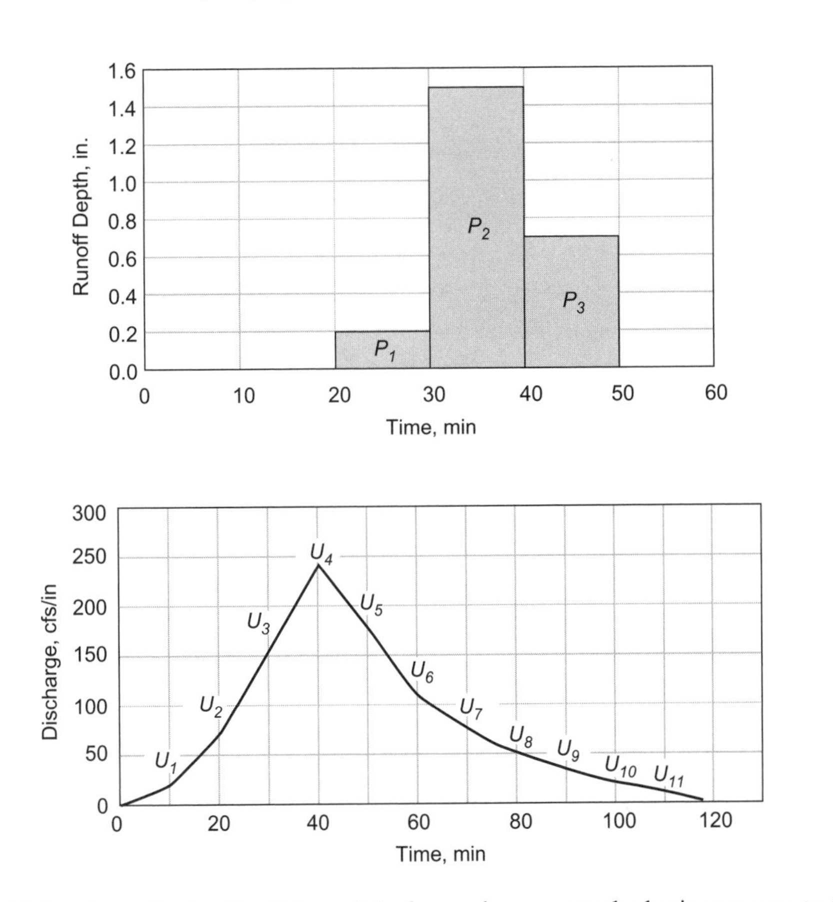

Figure 5.14 Effective rainfall (runoff) hyetograph

Figure 5.15 Unit hydrograph ordinates

Recall that the ordinates U_j of the unit hydrograph represent the basin response to 1 in. or 1 cm of direct runoff (effective rainfall) applied over a time of Δt time units. Because unit hydrograph theory assumes that drainage basins behave linearly, the ordinates Q_j of the runoff hydrograph produced by P_1 inches of effective rainfall must be equal to $P_1 U_j$. These ordinates are shown in the third column of Table 5.17 (the first and second columns of the table reproduce the information in Figure 5.15). These ordinates are calculated by multiplying each unit hydrograph ordinate by the runoff depth for pulse P_1 = 0.2 in. The values in column 3 are shifted down such that the time at which the runoff hydrograph in column 3 begins is the same as the time at which the first pulse of effective rainfall begins.

Columns 4 and 5 in Table 5.17 are computed by multiplying the unit hydrograph ordinates in the second column by the effective rainfall depths P_2 = 1.5 in. and P_3 = 0.7 in., respectively. Again, the time at which each runoff pulse's hydrograph begins corresponds to the times at which each runoff (effective rainfall) pulse begins. The sum of the runoff hydrograph ordinates in each row of the table (shown in column 6) is an ordinate of the direct runoff hydrograph caused by the complete rainfall event.

Table 5.17 Discrete convolution

(1)	(2)	(3)	(4)	(5)	(6)
t (min)	U (cfs/in)	P_1U (cfs)	P_2U (cfs)	P_3U (cfs)	Q (cfs)
0	0				0
10	20				0
20	70	0			0
30	160	4	0		4
40	240	14	30	0	44
50	180	32	105	14	151
60	110	48	240	49	337
70	75	36	360	112	508
80	50	22	270	168	460
90	33	15	165	126	306
100	21	10	113	77	200
110	10	7	75	53	134
120	0	4	50	35	89
130		2	32	23	57
140		0	15	15	30
150			0	7	7
160				0	0

Figure 5.16 shows the runoff hydrograph resulting from each rainfall pulse and the total runoff hydrograph for the entire rainfall event.

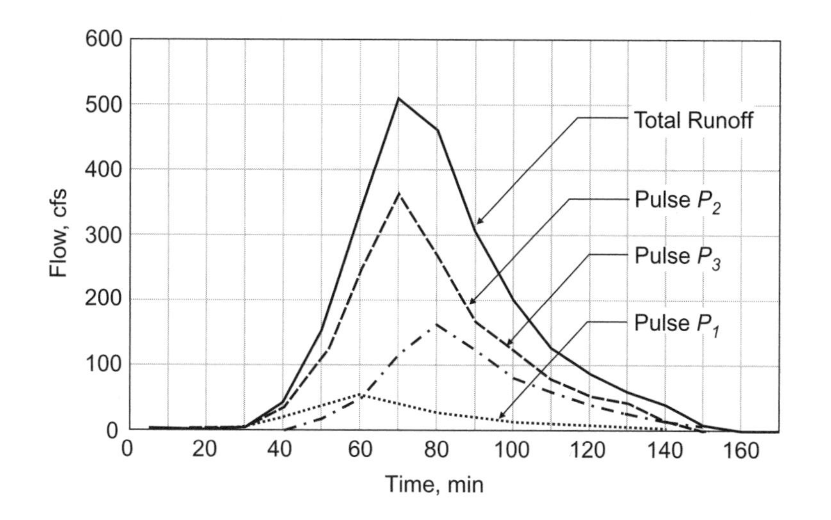

Figure 5.16 Summing of hydrographs from individual rainfall pulses

The tabular process of discrete convolution illustrated in Table 5.17 can be summarized in equation form for computation of the non-zero ordinates Q_k of the direct runoff hydrograph as

$$Q_k = \sum_{i=1}^{k} P_i U_{k-i+1} \tag{5.30}$$

where $k = 1, 2 \ldots n_q$, and $n_q = n_p + n_u - 1$

To clarify Equation 5.30, $Q_1 = P_1U_1$, $Q_2 = P_1U_2 + P_2U_1$, $Q_3 = P_1U_3 + P_2U_2 + P_3U_1$, etc. These expressions are easily rationalized with reference to Table 5.15.

Example 5.13 - Computing a Runoff Hydrograph. Compute the direct runoff hydrograph for the Memphis, Tennessee, drainage basin described in Example 5.12 on page 157. Use the unit hydrograph developed for the basin in that example, and use the runoff (effective rainfall) hyetograph developed in Example 5.6 on page 132.

Solution: There are a total of $n_p = 4$ runoff (effective rainfall) pulses, and a total of $n_u = 26$ non-zero unit hydrograph ordinates. Thus, there will be a total of $n_q = 29$ non-zero direct runoff hydrograph ordinates.

Table 5.18 illustrates the tabular calculation (discrete convolution). The first column is the time since the beginning of rainfall in $\Delta t = 10$ min increments (the duration of the unit hydrograph and the duration of the effective rainfall pulses). The second column contains the ordinates of the unit hydrograph from Example 5.12. The third through the sixth columns are the unit hydrograph ordinates multiplied by the effective rainfall depths from Example 5.6. Note that the first entry (the first zero) in each column corresponds to the time at which the corresponding effective rainfall pulse begins ($t = 20$ min for the first pulse, $t = 30$ min for the second pulse, and $t = 40$ min for the third pulse; see Figure E5.6.2). The seventh column is the sum of the previous four, and is the direct runoff hydrograph.

Table 5.18 Discrete convolution to obtain direct runoff hydrograph

(1)	(2)	(3)	(4)	(5)	(6)	(7)
t (min)	*U* (cfs/in.)	0.99*U*	0.45*U*	0.25*U*	0.09*U*	*Q* (cfs)
0	0					0
10	59					0
20	178	0				0
30	363	58	0			58
40	585	176	27	0		203
50	710	359	80	15	0	454
60	740	579	163	45	5	792
70	688	703	263	91	16	1073
80	599	733	320	146	33	1231
90	474	681	333	178	53	1244
100	348	593	310	185	64	1152
110	259	469	270	172	67	977
120	192	345	213	150	62	769
130	155	256	157	119	54	585
140	118	190	117	87	43	436
150	89	153	86	65	31	336
160	67	117	70	48	23	258
170	52	88	53	39	17	197
180	37	66	40	30	14	150
190	30	51	30	22	11	115
200	22	37	23	17	8	85
210	15	30	17	13	6	65
220	15	22	14	9	5	49
230	7	15	10	8	3	36
240	7	15	7	6	3	30
250	7	7	7	4	2	19
260	7	7	3	4	1	15
270	0	7	3	2	1	13
280		7	3	2	1	12
290		0	3	2	1	6
300			0	2	1	2
310				0	1	1
320					0	0

Kinematic Wave Model

Unit hydrograph-based approaches for runoff estimation tend to be straightforward mathematical procedures with only a limited physical basis (but see Rodríguez-Iturbe and Valdes, 1979, for a physical interpretation of them). A more physically based approach to runoff estimation is the *kinematic wave model*, which involves application of the fundamental laws of conservation of mass and momentum to overland and channelized flows. This approach is *hydraulic* (as opposed to *hydrologic*) in nature, as it applies conservation of mass and momentum to describe free-surface (open-channel) flows.

As mentioned previously, use of the kinematic wave model requires that the surface of a drainage basin be idealized as a set of overland flow planes and collector channels (see Figure 5.17). One-dimensional, free-surface flow on an overland flow plane or in a collector channel is then described using the Saint-Venant equations, which are a pair of partial differential equations describing conservation of mass and momentum (see Appendix C).

Occasionally, some of the terms appearing in the momentum equation can be eliminated because they are small in comparison to the remaining terms. When all terms are retained in the momentum equation, the Saint-Venant equations become what is known as the *dynamic wave* formulation. When the so-called inertial acceleration terms are dropped in the momentum equation, the *diffusion wave* formulation results. When only the gravitational and frictional terms are retained in the momentum equation, the *kinematic wave* formulation results (see also Appendix C).

The Saint-Venant equations, when the kinematic wave simplification is invoked, take the forms given in Equation 5.31 and Equation 5.32:

$$\frac{\partial A}{\partial t} + \frac{\partial Q}{\partial x} = q_o \qquad (5.31)$$

where A = cross-sectional area of flow (ft^2, m^2)
t = time (s)
Q = flow rate (ft^3/s, m^3/s)
x = distance along the flow path (ft, m)
q_o = lateral discharge added to the flow per unit length of the flow path (cfs/ft, m^3/s/m)

$$A = \alpha Q^{\beta} \qquad (5.32)$$

where α and β are coefficients representing the physical nature of the land surface or stream channel and the type of constitutive relationship applied.

The lateral discharge q_L in Equation 5.31 may be due to additional flow entering the channel from its sides. It may also be due to rainfall directly onto the water surface, or due to infiltration into the channel bottom (in which case q_L would be negative).

Equation 5.32, which is a simplified version of the momentum equation, may be invoked using a constitutive relationship for uniform flow. For example, Manning's equation can be expressed as

$$A = \left(\frac{QnP^{2/3}}{1.49S_o^{1/2}} \right)^{3/5} \qquad (5.33)$$

where n = Manning's roughness coefficient
P = wetted perimeter (ft, m)
S_o = longitudinal slope

Comparing this form to Equation 5.32 shows that the coefficients are given by $\alpha = (nP^{2/3}/1.49S_o{}^{1/2})^{3/5}$ and $\beta = 3/5$.

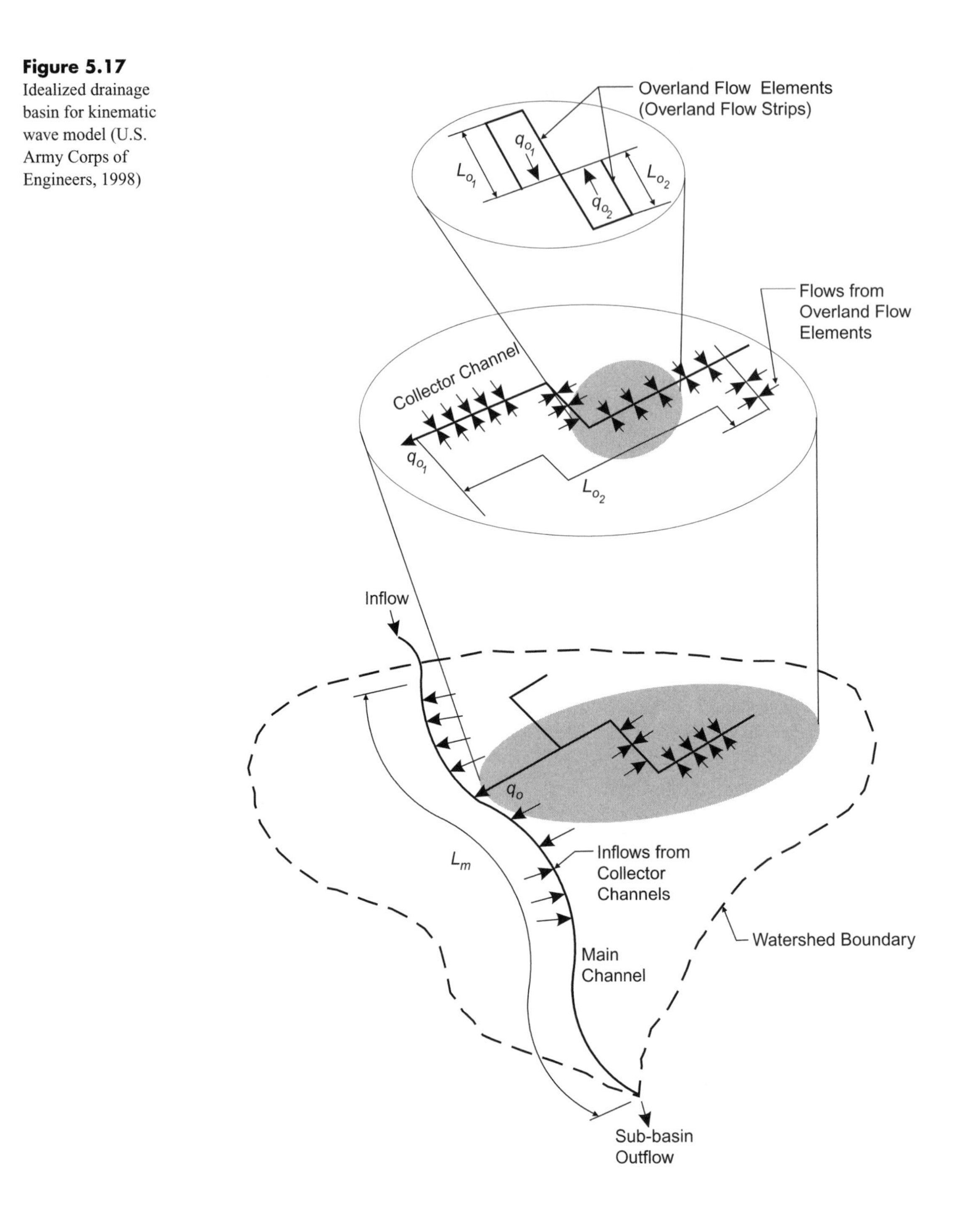

Figure 5.17
Idealized drainage basin for kinematic wave model (U.S. Army Corps of Engineers, 1998)

Substitution of Equation 5.32 into Equation 5.31 yields a single partial differential equation in Q:

$$\alpha\beta Q^{\beta-1}\frac{\partial Q}{\partial t}+\frac{\partial Q}{\partial x}=q_L \tag{5.34}$$

The solution of this equation, which expresses Q as a function of location and time within a flow path, may be approximated using a variety of numerical methods. The interested reader is referred to Chow, Maidment, and Mays (1988) for an introductory exposition, and to other references cited therein for more detail.

5.6 BASE FLOW

The total discharge in a stream channel is composed of *surface runoff* (direct runoff) and *subsurface flow* (base flow) from groundwater, as shown in the hydrograph in Figure 5.18. Hydrologists may also include interflow in the total flow of a stream, where *interflow* is defined as the runoff that occurs within near-surface soils. Interflow passes through the near-surface soils much faster than it does through deeper soils, and thus reaches streams more quickly than does base flow. In this discussion, it is assumed that interflow is lumped into surface runoff and therefore does not need to be treated separately.

Base flow may or may not exist in a stream channel, or it may exist only during certain times of the year. A stream in which base flow is always present is called a *perennial stream*, and one in which base flow either does not exist, or in which it exists only some of the time, is called an *ephemeral* or *intermittent stream*. Perennial streams are the norm in regions with an abundance of rainfall and where the elevation of the groundwater table is above stream channel elevations. Ephemeral streams are common in arid and semi-arid regions. Ephemeral streams may also exist in humid regions, particularly in locations where the contributing drainage basin is small and/or the groundwater table lies below the bottom of the stream channel.

When base flow is present in a stream, the total discharge for which a water conveyance system is designed should include the base flow as well as the surface runoff (direct runoff). For estimation of base flow to support the design of a stormwater conveyance system, the best course of action may be to periodically monitor and measure the flow in the stream of interest using a current meter. Base flow should be measured during dry-weather periods when enough time has elapsed following storm events for the discharge to become nearly constant with respect to time.

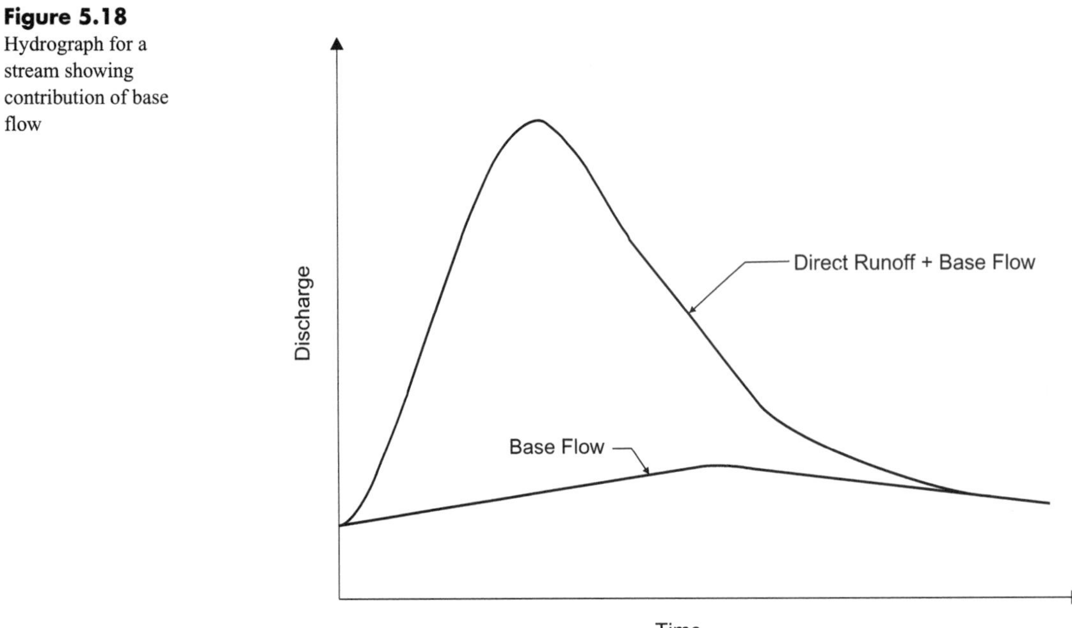

Figure 5.18
Hydrograph for a stream showing contribution of base flow

Base flow may change seasonally due to fluctuations in the groundwater table. Urbanization of an area may cause base flow to decrease, or it may initiate base flow in a stream where none existed previously. The latter is especially true in semi-arid regions, where the wastewater discharges, leaking water mains, irrigation, car washing, and so forth that accompany urbanization may cause local groundwater levels to rise or perched groundwater conditions to develop.

Once a base flow discharge has been estimated, that discharge may be simply added to each ordinate of a direct runoff hydrograph to determine the total runoff hydrograph, as in Example 5.14 on page 172.

5.7 SNOWMELT

In many parts of the world, snow is the dominant form of precipitation. Water produced by melting snow supplies reservoirs, lakes, and rivers, and infiltrating meltwater recharges soil moisture and groundwater (Gray and Prowse, 1993). Because snowmelt is an additional mechanism by which urban runoff may be generated, it affects facilities such as storm sewers, drainage channels, culverts, and detention ponds.

Although runoff flow rates from snowmelt are typically low, they may be sustained over several days. Rainfall events superimposed on snowmelt baseflow will produce higher runoff peak flows and volumes, and rainfall increases the melt rate of snow (Huber and Dickinson, 1988). Simulation of snowmelt effects is typically more critical for facilities significantly impacted by an increase in runoff volume, such as stormwater pump stations and detention facilities. Snowmelt can also have significant effects on water quality in urban areas because stored pollutants such as hydrocarbons, oil and grease, chlorides, sediment, and nutrients are flushed into surface waters (Oberts, 1994).

Most of the techniques for modeling runoff generated by melting snow were developed for mountainous, rural, and agricultural watersheds. These same models have recently been applied to urban watersheds. This section describes, in general terms, the models for forecasting runoff from snowmelt in urban environments.

Runoff Potential

The physical properties of interest to modelers are depth, density and water equivalent (Gray and Prowse, 1993). The potential quantity of runoff in snowpack is the *snow-water equivalent.* It is the equivalent depth of water in snow cover, and is calculated as

$$SWE = d_s \rho_s / \rho_w \tag{5.35}$$

where SWE = snow water equivalent (in, mm)
d_s = depth of snow (in, mm)
ρ_s = density of snow (lb/ft^3, kg/m^3)
ρ_w = density of water (62.4 lb/ft^3, 1,000 kg/m^3)

Note that density is expressed in force units when U.S. units are used and in mass units when SI units are used. The density of freshly fallen snow varies widely depending on the amount of air contained within the lattice of the snow crystals, but is commonly in the range of 3 to 7.5 lb/ft^3 (50 to 120 kg/m^3) (Gray and Prowse, 1993). Following deposition, the density of snow increases due to *metamorphism*, which is a change in the size, shape, and bonding of snow crystals due to temperature and water-vapor gradients, settlement, and wind packing. Typical densities for settled snow are 12 to 19 lb/ft^3 (200 to 300 kg/m^3), but densities may be as high as 22 lb/ft^3 (350 kg/m^3) for hard wind slab (Gray and Prowse, 1993).

Snowmelt Models

The temperature-index or degree-day method is described in numerous references, including the NEH-4 (Mockus, 1973), Huber and Dickinson (1988), and Semadeni-Davies (2000). The *meltwater rate, M,* is a function of daily average air temperature:

$$M = 0, \text{ if } T_a < T_m \tag{5.36}$$

$$M = C(T_a - T_m), \text{ if } T_a \geq T_m \tag{5.37}$$

where M = meltwater rate, (in/day, mm/day)
C = melt rate factor (in/°F/day, mm/°C/day)
T_a = daily average air temperature (°F, °C)
T_m = threshold temperature, (32°F, 0°C)

Values of C for rural areas range from 0.72 to 3.6 in/°F/day (1.4 to 6.9 mm/°C/day) and from 1.7 to 4.1 in/°F/day (3 to 8 mm/°C/day) for urban areas (Huber and Dickenson, 1988). Gray and Prowse (1993) presented an empirical equation for C:

$$C = K\rho_s \tag{5.38}$$

where C = melt rate factor, mm/°C/day
K = empirical coefficient equal to 0.00385 for U.S. Customary units or 0.011 for SI units
ρ_s = snow density (lb/ft^3, kg/m^3)

Although the degree-day method is convenient because it combines several complex processes into a single constant C, it can yield misleading results. An equation such as Equation 5.38 should only be used with great care. Ideally, developing the melt-rate factor should be accomplished by calibration.

Equation 5.37 calculates the meltwater rate on a daily basis. The air-temperature data typically available are the minimum and maximum values for each day, and T_a is commonly taken as the average of these values. Incorporation of meltwater into continuous-simulation hydrologic models requires temperature data for more frequent time steps. Procedures for developing hourly meltwater rates from this type of data are presented by Huber and Dickenson (1988).

More detailed models for forecasting snowmelt runoff are available. Most of these are based on the energy-balance method described by Anderson (1973). The units for each energy budget term are energy/area-time (for instance, langleys/day, where 1 langley = 1 cal/cm^2). The energy balance is presented in Equation 5.39 (Huber and Dickenson, 1988).

$$\Delta H = H_{rs} + H_g + H_{rl} + H_c + H_e + H_p \tag{5.39}$$

where ΔH=change in heat storage in snowpack
H_{rs}=net shortwave radiation entering snowpack
H_g=conduction of heat to snowpack from underlying ground
H_{rl}=net (incoming minus outgoing) longwave radiation entering snowpack
H_c=conductive exchange of sensible heat between air and snowpack
H_e=release of latent heat of vaporization by condensation of atmospheric water vapor
H_p=advection of heat to snowpack by rain

Equation 5.39 assumes that the *cold content* of the snow pack is zero; that is, none of the energy available to the snow pack is used to bring the snow pack up to the temperature at which any additional energy will result in melt. Procedures for estimating each term in Equation 5.39 are described by Huber and Dickenson (1988) and Gray and Prowse (1993).

The meltwater rate may then be determined with Equation 5.40:

$$M = \frac{(10,000)\Delta H}{L\rho_w} \tag{5.40}$$

where M = meltwater rate (mm/day)
ΔH = change in heat storage in snowpack (langley/day)
L = latent heat of fusion (80 cal/g water)

Eighty calories are required to melt 1 g of water, or 80 langley/cm (8 langley/mm) of snow depth. The rate of snowmelt is calculated by multiplying the meltwater rate times the area of the watershed. Production of meltwater is not necessarily equivalent to runoff, however, because snowpack is porous and free water moves downward through the snowpack before runoff is generated. Huber and Dickinson (1988) use a simple reservoir-type routing procedure to simulate this process. The free water holding capacity of snowpack is modeled as a fraction of the snow depth, usually between 0.02 and 0.05. In a continuous simulation, this depth is filled with meltwater before runoff occurs.

Snow tends to insulate the soil beneath it. If the ground is frozen before snow falls, it will usually remain frozen. Unfrozen soil tends to remain unfrozen underneath a snowpack. Therefore, the soil properties related to runoff (infiltration, detention storage, and so forth) are assumed to remain unchanged by snow.

The energy-balance method offers the advantage of accounting for factors like solar radiation, wind speed, and *albedo* (the fraction of sunlight reflected from the surface of the snowpack) that impact the meltwater rates. It can also account for melting that is induced by rain. However, the application of this method to urban environments has been limited. Part of the difficulty is that the energy balance method has extensive data requirements. Also, urban snow data are sparse, which limits the possibility of assessing or improving model performance.

Huber and Dickinson (1988) showed that the energy balance equation can be reduced to the degree-day equation by making appropriate substitutions and assumptions. Anderson (1976) reported that, under many conditions, results obtained using the energy-balance model were not significantly better than those obtained using the degree-day method. However, Matheussen and Thorolfsson (2001) stated that the energy fluxes related to snowmelt are significantly altered in the urban environment, and degree-day models are not suitable for urban areas. The HEC-1 model (U.S. Army Corps of Engineers, 1998) allows the modeler to use either the degree-day or energy-balance model. The energy-budget routine can be applied to heavily forested areas.

5.8 CHANNEL ROUTING

Channel routing is a type of unsteady flow analysis necessary when rainfall occurs in a subbasin some distance upstream of a point of interest, or when the runoff hydrographs from multiple subbasins must be combined and routed downstream to a point of interest. Channel routing is usually applied in cases where runoff hydrographs, as opposed to just peak flows, are being used. In storm sewer designs that use the rational method for peak discharge prediction, routing is not performed. Instead, the time of concentration of the entire contributing drainage basin area is determined for each point of interest in the system, and the peak discharge is computed as illustrated in Example 5.9 on page 143.

Channel routing may be done by using one of many available hydrologic or hydraulic methods. Hydrologic methods involve the principle of conservation of mass and a rather simple storage-discharge relationship (Chow, Maidment, and Mays, 1988). Hydraulic methods involve the simultaneous solution of the Saint-Venant equations (see Appendix C). The next section introduces the concepts of channel routing by demonstrating a hydrologic routing method called the *Muskingum method* (McCarthy, 1938). Additional information on unsteady flow modeling methods can be found in Appendix C.

Muskingum Routing

This method derives its name from the Muskingum River basin in Ohio, where it was first developed and applied. As described by Linsley, Kohler, and Paulhus (1982), an expression for the volume S of water stored within a channel reach may be expressed as

$$S = \frac{b}{a}\left[XI^{m/n} + (1-X)O^{m/n}\right] \quad (5.41)$$

where

S = volume stored in channel reach (ft^3, m^3)
a, n = constants in the stage-discharge relationship for the reach
b, m = constants in the stage-storage relationship for the reach
X = a dimensionless weighting factor
I = inflow discharge into the reach (cfs, m^3/s)
O = outflow discharge from the reach (cfs, m^3/s)

The Muskingum method assumes that $m/n = 1$ and assigns $K = b/a$ such that

$$S = K\left[XI + (1-X)O\right] \quad (5.42)$$

K is approximately equal to the travel time through the channel reach and, in the absence of additional information, is often estimated that way. For most streams and rivers, the constant X is in the range of about 0.1 to 0.4. Small values of X tend to be associated with rivers having wide floodplains and therefore large amounts of storage. Large values of X tend to be associated with narrow canyons where little storage is available as the water level rises.

If conservation of mass is approximated over a time interval $\Delta t = t_2 - t_1$, the following expression results:

$$S_2 - S_1 = \frac{I_1 + I_2}{2}\Delta t - \frac{O_1 + O_2}{2}\Delta t \qquad (5.43)$$

Substituting Equation 5.42 for each of the S terms in this expression and collecting like terms yields

$$O_2 = C_0 I_2 + C_1 I_1 + C_2 O_1 \qquad (5.44)$$

In Equation 5.44, the coefficients C_0, C_1, and C_2 are defined by Equations 5.45, 5.46, and 5.47, respectively. Note that $C_0 + C_1 + C_2 = 1$.

$$C_0 = \frac{0.5\Delta t - KX}{K(1-X) + 0.5\Delta t} \qquad (5.45)$$

$$C_1 = \frac{KX + 0.5\Delta t}{K(1-X) + 0.5\Delta t} \qquad (5.46)$$

$$C_2 = \frac{K(1-X) - 0.5\Delta t}{K(1-X) + 0.5\Delta t} \qquad (5.47)$$

K must be in the same time units as Δt. With K, X, and Δt established, these expressions may be used to estimate C_0, C_1, and C_2. Equation 5.44 can then be applied to predict the outflow rate from the channel reach at time t_2 on the basis of the current inflow and the previous inflow and outflow rates.

The number of routing reaches to which the Muskingum method is applied is designated as N_R. To ensure computational stability and accuracy of the computed outflow hydrograph, the U.S. Army Corps of Engineers (1960) recommends choosing Δt or N_R such that

$$\frac{1}{2(1-X)} \le \frac{K}{N_R \Delta t} \le \frac{1}{2X} \qquad (5.48)$$

If Δt is fixed by the time step of an available inflow hydrograph, a value for N_R can be chosen to ensure that Equation 5.48 is satisfied. If, for example, $N_R = 2$ satisfies Equation 5.48, the stream reach would be treated as a series of two reaches, each of which would be routed using the preceding expressions. The inflow hydrograph to the second (downstream) reach would be taken as the outflow hydrograph from the first (upstream) reach, and so on. As an alternative to fixing the time step based on the inflow hydrograph as just described, points on the inflow hydrograph can be interpolated for a time step other than that for which it was initially tabulated.

Example 5.14 – Routing a Channel Using the Muskingum Method. Route a runoff hydrograph through a channel reach for which $X = 0.2$ and $K = 20$ min. Use the hydrograph computed in Example 5.13 (page 161), but add 25 cfs to each hydrograph ordinate to account for base flow in the stream. Assume that flow in the channel is steady prior to the onset of runoff.

Note that $\Delta t = 10$ min is the time step at which the inflow hydrograph is tabulated. Using Equation 5.48, the ratio $K/N_R\Delta t$ must be between 0.28 to 2.50. For $K = 20$ min and $\Delta t = 10$ min, the relationship is satisfied for $N_R = 1$. Thus, the channel reach can be treated as a single reach, and the inflow hydrograph will not need to be interpolated for a different time step.

The values of the coefficients C_0, C_1, and C_2 are estimated to be

$$C_0 = [0.5(10) - 20(0.2)]/[20(1 - 0.2) + 0.5(10)] = 0.048$$

$$C_1 = [20(0.2) + 0.5(10)]/[20(1 - 0.2) + 0.5(10)] = 0.429$$

$$C_2 = [20(1 - 0.2) - 0.5(10)]/[20(1 - 0.2) + 0.5(10)] = 0.523$$

A check confirms that $C_0 + C_1 + C_2 = 1$.

Table 5.19 is set up for performing the hydrograph routing. The first column is the time, in minutes, since the beginning of rainfall. The second column is the inflow hydrograph to the reach, in cfs, and has been obtained by adding 25 cfs to each direct runoff hydrograph ordinate computed in Example 5.13. The first several entries in the third column (the outflow hydrograph column), up until a time of 20 min, are equal to 25 cfs because of the steady-state assumption.

At time $t = 30$ min, the discharge is computed as

$$Q_2 = 0.048I_2 + 0.429I_1 + 0.523Q_1 = 0.048(83) + 0.429(25) + 0.523(25) = 28 \text{ cfs}$$

At time $t = 40$ min, the discharge is computed as

$$Q_2 = 0.048(228) + 0.429(83) + 0.523(28) = 61 \text{ cfs}$$

Proceeding in this fashion, the table is completed. The inflow and outflow hydrographs are plotted in Figure E5.14.1.

Table 5.19 Muskingum channel routing

t (min)	*I* (cfs)	*Q* (cfs)	*t* (min)	*I* (cfs)	*Q* (cfs)
0	25	25	170	222	389
10	25	25	180	175	307
20	25	25	190	140	242
30	83	28	200	110	192
40	228	61	210	90	152
50	479	153	220	74	122
60	817	325	230	61	98
70	1098	573	240	55	80
80	1256	831	250	44	68
90	1269	1034	260	40	56
100	1177	1142	270	38	49
110	1002	1150	280	37	44
120	794	1070	290	31	40
130	610	930	300	27	35
140	461	770	310	26	32
150	361	618	320	25	29
160	283	492			

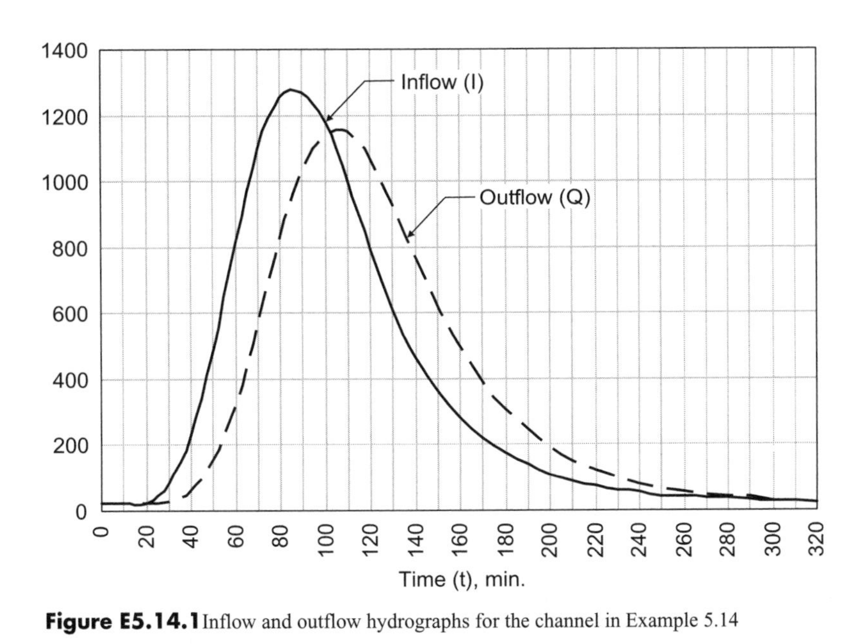

Figure E5.14.1 Inflow and outflow hydrographs for the channel in Example 5.14

When inflow and outflow hydrograph data are available for a channel reach, they may be used to estimate the parameters K and X for that reach. Either the graphical approach discussed by Chow, Maidment, and Mays (1988) may be applied for this, or a constrained least-squares method using a Lagrange multiplier can be applied.

5.9 RESERVOIR ROUTING

Reservoir routing is used when one needs to assess the effects of a natural pond or lake, or of a constructed facility such as a detention pond, on the characteristics of a hydrograph. As in channel routing, either hydrologic or hydraulic methods may be applied in reservoir routing. Level-pool hydrologic methods are most commonly used.

In the design and analysis of stormwater conveyance systems, reservoir routing is usually performed for facilities such as detention ponds. Therefore, descriptions of reservoir routing methods are presented with detention pond design in Chapter 12.

5.10 SUMMARY OF METHODS FOR COMPUTING FLOWS

Table 5.20 summarizes the basic characteristics of a number of methods for computing peak flows, developing runoff hydrographs, and routing hydrographs. The methods presented in this chapter are included, as well as a sampling of other popular models.

Table 5.20 Descriptions of Selected Methods for Computing Peak Flows

Type	Model	Description	Advantages	Disadvantages
Peak Flow These methods are generally statistically based or rely on an empirical formula to estimate the peak runoff from a catchment. Coefficients and parameters used in the formulas are lumped to express catchment properties such as surface detention and infiltration.	Rational	This method estimates the peak flow from rainfall events of a specific duration and frequency occurring over a given catchment area. Infiltration losses, surface detention, and antecedent conditions are accounted for by a runoff coefficient applied to the product of the rainfall intensity and the catchment area.	Peak flow models are easy to formulate because there are an abundance of data for determining the values of the parameters. With the rational method, multiple storms can be analyzed quickly and easily to determine the worse case scenario. Partial area effects can also be included to improve the quality of the analysis. The SCS model is able to account for storage within the catchment by using a pond and swamp adjustment factor.	The lumped system approach of peak flow models does not allow the analysis of spatial variation in rainfall and catchment properties. The rational method cannot be used on large catchments or those which experience large storage effects. Its runoff coefficient *C* is a lumped parameter taking into account infiltration, surface detention, antecedent conditions, etc.These factors cannot be effectively quantified by a single parameter, and variation in the value of the runoff coefficient has been shown to occur with more extreme events (Cordery and Pilgrim, 1983). The SCS method was developed for use with 24-hour storm rainfall depths and its use for storm durations other than this is not advised.
	SCS	The depth of effective precipitation during a storm is generated using the SCS Curve Number method and the peak discharge is then estimated as the product of this effective precipitation, the unit peak flow rate, and the drainage basin area. The unit peak flow rate is based on regression analysis of a large number of computer runs of SCS Curve Number method (TR-20). Consideration of storage effects in the catchment is also achieved through the introduction of a pond and swamp adjustment factor.		
Unit Hydrograph Unit hydrograph methods are based on the relationship of a pulse of rainfall (1 mm or 1 in.) applied to a catchment and the resulting response. With this relationship, the actual rainfall hyetograph is convolved to produce a runoff hydrograph. There are two forms of unit hydrograph: measured and synthetic. A measured hydrograph is the result of deconvolving an observed runoff hydrograph and the hyetograph. A synthetic hydrograph is developed from the catchment properties.	Clark	This method (Clark, 1945) uses the time-area approach to develop a unit hydrograph for the catchment. The unit hydrograph is convolved with the rainfall hyetograph to produce a runoff hydrograph, which is then routed through a linear storage concentrated at the catchment outlet so as to reproduce the effect of hydrograph attenuation and diffusion.	The application of the unit hydrograph to the rainfall hyetograph to estimate the runoff hydrograph is a simple process and takes account of all the catchment characteristics that influence the runoff hydrograph, such as infiltration, abstraction, storage, etc. With the Clark unit hydrograph, the characteristics of the catchment (shape, hydraulic length, etc.) are captured in the time-area histogram and therefore reflected in the unit hydrograph. Also with the Clark unit hydrograph, the use of the concentrated storage at the outlet addresses storage effects experienced in the catchment.	Unit hydrographs are not capable of estimating runoff hydrographs on catchments that experience a high level of nonlinearity caused by storages. Because they deal with lumped systems, it is not possible to analyze the effect of spatial rainfall. With the Snyder unit hydrograph, lag times may actually vary with the magnitude of the event and result in an underestimation of the peak of the runoff hydrograph.
	Snyder	Snyder (1938) developed this synthetic unit hydrograph from a study in the Appalachian Highlands. Catchment parameters, such as length along the main stream and length along the main stream from the outlet to the centroid, are used in an expression for the basin lag time. Once this lag has been determined, other parameters affecting the shape of the unit hydrograph can be estimated. The U.S. Army Corps of Engineers has developed empirical equations to further define the shape of the unit hydrograph.		
Time Area A series of isochrones are used to divide the catchment and produce a histogram of contributing area for each time interval. The partial flow from each of these time-based areas is generated by multiplying the area by the effective rainfall intensity for each time interval. Ordinates of the hydrograph at the outlet of the catchment are then developed by lagging and summing these partial flows.	TRRL	This approach is similar to the Clark method except that there is a nonlinear concentrated storage at the outlet instead of a linear storage. The time-area method is used to generate a hydrograph which is then routed through a concentrated nonlinear storage at the outlet. The model is also capable of performing simple routing of the runoff hydrograph through pipe networks.	It is possible to analyze spatial variations in the rainfall. The characteristics of the catchment (shape, hydraulic length, etc.) are captured in the time-area histogram and therefore reflected in the runoff hydrograph. An advantage of the TRRL method over the Clark method is that nonlinearity can also be modeled. Also with TRRL, diffusion and attenuation effects are addressed by including the concentrated storage at the catchment outlet	Storage effects within the catchment cannot be accounted for because this is a translation method and therefore the resulting hydrographs lack diffusion and attenuation.
	Modified RRL	Manning equation is used to develop a storage relationship for the main drainage pipe in the catchment. The hydrograph generated by the time-area method is then routed through this storage.		
	ILLUDAS-SA, ILLUDAS, and ILSAX	These models are derived from the TRRL implementation. The main difference is that the concentrated storage at the outlet is neglected, resulting in only translation of the hydrograph.		
Hydraulic Routing These methods generally involve the representation of the overland and stream flow within the catchment by the unsteady, one-dimensional, gradually varied open-channel flow equations (continuity and momentum).	Kinematic Wave	This model consists of various plane elements and channel sections arranged and connected in a manner that depicts the subcatchment properties and channel network within the catchment. A lumped or distributed approach can be applied depending on whether the parameters are kept constant or allowed to vary in space.	Since this is a deterministic method, it may be used on a catchment where no gauged data is available.	Although the model has a sound mathematical basis and physical attribute of the catchment are used to parameterize the equations, the assumptions of one-dimensional sheet flow on the plane surfaces and uniform flow in channel are approximations to reality.

Type	Model	Description	Advantages	Disadvantages
The full form of the equations is known as the Saint-Venant equations. The kinematic wave equation, which uses an approximation of the full Saint-Venant momentum equation, is more commonly used for runoff routing.	EPA Runoff	This method is similar to a kinematic wave. The main difference being in the formulation of the momentum equation, which is in a form that is very similar to Manning's equation.		
Storage Routing Storage routing models generally fall into two broad categories, distributed and lumped storage models, and can be either linear or nonlinear. Distributed models are capable of accounting for spatial variation in rainfall and catchment properties, while nonlinear models better reflect runoff variation that occur at different intensities.	RSWM	Regional Stormwater Model (Goyen & Aitken, 1979; Black and Codner, 1979) is a distributed storage model where the catchment is divided into 10 subareas using isochrones at equal time intervals. A concentrated nonlinear storage is placed at the outlet of each of these subareas, and flow is routed through these storages in series to produce a cascading storage routing scheme.	The arrangement of storage models within the catchment allows the physical characteristics of the catchment to be modeled with accuracy. In this way, the effects of urbanization can easily be incorporated into the design. The ability to include various loss models further improves the applicability of these models to the design of stormwater facilities. The ability of storage routing models to account for nonlinearity in storage is an added advantage for the analysis of urban catchments, which often contain storage facilities. With the WBNM model, as the storage characteristics and storage delay times of two different types of storages are different, the provision of these storages is physically realistic. The SBUH method directly computes a runoff hydrograph without going through an intermediate process such as developing a unit hydrograph. With the Tank Model, accurate results can be obtained as all the physical characteristics of the hydrological cycle are taken into consideration	The power function relationship that provides the nonlinear response is not always applicable to some catchments. Errors in establishment and calibration of nonlinear models can have a more pronounced effect on the results and can lead to underestimates for large events where data is not available. With the Nash model, there is a tendency for the model to average disparities within the catchment, since the collective runoff from the entire catchment is routed through the same storage. Also with the Nash model, the number of storages and the storage coefficient has been seen to vary from event to event, and it is difficult to reproduce the response of sharp peaks in observed hydrographs. The main problem with the Tank Model is that it requires a considerable amount of expertise to calibrate because the model structure and parameters must be determined subjectively.
	RORB	This model was developed by Laurenson and Mein (1983). A distributed nonlinear network model is developed by dividing the catchment into subcatchments, which are usually based on the major tributary network. Nodes are placed at the subcatchment centroids, the stream confluences, downstream of storages, etc. Catchment storage effects are represented by nonlinear concentrated storage placed midway between the nodes with special concentrated storages used to represent detention basins and reservoirs.		
	WBNM	The watershed bounded network model (WBNM) was developed by Boyd et al. (1979a, 1979b) and is similar to RORB, although it is based on more detailed consideration of geomorphological relations. The main difference is that WBRM has two different types of storages for types of subcatchments: ordered basins and interbasin areas. Ordered basins are complete subcatchments where no water flows into the area across the watershed boundary and their storage represents the transformation of rainfall excess into surface runoff at the downstream end of the catchment. Interbasin areas are subcatchments with a stream draining upstream areas that flow through them. Interbasins transform rainfall into runoff and contain transmission storages that route upstream runoff through the stream in the subcatchment.		
	Nash	This method is based on the average rainfall excess over the catchment being routed through a series of cascading linear storages concentrated at the catchment outlet.		
	SBUH	The SBUH method is based on the curve number (CN) approach and uses SCS equations for computing soil absorption and precipitation excess. This method converts the incremental runoff depths into instantaneous hydrographs that are then routed through a conceptual reservoir with a time delay equal to the basin time of concentration.		
	Tank Model	The tank model was developed by Sugawara and further refined as reported in Sugawara et al. (1984). The main concept behind this model is to represent the various components in the hydrologic cycle by linear or nonlinear storages ("tanks"). Each tank has one or more outlets located on the bottom or sides. The various parameters for each tank are estimated by considering catchment conditions such as evaporation, infiltration, and soil moisture, while calibrating to observed data.		

5.11 CHAPTER SUMMARY

Many methods and levels of analysis are available to quantify runoff from a watershed area. Several of these methods are described within this chapter.

The portion of the rainfall occurring over the area of interest that does not contribute to runoff is the rainfall abstraction. The total abstraction consists of several components: interception, depression storage, infiltration, and evapotranspiration. The first three components are typically accounted for in hydrologic analysis for stormwater management facilities. Evapotranspiration can usually be neglected. Two possible methods for modeling infiltration are Horton and Green-Ampt.

The portion of the total rainfall that contributes to runoff is called the effective precipitation. Effective precipitation, regardless of the method used for its determination, is used as the basis for practical estimation of the runoff volume, for peak runoff flow rate calculations, and for computing runoff flow rates over time (that is, a runoff hydrograph). Computation of the peak flow rate or a runoff hydrograph requires an estimate of the basin's time of concentration or lag time.

Methods used to compute the volume of effective precipitation include the runoff coefficient and the curve number method. The most accurate way to model effective precipitation is to create a runoff hyetograph by subtracting interception, infiltration, and depression storage from a rainfall hyetograph.

To compute the peak flow rate resulting from a rainfall event, the rational method, curve number method, or regression formula developed for a particular area and application may be used. For most approaches, it is necessary to compute the response time of the drainage basin (either the time of concentration or basin lag) to compute runoff flow rates.

A number of techniques for developing a runoff hydrograph exist; this text focuses on the unit hydrograph approach. With this method, an effective rainfall hyetograph is merged with the unit hydrograph for the basin through a process called discrete convolution. If applicable, base flow can be added to the resulting hydrograph. A more physically-based approach to hydrograph development is the kinematic wave model.

Runoff from snowmelt is an important consideration in designing stormwater facilities in many regions of the world. In determining the runoff potential for snowmelt, it is necessary to calculate the snow water equivalent, which is a function of the snow depth and density. The meltwater rate for snow is based on a melt rate factor and temperature.

Channel routing and reservoir routing are processes through which the effects of storage in a conveyance channel or basin within the watershed are accounted for. Reservoir routing is presented in more detail in Chapter 12.

REFERENCES

American Society of Civil Engineers (ASCE) 1992. *Design and Construction of Urban Storm Water Management Systems.* New York: American Society of Civil Engineers.

Anderson E. A. 1973. "National Weather Service River Forecast System – Snow Accumulation and Ablation Model." Report NWS 17. Washington, D.C.: U.S. Department of Commerce.

Anderson E. A. 1976. "A Point Energy and Mass Balance Model of Snow Cover." Report NWS 19. Washington, D.C.: U.S. Department of Commerce.

Bedient, P. B., and W. C. Huber. 1992. *Hydrology and Floodplain Analysis.* 2nd ed. Reading, Mass.: Addison-Wesley.

Black, D. C. and G. P. Codner. 1979. "Investigation of urban drainage strategy using simulation." Pp 199–204 in *Hydrology and Water Resources Symposium 1979.* Institute of Engineers Australia, National Conference Publication No. 79/10.

Boyd, M. J., B. C. Bates, D. H. Pilgrim, and I. Cordery. 1979a. "A storage routing model based on catchment geomorphology." *Journal of Hydrology* 42: 209-230.

Boyd, M. J., B. C. Bates, D. H. Pilgrim, and I. Cordery. 1979b. "An improved routing model based on geomorphology." Pp 189-193 in *Hydrology and Water Resources Symposium 1979*, Institute of Engineers Australia, National Conference Publication No. 79/10.

Boyd, M. J., B. C. Bates, D. H. Pilgrim, and I. Cordery. 1987. "WBNM: a general runoff routing model." Pp 137-141 in *4th National Local Government Eng. Conference*, Inst. Engrs Australia, National Conference Publication No. 87/9.

Brown, S. A., S. M. Stein, and J. C. Warner. 2001. *Urban Drainage Design Manual.* Hydraulic Engineering Circular No. 22, FHWA-SA-96-078. Washington, D.C.: U.S. Department of Transportation.

Chin, D. A. 2000. *Water-Resources Engineering.* Upper Saddle River, New Jersey: Prentice Hall.

Chow, V. T., D. R. Maidment, and L. W. Mays. 1988. *Applied Hydrology*, New York: McGraw-Hill.

Clark, C. O. 1945. "Storage and the unit hydrograph." *Transactions, American Society of Civil Engineering* 110: 1419-1446.

Cordery, I. and D. H. Pilgrim. 1983. "On the lack of dependences of losses from flood runoff on soil and cover characteristics." Pp. 187-195 in IAHR Publ. No. 140.

Federal Aviation Administration (FAA). 1970. *Advisory Circular on Airport Drainage.* Report A/C 150-5320-58. Washington, D.C.: U.S. Department of Transportation.

Goyen, A. J. and A. P. Aitken. 1979. "A stormwater drainage model." Pp 40-44 in Hydrology Symposium 1976. Institute of Engineers Australia, National Conference Publication No. 76/2.

Gray, D. M. and T. D. Prowse. 1993. "Snow and Floating Ice." In *Handbook of Hydrology*, D.R. Maidment, ed. New York: McGraw-Hill, Inc.

Green, W. H., and G. Ampt. 1911. "Studies of Soil Physics, Part I: The Flow of Air and Water Through Soils." *Journal of Agricultural Science* 4: 1–24.

Hathaway, G. A. 1945. "Design of Drainage Facilities." *Transactions, American Society of Civil Engineers* 110: 697–730.

Helvey, J. D. 1967. "Interception by Eastern White Pine." *Water Resources Research* 3: 723–730.

Helvey, J. D, and J. H. Patric. 1965. "Canopy and Litter Interception of Rainfall by Hardwoods of the Eastern United States." *Water Resources Research* 1: 193–206.

Henderson, F. M., and R. A. Wooding. 1964. "Overland Flow and Groundwater Flow from a Steady Rain of Finite Duration." *Journal of Geophysical Research* 69, no. 8: 1531–1540.

Horton, R. 1939. "Analysis of Runoff Plot Experiments with Varying Infiltration Capacity." *Transactions, American Geophysical Union* 20: 693–711.

Horton, R. 1919. "Rainfall Interception." *Monthly Weather Review* 47: 602–623.

Huber, W. C. and Dickinson, R. E. 1988. *Storm Water Management Model User's Manual, Version 4*. EPA-600/3-88-001a (NTIS PB88-236641/AS). Athens, Ga.: U.S. EPA.

Izzard, C. F. 1946. "Hydraulics of Runoff From Developed Surfaces." *Proceedings, Highway Research Board* 26: 129–150.

Jennings, M. E., W. O. Thomas, Jr., and H. C. Riggs. 1994. *Nationwide Summary of U.S. Geological Survey Regional Regression Equations for Estimating Magnitude and Frequency of Floods for Ungaged Sites.* Water Resources Investigations Report 94-4002. U.S. Geological Survey.

Johnstone, D., and W. P. Cross. 1949. *Elements of Applied Hydrology.* New York: Ronald Press.

Kent, K. 1972. "Travel Time, Time of Concentration, and Lag." *National Engineering Handbook,* Section 4: Hydrology, Chapter 15. Washington, D.C.: Soil Conservation Service, U.S. Department of Agriculture.

Kerby, W. S. 1959. "Time of Concentration of Overland Flow." *Civil Engineering* 60: 174.

Kibler, D. F. 1982. "Desk-Top Runoff Methods." Pp. 87–135 in *Urban Stormwater Hydrology*, ed. D.F. Kibler. Water Resources Monograph Series. Washington, D.C.: American Geophysical Union.

Kirpich, T. P. 1940. "Time of Concentration of Small Agricultural Watersheds." *Civil Engineering* 10, no. 6: 362.

Kuichling, E. 1889. "The Relation Between the Rainfall and the Discharge of Sewers in Populous Districts," *Transactions, American Society of Civil Engineers,* vol. 20, pp. 1-60.

Laurenson, E. M. and R. G. Mein. 1983. *RORB version 3: Runoff routing program-users manual*, 2d ed. Department of Civil Engineering, Monash University.

Linsley, R. K. 1943. "Application of the Synthetic Unit Graph in the Western Mountain States." *Transactions American Geophysical Union* 24:581–587.

Linsley, R. K., Jr., M. A. Kohler, and J. L. H. Paulhus. 1982. *Hydrology for Engineers*. 3d ed. New York: McGraw-Hill.

Matheussen, B. R. and Thorolfsson, S. T. 2001. Urban Snow Surveys in Risvollan, Norway, Urban Drainage Modeling. *Proceedings of the Specialty Symposium of the World and Water Environmental Resources Conference*. Alexandria, VA: American Society of Civil Engineers.

McCarthy, G. T. 1938. "The Unit Hydrograph and Flood Routing." North Atlantic Division Conference. U.S. Army Corps of Engineers.

McCuen, R. H., S. L. Wong, and W. J. Rawls. 1984. "Estimating Urban Time of Concentration." *Journal of Hydraulic Engineering* 110, no. 7: 887–904.

Merriam, R. A. 1960. "A Note on the Interception Loss Equation." *Journal of Geophysical Research* 65: 3850–3851.

Merriam, R. A. 1961. "Surface Water Storage on Annual Ryegrass." *Journal of Geophysical Research* 66: 1833–1838.

Mockus, V. 1969. "Hydrologic Soil-Cover Complexes." *National Engineering Handbook,* Section 4: Hydrology, Chapter 9. Washington, D.C.: Soil Conservation Service, U.S. Department of Agriculture.

Mockus, V. 1972. "Estimation of Direct Runoff from Storm Rainfall." *National Engineering Handbook,* Section 4: Hydrology, Chapter 10. Washington, D.C.: Soil Conservation Service, U.S. Department of Agriculture.

Mockus, V. 1973. "Estimation of Direct Runoff from Snowmelt." *National Engineering Handbook,* Section 4: Hydrology, Chapter 11. Washington, D.C.: Soil Conservation Service, U.S. Department of Agriculture.

Morel-Seytoux, H. J., and J. P. Verdin. 1981. *Extension of Soil Conservation Service Rainfall Runoff Methodology for Ungaged Watersheds.* Report FHWA/Rd-81/060. Washington, D.C.: U.S. Department of Transportation, Federal Highway Administration.

Natural Resources Conservation Service (NRCS). 1993. "Storm Rainfall Depth." *National Engineering Handbook,* Part 630: Hydrology, Chapter 4. Washington, D.C.: U.S. Department of Agriculture.

Oberts, G. 1994. "Influence of snowmelt dynamics on stormwater runoff quality." *Watershed Protection Techniques* 1 no. 2: 55–61.

Olin, D. A. 1984. *Magnitude and Frequency of Floods in Alabama.* Water Resources Investigations Report 84-4191. U.S. Geological Survey.

Overton, D. E., and M. E. Meadows 1976. *Stormwater Modeling.* New York: Academic Press.

Patric, J. H. 1966. "Rainfall Interception by Mature coniferous Forests of Southeast Alaska." *Journal of Soil and Water Conservation* 21: 229–231.

Rawls, W., P. Yates, and L. Asmusse. 1976. *Calibration of Selected Infiltration Equations for the Georgia Coastal Plain.* Technical Report ARS-S-113. Washington, D.C.: U.S. Dept. of Agriculture, Agricultural Research Service.

Rawls, W. J., L. R. Ahuja, D. L. Brakensiek, and A. Shirmohammadi. 1993. "Chapter 5: Infiltration and Soil Water Movement" in D.R. Maidment, ed. *Handbook of Hydrology.* New York: McGraw Hill.

Rodriguez-Iturbe, I., and J. B. Valdes. 1979. "The Geomorphologic Structure of Hydrologic Response." *Water Resources Research* 15, no. 6: 1409–1420.

Sauer, V. B., W. O. Thomas Jr., V. A. Stricker, and K. V. Wilson. 1983. *Flood Characteristics of Urban Watersheds in the United States.* Water Supply Paper 2207. U.S. Geological Survey.

Schaake, J. C., Jr., J. C. Geyer, and J. W. Knapp. 1967. "Experimental Examination of the Rational Method." *Journal of the Hydraulics Division,* American Society of Civil Engineers (ASCE) 93, no. HY6.

Semadeni-Davies, A. 2000. "Representation of snow in urban drainage models." *J. Hydrologic Engineering* 5 no. 4: 363–370.

Sherman, L. K. 1932. "Stream Flow from Rainfall by the Unit Graph Method." *Engineering News-Record* 108: 501–505.

Singh, V. P. 1988. *Hydrologic Systems, Rainfall-Runoff Modeling,* vol. I. Englewood Cliffs, New Jersey: Prentice Hall.

Singh, V. P. 1992. *Elementary Hydrology.* Englewood Cliffs, New Jersey: Prentice Hall.

Snider, D. 1972. "Hydrographs." *National Engineering Handbook,* Section 4: Hydrology, Chapter 16. Washington, D.C.: Soil Conservation Service (SCS), U.S. Department of Agriculture.

Snyder, F. F. 1938. "Synthetic Unit Graphs." *Transactions, American Geophysical Union* 19: 447.

Soil Conservation Service (SCS). 1969. Section 4: Hydrology in *National Engineering Handbook.* Washington, D.C.: U.S. Dept. of Agriculture.

Soil Conservation Service (SCS). 1986. *Urban Hydrology for Small Watersheds.* Technical Release 55. Washington, D.C.: U.S. Department of Agriculture.

Sugawara, M., I. Watanabe, E. Ozaka, Y. Katsuyama. 1984. *Tank model with snow component.* Research Notes of the National Research Center for Disaster Prevention No. 65. Ibaraki-Ken, Japan: Science and Technology Agency.

Taylor, A. B., and H. E. Schwarz. 1952. "A unit hydrograph lag and peak flow related to basin characteristics." *Trans. Am. Geophys. Union* 33: 235.

U.S. Army Corps of Engineers. 1960. *Routing of Floods Through River Channels.* Engineering and Design Manuals, EM 1110-2-1408. Washington, D.C.: U.S. Army Corps of Engineers.

U.S. Army Corps of Engineers Hydrologic Engineering Center (HEC). 1998. HEC-1 Flood Hydrograph Package User's Manual. Davis, Ca.: HEC.

Viessman, W., J. Knapp, and G. L. Lewis. 1977. *Introduction to Hydrology,* 2d ed. New York: Harper & Row.

Williams, G. B. 1922. "Flood Discharges and the Dimensions of Spillways in India." *Engineering* (London) 134: 321.

Zinke, P. J. 1967. "Forest Interception Studies in the United States." Pp. 137–160 in W. E. Sopper and H. W. Lull, eds. *Forest Hydrology*. Oxford: Pergamon Press.

PROBLEMS

5.1 Using the rational method, determine the composite runoff coefficient from a shopping center with the following characteristics:

Description	Area (ac)	C
Roadways and parking lots	8.5	0.90
Rooftops	3.4	0.85
Landscape areas	1.2	0.20

5.2 Use the data below to plot the incremental rainfall excess over the duration of the storm. What is the total effective precipitation, in inches?

Time (hr)	Rainfall Intensity (in/hr)	Infiltration Rate (in/hr)	Depression storage (in/hr)
0 to 1	0.41	0.2	0.2
1 to 2	0.49	0.16	0.14
2 to 3	0.32	0.13	0.04
3 to 4	0.31	0.09	0.02
4 to 5	0.22	0.08	0
5 to 6	0.08	0.07	0
6 to 7	0.06	0.06	0
7 to 8	0.04	0.04	0
8 to 9	0.04	0.03	0

5.3 Use the curve number method to calculate the effective precipitation from a 4-in. rainfall event over a watershed for each of the three antecedent moisture conditions. The watershed characteristics are given in the table.

Land Cover	Soil Group	Area (mi^2)
Pasture in good hydrologic condition	A	0.72
Residential (1/2 ac lots)	B	1.52
Streets (paved with curbs and gutters)	B	0.58

5.4 A subdivision has the following land uses: 160 ac of 1-ac-lot residential, 68 ac of good-condition open space, and 13 ac of gravel roads. Three percent of the subdivision is covered by swampy areas and the entire area overlays hydrologic soil type C. What are the effective rainfall and the peak discharge from a 5-in. rainfall event? The time of concentration is estimated to be 0.6 hr. Use the curve number method for a Type II rainfall distribution and assume that AMC-II applies.

5.5 A stormwater collection system is to be designed for a subdivision consisting of subbasins A, B, C, D, and E, each draining to an inlet with the same designation. The layout of the subbasins and collection system is shown. The characteristics of the subbasins, pipe flow times, and IDF data for the 25-

year design storm are given in the tables that follow. (Note that the IDF data is the same as that developed in Example 4.2.) Assume that every inlet captures 100 percent of approaching flows and use the rational method and the 25-year-storm IDF data from Example 4.2 to compute the design peak discharge to each inlet and within each pipe. Complete the results tables at the end of the problem.

Subbasin	Area (ac)	Runoff Coefficient	Inlet t_c (min)
A	1.5	0.50	7
B	1.7	0.60	8
C	1.1	0.75	5
D	2.7	0.55	10
E	1.1	0.70	5

Pipe	Flow Time (min)
Pipe 1	0.85
Pipe 2	0.55
Pipe 3	0.54
Pipe 4	0.50

Intensity Data for 25-Year Storm (in./hr)				
5 min	10 min	15 min	30 min	60 min
8.40	7.02	5.96	4.42	2.97

Inlet	Peak Flow (cfs)
A	
B	
C	
D	
E	

Pipe	Peak Flow (cfs)
Pipe 1	
Pipe 2	
Pipe 3	
Pipe 4	
Outlet Pipe	

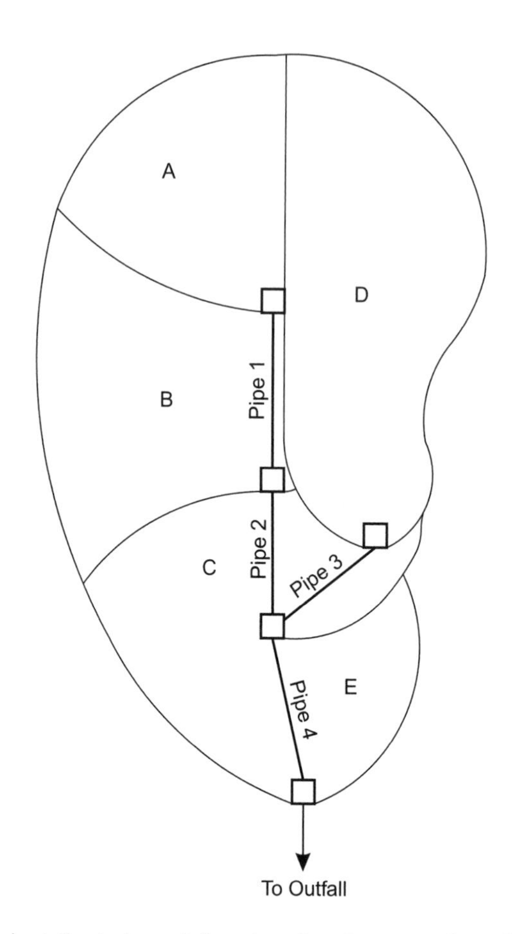

5.6 Given the following inflow hydrograph for a channel reach, compute the outflow at a section 3 mi downstream using the Muskingum method for $K = 11$ hr and $X = 0.13$.

Time (hr)	Inflow (cfs)	Outflow (cfs)
0	90	90
6	310	
12	690	
18	505	
24	395	
30	310	
36	220	
42	90	

5.7 Construct a synthetic unit hydrograph using the NRCS (SCS) method for a watershed with a drainage area of 250 ac. The time of concentration has been computed as 0.5 hr. Assume that Equation 5.26 is valid for this watershed.

Discharge in this force main is an example of closed-conduit flow. The brick structure is the pump station.

CHAPTER

6

Flow in Closed Conduits

Flow of water through a conduit is said to be *pipe flow,* or *closed-conduit flow*, if it occupies the entire cross-sectional area of the conduit. Conversely, if only a portion of the total cross-sectional area is occupied by the flow such that there is a free water surface at atmospheric pressure within the conduit, *open-channel flow* exists. Figure 6.1 shows a cross-sectional view of open-channel flow in a pipe.

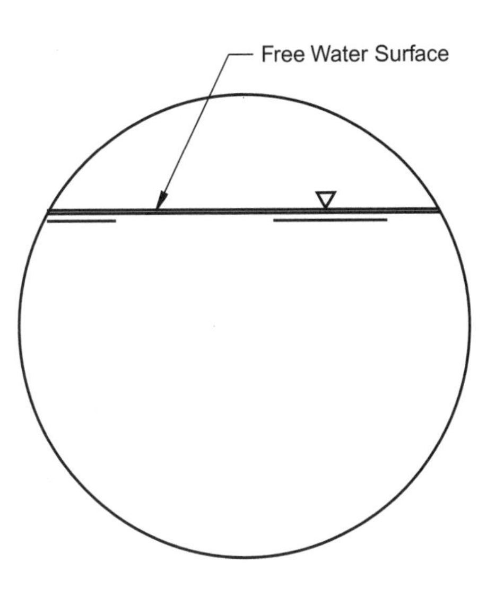

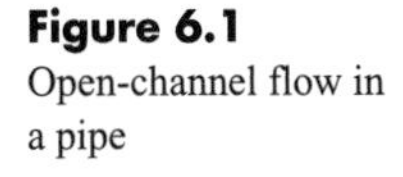

Figure 6.1
Open-channel flow in a pipe

Storm sewers and culverts are most commonly designed and constructed using commercially available pipe, which is manufactured in a number of different materials and cross-sectional shapes. Reinforced concrete and corrugated metal (steel or aluminum) are the most frequently used pipe materials, and circular is the most common cross-sectional shape. In some cases, notably for large culverts and sewers, concrete may be cast in place. For cast-in-place culverts, it is usually advantageous to use simple, rectangular cross-sectional shapes.

Although stormwater-related systems are primarily thought of as open-channel conveyances, closed-conduit flow occurs in a number of instances. The most obvious of these are stormwater pumping facilities; however, storm sewers, culverts, and detention pond outlet structures can also experience closed-conduit flow conditions when the hydraulic grade is sufficiently high.

In closed-conduit flow, the cross-sectional area of the flow is the same as that of the inside of the conduit. However, in open-channel flow, the cross-sectional area of the flow, and hence other flow parameters such as the velocity, depend not only on the size and shape of the conduit, but also on the depth of flow. As a consequence, open-channel flow is more difficult to treat from an analytical point of view than is pipe flow. This chapter, therefore, focuses on closed-conduit flow only; Chapter 7 addresses open-channel flow.

The hydraulics of pipe flow are examined in this chapter by first reviewing the energy equation from Section 3.2. Section 6.2 then covers the subject of flow resistance and presents several commonly used constitutive relationships for estimating the energy loss due to friction. Minor (or local) energy losses that occur at bends and other appurtenances in pipe systems are addressed in Section 6.3.

When flow momentum is changed, either by a change in flow direction (such as at a bend or manhole), or by a change in the magnitude of the velocity (due to a pipe diameter change, for instance), the change in momentum is accompanied by a corresponding hydraulic force that is exerted on the conduit or appurtenance. Prediction of the magnitudes and directions of these forces is presented in Section 6.4. Calculation of the predicted forces provides a basis for design of structural elements such as thrust blocks or other force restraints.

6.1 THE ENERGY EQUATION

In Section 3.2, the energy equation was introduced as a simplification of the most formal statement of conservation of energy, the First Law of Thermodynamics. The energy equation is

$$\frac{p_1}{\gamma} + Z_1 + \frac{\alpha_1 V_1^2}{2g} = \frac{p_2}{\gamma} + Z_2 + \frac{\alpha_2 V_2^2}{2g} + h_L - h_P + h_T \tag{6.1}$$

where p = fluid pressure (lb/ft^2, Pa)
γ = specific weight of fluid (lb/ft^3, N/m^3)
Z = elevation above an arbitrary datum plane (ft, m)
α = velocity distribution coefficient (see next section)
V = fluid velocity, averaged over a cross-section (ft/s, m/s)
g = acceleration of gravity (32.2 ft/s^2, 9.81 m/s^2)
h_L = energy loss between cross-sections 1 and 2 (ft, m)
h_P = fluid energy supplied by a pump between cross-sections 1 and 2 (ft, m)
h_T = energy lost to a turbine between cross-sections 1 and 2 (ft, m)

The first three terms on each side of Equation 6.1 are the *pressure*, *elevation*, and *velocity heads* at a given cross-section of the flow. The subscript 1 applies to an upstream cross-section, and the subscript 2 applies to a downstream cross-section. The last three terms on the right side of Equation 6.1 represent energy losses and gains that may occur along the flow path between cross-sections 1 and 2. In stormwater conveyance systems, it is rare for a *turbine* to be present; thus the term h_T will be assumed to equal zero throughout this text. *Pumps* are occasionally present in stormwater conveyance systems and are discussed in Chapter 13. However, in this chapter, it will be assumed that the term h_P equals zero.

Energy losses between any two cross-sections in pipe flow are composed of distributed losses due to pipe *friction* (distributed because they occur all along the length of a pipe, as opposed to occurring at one location) and localized or point losses (minor losses) due to the presence of bends, pipe entrances and exits, and other appurtenances. Thus, the h_L term in Equation 6.1 includes both frictional losees and minor losses. Frictional losses are discussed in detail in Section 6.2. Minor losses are discussed in Section 6.3.

The Velocity Distribution Coefficient

The velocity term, V, in Equation 6.1 refers to the *average* cross-sectional velocity of the fluid. This definition implies that velocity distribution is *non-uniform* over a cross section. In fact, due to friction, the velocity of the fluid is zero at the walls of the conduit.

Strictly speaking, the energy equation applies only to a particular streamline in a flow. The velocity distribution coefficient α, also known as the *kinetic energy correction factor*, applied to the velocity head terms in Equation 6.1 should be included when applying the equation to an averaged cross-sectional velocity ($V = Q/A$) rather than an individual streamline velocity. By definition, the value of α at any cross section is

$$\alpha = \frac{1}{AV^3} \int_A v^3 dA \qquad (6.2)$$

where A = total cross-sectional area of flow (ft^2, m^2)
V = cross-sectionally averaged velocity (ft/s, m/s)
v = velocity through an area dA of the entire cross-section (ft/s, m/s)

The integration is performed over the entire cross-sectional area.

An understanding of the basic types of flow that can exist in a conduit is essential to determining the appropriate value of α for use with the average velocity in Equation 6.1. With *turbulent flow*, eddies form in the flow and cause flow mixing. In the case of *laminar flow*, this intense mixing does not occur; in fact, the velocity for a steady flow will be constant at a given location in the cross section over time. A parameter called the *Reynolds number (Re)* (see page 192) is used in classifying flow as laminar or turbulent. Between these classifications is a transitional range in which flow may be either laminar or turbulent, depending on outside disturbances. The velocity distributions for typical turbulent and laminar flow conditions in a pipe are shown in Figure 6.2, as well as the idealized uniform velocity distribution in which the velocity is the same across the entire cross section.

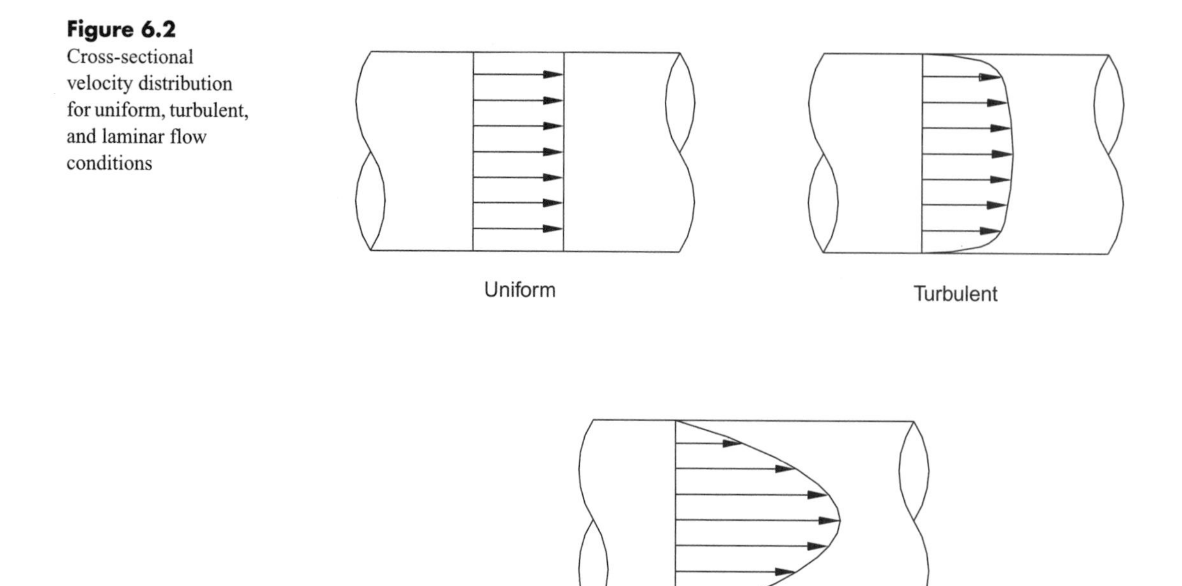

Figure 6.2
Cross-sectional velocity distribution for uniform, turbulent, and laminar flow conditions

Numerically, the velocity distribution coefficient α is greater than or equal to 1; it is equal to 1 when the velocity distribution is uniform. Because the velocity distribution in pipe flow is nonuniform, α is larger than 1 in pipe flow; however, it is often assumed to be equal to 1 as an approximation. Table 6.1 lists recommended values of α. Note that in fully turbulent flow, the flow distribution is approximately uniform (see Figure 6.2), and $\alpha \approx 1$. For laminar flow, which has a parabolic velocity distribution, $\alpha = 2$. (For an explanation of β, see Section 6.4.)

Table 6.1 Recommended values of α and β

Flow Description	Re	α	β
Uniform velocity distribution	N/A	1.0	1.00
Turbulent	> 4,000	1.04–1.06	1.03–1.07
Transitional	2,000–4,000	1.1–1.9	1.1–1.3
Laminar	< 2,000	2.0	1.33

Graphical Representation of the Energy Equation

Because each term in the energy equation has units of length and is referred to as *head*, the equation can be graphically represented on a profile view of a pipe system. Figure 6.3 illustrates such a profile with the various terms in the energy equation depicted graphically.

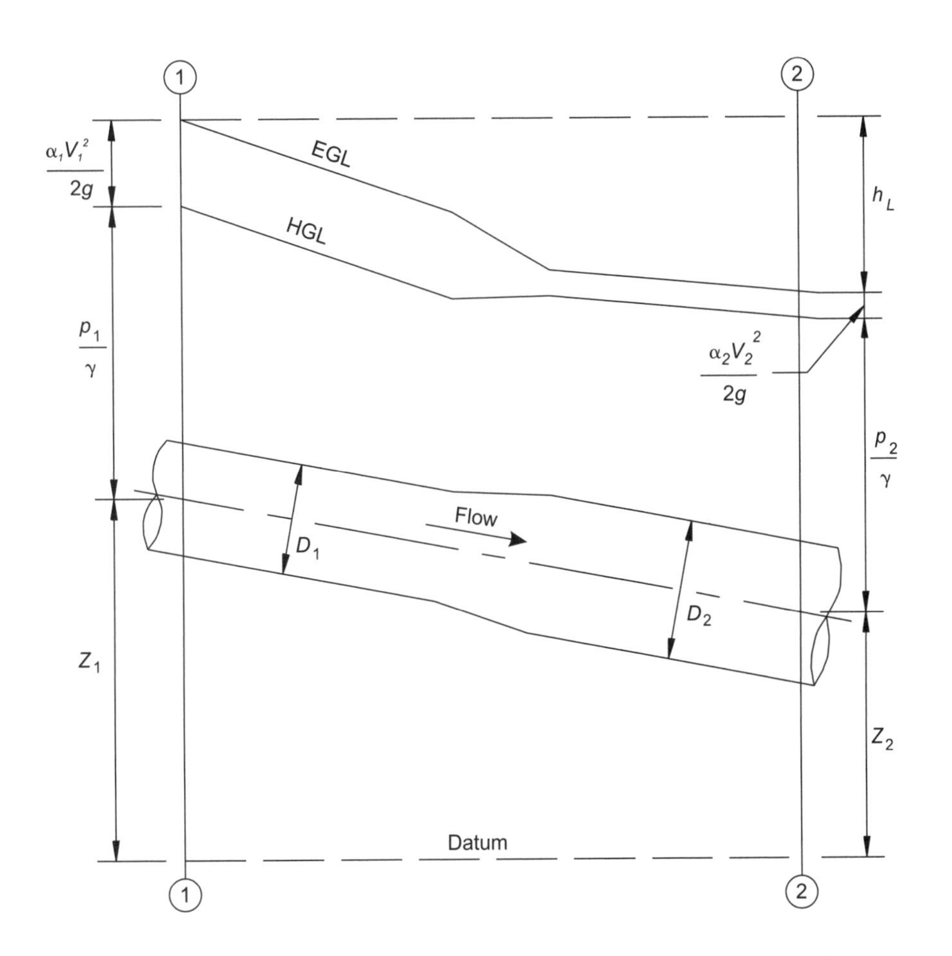

Figure 6.3
Graphical representation of the energy equation

Figure 6.3 shows two arbitrarily chosen flow cross sections, 1 and 2 (note that cross section 1 is upstream of cross section 2), as well as an elevation *datum*. In practice, cross sections are selected to correspond to locations where the flow behavior is of interest (such as the entrance and outlet of a culvert). The datum may be taken as mean sea level or any other convenient elevation.

At any cross section, the distance from the datum to the centerline of the pipe is known as the *elevation head* and is denoted by the term Z on each side of the energy equation. The *pressure head*, p/γ, is the distance from the pipe centerline to the *hydraulic grade line (HGL)*. The HGL elevation at a particular location represents the height to which water would rise, due to pressure, if a standpipe were there. The distance from the datum to the HGL, which is the sum of the pressure and elevation heads at any cross section, is called the *piezometric head*:

$$h = \frac{p}{\gamma} + Z \qquad (6.3)$$

where h = piezometric head (ft, m)

The HGL represents the head due to the potential and internal energies only. Kinetic energy is represented by the *velocity head* term, $\alpha V^2/2g$, which is the distance between the HGL and the *energy grade line (EGL)* in Figure 6.3. The EGL represents the total energy at any cross section. The distance from the datum to the EGL at any section is called the *total head:*

$$H = \frac{p}{\gamma} + Z + \frac{\alpha V^2}{2g} \tag{6.4}$$

where H = total head (ft, m)

Noting that H is equal to the sum of the first three terms on each side of the energy equation (Equation 6.1), it can be seen that a difference in H between any two cross sections is due to the head loss h_L (and the pump head, if a pump exists) between those cross sections. If no pump is present, H_1 is always larger than H_2 by the amount h_L because the total energy decreases due to frictional and minor losses as the flow progresses in a downstream direction. In Figure 6.3, the energy loss is due to a combination of the frictional losses in the two pipes and the minor loss at the pipe expansion.

In summary, the following statements about the graphical interpretation of the energy equation can be made:

- The EGL represents the total head at a cross section relative to the chosen datum, and always decreases in elevation as one moves in a downstream direction (unless a pump is present).
- The HGL represents the height to which water would rise in a standpipe. In a constant-diameter pipe, it is parallel to and below the EGL by the amount $\alpha V^2/2g$. At locations where the flow velocity changes, the HGL becomes either closer to or further from the EGL.
- The difference in the elevation of the EGL between any two cross sections is equal to the head loss (and/or the pump energy supplied) between those cross sections. For example, if the elevation of the EGL at one section is known, and if the head loss between that section and a second cross section is also known, then the elevation of the EGL at the second cross section can be determined.
- The difference in elevation between the HGL and the pipe centerline is the pressure head, p/γ. If this value is known, then the (gauge) fluid pressure in the pipe can be found as the product of γ and the pressure head. (To convert from gauge pressure to pressure head in feet if pressure is in psi, multiply the pressure by 2.31. To obtain pressure head in meters, multiply the pressure in kPa by 0.102.) The gauge pressure will be positive (that is, the absolute fluid pressure will be greater than atmospheric pressure) if the HGL is above the pipe centerline. Otherwise, the gauge pressure will be negative and the absolute fluid pressure will be lower than the atmospheric pressure.

- The EGL slopes downward (in the direction of flow) along pipe reaches, with the magnitude of the slope being dependent on the pipe diameter, friction factor, and flow velocity (see Section 6.2). At locations of minor losses, the elevation of the EGL is usually assumed to drop abruptly, with the magnitude of the drop being equal to the minor energy loss (see Section 6.3).

This set of observations is of great practical use for engineering calculations. Delineation of the HGL and EGL on design drawings illustrating the profile of a stormwater conveyance system will clearly show the adequacy of the design, at least from an energy point of view. Section 11.5 discusses this concept in detail.

Example 6.1 – Application of the Energy Equation. For the pipe flow situation shown in Figure 6.3, the following information is given:

$Z_1 = 50$ ft $Z_2 = 42$ ft $Q = 5$ cfs

$D_1 = 12$ in. $D_2 = 24$ in. $h_L = 25$ ft

$p_2/\gamma = 30$ ft

At both cross sections, determine the velocity head, the elevations of the EGL and HGL, and the fluid pressure in the pipe. Assume that $\alpha = 1$ at both cross sections.

Solution: The pipe areas are required to determine the velocity at each cross section, which in turn is necessary to compute the velocity head. At cross section 1, where the pipe diameter is 12 in., the area is $A_1 = 0.785$ ft^2. At cross section 2, the area is $A_2 = 3.14$ ft^2. The velocity at cross section 1 is

$$V_1 = Q/A_1 = 5/0.79 = 6.33 \text{ ft/s}$$

and the velocity at cross section 2 is $V_2 = 1.59$ ft/s. The velocity heads at the two cross sections are

$$(\alpha V^2/2g)_1 = 6.33^2/[2(32.2)] = 0.62 \text{ ft}$$

$$(\alpha V^2/2g)_2 = 1.59^2/[2(32.2)] = 0.04 \text{ ft}$$

At cross section 2, the HGL elevation is the sum of Z_2 and p_2/γ, or

$$HGL_2 = 42 \text{ ft} + 30 \text{ ft} = 72 \text{ ft}$$

The EGL at cross section 2 is determined as the sum of the HGL and the velocity head, or

$$EGL_2 = 72 \text{ ft} + 0.04 \text{ ft} = 72.04 \text{ ft}$$

The EGL elevation at cross section 1 is found by adding the EGL at cross section 2 and the head loss between the cross sections:

$$EGL_1 = 72.04 \text{ ft} + 25 \text{ ft} = 97.04 \text{ ft}$$

The HGL at cross section 1 is calculated by subtracting the velocity head from the EGL:

$$HGL_1 = 97.04 \text{ ft} - 0.62 \text{ ft} = 96.42 \text{ ft}$$

The pressure head at cross section 1 is found by subtracting the pipe centerline elevation from the HGL elevation:

$$p_1/\gamma = 96.41 \text{ ft} - 50 \text{ ft} = 46.42 \text{ ft}$$

The fluid pressure at each cross section is found by multiplying the pressure head by the specific weight of water:

$$p_1 = (46.42 \text{ ft})(62.4 \text{ lb/ft}^3) = 2{,}900 \text{ lb/ft}^2 = 20.1 \text{ psi}$$

$$p_2 = (30 \text{ ft})(62.4 \text{ lb/ft}^3) = 1{,}870 \text{ lb/ft}^2 = 13.0 \text{ psi}$$

6.2 RESISTANCE TO FLOW IN CLOSED CONDUITS

In most stormwater conveyance systems, the major cause of head losses is the friction exerted on the fluid by the pipe walls. Minor losses, as the name implies, are often small in comparison to frictional losses. This section discusses head losses caused by pipe friction, and describes several of the methods that have been developed to quantify pipe friction. Minor losses are discussed in Section 6.3.

The Reynolds Number

A set of hydraulic experiments carried out in the nineteenth century by Osborne Reynolds (1883) examined the onset of turbulence in pipe flow. Reynolds varied the velocity of flow through a glass tube into which a thread of dye was injected, and observed that the thread of dye would break up and disperse by mixing when the velocity became relatively large. By considering not only the velocity, but also the pipe diameter and fluid viscosity, Reynolds was able to summarize his observations in terms of a dimensionless parameter known as the Reynolds number, which is an indicator of the relative importance of inertial forces to viscous forces. The Reynolds number is computed as:

$$\mathrm{Re} = \frac{\rho V D}{\mu} = \frac{VD}{\nu} \qquad (6.5)$$

where Re = the Reynolds number (dimensionless)
ρ = fluid density (slugs/ft^3, kg/m^3)
V = flow velocity (ft/s, m/s)
D = pipe diameter (ft, m)
μ = dynamic viscosity of the fluid (lb-s/ft^2, N-s/m^2)
ν = kinematic viscosity of the fluid (ft^2/s, m^2/s)

In his experiments, Reynolds found that flow became *turbulent* when Re > 12,000. However, in the commercially available pipes used for construction of stormwater conveyance systems, flow is generally turbulent when the Reynolds number exceeds 4,000. Flow in a pipe is *laminar* when the Reynolds number is less than 2,000, and is *transitional* (that is, between laminar and fully turbulent) when the value of Re is between 2,000 and 4,000. Note that the Reynolds number of a flow increases when either the pipe diameter or fluid velocity increases, or when the fluid viscosity decreases.

Calculation of the Reynolds number is necessary to determine the friction factor to be used in the Darcy-Weisbach equation for computing head loss, as described in the following subsection.

The Darcy-Weisbach Equation

The *Darcy-Weisbach equation* can be derived several ways and is an expression for the head loss in a conduit due to friction. The head loss depends on both conduit and flow properties and is based on the fact that, for steady flow, the frictional forces at the pipe wall must be equal to the driving forces of pressure and gravity. Solving for head loss due to friction, the Darcy Weisbach equation can be written as

$$h_L = f \frac{L}{D} \frac{V^2}{2g} \tag{6.6}$$

where h_L = head loss due to friction (ft, m)
f = a friction factor depending on the Reynolds number of the flow and the *relative roughness* of the pipe
L = pipe length (ft, m)
D = pipe diameter (ft, m)
V = cross-sectionally averaged velocity of the flow (ft/s, m/s)
g = acceleration of gravity (32.2 ft/s^2, 9.81 m/s^2)

If both sides of Equation 6.6 are divided by the pipe length, the slope of the energy grade line S_f (that is, the *energy gradient* or *friction slope*) can be expressed as

$$S_f = \frac{h_L}{L} = f \frac{1}{D} \frac{V^2}{2g} \tag{6.7}$$

This form of the Darcy-Weisbach equation presents the relationship between the energy gradient and the flow velocity (or vice versa, by a simple algebraic rearrangement). Equations of this nature, relating the energy gradient and flow velocity, are called *constitutive relationships*. The Darcy-Weisbach equation can be argued to be the most fundamentally sound constitutive relationship for pipe flow although many others, such as the Manning equation, are also available and widely used.

As already noted, the friction factor f in the Darcy-Weisbach equation depends on the Reynolds number and on the relative roughness of the pipe. The relative roughness is defined as the ratio ε/D. Here, ε is the *equivalent sand grain roughness* of the pipe, which must be experimentally determined. Table 6.2 presents recommended values of ε for several common pipe materials. The values given are for new pipe. As pipe ages, corrosion, pitting, scale build-up, and/or deposition usually occur, causing the equivalent sand grain roughness to increase. Values for old pipes may be several orders of magnitude larger than those shown in Table 6.2.

Table 6.2 Equivalent sand grain roughness for commercial pipes

Pipe Material	ε (ft)	ε (mm)
Corrugated metal	0.1–0.2	30–61
Riveted steel	0.003–0.03	0.91–9.1
Concrete	0.001–0.01	0.3–3.0
Wood stave	0.0006–0.003	0.18–0.91
Cast iron	0.0008–0.018	0.2–5.5
Galvanized iron	0.00033–0.015	0.102–4.6
Asphalted cast iron	0.0004–0.007	0.1–2.1
Steel, wrought iron	0.00015	0.046
Drawn tubing	0.000005	0.0015

With the Reynolds number and relative roughness known, the *Moody diagram* (Moody, 1944), illustrated in Figure 6.4 on page 196, can be employed to obtain the friction factor f. Note that for Reynolds numbers below 2,000, where the flow is laminar, f depends only on the Reynolds number and is calculated as $f = 64/Re$. For large Reynolds numbers where the curves on the Moody diagram become horizontal, it can be seen that f depends only on the relative roughness of the pipe. Flow in this region is considered to be fully turbulent or completely rough flow. In the transitional region between laminar and fully turbulent flow, f depends on both Re and D.

Example 6.2 – Application of the Darcy-Weisbach Equation to Pipe Flow. Determine the head loss in 75 m of 610-mm diameter concrete pipe conveying a discharge of 0.70 m^3/s. Assume that the water temperature is 15° C.

Solution: Referring to Table 6.2, the equivalent sand grain roughness for the pipe can be taken as ε= 3 mm. The relative roughness is therefore $\varepsilon/D = 0.005$. Because the adopted value of ε is at the high end of the range of recommended values, the computed head loss will also be high. It is often wise in practical design to adopt high values of ε for head loss computations and to adopt low values of ε for velocity calculations. When this is done, both computed head losses and computed velocities will tend to be on the high side and thus conservative estimates.

The cross-sectional area of the 610-mm pipe is $A = 0.292$ m^2, and the velocity of flow is $V = Q/A$ = 2.40 m/s. Table A.1 (Appendix A) provides the kinematic viscosity for the water as $\nu = 1.13 \times 10^{-6}$ m^2/s. The Reynolds number is computed as

$$\text{Re} = \frac{VD}{\nu} = \frac{2.4(0.61)}{1.13 \times 10^{-6}} = 1.3 \times 10^6$$

From the Moody diagram, $f = 0.030$. Applying the Darcy-Weisbach equation, the head loss in the pipe is

$$h_L = 0.030\left(\frac{75}{0.61}\right)\frac{2.4^2}{2(9.8)} = 1.08 \text{ m}$$

Example 6.3 – Determining Required Pipe Size Using Darcy-Weisbach. The energy gradient available to drive the flow in a pipe is $S_f = 0.015$. If the required flow rate is $Q = 50$ cfs, and if the equivalent sand grain roughness of the pipe is $\varepsilon = 0.001$ ft, determine the required pipe diameter.

Solution: Because the pipe diameter is unknown, the relative roughness of the pipe is also unknown. The pipe cross-sectional area and flow velocity are unknown, and hence the Reynolds number and the friction factor f are also unknown. Because of all the unknowns, a trial and error solution is required.

The solution proceeds by first assuming an $f = 0.020$. The pipe diameter is then computed by rearranging Equation 6.7 to

$$S_f = f\frac{1}{D}\frac{Q^2}{2gA^2} = \frac{8fQ^2}{\pi^2 gD^5}$$

or

$$D = \left\{\frac{8fQ^2}{\pi^2 gS_f}\right\}^{1/5} = \left\{\frac{8(0.020)(50)^2}{32.2\pi^2(0.015)}\right\}^{1/5} = 2.43 \text{ ft}$$

With this initial estimate of the pipe diameter, the relative roughness and Reynolds number can be computed to obtain a more accurate estimate of f:

$$\frac{\varepsilon}{D} = \frac{0.001}{2.43} = 0.0004$$

$$V = \frac{Q}{A} = \frac{50}{\pi(2.43)^2/4} = 10.8 \text{ ft/s}$$

$$\text{Re} = \frac{VD}{\nu} = \frac{10.8(2.43)}{1.217\times10^{-5}} = 2.2\times10^6$$

$f = 0.016$

With this improved estimate of f, the pipe diameter can be recomputed as

$$D = \left\{\frac{8(0.016)(50)^2}{32.2\pi^2(0.015)}\right\}^{1/5} = 2.32 \text{ ft}$$

An additional iteration does not significantly change this diameter and thus the minimum required diameter is 2.32 ft or 28 in. Rounding up to the next commercially available size, a 30-in. diameter pipe should be selected.

Explicit Equations for *f*, *Q*, and *D*. Use of the Darcy-Weisbach equation for pipe flow calculations can be inconvenient because of the need to consult the Moody diagram. Further, as seen in the previous example, calculations are often iterative in nature. Because of this, a number of efforts have been made to express the relationship between the important variables in explicit mathematical form. This is especially necessary if solutions are to be computerized.

Prandtl and von Kármán (Prandtl, 1952) used Nikuradse's experimental results to express the friction factor for flow in smooth pipes as

$$\frac{1}{\sqrt{f}} = -2\log\left(\frac{2.51}{\text{Re}\sqrt{f}}\right) \qquad (6.8)$$

A *smooth pipe* is one in which the relative roughness ε/D is very small (see Figure 6.4). Prandtl and von Kármán also expressed the friction factor for *completely rough flow* in pipes (where f depends only on ε/D) as

$$\frac{1}{\sqrt{f}} = -2\log\left(\frac{\varepsilon/D}{3.7}\right) \qquad (6.9)$$

For the *transitional flow* region where f depends on both Re and ε/D, Colebrook (1939) developed the equation

$$\frac{1}{\sqrt{f}} = -2\log\left(\frac{\varepsilon/D}{3.7} + \frac{2.51}{\text{Re}\sqrt{f}}\right) \qquad (6.10)$$

Figure 6.4
The Moody diagram

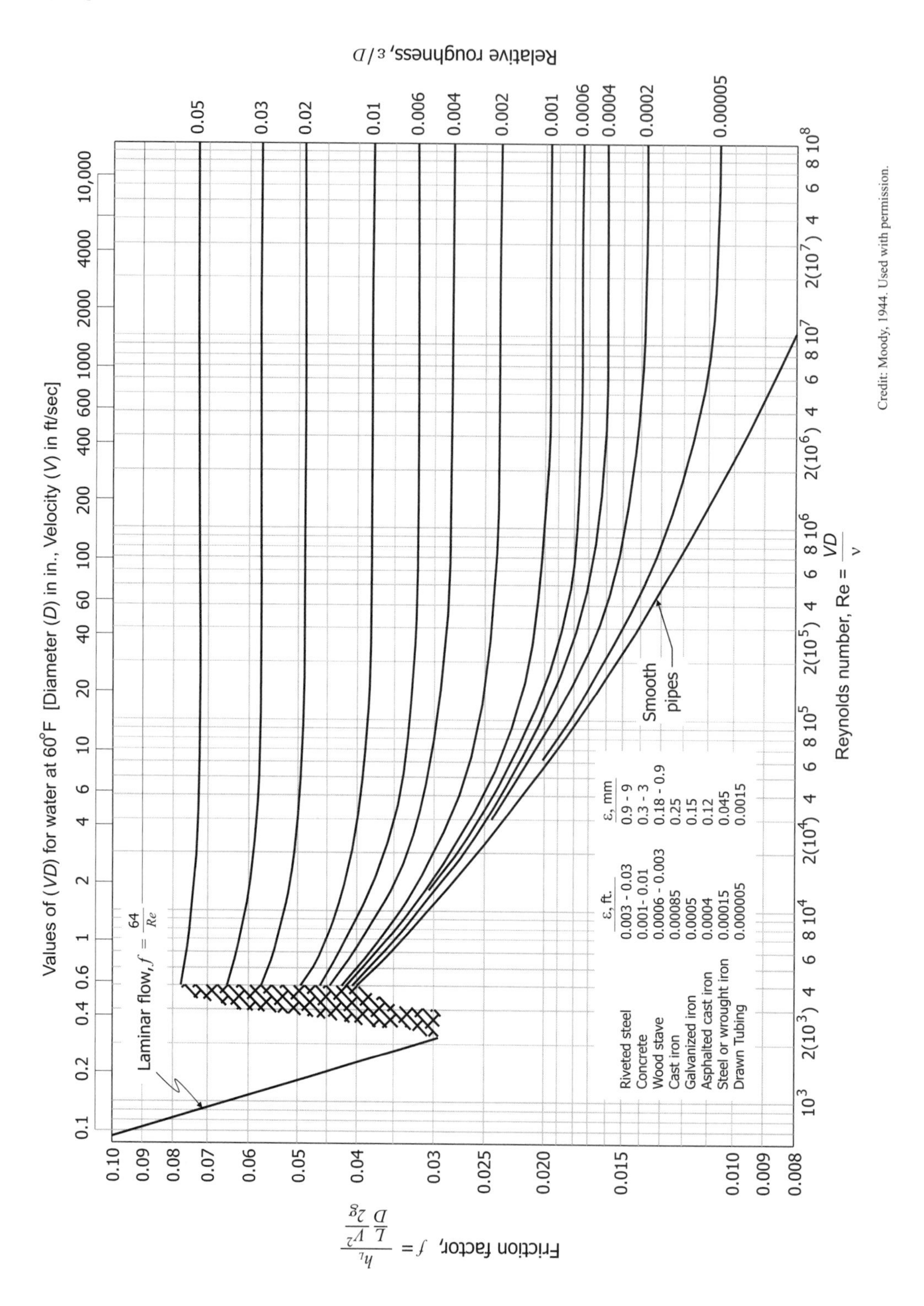

Credit: Moody, 1944. Used with permission.

Note that Equations 6.8 and 6.9 are asymptotes of this Equation 6.10. Equation 6.10 reduces to Equation 6.8 when ε/D is small (smooth pipe), and to Equation 6.9 when Re is large (fully turbulent flow).

The *Colebrook equation*, which was developed by Cyril Colebrook and Cedric White, preceded the Moody diagram. It is useful but complicated by its implicit form. Because f appears on both sides of the equation, a trial and error solution is required. Recognizing this, Jain (1976) developed an explicit approximation to the Colebrook equation as

$$\frac{1}{\sqrt{f}} = -2\log\left(\frac{\varepsilon / D}{3.7} + \frac{5.74}{\mathrm{Re}^{0.9}}\right) \tag{6.11}$$

This expression is valid for $10^{-6} \leq \varepsilon/D \leq 10^{-2}$ and $5{,}000 \leq \mathrm{Re} \leq 10^8$.

Swamee and Jain (1976) developed similar explicit equations for Q and D, which do not require the iterative trial-and-error solutions of the Darcy-Weisbach equation. Their equation for Q is

$$Q = -2.22D^2\sqrt{\frac{gDh_L}{L}} \cdot \log\left(\frac{\varepsilon / D}{3.7} + \frac{1.784\nu}{D\sqrt{gDh_L / L}}\right) \tag{6.12}$$

A modified version of their equation for D is (Streeter and Wylie, 1979)

$$D = 0.66\left\{\varepsilon^{1.25}\left(\frac{LQ^2}{gh_L}\right)^{4.75} + \nu Q^{9.4}\left(\frac{L}{gh_L}\right)^{5.2}\right\}^{0.04} \tag{6.13}$$

These expressions are valid for $10^{-6} \leq \varepsilon/D \leq 10^{-2}$ and $5000 \leq \mathrm{Re} \leq 10^8$.

Noncircular Conduits. It is sometimes necessary in design and construction of stormwater conveyance systems to employ conduits of noncircular shapes. The most common of these shapes are elliptical, arch, and rectangular, although others exist as well. Use of these shapes is most common in cases where the burial depth of the stormwater conduit is limited. Rectangular shapes are often used in large cast-in-place or precast concrete structures.

When the Darcy-Weisbach equation is applied to the analysis of flow in a noncircular conduit, the pipe diameter D should be replaced by four times the hydraulic radius, or

$$D = 4R \tag{6.14}$$

The *hydraulic radius* is defined as the ratio of the cross-sectional area to the wetted perimeter of the flow, or $R = A/P$. Note that for full flow in a circular cross-section, the area is simply the area of the circle ($D^2/4$), and the wetted perimeter is the circumference of the circle ($P = 2\pi r = \pi D$); thus, $R = A/P = D/4$.

"Janet, we have to stop meeting like this. It has become a source of increasing friction in my marriage to Elizabeth."

Use of the substitution $D = 4R$ means that the head loss, Reynolds number, and relative roughness expressions should all be modified to the following forms:

$$h_L = f\frac{L}{4R}\frac{V^2}{2g} \tag{6.15}$$

$$\mathrm{Re} = \frac{4\rho VR}{\mu} = \frac{4VR}{\nu} \tag{6.16}$$

$$\textit{Relative roughness} = \frac{\varepsilon}{4R} \tag{6.17}$$

This substitution is an approximation, but the error is small if the ratio of the maximum to minimum cross-sectional dimensions of the conduit is less than about 4.0 (Potter and Wiggert, 1991).

The Manning Equation

Experiments conducted by the Irish engineer Robert Manning on flows in open channels and pipes led to the formulation of a constitutive relationship of the form (Manning, 1891)

$$V = \frac{C_f}{n}R^{2/3}S_0^{1/2} \tag{6.18}$$

where V = flow velocity (ft/s, m/s)
C_f = unit conversion factor (1.49 for U.S. Customary units; 1.00 for SI units)
n = friction factor (also known as Manning's n)
R = hydraulic radius (ft, m)
S_o = channel slope (ft/ft, m/m)

Although it was primarily developed for open-channel flow problems, the Manning equation has been used extensively for analysis of pipe flow. The equation was originally developed for *uniform flow*, where the pipe or channel bed slope is equal to the energy gradient. However, it is also widely used for applications where S_o and S_f are different from one another. When the energy gradient, S_f, is used for the slope term, the equation can be rearranged as

$$S_f = \frac{(nV)^2}{C_f^2 R^{4/3}} \tag{6.19}$$

In pipe flow applications, it is usually assumed that the value of Manning's n depends only on the pipe material, and that it is independent of the Reynolds number and the pipe diameter. Therefore, the Manning equation should be considered to be valid only for fully turbulent flow considering earlier discussions regarding the friction factor f in the Darcy-Weisbach equation. It would also be expected that, in reality, the n value should be different for large pipes than for small ones, even when the pipe materials are the same. However, the same values of n are usually used in practice. Table 6.3 contains typical values of n for common pipe materials. Additional information for selecting n values for corrugated steel pipe is provided in Section 11.1.

Table 6.3 Typical Manning's n values

Pipe Material	n
Corrugated metal	0.019–0.032
Concrete pipe	0.011–0.013
Concrete (cast-in-place)	0.012–0.017
Plastic (smooth)	0.011–0.015
Plastic (corrugated)	0.022–0.025
Steel	0.009–0.012
Cast iron	0.012–0.014
Brick	0.014–0.017
Clay	0.012–0.014

Example 6.4 – Application of the Manning Equation to a Culvert Flowing Full. A 1200-mm diameter reinforced concrete pipe culvert conveys a discharge of 3.5 m^3/s under a highway. The culvert has a length of 60 m. If $n = 0.013$ and the culvert is flowing full, determine the head loss through the culvert. Neglect minor losses at the culvert entrance and outlet.

Solution: The cross-sectional area of the culvert is 1.13 m^2 and the flow velocity is

$$V = Q/A = 3.10 \text{ m/s}$$

The hydraulic radius of a circular cross section when flowing full is $R = D/4$, and hence $R = 0.3$ m.

Applying Equation 6.19, the energy gradient is

$$S_f = \frac{(0.013 \times 3.10)^2}{0.3^{4/3}} = 0.0081$$

Multiplying the energy gradient by the culvert length yields the head loss as

$h_L = S_f L = 0.0081(60) = 0.486$ m

The Chézy Equation

Around 1769, Antoine Chézy, a French engineer, developed what is probably the oldest constitutive relationship for uniform flow in an open channel (Herschel, 1897). The same relationship has also been used for pipe flow. The *Chézy equation* takes the form

$$V = C\sqrt{RS_o} \tag{6.20}$$

where V = flow velocity (ft/s, m/s)
C = a resistance coefficient known as Chézy's C ($ft^{0.5}/s$, $m^{0.5}/s$)
R = hydraulic radius (ft, m)
S_o = channel slope (ft/ft, m/m)

If one assumes that the slope can be replaced by the energy gradient, and the equation is rearranged, one obtains a constitutive relationship of the form

$$S_f = \frac{V^2}{C^2 R} \tag{6.21}$$

As is the case with the Manning equation, it is usually assumed in practice that the resistance coefficient (C in this case) is a constant and is independent of the Reynolds number and the relative roughness of the conduit. Chézy's C can be related to Manning's n by equating the right sides of Equations 6.19 and 6.21 to obtain

$$C = \frac{C_f R^{1/6}}{n} \tag{6.22}$$

where C_f = unit conversion factor (1.49 for U.S. Customary units; 1.00 for SI units)

Example 6.5 – Application of the Chézy Equation to Pipe Flow. The head loss along a 107-m long pipe is 1.2 m. If the pipe is 460 mm in diameter and the flow rate is 0.6 m^3/s, determine Chézy's C for the pipe.

Solution: The energy gradient is the ratio of the head loss to the pipe length, or

$S_f = 1.2 \text{ m}/107 \text{ m} = 0.01121$

The hydraulic radius of the pipe is $R = D/4 = 0.115$ m, and the flow velocity is $V = Q/A = 0.6/0.166 = 3.61$ m/s. Rearranging Equation 6.21, Chézy's C is determined as

$$C = \frac{V}{\sqrt{RS_f}} = \frac{3.61}{\sqrt{0.115(0.01121)}} = 100.5 \text{ m}^{1/2}/\text{s}$$

The Hazen-Williams Equation

The *Hazen-Williams equation* was developed primarily for use in water distribution design and has rarely been used for other applications. The equation is (Williams and Hazen, 1933):

$$V = C_f C_h R^{0.63} S_f^{0.54} \tag{6.23}$$

where V = flow velocity (ft/s, m/s)
C_f = a unit conversion factor (1.318 for U.S. Customary units; 0.849 for SI units)
C_h = Hazen-Williams resistance coefficient
R = hydraulic radius (ft, m)
S_f = energy gradient (ft/ft, m/m)

Some typical values of the resistance coefficient are given in Table 6.4.

Table 6.4 Typical values of the Hazen-Williams resistance coefficient (Viessman and Hammer, 1993)

Pipe Material	C_h
Cast iron (new)	130
Cast iron (20 yr old)	100
Concrete (average)	130
New welded steel	120
Asbestos cement	140
Plastic (smooth)	150

Chin (2000), in a discussion of the Hazen-Williams formula, points out that its use should be limited to flows of water at 60°F (16°C), to pipes with diameters from 2 to 72 in. (50 to 1800 mm), and to flow velocities less than 10 ft/s (3 m/s). Further, the resistance coefficient is dependent on the Reynolds number, which means that significant errors can arise from use of the formula.

For high velocities in rough pipes, Walski (1984) gives a correction factor for C as

$$C_h = C_o \left(\frac{V_o}{V}\right)^{0.081} \tag{6.24}$$

where C_o = measured Hazen-Williams resistance coefficient
V_o = velocity at which C was measured
V = velocity for which C_h is to be calculated

6.3 MINOR LOSSES AT BENDS AND APPURTENANCES

The previous section presented a number of methods for estimating the head loss caused by friction as flow moves along the length of a pipe. This section discusses an additional type of head loss, called *minor loss*, which occurs at locations of manholes, pipe entrances and exits, and similar pipe appurtenances. As stated in Section 6.1, the total head loss along a pipe is equal to the sum of the frictional and minor losses.

There are a number of methods by which minor losses can be predicted. The classical approach, which is presented in this section, is to express the head loss as the product of a minor loss coefficient and the absolute difference between the velocity heads upstream and downstream of the appurtenance. This approach is reasonable for many types of minor loss calculations. However, at stormwater inlets, manholes, and junction structures, where flow geometries are often complex, additional methods have been developed for minor loss prediction and are presented by Brown, Stein, and Warner (1996). These approaches are known as the energy-loss and power-loss methods, and they are covered in detail in Section 11.4.

Classical Method

As already noted, the classical approach to predicting minor losses at pipe entrances, exits, and other appurtenances is to express the head loss as the product of a *minor loss coefficient*, *K*, and the absolute difference between the upstream and downstream velocity heads:

$$h_L = K\left|\frac{V_2^2 - V_1^2}{2g}\right| \qquad (6.25)$$

where h_L = head loss due to minor losses (ft, m)
K = minor loss coefficient
V = velocity of flow (ft/s, m/s)
g = gravitational acceleration constant (32.2 ft/s^2, 9.81 m/s^2)

In cases where the upstream and downstream velocities are equal in magnitude, the head loss can be expressed as the product of a loss coefficient and the velocity head:

$$h_L = K\frac{V^2}{2g} \qquad (6.26)$$

Equation 6.25 reduces to Equation 6.26 when either V_1 or V_2 is small in comparison to the other. An assumption that either V_1 or V_2 is small in comparison to the other is often quite reasonable at, for example, culvert entrances and outlets.

Regardless of whether Equation 6.25 or Equation 6.26 is used, the numerical value of the minor loss coefficient depends on the type of the appurtenance where the loss occurs, as well as on the geometry of the appurtenance. The following subsections summarize several appurtenance types and their associated minor loss coefficients.

Pipe Entrances. The minor loss that occurs at a pipe entrance can be predicted using Equation 6.25, where V_1 is the velocity upstream of the entrance and V_2 is the velocity in the pipe. It is often reasonable to assume that V_1 is small compared to V_2; thus, Equation 6.26 can be used. The minor loss coefficient at a pipe entrance depends on the geometry of the pipe entrance, and particularly on conditions such as whether the edges of the entrance are square-edged or rounded. The loss coefficient also depends on whether the entrance is fitted with headwalls, wingwalls, or a warped transition, and on whether the entrance to the pipe is skewed.

Table 6.5 provides a listing of entrance loss coefficients for culverts operating under outlet control (see Section 9.5, page 338 for an explanation of outlet control). These values are also applicable to entrances to other pipes, such as storm sewers, where the water depth at the pipe entrance is controlled by downstream conditions. In cases where inlet control governs (usually in cases where the pipe is steep and/or downstream flow depths are small), the water surface elevation upstream of a pipe entrance should be determined using inlet control calculation procedures for culverts (see Section 9.5).

Table 6.5 Entrance loss coefficients for pipes and culverts operating under outlet control (Normann, Houghtalen, and Johnston, 2001)

Structure Type and Entrance Condition	K
Concrete Pipe	
Projecting from fill, socket or groove end	0.2
Projecting from fill, square edge	0.5
Headwall or headwall and wingwalls	
Socket or groove end	0.2
Square edge	0.5
Rounded (radius = $D/12$)	0.2
Mitered to conform to fill slope	0.7
End section conforming to fill slope	0.5
Beveled edges (33.7° or 45° bevels)	0.2
Side- or slope-tapered inlet	0.2
Corrugated Metal Pipe or Pipe-Arch	
Projecting from fill (no headwall)	0.9
Headwall or headwall and wingwalls (square edge)	0.5
Mitered to conform to fill slope	0.7
End section conforming to fill slope	0.5
Beveled edges (33.7° or 45° bevels)	0.2
Side- or slope-tapered inlet	0.2
Reinforced Concrete Box	
Headwall parallel to embankment (no wingwalls)	
Square-edged on 3 sides	0.5
Rounded or beveled on 3 sides	0.2
Wingwalls at 30° to 75° from barrel	
Square-edged at crown	0.4
Crown edge rounded or beveled	0.2
Wingwalls at 10° to 25° from barrel	
Square-edged at crown	0.5
Wingwalls parallel (extensions of box sides)	
Square-edged at crown	0.7
Side- or slope-tapered inlet	0.2

Pipe Outlets. The exit loss (or outlet loss) from a storm sewer or culvert can be predicted by using Equation 6.25, although it is often assumed that V_2 is small compared to V_1 and so Equation 6.26 is used instead. The minor loss coefficient at a pipe outlet is nearly always assumed to be equal to 1.0, although it may be as small as 0.8 if a warped transition structure is provided at the pipe outlet. If there is no change in the velocity of flow at a pipe outlet, the minor loss is equal to zero.

Pipe Bends. The head loss at a mitered bend in a pipe, or at a location where the alignment of the pipe is deflected at a pipe joint, can be predicted using Equation 6.26. The U.S. Bureau of Reclamation (USBR, 1983) indicates that the minor loss coefficient in this case depends on the pipe deflection angle, as shown in Figure 6.5.

Figure 6.5
The minor loss coefficient at a pipe bend depends on the pipe deflection angle (USBR, 1983)

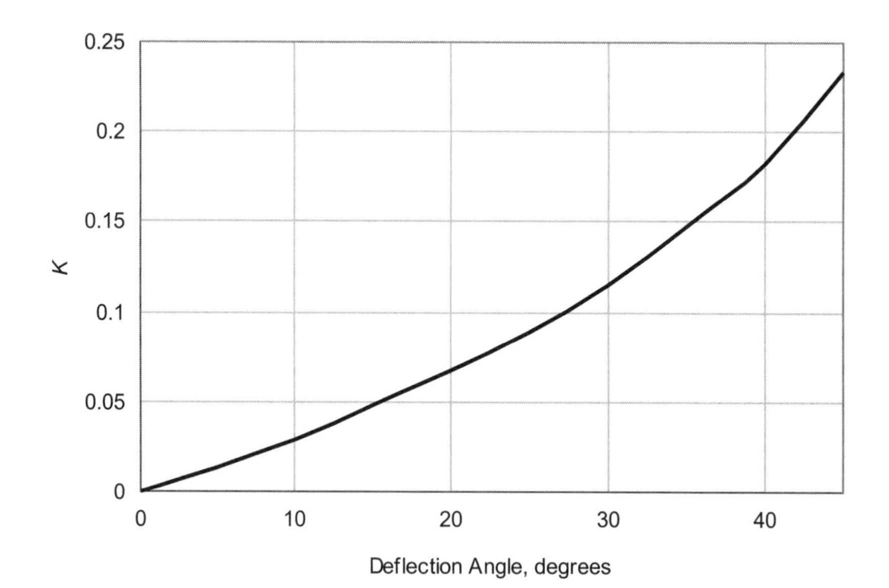

The energy-loss and composite-energy-loss methods for computing losses in manhole structures are presented in Section 11.4; however, the American Association of State Highway and Transportation Officials (AASHTO, 1991) indicates that minor losses at manholes also can be estimated using Equation 6.26. The value of K depends on the pipe deflection angle Δ (in degrees) as

$$K = 0.0033\Delta \qquad (6.27)$$

where Δ = pipe deflection angle (degrees)

This expression is limited to use when $\Delta \leq 90°$, when there is only one pipe flowing into the manhole and one pipe flowing out, and when the pipes are of equal diameter. When these conditions are not met, the more complex prediction methods described in Section 11.4 should be used.

Pipe Transitions. A *pipe transition* is a structure where the pipe cross-sectional area (that is, the diameter) changes. A transition is called an *expansion* if the area increases, and a *contraction* if the area decreases.

The minor loss coefficient for an expansion or contraction depends on the relative sizes of the upstream and downstream pipes, and on the abruptness (as measured in terms of the cone angle) of the transition (see Figure 6.6). Expansion and contraction loss coefficients also depend on whether the flow is pipe flow or open-channel flow. This section discusses minor loss coefficients for pipe flow; those for open-channel flow are discussed in Section 7.2.

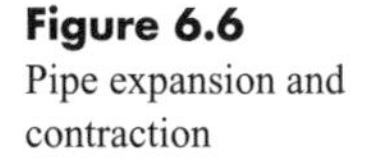

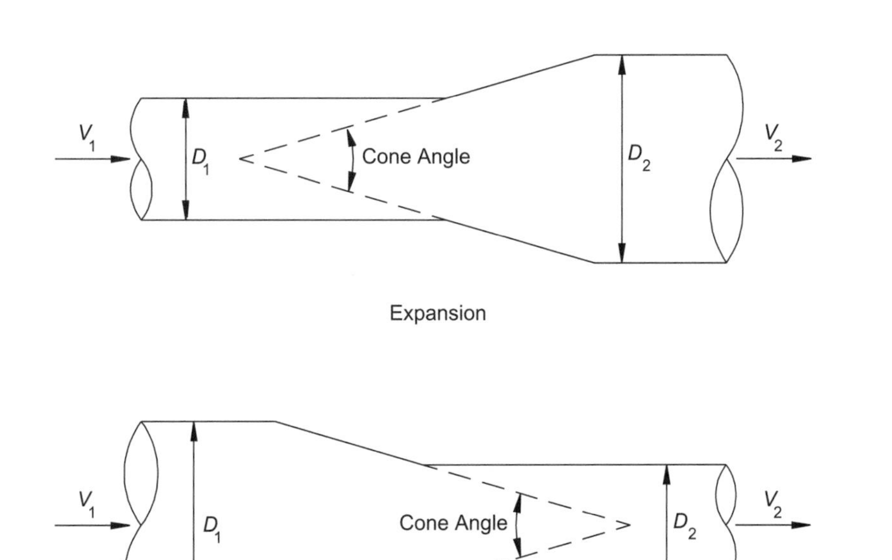

Figure 6.6 Pipe expansion and contraction

The head loss in a sudden or gradual pipe expansion can be predicted using Equation 6.26 if the upstream (higher) velocity is used to compute the velocity head. Minor loss coefficients are presented in Table 6.6 for *sudden expansions* (cone angle greater than 60 degrees) and in Table 6.7 for *gradual expansions* (cone angle 60 degrees or smaller).

The head loss in a sudden contraction can also be estimated using Equation 6.26, but the downstream (higher) velocity should be used to compute the velocity head. The minor loss coefficient for a sudden contraction may be obtained from Table 6.8. The head loss due to a gradual contraction may be estimated as one-half the head loss for a similar pipe expansion or, alternatively, by using Equation 6.25 and the minor loss coefficients presented in Table 6.9.

Table 6.6 Minor loss coefficients for sudden expansions in pipe flow [American Society of Civil Engineers (ASCE), 1992]

	D_2/D_1										
V_1 (ft/s)	1.2	1.4	1.6	1.8	2.0	2.5	3.0	4.0	5.0	10	∞
2	0.11	0.26	0.40	0.51	0.60	0.74	0.83	0.92	0.96	1.00	1.00
3	0.10	0.26	0.39	0.49	0.58	0.72	0.80	0.89	0.93	0.99	1.00
4	0.10	0.25	0.38	0.48	0.56	0.70	0.78	0.87	0.91	0.96	0.98
5	0.10	0.24	0.37	0.47	0.55	0.69	0.77	0.85	0.89	0.95	0.96
6	0.10	0.24	0.37	0.47	0.55	0.68	0.76	0.84	0.88	0.93	0.95
7	0.10	0.24	0.36	0.46	0.54	0.67	0.75	0.83	0.87	0.92	0.94
8	0.10	0.24	0.36	0.46	0.53	0.66	0.74	0.82	0.86	0.91	0.93
10	0.09	0.23	0.35	0.45	0.52	0.65	0.73	0.80	0.84	0.89	0.91
12	0.09	0.23	0.35	0.44	0.52	0.64	0.72	0.79	0.83	0.88	0.90
15	0.09	0.22	0.34	0.43	0.51	0.63	0.70	0.78	0.82	0.86	0.88
20	0.09	0.22	0.33	0.42	0.50	0.62	0.69	0.76	0.80	0.84	0.86
30	0.09	0.21	0.32	0.41	0.48	0.60	0.67	0.74	0.77	0.82	0.83
40	0.08	0.20	0.32	0.40	0.47	0.58	0.65	0.72	0.75	0.80	0.81

Table 6.7 Minor loss coefficients for gradual expansions in pipe flow (ASCE, 1992)

Cone	D_2/D_1								
Angle	1.1	1.2	1.4	1.6	1.8	2.0	2.5	3.0	∞
2°	0.01	0.02	0.02	0.03	0.03	0.03	0.03	0.03	0.03
6°	0.01	0.02	0.03	0.04	0.04	0.04	0.04	0.04	0.05
10°	0.03	0.04	0.06	0.07	0.07	0.07	0.08	0.08	0.08
15°	0.05	0.09	0.12	0.14	0.15	0.16	0.16	0.16	0.16
20°	0.10	0.16	0.23	0.26	0.28	0.29	0.30	0.31	0.31
25°	0.13	0.21	0.30	0.35	0.37	0.38	0.39	0.40	0.40
30°	0.16	0.25	0.36	0.42	0.44	0.46	0.48	0.48	0.49
35°	0.18	0.29	0.41	0.47	0.50	0.52	0.54	0.55	0.56
40°	0.19	0.31	0.44	0.51	0.54	0.56	0.58	0.59	0.60
50°	0.21	0.35	0.50	0.57	0.61	0.63	0.65	0.66	0.67
60°	0.23	0.37	0.53	0.61	0.65	0.68	0.70	0.71	0.72

Table 6.8 Minor loss coefficients for sudden contractions in pipe flow (ASCE, 1992)

	D_2/D_1												
V_2 (ft/s)	1.1	1.2	1.4	1.6	1.8	2.0	2.2	2.5	3.0	4.0	5.0	10	∞
2	0.03	0.07	0.17	0.26	0.34	0.38	0.40	0.42	0.44	0.47	0.48	0.49	0.49
3	0.04	0.07	0.17	0.26	0.34	0.38	0.40	0.42	0.44	0.46	0.48	0.48	0.49
4	0.04	0.07	0.17	0.26	0.34	0.37	0.40	0.42	0.44	0.46	0.47	0.48	0.48
5	0.04	0.07	0.17	0.26	0.34	0.37	0.39	0.41	0.43	0.46	0.47	0.48	0.48
6	0.04	0.07	0.17	0.26	0.34	0.37	0.39	0.41	0.43	0.45	0.47	0.48	0.48
7	0.04	0.07	0.17	0.26	0.34	0.37	0.39	0.41	0.43	0.45	0.46	0.47	0.47
8	0.04	0.07	0.17	0.26	0.33	0.36	0.39	0.40	0.42	0.45	0.46	0.47	0.47
10	0.04	0.08	0.18	0.26	0.33	0.36	0.38	0.40	0.42	0.44	0.45	0.46	0.47
12	0.04	0.08	0.18	0.26	0.32	0.35	0.37	0.39	0.41	0.43	0.45	0.46	0.46
15	0.04	0.08	0.18	0.26	0.32	0.34	0.37	0.38	0.40	0.42	0.44	0.45	0.45
20	0.05	0.09	0.18	0.25	0.32	0.33	0.35	0.37	0.39	0.41	0.42	0.43	0.44
30	0.05	0.10	0.19	0.25	0.29	0.31	0.33	0.34	0.36	0.37	0.38	0.40	0.41
40	0.06	0.11	0.20	0.24	0.27	0.29	0.30	0.31	0.33	0.34	0.35	0.36	0.38

Table 6.9 Minor loss coefficients for gradual contractions in pipe flow (Simon, 1981)

Cone Angle	K
20°	0.20
40°	0.28
60°	0.32
80°	0.35

Example 6.6 – Computing Friction, Entrance, and Exit Losses for a Culvert Flowing Full. Consider again the culvert described in Example 6.4 with a diameter of 1200 mm, a length of 60 m, and n = 0.013 (reinforced concrete pipe). The discharge is $Q = 3.5\ m^3/s$. The invert elevation at the culvert outlet is 80 m, and the culvert slope is $S_o = 0.002$. The culvert entrance consists of the groove end of a pipe joint at a headwall. No special transition is provided at the culvert outlet. If the depth of flow just downstream of the culvert outlet is 1.5 m, determine the depth of water and the water surface elevation just upstream of the culvert entrance. Assume that the velocities in the channels upstream and downstream of the culvert are small and approximately equal to one another.

Solution: As determined in Example 6.4, the head loss due to friction in the culvert is 0.486 m. Because the channel velocities upstream and downstream of the culvert are small, the minor losses due to the culvert entrance and outlet can be determined using Equation 6.26, where V is taken as the culvert barrel velocity (V = 3.10 m/s). Referring to Table 6.5, the entrance minor loss coefficient is K = 0.2. For the outlet, K = 1.0. Thus, the total head loss through the culvert is

$$h_L = 0.486 + 0.2\frac{3.10^2}{2(9.8)} + 1.0\frac{3.10^2}{2(9.8)} = 1.07 \text{ m}$$

In open-channel flow, which exists both upstream and downstream of the culvert, the water surface elevation is the hydraulic grade line (HGL), and the total head or energy grade line (EGL) lies above the water surface by an amount equal to the velocity head (see Section 7.2). Because the velocity heads upstream and downstream of the culvert in this example are approximately equal, the head loss can be taken as the difference between the water surface elevations upstream and downstream of the culvert.

The downstream water surface elevation, HGL_2, is the sum of the channel invert elevation and the water depth, or

$$HGL_2 = 80.0 + 1.50 = 81.5 \text{ m}$$

The upstream water surface elevation, HGL_1, is the sum of the downstream water surface elevation and the head loss in the culvert:

$$HGL_1 = 81.5 + 1.07 = 82.57 \text{ m}$$

The channel invert elevation, Z_1, at the culvert entrance is equal to the invert elevation at the outlet plus the change in elevation due to the culvert slope, or

$$z_1 = z_2 + S_o L = 80.0 + 0.002\,(60) = 80.12 \text{ m}$$

Thus, the water depth, y_1, at the culvert entrance is

$$y_1 = 82.57 - 80.12 = 2.45 \text{ m}$$

6.4 MOMENTUM AND FORCES IN PIPE FLOW

Section 3.3 presented the momentum equation as one of three fundamental physical laws applied to flows in stormwater conveyance systems. The discussion pointed out that analyses based on the principle of conservation of momentum are more complex than are analyses based on conservation of mass and/or energy because momentum is a vector-valued quantity. Thus, a separate equation must be written for each coordinate direction.

The momentum equation must be applied in cases where one needs to estimate the hydraulic forces acting on a component of a stormwater conveyance system. For example, knowledge of hydraulic forces may be required to determine the need for *thrust blocks* or *restraints* at pipe bends or other appurtenances. These types of hydraulic forces occur whenever there is a change in the velocity of a flow. Since velocity is a vector, a change in velocity means either a change in its magnitude or direction, or both.

Conservation of momentum also plays an important role in the analysis of time-variable pressures within a piping system as a consequence of *hydraulic transients*. These transients, often called *water hammer*, can arise as a result from starting or stopping of pumps. Hydraulic transients caused by pumps are discussed briefly in Chapter 11.

In its most general form, conservation of momentum is expressed as

$$\sum F_x = \frac{\partial}{\partial t}\int_{cv} \rho V_x dS + \sum (M_{out})_x - \sum (M_{in})_x \tag{6.28}$$

"Frankly, I hate weekends. They break my momentum."

where F_x = force in x-direction (lb, N)
t = time (s)
ρ = fluid density (slugs/ft^3, kg/m^3)
V_x = x-component of the velocity of fluid in the control volume (ft/s, m/s)
dS = incremental volume of fluid (ft^3, m^3)
$(M_{out})_x$ = momentum outflow rate from the control volume in the x direction (lb, N)
$(M_{in})_x$ = momentum inflow rate into the control volume in the x direction (lb, N)

This expression, written for the coordinate direction x, also can be written for other coordinate directions as needed. It states that the sum of the external forces acting on the water in a control volume is equal to the time rate of change of momentum within the control volume, plus the net momentum outflow rate (in the x direction) from the control volume. The sums of $\mathbf{M}_{out}$ and $\mathbf{M}_{in}$ account for the possibility of more than one inflow and/or outflow pipe to and from the control volume. A momentum flow rate (in the x direction) can be written as

$$M_x = \beta \rho Q V_x \tag{6.29}$$

where β = a coefficient defined by Equation 6.30
Q = discharge (cfs, m^3/s)

The term β, like α in the energy equation (Equation 6.1), is a coefficient accounting for the effect of a nonuniform velocity distribution. The definition of β is

$$\beta = \frac{1}{AV^2} \int_A v^2 dA \tag{6.30}$$

where A = total cross-sectional area of flow (ft^2, m^2)
V = averaged cross-sectional velocity (ft/s, m/s)
v = velocity through an area dA of the entire cross section (ft/s, m/s)

Typical values of β are provided in Table 6.1. Note that β for turbulent flow in a pipe is near 1, and thus it is often taken to be 1 in practice.

Evaluating Forces on Hydraulic Structures

For steady-state flow, the time derivative is equal to zero and Equation 6.28 reduces to the simpler expression given by Equation 3.12 (repeated here):

$$\sum F_x = \sum (M_{out})_x - \sum (M_{in})_x \qquad (6.31)$$

This steady-state case is adequate for use in most practical applications, although the more complicated Equation 6.28 must be used in the analysis of hydraulic transients.

In analyzing the forces acting on a structure such as a manhole, the forces due to fluid pressure and change in momentum must be computed and used to determine the reaction forces, R_x and R_y, on the water in the structure. The pressure and reaction forces are placed on the left side of Equation 6.31 for each coordinate direction, and the difference in the momentum of flows entering and leaving the structure is computed on the right side of the equation.

The hydraulic forces exerted by the water on the manhole are equal, but opposite in direction, to the computed reaction forces. Example 6.7 demonstrates the calculation of the hydraulic forces exerted on a manhole.

Example 6.7 – Hydraulic Forces. Determine the hydraulic forces exerted on the manhole in Figure E6.7.1. The hydraulic grade line elevations given in the figure should be used in computing the pressure forces. (Note that these HGL elevations have already been computed using the energy-loss method; these calculations will be presented in Example 11.3.) Assume that β = 1 for all pipes.

Solution: Figure E6.7.1 shows the various forces exerted on the water in the control volume. The control volume is taken as the circle in the figure, which represents the manhole. The *x*- and *y*-coordinate directions assumed for this example are also shown in the figure. Note that there are two momentum flows into the control volume, and one momentum flow out of the control volume.

As shown in Figure E6.7.1, the forces acting on the water in the control volume in the *x*-direction are due to the pressures in Pipe 1 and the outflow pipe, plus the reaction force R_x on the water in the control volume. The pressure in Pipe 1 is determined by multiplying the difference in elevation between HGL_1 and the centerline elevation of Pipe 1 by the specific weight of water. The pressure force exerted on the control volume by Pipe 1 is the product of its cross-sectional area and the pressure. Thus, the force is

$$P_{x1} = \gamma(HGL_1 - Z_1)A_1 = 62.4(104.12 - 102.00)(3.14) = 415 \text{ lb}$$

Similarly,

$$P_{xo} = -\gamma(HGL_o - Z_o)A_o = -62.4(106.00 - 101.50)(7.07) = -1{,}985 \text{ lb}$$

Note that P_{xo} is negative because it acts in the negative *x*-direction.

Forces acting on the water in the control volume in the *y*-direction are P_{y2} and the reaction force R_y. The pressure force is

$$P_{y2} = \gamma(HGL_2 - Z_2)A_2 = 62.4(106.57 - 102.25)(1.77) = 477 \text{ lb}$$

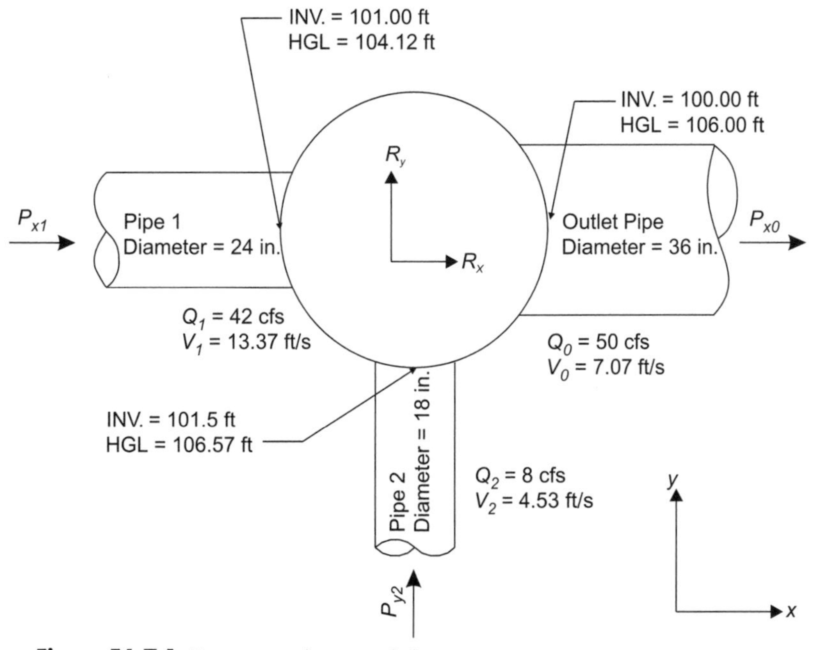

Figure E6.7.1 Forces exerted on a manhole

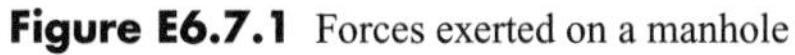

Momentum flows in the x-direction are

$$(M_{in})_x = \rho Q_1 V_{1x} = 1.94(42)(13.37) = 1{,}089 \text{ lb}$$

$$(M_{out})_x = \rho Q_o V_{ox} = 1.94(50)(7.07) = 686 \text{ lb}$$

Momentum flows in the y-direction are

$$(M_{in})_y = \rho Q_2 V_{2y} = 1.94(8)(4.53) = 70 \text{ lb}$$

$$(M_{out})_y = 0$$

Applying Equation 6.28, and assuming steady-state conditions (that is, the time derivative is zero),

$$415 - 1985 + R_x = 686 - 1{,}089$$

or

$$R_x = 1{,}167 \text{ lb}$$

Applying Equation 6.28 again, but writing it for the y-direction, leads to

$$477 + R_y = 0 - 70$$

or

$$R_y = -547 \text{ lb}$$

Thus, the actual reaction force R_y has a direction opposite to that assumed and shown in Figure E6.7.1.

It should be noted that the reaction forces R_x and R_y computed using the momentum equation are the forces acting on the water in the control volume. As stated previously, the hydraulic forces exerted by the water on the manhole are equal, but opposite in direction, to the computed reaction forces.

6.5 CHAPTER SUMMARY

A pipe or closed-conduit flow condition exists when the flowing water occupies the entire cross-sectional area of a closed conduit. The energy equation in the form

$$\frac{p_1}{\gamma} + Z_1 + \frac{\alpha_1 V_1^2}{2g} = \frac{p_2}{\gamma} + Z_2 + \frac{\alpha_2 V_2^2}{2g} + h_L - h_P + h_T \tag{6.32}$$

can be used to describe flow in closed conduits.

The hydraulic grade line elevation at a cross section is given by the sum of the pressure head (p/γ) and elevation head (Z) terms; this quantity is also known as the piezometric head. The energy grade line elevation or total head is given by the sum of the piezometric head and the velocity head ($V^2/2g$).

The primary cause of head loss in stormwater conveyance systems is pipe friction. Several methods for computing energy losses are available, including the Darcy-Weisbach, Manning, Chézy, and Hazen-Williams formulas.

In addition to friction losses, energy losses occur due to changes in flow velocity through pipe entrances, outlets, transitions, bends, and other appurtenance. These losses, known as minor losses, are computed by multiplying a coefficient K by the velocity head or change in velocity head. Typical values for K under many conditions are available.

In cases where forces on hydraulic structures need to be computed to determine whether thrust blocks or restraints are needed, the momentum equation is applied. Another application of the momentum equation is in the analysis of pressures due to hydraulic transients.

"I think you should be more explicit here in step two."

REFERENCES

American Association of State Highway and Transportation Officials (AASHTO). 1991. *Model Drainage Manual*, Washington, D.C.: American Association of State Highway and Transportation Officials.

American Society of Civil Engineers (ASCE). 1992. *Design and Construction of Urban Storm Water Management Systems,* New York: American Society of Civil Engineers.

Brown, S. A., S. M. Stein, and J. C. Warner. 1996. *Urban Drainage Design Manual.* Hydraulic Engineering Circular No. 22, FHWA-SA-96-078. Washington, D.C.: U.S. Department of Transportation., Federal Highway Administration.

Chin, D.A. 2000. *Water-Resources Engineering.* Upper Saddle River, N. J.: Prentice Hall.

Colebrook, C. F. 1939. "Turbulent Flow in Pipes with Particular Reference to the Transition Between Smooth and Rough Pipe Laws." *Journal of the Institution of Civil Engineers of London* 11.

Herschel, C. 1897. "On the Origin of the Chézy Formula." *Journal of the Association of Engineering Societies* 18: 363–368.

Jain, A. K. 1976. "Accurate Explicit Equation for Friction Factor." *Journal of the Hydraulics Division, ASCE* 102, no. HY5: 674–677.

Manning, R. 1891. "On the Flow of Water in Open Channels and Pipes. *Transactions, Institution of Civil Engineers of Ireland* 20: 161–207.

Moody, L. F. 1944. "Friction Factors for Pipe Flow." *Transactions, American Society of Mechanical Engineers* (ASME) 66.

Normann, J. M., R. J. Houghtalen, and W. J. Johnston. 2001. *Hydraulic Design of Highway Culverts, Second Edition.* Hydraulic Design Series No. 5. Washington, D.C.: Federal Highway Administration (FHWA).

Potter, M. C., and D. C. Wiggert. 1991. *Mechanics of Fluids.* Englewood Cliffs, NJ: Prentice Hall.

Prandtl, L. 1952. *Essentials of Fluid Dynamics.* New York: Hafner Publ.

Reynolds, O. 1883. "An Experimental Investigation of the Circumstances Which Determine Whether the Motion of Water Shall be Direct or Sinuous and of the Law of Resistance in Parallel Channels." *Philosophical Transactions of the Royal Society* 174: 935.

Simon, A. L. 1981. *Practical Hydraulics.* 2d ed. New York: John Wiley & Sons.

Streeter, V. L., and Wylie, E. B. 1979. *Fluid Mechanics.* 7th ed. New York: McGraw Hill.

Swamee, P. K., and A. K. Jain. 1976. "Explicit Equations for Pipe Flow Problems." *Journal of the Hydraulics Division, ASCE* 102, no. HY5: 657–664.

U.S. Bureau of Reclamation (USBR). 1983. *Design of Small Canal Structures.* Lakewood, Colorado: USBR.

Viessman, W., Jr., and M. J. Hammer. 1993. *Water Supply and Pollution Control.* 5th ed. New York: Harper Collins.

Walski, T. M. 1984. *Analysis of Water Distribution Systems.* New York: VanNostrand Reinhold.

Williams, G. S., and A. H. Hazen. 1933. *Hydraulic Tables.* 3d ed. New York: John Wiley and Sons.

PROBLEMS

6.1 The bottom of a 24-in. diameter pipe lies 8 ft below the ground surface. It is flowing full at a rate of 15 cfs. The pressure in the pipe is 35 psi. What is the total hydraulic energy at the centerline of the pipe with respect to the ground?

6.2 A 30-in. pipe lies 10 ft above a reference level. The flow is 30 cfs and the pressure is 45 psi. What is the total energy in the pipe with respect to the reference level?

6.3 A 300-mm pipe drains a small reservoir. The outlet to the pipe is 6.1 m below the surface of the reservoir and it discharges freely to the atmosphere. Assuming no losses through the system, what is the flow through the pipe and the pipe's exit velocity?

6.4 Find the total head loss in 305 m of 460-mm diameter cast-iron pipe carrying a flow of 0.43 m^3/s. Use the Hazen-Williams formula to compute head loss assuming $C = 130$.

6.5 Water flows through a 12-in. diameter riveted steel pipe with $\varepsilon = 0.01$ ft and a head loss of 20 ft over 1,000 ft. Determine the flow.

6.6 What size pipe is required to transport 10 ft^3/s of water a distance of 1 mile with a head loss of 6 ft? Use new cast-iron pipe with an equivalent sand roughness of 0.00085 ft.

6.7 Water is to be pumped at a rate of 0.057 m^3/s over a distance of 610 m. The pipe is level and the pump head cannot exceed 18.3 m. Determine the required concrete pipe diameter. Use $\varepsilon = 0.005$ ft (1.5 mm).

6.8 For the following figure, sketch the energy grade line between points 1 and 2.

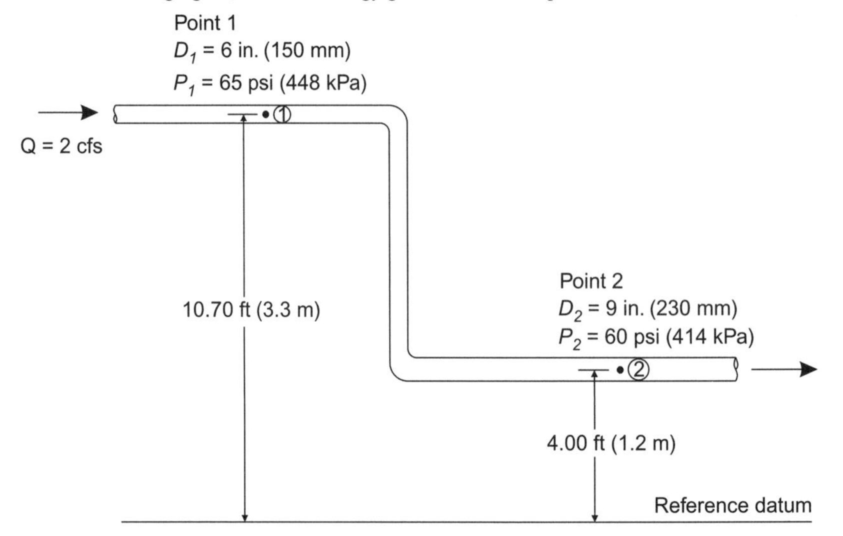

6.9 An 8-in. concrete pipe is carrying 2.25 ft^3/s and has a length of 200 ft. Calculate the head loss using

a) Darcy Weisbach equation with $\varepsilon = 0.01$ ft

b) Hazen-Williams formula with $C = 100$

c) Manning's equation with $n = 0.011$

6.10 Revisit the circular concrete culvert from Example 6.6 and assume that the headwater rises to a level of 3.0 m above the upstream invert of the culvert to an elevation 83.12 m. Determine the discharge through the culvert given the characteristics in the following table. The velocities in the channels upstream and downstream of the culvert are low and approximately equal.

Parameter	Value
Diameter	1200 mm
Length	60 m
Manning's *n*	0.013
Outlet invert elev.	80.00 m
Slope	0.002
Entrance *K*	0.2
Exit *K*	1.0
Tailwater depth	1.5 m

6.11 A 380-mm diameter new cast iron pipe connecting two reservoirs as shown carries water at 50° F. The pipe is 45.7 m long and the discharge is 0.57 m^3/s. Determine the elevation difference between the water surfaces in the two reservoirs.

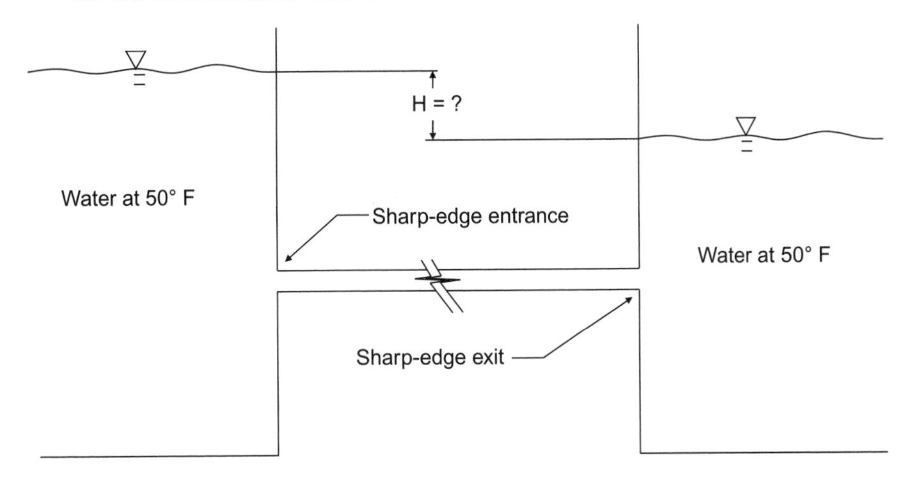

Overflows from this reservoir exit over a spillway and enter the rock channel below.

CHAPTER

7

Flow in Open Channels

Open channels are ubiquitous in both natural and constructed environments. They consist of natural stream channels as well as natural swales and depressions along which water runs following rainfall events. In the constructed or urbanized environment, an open channel may take the form of a street gutter, drainage ditch, pipe (if there is a free water surface at atmospheric pressure), lined or unlined stormwater collection channel, or natural or modified stream channel.

In open-channel flow, unlike closed-conduit flow, the cross-sectional area of flow is not a constant that is governed by the shape and size of the conveyance. Rather, the cross-sectional area and associated flow descriptors, such as the average flow velocity, are unknowns and must be determined as part of the solution to a problem. Further complicating analyses of open-channel flow is the fact that there is not, in general, a unique or one-to-one relationship between discharge Q and depth of flow y in an open channel. In most instances of open-channel flow, the depth of flow for any specified channel discharge depends on channel shape and roughness characteristics, as well as on specified boundary conditions at *flow control* sections (that is, cross sections where a known relationship between depth and discharge exists). Thus, open-channel flow is analytically much more difficult to treat than is pressure flow, but the challenge imposed by the additional difficulty also makes it more interesting.

This chapter presents methods for analysis of the hydraulics of steady flows in open channels. Unsteady flows are viewed as an advanced topic and are introduced in Appendix C. The reader may consult Chow (1959); Abbott (1979); Cunge, Holly, and Verwey (1980); French (1985); Fread (1993); and Chaudhry (1993) for more detailed coverage of unsteady flows in open channels.

Section 7.1 of this chapter describes the classification of the types of open channels and gives an overview of their geometric properties. Energy and constitutive relationships for the estimation of energy losses in open-channel flow are presented in Section 7.2.

Sections 7.3 and 7.4, respectively, deal with the fundamentally important concepts of uniform flow and critical flow in open channels—the only types of flow for which there is a one-to-one relationship between channel discharge and flow depth. Uniform flow is the basis for many types of channel design; the subject of design is taken up in detail in Chapter 8. The concept of critical flow is fundamental to the analysis of varied flows in open channels and to understanding many types of flow controls. Establishing critical flow in an open channel can also provide a basis for flow measurement.

The momentum principle, as applied in the analysis of open-channel flow, is taken up in Section 7.5. As in pipe flow, the momentum principle is applied in situations where the engineer desires an estimate of the hydraulic forces acting on a flow obstruction such as a bridge pier. The momentum principle is also fundamental to the analysis of the rapidly varied flow phenomenon known as the *hydraulic jump*.

Sections 7.6 and 7.7 address the topics of weirs and orifices, respectively, which are commonly used structures for flow control and measurement. Weirs and orifices may be used as outlets for stormwater detention facilities. The weir and orifice equations also play a role in describe the hydraulic performance of stormwater inlets and culvert entrances.

Gradually varied flow is discussed in Section 7.8. Gradually varied flows are those in which the depth of flow changes along the length of a channel as a consequence of convective accelerations of the flow. These accelerations occur where flow controls are such that the flow depth at the control is different from the depth of uniform flow and forces on the water are not balanced. Such situations are common upstream of culvert entrances, at channel entrances and exits, and at locations where channel geometric properties change.

Spatially varied flows—flows for which the discharge is not a constant along the length of a channel—are discussed briefly in Section 7.9. These flows may have either an increasing or a decreasing discharge in the direction of flow, and different methods exist for the analysis of each flow type. Spatially varied flows with a decreasing discharge occur, for example, in gutters at stormwater inlet locations. Spatially varied flows with an increasing discharge may occur in side overflow spillway channels. Flow in a natural stream channel is usually spatially varied because of the increase in contributing drainage basin area as one proceeds downstream.

7.1 CLASSIFICATION AND PROPERTIES OF OPEN CHANNELS

Open channels may be natural or man-made and come in a wide variety of cross-sectional shapes. For analytical purposes, channels may be classified as *prismatic* or *non-prismatic*. In a prismatic open channel, the descriptive geometric variables such as its width, side slopes, and longitudinal slope, do not change along the length of the channel. Practically speaking, one would typically not expect to encounter a perfectly prismatic channel outside of the laboratory, but many constructed channels can, for the purposes of analysis, be considered prismatic. A concrete curb and gutter along the edge of a paved roadway with a constant slope is one example of a prismatic channel. Another example is shown in Figure 7.1.

Figure 7.1
A riprap-lined prismatic channel

In a nonprismatic channel the descriptive geometric variables (shape, roughness, etc.) change with location along the channel length. Natural stream channels, which typically have variations in width and other properties, are examples of nonprismatic open channels (see Figure 7.2). Coverage of open-channel flow in this text deals primarily with prismatic channels. Nevertheless, certain principles associated with the hydraulics of flow in nonprismatic channels will be addressed as necessary.

Open channels can also be classified on the basis of whether they are lined or unlined. In a *lined channel*, the channel bottom and sides have been paved with an erosion-resistant material, such as concrete, asphalt, metal, riprap, or vegetation. An *unlined channel* is one in which the channel bottom and sides consist of natural geologic materials such as earth, gravel, or rock. As shown in Chapter 8, the lining material (or lack thereof) has practical significance in the design of the open channel.

Geometric Properties of Open Channels

The *longitudinal slope* of an open channel, denoted by S_o, is a critically important geometric variable. An alternative way of describing the channel slope is by use of its *angle of inclination* above the horizontal, denoted by θ (see Figure 7.3). Note that $S_o = \tan\theta$. Note also that if θ is small (less than about 5 degrees), then $S_o \approx \sin\theta$. A *horizontal open channel* has a longitudinal slope of zero, and an *adverse open channel* has a negative slope (that is, the channel bottom elevation increases in the downstream direction).

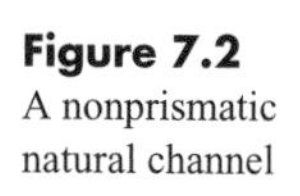

Figure 7.2
A nonprismatic natural channel

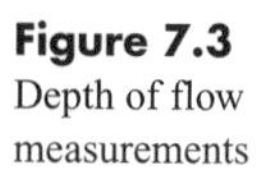

Figure 7.3
Depth of flow measurements

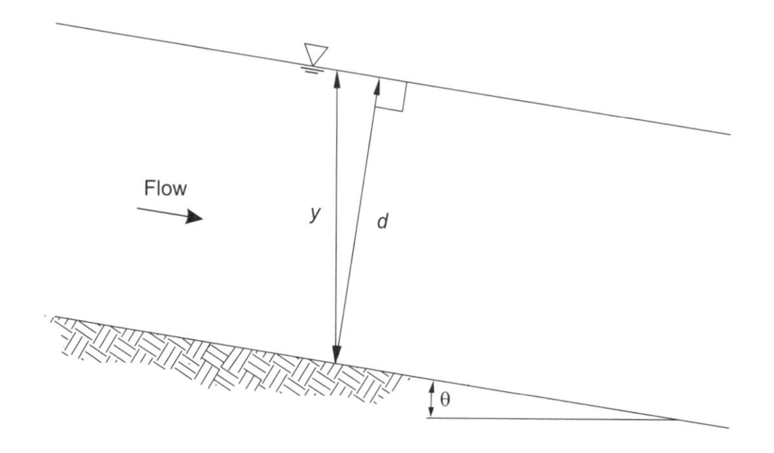

Most geometric properties of open channels are those related to a channel's cross section. Strictly speaking, channel cross sections should be taken perpendicular to the channel bottom (that is, normal to the direction of flow). The *depth of flow* measured perpendicular to the channel bottom is denoted by d in Figure 7.3. However, in most open channels, the slope is small, and vertical channel cross sections can be employed with little error. The depth of flow measured vertically is denoted by y (see Figure 7.3). The relationship between these depths is $d = y\cos\theta$.

The most common cross-sectional shapes of prismatic channels are *rectangular*, *triangular*, *trapezoidal*, and *circular*. Other shapes are those associated with *curbs and gutters*, which are discussed in Section 10.1. Figure 7.4 illustrates the dimensions used to describe common channel shapes. B denotes the channel bottom width, y denotes the depth of flow, D denotes the diameter of the channel (if a circular pipe), T denotes the top width of flow, and z represents a measure of steepness of the channel sides. For circular channels, ϕ represents the angle subtended by the intersections of the water surface with the channel or conduit surface, as shown in the figure. Note that rectangular and triangular channels are merely special cases of a trapezoidal channel, the former in the case of $z = 0$, and the latter in the case of $B = 0$.

The cross-sectional area, wetted perimeter, hydraulic radius, top width T, and hydraulic depth can be determined on the basis of the depth of flow y and the properties shown in Figure 7.4. The *hydraulic radius* is defined as

$$R = A/P \qquad (7.1)$$

where R = hydraulic radius (ft, m)
A = cross-sectional area (ft^2, m^2)
P = wetted perimeter (ft, m)

The definition of *hydraulic depth* is

$$D_h = A/T \qquad (7.2)$$

where D_h = hydraulic depth (ft, m)
T = top width (ft, m)

Table 7.1 summarizes the relationships for each of the channel cross-sectional shapes shown in Figure 7.4. Note that the formulas from triangular and trapezoidal channels are only applicable to cross sections with equal side slopes, and the depth y in an open channel is always taken as the greatest depth at any point in a cross section. The hydraulic depth can be thought of as the average depth of flow in a cross section. Use of the hydraulic depth can also be viewed as a type of shorthand, as the ratio $D_h = A/T$ occurs frequently in equations describing open-channel flow.

The cross-sectional properties of irregularly shaped channels can be determined by dividing a cross section into a set of rectangles and/or triangles and applying basic geometric formulae related to areas, perimeters, and slopes. Clearly, this process is quite tedious and time-consuming; for that reason, computer-based methods for solving open-channel flow problems are especially useful for irregularly shaped channels.

Figure 7.4 Dimensions of common channel shapes

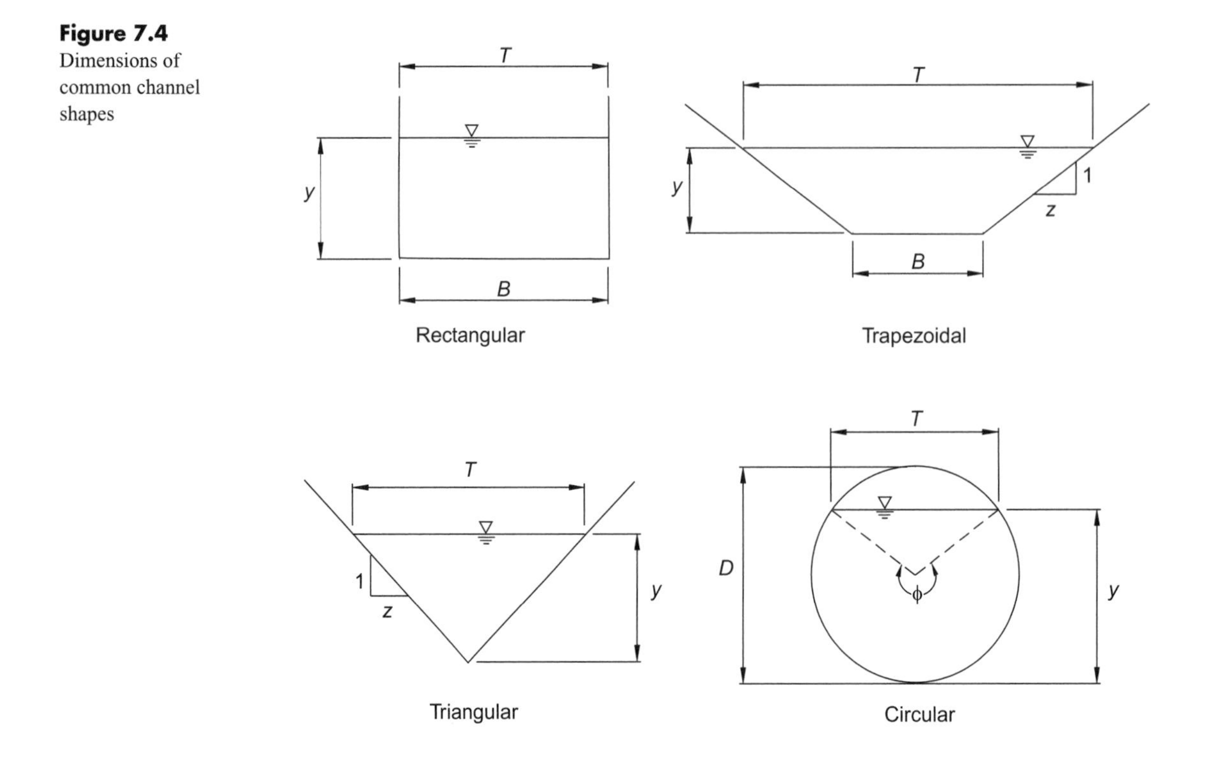

Table 7.1 Cross-sectional properties of prismatic open channels

Channel Shape	Area *(A)*	Wetted Perimeter *(P)*	Hydraulic Radius *(R)*	Top Width *(T)*	Hydraulic Depth (D_h)
Rectangular	By	$B+2y$	$\dfrac{By}{B+2y}$	B	y
Trapezoidal	$By+zy^2$	$B+2y\sqrt{1+z^2}$	$B+2y\sqrt{1+z^2}$	$B+2zy$	$\dfrac{By+zy^2}{B+2zy}$
Triangular	zy^2	$2y\sqrt{1+z^2}$	$\dfrac{zy}{2\sqrt{1+z^2}}$	$2zy$	$\dfrac{y}{2}$
Circular[a]	$\dfrac{D^2(\phi-\sin\phi)}{8}$	$\dfrac{D\phi}{2}$	$\dfrac{D}{4}\left(1-\dfrac{\sin\phi}{\phi}\right)$	$D\sin\left(\dfrac{\phi}{2}\right)$	$\dfrac{D}{8}\left(\dfrac{\phi-\sin\phi}{\sin(\phi/2)}\right)$

a. Angle ϕ is measured in radians. 1 radian ≈ 57.3 degrees.

Example 7.1 – Geometric Properties for a Trapezoidal Channel. An open channel has a cross-sectional shape as shown in Figure E7.1.1. If the depth of flow is 4 ft, determine the cross-sectional area, wetted perimeter, hydraulic radius, top width, and hydraulic depth.

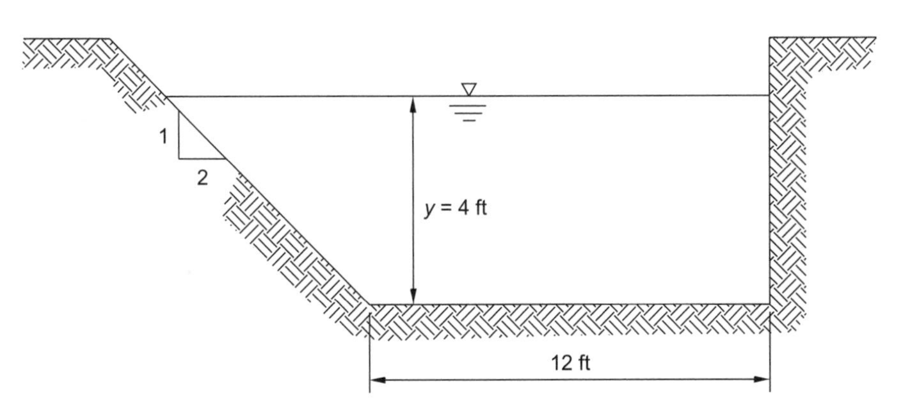

Figure E7.1.1 Open channel for Example 7.1

Solution: Because the cross-sectional shape of the channel does not coincide with any of those shown in Figure 7.4, the properties of the channel section must be derived on the basis of its geometric properties.

The cross-sectional area can be constructed as the sum of the areas of a rectangle and a triangle, and is

$$A = 12(4) + 0.5(4)[2(4)] = 64 \text{ ft}^2$$

The wetted perimeter can be computed as the sum of the water depth against the right-hand side of the channel, the channel bottom width, and the length along the sloping side of the channel:

$$P = 4 + 12 + \sqrt{4^2 + 8^2} = 24.94 \text{ ft}$$

The hydraulic radius is

$$R = A/P = 64/24.94 = 2.57 \text{ ft}$$

The top width is equal to

$$T = 12 + 2(4) = 20 \text{ ft}$$

and the hydraulic depth is

$$D_h = A/T = 64/20 = 3.20 \text{ ft}$$

7.2 ENERGY AND CONSTITUTIVE RELATIONSHIPS

In open-channel flow problems, the energy equation takes a somewhat different form than that presented in Equations 3.9 and 6.1. The difference arises in the pressure head term, which, instead of being written as p/γ, is written as $d\cos\theta$, where d is the depth of flow normal to the channel bottom and θ is the angle of inclination of the channel. Thus, the energy equation for flow in an open channel is

$$Z_1 + d_1 \cos\theta + \frac{\alpha_1 V_1^2}{2g} = Z_2 + d_2 \cos\theta + \frac{\alpha_2 V_2^2}{2g} + h_L \tag{7.3}$$

where Z = channel bottom elevation (ft, m)
d = depth of flow normal to the channel bottom (ft, m)
θ = channel slope angle

α = a velocity distribution coefficient defined by Equation 6.3
V = average flow velocity (ft/s, m/s)
g = gravitational acceleration constant (ft/s^2, m/s^2)
h_L = energy loss between section 1 and section 2 (ft, m)

Figure 7.5 depicts the terms in Equation 7.3, as well as the next two equations. Table 7.2 gives values of α and β (for an explanation of β, see Section 7.5) for open-channel flows. (Estimation of α for compound channel sections is presented in Section 7.3.)

Because $d = y\cos\theta$ (see Figure 7.3), an alternative expression for the energy equation for open-channel flow is

$$Z_1 + y_1 \cos^2\theta + \frac{\alpha_1 V_1^2}{2g} = Z_2 + y_2 \cos^2\theta + \frac{\alpha_2 V_2^2}{2g} + h_L \tag{7.4}$$

where y_1 and y_2 are the flow depths (measured vertically) at sections 1 and 2, respectively (ft, m).

When the angle of inclination, θ, is small, this equation simplifies to the version most commonly applied in practice:

$$Z_1 + y_1 + \frac{\alpha_1 V_1^2}{2g} = Z_2 + y_2 + \frac{\alpha_2 V_2^2}{2g} + h_L \tag{7.5}$$

Figure 7.5
Definition figure for energy equation for an open channel

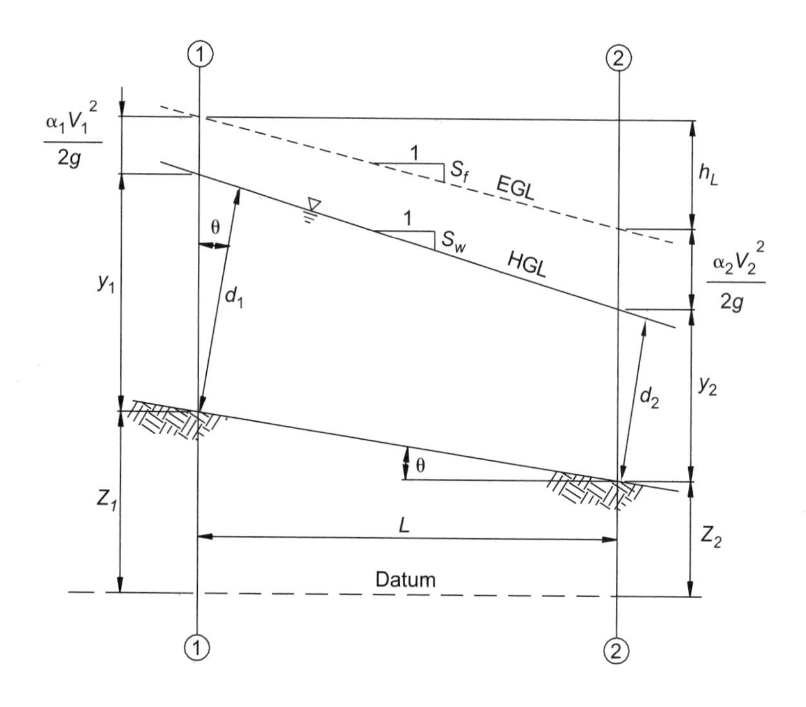

Table 7.2 Typical values of α and β in open-channel flow (Chow, 1959)

Channel Description	α			β		
	Min	Avg	Max	Min	Avg	Max
Regular channels, flumes, spillways	1.10	1.15	1.20	1.03	1.05	1.07
Natural streams and torrents	1.15	1.30	1.50	1.05	1.10	1.17
Rivers under ice cover	1.20	1.50	2.00	1.07	1.17	1.33
River valleys, over flooded	1.50	1.75	2.00	1.17	1.25	1.33

Pressure Variations in Open-Channel Flow

Replacement of the pressure head term, p/γ, with $d\cos\theta$ in Equation 7.3 means that the fluid pressure at depth d in open-channel flow is equal to

$$p = \gamma d \cos\theta \tag{7.6}$$

where p = fluid pressure at depth d (lb/ft^2, N/m^2)
d = depth measured normal to the flow streamlines (ft, m)
γ = fluid specific weight (lb/ft^3, N/m^3)

If one desires instead to use the vertical depth, y, the pressure is found to be

$$p = \gamma y \cos^2\theta \tag{7.7}$$

Because $\cos\theta$ is less than unity in all nonhorizontal channels, this expression indicates that the pressure at depth y in an open-channel flow is less than the pressure at the same depth in a static fluid.

Example 7.2 – Pressure Variation over a Spillway. Flow occurs on a spillway with a depth d = 1.00 m. The slope of the spillway is 15 percent. Determine the fluid pressure on the face of the spillway and compare that pressure to the pressure existing at a depth of 1.00 m in standing water.

Solution: At a depth of 1.00 m in standing water, the pressure is found to be

$$p = \gamma d = 1{,}000(1.00) = 1{,}000 \text{ kg/m}^2$$

The pressure at the same depth on the face of the spillway can be evaluated using Equation 7.6, where $\theta = \arctan S_o = \arctan(0.15) = 8.53°$.

$$p = 1{,}000(1.0)(\cos 8.53°) = 988.9 \text{ kg/m}^2$$

The pressure reduction caused by the slope of the channel is slightly more than 1 percent in this case.

Constitutive Relationships for Open-Channel Flow

Energy losses represented by the h_L term in the energy equation can be viewed as consisting of localized and distributed losses. *Local losses*, often referred to as *minor losses*, occur at locations where the flow velocity changes in either magnitude or direction. They are described later in this section on page 230. *Distributed losses* are due to friction exerted on the flow by the channel bottom and sides. Distributed losses may be described by expressing the head loss for a channel reach as the product of the energy gradient, S_f, and the length, L, of the reach:

$$h_L = S_f L \tag{7.8}$$

where S_f = energy gradient (ft/ft, m/m)
L = reach length (ft, m)

A number of constitutive relationships are available to express the energy gradient as a function of flow and channel characteristics. The most commonly used relationships are described in the subsections that follow.

The Manning Equation. The *Manning equation* (Manning, 1891) is an empirically derived expression obtained from observations of flows in laboratory channels. It was developed as a formula for uniform flow but has also been successfully applied for analyses of gradually and spatially varied flows. It is the most widely used of all constitutive relationships for open-channel–flow analysis.

The Manning equation can be written as

$$V = \frac{C_f}{n} R^{2/3} S_f^{1/2} \tag{7.9}$$

where V = cross-sectionally averaged flow velocity (ft/s, m/s)
C_f = unit conversion factor (1.49 for US customary units, 1.00 for SI)
n = Manning's roughness coefficient
R = hydraulic radius of the flow (ft, m)

Note that S_f is equal to the channel slope S_o in the case of *uniform flow*, and is different from S_o in the case of *varied flow*. The numerical value of Manning's n is independent of the set of units employed.

Table 7.3 provides a listing of typical values of Manning's n for various channel types and surfaces. Note that n is not a constant for a particular channel; it depends on a number of factors, including flow depth, channel alignment and irregularity, the presence of vegetation and flow obstructions, and the amount of sediment being transported by the stream. Chow (1959) provides a detailed discussion of the effects of these factors.

Because computations of flow velocities and energy gradients via the Manning equation are fairly sensitive to the assumed value of n, an understanding of influential factors should enable one to make more informed choices regarding this important parameter. For example, Chow (1959), in a theoretical discussion of channel roughness and uniform flow formulas, showed that the value of Manning's n is related to the channel roughness height as:

$$n = \varphi\left(\frac{R}{k}\right) k^{1/6} \tag{7.10}$$

where $\varphi(R/k)$ = a value that is a function of the dimensionless ratio of the hydraulic radius R to the roughness height k

k = channel roughness height (ft, m)

Table 7.3 Typical values of Manning's n (adapted from Chow, 1959)

Channel Description	Minimum n	Average n	Maximum n
Closed conduits flowing partly full			
Brickwork	0.011	0.014	0.017
Cast iron, coated	0.010	0.013	0.014
Cast iron, uncoated	0.011	0.014	0.016
Clay, vitrified	0.011	0.014	0.017
Concrete, finished	0.011	0.012	0.014
Concrete, unfinished	0.012	0.014	0.020
Corrugated metal	0.021	0.024	0.030
Rubble masonry	0.018	0.025	0.030
Steel, welded	0.010	0.012	0.014
Steel, riveted and spiral	0.013	0.016	0.017
Wood stave	0.010	0.012	0.014
Lined or built-up channels			
Asphalt	0.013	0.016	—
Brickwork	0.011	0.014	0.018
Concrete, troweled	0.011	0.013	0.015
Concrete, unfinished	0.014	0.017	0.020
Corrugated metal	0.021	0.025	0.030
Rubble masonry	0.017	0.029	0.035
Steel, smooth	0.011	0.013	0.017
Wood, planed and creosoted	0.011	0.012	0.015
Wood, planed and untreated	0.010	0.012	0.014
Excavated or dredged channels			
Earth, straight and uniform	0.016	0.022	0.033
Earth, winding and sluggish	0.023	0.035	0.050
Overgrown with vegetation	0.040	0.080	0.140
Rock cuts, smooth and uniform	0.025	0.035	0.040
Rock cuts, jagged and irregular	0.035	0.040	0.050
Natural streams			
Flood plains, pasture	0.025	0.033	0.050
Flood plains, cultivated	0.020	0.035	0.160
Flood plains, trees	0.050	0.080	0.200
Major streams (width > 100 ft)	0.025	—	0.100
Minor streams on plain areas	0.025	0.045	0.150
Mountain streams	0.030	0.045	0.070

The value of the function $\varphi(R/k)$ does not vary much over a wide range of R/k. Strickler (1923) showed that, on average, $\varphi(R/k) = 0.0342$ when the roughness height (in ft) is taken as the median grain size of the channel bed material. (When the median grain size is expressed in units of meters, $\varphi(R/k) = 0.0281$.) This relationship indicates that Manning's n is proportional to the one-sixth power of the median sediment size in a stream.

Example 7.3 – Determining the Energy Gradient with the Manning Equation.

A discharge of 300 cfs occurs at a depth of 2.50 ft in a concrete-lined rectangular open channel with a width of 10 ft and $n = 0.013$. What is the energy gradient S_f for this flow? If the channel bottom slope is $S_o = 0.001$, is this flow uniform or varied?

Solution: The cross-sectional area, wetted perimeter, and hydraulic radius of the flow are

$$A = 10(2.50) = 25 \text{ ft}^2$$

$$P = 10 + 2(2.50) = 15 \text{ ft}$$

$$R = A/P = 25/15 = 1.67 \text{ ft}$$

$$V = Q/A = 300/25 = 12 \text{ ft/s}$$

Substituting these values into the Manning equation and rearranging yields the energy gradient:

$$S_f = \frac{(nV)^2}{1.49^2 R^{4/3}} = \frac{[0.013(12)]^2}{1.49^2 (1.67)^{4/3}} = 0.00555$$

Because the energy gradient is not equal to the channel slope, the flow must be varied.

The Chézy Equation. The *Chézy equation*, developed around 1769 by the French engineer Antoine Chézy, is probably the oldest constitutive relationship for open-channel flow (Herschel, 1897). Unlike the Manning equation, the Chézy equation can be derived theoretically by equating the force resisting a flow to the gravitational force driving the flow (a condition that must be true for uniform flow). Because the resisting force is proportional to the square of the flow velocity, the net result is the expression

$$V = C\sqrt{RS_f} \qquad (7.11)$$

where C = a roughness coefficient known as *Chézy's C* ($\text{ft}^{0.5}/\text{s}$, $\text{m}^{0.5}/\text{s}$)

Like Manning's n, Chézy's C depends on a number of factors and thus is not a constant. In 1869, the Swiss engineers Ganguillet and Kutter (1869) published a formula relating C to the roughness, slope, and hydraulic radius of a channel as

$$C = \frac{41.65 + \dfrac{0.00281}{S} + \dfrac{1.811}{n}}{k + \left(41.65 + \dfrac{0.00281}{S}\right)\dfrac{n}{\sqrt{R}}} \qquad (7.12)$$

where S = channel slope (ft/ft, m/m)
n = Kutter's n
k = a constant equal to 1 for U.S. Customary units or 1.811 for SI units

For practical purposes, Kutter's n is the same as Manning's n when the channel slope is larger than 0.0001 and the hydraulic radius is between 1 and 30 feet (0.3 and 9 m).

Bazin (1897) published a formula for C in which slope is not considered. His equation is

$$C = \frac{157.6}{k + \frac{m}{\sqrt{R}}} \quad (7.13)$$

where m = a roughness coefficient (see Table 7.4)

Table 7.4 Proposed values of Bazin's m (Chow, 1959)

Channel Description	m
Very smooth cement or planed wood	0.11
Unplaned wood, concrete or brick	0.21
Rubble masonry or poor brickwork	0.83
Earth channels in good condition	1.54
Earth channels in ordinary condition	2.36
Earth channels in rough condition	3.17

Although considerably simpler than the Ganguillet and Kutter (1869) formula, Bazin's is considered to be less reliable.

Powell (1950) suggested that Chézy's C can be expressed as

$$C = -42 \log_{10}\left(\frac{C}{4\mathrm{Re}} + \frac{\varepsilon}{R}\right) \quad (7.14)$$

where C = Chézy's C ($ft^{0.5}/s$)
Re = Reynolds number
ε = measure of roughness having the tentative values shown in Table 7.5
R = hydraulic radius (ft)

For open-channel flow, the Reynolds number can be computed as

$$\mathrm{Re} = \frac{VR}{\nu} = \frac{\rho VR}{\mu} \quad (7.15)$$

where ν = fluid kinematic viscosity (ft^2/s, m^2/s)
ρ = fluid density ($slugs/ft^3$, kg/m^3)
μ = fluid dynamic viscosity ($lbf\text{-}s/ft^2$, $N\text{-}s/m^2$)

The reader may want to compare Equation 7.15 to Equation 6.5, which is used to compute the Reynolds number in the case of a circular pipe.

Table 7.5 Tentative values of Powell's ε (Chow, 1959)

Channel Description	ε
Neat cement	0.0002–0.0004
Unplaned plank flumes	0.0010–0.0017
Concrete-lined channels	0.004–0.006
Earth, straight and uniform	0.04
Dredged earth channels	0.10

The Powell formula, being an implicit function of C, requires a trial-and-error solution and therefore is less convenient than other approaches. It should be used with caution, as further research is required for determination of proper values of the roughness parameter ε.

Example 7.4 – Determining the Energy Gradient with Chézy's Equation. Repeat Example 7.3 using the Chézy equation to compute the energy gradient. Use the Ganguillet and Kutter formula to determine the value of Chézy's C. In Example 7.3, a discharge of 300 cfs occurs at a depth of 2.50 ft in a concrete-lined rectangular open channel with a width of 10 ft and $n = 0.013$.

Solution: Using $n = 0.013$ and $R = 1.67$ ft from Example 7.3, the Ganguillet and Kutter formula may be stated as

$$C = \frac{180.96 + \frac{0.00281}{S_f}}{1.42 + \frac{0.0000283}{S_f}}$$

Substitution of this expression, $V = 12$ ft/s, and $R = 1.67$ ft into Equation 7.11 leads to an expression that is implicit in S_f. Solving by trial and error,

$S_f = 0.00534$

Again, because the energy gradient is not equal to the channel slope, the flow must be varied.

Minor Losses in Open-Channel Flow

As indicated in Section 6.3, the classical method of predicting minor energy losses caused by flow expansions and contractions is to express the energy loss as being proportional to the absolute difference between the upstream and downstream velocity heads:

$$h_L = K\left|\frac{V_2^2 - V_1^2}{2g}\right| \tag{7.16}$$

where h_L = energy loss (ft, m)
K = minor loss coefficient

The *minor loss coefficient, K*, depends on the geometry of the flow contraction or expansion. For channel transitions such as entrances and exits to and from flumes, tunnels, or siphons, Chow (1959) presents the recommended design values shown in Table 7.6. Note that the loss coefficient for a flow expansion is always greater than the coefficient for a flow contraction.

Table 7.6 Minor loss coefficients for open-channel flow (Chow, 1959)

Type of Channel Transition	Contraction	Expansion
Cylinder quadrant	0.15	0.25
Square-ended	0.30 or more	0.75
Straight line	0.30	0.50
Warped	0.10	0.20

In nonprismatic channels, where the cross-sectional area and hence the flow velocity may change from one cross section to another, *eddy losses* occur and can be predicted using Equation 7.16. For gradually converging and diverging reaches, respectively, recommended values of K are about 0.1 and 0.3 [Chow, 1959; U.S. Army Corps of Engineers (USACE), 1991].

7.3 UNIFORM FLOW AND NORMAL DEPTH

Uniform flow in an open channel occurs when the depth, velocity, cross-sectional area, and discharge are constant along the channel length. As a practical matter, this condition can exist only in a prismatic channel, and only if the flow is steady. When uniform flow does exist, the constancy of the flow variables (such as velocity and depth) can be used to recognize that the slopes of the channel bottom (S_o), water surface (S_w), and energy grade line (S_f) are all equal.

Uniform flow occurs in a channel reach between two cross sections when all forces acting on the water in the reach are in balance and there is no flow acceleration. With reference to Figure 7.6, these forces are the *hydrostatic pressure forces,* P_1 and P_2, acting at the upstream and downstream ends of the reach, the *frictional force,* F_f, opposing the flow, and the *gravitational force,* $W\sin\theta$ (where W is the weight of the water), that is driving the flow. Because the flow is uniform, P_1 and P_2 must be equal and the frictional force must be equal to the driving force, or $F_f = W\sin\theta$.

If the frictional force is expressed as the product of a *shear stress* and the area of the channel bottom and sides to which that stress is applied, one can write $F_f = \tau PL$, where τ is the shear stress, P is the wetted perimeter of the flow, and L is the length of the channel reach (Figure 7.6). Similarly, if the weight of the water is expressed as $W = \gamma AL$, where A is the cross-sectional flow area, and if the slope S_o is approximated by $\sin\theta$, then $W\sin\theta = \gamma ALS_o$. Equating the frictional and driving forces leads to Equation 7.17:

$$\tau = \gamma RS_o \tag{7.17}$$

where τ = average shear stress (lb/ft^2, N/m^2)
γ = fluid specific weight (lb/ft^3, N/m^3)
R = hydraulic radius (ft, m)
S_o = channel slope (ft/ft, m/m)

Thus, when flow is uniform, Equation 7.17 gives the average shear stress acting on the channel bed and sides. This parameter is important in the design of unlined channels, as it has a direct bearing on whether the flow will mobilize soil particles on the channel boundaries.

Other than its use for the derivation of Equation 7.17, the concept of the force balance that must exist in uniform flow illustrates that uniform flow represents a condition of *equilibrium*. The depth at which this equilibrium condition is reached is called the *normal depth* and is denoted by y_n. The normal depth depends on the discharge, channel cross-sectional geometry, channel roughness, and channel slope.

If a flow control exists that causes the actual depth of flow at a cross section to be different from the normal depth, then a *varied flow profile* will exist as a transition between the depth at the flow control and the normal depth in the channel. In other words, despite the existence of the flow control, there will be a region of the channel in which the flow adjusts itself to seek out its equilibrium condition. Depending on the length of the channel, the flow may or may not have the opportunity to actually attain its normal depth condition.

Figure 7.6
Forces on a fluid element for a uniform flow condition

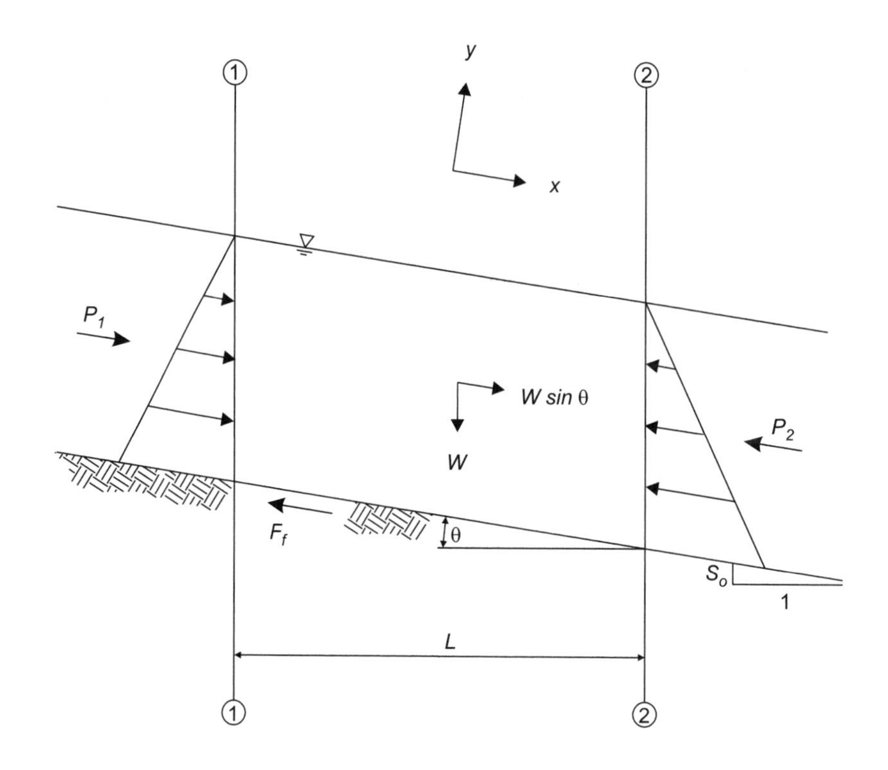

Uniform Flow in Simple Channels

A *simple open channel* is one with constant roughness and without overbank (or floodplain) areas. For the purpose of this text, a simple channel will also be taken as one without a gradually closing top, such as occurs in a pipe. In a simple open channel, the relationships between flow variables under conditions of uniform flow are given by a constitutive relationship for which it is assumed that the energy gradient S_f is equal to the channel bottom slope S_o. As discussed in Section 7.2, the most common constitutive relationships are the Manning and Chézy equations.

If either the Manning or the Chézy equation is multiplied by the cross-sectional area A of flow, it can be expressed in terms of the discharge rather than the velocity of the flow. That is,

$$Q = \frac{C_f}{n} A R^{2/3} S_o^{1/2} \tag{7.18}$$

where Q = discharge (cfs, m^3/s)
C_f = is a conversion factor equal to 1.49 for U.S. Customary units or 1.00 for SI units
A = cross-sectional area of flow (ft^2, m^2)
n = Manning's roughness coefficient

or

$$Q = CA\sqrt{RS_o} \tag{7.19}$$

where C = Chézy's C ($ft^{0.5}/s$, $m^{0.5}/s$)

It can be seen from these expressions that, for a simple channel of fixed slope and discharge, there is a unique value of flow depth (recall that R and A depend on depth) that will balance each equation. That unique depth is the normal depth y_n. Similarly, if the slope and normal depth are fixed, there is a unique discharge that results. Finally, if the discharge and normal depth are fixed, there is a unique channel slope that results (recognizing that the slope represented by the negative square root can be disregarded). That slope is called the *normal slope* of the channel for the given discharge and depth, and is denoted by S_n.

Computation of uniform flow involves determining the depth, discharge, or slope that satisfies Equation 7.18 or 7.19 when the other variables are specified. As illustrated in the following examples, determination of the normal depth generally requires a trial-and-error solution. Determination of either the discharge or the normal slope can be accomplished without iterating.

Example 7.5 – Determining Discharge with the Manning Equation. A trapezoidal open channel has a bottom width B = 10 ft, side slopes with z = 3 (see Figure E7.5.1), a slope S_o = 0.002, and n = 0.035. Determine the discharge in the channel if the flow is uniform and its depth is 3.20 ft. Use the Manning equation.

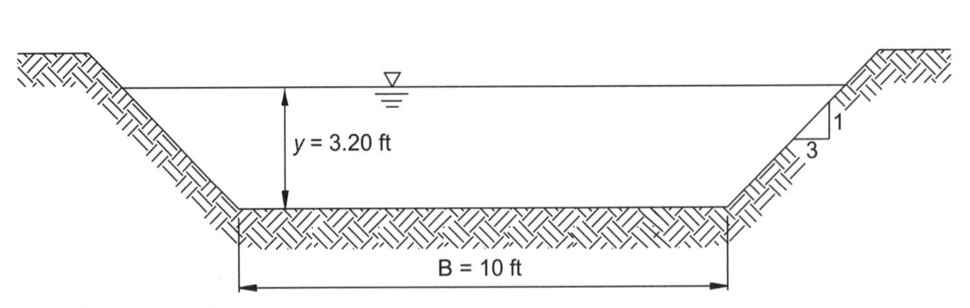

Figure E7.5.1 Trapezoidal channel for Example 7.5

Solution: From the expressions given in Table 7.1, the cross-sectional area and wetted perimeter of the channel can be computed as

$$A = 10(3.20) + 3(3.20)^2 = 62.72 \text{ ft}^2$$

$$P = 10 + 2(3.20)(1 + 3^2)^{1/2} = 30.24 \text{ ft}$$

The hydraulic radius is therefore $R = 62.72/30.24 = 2.07$ ft. Substitution into Equation 7.18 yields

$$Q = \frac{1.49}{0.035}(62.72)(2.07)^{2/3}(0.002)^{1/2} = 194 \text{ cfs}$$

Example 7.6 – Determining Normal Slope with the Manning Equation. Determine the normal slope of the channel in Example 7.5 if the discharge is $Q = 350$ cfs.

Solution: By algebraic rearrangement, Equation 7.18 can be expressed in the form

$$S_n = \left(\frac{nQ}{1.49AR^{2/3}}\right)^2$$

Substituting the given values, the normal slope is

$$S_n = 0.00651$$

Example 7.7 – Determining Normal Depth with the Manning Equation. A rectangular open channel has a width of $B = 3.6$ m, $n = 0.013$, and $S_o = 0.015$. Determine the normal depth of flow if the discharge is $Q = 7.0 \text{ m}^3/\text{s}$.

Solution: The cross-sectional area of flow can be written as

$$A = 3.6y$$

The wetted perimeter can be written similarly as

$$P = 3.6 + 2y$$

Substitution of these expressions into Equation 7.18 yields

$$Q = 7.0 = \frac{1.00}{0.013}(3.6y)^{5/3}(3.6 + 2y)^{-2/3}0.015^{1/2}$$

or

$$(3.6y)^{5/3}(3.6 + 2y)^{-2/3} = 0.74$$

By trial and error, the value of y that balances this equation is $y_n = 0.42$ m.

Uniform Flow in a Channel with a Gradually Closing Top

A pipe, whether it is circular, elliptical, or another shape, is an example of a channel with a *gradually closing top*. In a channel with a gradually closing top, there may be more than one depth of flow that satisfies the normal depth criterion of equilibrium. Furthermore, the maximum discharge that can be conveyed by the channel under uniform flow conditions occurs at a depth of flow smaller than that corresponding to full flow in the conduit.

Figure 7.4 on page 222 illustrates a circular pipe of diameter D in which the depth of flow is denoted by y. The angle ϕ defined by the edges of the free water surface is also shown. Note that $\phi = 0$ when $y = 0$, and that $\phi = 2\pi$ radians when $y = D$. The relationship between y, D, and ϕ (in radians) is given by

$$y = \frac{D}{2}\left[1 - \cos\left(\frac{\phi}{2}\right)\right] \qquad (7.20)$$

From Table 7.1, the cross-sectional area and wetted perimeter are given by

$$A = \frac{D^2}{8}(\phi - \sin\phi) \qquad (7.21)$$

$$P = \frac{D\phi}{2} \qquad (7.22)$$

Substitution of these expressions into the Manning equation yields

$$Q = \frac{C_f}{n}\left[\frac{D^2}{8}(\phi - \sin\phi)\right]^{5/3}\left(\frac{D\phi}{2}\right)^{-2/3} S_o^{1/2} \qquad (7.23)$$

where Q = discharge (cfs, m^3/s)
C_f = unit conversion factor (1.49 for US customary units, 1.00 for SI)
n = Manning's n
D = diameter of pipe (ft, m)
ϕ = angle defined by edges of free water surface of circular channel (see Figure 7.4)

For chosen values of D, n, and S_o, Equation 7.23 may be applied to develop a graph of Q as a function of ϕ (or as a function of y by rearranging Equation 7.20 to solve for ϕ and substituting this expression into Equation 7.23). Thus, ϕ (or y) could be obtained from the graph if Q were specified. As shown in Figure 7.7, which illustrates such a graph, two possible depths of flow (y-values) exist for any possible discharge greater than or equal to the *full-flow discharge* (that is, the discharge when $y = D$). It can also be seen from the graph that the maximum discharge occurs when $y \approx 0.94D$.

Although Figure 7.7 is useful for demonstrating that there may be more than one possible depth of uniform flow in a channel with a gradually closing top, it was developed for the case of a specific pipe diameter, roughness, and slope. Figure 7.8 depicts a more useful dimensionless graph that can be used for any combination of pipe geometric variables. In this figure, the relative depth of flow, y/D, is shown as a function of the ratio of actual flow to full flow and the ratio of actual velocity to full velocity. Example 7.8 illustrates the use of Figure 7.8.

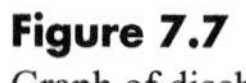

Figure 7.7
Graph of discharge versus normal depth for a representative 1-ft diameter pipe

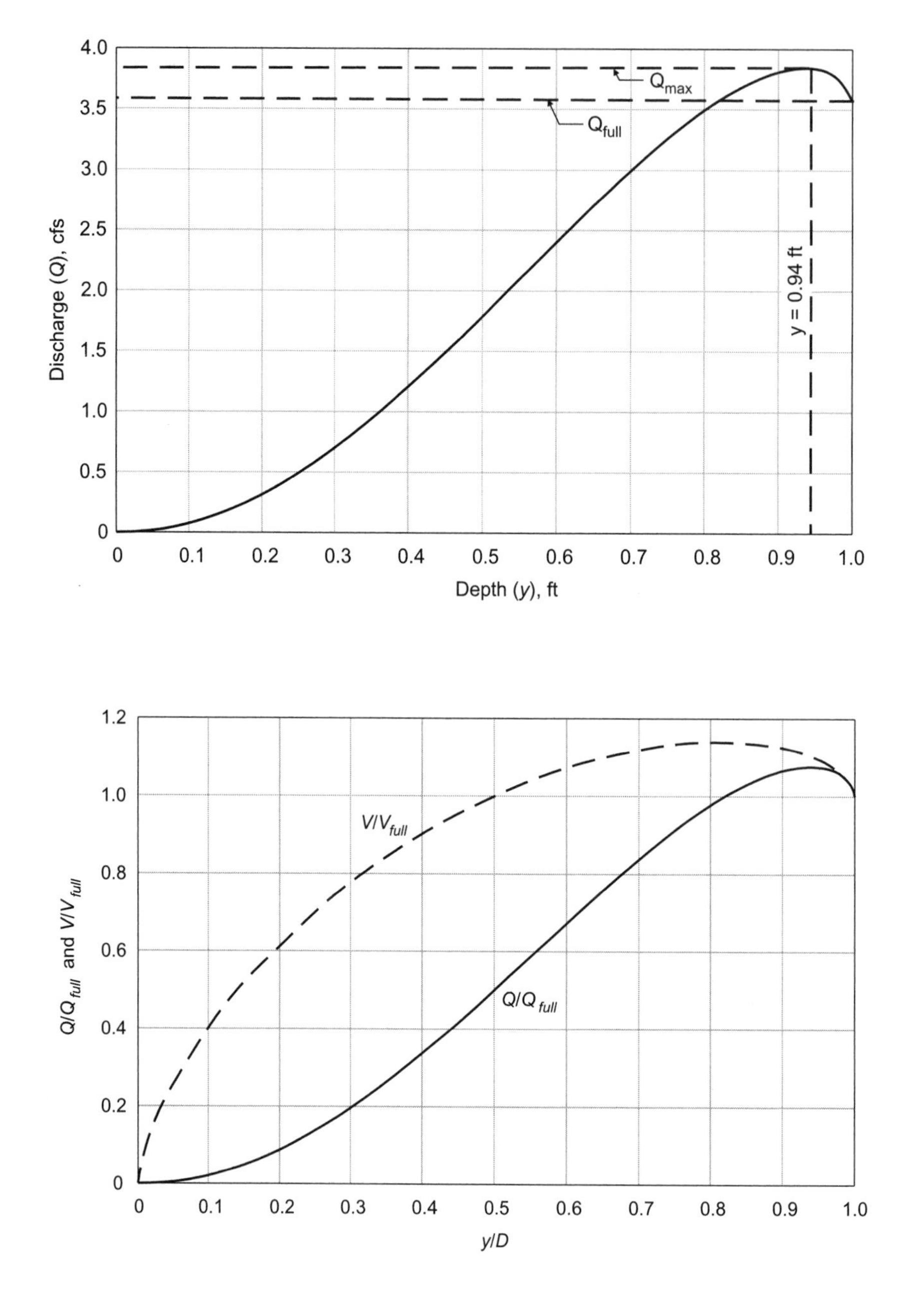

Figure 7.8
Dimensionless graph of discharge and velocity fractions versus depth fraction for circular pipe

---

Example 7.8 – Depth and Velocity in a Circular Pipe. A 36-in. diameter pipe with $n = 0.013$ and $S_o = 0.01$ conveys a discharge of 45 cfs. Determine the depth and velocity of flow in the pipe. If the pipe flows at its maximum capacity ($y \approx 0.94D$), what is the discharge?

Solution: For full-flow conditions, the pipe's cross-sectional area and wetted perimeter are, respectively, 7.07 ft^2 and 9.42 ft. Applying the Manning equation, the full-flow discharge is $Q_f = 67$ cfs. The full-flow velocity is

$$V_f = 67/7.07 = 9.46 \text{ ft/s}$$

The actual discharge in the pipe is $Q = 45$ cfs, so

$$Q/Q_f = 45/67 = 0.67$$

From Figure 7.8, the relative depth of flow is $y/D = 0.60$. Thus, the actual depth of flow is

$$D(y/D) = 3(0.60) = 1.80 \text{ ft.}$$

Using $y/D = 0.60$, Figure 7.8 shows that $V/V_f = 1.09$. Thus, the actual flow velocity is

$$V_f(V/V_f) = 9.46(1.09) = 10.3 \text{ ft/s}$$

At maximum capacity, the actual discharge is about 1.09 times the full-flow discharge (see Figure 7.8). Therefore, the maximum discharge is

$$1.09(67) = 73 \text{ cfs}$$

Dimensionless graphs of geometric elements similar to the one shown in Figure 7.8 can be developed for pipe shapes other than circular. The *Concrete Pipe Design Manual* (American Concrete Pipe Association, 2000) and similar publications developed by associations of pipe manufacturers may be consulted to obtain graphs for other common pipe shapes.

Uniform Flow in a Channel with Composite Roughness

In a simple channel with *composite roughness,* the roughness characteristics vary from one part of the channel boundary to another. For example, a constructed prismatic channel might have an earthen bottom and riprap side slopes. A laboratory channel might have a steel bottom and glass sides. In such channels, a uniform flow formula such as the Manning equation can be applied if the n-value used is an *equivalent n-value* computed to represent an average measure of the channel roughness.

There are several ways in which an equivalent n value may be determined. As shown in Figure 7.9, a channel cross section may be subdivided into a number of subareas N, each of which has its own roughness n_i, area A_i, wetted perimeter P_i, hydraulic radius R_i, and average depth y_i, where $i = 1, 2 \ldots N$. The wetted perimeter of each subarea is computed using only that part of the perimeter that coincides with the channel bottom or sides.

Figure 7.9
Division of channel for computation of the equivalent roughness coefficient

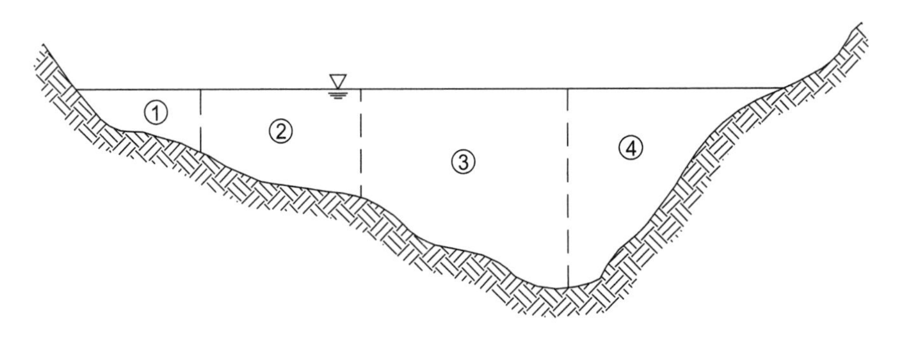

If one assumes that the velocity in each subarea is equal to the mean velocity for the cross section as a whole, the equivalent n value may be determined as (Horton, 1933; Einstein, 1934)

$$n_e = \left\{ \frac{\sum P_i n_i^{3/2}}{\sum P_i} \right\}^{2/3} \tag{7.24}$$

where n_e = equivalent Manning's n
P_i = wetted perimeter for the i-th channel subarea (ft, m)
n_i = Manning's n for the i-th channel subarea

If one assumes that the total force resisting the flow is equal to the sum of the resisting forces in each of the subareas, the equivalent n value is (Pavlovskii, 1931; Muhlhofer, 1933; Einstein and Banks, 1951)

$$n_e = \left\{ \frac{\sum P_i n_i^2}{\sum P_i} \right\}^{1/2} \tag{7.25}$$

Lotter (1933) showed that the equivalent n value could be expressed as the following, assuming that the total discharge is equal to the sum of the discharges in the individual subareas:

$$n_e = \frac{PR^{5/3}}{\sum\left(\frac{P_i R_i^{5/3}}{n_i}\right)} \tag{7.26}$$

Finally, if the velocity distribution in the cross section is assumed to be logarithmic, the equivalent n value is (Krishnamurthy and Christensen, 1972)

$$\ln n_e = \frac{\sum P_i y_i^{3/2} \ln n_i}{\sum P_i y_i^{3/2}} \tag{7.27}$$

where y_i = average depth of the i-th subarea

In an evaluation of these alternative expressions based on data from 36 actual channel cross sections, Motayed and Krishnamurthy (1980) concluded that Equation 7.24 performed the best.

Example 7.9 – Composite Roughness in a Rehabilitated Channel. Figure E7.9.1 illustrates the cross section of an old masonry tunnel whose hydraulic performance is to be improved by paving the invert with finished concrete. If the depth of flow in the tunnel is 4 ft both before and after the rehabilitation, determine the amount by which the tunnel discharge capacity increases following the change. The tunnel slope is $S_o = 0.002$, and the n values for rough masonry and finished concrete are 0.030 and 0.013, respectively. Use Equation 7.24 to determine the equivalent n value following the rehabilitation.

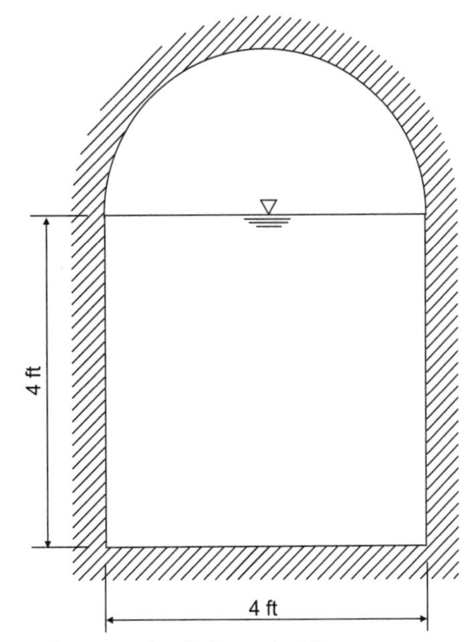

Figure E7.9.1 Tunnel cross section for Example 7.9

Solution: In both cases, the cross-sectional area of flow is $A = 16\ \text{ft}^2$ and the wetted perimeter is $P = 12$ ft. The tunnel capacity prior to the change is

$$Q = \frac{1.49}{0.030}(16)^{5/3}(12)^{-2/3}(0.002)^{1/2} = 43 \text{ cfs}$$

The equivalent n value following the rehabilitation is

$$n_e = \frac{\left\{4(0.030)^{3/2} + 4(0.013)^{3/2} + 4(0.030)^{3/2}\right\}^{2/3}}{12^{2/3}} = 0.025$$

Using this equivalent value, the capacity following the rehabilitation is

$$Q = \frac{1.49}{0.025}(16)^{5/3}(12)^{-2/3}(0.002)^{1/2} = 52 \text{ cfs}$$

In this instance, the capacity of the tunnel is increased by over 20 percent.

Uniform Flow in Compound Cross Sections

Figure 7.10, which is a simplified representation of a flooded river valley, illustrates a *compound channel* cross section. It consists of a main channel and *overbank* (or *floodplain*) areas on either side of the main channel that are inundated by high flows. Roughness is typically higher in the overbank areas than in the main channel, with the effect that flow velocities are typically lower in the overbank areas than in the main channel. This condition causes either of two possible problems to arise in the analysis of the flow.

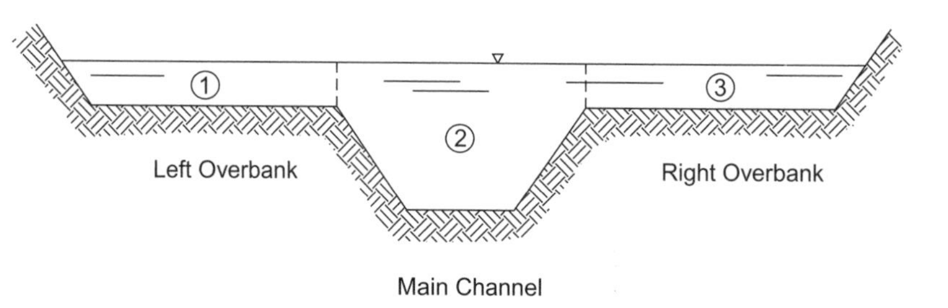

Figure 7.10
A compound channel cross section

To illustrate the first type of problem, assume that at a given cross section the water surface elevations are the same in both the main channel and the overbank areas. Under this assumption, there must be three different energy grade line elevations within the cross section: one for the main channel and one for each overbank area. The multiple energy grade lines occur because the velocities, and hence the velocity heads, are different in each area of the cross section.

Alternatively, one might assume that the energy grade line elevation at a cross section is the same for all three parts of the flow area. In this case, again because of velocity differences, one comes to realize that the water surface elevations must be different in the main channel and overbank areas.

To resolve these issues and arrive at both a single water surface elevation for a cross section and a single energy grade line elevation, the velocity distribution in the cross section must be accounted for. The velocity distribution coefficient α is used for this purpose.

If a channel cross section is subdivided into N subareas, each of which has an area A_i and average velocity V_i, where $i = 1, 2 \ldots N$, then α can be computed using an approximation of Equation 6.3:

$$\alpha = \frac{\sum A_i V_i^3}{A V_m^3} \tag{7.28}$$

where α = velocity distribution coefficient
A_i = cross-sectional area of the i-th subarea (ft^2, m^2)
V_i = average velocity of the i-th subarea (ft/s, m/s)
A = total cross-sectional area (ft^2, m^2)
V_m = the mean velocity in the entire cross section (ft/s, m/s)

V_m is computed as

$$V_m = \frac{1}{A} \sum A_i V_i \tag{7.29}$$

Using these expressions, the velocity head at a cross section may be expressed as $\alpha V_m^2/2g$.

The total discharge at a cross section may be computed as

$$Q = \sum Q_i = \sum A_i V_i \tag{7.30}$$

where Q = total discharge at a channel cross section (cfs, m^3/s)
Q_i = discharge at a cross-sectional subarea (cfs, m^3/s)

If one assumes that the energy gradient S_f is equal for each subarea, it may be shown that

$$Q_i = \kappa_i S_f^{1/2} \tag{7.31}$$

where κ_i = the *conveyance* of the i-th subarea (cfs, m^3/s)
S_f = energy gradient (ft/ft, m/m)

and

$$Q = \kappa S_f^{1/2} \tag{7.32}$$

where

$$\kappa = \sum \kappa_i \tag{7.33}$$

In these expressions, κ_i denotes the *conveyance* of the i-th subarea of the cross section. If the Manning equation (Equation 7.9) is employed as the constitutive relationship for the flow, then the conveyance can be written as

$$\kappa_i = \frac{C_f}{n_i} A_i R_i^{2/3} \tag{7.34}$$

where C_f = unit conversion factor (1.49 for U.S. customary units, 1.00 for SI)

Dividing Equation 7.31 by 7.32, one obtains

$$Q_i = Q \frac{\kappa_i}{\sum \kappa_i} \tag{7.35}$$

This expression shows that the discharge in the i-th subarea is a fraction of the total discharge, where the fraction is computed as the ratio of the conveyance for the subarea in question to the total conveyance for the cross section.

Use of these equations is typically made in either of two ways. If one desires to find the depth of uniform flow for a specified set of discharge and channel characteristics, the solution involves iterative calculations using assumed water surface elevations. For an assumed water surface elevation, the velocity and area of each subarea can be determined, and a total discharge Q can be computed using Equations 7.30 and 7.31. If the computed flow rate is different from the known discharge, then a new water surface elevation is assumed, and the procedure is repeated until agreement is attained. After the water surface elevation is found, the mean velocity of flow can be determined as $V_m = Q/A$, and α can be determined by using Equation 7.28. Finally, the velocity head is computed as $\alpha V_m^2/2g$ and added to the water surface elevation to obtain the energy grade line elevation.

A second way in which these equations can be applied occurs when the discharge, channel characteristics, and energy grade line elevation are known at a cross section. This situation occurs in gradually varied flow profile analysis via the standard-step method (see Section 7.8). As in the previous case, iterations are made on assumed water surface elevations. For an assumed water surface elevation, one can determine the areas and conveyances of each of the subareas of the channel section. The discharge in each subarea can then be determined using Equation 7.35, and the velocity in each subarea can be found as the ratio of the discharge to the area. The mean velocity can be found by using Equation 7.29, and α can be found by using Equation 7.28. Finally, the velocity head can be computed as $\alpha V_m^2/2g$ and added to the assumed water surface elevation to obtain a computed energy grade line elevation. If the computed energy grade line elevation is not the same as the known elevation, a new water surface elevation is assumed, and the procedure is repeated.

Example 7.10 – Uniform Flow in a Compound Cross Section. Figure E7.10.1 illustrates a compound channel section consisting of two rectangular elements. Determine the depth of uniform flow in this channel section if the total discharge is $Q = 450$ cfs and the channel slope is $S_o = 0.005$. Also, determine the velocity distribution coefficient α for this uniform flow. Manning's n values for the main channel and overbank area are 0.020 and 0.030, respectively.

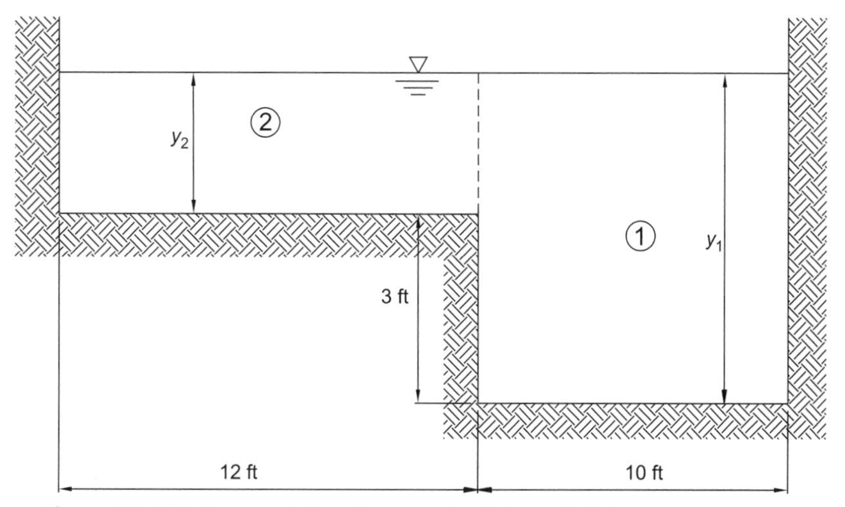

Figure E7.10.1 Compound channel cross section for Example 7.10

Solution: For any assumed depth of flow y_1 greater than 3 ft, the areas and wetted perimeters of the two subareas can be found as

$$A_1 = 10y_1$$

$$A_2 = 12y_2 = 12(y_1 - 3)$$

$$P_1 = 13 + y_1$$

$$P_2 = 12 + y_2 = 12 + (y_1 - 3)$$

where the subscripts 1 and 2 refer to the main channel and overbank, respectively.

These expressions can be used with Equations 7.31 and 7.34 to find the discharge in each subarea, and then a total discharge as the sum of the two subarea discharges. Because the flow is uniform, S_o is used in place of S_f in Equation 7.31. By trial and error, the correct depth is found when the computed discharge is equal to the known discharge.

If the initial assumption for depth is $y_1 = 4$ ft, then $A_1 = 40$ ft^2, $A_2 = 12$ ft^2, $P_1 = 17$ ft$_2$, and $P_2 = 13$ ft^2. Therefore, from Equations 7.31 and 7.34,

$$\kappa_1 = \frac{C_f}{n_1} A_1 R_1^{2/3} = \frac{1.49}{0.020}(40)\left(\frac{40}{17}\right)^{2/3} = 5{,}272 \text{ cfs}$$

$$Q_1 = \kappa_1 S_o^{1/2} = (5272)(0.005)^{1/2} = 373 \text{ cfs}$$

$$\kappa_2 = \frac{C_f}{n_2} A_2 R_2^{2/3} = \frac{1.49}{0.030}(12)\left(\frac{12}{13}\right)^{2/3} = 565 \text{ cfs}$$

$$Q_2 = \kappa_2 S_o^{1/2} = (565)(0.005)^{1/2} = 40 \text{ cfs}$$

So, according to Equation 7.30,

$$Q = \sum Q_i = 373 + 40 = 413 \text{ cfs}$$

Because 413 cfs is less than the given flow of 450 cfs, a slightly higher water surface elevation must be assumed and the flow computed again until the correct elevation is found. For this example, trial and error leads to $y_n = 4.18$ ft.

For the depth of 4.18 ft, the individual subareas and velocities are

$$A_1 = 41.8 \text{ ft}^2$$

$$A_2 = 14.16 \text{ ft}^2$$

$$V_1 = 9.52 \text{ ft/s}$$

$$V_2 = 3.67 \text{ ft/s}$$

Therefore, from Equations 7.28 and 7.29:

$$V_m = \frac{1}{A}\sum A_i V_i = \frac{1}{41.8 + 14.16}\left[(41.8)(9.52) + (14.16)(3.67)\right] = 8.04 \text{ ft/s}$$

$$\alpha = \frac{\sum A_i V_i^3}{A V_m^3} = \frac{(41.8)(9.52)^3 + (14.16)(3.67)^3}{(41.8 + 14.16)(8.04)^3} = 1.26$$

As shown, the velocity distribution coefficient is 1.26.

7.4 SPECIFIC ENERGY AND CRITICAL FLOW

Critical flow, like uniform flow, is a case in open-channel flow for which a unique relationship between depth and discharge exists. The occurrence of critical flow corresponds to the point of minimum specific energy for a given discharge and channel geometry; thus, the study of critical flow is facilitated by an understanding of specific energy concepts.

Specific Energy

Specific energy is the energy head relative to the channel bottom elevation. For a channel of small slope, the specific energy E is

$$E = y + \frac{\alpha V^2}{2g} = y + \frac{\alpha Q^2}{2gA^2} \tag{7.36}$$

where E = specific energy (ft, m)
y = depth of flow (ft, m)
α = velocity distribution coefficient
V = cross-sectionally averaged flow velocity (ft/s, m/s)
g = gravitational acceleration (32.2 ft^2/s, 9.81 m^2/s)
Q = discharge (cfs, m^3/s)
A = cross-sectional area (ft^2, m^2)

Clearly, the specific energy at a channel cross section consists of the sum of the pressure head (the depth of flow) and velocity head terms from the energy equation (Equation 7.25).

For a given discharge Q and channel cross-sectional geometry, inspection of Equation 7.36 reveals that E depends only on the depth of flow. A graph showing the dependence of E on y for a given discharge and channel geometry is called a *specific energy curve*. Qualitatively, the characteristics of a specific energy curve can be deduced from Equation 7.36. For small values of y, the velocity will be large and the second term on the right side of the equation will be much larger than the first. Thus, the specific energy curve will approach the asymptote $E = \alpha V^2/2g$. Conversely, for large values of y, the velocity will be small and the first term will be much larger than the second. In that case, the curve will asymptotically approach $E = y$. Figure 7.11 illustrates the characteristics of a typical specific energy curve.

Because a specific energy curve shows the dependence of E on y for a fixed discharge and channel cross-sectional geometry, a different curve must arise if either the discharge or geometry changes. (Note, however, that changing only the slope simply means moving another point on the same curve.) It is difficult (if not impossible) to state any generally applicable effects of geometric changes, but Equation 7.36 shows that E increases when the discharge increases. That is, if two specific energy curves for a given channel geometry were plotted on the same graph, with one curve corresponding to a discharge Q_1 and the second corresponding to a discharge $Q_2 > Q_1$, the result would be a shift of the curve for Q_2 to higher values of E. This shift is shown in Figure 7.11.

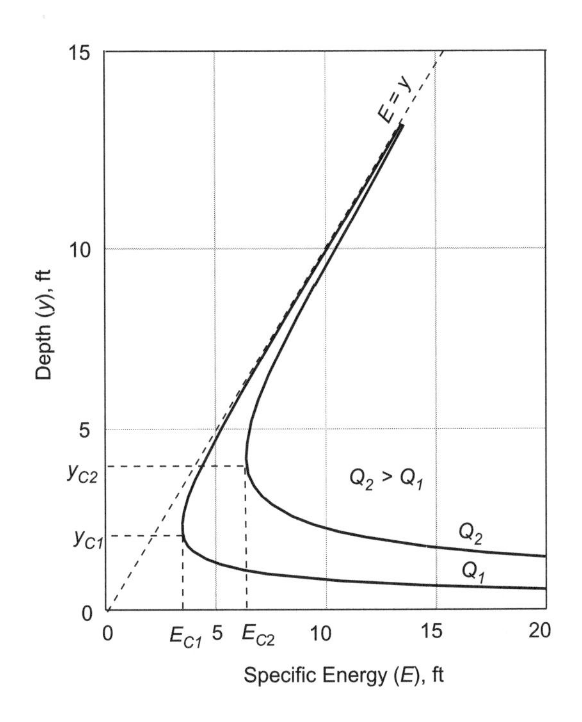

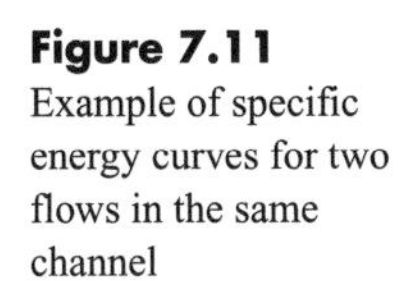
Figure 7.11
Example of specific energy curves for two flows in the same channel

Inspection of a specific energy curve reveals that there are two possible depths of flow corresponding to any specific energy greater than the minimum specific energy level denoted by E_c. These two depths are called *alternate depths*. The smaller of the two depths corresponds to a shallow and rapid flow known as *supercritical flow*. The larger of the two depths corresponds to a deep and tranquil flow known as *subcritical flow*. The two limbs of a specific energy curve join at a point where the specific energy is E_c and there is a single corresponding value of depth denoted by y_c (see Figure 7.11). The depth y_c is called the *critical depth*, and a flow at this depth is correspondingly called a *critical flow*.

Example 7.11 – Specific Energy in a Rectangular Channel. Compute and graph a specific energy curve for a rectangular open channel with $B = 10$ ft and $Q = 200$ cfs.

Solution: To graph the curve, one may assume depths of flow and compute the corresponding E values. For an assumed depth y, the cross-sectional area is $A = 10y$ and the velocity is

$$V = Q/A = 20/y$$

The specific energy (assuming $\alpha = 1$) is therefore

$$E = y + \frac{(20/y)^2}{2g}$$

Figure E7.11.1 shows this specific energy curve for this example.

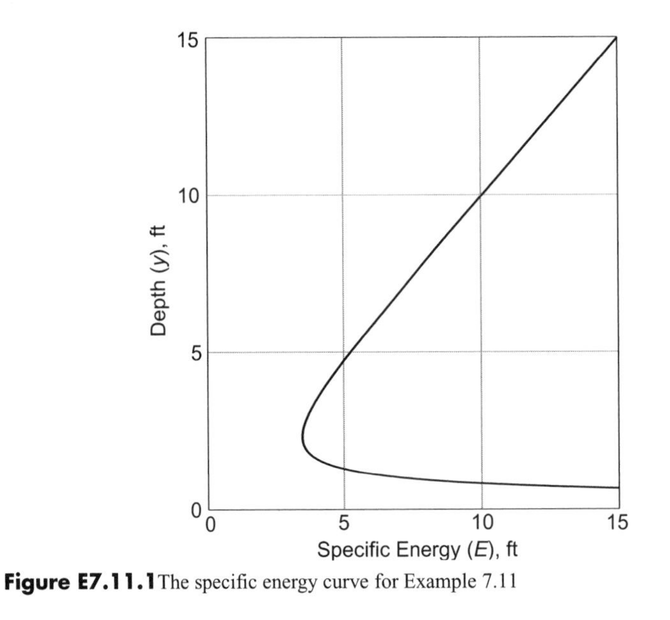

Figure E7.11.1 The specific energy curve for Example 7.11

Critical Depth and Critical Flow

As shown in the preceding section, critical depth and the corresponding critical flow occur when the specific energy is a minimum for a fixed discharge and channel cross-sectional geometry. Computation of the critical depth y_c can be accomplished in a trial-and-error manner to minimize E. An alternative (and more enlightening) way of computing critical depth is to recognize that it occurs when dE/dy is equal to zero (see Figure 7.11). Differentiating Equation 7.36 and assuming $\alpha = 1$, one obtains

$$\frac{dE}{dy} = 1 - \frac{Q^2}{gA^3}\frac{dA}{dy} = 1 - \frac{Q^2 T}{gA^3} \tag{7.37}$$

where $T = dA/dy =$ top width of the flow

Equating this expression to zero, critical depth occurs when

$$\frac{Q^2 T}{gA^3} = 1 \tag{7.38}$$

The *Froude number* of an open-channel flow is a dimensionless quantity defined as

$$\mathrm{Fr} = \left(\frac{Q^2 T}{gA^3}\right)^{1/2} = \frac{V}{\sqrt{gD_h}} \tag{7.39}$$

where Fr = the Froude number
D_h = hydraulic depth (see Equation 7.2) (ft, m)

Comparing Equations 7.38 and 7.39, it can be seen that critical depth and critical flow occur when the Froude number is equal to unity. Subcritical flow occurs when $\mathrm{Fr} < 1$, and supercritical flow occurs when $\mathrm{Fr} > 1$.

An additional fact that can be deduced from Equations 7.38 and 7.39 is that when a flow is at critical depth, the velocity head is equal to one-half of the hydraulic depth:

$$\frac{V^2}{2g} = \frac{D_h}{2} \tag{7.40}$$

This relationship can be quite useful for practical problem solving, particularly when a channel is rectangular in shape.

The reader should note that the hydraulic depth D_h of a channel is generally not the same as the depth of flow y. For common prismatic channel shapes, the hydraulic depth is smaller than the depth of flow, and is equal to the depth of flow only in rectangular channels.

Example 7.12 – Finding Critical Depth and Minimum Specific Energy in a Rectangular Channel. Determine the critical depth and minimum specific energy for the channel in Example 7.11.

Solution: The criterion for critical depth is that Equation 7.38 must be satisfied. For the channel of interest, $B = T = 10$ ft, $Q = 200$ cfs, and $A = 10y$. Substituting and solving for y, one obtains the critical depth as

$$y_c = \sqrt[3]{\frac{200^2}{32.2(10)^2}} = 2.32 \text{ ft}$$

From Equation 7.40, the velocity head for critical flow is one-half the hydraulic depth, or 1.16 ft in this example. Thus, the minimum specific energy is

$$E_c = 2.32 + 1.16 = 3.48 \text{ ft}$$

Example 7.13 – Finding the Critical Depth in a Trapezoidal Channel. A trapezoidal open channel has a bottom width of $B = 3$ m and side slopes with $z = 2$. Determine the critical depth if $Q = 7$ m^3/s.

Solution: The cross-sectional area of flow, in terms of the depth, is

$$A = 3y + 2y^2$$

The top width is

$$T = 3 + 4y$$

Substituting into Equation 7.38, one obtains

$$\frac{Q^2 T}{gA^3} = \frac{7^2(3+4y)}{9.81(3y+2y^2)^3} = 1$$

By trial and error, the depth that satisfies this expression is $y_c = 0.70$ m.

Compound Channel Sections

In a compound channel section, consisting of a main channel and overbank areas, there may be more than one critical depth for a given discharge. Further, determination of the number of critical depths and their values is more complex than simply locating all local minima on a specific energy curve, because a critical flow condition may exist even in situations for which the specific energy is not at a local minimum (Chaudhry and Bhallamudi, 1988).

The procedure described by Chaudhry and Bhallamudi (1988) for determining the critical depths in a compound channel is presented here. The procedure first determines the number of critical depths that occur within a channel section. It then determines each of the critical depths. It assumes that the channel cross section is symmetrical about its centerline, with dimensional terminology as shown in Figure 7.12.

Based on these dimensions, one can define a number of dimensionless parameters:

$$y_r = \frac{y}{y_f} \tag{7.41}$$

where y_r = ratio of the depth y to the vertical distance from the bottom of the overbank to the bottom of the main channel y_f

$$b_r = \frac{B_f}{B_m} \tag{7.42}$$

where b_r = ratio of the width of an overbank B_f to the width of the main channel B_m

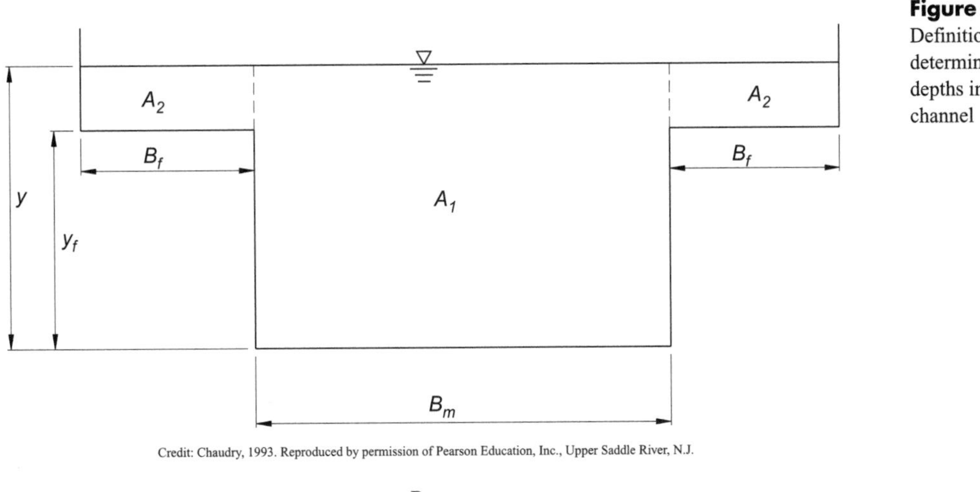

Credit: Chaudry, 1993. Reproduced by permission of Pearson Education, Inc., Upper Saddle River, N.J.

Figure 7.12 Definition figure for determining critical depths in a compound channel

$$b_f = \frac{B_f}{y_f} \tag{7.43}$$

where b_f = ratio of the width of an overbank B_f to the vertical distance from the bottom of the overbank to the bottom of the main channel y_f

$$n_r = \frac{n_m}{n_f} \tag{7.44}$$

where n_r = ratio of Manning's n for the main channel n_m to Manning's n for an overbank n_f

As described by Chaudhry and Bhallamudi (1988), one can use these dimensions and dimensionless ratios to write an expression of the form

$$C = \frac{gB_m^2 y_f^3}{Q^2} \tag{7.45}$$

where

$$C = c_1 + c_2 \tag{7.46}$$

and

$$c_1 = \frac{1}{y_r + 2b_r(y_r - 1)}\left\{\left(\frac{m}{y_r}\right)^2 + \left(\frac{1-m}{y_r - 1}\right)^2 \frac{1}{2b_r}\right\} \tag{7.47}$$

$$c_2 = \frac{2m(1-m)}{3[y_r + 2b_r(y_r - 1)]}\left\{\frac{5}{y_r(y_r-1)} - \frac{2}{b_f + y_r - 1}\right\}\left\{\frac{m}{y_r} - \frac{1}{2b_r}\left(\frac{1-m}{y_r-1}\right)\right\} \tag{7.48}$$

$$m = \frac{1}{1 + 2n_r(A_2/A_1)^{5/3}(P_1/P_2)^{2/3}} \tag{7.49}$$

where A_1 = flow area for the main channel
A_2 = flow area for one of the overbanks
P_1 = wetted perimeter for the main channel ($2y_f + B_m$)
P_2 = wetted perimeter for one of the overbanks ($y - y_f + B_f$)

For specified channel dimensions, roughness values, and a discharge, one can determine a target value of C using Equation 7.45. Various flow depths (and hence values of y_r) can be assumed so that Equations 7.46 through 7.49 can be applied and C graphed as a function of y_r. The values of y_r for which C is equal to the target value (the right side of Equation 7.45) are the critical depths for flows over the floodplain. An additional critical depth, depending on the discharge, may also occur at a depth smaller than $y = y_f$. Example 7.14, adapted from Chaudhry (1993), illustrates the procedure for computing critical depths in a compound channel.

Example 7.14 – Determining the Critical Depths in a Compound Channel.

Determine the critical depths for $Q = 1.7$ and $2.5\ m^3/s$ in a channel having the following dimensions: $B_m = 1.0$ m, $B_f = 3.0$ m, $y_f = 1.0$ m, and $n_r = 0.9$ (see Figure 7.12).

Solution: Consider first the possibility that a critical depth might occur within the main channel (i.e., at a depth smaller than y_f). Applying Equation 7.38 with $T = B_m$ and $Q = 1.7\ m^3/s$, one finds a critical depth of $y_c = 0.67$ m. Applying the same equation, but with $Q = 2.5\ m^3/s$, one finds a critical depth of $y_c = 0.86$ m. Thus, for both of these discharges, a critical depth occurs within the main channel. It must now be determined if additional critical depths occur once the flow rises to a depth greater than y_f.

Figure E7.14.1 is a graph of C versus y_r obtained by applying Equations 7.46 through 7.49. Note that the minimum value of y_r employed is 1.00, which corresponds to an actual depth of $y = y_f$. Note also that C approaches 1 when y_r approaches 1 m, and that C attains a maximum value of $C_{max} = 2.88$ when $y_r = 1.033$ m.

Using Equation 7.45 with $Q = 1.7\ m^3/s$ yields $C = 3.39$. This is shown as the upper of the two horizontal lines in Figure E7.14.1. Because it does not intersect the C- y_r curve in that figure, there is no critical depth larger than y_f when the discharge in the channel is $1.7\ m^3/s$.

Using Equation 7.45 with $Q = 2.5\ m^3/s$ yields $C = 1.57$. This is shown as the lower of the two horizontal lines in Figure E7.14.1. Because it intersects the C- y_r curve in the figure twice, two critical depths larger than y_f exist when the discharge in the channel is $2.5\ m^3/s$. Those critical depths are 1.002 m and 1.12 m.

Figure E7.14.2 shows specific energy curves computed for each of the two discharges in the channel. Note that when the discharge is $Q = 1.7\ m^3/s$, only one critical depth exists: 0.67 m. However, when the discharge is $Q = 2.5\ m^3/s$, there are three critical depths, at values of 0.86, 1.002, and 1.12 m. As shown in the figure, not all of the critical depths correspond to local minima on the specific energy curve.

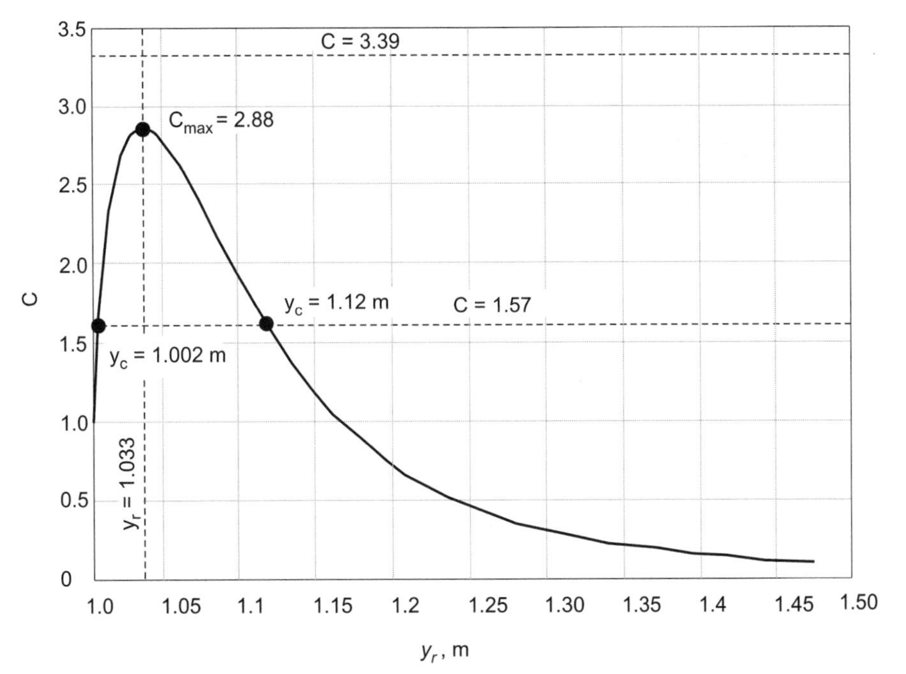

Figure E7.14.1 C versus y_r for the compound channel in Example 7.14

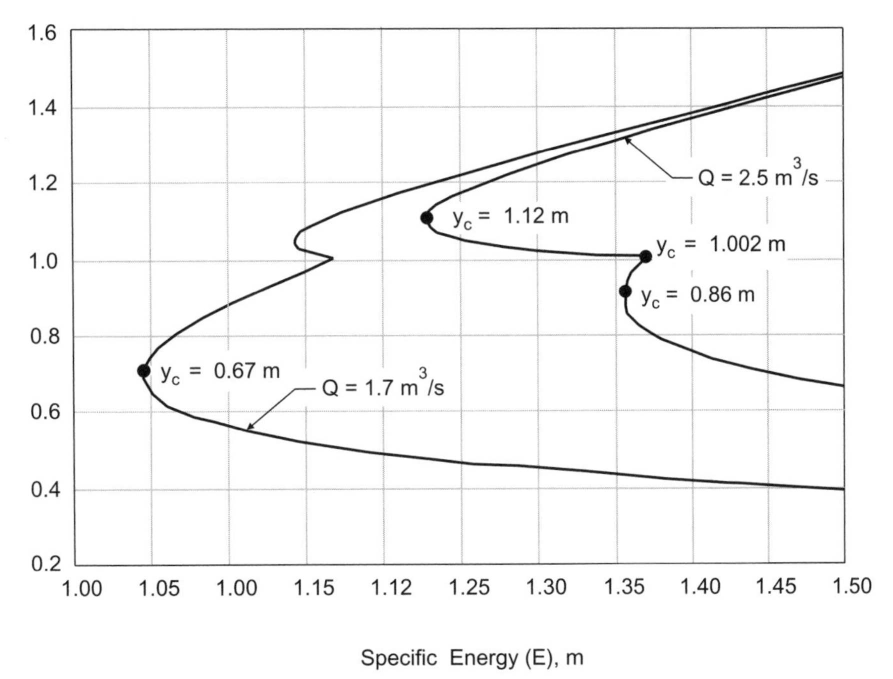

Figure E7.14.2 Specific energy curves for two discharges in a compound channel for Example 7.14

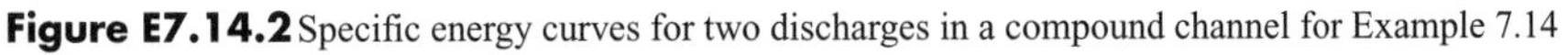

Flow Control

A *control section* in open-channel flow is said to occur at a channel location for which there is a known or unique value of the depth of flow for a specified discharge. The depth of flow at a control section is essentially a *boundary condition* that provides a unique solution to the differential equation describing a varied flow in a channel. This concept is discussed further in subsequent sections of this chapter.

Because a unique relationship exists between discharge and flow depth for a critical flow, a channel cross section at which critical flow occurs is necessarily a flow control section. Common occurrences of such *critical flow control sections* occur at free overfalls and at locations where a mildly sloping channel ends and becomes a steeply sloping channel. Rouse (1936) has shown that, at a free overfall, critical depth actually occurs at a distance of about $3y_c$ to $4y_c$ upstream of the brink of the overfall.

In open-channel flow, flow control sections can exist at locations other than critical flow sections. For example, the flow depth created at the upstream end of a culvert is often a flow control section, as is the depth of a supercritical flow issuing from beneath a sluice gate. Many other examples of flow control sections can be found.

Flow Measurement

Critical flow is an instance in which there is a unique definitive relationship between the depth of flow and discharge in an open channel. Thus, if one can cause a flow to pass through its critical depth, measurement of that depth can be used to deduce the discharge.

There are a number of ways that one can force a flow to pass through critical depth. One way is to install a *weir* in a channel (see Section 7.6). Weirs, however, tend to cause large head losses in the flow, which may not be acceptable in some circumstances. Critical depth can be caused by a flow contraction, and it occurs naturally at a free overfall or at a location where a mild and steeply sloping channel join. A *Parshall flume* is another widely used device for measurement of open-channel flow, though critical depth may or may not occur in such a flume.

Analysis of Flow in Channel Transitions

A *transition* occurs in an open channel at a location where the cross-sectional geometry of the channel changes. For example, a transition might correspond to a change in the width of a rectangular channel, or in the diameter of a pipe (such as at a manhole). A transition may also be as simple as a step or drop in the bottom elevation of the channel, with no corresponding change in cross-sectional geometry.

When there is a reduction in cross-sectional area and a corresponding increase in velocity at a transition, the flow is *contracted*. A flow *expansion* occurs when the cross-sectional area increases and the velocity decreases.

Consider a transition consisting of a simple step in the bottom elevation, as shown in Figure 7.13. The discharge, Q, is constant through the transition, and the cross-sectional geometry of the channel is the same both upstream and downstream of the step. If the depth of flow at a cross section downstream of the step is known and designated by y_2, the corresponding specific energy E_2 can easily be found. If one assumes that there is no energy loss in the transition (that is, that the energy grade line has a negligible slope in the transition), the specific energy E_1 at a cross section upstream of the step can be determined as $E_1 = E_2 + \Delta Z$, where ΔZ is the height of the step. A specific energy relationship can then be applied to determine the two alternate depths of flow at the upstream section.

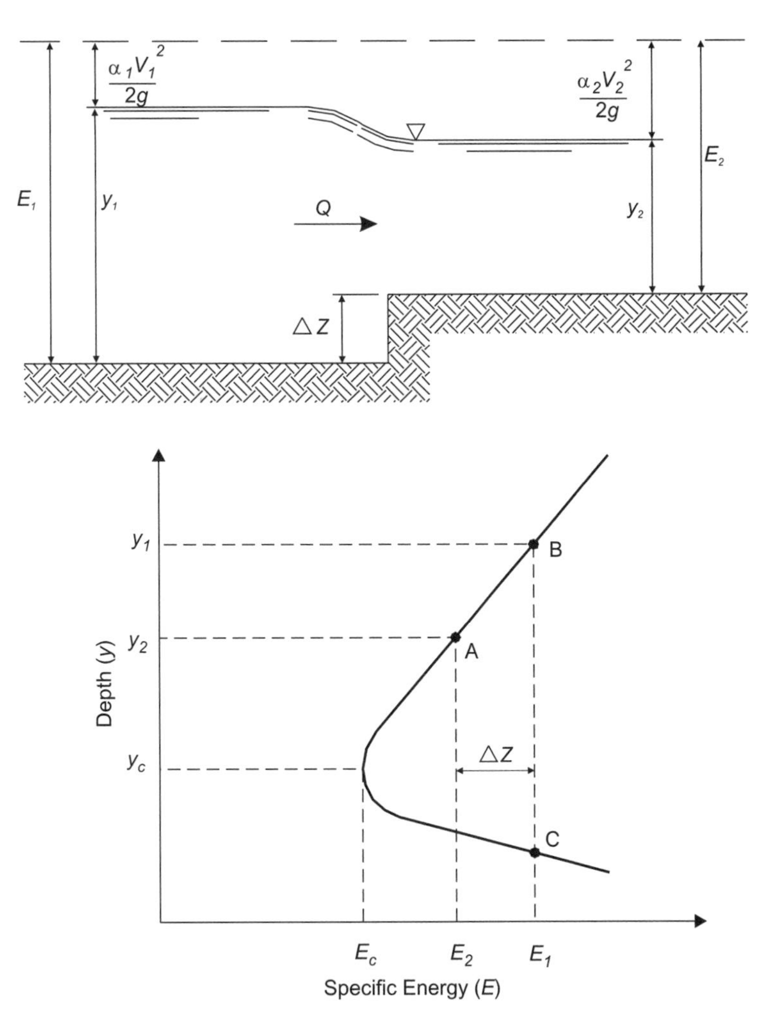

Figure 7.13
A simple step channel transition and the resulting specific energy curve

Determining which of the two alternate depths is the correct upstream depth requires one to consider the concept of *attainability*. Because neither the channel geometry nor the discharge changes in the transition, the single specific energy curve shown in Figure 7.13 applies to both the upstream and downstream portions of the transition. To move from one location on the curve (represented by a particular combination of depth and specific energy) to another location on the curve requires that one traverse along the path of the curve. Point A on the curve represents the known depth and specific energy downstream of the transition, and points B and C represent the alternate depths upstream of the transition. Note that movement along the curve from point A to point B requires an increase in specific energy, which occurs in this situation since E_1 is larger than E_2 by the amount ΔZ. On the other hand, movement along the curve from point A to point C requires that the specific energy first decrease to E_c, and then increase to the amount E_1.

Noting that specific energy may be viewed graphically as the distance from the energy grade line to the channel bottom, it is seen in Figure 7.13 that in moving from the downstream to the upstream cross section, the specific energy only increases. It does not decrease and then increase, and thus point C on the specific energy curve is not attainable from point A in this instance. Thus, the correct upstream depth is the larger of the two alternate depths. Had there been a hump or sill in the channel bottom between the two cross sections such that the specific energy at the hump were equal to E_c, then the smaller of the two alternate depths would have been the correct solution.

This concept of attainability is applicable to any type of flow in a channel transition, whether it is subcritical or supercritical, provided that neither the discharge nor the channel geometry changes in the transition. When the geometry changes, the concept of attainability remains valid, but its analysis becomes more complicated. The complication arises because the specific energy relationships upstream and downstream of the transition are different from one another, and there is generally a continuum of specific energy relationships due to the gradually changing geometry of the transition along its length.

In the preceding discussion, it was assumed that energy losses in a transition were negligible. In truth, the energy losses occurring in a channel transition consist of frictional losses along the length of the transition and minor losses caused by velocity changes. Because channel transitions are often short in length, the frictional losses are usually small compared to the minor losses and can be safely ignored. Minor loss coefficients applicable to channel transitions were presented in Section 7.2 (page 230).

Incorporation of energy losses into an analysis of flow through a transition requires a trial-and-error approach. To illustrate, if the downstream depth and specific energy are known, a first estimate of the upstream depth and specific energy can be made using the procedures discussed previously. With an initial estimate of the upstream depth, the velocities at the upstream and downstream cross sections can be determined, and the minor energy loss can be computed using Equation 7.16. A refined estimate of the upstream specific energy is then determined by recognizing that the minor loss causes the energy grade line elevation to increase by the amount h_L. This refined estimate of the upstream specific energy can be used to obtain a refined estimate of the upstream depth, and the procedure is repeated until convergence is obtained.

Example 7.15 – Bottom Transition in a Rectangular Channel. A sudden drop of 0.50 ft occurs in the channel described in Examples 7.11 and 7.12. Recall that the channel is rectangular with B = 10 ft and Q = 200 cfs. The depth of flow downstream of the drop is y_2 = 4.80 ft. Determine the depth upstream of the drop assuming that energy losses are negligible.

Solution: The specific energy downstream of the drop is

$$E_2 = y_2 + \frac{V_2^2}{2g} = 4.8 + \frac{\{200/[(10)(4.8)]\}^2}{2(32.2)} = 5.07 \text{ ft}$$

Based on the assumption of zero energy loss, the specific energy upstream of the 0.5-ft drop must therefore be

$$E_1 = 5.07 - 0.5 = 4.57 \text{ ft}$$

The alternate depths upstream of the drop are the two positive solutions of the following expression:

$$E_1 = y_1 + \frac{(200/10y_1)^2}{2(32.2)} = 4.57 \text{ ft}$$

The two alternate depths are found, by trial and error, to be y_1 = 4.22 ft and y_1 = 1.40 ft.

From Example 7.12, the critical depth in this channel is 2.32 ft, and the corresponding minimum specific energy is 3.48 ft. Because the specific energy in the channel decreases from 5.07 to 4.57 ft, but never drops to a value as small as 3.48 ft, the larger of the two alternate depths is the only attainable one. Therefore, the upstream depth of flow is y_1 = 4.22 ft.

7.5 MOMENTUM IN OPEN-CHANNEL FLOW

Equation 3.12, the momentum principle, as applied to a coordinate direction x for a steady flow, states that the sum of the external forces acting on the water within a control volume is equal to the net rate of momentum outflow from the control volume. This principle is applied when determining the upstream and downstream depths for a hydraulic jump and when evaluating forces on submerged structures.

Figure 7.14(a) illustrates a reach of channel (a control volume) of length L between an upstream cross section 1 and a downstream cross section 2. The coordinate direction x is taken as positive in the downstream direction and parallel to the channel bottom. External forces acting on the water in the control volume consist of the pressure force P_1 acting at cross section 1, the pressure force P_2 acting at cross section 2, the frictional force F_f opposing the flow, and the gravitational force $W\sin\theta$ driving the flow.

Unlike the case in uniform flow, the pressure forces P_1 and P_2 shown in Figure 7.14 are not necessarily equal. The sum of the forces is equal to the change in momentum across the control volume.

Figure 7.14
Definition figure for forces acting on a fluid element with non-uniform flow

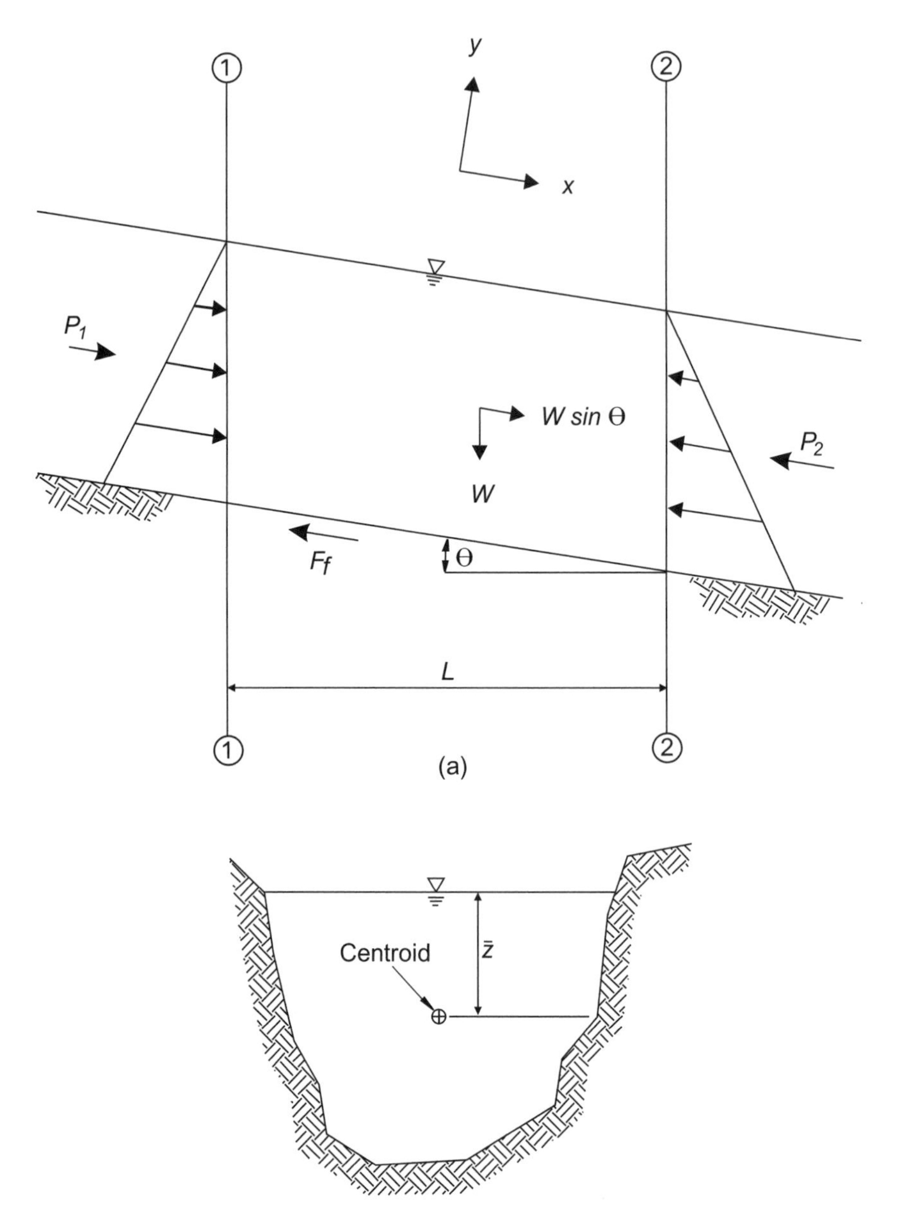

Substituting the sum of the external forces into Equation 3.12 and equating the sum to the net momentum outflow rate, one obtains

$$\sum F_x = P_1 - P_2 + W\sin\theta - F_f = \rho Q(\beta_2 V_2 - \beta_1 V_1) \tag{7.50}$$

where $\sum F_x$ = sum of external forces in the x direction
P_1 and P_2 = forces exerted on cross section due to hydrostatic pressure (see Equation 7.53) (lb, N)
W = weight of fluid element (lb, N)
θ = angle that channel bed makes with horizontal
F_f = frictional force (lb, N)
ρ = fluid density (slugs/ft^3, kg/m^3)
Q = discharge (cfs, m^3/s)
β_1 and β_2 = velocity distribution coefficients as defined by Equation 6.30
V_1 and V_2 = average flow velocities (ft/s, m/s)

Typical values of β for open-channel flow are presented in Table 7.2. In compound channels, where the velocity may vary greatly from one part of the cross section to another, an estimate of β may be obtained as

$$\beta = \frac{\sum_{i=1}^{n} A_i V_i^2}{A V^2} \tag{7.51}$$

where β = velocity distribution coefficient
A_i = cross-sectional area of the i-th subarea (ft^2, m^2)
V_i = average velocity of the i-th subarea (ft/s, m/s)
n = number of subareas
A = total cross-sectional area (ft^2, m^2)
V = average velocity for the entire cross section (ft/s, m/s)

The *pressure force* P exerted at a cross section in an open-channel flow (see Equation 7.50) can be expressed as

$$P = \gamma \bar{z} A \tag{7.52}$$

where γ = fluid specific weight (lb/ft^3, N/m^3)
$\bar{z}$ = distance from the free water surface to centroid of cross-sectional area [see Figure 7.14(b)] (ft, m)
A = cross-sectional area of flow (ft^2, m^2)

Table 7.7 summarizes expressions for $\bar{z}$ for rectangular, trapezoidal, triangular, and circular cross-sectional shapes (for dimension definitions, see Figure 7.4). (Note that both side slopes must be equal for triangular and trapezoidal channels if these equations are used.) For other channel shapes, $\bar{z}$ can be computed as

$$\bar{z} = \frac{\sum_{i=1}^{n} A_i \bar{z}_i}{A} \tag{7.53}$$

where $\bar{z}_i$ = vertical distance from free water surface to centroid of cross section of i-th subarea (ft, m)

If other forces are exerted on the water in the control volume, such as by a step in the channel bottom or by a bridge pier, then those forces should be included in Equation 7.50. This point is illustrated in the "Forces on Submerged Structures" discussion later in this section.

Table 7.7 Centroidal distance $\bar{z}$ for prismatic channels[a]

Channel shape	Area	$\bar{z}$
Rectangular	By	$\frac{y}{2}$
Triangular	zy^2	$\frac{y}{3}$
Trapezoidal	$By + zy^2$	$\frac{y^3(3B - 2zy)}{6(By + zy^2)}$
Circular[a]	$\frac{D^2(\phi - \sin\phi)}{8}$	$\frac{2D\sin\left(\frac{\phi}{2}\right) - D\sin\phi\cos\left(\frac{\phi}{2}\right)}{3(\phi - \sin\phi)} - \frac{D\cos\left(\frac{\phi}{2}\right)}{2}$

a. Angle ϕ is measured in radians. 1 radian ≈ 57.3 degrees.

Specific Force

If Equations 7.50 and 7.52 are applied to a short reach of a horizontal channel, the frictional force F_f will be small and $W\sin\theta$ will be zero. Thus, if β_1 and β_2 are assumed to be 1, the momentum equation simplifies to

$$\gamma\left(\bar{z}_1 A_1 + \frac{Q^2}{gA_1}\right) = \gamma\left(\bar{z}_2 A_2 + \frac{Q^2}{gA_2}\right) \tag{7.54}$$

or

$$F_{s1} = F_{s2} \tag{7.55}$$

where F_s = specific force (lb, N), given by

$$F_s = \gamma\left(\bar{z}A + \frac{Q^2}{gA}\right) \tag{7.56}$$

The quantity F_s expressed by Equation 7.56 is known as the *specific force* in an open-channel flow. It consists of two terms, one representing the hydrostatic pressure force, and the second representing the additional dynamic force caused by the flow momentum.

Specific force, like specific energy, depends on the flow depth, channel geometry, and discharge. For a specified discharge and channel shape, a graph of specific force can be developed as a function of flow depth. This graph is known as a *specific force curve*.

Like a specific energy curve, a specific force curve contains two limbs: the first corresponds to subcritical flow, and the second corresponds to supercritical flow. The specific force curve also has a minimum force given by F_{sc}, which occurs at the critical depth y_c. Note that for any value of the specific force greater than F_{sc}, there are two possible depths of flow. Again, this is exactly analogous to the case of a specific energy curve. However, in the case of a specific force curve, the two depths are called *sequent depths* (some texts refer to them as *conjugate depths*).

Example 7.16 – Specific Force in a Trapezoidal Channel. Develop a specific force curve for a trapezoidal open channel with $B = 8$ ft, $z = 2$, and $Q = 250$ cfs.

Solution: Using expressions for A and $\bar{z}$ from Table 7.7, one can substitute into Equation 7.56 to obtain an expression for F_s as a function of flow depth y:

$$F_s = \gamma\left(\frac{By^2}{2} + \frac{zy^3}{3} + \frac{Q^2}{g(By + zy^2)}\right)$$

A graph of the resulting relationship is shown in Figure E7.16.1, which is the specific force curve. Note that the specific force has a minimum at the critical depth of 2.51 ft.

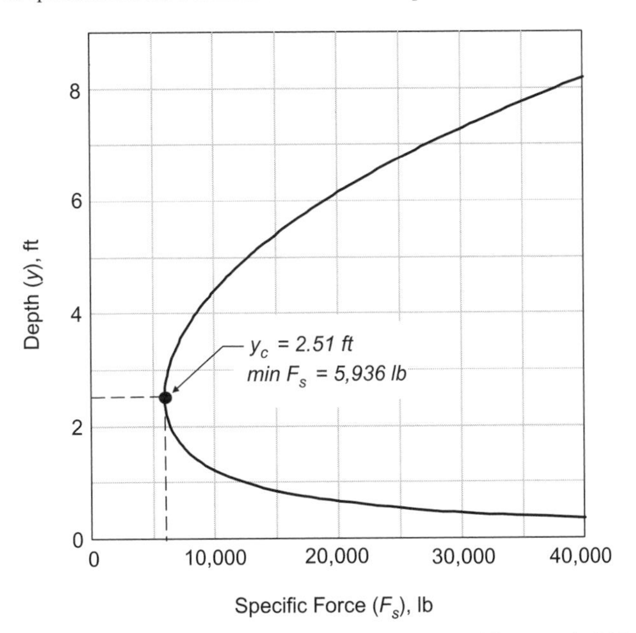

Figure E7.16.1 Graph of specific force versus depth for channel in Example 7.16

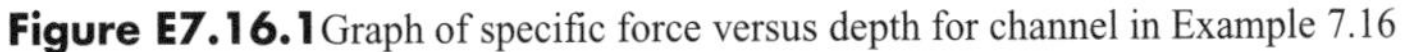

The Hydraulic Jump

The *hydraulic jump* is a phenomenon that occurs when flow in an open channel increases from a small supercritical depth to a larger subcritical depth with a corresponding decrease in energy caused by turbulent eddies. This loss of energy can be considerable. For this reason, hydraulic jumps are often employed to dissipate otherwise harmful energy in a channel.

To illustrate the characteristics of a hydraulic jump, consider a jump in a horizontal trapezoidal channel with dimensions and discharge as specified in Example 7.16 on page 259. Assume that the depth of flow upstream of the jump is y_1 = 1.00 ft (0.3 m), as shown in Figure 7.15, and that the length of the jump is short. Because the channel is horizontal and the jump is short, conditions assumed at the beginning of the preceding subsection (that is, F_f is small and $W\sin\theta$ is 0) are applicable, and one can conclude that $F_{s1} = F_{s2}$. Using the information in Figure E7.16.1 with y_1 = 1.00 ft (0.3 m) yields $Fs_1 = Fs_2$ = 12,403 lb (5,625 kg) and y_2 = 4.93 ft (1.5 m). Thus, for a hydraulic jump in a horizontal channel, the relationship between the depths upstream and downstream of the jump is given by the sequent depths on a specific force curve.

The energy loss in the hydraulic jump can be assessed by computing the specific energy upstream and downstream of the jump. Upstream of the jump where the depth is y_1 = 1.00 ft (0.3 m), the corresponding specific energy (from Equation 7.36) is E_1 = 10.70 ft (3.3 m). Downstream of the jump, where the depth is y_2 = 4.93 ft (1.5 m), the specific energy is E_2 = 5.06 ft (1.54 m). The difference between the two specific energies is the head loss in the jump, which is h_L = 5.64 ft or 1.72 m (see Figure 7.15). Note that the alternate depths do not play a role in the analysis of a hydraulic jump, but the specific energy curve does provide information relevant to the head loss that occurs in the jump.

Figure 7.15
Definition figure for a hydraulic jump

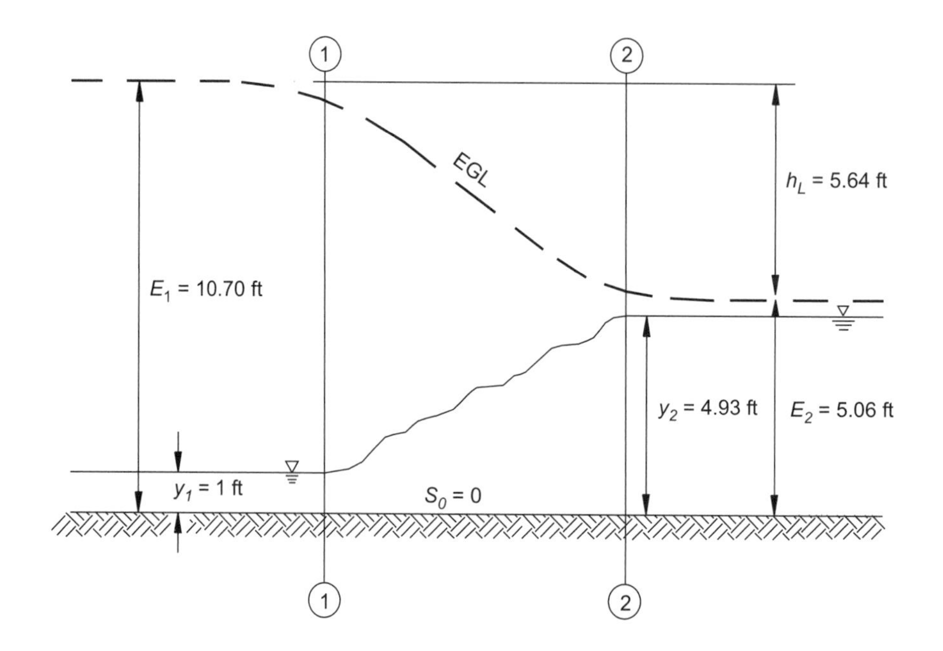

As described in the previous paragraphs, the characteristics of a hydraulic jump in a horizontal channel of any cross-sectional shape can be determined by use of specific force and specific energy curves. However, in the special case of a horizontal, rectangular channel, one can show that the sequent depths are related as

$$\frac{y_2}{y_1} = \frac{1}{2}\left(\sqrt{1+8\text{Fr}_1^2} - 1\right) \tag{7.57}$$

where y_1, y_2 = flow depths at sections 1 and 2, respectively (ft, m)
Fr = Froude number

and

$$\text{Fr}_1^2 = \frac{V_1^2}{gy_1} \tag{7.58}$$

where g = gravitational acceleration constant (32.2 ft/s^2, 9.81 m/s^2)

In such a channel, one can also show that the energy loss is

$$h_L = E_1 - E_2 = \frac{(y_2 - y_1)^3}{4y_1y_2} \tag{7.59}$$

where h_L = head loss in hydraulic jump (ft, m)
E_1, E_2 = total energy (ft, m)

If a hydraulic jump occurs in a sloping channel such that the term $W\sin\theta$ in the momentum equation cannot be neglected, the specific force approach to estimating the relationship between the depths upstream and downstream of the jump cannot be applied. Instead, one must rely on an experimental relationship, such as that illustrated in Figure 7.16. The figure illustrates the ratio of the downstream depth to the upstream depth (y_2/y_1 measured vertically or d_2/d_1 measured normal to the channel bottom) as a function of the channel slope (S_o) and the Froude number of the upstream flow (Fr_1). For this figure, a rectangular cross-section shape is assumed.

The length of a hydraulic jump is a characteristic that must be evaluated on the basis of experimental data. Figure 7.17, which is based on data collected by the U.S. Bureau of Reclamation (1955), illustrates the length L of a jump as a function of the downstream depth, the channel slope, and the Froude number of the upstream flow. This figure was developed for rectangular channels; however, in the absence of other data, it can be used to approximate jump length in nonrectangular channels.

Figure 7.16
Ratio of downstream to upstream depths for a hydraulic jump in a rectangular channel as a function of the channel slope and upstream Froude number

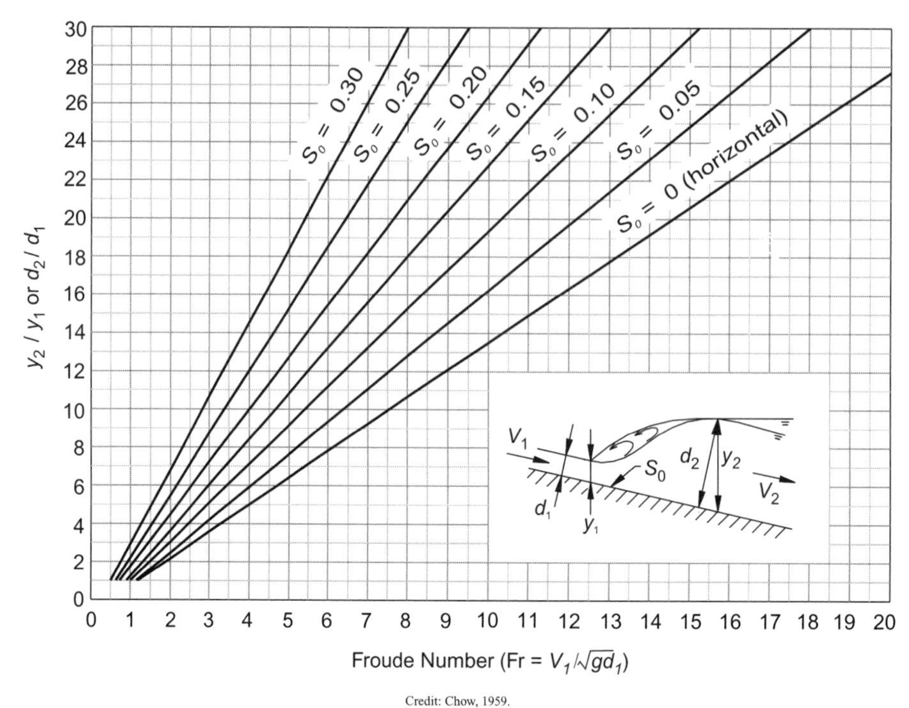

Credit: Chow, 1959.

Figure 7.17
Graph for determining the length of a hydraulic jump

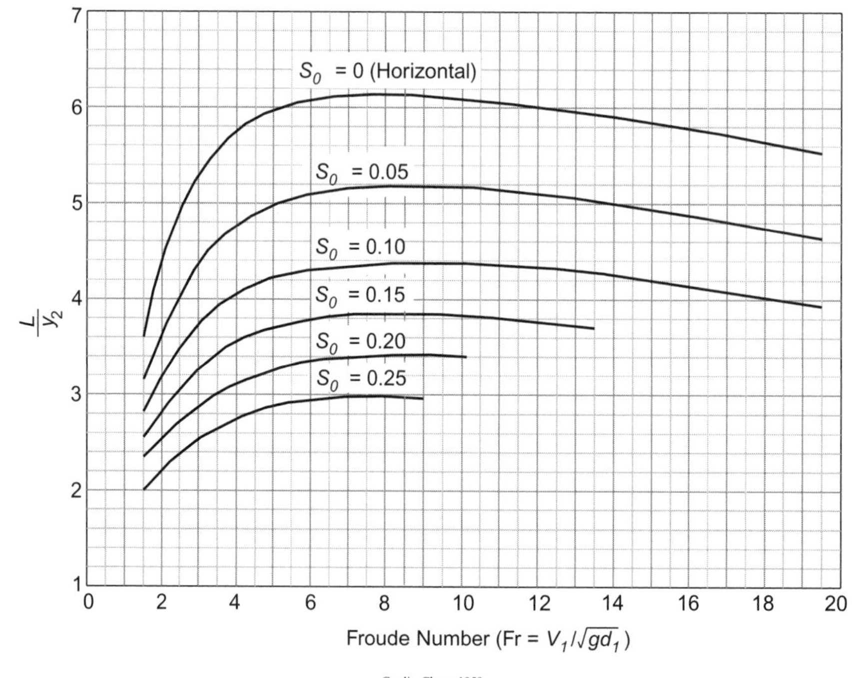

Credit: Chow, 1959.

Example 7.17 – Hydraulic Jump in a Rectangular Channel. A hydraulic jump forms in a horizontal rectangular open channel with $B = 3$ m and $Q = 6\ \text{m}^3/\text{s}$. The depth of the approaching flow is $y_1 = 0.40$ m. Determine the depth downstream of the jump, the energy loss in the jump, and the length of the jump.

Solution: Because the channel is horizontal and rectangular, Equation 7.57 may be used to determine the downstream depth. The Froude number of the approaching flow is

$$\text{Fr} = \frac{V_1}{\sqrt{gy_1}} = \frac{6(3 \times 0.04)}{\sqrt{9.81 \times 0.4}} = 2.52$$

Therefore, applying Equation 7.57, the downstream depth is

$$y_2 = \frac{y_1}{2}(\sqrt{1 + 8\text{Fr}^2} - 1) = 1.24 \text{ m}$$

The energy loss in the jump may be determined using Equation 7.59, and is

$$h_L = \frac{(y_2 - y_1)^3}{4y_1y_2} = \frac{(1.24 - 0.40)^3}{4 \times 1.24 \times 0.40} = 0.30 \text{ m}$$

The length of the jump can be estimated using Figure 7.17. From that figure, the dimensionless ratio L/y_2 is estimated to be about 4.8. Therefore,

$$L = 4.8y_2 = 4.8(1.24) = 5.95 \text{ m}$$

Forces on Submerged Structures

When an open-channel flow occurs past a submerged structure, such as a bridge pier, the flow exerts a hydraulic force on that structure. The structure in turn exerts an equal and opposite force on the flow. The force exerted on the flow by the structure must be considered as an external force on the flow when applying the momentum equation.

Consider flow in an open channel of small slope and short length where there is a submerged structure (such as a bridge pier) in the flow path. In this case, the weight of the water and frictional forces may be neglected, and the momentum equation becomes

$$\gamma \bar{z}_1 A_1 - \gamma \bar{z}_2 A_2 - F_p = \rho Q(\beta_2 V_2 - \beta_1 V_1) \tag{7.60}$$

where F_p = force exerted on the flow by the pier (lb, N)

Note that it is assumed in this expression that the force exerted by the pier opposes the flow direction (that is, it is in the negative x-direction).

A comparison of Equation 7.60 to Equations 7.54 and 7.55 indicates that

$$F_{s1} = F_{s2} + F_p \tag{7.61}$$

Thus, the pier force F_p can be determined graphically from a specific force curve if the depths of flow y_1 and y_2 upstream and downstream of the pier are known. The depth change due to the pier may be found through an application of the energy equation based on an estimate of the energy loss due to the pier.

Example 7.18 – Force in a Channel Transition. Figure E7.18.1 illustrates a channel transition consisting of a step of height ΔZ. Determine the force exerted on the step by the water if the channel is rectangular with $B = 6$ m, $y_1 = 2$ m, $Q = 12$ m^3/s, and $\Delta Z = 0.25$ m. Assume that the energy loss in the transition is negligible.

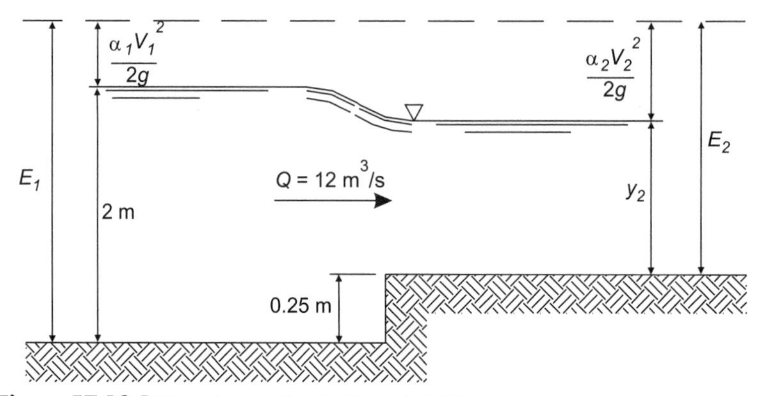

Figure E7.18.1 Channel transistion for Example 7.18

Solution: First, the downstream depth y_2 must be determined. The specific energy upstream of the step is

$$E_1 = y_1 + \frac{V_1^2}{2g} = 2 + \frac{\left[12/(2)(6)\right]^2}{2(9.81)} = 2.05 \text{ ft}$$

and the specific energy downstream of the step is

$$E_2 = E_1 - \Delta Z = 2.05 - 0.25 = 1.80 \text{ m}$$

The equation for E_2 can then be written as

$$1.8 = y_2 + \frac{\left[12/(6y_2)\right]^2}{2(9.81)}$$

By trial and error, the downstream depth is $y_2 = 1.74$ m.

Assuming that the step exerts a force on the flow in the upstream direction (the negative x-direction), that force may be determined from Equation 7.61 as

$$F_p = F_{s1} - F_{s2}$$

or, substituting from Equation 7.56,

$$F_p = \gamma\left(\bar{z}_1 A_1 + \frac{Q^2}{gA_1}\right) - \gamma\left(\bar{z}_2 A_2 + \frac{Q^2}{gA_2}\right)$$

$$F_p = 9{,}810\left\{\left[1(12) + \frac{12^2}{9.81(12)}\right] - \left[\frac{1.74}{2}(1.74)(6) + \frac{12^2}{9.81(1.74)(6)}\right]\right\} = 26{,}800 \text{ N}$$

Because the sign of this result is positive, the assumed direction of the force exerted by the step is correct. Therefore, the force exerted on the step by the water is in the downstream direction and has a magnitude of 26,800 N.

7.6 WEIRS

A *weir* in an open channel is a structure that may be employed for flow measurement or as a flow control structure. Weirs may be used as outlet structures for water storage facilities such as detention ponds or reservoirs. Weir flows over roadway embankments may also occur during high flow events at culvert crossings. Finally, some hydraulic structures, notably stormwater inlets, behave as weirs during some types of flow conditions.

Weir Characteristics

Weirs are classified as being either *sharp-crested* or *broad-crested*. A sharp-crested weir is one in which a relatively thin metal, concrete, or plastic plate forms the crest of the weir. For a broad-crested weir, the weir crest thickness is relatively large, and flow over the weir crest is critical.

As illustrated in Figure 7.18, additional classifications for weirs concern the attributes of the crest cross-sectional *shape* (such as rectangular, triangular, or trapezoidal), and whether there are contractions of the flow at the ends of the weir. A *contracted* weir [Figure 7.18(b)] is a weir type that causes the flow to contract. Flow over a *suppressed* or uncontracted weir [Figure 7.18(a)], however, has no end contractions.

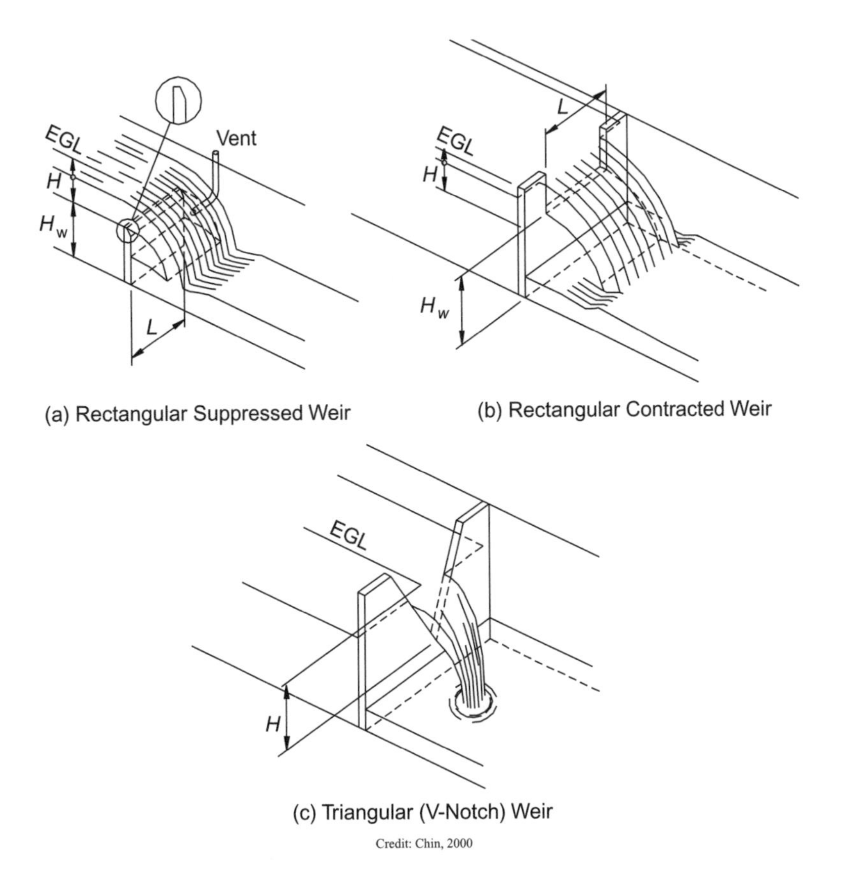

Figure 7.18 Definition figures for rectangular and triangular weirs

The jet of water that issues in a downstream direction from the crest of a sharp-crested weir is called the *nappe*. Ideally, the nappe should spring from the crest and should have atmospheric pressures prevailing at both its upper and lower surfaces. If the lower face of the nappe is at a pressure below atmospheric, the nappe may cling to the downstream face of the weir and change its hydraulic performance. If the possibility of a clinging nappe presents itself in a particular application (usually it is only a problem with suppressed weirs), vents should be installed so as to enable atmospheric pressure to exist on the underside of the nappe [Figure 7.18(a)].

Sharp-Crested Weirs

Theoretically, the discharge over a sharp-crested weir may be determined via the energy equation. The procedure amounts to equating the total energy head upstream of the weir to the total energy head on the weir crest. The resulting equation for discharge in terms of the upstream energy level depends on the cross-sectional shape of the weir, and inevitably also involves some simplifying assumptions. An empirical weir *discharge coefficient*, C_d, is introduced to account for the simplifying assumptions and to correct the theoretical discharge. (Note that a "weir coefficient" is sometimes used instead of the discharge coefficient C_d described here. The weir coefficient is equivalent to $C_d(2g)^{1/2}$. Whereas C_d is unitless, the weir coefficient value depends on the units being used.)

Two types of sharp-crested weirs, rectangular and triangular, are discussed in this section.

Rectangular Weirs. The discharge Q over a suppressed rectangular sharp-crested weir [Figures 7.18(a) and 7.19] may be expressed in terms of the head H on the weir as

$$Q = \frac{2}{3} C_d \sqrt{2g} L H^{3/2} \tag{7.62}$$

where Q = discharge (cfs, m^3/s)
C_d = discharge coefficient determined using Equation 7.63
g = gravitational acceleration constant (32.2 ft/s^2, 9.81 m/s^2)
L = length of the weir crest (ft, m)
H = height of the energy grade above weir crest (ft, m)

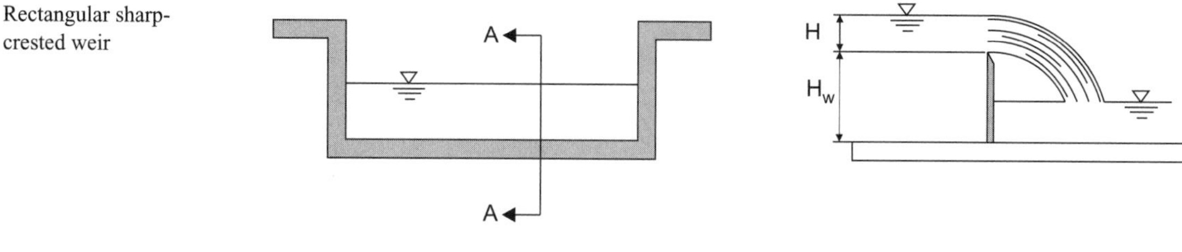

Figure 7.19
Rectangular sharp-crested weir

If the velocity upstream of the weir is low, as in the case of a pond spillway, the velocity head is often considered negligible, and H is taken as the difference in elevation between the upstream water surface and the weir crest. H should be measured far enough upstream of the weir that drawdown effects are avoided.

Fundamentally, the discharge coefficient in Equation 7.62 depends on the Reynolds and Weber numbers of the flow, and on the dimensionless head H/H_w, where H_w is the height of the weir crest above the bottom of the channel (Figure 7.18). However, for practical purposes, experiments have shown that the discharge coefficient can be represented as (Rouse, 1946; Blevins, 1984)

$$C_d = 0.611 + 0.075\frac{H}{H_w} \qquad (7.63)$$

where H_w = height of weir crest above bottom of channel (ft, m)

Equation 7.63 is valid for H/H_w less than 5 and is approximate up to $H/H_w = 10$. For H/H_w greater than 15, the discharge can be computed from the critical flow equation (see Section 7.4) by assuming that $y_c = H$ (Chaudhry, 1993).

When flow occurs over a contracted rectangular weir [see Figure 7.18(b)], the end contractions cause the effective crest length of the weir to be reduced from the actual crest length. In this instance, the head-discharge relationship for the weir may be expressed as

$$Q = \frac{2}{3}C_d\sqrt{2g}(L - 0.1NH)H^{3/2} \qquad (7.64)$$

where N = number of weir end contractions ($N = 1$ or 2; usually, $N = 2$)

This equation assumes that L is greater than $3H$. Note that the effective length of the weir is $L - 0.1NH$, which is $0.1NH$ shorter than the actual length of the weir crest.

A *Cipolletti* weir is a contracted weir that, rather than a rectangular cross-section, has a trapezoidal cross section with 1:4 (horizontal:vertical) side slopes to compensate for the flow contraction. Denoting the base width of the trapezoid (that is, the weir crest length) as L, Equations 7.62 and 7.63 can be used to determine the head-discharge relationship. (Note that Equation 7.64 is not used in this case because the sloped sides of the trapezoid offset the effective crest length reduction that otherwise occurs with a contracted weir.)

Example 7.19 – Discharge over a Contracted Rectangular Weir. A rectangular weir with contractions on both sides ($N = 2$) has a crest length $L = 3$ m, $H_w = 2.5$ m, and $H = 0.30$ m. Determine the discharge over the weir.

Solution: Using Equation 7.63, the discharge coefficient is

$$C_d = 0.611 + 0.075\frac{0.30}{2.5} = 0.62$$

Applying Equation 7.64, the discharge is

$$Q = \frac{2}{3}(0.62)\sqrt{2(9.81)}\,\{3 - 0.1(2)(0.30)\}\,(0.30)^{3/2} = 0.88\ \text{m}^3/\text{s}$$

Triangular Weirs. A triangular, or v-notch, weir has a v-shaped cross section [see Figures 7.18(c) and 7.20]. This type of weir can control flow more accurately than a rectangular weir when the discharge and head on the weir are small. Because the discharge over a triangular weir for any given weir head H is much smaller than that for a rectangular weir with the same head, triangular weirs are usually used only in instances where the discharge is small.

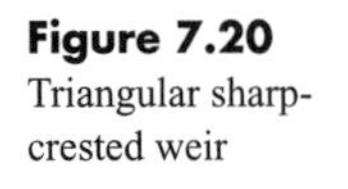

Figure 7.20
Triangular sharp-crested weir

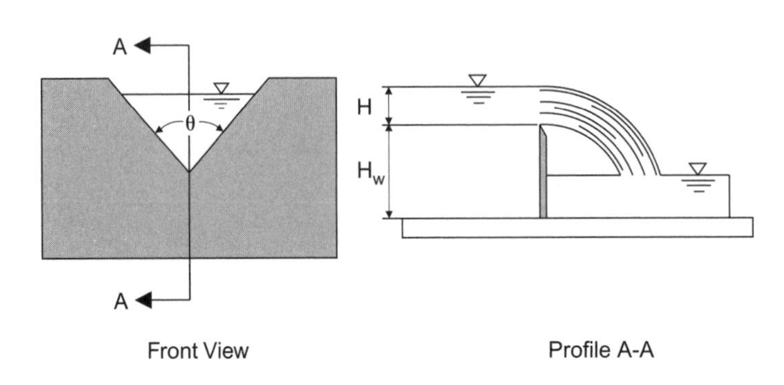

The discharge over a sharp-crested triangular weir can be shown to be

$$Q = \frac{8}{15} C_d \sqrt{2g} \tan\left(\frac{\theta}{2}\right) H^{5/2} \tag{7.65}$$

where θ = angle formed by vertex of v-shaped weir cross section (radians)

Values of the discharge coefficient for use in this expression are shown in Figure 7.21. The solid curves on the graph should be used for nearly all design purposes, especially when the head on the weir is 2 in. (5 cm) or more. For smaller heads, both viscous and surface-tension effects are important determinants of weir behavior, and the dashed lines reflect this combination of effects. Although the dashed lines extend above $H = 0.2$ ft in the graph, it is recommended that the solid lines be used for larger values of H. As a practical matter, the difference in using the dashed or solid curves will be small, as changes in C_d are typically less than 0.01.

Broad-Crested Weirs

Broad-crested weirs, as the name implies, have weir crests that are broad as opposed to sharp-edged. For example, a roadway embankment containing a culvert behaves as a broad-crested weir if the water is high enough to flow over the roadway surface.

Theoretically, the flow profile over a broad-crested weir can be evaluated through the concepts of specific energy and critical depth. Figure 7.22(a) illustrates a broad-crested weir of height $H_w = \Delta Z$ with an approaching flow of discharge Q and depth y_1. The thickness of the weir crest in the direction of flow is Δx, and the cross section is rectangular with a weir crest length of L normal to the plane of the page.

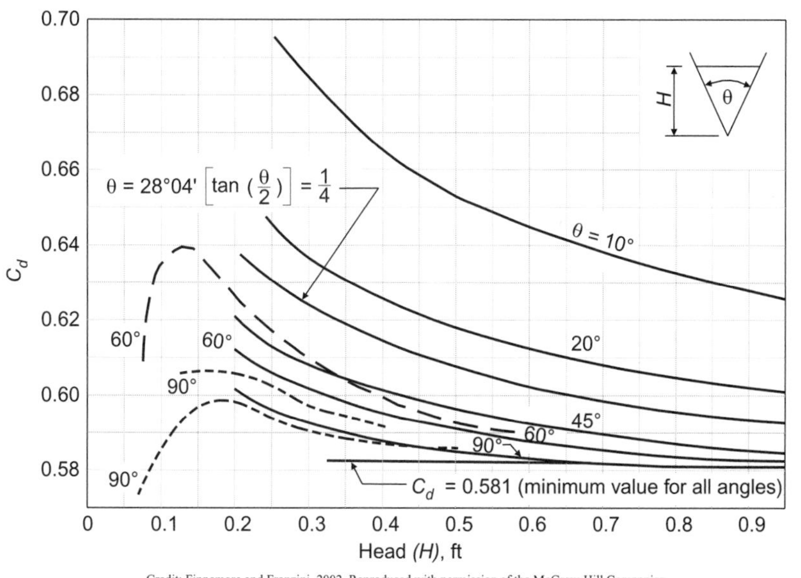

Figure 7.21 Discharge coefficient values for triangular weirs (Franzini and Finnemore, 1997)

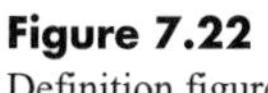

Figure 7.22 Definition figure and specific energy curve for a broad-crested weir

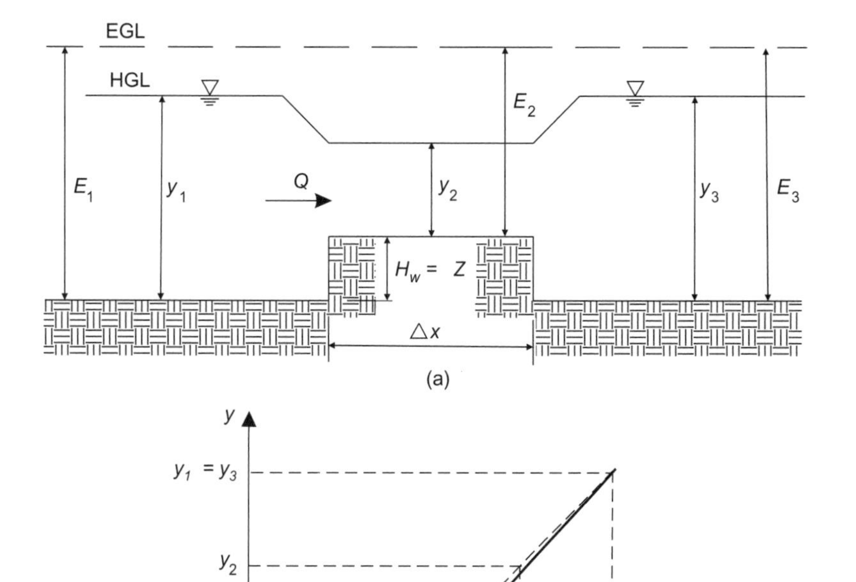

The specific energy upstream of the weir is E_1. Assuming no energy loss, the specific energy on the crest of the weir is $(E_1 - \Delta Z) = E_2$, from which the depth y_2 may be determined using the specific energy curve [Figure 7.22(b)]. Thus, if the weir crest height $H_w = \Delta Z$ is not excessively large and the assumption of zero energy loss is made, then the flow profile over the weir may be evaluated using a specific energy curve. Further, the specific energy E_3 and depth y_3 downstream of the weir will be equal to E_1 and y_1, respectively, if the channel bottom elevation is the same upstream and downstream of the weir. If desired, a minor loss can be introduced to account for the energy loss that actually takes place.

If, assuming fixed values for E_1, y_1, and $Hw = \Delta Z$, E_2 is found to be less than E_c, the assumptions described in the preceding paragraph are not valid. This situation is impossible because E_c represents the lowest energy on the specific energy curve [see Figure 7.22(b)]. What actually occurs is that the flow is *choked*, meaning that a backwater effect is created by the weir. In such a case, one must set $y_2 = y_c$ and $E_2 = E_c$. The specific energy upstream of the weir is then found as $E_1 = E_c + H_w$, and the depth y_1 is determined from the specific energy curve.

When a flow is choked (that is, when a structure behaves as a broad-crested weir), the discharge may be computed as a function of the head $H = E_1 - H_w$ as

$$Q = C_d \sqrt{g} L \left(\frac{2H}{3} \right)^{3/2} \tag{7.66}$$

where values of the discharge coefficient can be estimated from

$$C_d = \frac{0.65}{\left(1 + H / H_w\right)^{1/2}} \tag{7.67}$$

These expressions are valid provided that $0.08 < (y_1 - H_w)/\Delta x < 0.50$. Head losses across the crest of the weir cannot be neglected if $(y_1 - H_w)/\Delta x < 0.08$. If $(y_1 - H_w)/\Delta x > 0.50$, streamlines over the weir crest are not horizontal. In these cases, the flow profile should be evaluated using gradually varied flow profile analysis methods (see Section 7.8).

7.7 ORIFICES

Like a weir, an orifice functions as a hydraulic control. For example, in stormwater systems, an orifice may be part of a detention pond outlet structure configuration in the form of a restrictor plate or an opening in a riser pipe or inlet box (see Chapter 12). Culvert entrances and stormwater inlets may also function hydraulically as orifices.

The discharge through a single orifice opening depends on the area of the opening, the effective head on the orifice, and the type of edge around the opening. The relationship between these variables, which are shown in Figure 7.23, is expressed as:

$$Q = C_d A_0 (2gh_0)^{1/2} \tag{7.68}$$

where Q = discharge (cfs, m^3/s)
C_d = discharge coefficient depending on the orifice edges
A_0 = area of the orifice opening (ft^2, m^2)
g = acceleration of gravity (ft/s^2, m/s^2)
h_0 = effective head on the orifice (ft, m)

Brown et al. (1996) indicate that C_d can be taken as 0.60 if the edges of the orifice opening are uniform and square-edged. For ragged edges, such as those resulting from use of an acetylene torch to cut an opening, a value of 0.40 should be used.

The effective head on an orifice depends on whether it has a free or submerged discharge. If the discharge is a free outfall, the effective head is the difference between the upstream water surface elevation and the elevation at the centroid of the orifice opening. If tailwater submerges the outlet of an orifice, the effective head is equal to the difference between the water surface elevations immediately upstream and downstream of the orifice.

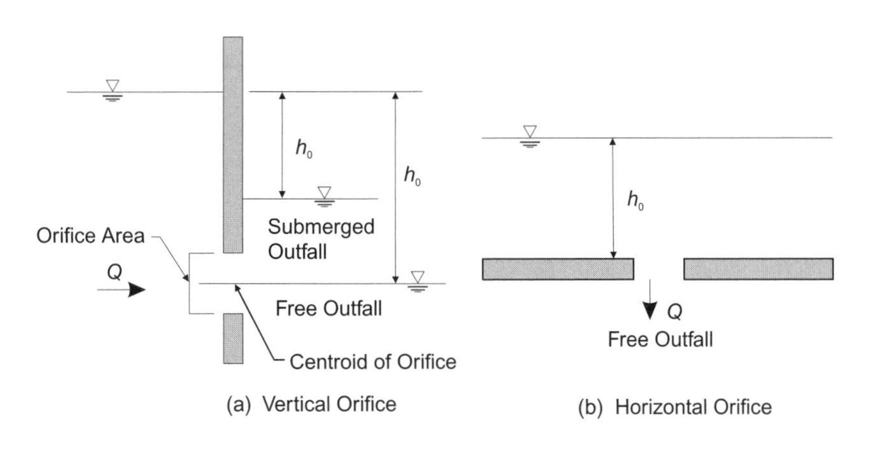

Figure 7.23
Orifice flow diagram

7.8 GRADUALLY VARIED FLOW

For *gradually varied flow* in an open channel, the depth and velocity of a steady flow change along the length of the channel. As the name implies, the rates of change of depth and velocity are gradual and there are no abrupt changes such as hydraulic jumps. Because the curvatures of streamlines are small, hydrostatic pressures can be assumed to exist in the flow (that is, the pressure head can be taken as the depth of flow), and the constitutive relationships developed for uniform flow, such as the Manning equation, can be assumed to be valid for analyses of gradually varied flow.

Gradually varied flow occurs in an open channel when a flow control causes the depth of flow at a cross section to be different from normal depth in the channel. This situation can occur, for instance, upstream of culverts, at locations where the slope of a channel changes, and at channel entrances. Depending on the depth at the control section relative to the normal depth, the flow may be accelerated or decelerated in the channel.

The evaluation of gradually varied flow requires computations to determine flow profile type and flow depths at multiple locations along the length of the channel. This section presents an example in which a gradually varied flow profile is evaluated. However, because of the number of calculations involved, engineers typically use spreadsheets or other computer programs (such as StormCAD or HEC-RAS) when performing this type of analysis.

Water Surface Profile Equation

For a channel of small slope, the total energy head H at a cross section consists of the sum of the channel bottom elevation Z, the depth of flow y, and the velocity head $\alpha V^2/2g$. Taking the x-direction to be horizontal and positive in the downstream direction, the derivative dH/dx is

$$\frac{dH}{dx} = \frac{dZ}{dx} + \frac{dy}{dx} + \frac{d}{dx}\left(\frac{\alpha Q^2}{2gA^2}\right) \tag{7.69}$$

where H = total energy head at a cross section (ft, m)
x = distance along channel (ft, m)
Z = channel bottom elevation (ft, m)
y = vertical flow depth (ft, m)
α = velocity distribution coefficient
Q = discharge (cfs, m^3/s)
g = gravitational constant (32.2 ft^2/s, 9.81 m^2/s)
A = area of cross section (ft^2, m^2)

Noting that the cross-sectional area A depends on the depth of flow y, the chain rule can be applied to rewrite this expression as

$$\frac{dH}{dx} = \frac{dZ}{dx} + \frac{dy}{dx}\left(1 - \alpha\frac{Q^2T}{gA^3}\right) = \frac{dZ}{dx} + \frac{dy}{dx}\left(1 - \alpha \mathrm{Fr}^2\right) \tag{7.70}$$

where Fr = Froude number

Because $dH/dx = -S_f$ is the slope of the energy grade line, and because $dZ/dx = -S_o$ is the channel bottom slope, this equation can be further simplified to read

$$\frac{dy}{dx} = \frac{S_o - S_f}{1 - \alpha \mathrm{Fr}^2} \tag{7.71}$$

where S_o = channel bottom slope (ft/ft, m/m)
S_f = energy gradient (ft/ft, m/m)

This equation is the *governing equation of gradually varied flow* in an open channel. It describes how the depth changes with distance along the channel as a function of channel and flow characteristics. A unique solution to this differential equation that describes the flow profile for a particular case of interest requires knowledge of a *boundary condition* (that is, a known depth of flow at a particular channel cross section).

Classification of Flow Profiles

For a specified discharge Q in a prismatic open channel of known dimensions, one can apply the procedures described in Section 7.3 and Section 7.4 to determine the normal and critical depths y_n and y_c. Knowledge of these two depths, as well as the depth at a flow control section, enables one not only to classify the channel and flow type, but also to determine the direction in which numerical calculations of the flow profile must proceed.

When one compares the normal and critical depths for a particular channel geometry and discharge, several possibilities arise. If one finds that y_n is greater than y_c, then the channel is said to have a *hydraulically mild* (M) slope. If y_n is equal to y_c, the channel has a *critical* (C) slope. If y_n is less than y_c, the channel is said to have a *hydraulically steep* (S) slope. If the channel has a slope of zero, then normal depth does not exist, and the channel is said to be *horizontal* (H). Finally, if the channel bottom elevation increases as one moves in a downstream direction, the channel is said to be *adverse* (A), and the normal depth is imaginary (that is, it involves the square root of a negative quantity).

Generally, the range of possible flow depths in an open channel can be subdivided into three subranges, called zones. *Zone 1* corresponds to all flow depths that are larger than both the normal and critical depths. *Zone 2* corresponds to flow depths between normal and critical depths, and *Zone 3* corresponds to flow depths smaller than both normal and critical depths. For example, in a hydraulically mild (M) channel, where y_n is greater than y_c, Zone 1 corresponds to $y > y_n$, Zone 2 corresponds to $y_c < y < y_n$, and Zone 3 corresponds to $y < y_c$. In a critical (C) channel, where y_n equals y_c, Zone 2 does not exist. Similarly, in horizontal (H) and adverse (A) channels, where normal depth is either non-existent or imaginary, Zone 1 does not exist.

By comparing a known depth of flow at a cross section with the normal and critical depths for that discharge, one can conclude the flow profile type. For example, if the known depth at a particular cross section in a mild (M) channel lies in Zone 1, the flow profile is of the M1 type. If the known depth of flow at a cross section in a steep (S) channel lies in Zone 3, the flow profile is of the S3 type. Figure 7.24 illustrates the various flow profile types that may exist in prismatic channels and indicates the shapes of the water surface profiles in terms of their curvatures.

Classification of the flow profile type in a channel for a particular discharge is an important first step that must take place before the actual computation of the water surface elevations along the length of the channel. The classification aids in the understanding of what the profile should look like, and hence provides a check on subsequent calculations. Understanding the flow profile type also helps in terms of classifying the flow as being subcritical or supercritical. Regardless of the channel slope classification (M, C, S, H, or A), a flow is subcritical when y is greater than y_c and supercritical when y is less than y_c.

Knowledge of whether a flow profile is subcritical or supercritical has implications for profile computations. Computation of a subcritical profile should always proceed in an upstream direction from a known depth at a downstream control section. Conversely, computation of a supercritical profile should always proceed in a downstream direction from a known depth at an upstream control section. This distinction arises because of fundamental differences in the ranges of influences in open-channel flow, particularly with reference to the *celerities* (velocities) of small gravity waves caused by flow disturbances.

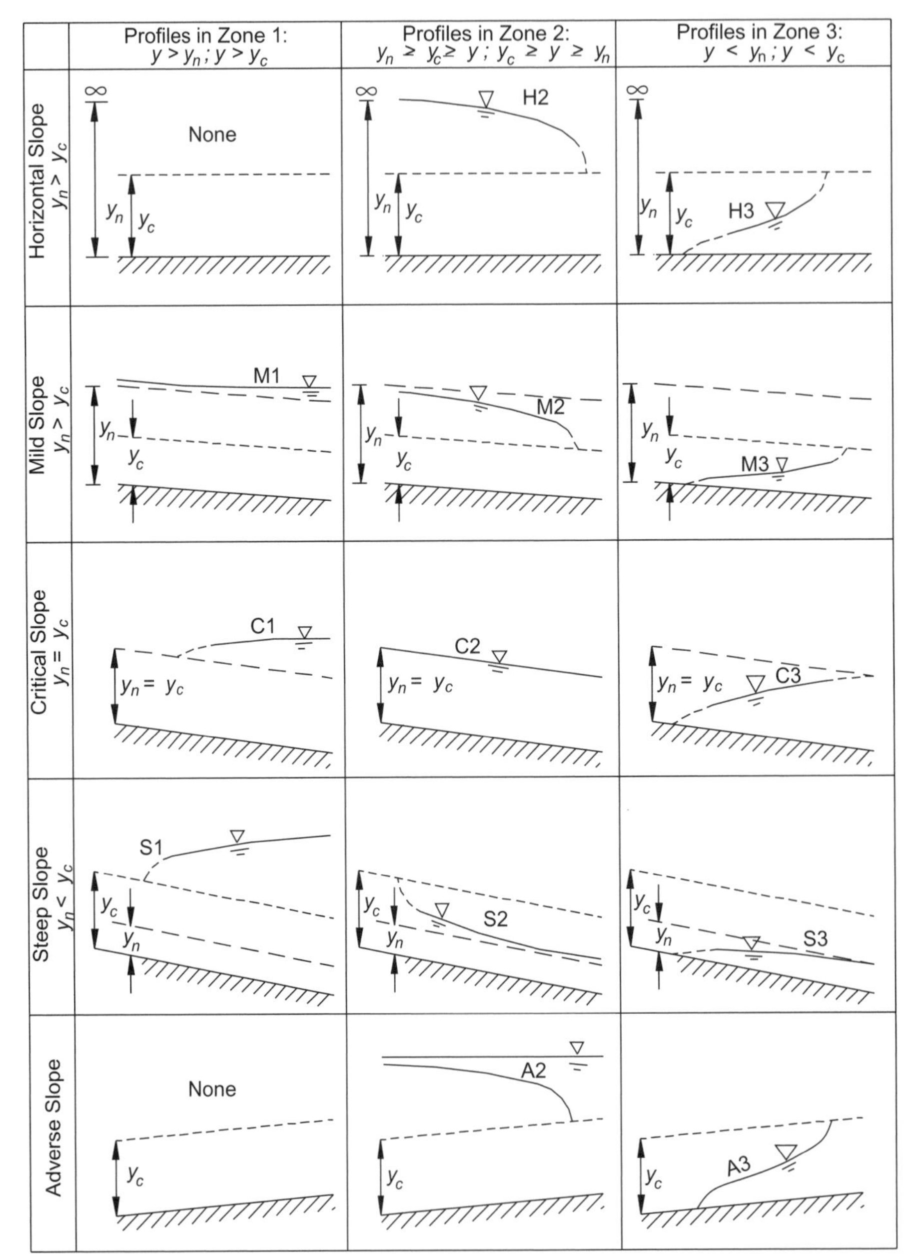

Credit: Chow, 1959.

Figure 7.24
Flow profile types for prismatic channels

Profile Calculations

Equation 7.71 is the fundamental equation describing a gradually varied flow profile. Given a known flow depth at a control section (that is, a boundary condition), a solution of that equation to define depths at additional locations (cross sections) of interest can be accomplished in a number of ways. Chow (1959) provides a discussion of many of the historically used methods, as well as of methods that continue to be used today. More modern methods involving numerical integration of the governing equation can also be developed.

There are two basic classifications of methods for solving Equation 7.71. In the first, one solves for values of x—the cross section locations—corresponding to assumed flow depths y. This type of method is generally useful only for prismatic channels because the cross-sectional geometry is independent of location. The *direct-step method* is an example of this type of procedure.

The second solution method classification is useful for non-prismatic channels. In this instance, methods for solution of Equation 7.71 are designed to determine the flow depth y at each specified channel location x. The most widely used of these approaches is called the *standard-step method*. Further discussion of this method can be found in texts on the subject of open-channel flow such as those by Chow (1959) and Chaudhry (1993).

Because the focus of this text is prismatic channels, the remainder of the discussion is limited to the direct-step method. To develop the direct-step method, the form of the energy equation given in Equation 7.5 is written for a reach of channel as

$$Z_1 + E_1 = Z_2 + E_2 + h_L \tag{7.72}$$

whereZ_1 and Z_2=channel bed elevation at sections 1 and 2, respectively (ft, m)
E_1 and E_2 = total energy at sections 1 and 2, respectively (ft, m)
h_L = head loss due to friction (ft, m)

If the length of the channel reach between the two cross-sections is denoted by Δx, then

$$h_L = S_f \Delta x \tag{7.73}$$

where S_f = friction slope
Δx = distance between cross sections 1 and 2

and

$$Z_1 = Z_2 + S_o \Delta x \tag{7.74}$$

where S_o = channel bed slope

Making these substitutions, Equation 7.72 can be rewritten in terms of Δx as

$$\Delta x = \frac{E_2 - E_1}{S_o - S_f} \tag{7.75}$$

This last expression, while not resembling the differential Equation 7.70, provides the basis for the flow profile computation. For a given discharge Q and channel cross-sectional geometry, and for known or assumed flow depths y_1 and y_2, one can determine the specific energies E_1 and E_2. Using a constitutive relationship such as the Manning equation, one can also determine the friction slopes S_{f1} and S_{f2} at each of the two cross sections. If the average of those two slopes is used in Equation 7.74, one can finally find the distance Δx between the two cross sections. The procedure is best explained through an example.

Example 7.20 – Gradually Varied Flow Analysis of a Channel Slope Transition. A slope change occurs in a trapezoidal open channel with $B = 5$ m and $z = 2$. Upstream of the break point, the channel slope is $S_o = 0.001$, and downstream of the break point the slope is $S_o = 0.020$. Figure E7.20.1 illustrates the channel profile. If $n = 0.025$ and $Q = 15$ m^3/s, compute the flow profile in the vicinity of the slope change. Assume that the channel lengths upstream and downstream of the slope change are long.

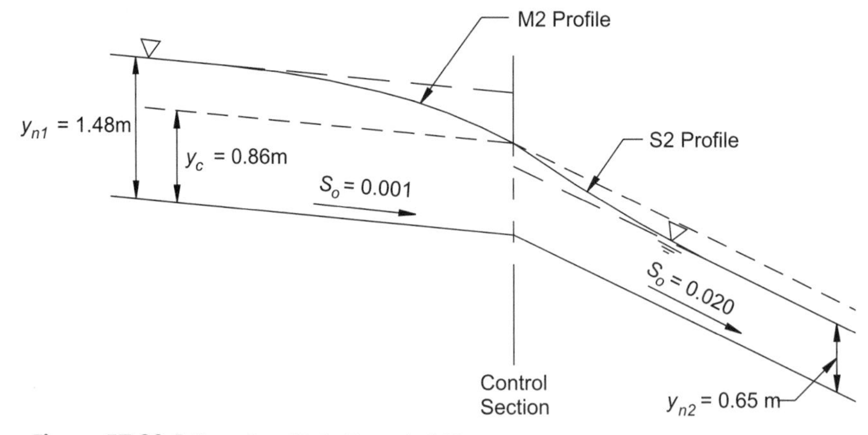

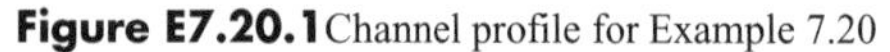
Figure E7.20.1 Channel profile for Example 7.20

Solution: The critical depth in the channel is found, by trial and error, to be $y_c = 0.86$ m. Upstream of the slope break point, the normal depth is $y_{n1} = 1.48$ m, and downstream of the slope break point the normal depth is $y_{n2} = 0.65$ m. Thus, upstream of the break point the channel is mild (M), and downstream of the break point it is steep (S).

In this situation, the control section occurs at the break point and the depth at the control section is the critical depth of 0.86 m. Moving in an upstream direction from the break point, the flow profile depth should converge toward the normal depth, and the flow profile is of the M2 type. Downstream of the break point, the flow profile is of the S2 type. Because an M2 profile is subcritical and an S2 profile is supercritical, the entire flow profile must be computed in two parts. The M2 profile is computed in an upstream direction from the control section, and the S2 profile is computed in a downstream direction from the control section.

Calculations for the M2 profile are illustrated in Table 7.8. The first column contains assumed flow depths, beginning with the known depth at the flow control section and proceeding in an upstream direction. [Note that as one moves upstream, the depth increases until it becomes the normal depth (see Figure 7.21). Assumed flow depths were developed by subtracting from the normal depth in 0.1-m increments.] The second through fourth columns are computed on the basis of the depth and channel geometry. The fifth column, velocity, is computed as the ratio of the discharge to the cross-sectional area, and the specific energy in the sixth column is computed using Equation 7.36. Friction slopes, in the seventh column, have been computed for this example using the Manning equation.

Table 7.8 Direct-step flow profile calculations

(1)	(2)	(3)	(4)	(5)	(6)	(7)	(8)	(9)	(10)
y (m)	A (m^2)	P (m)	R (m)	V (m/s)	E (m)	S_f	Avg S_f	Δx (m)	$\Sigma\Delta x$ (m)
0.86	5.78	8.85	0.65	2.60	1.203	0.007427			0
0.98	6.82	9.38	0.73	2.20	1.226	0.004624	0.006026	4.6	4.6
1.08	7.73	9.83	0.79	1.94	1.272	0.003238	0.003931	15.4	20.1
1.18	8.68	10.28	0.85	1.73	1.332	0.002334	0.002786	33.7	53.8
1.28	9.68	10.72	0.90	1.55	1.402	0.001722	0.002028	68.5	122.3
1.38	10.71	11.17	0.96	1.40	1.480	0.001297	0.001510	152.1	274.4
1.48	11.78	11.62	1.01	1.27	1.563	0.000995	0.001146	565.7	840.1

The average friction slope for successive rows in the table is shown in column eight, and the corresponding Δx, determined from Equation 7.75, is shown in column nine. A running summation of Δx values is shown in the last column of the table. Those values represent the distances upstream of the slope break at which the corresponding flow depths occur.

Table 7.9 is identical to Table 7.8, except that it pertains to the steep reach of the channel downstream of the slope break point. (Assumed flow depths were developed by adding to the normal depth in 0.1-m increments.) Again, the summed Δx values shown in the last column of the table represent locations at which the corresponding depths occur. In this case, however, the distances are measured downstream rather than upstream from the slope break point. The computed water surface profile is illustrated in Figure E7.20.1.

Table 7.9 Direct-step flow profile calculations

(1)	(2)	(3)	(4)	(5)	(6)	(7)	(8)	(9)	(10)
y (m)	A (m^2)	P (m)	R (m)	V (m/s)	E (m)	S_f	Avg S_f	Δx (m)	$\Sigma\Delta x$ (m)
0.86	5.78	8.85	0.65	2.60	1.203	0.007427			0
0.80	5.28	8.58	0.62	2.84	1.211	0.009633	0.008530	0.7	0.7
0.75	4.88	8.35	0.58	3.08	1.233	0.012134	0.010884	2.3	3.0
0.70	4.48	8.13	0.55	3.35	1.271	0.015510	0.013822	6.3	9.3
0.65	4.10	7.91	0.52	3.66	1.334	0.020163	0.017837	28.9	38.2

7.9 SPATIALLY VARIED FLOW

With *spatially varied flow* in an open channel, the discharge varies along the channel length, even in otherwise steady flow conditions. This situation occurs when flow is either added to or diverted from the channel along its length. Flow may be added, for example, in a side-channel spillway. A stormwater inlet on a continuous grade (discussed in Section 10.2) is an example where flow is diverted from the channel as it passes the inlet opening.

Methods for dealing with spatially varied flow are distinguished in terms of whether flow is added to or diverted from the channel. When flow is added to a channel, as in the case of a side channel spillway, turbulence generated in the flow is such that energy losses are greater than those due to friction alone, and the energy equation is not easily applied. Therefore, in a channel with an increasing discharge, momentum rather than energy should be used in computing the flow profile. In the case of a channel with a decreasing discharge, experience has shown that the energy equation can be applied for flow profile computations because of the general lack of turbulence generated in this sort of flow.

Spatially Varied Flow with an Increasing Discharge

A derivation of the fundamental differential equation describing a spatially varied flow with an increasing discharge is accomplished by applying the momentum principle to the flow in a channel reach. Assuming that the coefficients α and β are unity, the resulting equation is (Chow, 1959)

$$\frac{dy}{dx} = \frac{S_o - S_f - \dfrac{2Qq}{gA^2}}{1 - \mathrm{Fr}^2} \tag{7.76}$$

where y = vertical depth of flow (ft, m)
x = distance along channel bottom (ft, m)
S_o = channel bottom slope (ft/ft, m/m)
S_f = energy gradient (ft/ft, m/m)
Q = discharge (cfs, m^3/s)
q = dQ/dx (cfs/ft, $m^3/s/m$)
g = gravitational acceleration (32.2 ft/s^2, 9.81 m/s^2)
A = cross-sectional area of channel (ft^2, m^2)
Fr = Froude number

Note that this equation is identical to Equation 7.71 with the exception of the third term in the numerator. In the third term, Q, is the discharge, which depends on the location x along the length of the channel, and $q = dQ/dx$ is the rate at which flow is added to the channel along its length. Note that q has units of cfs per foot of channel length, or of m^3/s per meter of channel length. The friction slope S_f appearing in the equation, which accounts for friction along the channel bottom and sides, may be evaluated using a constitutive relationship such as the Manning equation.

If the depth of flow is known at a control section, a numerical solution of Equation 7.76 may be used to determine the water surface profile in a channel. French (1985) has shown that the location x_c of the control section in a channel (measured from upstream end of the reach) with $Q = 0$ when $x = 0$ is

$$x_c = \frac{8q^2}{gT_c^2\left(S_o - \dfrac{gn^2 P_c^{4/3}}{C_f T_c A_c^{1/3}}\right)^3} \tag{7.77}$$

where x_c = location of control section (ft, m)
T_c = top width of flow at control section (ft, m)
n = Manning's roughness coefficient
P_c = wetted perimeter at control section (ft, m)
C_f = unit conversion factor (2.21 U.S. Customary units, 1.0 SI units)
A_c = cross-sectional area of flow at control section (ft, m)

Critical depth occurs at the control section location determined by Equation 7.77.

Solution of Equation 7.77 must be accomplished in an iterative, trial-and-error manner. One begins by assuming a value x_c, from which $Q = qx_c$. For this discharge, one determines the corresponding critical depth y_c, and then the values of A_c, P_c, and T_c. Finally, Equation 7.76 is applied to compute a new value of x_c. If the computed value of x_c agrees with the assumed one, convergence is attained. Otherwise, the procedure is repeated with a new trial value.

If the location x_c of the control section lies within the length L of the channel that is subjected to lateral inflow, then flow upstream of that point will be subcritical, and flow downstream of that point will be supercritical. If x_c is greater than L, then flow in the channel will be subcritical throughout and will be controlled by conditions existing at the location $x = L$. Even if x_c is less than or equal to L, the critical section may be "drowned" by downstream conditions.

Spatially Varied Flow with a Decreasing Discharge

The equation describing spatially varied flow with a decreasing discharge is derived using the energy principle, as shown by Chow (1959). Assuming that the coefficients α and β are unity, the equation is

$$\frac{dy}{dx} = \frac{S_o - S_f - \dfrac{Qq}{gA^2}}{1 - \text{Fr}^2} \tag{7.78}$$

Note that this expression is identical to Equation 7.76, except for the coefficient on the last term in the numerator. Note also that in the case of spatially varied flow with a decreasing discharge, the quantity $q = dQ/dx$ is negative.

As in the case of Equation 7.75, a numerical method may be employed to solve Equation 7.78 to yield a flow profile.

7.10 CHAPTER SUMMARY

Open channels are extremely common in both natural and constructed environments, and include streams, swales, ditches, gutters, and gravity-flow pipes. Unlike closed-conduit flow analysis, descriptors such as cross-sectional flow area and average velocity are unknown variables when evaluating flow in open channels. Their values depend on channel shape, slope, and roughness characteristics, and on channel boundary conditions.

Open channels may be classified as either prismatic or non-prismatic. Common cross-sectional shapes for prismatic channels are rectangular, trapezoidal, triangular, and circular. Channels may also be classified on the basis of whether they are lined or unlined.

When evaluating open-channel flow, the energy equation takes the form

$$Z_1 + d_1 \cos\theta + \frac{\alpha_1 V_1^2}{2g} = Z_2 + d_2 \cos\theta + \frac{\alpha_2 V_2^2}{2g} + h_L \tag{7.79}$$

Uniform flow in an open channel means that the flow depth, discharge, and other flow properties are constant. This condition can occur in a sufficiently long prismatic channel. The depth at which uniform flow occurs is called *normal depth.* Constitutive relationships such as the Manning and Chézy equations can, in addition to being used to compute friction losses, be used to predict the normal depth for a given channel and discharge.

Critical flow is another condition for which a unique relationship exists between depth and discharge. It also corresponds to the point of minimum specific energy (the energy relative to the channel bottom) for a given channel geometry and discharge. Critical flow often occurs at channel control sections, such as at free overfall locations. The concept of specific energy of flow can also be applied to the evaluation of flow properties at channel transitions such as expansions, contractions, and steps.

Momentum and specific force concepts are used in evaluating the depths upstream and downstream of hydraulic jumps, and in computing the forces on hydraulic structures such as channel transitions and bridge piers. The momentum equation for open-channel flow is written as

$$\sum F_x = P_1 - P_2 + W \sin\theta - F_f = \rho Q(\beta_2 V_2 - \beta_1 V_1) \tag{7.80}$$

"GENTLEMEN, I'LL SUM THIS UP IN A NUTSHELL."

For a short, horizontal channel reach, specific force can be computed as

$$F_s = \gamma\left(\bar{z}A + \frac{Q^2}{gA}\right) \tag{7.81}$$

Weirs are used to control or measure flow in open channels. Weirs may be classified as either sharp- or broad-crested. They may also be classified by shape; common shapes are rectangular and triangular. Finally, weirs can be classified as to whether the weir flow is contracted or suppressed. Equations and coefficients used in computing the flow or head for some types of weirs were presented in the chapter.

When depth and velocity gradually change along the length of a channel, the condition is termed *gradually varied flow.* Evaluation of gradually varied flow involves classifying the type of profile based on the relative values of normal depth and critical depth for the channel, and on the location of the water surface relative to these depths. Once the profile type is known, corresponding flow depths and locations along the channel reach are determined.

If discharge varies along a channel reach, *spatially varied flow* exists. If the discharge increases along the channel reach, momentum concepts (rather than energy) should be used to evaluate the flow with a trial-and-error procedure. If the discharge is decreasing, energy concepts can be used.

REFERENCES

Abbott, M. B. 1979. *Computational Hydraulics: Elements of the Theory of Free Surface Flows.* London: Pitman.

American Concrete Pipe Association (ACPA). 2000. *Concrete Pipe Design Manual.* 13th Printing (revised). Irving, Texas: ACPA.

Bazin, H. 1897. "A New Formula for the Calculation of Discharge in Open Channels." *Mémoire No. 41, Annales des Ponts et Chausées,* Paris 14, Ser. 7, 4th trimester: 20–70.

Blevins, R. D. 1984. *Applied Fluid Dynamics Handbook.* New York: Van Nostrand Reinhold.

Chow, V. T. 1959. *Open-Channel Hydraulics.* New York: McGraw Hill.

Chaudhry, M. H. 1993. *Open-Channel Flow.* Englewood Cliffs, New Jersey: Prentice Hall.

Chaudhry, M. H., and S. M. Bhallamudi. 1988. "Computation of Critical Depth in Compound Channels." *Journal of Hydraulic Research* 26, no. 4: 377–395.

Cunge, J. A., F. M. Holly, and A. Verwey. 1980. *Practical Aspects of Computational River Hydraulics.* London: Pitman.

Einstein, H. A. 1934. "The Hydraulic or Cross Section Radius." *Schweizerische Bauzeitung* 103, no. 8: 89–91.

Einstein, H. A., and R. B. Banks. 1951. "Fluid Resistance of Composite Roughness." *Transactions of the American Geophysical Union* 31, no. 4: 603–610.

Finnemore, E. J. and J. Franzini. 2000. *Fluid Mechanics with Engineering Applications.* New York: McGraw-Hill.

Franzini, J. and E. J. Finnemore. 1997. *Fluid Mechanics,* 9th Ed. New York: McGraw-Hill.

Fread, D. L 1993. "Flow Routing" in Maidment, D. R., ed. *Handbook of Hydrology*. New York: McGraw-Hill.

French, R. H. 1985. *Open-Channel Hydraulics*. New York: McGraw-Hill.

Ganguillet, E., and W. R. Kutter. 1869. "An Investigation to Establish a New General Formula for Uniform Flow of Water in Canals and Rivers" (in German), *Zeitschrift des Oesterreichischen Ingenieur und Architekten Vereines* vol. 21, no. 1: 6–25; no. 2: 46–59. Translated into English as a book by R. Hering and J.C. Trautwine, Jr., *A General Formula for the Uniform Flow of Water in Rivers and* Other Channels, John Wiley and Sons, New York, 1888.

Herschel, C. 1897. "On the Origin of the Chézy Formula." *Journal of the Association of Engineering Societies* 18: 363–368.

Horton, R. 1933. "Separate Roughness Coefficients for Channel Bottom and Sides." *Engineering News-Record* 111, no. 22: 652–653.

Krishnamurthy, M., and B. A. Christensen. 1972. "Equivalent Roughness for Shallow Channels." *Journal of the Hydraulics Division,* ASCE 98, no. 12: 2257–2263.

Lotter, G. K. 1933. "Considerations of Hydraulic Design of Channels with Different Roughness of Walls." *Transactions All Union Scientific Research, Institute of Hydraulic Engineering* 9: 238–241.

Manning, R. 1891. "On the Flow of Water in Open Channels and Pipes." *Transactions, Institution of Civil Engineers of Ireland* 20: 161–207.

Motayed, A. K., and M. Krishnamurthy. 1980. "Composite Roughness of Natural Channels." *Journal of the Hydraulics Division,* ASCE 106, no. 6: 1111–1116.

Muhlhofer, L. 1933. "Rauhigkeitsunteruchungen in einem stollen mit betonierter sohle und unverkleideten wanden." *Wasserkraft und Wasserwirtschaft* 28(8): 85–88.

Pavlovskii, N. N. 1931. "On a Design Formula for Uniform Movement in Channels with Nonhomogeneous Walls" (in Russian), *Izvestiia Vsesoiuznogo Nauchno-Issledovatel'skogo Insituta Gidrotekhniki* 3: 157–164.

Powell, R. W. 1950. "Resistance to Flow in Rough Channels." *Transactions, American Geophysical Union* 31, no. 4: 575–582.

Rouse, H. 1936. "Discharge Characteristics of the Free Overfall." *Civil Engineering* 6(7): 257.

Rouse, H. 1946. *Elementary Fluid Mechanics*. New York: Wiley.

Strickler, A. 1923. "Some Contributions to the Problem of the Velocity Formula and Roughness Factors for Rivers, Canals and Closed Conduits" (in German). *Mitteilungen des Eidgenossischen Amtes fur Wasserwirtschaft* 16.

U.S. Army Corps of Engineers (USACE). 1991. *HEC-2, Water Surface Profiles, User's Manual*. Davis, California: USACE Hydrologic Engineering Center.

U.S. Bureau of Reclamation (USBR). 1955. *Research Studies on Stilling Basins, Energy Dissipators, and Associated Appurtenances,* Hydraulic Laboratory Report No. Hyd-399. Lakewood, Colorado: USBR.

PROBLEMS

7.1 From the natural channel cross-section data given in the table, determine the area, wetted perimeter, hydraulic radius, top width, and hydraulic depth if the depth of flow is 4 ft.

Station (ft)	Elevation (ft)
0	10.1
5	7.8
15	5.5
18	6.0
22	8.2
26	9.8
30	10.7

7.2 Determine the area of flow, wetted perimeter, hydraulic radius, top width, and hydraulic depth for each channel described. The flow depth for all channels is 1.22 m.

a) Rectangular channel with a bottom width of 1.5 m

b) Triangular channel with side slopes of 2 horizontal to 1 vertical

c) Semicircular channel with a radius of 2.1 m

d) Trapezoidal channel with a bottom width of 3.0 m and side slopes of 3 horizontal to 1 vertical

7.3 Demonstrate that the hydraulic depth D_h in a rectangular channel is equal to the actual depth of flow. Also demonstrate that the hydraulic depth is smaller than the actual depth of flow in trapezoidal and triangular channels.

7.4 Determine the discharge in a 3-ft diameter sewer pipe if the depth of flow is 1 ft and the slope of the pipe is 0.0019. Use a value of 0.012 for the Manning roughness coefficient.

7.5 An asphalt-lined trapezoidal channel has the dimensions indicated in the figure. The channel slope is 0.001. Use the Chézy equation to find the discharge if $B = 4.9$ m and $y = 1.4$ m.

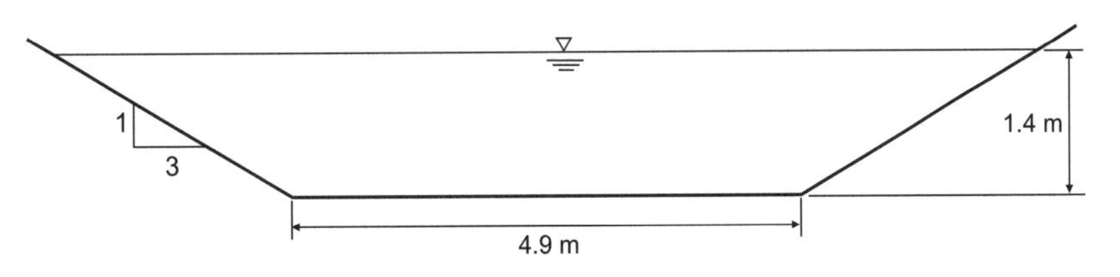

7.6 By comparing the Manning and Chézy equations, show that the relationship between n and C is $C = 1.49R^{1/6}/n$ if R is in feet.

7.7 Explain why uniform flow cannot occur in frictionless and horizontal channels.

7.8 Determine the normal and critical depths in a triangular channel if $Q = 5$ cfs, $n = 0.045$, and $S_o = 0.008$. The channel has side slopes of 3 horizontal to 1 vertical. Is uniform flow in this channel subcritical or supercritical?

7.9 A 36-in. diameter pipe with $n = 0.013$ and $S_o = 0.01$ conveys a discharge of 25 cfs. Determine the depth and velocity of flow in the pipe.

7.10 Determine the critical depth in a trapezoidal channel if the discharge is 14.2 m^3/s. The width of the channel bottom is 6.1 m and the side slopes are 1:1.

7.11 In the triangular section shown, the flow is 350 cfs. Using $n = 0.012$, calculate the critical depth and the critical velocity.

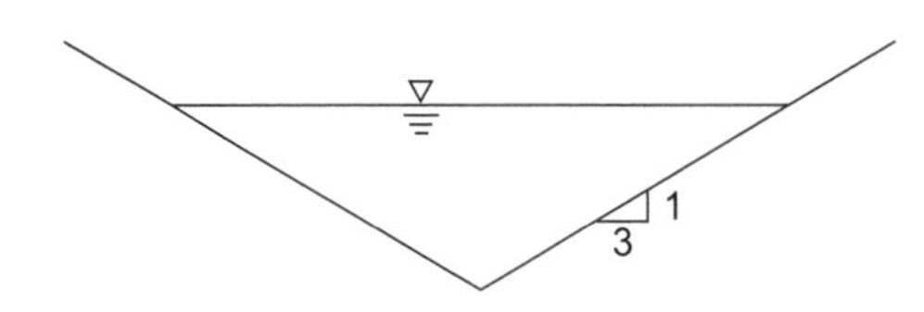

7.12 The discharge is 7.1 m^3/s in a 3-m-wide rectangular section. Plot the specific energy as a function of depth of flow.

7.13 Calculate the critical depth of the channel in Problem 7.12 using Equation 7.38.

7.14 Determine the sequent depth and length of a hydraulic jump in a rectangular channel with $B = 5$ ft, $Q = 150$ cfs, $n = 0.013$, and $S_o = 0.05$. Flow upstream of the jump is uniform. What is the energy loss in the jump?

7.15 Water flows through a rectangular section at a rate of 14.2 m^3/s. The channel is 3.05 m wide and the upstream depth is 0.94 m. The flow undergoes a hydraulic jump as indicated in the diagram.

a) Find the downstream depth.

b) Find the energy loss in the hydraulic jump.

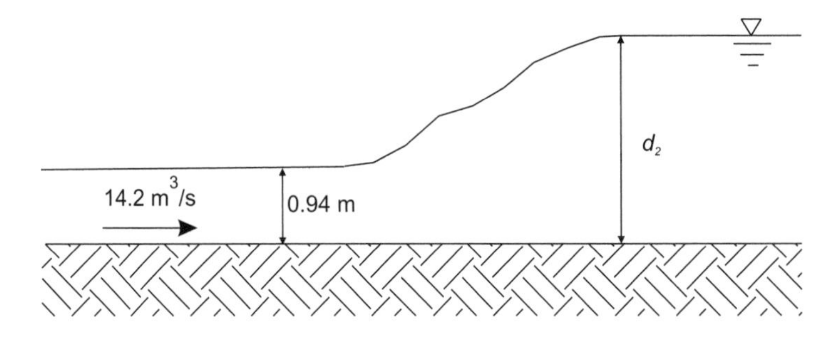

7.16 A sharp-crested, contracted, rectangular weir ($N = 2$) is 20 ft long. The height of the weir crest above the channel is 8 ft. Develop and plot a rating curve (that is, a relationship between Q and H) for this weir. Use H values ranging from 0 to 3 ft to develop the rating curve.

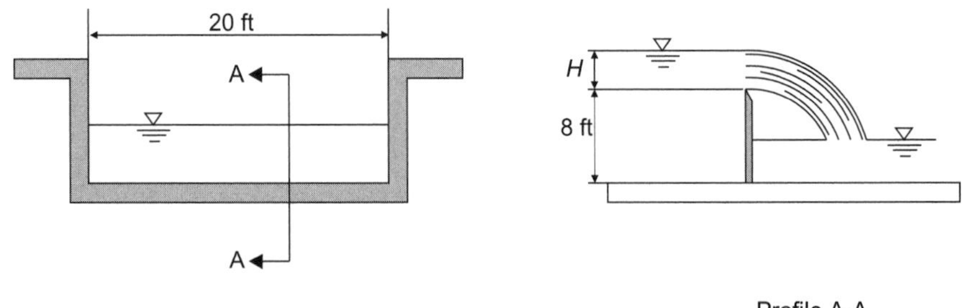

7.17 Water flows in a 6.1-m-wide rectangular concrete channel at a rate of 14.2 m^3/s. The slope changes from 0.0125 to 0.05. Plot the water surface profile through the transition.

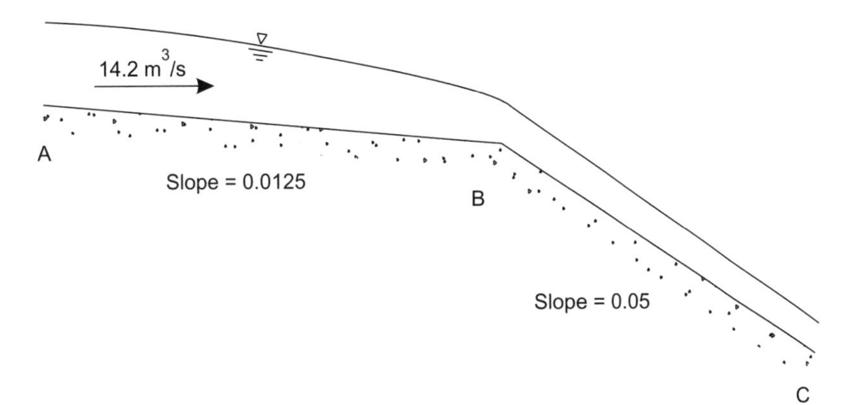

Stormwater is conveyed through this rectangular concrete channel, which is fenced for safety reasons.

CHAPTER

8

Design of Open Channels

Chapter 7 presented methods of open-channel flow *analysis*. It focused on fundamental hydraulic principles and how they could be applied to analyze the characteristics of a flow in a specified channel. This chapter focuses on channel *design*, with a corresponding emphasis on determining appropriate channel geometry and materials. The design methods presented in this chapter are applicable to open channels such as ditches, gullies, and streams. (Chapter 9 covers culvert design, and Chapters 10 and 11 cover the design of storm sewer systems.) The techniques presented in this chapter are further limited to uniform flow in prismatic channels. For instances in which the flow is varied (non-uniform), design modifications, such as increasing the channel depth, must be made in order to ensure adequate channel capacity.

Section 8.1 of this chapter discusses the various issues that should be considered when designing an open channel. Some of the issues relate to channel hydraulics, and others relate to public safety, right-of-way requirements, maintenance, and channel alignment.

The concept of the *best hydraulic section* as a basis for channel design is briefly discussed in Section 8.2. Channel design using this approach provides optimal hydraulic performance in that the cross-sectional area of the channel is maximized for a fixed wetted perimeter, thus minimizing the resistance to flow. Channels designed using this method, however, must be carefully examined for feasibility in light of other design considerations.

For design purposes, a channel may be classified according to three basic types: lined channels, unlined channels, and grassed channels. *Lined channels* have erosion-resistant materials lining their bottom and sides. Most commonly, they are constructed using concrete or asphalt as the lining material. In *unlined channels*, also referred to as *erodible channels*, the channel bottom and sides consist of unprotected, natural soil materials. *Grassed channels* have intermittent flows such that grass or similar vegetation can establish itself in the channel during no-flow periods. Various combinations of these three basic types also exist; for instance, a channel may have a concrete bottom for frequent, lower flows and, above the concrete, grassed sides for higher flows. Design methods for lined, erodible, and grassed channels are covered in Sections 8.3, 8.4, and 8.5, respectively. Design of riprap lined channels is discussed in Section 8.6

Frequently, open-channel design involved modifying an existing channel rather than creating a completely new drainageway. Most often, the purpose of the modifications is to increase channel capacity. Information on designing modifications to existing channels is presented in Section 8.7.

8.1 DESIGN CONSIDERATIONS

Proper design of an open channel involves careful hydraulic analysis, as well as an understanding that the constructed channel will be only a part of the larger urban infrastructure. Therefore, the design of an open channel should involve not only consideration of technical criteria related to flow velocities and the like, but also issues such as public safety, aesthetics, and maintainability. Brief discussions of a number of design considerations are provided in the following paragraphs.

Velocity of Flow

The velocity at which various flows will travel in an open channel is an important design consideration for a number of reasons. Excessively high velocities pose a public safety hazard. They also tend to cause channel erosion even in lined channels, particularly if the water is transporting a significant amount of sediment or debris. If velocities are too small [less than about 2 to 3 ft/s (0.6 to 0.9 m/s)], sediments and debris may settle out of the water column, with the net effect being that the channel will require regular cleaning. Biological slime and/or vegetation growth may also become problematic if flow velocities are too small.

"Hey, buddy, wouldn't the Grand Canal be faster?"

Froude Number

The Froude number of a flow should always be checked when designing an open channel. Except for cases in which drop structures, spillways, or other grade-change facilities are being addressed, channel design generally seeks to establish a subcritical flow regime where $Fr < 1$. Also, because critical or near-critical flow tends to be unstable and may be accompanied by a wavy water surface, the Froude number corresponding to a particular design should not be close to unity (that is, it should not be between approximately 0.9 and 1.1–1.2).

Side Slopes

The side slopes of a channel should be selected with regard to maintainability, safety, and slope stability. Channels that require periodic mowing or other maintenance must have side slopes that are flat enough to accommodate equipment ingress and egress. In many areas, side slopes of 3 horizontal to 1 vertical are used as a rule of thumb for channel design to facilitate mower operation. If side slopes are too steep, and especially if flow velocities are also high, it may be difficult for someone who falls into the channel to get out. A 3 to 1 slope is fairly mild and fairly easy to navigate when it is dry, but may present more of an obstacle when it is wet or, worse still, when water is flowing.

Soil materials have a natural angle of repose (the maximum slope for which they are stable) that should not be exceeded. This angle may even have to be reduced when designing channel side slopes due to seepage forces that develop as the soils drain following a large flow event.

Longitudinal Slope

The longitudinal slope of a channel is governed in large part by local topography, but there are other considerations to bear in mind as well. For example, if an existing natural channel is straightened and/or reshaped to improve its hydraulic efficiency, the velocity of flow usually increases and may pose erosion problems. In such instances, a designer may wish to decrease the channel slope and provide one or more drop structures where necessary to make up required channel bed elevation changes.

The engineer must also consider that a channel conveys a range of flows, not just a single design discharge. Even if a channel is designed to convey the design discharge with a subcritical flow regime, supercritical flow may still occur at other discharges. The longitudinal slope should be selected such that all possible flows are subcritical.

Freeboard

The total depth of an open channel from the bed to the top of either bank should, of course, be greater than the depth of the design flow. *Freeboard* is difference in height between the lower of the banks and the free water surface, and is essentially a factor of safety. While various sources may provide some guidance on freeboard requirements, the selection of a freeboard height remains largely a professional judgment issue. Freeboard heights are usually at least 1 ft (0.3 m). This value often increases when flows have large velocities or depths, or if there is a curve in the horizontal alignment of the channel that causes the water surface to superelevate (that is, to be higher on one side of a cross section).

Lining Material

Selection of the channel lining material or type of grass cover, if any, is a major consideration in channel design. Channels lined with concrete, asphalt, or other impermeable materials are hydraulically efficient, but are also more costly to construct, are frequently less aesthetically pleasing, and have little value as habitat. They may also require consideration of uplift forces such as buoyancy and frost heave. In grass-lined channels, selection of the grass species is a design consideration. Sod-forming grasses, once established, provide uniform protection to the channel bottom and sides. Bunch-forming grasses such as alfalfa may channelize flows between grass bunches, causing localized erosion.

Cost and Maintenance

The life-cycle cost of a channel design includes not only the capital costs associated with engineering and construction, but also long-term maintenance costs. Evaluations of alternative designs should seek to minimize the total life-cycle cost, but design objectives and/or constraints other than cost may ultimately dictate the final design selection.

Safety

A design issue of paramount importance, especially in urban and suburban settings, is public safety. Channels (and water in general) often become magnets for leisure and recreational activities, with the net effect being that humans often come into contact with flowing waters. Children especially are attracted to open streams, and are more vulnerable than adults to the hazards posed by large flow velocities and/or depths. As mentioned previously, whenever possible, channel sides slopes should be flat enough for a person to climb out. Steps or ladder rungs similar to those in sewer manholes can be installed if necessary to facilitate egress. It may be appropriate to use fencing along larger channels, channels that experience high velocities, and channels with steep sides in order to keep people away, especially in urban areas. The U.S. Bureau of Reclamation (1974) and Aisenbrey et al. (1978) provide some safety-related guidance.

Aesthetics

More and more frequently, open channels and other types of water bodies are becoming centerpieces for parks and similar cultural centers. Greenways with trails and bike paths often follow streams. In such instances, the aesthetics of a design are typically among the most significant evaluation criteria, and other issues such as hydraulic efficiency are of less concern.

The aesthetics of a channel can be enhanced to make it more pleasing and natural in appearance. Materials such as stones and boulders can be placed in the bottom and on the sides. Where possible, the vertical alignment can be designed to allow for drops such as waterfalls, and a naturally meandering horizontal alignment can be used. A liberal amount of vegetation along the channel edges adds to its natural character. A landscape architect can assist in the design of a channel for which aesthetic considerations are important.

Right-of-Way Requirements

In urban and suburban areas where land values may be high and space at a premium, right-of-way requirements can be a major consideration in channel design. This is true both for areas that are just becoming urbanized and for already-established areas where retrofits or upgrades are being made to the stormwater conveyance system.

For areas where available right-of-way is limited, one should clearly seek to develop a channel design with a small top width. A rectangular cross section may be used, but typically the channel must then be lined with concrete, which can increase the channel construction cost considerably. This type of channel may also give rise to problems with maintainability and safety.

Dry-Weather Flows

Open channels must be capable of conveying large design storm discharges, but they also should be designed with due regard for relatively small dry-weather flows (also called *base flows*). Particularly in park-like or greenbelt settings where grassed channels are often used for conveyance, it may be desirable to construct a small low-flow channel or *trickle channel* using riprap or other lining material. This practice helps to define the path of low flows, preventing them from meandering and shifting with time across the bottom of the larger channel in which they reside.

Clearly, not all of the considerations described above will arise in every channel design. In some cases, there may be additional considerations not cited here. The important point is that there are many issues that must be considered in design, and tradeoffs occur regularly. A critical aspect of the design process is to understand and weigh the tradeoffs and associated consequences so that a well-balanced design may be selected for implementation.

8.2 BEST HYDRAULIC SECTION

If a channel's longitudinal slope and roughness characteristics are fixed, maximization of its hydraulic efficiency amounts to either maximizing the cross-sectional area for a fixed wetted perimeter or, equivalently, minimizing its wetted perimeter for a fixed cross-sectional area. This is observed by noting that, with all other things equal, a large cross-sectional area yields a greater flow capacity than does a small area. Similarly, a large wetted perimeter exerts a greater frictional resistance to flow than does a small wetted perimeter. The *best hydraulic section* is defined as a cross section having the highest possible ratio of cross-sectional flow area to wetted perimeter.

Of all possible geometric shapes that could be selected as the cross section for an open channel, a semicircle with the flat face corresponding to the water surface is the "best" hydraulically. The semicircular shape results in the largest possible cross-sectional area for a fixed wetted perimeter, or the smallest possible wetted perimeter for a fixed cross-sectional area.

When one considers non-semicircular shapes such as rectangles, triangles, and trapezoids, one finds that the best dimensions for these shapes are those that most closely approximate a semi-circle. For example, the best rectangular cross section is one that is the bottom half of a square (that is, one for which the depth of flow is one-half the channel width). The best triangular cross section is the bottom half of a square tilted to resemble a diamond shape (that is, one with side slopes inclined at 45 degrees). The best trapezoidal cross section is the bottom half of a hexagon, such that its side slopes are inclined 60 degrees from the horizontal. Figure 8.1 illustrates these cross sections.

Figure 8.1
Best hydraulic cross-sections for various channel shapes

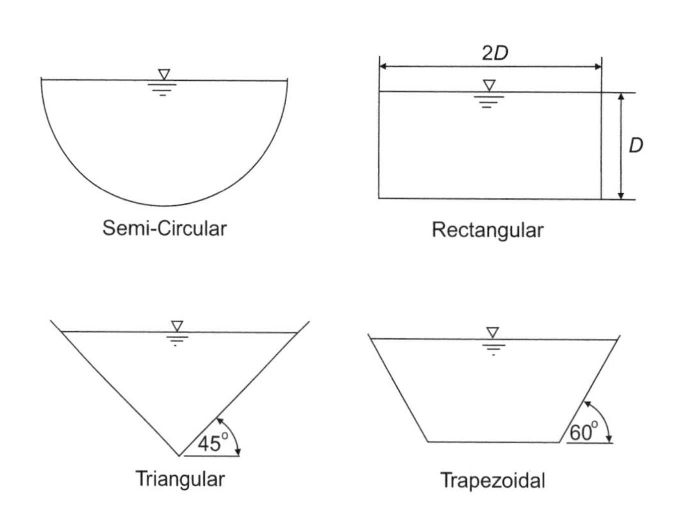

Because the best hydraulic section approximates a semicircle, its side slopes are quite steep, if not vertical. Unfortunately, this geometry limits the practical use of the best hydraulic section to instances in which the channel is either to be lined or cut through solid rock. Even in instances where the side slopes can be steep or vertical, design considerations such as safety and maintainability may limit the practicality of the best hydraulic section.

Example 8.1 – Computing the Best Hydraulic Section. Use the concept of the best hydraulic section and the Manning equation (see Section 7.2, page 226) to design a concrete-lined trapezoidal channel to convey $Q = 3.0\ \mathrm{m^3/s}$ with $n = 0.013$ and $S_o = 0.0015$. Check the velocity and Froude number of the flow, and add 20 percent of the computed depth of flow as freeboard.

Solution: The best trapezoidal cross section is half of a hexagon. Letting B denote the cross-section bottom width (that is, the length of one side of the hexagon), y the flow depth, and P the wetted perimeter, it can be observed from Figure E8.1.1 that

$$P = B + B + B = 3B$$

$$y = B \sin 60° = 0.866B$$

$$A = By + (B/2)y = 1.299B^2$$

$$T = B + 2(B/2) = 2B$$

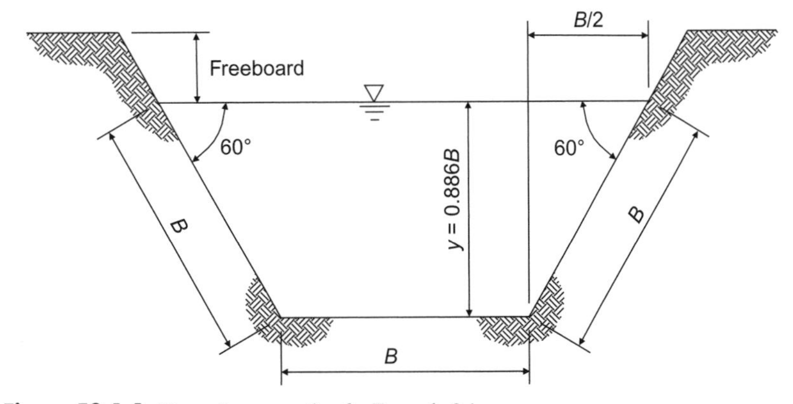

Figure E8.1.1 Channel cross section for Example 8.1

Substituting into the Manning equation (Equation 7.9), the discharge can be expressed as

$$Q = 3 = \frac{1}{n} A R^{2/3} S_o^{1/2} = \frac{1}{0.013}(1.299B^2)^{5/3}(3B)^{-2/3}(0.0015)^{1/2} = 2.215B^{8/3}$$

Solving this equation,

$$B = (3/2.215)^{3/8} = 1.12 \text{ m}$$

The other hydraulic dimensions are

$$P = 3.36 \text{ m}$$

$$y = 0.97 \text{ m}$$

$$A = 1.63 \text{ m}^2$$

$$T = 2.24 \text{ m}$$

The average flow velocity is

$$V = 3/1.63 = 1.84 \text{ m/s}$$

This velocity is not excessively high, and hence the design would likely be considered adequate on the basis of this consideration.

The Froude number is computed as (from Equation 7.39):

$$\text{Fr} = \sqrt{\frac{Q^2 T}{g A^3}} = \sqrt{\frac{3^2(2.24)}{9.81(1.63)^3}} = 0.69$$

This is well into the subcritical range and again leads to a positive evaluation of this design.

Using 20 percent of the depth as the freeboard criterion, the freeboard height is computed to be

$$0.2 \times 0.97 = 0.19 \text{ m}$$

Thus, the total channel depth is

$$0.97 + 0.19 = 1.16 \text{ m}$$

8.3 Design of Lined Channels

From the point of view of determining channel cross-section dimensions, the design of a lined channel is the most straightforward. The steps that follow outline a design procedure for a specified design discharge Q.

1 Determine Manning's n and the channel slope S_o. These selections will depend on the lining material chosen, as well as on topography.

2 Using the design discharge and the values determined in step 1, rearrange the Manning equation (Equation 7.9) such that the unknown variables describing channel geometry are on the left side and known variables are on the right. Compute the channel section factor $AR^{2/3}$.

$$AR^{2/3} = \frac{nQ}{C_f \sqrt{S_o}} \tag{8.1}$$

where A = cross-sectional area of channel (ft^2, m^2)
R = hydraulic radius (ft, m)
n = Manning's roughness coefficient
Q = discharge (cfs, m^3/s)
C_f = unit conversion factor (1.49 for U.S. customary units; 1.00 for SI)
S_o = channel bottom slope (ft/ft, m/m)

3 Select the channel dimensions B and/or z and write an expression for the section factor $AR^{2/3}$ in terms of those dimensions and the depth of flow. Figure 8.2 (repeat of Figure 7.4) and Table 8.1 (repeat of Table 7.1) may be consulted to aid in this step.

4 Equate the value of the section factor determined in step 2 to the expression written in step 3, and solve for the normal depth of flow, y_n. This solution usually must be accomplished by trial and error.

5 Check the flow velocity and Froude number for acceptability and add freeboard to obtain the required channel depth.

6 Develop rating curves or tables to check velocity and flow depth for a wide range of possible flow rates, and consider whether a low-flow channel is needed.

For lined channels constructed using unreinforced concrete, ASCE (1992) has recommended that the flow velocity not exceed 7 ft/s (2.1 m/s). In reinforced concrete channels, the flow velocity should not exceed 18 ft/s (5.5 m/s). In either case, the velocity should not be less than about 2 to 3 ft/s (0.6 to 0.9 m/s). The Froude number should be in the subcritical range, and should not be close to unity (1.0).

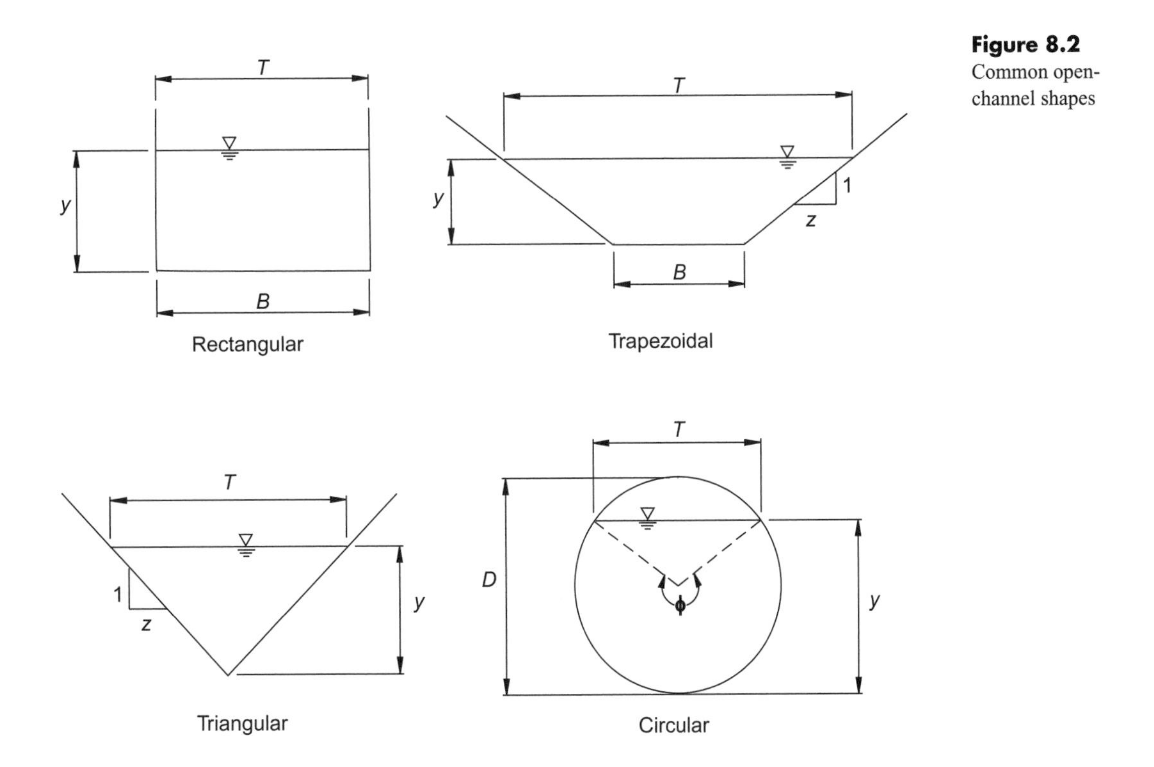

Figure 8.2
Common open-channel shapes

Table 8.1 Cross-sectional properties of prismatic open channels

Channel Shape	Area *(A)*	Wetted Perimeter *(P)*	Hydraulic Radius *(R)*	Top Width *(T)*	Hydraulic Depth *(D_h)*
Rectangular	By	$B+2y$	$\frac{By}{B+2y}$	B	y
Trapezoidal	$By+zy^2$	$B+2y\sqrt{1+z^2}$	$B+2y\sqrt{1+z^2}$	$B+2zy$	$\frac{By+zy^2}{B+2zy}$
Triangular	zy^2	$2y\sqrt{1+z^2}$	$\frac{zy}{2\sqrt{1+z^2}}$	$2zy$	$\frac{y}{2}$
Circular[a]	$\frac{D^2(\phi-\sin\phi)}{8}$	$\frac{D\phi}{2}$	$\frac{D}{4}\left(1-\frac{\sin\phi}{\phi}\right)$	$D\sin\left(\frac{\phi}{2}\right)$	$\frac{D}{8}\left(\frac{\phi-\sin\phi}{\sin(\phi/2)}\right)$

a.Angle ϕ is measured in radians. 1 radian ≈ 57.3 degrees.

Although a general expression for required freeboard height has not been developed, Chow (1959) indicates that the U.S. Bureau of Reclamation has suggested that an estimate may be made using the expression

$$F = \sqrt{Cy} \tag{8.2}$$

where F = required freeboard height (ft)
C = coefficient based on discharge
y = depth of flow (ft)

The coefficient C ranges from 1.5 for a discharge of about 20 cfs (0.6 m^3/s) to 2.5 for discharges in excess of 3,000 cfs (85 m^3/s). This estimate is based on average U.S. Bureau of Reclamation practice and should not be expected to apply to all conditions. Freeboard requirements for lined channels are usually at least 1 ft (0.3 m) (ASCE, 1992).

At channel bends, where superelevation of the free water surface occurs, freeboard usually needs to be increased. For subcritical flow, the superelevation height of the water surface can be approximated as (Chow, 1959)

$$h_s = \frac{V^2 T}{g r_c} \tag{8.3}$$

where h_s = superelevation height (ft, m)
V = flow velocity (ft/s, m/s)
T = top width (ft, m)
g = gravitational acceleration (32.2 ft/s^2, 9.81 m/s^2)
r_c = radius of curvature of the channel center line (ft, m)

Example 8.2 – Lined Channel Design. Design a concrete-lined open channel to convey a maximum design discharge of 500 cfs. The channel is to be trapezoidal in shape with $z = 2$, and its slope will be $S_o = 0.001$.

Solution: Referring to Table 7.3, a typical n value for concrete of 0.013 is selected. This value, along with the stated channel slope and discharge, are used to compute the section factor as

$$AR^{2/3} = \frac{0.013(500)}{1.49\sqrt{0.001}} = 138$$

Using a trial channel bottom width of $B = 10$ ft, one can use the expressions in Table 8.1 to write the section factor as

$$AR^{2/3} = 138 = (10y + 2y^2)\left(\frac{10y + 2y^2}{10 + 2y\sqrt{1+2^2}}\right)^{2/3} = (10y + 2y^2)^{5/3}(10 + 2y\sqrt{5})^{-2/3}$$

Solving this expression by trial and error, one obtains

$$y_n = 4.04 \text{ ft}$$

For this depth of flow, other hydraulic dimensions are

$$A = 73.0 \text{ ft}^2$$

$$T = 26.2 \text{ ft}$$

The flow velocity is

$$V = 500/73.0 = 6.85 \text{ ft/s}$$

and the Froude number is

$$\mathrm{Fr} = \sqrt{\frac{Q^2 T}{gA^3}} = \sqrt{\frac{500^2(26.2)}{32.2(73.0)^3}} = 0.72$$

Both of these values are within acceptable ranges. Applying linear interpolation to the range of values for C provided in the text (for Equation 8.2), C is computed to be 1.7, and the required freeboard height is estimated as

$$F = [1.7(4.04)]^{1/2} = 2.62 \text{ ft}$$

Thus, the total channel depth should be about 6.7ft from its bottom to the tops of the banks. Figure E8.2.1 illustrates the designed cross section.

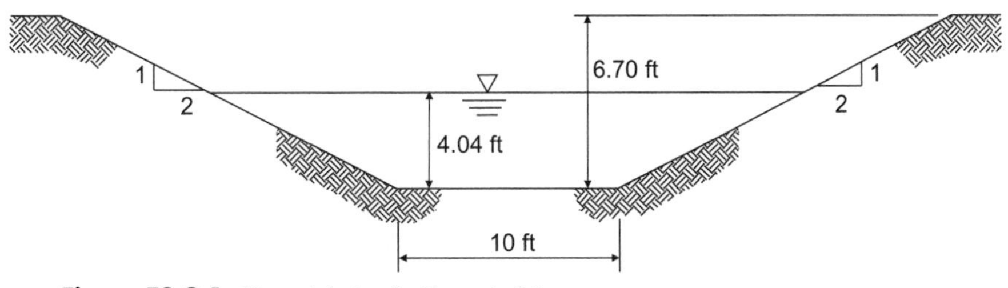

Figure E8.2.1 Channel design for Example 8.2

The step-by-step procedure for a lined channel design may be repeated for a number of alternatives using different lining materials and different channel dimensions and shapes. The engineer evaluates the alternative designs, and then chooses the best one for implementation.

8.4 Design of Erodible Channels

The bottom and sides of an *erodible channel* consist of natural soil materials that are not protected by a rigid lining or vegetation. Depending on flow characteristics, sediments may be either scoured from or deposited onto the channel boundaries. One can therefore speak of design of channels that scour but do not silt, channels that silt but do not scour, or channels that silt and scour simultaneously. Only the first of these three types is considered here. The latter two types require that sediment transport issues be addressed, which is beyond the scope of this text.

Two approaches for designing erodible channels are used by hydraulic engineers. The first approach is based on the idea that excessive velocities are the cause of scour and seeks to design a channel on the basis of a *maximum permissible velocity*. The second approach, which is based on a more mechanistic understanding of sediment transport processes, seeks to design a channel on the basis of a *maximum permissible shear stress, tractive force,* or *tractive tension*. Each of these design approaches is discussed in the subsections that follow.

Method of Permissible Velocity

The *permissible velocity* in an open channel is defined as the maximum *average* velocity for which soil particles on the channel boundaries will not be set into motion. The permissible velocity depends on many factors and therefore has significant uncertainty associated with it. It tends to be higher for old channels than for new ones, and higher for deep channels than for shallow ones.

The permissible velocity depends strongly on the characteristics of the soil making up the channel boundaries, and on whether the water is clear or is transporting colloidal silts. For noncohesive soils such as sands, the permissible velocity depends mainly on the median grain size. On the other hand, the permissible velocity for a cohesive soil depends on the void ratio of the material, as well as on cohesive and/or electrostatic forces within the soil material.

Table 8.2 lists permissible velocities assembled by Fortier and Scobey (1926) using data from canals. Values given in the table are applicable to aged channels with small slopes. Also shown in the table are suitable Manning's n values for the soil materials listed. Note that the permissible velocities for water transporting colloidal silts are greater than the corresponding velocities for clear water. Figure 8.3 illustrates U.S.S.R. data on permissible velocities for cohesive soils (Chow, 1959). It can be seen that the soil type itself has some influence on the permissible velocity, but the voids ratio (or degree of soil compaction) is the most important parameter.

An alternative to Table 8.2 and Figure 8.3 is to employ Figure 8.4, which shows the permissible velocity as a function of mean sediment size. This latter figure is based mainly on the work of Hjulstrom (1935) and has been suggested for use by the ASCE Sedimentation Task Committee (ASCE, 1977). Note that the permissible velocity is at a minimum for a mean sediment size of approximately 0.2 mm. Sediments smaller than 0.2 mm tend to be associated with cohesive materials, which have cohesive and/or electrostatic forces that tend to resist particle motion. Larger sediments tend to correspond to larger permissible velocities because, by virtue of their mass, they are more resistant to motion.

Table 8.2 Permissible velocities in open channels (Fortier and Scobey, 1926)

		Permissible Velocity			
		Clear Water		Water with Colloidal Silts	
Material	n	**ft/s**	**m/s**	**ft/s**	**m/s**
Fine sand, colloidal	0.02	1.50	0.46	2.50	0.76
Sandy loam, noncolloidal	0.02	1.75	0.53	2.50	0.76
Silt loam, noncolloidal	0.02	2.00	0.61	3.00	0.91
Alluvial silts, noncolloidal	0.02	2.00	0.61	3.50	1.07
Ordinary firm loam	0.02	2.50	0.76	3.50	1.07
Volcanic ash	0.02	2.50	0.76	3.50	1.07
Stiff clay, very colloidal	0.025	3.75	1.14	5.00	1.52
Alluvial silts, colloidal	0.025	3.75	1.14	5.00	1.52
Shales and hardpans	0.025	6.00	1.83	6.00	1.83
Fine gravel	0.02	2.50	0.76	5.00	1.52
Graded loam to cobbles, noncolloidal	0.03	3.75	1.14	5.00	1.52
Graded silts to cobbles, colloidal	0.03	4.00	1.22	5.50	1.68
Coarse gravel, noncolloidal	0.025	4.00	1.22	6.00	1.83
Cobbles and shingles	0.035	5.00	1.52	5.50	1.68

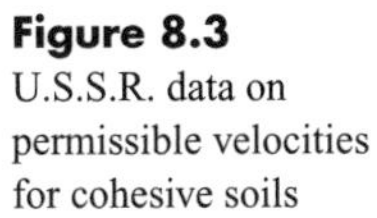

Figure 8.3
U.S.S.R. data on permissible velocities for cohesive soils

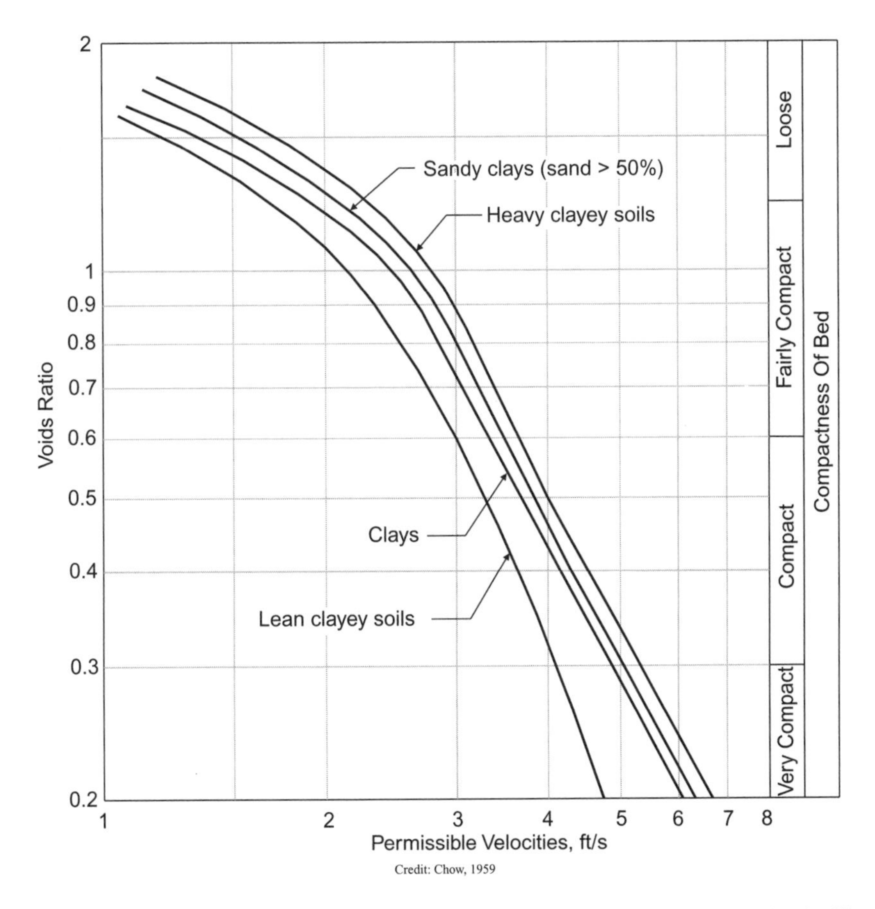

Credit: Chow, 1959

The scatter of the data points in Figure 8.4, combined with the differences in the illustrated curves, reinforce the fact that the permissible velocity is highly uncertain and is affected by variables other than just sediment size. For design purposes, one may reasonably use the mean Hjulstrom curve shown in the figure.

The permissible velocities indicated in Table 8.2 and in Figures 8.3 and 8.4 are applicable to depths of flow equal to about 3 ft (0.9 m). For depths greater or smaller than this, the permissible velocity should be modified using a correction factor such that

$$V_c = kV \qquad (8.4)$$

where V_c = corrected permissible velocity (ft/s, m/s)
k = correction factor
V = permissible velocity for flow depth of 3 ft (0.9 m)

Figure 8.5, from U.S.S.R. data (Chow, 1959), shows values for the correction factor k as a function of average depth. These correction factors are applicable to both cohesive and noncohesive soil materials. It can be seen that the correction factor is greater than unity for depths of more than 3 ft (0.9 m), and less than unity for depths less than 3 ft. Therefore, depths shallower than 3 ft have lower permissible average velocities, and those deeper than 3 ft have higher permissible average velocities. This can be rationalized in terms of the typical shape of the velocity distribution in an open channel; other factors being equal, the average velocity will be higher in a deeper channel.

Mehrota (1983) indicated that the correction factor, denoted by k, can be computed as

$$k = y^{1/6} \tag{8.5}$$

where y = depth of flow (m)

Figure 8.4
Permissible velocity as a function of mean sediment size

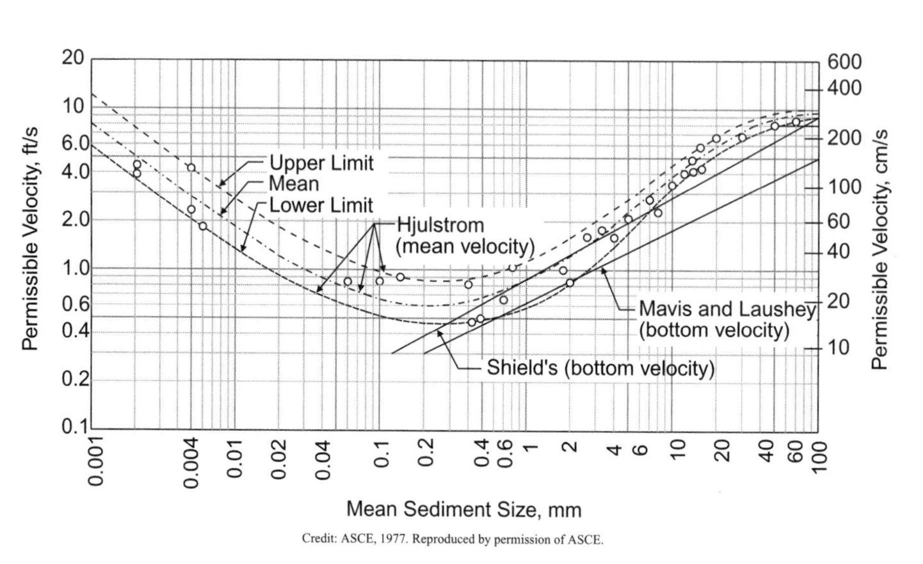

Credit: ASCE, 1977. Reproduced by permission of ASCE.

Figure 8.5
U.S.S.R. data on correction factor (k) that may be applied to permissible velocity

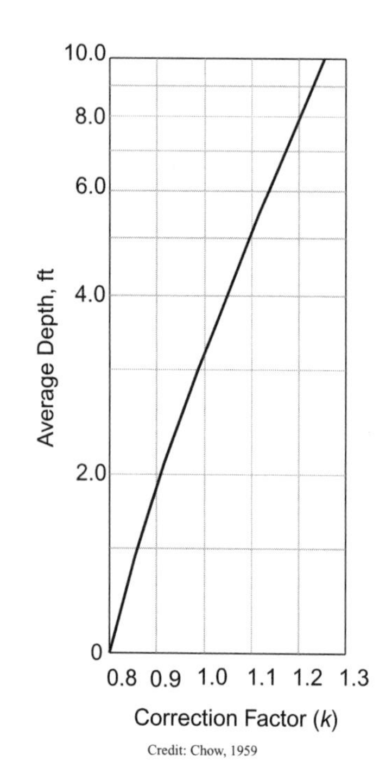

Credit: Chow, 1959

The permissible velocities described previously are for straight channels. In sinuous channels, flow can impinge on the outside bank in bends, and thus the permissible velocity should be reduced. Lane (1955) suggested a 5-percent reduction for mildly sinuous channels, a 13-percent reduction for moderately sinuous channels, and a 22-percent reduction for highly sinuous channels.

Based on the estimate of permissible velocity described previously, the design of a trapezoidal erodible channel can proceed through the following steps:

1 Estimate Manning's n and the channel side slope z to be used in the design. Table 8.2 and/or Table 7.3 may be used to guide a selection of Manning's n. Recommended maximum side slopes for various soil materials are presented in Table 8.3.

2 Using the estimated permissible velocity V and the values of n and z determined in step 1, apply the Manning equation (Equation 7.9) to compute the required hydraulic radius:

$$R = \left(\frac{nV}{C_f \sqrt{S_o}} \right)^{3/2} \tag{8.6}$$

where R = hydraulic radius (ft, m)
n = Manning's roughness coefficient
V = average velocity (ft/s, m/s)
C_f = unit conversion factor (1.49 for U.S. customary units; 1.00 for SI units)
S_o = channel bottom slope (ft/ft, m/m)

3 Compute the values of the required cross-sectional area and wetted perimeter as $A = Q/V$ and $P = A/R$, where Q is the design discharge.

4 Write expressions for A and P using Table 8.1 and substitute the previously determined values of A, P, and z into these expressions. Solve the resulting two expressions simultaneously for the values of B and y_n.

5 Check the Froude number, add freeboard, and modify the channel bottom width B for practicality.

Example 8.3 illustrates this procedure.

Table 8.3 Recommended maximum side slope, z, in various materials (Chow, 1959)

Material	Maximum side slope (z)
Rock	Nearly vertical
Muck and peat soils	0.25
Stiff clay or earth with concrete lining	0.5–1
Earth with stone lining, or earth for large channels	1
Firm clay or earth for small ditches	1.5
Loose sandy earth	2
Sandy loam or porous clay	3

Example 8.3 - Erodible Channel Design with the Permissible Velocity Method. Using the method of permissible velocity, design a trapezoidal channel to convey a discharge of $Q = 400$ cfs. The channel bottom and sides will be composed of non-colloidal coarse gravel, and the channel slope will be $S_o = 0.0015$.

Solution: Because coarse gravel is similar to a sand, the maximum allowable side slope is estimated from Table 8.3 as $z = 2$. Manning's n and the permissible velocity are estimated from Table 8.2 as $n = 0.025$ and $V = 4$ ft/s. In the interest of developing a conservative design, the water is assumed to be clear.

The required hydraulic radius is computed from Equation 8.6 and is

$$R = \left(\frac{0.025(4)}{1.49\sqrt{0.0015}}\right)^{3/2} = 2.28 \text{ ft}$$

The required area and wetted perimeter are

$A = 400/4 = 100 \text{ ft}^2$

$P = 100/2.28 = 43.8 \text{ ft}$

Using the expressions for a trapezoidal channel in Table 8.1, the following equations may be written:

$$A = 100 = By + 2y^2$$

$$Py = 43.8y = By + 2y^2\sqrt{5}$$

Note that the value of $z = 2$, determined in step 1, has been used in these equations. Note also that the equation for wetted perimeter has been multiplied by y. Simultaneous solution of these equations can be accomplished by subtracting the second from the first to obtain

$$A - Py = 100 - 43.8y = 2y^2(1 - \sqrt{5}) = -2.47y^2$$

or

$$2.47y^2 - 43.8y + 100 = 0$$

The two roots of this quadratic equation are $y = 2.69$ ft and $y = 15.04$ ft. Using each of these two values of y, and knowing that the cross-sectional area should equal 100 ft^2, one can determine the channel bottom width B. For the depth $y = 2.69$ ft, the required value of B is

$$B = \frac{A - zy^2}{y} = \frac{100 - 2(2.69)^2}{2.69} = 31.8 \text{ ft}$$

For the depth $y = 15.04$ ft, the corresponding value of B is –23.4 ft, clearly an invalid result. Therefore, one concludes that the channel bottom width should be approximately 32 ft and the depth of uniform flow will be about $y_n = 2.69$ ft. The Froude number of this flow is approximately 0.46.

From Equation 8.2, the required freeboard for this channel is estimated to be about 2 ft; thus, the total channel depth should be about 4.7 ft. Figure E8.3.1 illustrates the completed design.

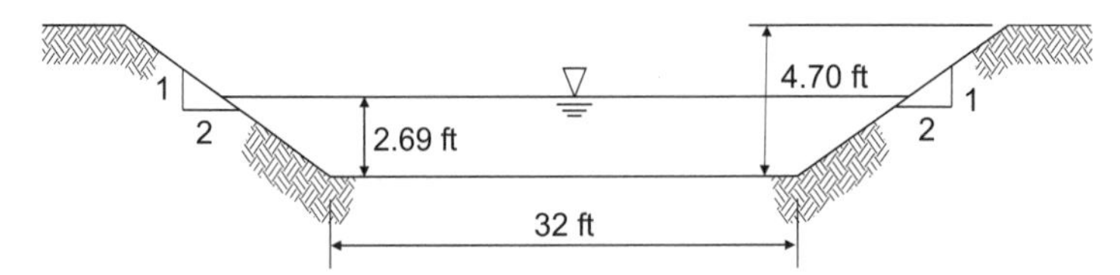

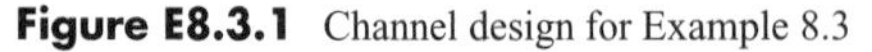

Figure E8.3.1 Channel design for Example 8.3

Note that because the depth of flow in this example is fairly close to 3 ft (0.9 m), there is no need to correct the permissible velocity using Figure 8.5 or Equation 8.4. However, had one decided to correct the permissible velocity, a second iteration of the illustrated calculations would yield a refined design.

Method of Tractive Force

Equation 7.17 showed that the average shear stress acting on the boundary of an open channel can be expressed as

$$\tau = \gamma R S_o \tag{8.7}$$

where τ = average shear stress (lb/ft^2, N/m^2)
γ = specific weight of water (lb/ft^3, N/m^3)

The shear stress, also called the *tractive force*, is the force applied per unit area of the channel bottom and sides. This force will mobilize individual soil particles and lead to scour if it is sufficiently large. As channel width increases (relative to depth), the value of the hydraulic radius approaches that of the flow depth, and Equation 8.7 can be rewritten as

$$\tau = \gamma y S_o \tag{8.8}$$

where y = vertical flow depth (ft, m)

The distribution of shear stress is not uniform on the bottom and sides of a channel. Olsen and Florey (1952) and Lane (1955) found that the shear stress distribution in trapezoidal channels resembles that shown in Figure 8.6. Further, they related the maximum shear stresses on the channel bottom and sides to the shear stress for an infinitely wide channel having the same depth and slope, $\gamma y S_o$ (see Figure 8.7). Figure 8.6 also illustrates the calculation of the maximum values for shear stress on the bottom and sides of the representative channel given the data presented in Figure 8.7.

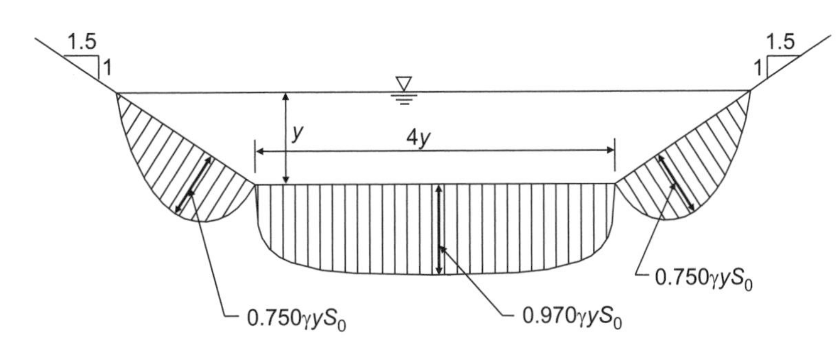

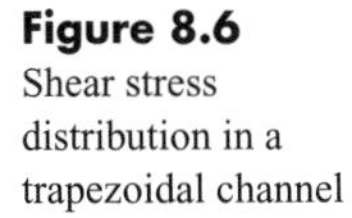

Figure 8.6 Shear stress distribution in a trapezoidal channel (Chow, 1959)

The *maximum permissible shear stress* acting on the bottom of a channel is the critical shear stress that causes incipient motion of a soil particle. Its value depends on sediment size and shape, sediment particle density, fluid properties, and the suspended sediment concentration in the water column. Based on data from a number of sources, the U.S. Bureau of Reclamation (1987) recommended use of Figure 8.8 for determining the permissible shear stress on the bottom of a straight channel with non-cohesive soil. In this figure, the permissible shear stress is referred to as "critical tractive force" and is given in units of g/m^2 (1 g/m^2 = 0.00981 N/m^2 = 0.000205 lb/ft^2). The large number of curves shown in Figure 8.8, and the corresponding wide range of permissible shear stresses for any mean sediment size, again reinforces the fact that many variables affect the permissible stress. The engineer must apply his or her best judgment in selecting which of the many cases to use.

Figure 8.7
Maximum shear stress (unit tractive force) on the bottom and sides of a trapezoidal or rectangular channel as a function of channel geometry (Chow, 1959)

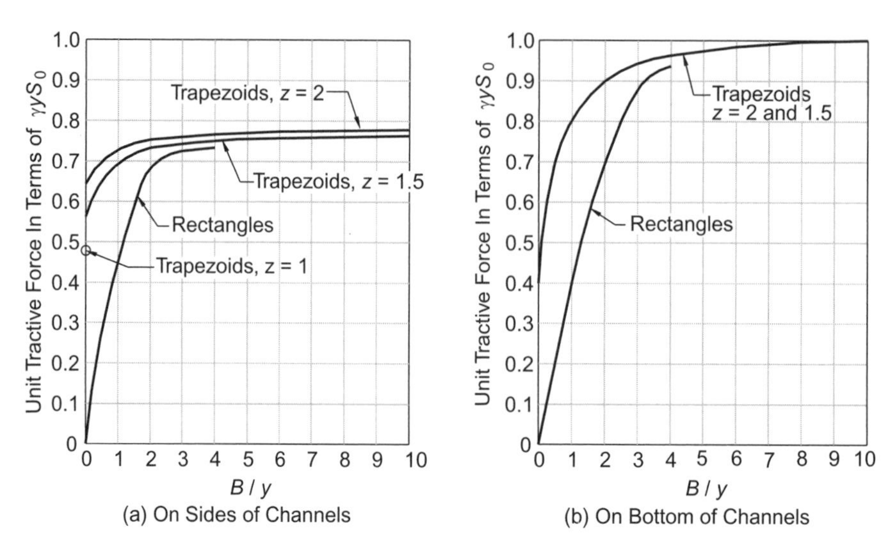

Credit: Chow, 1959.

Figure 8.8
Figure for determining the permissible shear stress (critical tractive force) on the bottom of a straight channel with noncohesive soil (USBR, 1987)

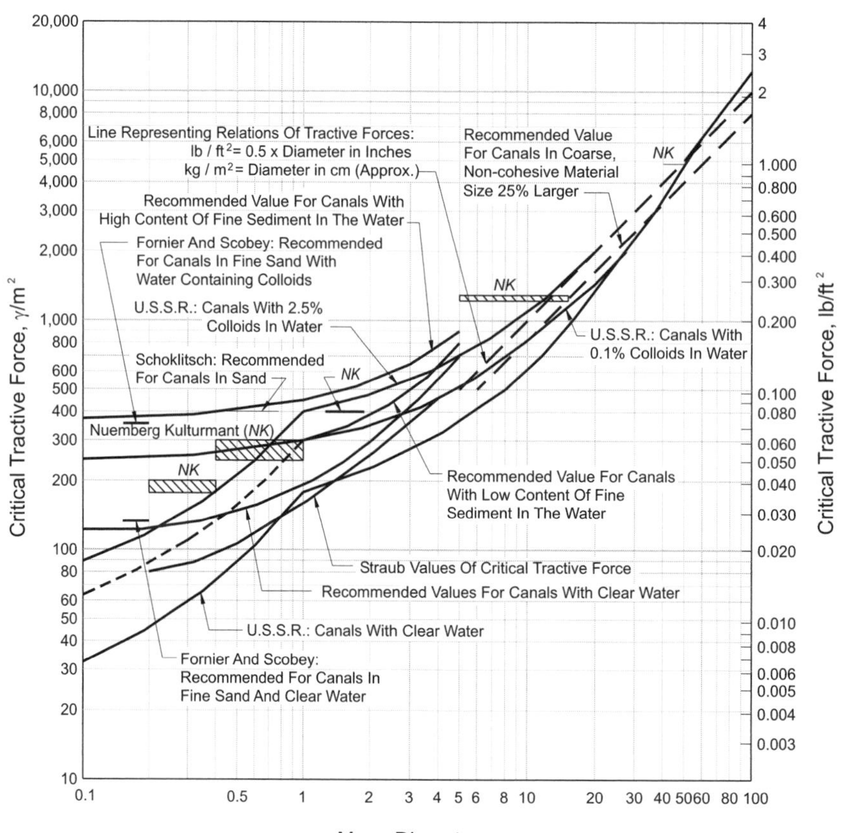

The permissible shear stress on the bottom of a straight channel in a cohesive soil can be obtained from Figure 8.9. It is called "unit tractive force" in this figure and is given in lb/ft^2 (1 lb/ft^2 = 47.9 N/m^2). Lane (1955) suggested that the permissible values should be reduced by about 10 percent for slightly sinuous channels, by 25 percent for moderately sinuous channels, and by 40 percent for very sinuous channels.

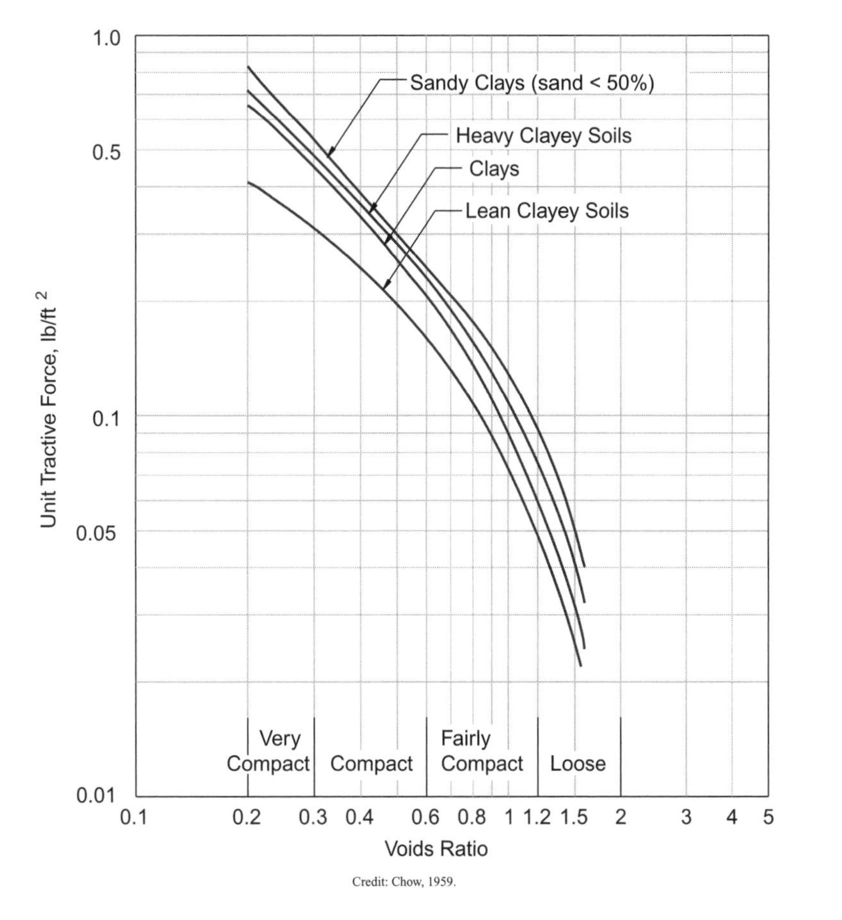

Figure 8.9 Determining the permissible shear stress (unit tractive force) on the bottom of a straight channel with cohesive soil (Chow, 1959)

The permissible shear stress on the sides of a channel is smaller than the permissible stress on the channel bottom. This difference is due to the addition of the gravitational force that acts on a soil particle and tends to roll it down the channel side slope. The *tractive force ratio* expresses the relationship between permissible stresses for the channel bottom and sides and can be used to find the permissible shear stress on the sides of the channel if the permissible shear stress for the bottom of the channel is known. It is defined as

$$K = \frac{\tau_s}{\tau_b} \qquad (8.9)$$

where K = tractive force ratio

τ_s = permissible stress for the channel sides (e.g., lb/ft^2, kg/m^2)

τ_b = permissible stress for the channel bottom (e.g., lb/ft^2, kg/m^2)

The value of the tractive force ratio can be computed as

$$K = \cos\xi\left(1 - \frac{\tan^2\xi}{\tan^2\varsigma}\right) = \left(1 - \frac{\sin^2\xi}{\sin^2\varsigma}\right) \tag{8.10}$$

where ξ = angle of inclination of the channel side slopes = arctan($1/z$) (z is the H:V ratio of the side slope)

ζ = angle of repose for the soil material

The *angle of repose* is the maximum angle of rest for dry sediment. This angle is a function of soil particle size and shape; in general, it increases with size and angularity. Figure 8.10 can be used to find the angle of repose for a particular soil (Chow, 1959).

In summary, design of a channel by the method of tractive force involves a determination of the permissible stresses on the channel bottom and sides, followed by the dimensioning of the channel such that the maximum actual stresses on the bottom and sides do not exceed the permissible values. Stresses on the sides are usually the controlling factor in a design, but this should always be checked. Example 8.4, which is adapted from Yang (1996), illustrates the procedure.

"And that's why we need a computer."

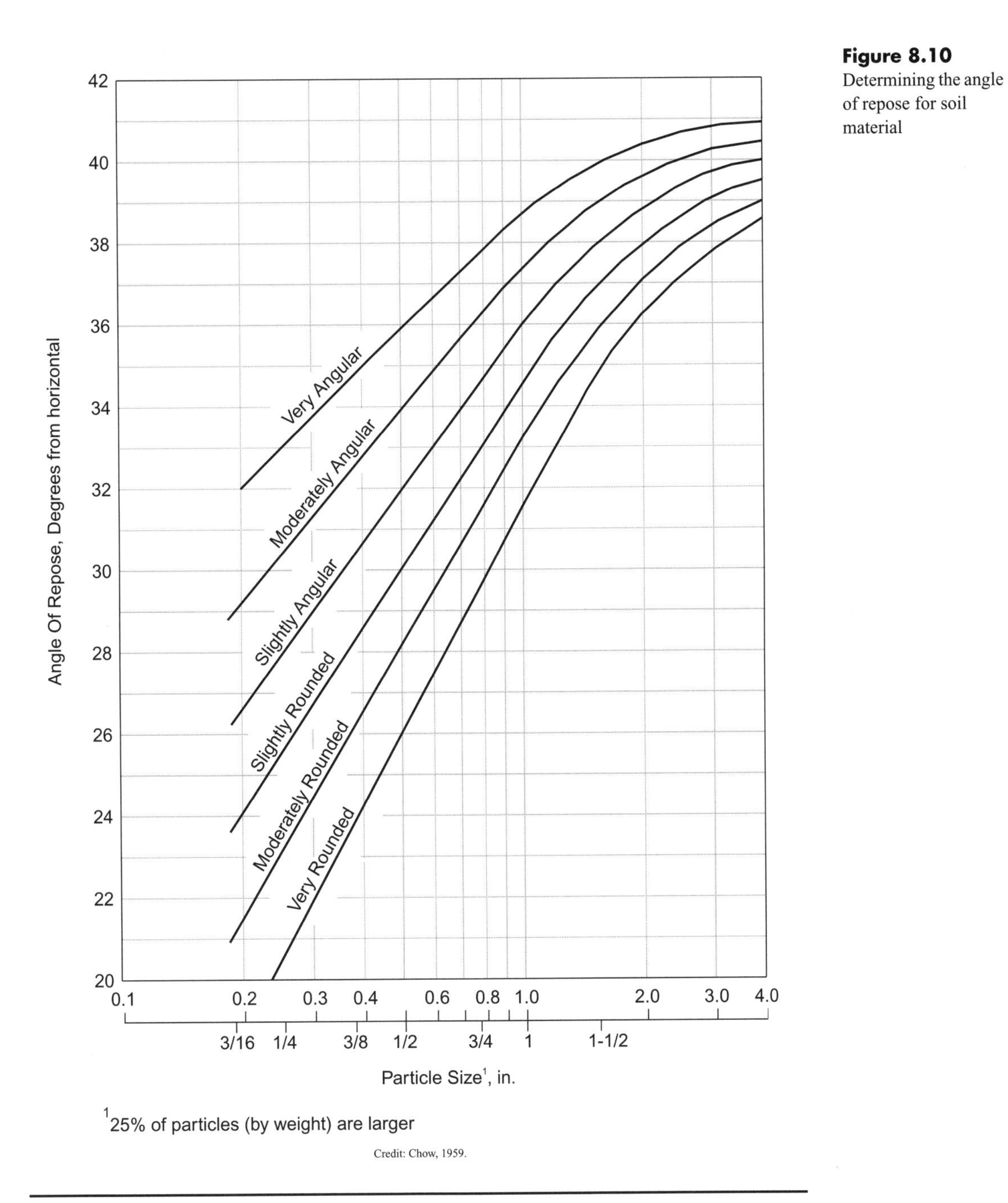

Figure 8.10
Determining the angle of repose for soil material

Example 8.4 – Erodible Channel Design with the Tractive Force Method. Design a straight trapezoidal channel using the tractive force method. The discharge is 60 m^3/s, the channel slope is 0.001, and Manning's $n = 0.015$. The soil material in which the channel is to be constructed is slightly rounded gravel with a median diameter of 40 mm.

Solution: The soil size of 40 mm is equivalent to 1.57 in. From Figure 8.10, the angle of repose is $\zeta = 37°$. The channel side slope must be flatter than this, and so it is chosen as 1.5:1 (H:V). The side slope angle is therefore $\xi = \arctan(1/1.5) = 33.7°$.

The permissible tractive force on the channel bottom is obtained from Figure 8.8 as $\tau_b = 4$ kg/m^2. The tractive force ratio is computed using Equation 8.10 as

$$K = \cos 33.7^\circ \left(1 - \frac{\tan^2 33.7^\circ}{\tan^2 37^\circ}\right)^{1/2} = 0.388$$

Therefore, the permissible tractive force on the channel sides is

$\tau_s = K \times \tau_b = 0.388 \times 4 = 1.55$ kg/m^2.

Determination of the actual channel cross-section dimensions requires an assumption as to whether the channel sides or bottom will control the design. One also needs to assume a channel width to depth ratio, B/y, for use with Figure 8.7. Once the dimensions have been determined on the basis of these assumptions, the reasonableness of the assumptions can be checked.

Assuming that the sides will control the design, and that the ratio $B/y = 4$, Figure 8.7 can be used to express the maximum stress on the channel sides as

$$\tau_s = 0.76\gamma y S_o = 0.76(1000)(0.001)y = 0.76y$$

Note that γ in this case is expressed in kg/m^3 rather than N/m^3 because the permissible shear stress is expressed in kg/m^3. Equating this expression to the permissible stress on the channel sides, determined earlier to be 1.55 kg/m^2, yields $y = 2.04$ m. Further, using the assumed width to depth ratio,

$$B = 4y = 8.2 \text{ m}$$

To check the validity of the assumption that the sides control the design, the maximum shear stress on the channel bottom is determined using Figure 8.7 as

$$\tau_b = 0.96\gamma y So = 0.96(1000)(2.04)(0.001) = 1.96 \text{ kg/m}^2$$

Because this value is smaller than the permissible tractive force on the bottom, the validity of the assumption is established.

A check of the validity of the width to depth ratio assumption requires that the channel discharge be computed on the basis of the determined dimensions and compared to the actual design discharge. For the determined channel dimensions of $z = 1.5$, $y = 2.04$ m, and $B = 8.2$ m, an application of the Manning equation yields $Q = 62.8$ m^3/s, which is very close to the design discharge of 60 m^3/s. If more accuracy is needed, the B/y ratio can be changed and the calculations repeated.

Completion of the design in this example would include checking the velocity and Froude number and adding a suitable amount of freeboard to the channel depth.

8.5 DESIGN OF GRASSED CHANNELS

Grassed channels are often preferable to lined or erodible channels in areas where aesthetics are a concern and for cases in which flow in the channel is intermittent. They are widely used as swales, in park settings, and for borrow and median ditches along highway rights-of-way. Grassed channels are generally not appropriate for use in channels that convey continuous flows.

Resistance to flow in a grassed channel depends on grass type and length, on the season of the year (because of its influence on the grass), and on the depth of flow relative to the grass height. Experiments conducted by the U.S. Soil Conservation Service

(Ree, 1949; Ree and Palmer, 1949; Stillwater Outdoor Hydraulic Laboratory, 1954) showed that, for a specific type and condition of grass, Manning's n could be related to VR, the product of the mean flow velocity and the hydraulic radius. The relationship between n and VR was found to be different for different types of grasses, and even for the same type of grass under different growing conditions. The concept of *retardance* of a grass was defined and introduced to account for these differences, and different grass species and growing conditions were classified into retardance categories from A (very high retardance) to E (very low retardance).

High retardances are generally associated with long and stiff grasses that are not easily flattened by flowing water. Low retardances, on the other hand, are associated with short grasses and those that are easily flattened out. Table 8.4 provides a listing of a number of common grass species with their retardance classifications for various growing conditions. Figure 8.11 shows the relationship between n and VR for each retardance category. The curves shown in Figure 8.11 are average relationships for all of the grasses within each category. Chow (1959) provides more detailed curves showing the variations from one grass species to another.

Permissible velocities in grassed channels depend on the grass species, growing condition, channel slope, and soil erodibility. Well-maintained grasses with high densities can generally tolerate higher velocities than can poorly maintained grasses with poor ground cover. Permissible velocities suggested by the Stillwater Outdoor Hydraulic Laboratory (1954) are presented in Table 8.5.

Recommended freeboard requirements for grassed channels are different than for lined and erodible channels. The 1-ft (30-cm) minimum recommended by ASCE (1992) should be maintained, but ASCE has also suggested that the minimum freeboard should be 6 in. (152 mm) plus the velocity head of the flowing water. Additional freeboard should be provided in curves where superelevation of the water surface occurs.

Design of a grassed channel begins with selection of the grass species to be used. This selection depends in large part on the climate and soil conditions where the channel is to be constructed. Sod-forming grasses, such as Kentucky bluegrass, centipede, and Bermuda grass, are generally preferable to bunch-forming grasses, particularly where discharges are high and/or channel slopes are steep. In newly constructed channels, consideration may be given to using annual grasses to provide temporary erosion protection until the permanent grass species is well established.

The second issue to be addressed in grassed channel design is selection of a channel side slope, z. Grassed channels typically require periodic mowing and other maintenance such as fertilization, and thus must be accessible by mowers and similar machinery. Slopes steeper than about 3 or 4:1 (horizontal:vertical) may be difficult to maintain.

Table 8.4 Degree of retardance for various grasses (Stillwater Outdoor Hydraulic Laboratory, 1954)

Retardance	Grass Type	Condition
A: Very high	Weeping love grass	Excellent stand, tall [avg. 30 in. (76 cm)]
	Yellow bluestem ischaemum	Excellent stand, tall [avg. 36 in. (91 cm)]
B: High	Kudzu	Very dense growth, uncut
	Bermuda grass	Good stand, tall [avg. 12 in. (30 cm)]
	Native grass mixture (little bluestem, blue grama, and other long and short Midwest U.S. grasses)	Good stand, unmowed
	Weeping love grass	Good stand, tall [avg. 24 in. (61 cm)]
	Lespedeza sericea	Good stand, not woody, tall [avg. 19 in. (48 cm)]
	Alfalfa	Good stand, uncut [avg. 11 in. (28 cm)]
	Weeping love grass	Good stand, mowed [avg. 13 in. (33 cm)]
	Kudzu	Dense growth, uncut
	Blue grama	Good stand, uncut [avg. 13 in. (33 cm)]
C: Moderate	Crabgrass	Fair stand, uncut [10 to 48 in. (25 to 122 cm)]
	Bermuda grass	Good stand, mowed [avg. 6 in. (15 cm)]
	Common lespedeza	Good stand, uncut [avg. 11 in. (28 cm)]
	Grass-legume mixture – summer (orchard grass, redtop, Italian rye grass, and common lespedeza)	Good stand, uncut [6 to 8 in. (15 to 20 cm)]
	Centipede grass	Very dense cover [avg. 6 in. (15 cm)]
	Kentucky bluegrass	Good stand, headed [6 to 12 in. (15 to 30 cm)]
D: Low	Bermuda grass	Good stand, cut to 2.5-in. (6-cm) height
	Common lespedeza	Excellent stand, uncut [avg. 4.5 in. (11 cm)]
	Buffalo grass	Good stand, uncut [3 to 6 in. (8 to 15 cm)]
	Grass-legume mixture – fall to spring (orchard grass, redtop, Italian rye grass, and common lespedeza)	Good stand, uncut [4 to 5 in. (10 to 13 cm)]
	Lespedeza sericea	After cutting to 2-in. (5-cm) height; very good stand before cutting
E: Very low	Bermuda grass	Good stand, cut to 1.5 in. (4-cm) height
	Bermuda grass	Burned stubble

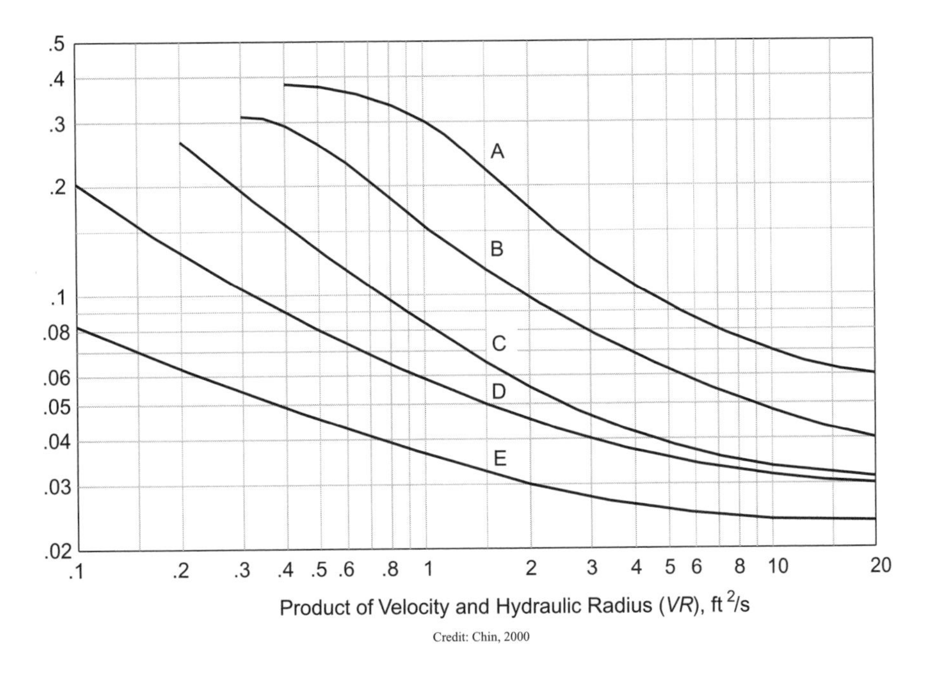

Credit: Chin, 2000

Figure 8.11 Relationship between Manning's n and VR for each retardance category

Table 8.5 Permissible velocities in grassed channels (Stillwater Outdoor Hydraulic Laboratory, 1954)

Grass Type	Slope (%)	Erosion-Resistant Soils ft/s	Erosion-Resistant Soils m/s	Easily Eroded Soils ft/s	Easily Eroded Soils m/s
Bermuda grass	0–5	8	2.4	6	1.8
	5–10	7	2.1	5	1.5
	> 10	6	1.8	4	1.2
Buffalo grass, Kentucky blue-grass, smooth brome, blue grama	0–5	7	2.1	5	1.5
	5–10	6	1.8	4	1.2
	> 10	5	1.5	3	0.9
Grass mixture	0–5	5	1.5	4	1.2
	5–10	4	1.2	3	0.9
	Do not use on slopes > 10%				
Lespedeza sericea, weeping love grass, ischaemum (yellow bluestem), kudzu, alfalfa, crabgrass	0–5	3.5	1.1	2.5	0.8
	Do not use on slopes > 5%, except for side slopes in a compound channel				
Annuals: used on mild slopes or as temporary protection until permanent covers are established – common lespedeza, Sudan grass	0–5	3.5	1.1	2.5	0.8
	Use on slopes > 5% is not recommended				

Procedure for Design of a Grassed Channel

Once the grass species and side slope have been selected, design of a grassed channel is usually treated as a two-stage procedure. In the first stage, it is assumed that the retardance category for the grass is relatively low (for example, due to winter conditions), with the net result being that the velocity and erosion potential in the channel will be relatively high. Thus, the intent of the first stage is to develop a channel cross section such that the permissible velocity will not be exceeded. In the second stage of the process, the retardance is assumed to be relatively high, with a corresponding low velocity and large depth of flow. The intent of this stage is to ensure that the channel depth is adequate to convey the design discharge in conditions where, for example, the grass has not been mowed in some time. Following is a description of the steps involved in each of the two design stages.

Stage 1.

1 Determine the design discharge and channel slope, and select the grass species and channel side slope to be used.

2 Assume a value for n and determine the corresponding value of VR from Figure 8.11. Use the curve indicative of a relatively low degree of retardance for the selected grass species.

3 Determine the permissible channel velocity from Table 8.5, and use it with the value of VR obtained in step 2 to determine the required hydraulic radius R.

4 Using the Manning equation, compute the product VR as

$$VR = \frac{1.49}{n} R^{2/3} S_o^{1/2} \qquad (8.11)$$

Compare this computed value to the value read from Figure 8.11 in step 2.

5 If the values of VR obtained in steps 2 and 4 do not agree, repeat steps 2 through 4 with different assumed n values until agreement is obtained.

6 Compute the required cross-sectional area as $A = Q/V$.

7 Use the values of A and R determined in steps 3 and 6 to determine the required cross-sectional dimensions.

Stage 2.

1 Using the cross-sectional dimensions determined in Stage 1 and an assumed depth of flow y, compute the cross-sectional area A and hydraulic radius R.

2 Compute the values $V = Q/A$ and VR.

3 Using the curve in Figure 8.11 for a relatively high retardance for the chosen grass, determine the value of Manning's n.

4 Compute the flow velocity using Manning's equation and compare it to the value determined in step 2.

5 If the velocities in steps 2 and 4 do not agree, repeat steps 1 through 4 with different assumed flow depths until agreement is obtained.

6 Add freeboard to the computed depth to obtain the total required depth of the channel.

Example 8.5, from Chow (1959), illustrates the procedure.

Example 8.5 – Grassed Channel Design. Design a grassed channel to convey a discharge of 50 cfs on a 4-percent slope. The soil is erosion-resistant, and the grass selected is a grass-legume mixture. Assume that the channel will be trapezoidal with side slopes of 3:1 (H:V).

Solution: Referring to Table 8.4, the grass-legume mixture has a low (D) retardance in the fall, winter, and spring, and a moderate (C) retardance during summer months. Thus, the D curve will be used in the first stage of the design, and the C curve will be used in the second stage. For the first stage, the permissible velocity is found from Table 8.5 to be 5 ft/s.

The trial-and-error steps in the first stage of the design are summarized in the following table.

Trial	n	VR	R	$VR = (1.49/n)R^{5/3}S^{1/2}$
1	0.04	3.0	0.60	3.2
2	0.038	3.5	0.70	4.3
3	0.042	2.5	0.50	2.2
4	0.041	2.75	0.55	2.7

The required cross-sectional area and hydraulic radius are therefore $A = 50/5 = 10$ ft^2 and $R = 0.55$ ft. Thus, using Table 8.1, one can write the equations

$$A = 10 \text{ ft}^2 = By + 3y^2$$

$$R = A/P = 0.55 \text{ ft} = (By + 3y^2)/[(B + 2y(1+3^2)^{1/2}]$$

Solving these equations simultaneously yields $B = 14$ ft and $y = 0.62$ ft. Thus, the channel cross section will have a bottom width of 14 ft, side slopes of 3:1, and a low-retardance depth of 0.62 ft.

The intent of Stage 2 is to determine the depth of flow that will occur under high-retardance (summer) conditions. The trial-and-error calculations are shown in the following table:

Trial	y	A	R	V	VR	n	$V = (1.49/n)R^{2/3}S^{1/2}$
1	0.65	10.4	0.57	4.81	2.74	0.049	4.18
2	0.70	11.3	0.61	4.44	2.70	0.049	4.37

The correct depth is therefore $y = 0.70$ ft.

The minimum amount of freeboard provided should be the larger of either 1 ft or 6 in. plus the velocity head. In Stage 2, the velocity was found to be 4.4 ft/s, so the velocity head plus 6 in. is 0.80 ft. Therefore, the required freeboard is determined to be 1 ft, and the total channel depth should be about 1.7 ft.

Open-Channel Stabilization Options

Many channel lining options are available for consideration in the design of open channels. The choice of lining primarily depends on regulations and site-specific issues. Applicable regulations may include highway design standards, environmental regulations related to concerns such as erosion and sediment control and stream stabilization. The features of a site that must be considered include slopes, other drainage systems, soil erodibility, aesthetics, and existing riparian vegetation.

States and localities typically have design handbooks or guidelines to make the process of selecting the channel lining more straightforward. For example, the Virginia Erosion and Sediment Control Handbook (Virginia Dept. of Conservation and Recreation, 1992) details applicable regulations, channel design criteria, and lining types that satisfy Virginia erosion and sediment control regulations, primarily for site construction. The FHWA also provides guidance on designing channels with flexible linings (Chen and Cotton, 2000).

Channel linings range from simple grass vegetation to heavily armored, concrete, engineered structures. The table below shows approximate ranges for shear stress for different channel lining types. The second table provides descriptions of channel lining materials and suggested references for further information.

Design standards for all of these lining types are very product specific. Although shear stress is one consideration, a factor such as permissible velocity, side slope, soil erodibility, or lining decomposition time may be of greater concern for some lining materials.

Maximum Allowable Shear Stress Ranges (Burchett, 1998)

Lining Material	Shear Stress (lb/ft^2)	Shear Stress (Pa)
Bare soil	0.1–2	0.15–3.0
Erosion control blanket over bare soil	1–5	1.5–7.4
Turf reinforcement mat over bare soil	3–8	4.5–11.9
Unreinforced Vegetation	0.5–4	0.7–11.9
Turf reinforcement mat, reinforced vegetation	4–10	6.0–14.9
Rock rip-rap	8–20	11.9–29.8
Gabions	Design Dependent	

Lining Type	Lining	Description
Erosion control blankets: Degradable blankets used to provide channel protection until permanent vegetation is established	Excelsior mat	Excelsior (dried, shredded wood) covered with fine paper net covering
	Jute mesh	Mat lining woven of jute yarn that varies in diameter
	Combination mats or blankets	Photo-degradable plastic netting that covers and is entwined with a natural organic or man-made mulching material such as straw or coconut
Permanent turf reinforcement mats: Mats designed to stay in place that provide a matrix to help support roots and vegetation; permanent vegetation will grow through the lining openings	Fiberglass mat	Fine, loosely woven glass fiber mat similar to furnace air-filter material; different fiber densities allow or inhibit underlying particular plant growth
	Fiberglass roving	A lightly bound ribbon of continuous glass fibers, applied using a special nozzle attached to an air compressor
Permanent armoring: Erosion-resistant structures composed of hard materials	Rip-rap	Permanent, erosion-resistant ground cover of large, loose, angular stone; graded rip rap contains stones of various sizes, while uniform riprap contains stones that are fairly close in size
	Gabions	Rectangular, rock-filled wire baskets; the amount of channel armoring needed affects rock and wire specifications
	Interlocking concrete blocks	Interlocking concrete blocks that may have extra space for soil for the establishment of permanent vegetation

8.6 RIPRAP-LINED CHANNELS

A special case of lined channels has the side slopes (and possibly the bottom also) of the channel lined with riprap. Construction of channels in this manner is a common practice, and may be viewed as a compromise between a concrete-lined channel and an erodible or grassed channel. In some cases, a trickle channel may be lined with riprap, while the main body of the conveyance may be vegetated or unlined.

While a riprap-lined channel may be attractive in terms of its capital cost, the designer and review agency should bear in mind that such channels are more difficult to maintain and may require more frequent inspections than other channel types. Design of riprap-lined channels requires detailed attention to required rock dimensions, not only to withstand the erosive forces of moving water, but also the "erosive" effect of neighborhood children who revel in the splashes caused by throwing the rocks into the stream. The availability or cost of obtaining suitably sized rock may make a riprap-lined channel an unsuitable alternative. It is foolhardy to use rock that is too small for a particular application simply because larger rock is not available or too expensive, as the riprap lining will fail in short order.

Design of riprap-lined channels is a rather specialized topic and is beyond the scope of this text. The reader desiring more information on this topic is referred to Volume 2 of Denver's *Urban Storm Drainage Criteria Manual* (UDFCD, 2000) and to other references cited therein.

8.7 MODIFYING EXISTING CHANNELS

Many open channel design projects involve making modifications to existing natural channels, typically to increase channel capacity. These modifications may be necessary because of increased runoff from new impervious areas, or because of flooding of existing developments. Other options for dealing with these types of issues, such as stormwater detention and floodproofing structures in the floodplain, may be viable and should be considered prior to undertaking a channelization project.

Channelization measures that increase channel capacity are the focus of this section. The basic ways in which capacity can be increased are:

- Increasing the cross-sectional area of the channel be making it deeper or wider, or by installing levees or floodwalls
- Reducing the roughness of the channel by removing debris and woody vegetation, straightening the channel, or installing a smooth lining such as concrete
- Increasing the slope of the channel, which is often accomplished in concert with channel straightening
- Removing or enlarging downstream controls that cause flow to back up in the channel

In addition to hydraulic capacity, an engineer designing channel improvements should consider factors such as habitat development and preservation, aesthetics, recreation, and sediment transport. Channel stability is another important issue, except in special cases such as a stream flowing over bedrock or a channel lined with paving material or large boulders.

Traditional channelization projects have often led to undesirable responses, such as increased erosion and/or sedimentation (McCarley et al., 1990; Brookes and Shields, 1996). Designing an open channel involves more than simply drawing a straight line between two points of known elevation and applying the Manning equation using the design flow rate. Engineers designing projects dealing with alluvial conditions need to understand *fluvial geomorphology*—the science that assesses the shape and form of a watercourse and the contributing physical processes—or have someone on the design team with expertise in channel erosion and sedimentation (Knighton, 1998; Nunnally, 1978).

A current trend is to move away from designing a simple channel shape, such as a trapezoid, to carry peak flow. Instead, there is an increasing use of more complex geometries that perform better over a wide range of flows. Various governmental guidelines reflect this trend (Federal Interagency Stream Corridor Restoration Working Group, 2001; U.S. Army Corps of Engineers, 1989). Stream restoration has also received considerable attention from engineers in recent years, including ASCE conferences such as the Joint Conference on Water Resources Engineering and Water Resources Planning and Management in 2000 (ASCE, 2000) and the Wetlands Engineering and River Restoration conference in 2001 (Hayes, 2001).

An improved cross section often consists of a main channel to handle smaller flows and overbank areas that convey flows from larger storm events. This approach is being used in the construction of new channels, improvements to existing channels, and restoration of channels previously impacted by development. Figure 8.12 contrasts a simple channel design that considers only peak flow with a more complex cross section that includes a floodway/overbank area to be used for recreational purposes or as habitat. In some cases, this overbank area may be large enough for hiking and bike paths or even ball fields, as with Cherry Creek in Denver, Colorado.

Figure 8.12
A simple channel design versus a complex channel design with a smaller main channel and floodway

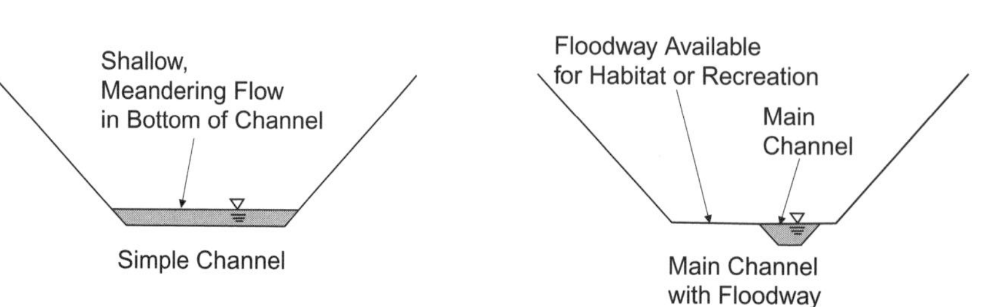

For natural channels, it is important to evaluate the channel at the peak design flow as well as at other important flow rates. An important flow is the *dominant discharge,* or *channel-forming flow,* that corresponds to a one- or two-year return interval flow for humid areas (Richards, 1982) and slightly longer periods in arid areas (U.S. Army Corps of Engineers, 1994). The channel-forming flow is not necessarily the same as the *bankfull discharge,* which is the discharge at which the channel begins overflowing onto the floodplain (Copeland et al., 2001). If a channel remains stable for long periods of the time, the bankfull discharge should be close to the dominant discharge because the channel has adjusted to hydraulic and sediment loads (Shields et al., 2003). Computer programs can assist the engineer in creating a rating curve for evaluation of a channel over a range of flows.

Modeling Focus: Channel Design

Computer programs can assist in the design of open channels, but there is great variation in the level of analysis among available programs. In the design of smaller prismatic channels, it is frequently reasonable to assume that the channel operates under normal flow conditions. FlowMaster is an example of a program that can be used in the design of open channels where a normal depth assumption is appropriate.

This example outlines a design for a proposed channel to convey the peak discharge of 125 cfs for the 25-year storm. It is assumed that the flow in the channel will be close to normal depth. The lower portion of the channel will be designed to handle a low flow of 31 cfs. This portion of the channel will be lined with concrete (Manning's $n = 0.013$) and will be trapezoidal with a 3-ft-wide bottom and 1:1 side slopes. The banks of the channel above the elevation of the concrete will be Bermuda grass on a 3:1 slope. The longitudinal slope of the channel is 0.25 percent. A freeboard of 1 ft should be provided above the 25-year storm flow depth.

The engineer can start by designing the trapezoidal, concrete, low-flow portion of the channel. The following data can be entered into a program such as FlowMaster to solve for depth:

Manning's *n*	0.013	Bottom width	3 ft
Long. Slope	0.0025	Discharge	31 cfs
Side slopes	1:1		

The normal depth is computed to be 1.37 ft. The flow is subcritical, with an average velocity of 5.2 ft/s. Based on this computed depth, the engineer decides the make the concrete portion of the channel 1.5 ft deep.

Next, the engineer must design the upper (grass) portion of the channel. A value for *n* corresponding to a low retardance for Bermuda grass must be selected. The lowest retardance category for Bermuda grass is category E. From Figure 8.11, an *n* value of 0.03 is assumed for the initial calculation. The corresponding value of *VR* is 2 ft^2/s.

The engineer assumes a conservative maximum depth and enters the geometry of the full cross-section into the computer program being used. For this example, the following cross-section points were used:

Station (ft)	Elevation (ft)	Station (ft)	Elevation (ft)
0+00.0	5.5	0+16.5	0.0
0+12.0	1.5	0+18.0	1.5
0+13.5	0.0	0+30.0	5.5

The assumed *n* for grass of 0.03 was used for stations 0+00.0 to 0+12.0 and 0+18.0 to 0+30.0. The *n* value for concrete, 0.013, is used between stations 0+12.0 and 0+18.0. The engineer can then solve for the flow characteristics in the channel for a flow of 125 ft^3/s.

Note: To apply Manning's equation to a channel such as this one, which has a composite roughness, a method for computing a weighted *n* value must be employed. In this example, the method proposed by Pavlovskii, by Muhlhofer, and by Einstein and Banks (Chow, 1959) is used. Software programs may vary in the method or methods available for weighting Manning's *n*.

The program computes the following:

Flow depth	3.54 ft	Wetted perim.	20.16 ft
Flow area	31.5 ft^2	Velocity	3.96 ft/s

Checking the initial assumption of $n = 0.03$ for the grass, *VR* is computed as (3.96)(31.5)/20.16 = 6.19 ft^2/s. From Figure 8.11, $n = 0.026$. With this *n* for the grass portions of the channel, the following output is obtained:

Flow depth	3.36 ft	Wetted perim.	19.03 ft
Flow area	28.3 ft^2	Velocity	4.41 ft/s

VR is now found to be 6.6 ft^2/s, and our assumed $n = 0.026$ is acceptable. Checking the velocity with the permissible velocities for Bermuda grass on slopes less than 5 percent, the velocity of 4.41 ft/s is acceptable, even for easily eroded soils.

Next, the design must be checked assuming a high retardance. The highest retardance for Bermuda grass is category B. The initial value for *n* is selected as 0.07 after consulting Figure 8.11, and this value corresponds to *VR* = 3 ft^2/s. Updating the input data for this *n*, the results are as follows:

Flow depth	4.89 ft	Wetted perim.	28.68 ft
Flow area	61.6 ft^2	Velocity	2.03 ft/s

Thus, *VR* is computed to be 4.36. Referring again to Figure 8.11, the new value for n is read as 0.065. Updating this in the input data and solving, the results are as follows:

Flow depth	4.75 ft	Wetted perim.	27.82 ft
Flow area	58.0 ft^2	Velocity	2.15 ft/s

In this iteration, *VR* is computed as 4.48 ft^2/s, and our assumed value of *n* is acceptable. For the final design, the engineer adds an additional 1 ft to the flow depth for freeboard, making the total depth of the channel 5.75 ft. A plot of the cross-section for the final channel design is shown below.

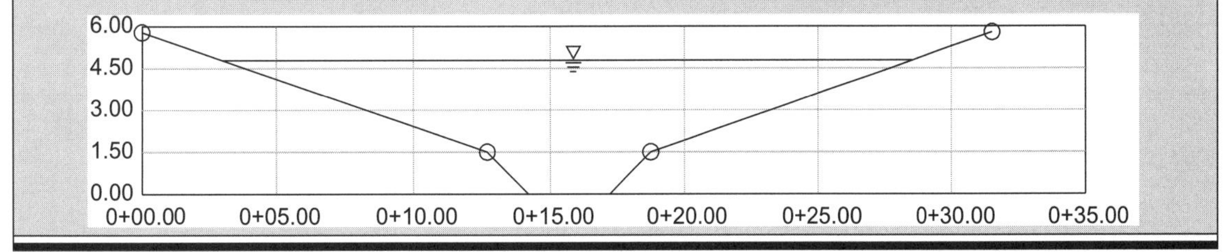

In natural channels, the roughness, particularly in the overbanks, may change over time due to the growth of vegetation, the addition of obstructions such as buildings and fences, and maintenance practices or lack of maintenance. The extent of the effects of such changes can be predicted by performing additional analyses to investigate the sensitivity of the channel to changes in the roughness factor.

The channel geometry of bends differs from straight sections of channels. The channel is deeper at the outside of the bend where the maximum erosion occurs, and the channel is shallower with slower velocity on the inside of the bend where a gravel or sand bar usually forms. Special protection measures may be necessary on the outside bank to prevent erosion.

Projects that involve straightening of streams have the effect of increasing velocity and therefore erosion. Armored riffle sections or drop structures can enable the flow to remain subcritical through most of the reach, and may be preferable to lining the entire length of the channel.

8.8 CHAPTER SUMMARY

Many factors must be considered in the design of open channels. The most obvious is the hydraulic capacity of the channel, but other factors such as flow velocity, Froude number, side slopes, longitudinal slope, freeboard, lining material, construction cost, maintenance cost, aesthetics, right-of-way requirements, and dry-weather flow performance should be considered as well.

The best hydraulic section is a channel section for which the wetted perimeter is minimized for a given cross-sectional area. However, the steepness of this section's side slopes limit its practical application.

Open channels may be classified as lined, erodible, or grassed. Some channels have characteristics of more than one of these types. The design of erosion-resistant lined channels is the most straightforward, and is primarily concerned with conveyance capacity.

When designing erodible channels, one must pay particular attention to the forces that could cause the channel to scour. Two methods for designing channels in order to prevent scour are described in the text. The first method uses a maximum permissible velocity, and the second uses a maximum permissible shear stress.

Grassed channels are often used in instances where there is intermittent flow. The flow retardance properties of various grass types in different conditions are considered, as well as the maximum allowable velocity and the suitability of different grass species for particular climates and soil conditions.

Often, channel design involves planning modifications to an existing channel rather than construction of an entirely new conveyance. In addition to ensuring that the channel will have adequate capacity for the design flow, the engineer should evaluate flow levels that will occur more frequently and should consider other factors such as the effect on aquatic habitat, channel stability, and the potential for erosion.

REFERENCES

Aisenbrey, A. J., Jr., R. B. Hayes, H. J. Warren, D. L. Winsett and R. B. Young. 1978. *Design of Small Canal Structures.* U.S. Department of Interior, Bureau of Reclamation.

American Society of Civil Engineers (ASCE). 1977. *Sedimentation Engineering.* ASCE Manuals and Reports on Engineering Practice, No. 54. New York: ASCE.

American Society of Civil Engineers (ASCE). 1992. *Design and Construction of Urban Storm Water Management Systems.* New York: ASCE.

American Society of Civil Engineers (ASCE). 2000. *Joint Conference on Water Resources Engineering and Water Resources Planning and Management.* Minneapolis, Minn.: ASCE.

Armortec. 2003. Product literature for Armorflex concrete mats. (www.armortec.com).

Brookes, A. and F. D. Shields. 1996. *River Channel Restoration: Guiding Principles for Sustainable Projects.* New York: John Wiley & Sons.

Burchett, C. 1998. "Open Channel Linings: Choosing the Right Erosion Control Proudcts." Burns & McDonnell Tech Briefs 1998, no. 3. Kansas City, Missouri: Burns & McDonnell (www.burnsmcd.com).

Chen, Y. H. and B. A. Cotton. 2000. *Design of Roadside Channels with Flexible Linings,* Hydraulic Engineering Circular no. 15 (HEC-15). McLean, Virginia: Federal Highway Administration (FHWA).

Chow, V. T. 1959. *Open-Channel Hydraulics.* New York: McGraw-Hill.

Copeland, R. R., D. N. McComas, C. R. Thorne, P. J. Soar, M. M. Jonas, and J. B. Fripp. 2001. *Hydraulic Design of Stream Restoration Projects.* Technical Report ERDC/CHL TR-01-28. Vicksburg, Mississippi: U.S. Army Corps of Engineers, Engineer Research and Development Center.

Federal Interagency Stream Corridor Restoration Working Group. 2001. *Stream Corridor Restoration: Principles, Processes and Practices.* GPO 0120-A. (http://www.usda.gov/stream_restoration/).

Fortier, S., and F. C. Scobey. 1926. "Permissible Canal Velocities." *Transactions, American Society of Civil Engineers* 89: 940–956.

Hayes, D. F., ed. 2001. "Wetlands Engineering and River Restoration." ASCE Conference Proceeding. Reno, Nevada: ASCE.

Hjulstrom, F. 1935. "The Morphological Activity of Rivers as Illustrated by River Fyris," *Bulletin of the Geological Institute* 25, chap. 3.

Knighton, D. 1998. *Fluvial Forms and Processes.* Arnold.

Lane, E. W. 1955. "Design of Stable Channels." *Transactions, American Society of Civil Engineers* 120: 1234–1260.

McCarley, R. W., J. J. Ingram, B. J. Brown, and A. J. Reese. 1990. *Flood Control Channel National Inventory.* MP HL-90-10. Vicksburg, Mississippi: US Army Engineer Waterways Experiment Station.

McWhorter, J. C., T. G. Carpenter, and R.N. Clark. 1968. "Erosion control criteria for drainage channels." Study conducted for the Mississippi State Highway Department and the Federal Highway Administration. State College, Mississippi: Department of Agricultural Engineering, Mississippi State University.

Mehrota, S. C. 1983. "Permissible Velocity Correction Factor." *Journal of the Hydraulics Division, ASCE* 109, no. HY2: 305–308.

Nunnally, N. R. 1978. "Stream Renovation: An Alternative to Channelization." *Environmental Management* 2(5): 403.

Olsen, O. J. and Q. L. Florey. 1952. *Sedimentation Studies in Open Channels: Boundary Shear and Velocity Distribution by Membrane Analogy, Analytical, and Finite-Difference Methods.* U.S. Bureau of Reclamation, Laboratory Report Sp-34, August.

Petratech. 2003. Product literature for Petraflex interlocking blocks. (www.petraflex.com).

Ree, W. O. 1949. "Hydraulic Characteristics of Vegetation for Vegetated Waterways." *Agricultural Engineering* 30, no. 4: 184–187 and 189.

Ree, W. O. and V. J. Palmer. 1949. *Flow of Water in Channels Protected by Vegetative Lining.* U.S. Soil Conservation Service, Technical Bulletin No. 967.

Richards, K. 1982. *Rivers: Form and Processes in Alluvial Channels. New York:* Metheun and Co.

Shields, F. D., R. R. Copeland, P. C. Klingeman, M. W. Doyle, and A. Simon. 2003. "Design for Stream Restoration." Accepted for ASCE *Journal of Hydraulic Engineering.*

Stillwater Outdoor Hydraulic Laboratory. 1954. *Handbook of Channel Design for Soil and Water Conservation.* U.S. Soil Conservation Service, SCS-TP-61.

Urban Drainage and Flood Control District (UDFCD). 2001. *Urban Storm Drainage Criteria Manual, Vol. 2, Major Drainage.* Denver, Colorado: UDFCD.

U.S. Army Corps of Engineers. 1989. *Environmental Engineering for Local Flood Control Channels.* Engineering Manual 1110-21601. Washington, D.C.: USACE.

U.S. Army Corps of Engineers. 1994. *Hydraulic Design of Flood Control Channels.* Engineering Manual 1110-21601. Washington, D.C.: USACE.

U.S. Bureau of Reclamation (USBR). 1987. *Design of Small Dams.* Lakewood, Colorado: USBR.

Virginia Department of Conservation and Recreation (DCR). 1992. *Virginia Erosion and Sediment Control Handbook,* 3rd ed. Richmond, Virginia: DCR.

Yang, C. T. 1976. "Minimum Unit Stream Power and Fluvial Hydraulics." *Journal of Hydraulic Engineering.* 105 (HY7): 769.

Yang, C. T. 1996. *Sediment Transport: Theory and Practice.* New York: McGraw-Hill.

PROBLEMS

8.1 Determine the discharge in a trapezoidal concrete channel with a bottom width of 2.4 m and side slopes of 1 to 1. Uniform depth is 1.8 m, the bottom slope is 0.009, and Manning's $n = 0.013$.

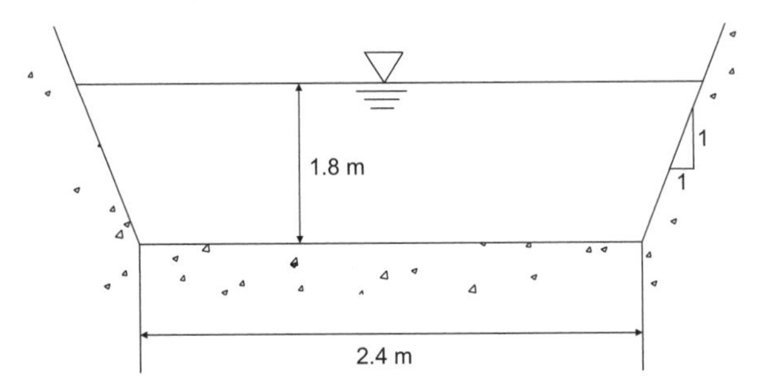

8.2 A concrete-lined channel with a triangular cross-section conveys a flow of 250 cfs. The slope is 0.005. The side slopes are 2H:1V and Manning's n is 0.012. Find the uniform flow depth.

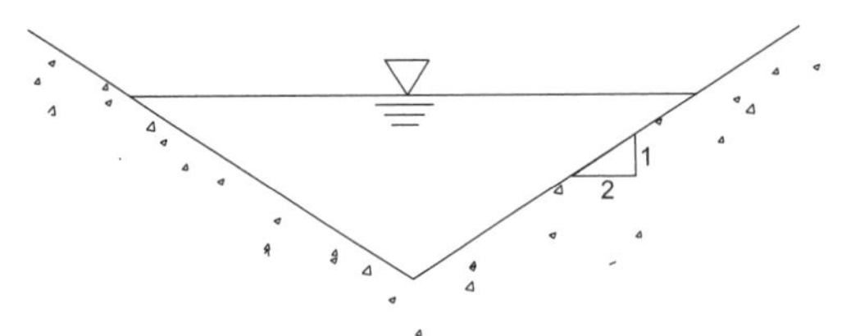

8.3 What is the uniform depth for a flow of 4.25 m^3/s in a 1.8-m wide, rectangular wood-planed channel ($n = 0.012$) with a bottom slope of 0.002?

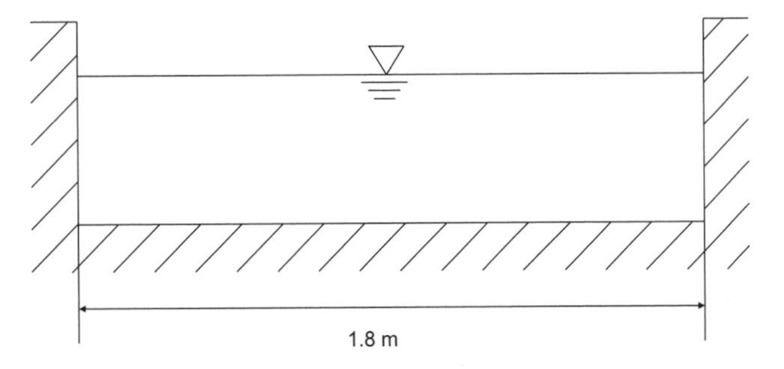

8.4 Use the maximum permissible velocity method to design an erodible channel to transport 250 cfs. The channel bottom and the side slopes are to be constructed from sandy loam. Use a trapezoidal cross section and a slope of 0.00033, and design for water with colloidal silts. Sketch your design cross section.

8.5 Use the tractive force method to design a channel lined with 2-in., slightly angular gravel. The design discharge is 500 cfs and the channel slope will be 0.002. Use a Manning's $n = 0.020$. Sketch your design cross section.

8.6 A trapezoidal channel is lined with a good stand of uncut buffalo grass that is 75 to 150 mm high. The channel slope is 0.01, the bottom width is 0.8 m, and the side slopes are 3H:1V. Find the channel capacity and the flow velocity for a depth of 0.5 m.

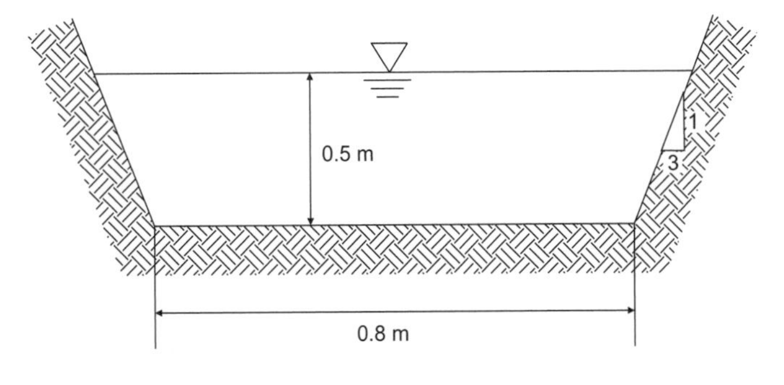

A triple-barrel box culvert with small wingwalls in a riprap channel conveys flow under a roadway.

CHAPTER

9

Culvert Design

A culvert is a hydraulically short underground water conveyance conduit that allows water to flow through an obstruction such as a highway or railway embankment (see the preceding page). The orientation of a culvert's longitudinal axis is usually at a right angle, or nearly so, to the centerline of the obstruction under which it passes. In some instances, structures constructed as culverts may be considered bridges if their spans exceed a certain dimension. Transportation departments in the United States usually consider any culvert with a total span (counting all culvert barrels, if there are more than one) greater than 20 ft (6 m) to be a bridge. It may sometimes be appropriate to hydraulically model a structure classified as a "bridge" by using culvert analysis methods; however, the techniques used to model bridge hydraulics generally differ from those used with culverts and are beyond the scope of this book. [The reader is referred to *Floodplain Modeling Using HEC-RAS* (Dyhouse et al., 2003) for information on bridge modeling.]

Typically, a culvert is designed to convey the peak flow from a specified design storm, and its performance is checked for a storm having a longer recurrence interval. The purpose of this check is to assess the magnitude of roadway or embankment overtopping expected to occur for the larger event. A local or regional review agency generally specifies the recurrence interval for a design. Chapters 4 and 5 present hydrologic procedures for calculating discharges for both the design event and check event, and Section 9.1 of this chapter presents a discussion of design flow estimation as it relates specifically to culverts.

As stated in Section 2.2 (page 22), culverts are available in various shapes, sizes, and materials. Materials and shapes are described in detail in Section 9.2. Section 9.3 provides a discussion of various types of culvert end treatments and their hydraulic performance characteristics and describes special inlet configurations that can be used when allowable headwater depths are limited.

The horizontal and vertical alignments of a culvert should closely follow those of the channel or stream. This typical alignment is sometimes modified for reasons relating to hydraulic performance, cost, or other constraints. Possible modifications include the use of a doglegged alignment or the use of a drop at the culvert entrance to reduce the barrel slope. Issues associated with such modifications are presented in Section 9.4.

Section 9.5 presents the detailed hydraulic performance characteristics of a culvert. The discussion includes the controls that affect hydraulic performance, the potential effects of tailwater elevation, and detailed computational procedures for flow profile estimation. Also discussed are methods for estimating outlet velocity, which is necessary for design of energy dissipation measures.

During storm events having recurrence intervals longer than that of the culvert design storm, or in cases where the culvert becomes clogged with debris, overtopping of the roadway or embankment is expected to occur. Prediction of the weir flow characteristics of roadway overtopping is presented in Section 9.6. Section 9.7 describes the development of culvert performance curves, which provide a convenient way of assessing the headwater elevation at a culvert for any discharge of interest.

Because the flow exiting the downstream end of a culvert often has a velocity higher than what can be tolerated in the downstream channel, or because of the acceleration of flow at the entrance, erosion protection measures are generally needed at the outlet and/or inlet of a culvert. This protection may consist of a concrete apron, protective riprap, or a geosynthetic material for reinforcement of soil and vegetation. Procedures for riprap erosion protection design at storm sewer outlets are presented in Section 11.7, and these procedures can be applied to culverts as well. Section 9.8 discusses outlet protection considerations specific to culvert design.

Because culverts often convey water in natural streams that may support fish and other aquatic wildlife, minimizing the disruption of migration pathways may be a design consideration. A discussion of designing for some of these environmental factors is provided in Section 9.9.

9.1 CULVERT DESIGN FLOWS

In most cases, culverts are designed to convey a discharge corresponding to a chosen recurrence interval. In such instances, one must estimate only the peak runoff rate. However, for cases in which a culvert causes significant ponding upstream of its entrance, the reduction of the peak discharge caused by detention effects may need to be considered. In such cases, the engineer should develop and route a complete runoff hydrograph. Chapter 5 outlines procedures for peak flow and hydrograph estimation. Section 12.7 discusses flow routing procedures.

A government review agency usually specifies the recurrence interval for which a design discharge or hydrograph should be determined. This criterion and others are provided in the interest of promoting consistency from one design to another. In highway design applications, culvert design flows are often estimated for a 50-year event based on FHWA design frequency recommendations (Brown, Stein, and Warner, 2001). Lower or higher recurrence intervals may be justified in other applications.

In addition to developing the design discharge, the engineer should estimate a *check flow*. The recurrence interval associated with the check flow is longer (and thus the discharge greater) than that of the design flow, because the check flow is used to assess performance when embankment overtopping is expected to occur. Again, local or state review agencies usually specify the check flow recurrence interval that should be used; in highway design applications, it is often the 100-year event.

For cases in which a performance curve is needed for a culvert design, a range of discharges corresponding to various recurrence intervals should be estimated. The low end of the discharge range should be lower than the design discharge. The high end of the range should be equal to or greater than the check flow discharge. In performance curve development (Section 9.7), the headwater elevation is estimated and graphed for each of these estimated discharges.

An alternative to basing design discharge on recurrence interval criteria is to base it on an economic analysis. Figure 9.1 shows the costs associated with a circular culvert as a function of its diameter. All costs are expressed as a present worth using appropriate discounting formulas. The capitalized cost curve represents the costs of design, construction, and maintenance of the culvert over its lifetime. Culverts with large diameters are clearly more costly than culverts with small diameters, and thus the capitalized cost increases with the diameter. The damage curve represents the present value of flood damages anticipated to occur when the culvert's capacity is exceeded. Because the capacity of a smaller culvert will be exceeded more frequently, the resulting damages will be higher than those expected for a larger culvert. Thus, the damage curve has a negative slope. The total cost associated with a culvert is the sum of the capitalized cost and the damages. It is shown as a third curve in Figure 9.1. The diameter corresponding to the minimum cost on the total cost curve represents the least-cost culvert design.

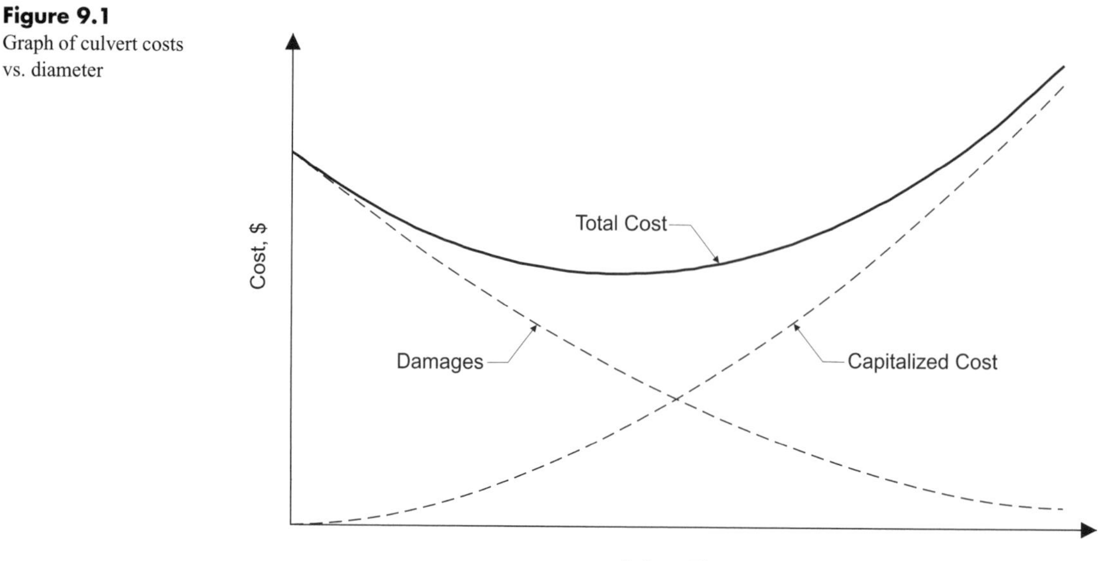

Figure 9.1
Graph of culvert costs vs. diameter

An economic analysis such as that just described is usually not accomplished in practice because of difficulties associated with estimating the damage curve. The analysis for an area can be further complicated by the presence of multiple culverts providing differing levels of protection. Nevertheless, performing an economic analysis is an enlightening exercise, as the true total cost of a culvert becomes evident. Such an analysis also shows that most culvert designs are likely less than optimal in economic terms.

9.2 CULVERT TYPES AND MATERIALS

Culverts, like storm sewers, come in a variety of shapes, sizes, and materials. Cross-sectional shapes associated with pipe culverts are circular, elliptical (horizontal or vertical), and arched. Culverts may also be constructed from precast or cast-in-place rectangular box sections, or from field-assembled structural-plate-arch materials. These latter shapes are common for large-capacity culverts. Materials used for culvert design are most commonly reinforced precast concrete, cast-in-place concrete, and corrugated steel. Section 11.1 discusses available materials and sizes. Information that is more detailed can be obtained from pipe manufacturer associations such as the American Concrete Pipe Association (ACPA) and the American Iron and Steel Institute (AISI). Other materials used for culverts include corrugated aluminum, polyethylene, and polyvinyl chloride (PVC). Aluminum and plastic materials offer advantages over steel in terms of corrosion, but are more limited in size availability. They also do not offer the structural strength of concrete or steel.

In many cases, to address a situation where there is limited cover over the top of a culvert, or to approximate the full width of a stream channel, multiple culvert barrels may be placed side by side. Such a configuration can minimize flow contractions at the culvert entrance and make the installation more hydraulically efficient. For instances in which multiple culvert barrels are used, it is usually assumed during design that the total discharge is divided equally among the individual barrels. Exceptions to this practice are cases in which the individual barrels have different sizes, or where they have different invert elevations or slopes. In these latter cases, a trial-and-error procedure is required to determine the discharge through each barrel. Computer-based modeling tools are particularly useful in such instances.

In the absence of review agency criteria, which would otherwise govern, the minimum recommended diameter for a culvert is 12 in. (300 mm). A higher minimum may be desirable for relatively long culverts, which are more difficult to maintain than short ones. Culverts are susceptible to clogging, especially in areas where sediment transport is high and where debris such as fallen leaves and branches is significant. When non-circular materials are used, it is recommended that the shortest cross-sectional dimension be at least 12 in. (300 mm).

9.3 CULVERT END TREATMENTS

The entrance and exit of a culvert are locations where flow is accelerated or decelerated, often causing erosion of the channel bed and embankment side slopes. Erosion protection measures are generally required in these locations and usually take the form of a riprap blanket and/or special culvert end treatments. The section focuses on the various types of end treatments that are used; Section 9.8 and Section 11.7 discuss riprap protective measures.

Culvert *end treatments* do more than provide protection against erosion; they help to stabilize embankment slopes around culvert entrances and exits and blend the culvert ends into the embankment slope. The type of culvert end treatment used in a design also has hydraulic ramifications, as some types may actually enhance the hydraulic conveyance capacity of the structure. For instance, all other factors being equal (culvert dimensions, material, discharge, tailwater elevation, etc.), some culvert entrance types will result in lower upstream water surface elevations (that is, lower headwater depths); these entrance types are considered to be more hydraulically efficient.

Figure 9.2 illustrates the most commonly used types of culvert end treatments. Each of these types can be used at either end (entrance or exit) of a culvert. A plain or *projecting* culvert [Figure 9.2(a)] has no special treatment. The culvert barrel simply projects from the embankment slope. This type of end treatment offers no protection against erosion of the embankment slope; its primary advantage is that it is inexpensive. Hydraulically, a projecting end is neither efficient nor inefficient if the wall thickness of the culvert pipe is significant (as in the case of concrete pipe); it may be viewed as a "middle-of-the-road" treatment. If corrugated metal pipe (which has a small wall thickness) is used, a projecting end is hydraulically inefficient.

A *mitered* end [Figure 9.2(b)] provides a visual improvement over the projecting end and usually incorporates an erosion protection blanket or apron around the mitered end. However, from a hydraulic capacity point of view, a mitered-end configuration is inefficient. The culvert entrance loss coefficient associated with a mitered end is often higher than that of other types of end treatments (see Table 6.5 on page 203).

Figure 9.2
Typical culvert end treatments

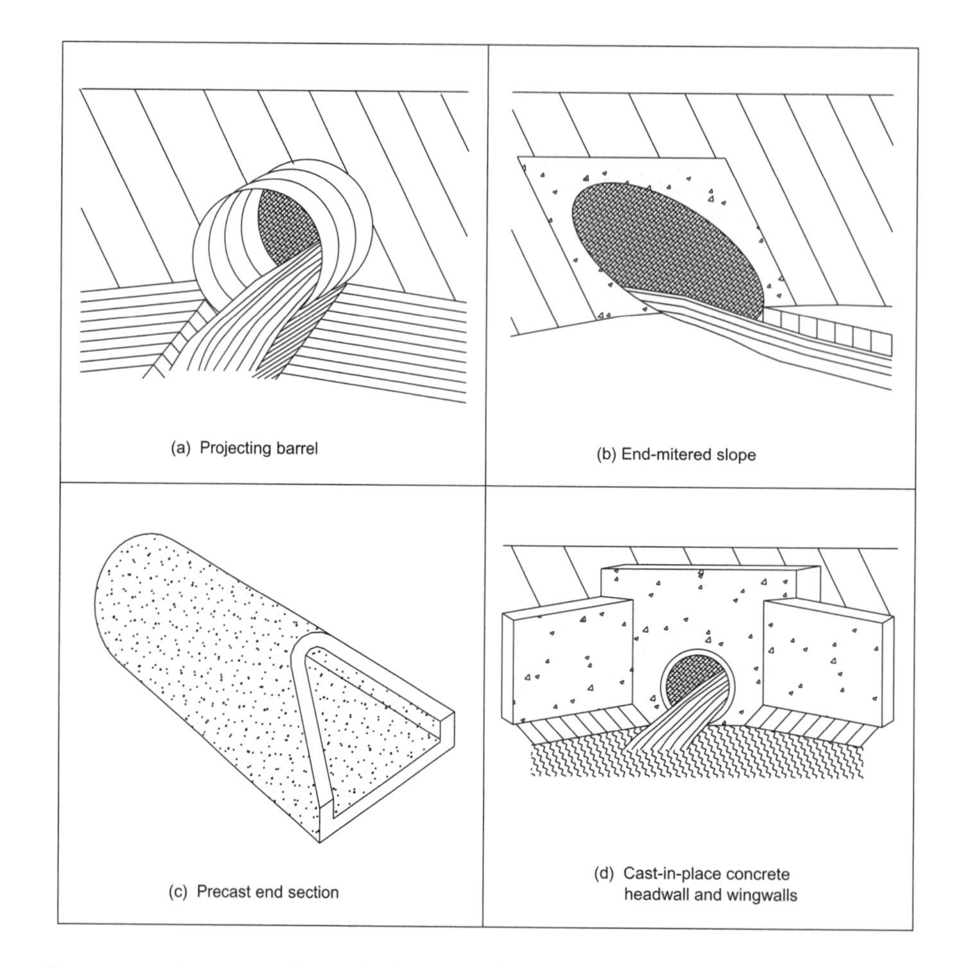

Concrete and corrugated metal pipe manufacturers can supply, in addition to pipe materials, specially formed *end sections* such as the precast end section shown in Figure 9.2(c). These end sections are usually tapered to gradually accelerate or decelerate the flow, and are shaped to facilitate the blending of the culvert end into the embankment slope. Entrance loss coefficients shown in Table 6.5 indicate that special end sections have approximately the same efficiency as a square-edged headwall or square-edged projecting end.

Headwalls and *wingwalls* [see Figure 9.2(d)] offer perhaps the best slope stability and erosion protection, but are generally the most expensive of treatment types of those shown in Figure 9.2. A *headwall* is a concrete wall at the inlet or outlet of a culvert through which the pipe projects. A headwall located at the outlet end of a culvert is sometimes referred to as an endwall. Wingwalls may be attached to either side of a headwall to provide additional stability or improve hydraulic performance. Headwalls and wingwalls, which may be precast or cast-in-place, are essentially retaining walls and should be designed as such by considering factors such as earth pressure, loading, and soil properties. The engineer should consider requiring a pipe joint near the headwall to allow for possible differential settlement or movement between the headwall and pipe. Headwalls and wingwalls should be provided with weep holes to allow seepage water trapped behind them to escape.

Any type of end treatment illustrated in Figure 9.2 can be enhanced hydraulically by *rounding* or *mitering* the edges of the entrance to the culvert barrel. Placing the socket or *groove end* of a pipe joint at the entrance has essentially the same effect as rounding or mitering. The hydraulic enhancement derives from a reduction in flow contraction effects as the water enters the culvert. A square-edged entrance causes a large contraction of the flow streamlines, resulting in an effective reduction of cross-sectional flow area. Rounding or mitering the entrance edges as shown in Figure 9.3 reduces the contraction. Table 6.5 shows that loss coefficients for rounded, mitered, or socket-end edges are always less than those for square-edged entrances.

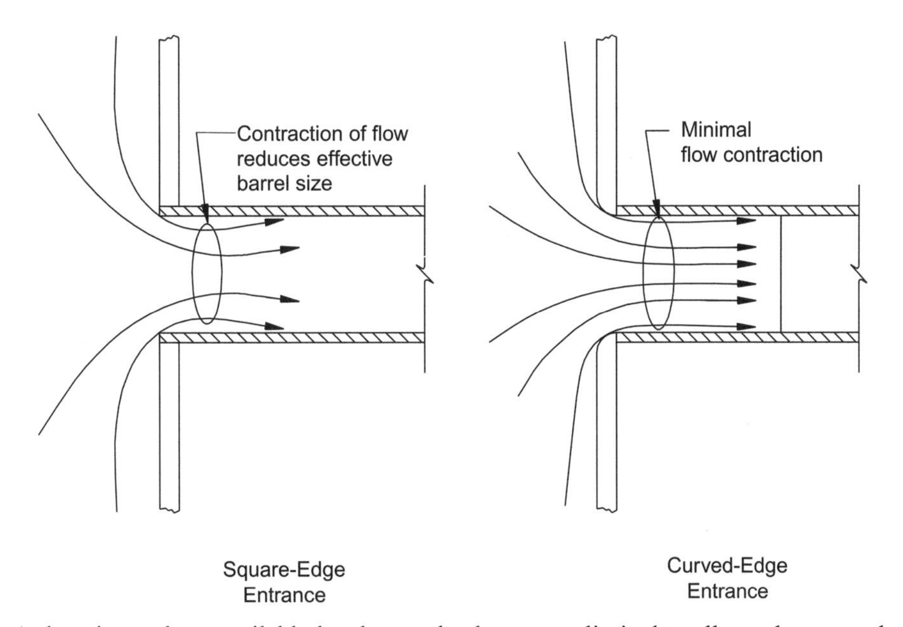

Figure 9.3
Square-edge and curved-edge culvert entrances (Norman, Houghtalen, and Johnston, 2001)

At locations where available headwater depths are too limited to allow adequate culvert performance, special culvert entrance configurations may be used to increase the hydraulic efficiency of the culvert inlet. *Side-tapered* or *slope-tapered inlets*, as illustrated in Figure 9.4, may be employed to fulfill this need. (Note that the slope-tapered inlet in this figure also has a side taper.) These special inlet configurations are more costly to construct but can be advantageous in locations where other alternatives fail to provide the required performance. Harrison et al. (1972) provide further information on the hydraulic performance for these types of culverts.

Regardless of the type of end treatment used for a culvert, it is recommended that a concrete *cradle* or *cutoff wall* (see Figure 9.5) be installed around and under the ends of the pipe. The cradle is especially important at the culvert outlet to prevent undercutting of the pipe and subsequent collapse. Of course, if a headwall end treatment is used, the headwall itself performs this function. If a concrete apron is used at the entrance or exit of a culvert, a cutoff wall should be integrated into it to prevent undermining.

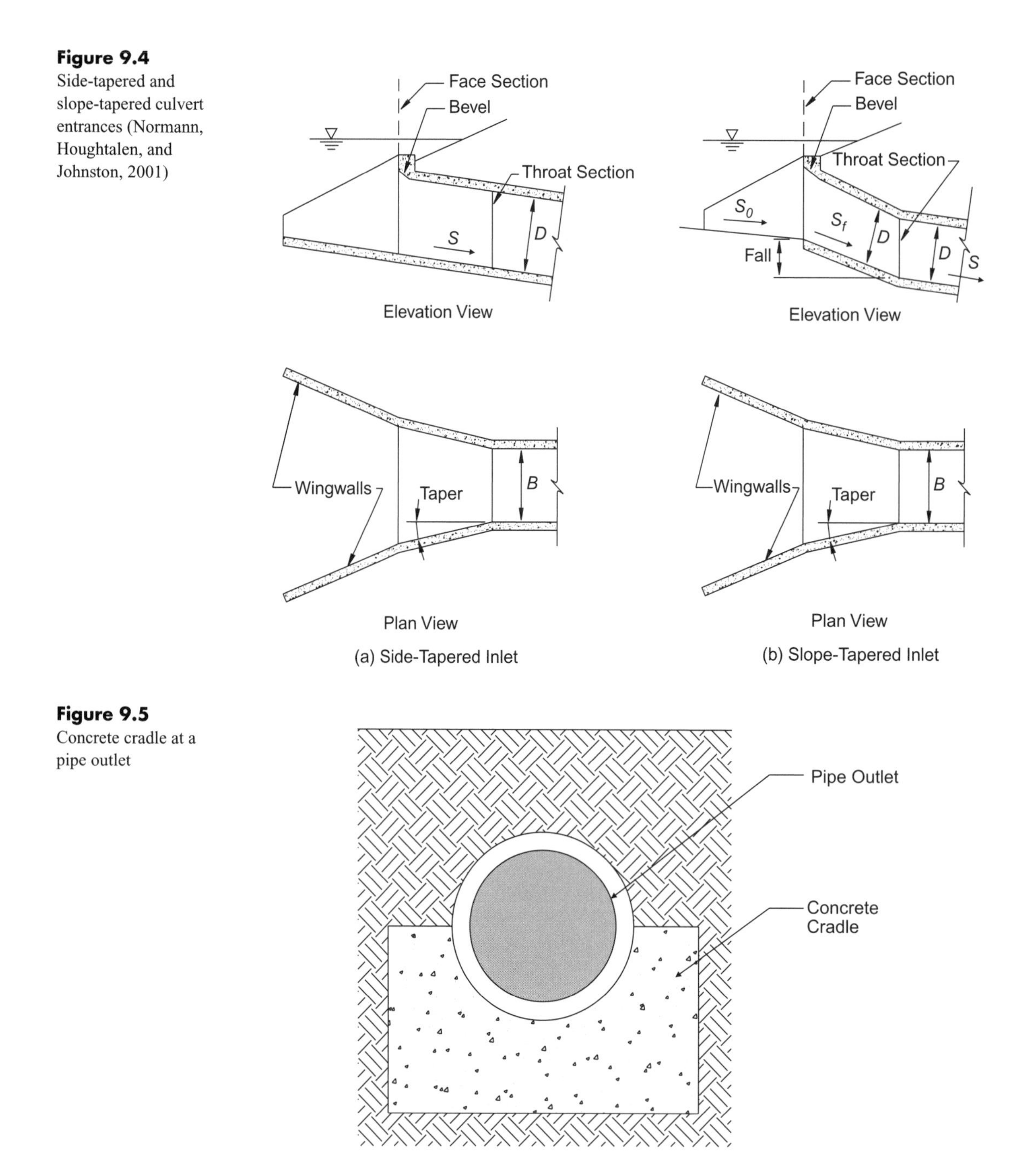

Figure 9.4
Side-tapered and slope-tapered culvert entrances (Normann, Houghtalen, and Johnston, 2001)

Figure 9.5
Concrete cradle at a pipe outlet

It is frequently desirable to provide *trash racks* (sometimes called *safety racks*) secured to the wingwalls at culvert entrances and/or exits to prevent unauthorized entry and/or clogging by debris. Trash racks should have clear openings between bars no greater than about 6 in. (150 mm) and should be constructed in modular sections to facilitate removal for maintenance of the culvert. The rack itself should receive periodic maintenance to clear trapped debris. Installing the rack so that the bars are inclined from the vertical can help keep the rack clear because the flowing water will push floating debris up the slope of the rack. Inclining the rack also makes it easier for a person trapped against the rack to climb out.

Although guidelines and recommendations vary, a trash rack should be structurally designed to withstand the hydraulic loads that would occur under partial or full clogging of the rack. Hydraulically, a trash rack causes a head loss, but the loss essentially can be neglected if the total clear opening area of the rack is much greater (about four times) than that of the cross-sectional area of the culvert barrel. Racks should be placed far enough upstream of the culvert entrance to ensure that the velocity of the accelerating flow is low enough for a person to escape.

9.4 ALIGNMENTS OF CULVERTS

Generally, the horizontal and vertical alignments of a culvert should conform as closely as possible to those of the channel. However, it may be necessary or desirable in some cases to depart from this ideal. Discussions of alignment problems and solutions are presented in the subsections that follow.

Horizontal Alignment

If a stream channel is very sinuous, it can be difficult or impossible to develop a culvert alignment that provides a smooth transition of the flow at both the upstream and downstream ends of a culvert. In such instances, it may be desirable to realign the channel to provide for smooth transitions. However, such efforts should be undertaken with great care because the newly aligned channel may be unstable, resulting in sedimentation and/or erosion problems upstream or downstream of the culvert. Brice (1981) concluded that channel realignments are generally successful, except in instances where the channel was already unstable.

An alternative to channel realignment is to use a *curved* or *doglegged alignment* for the culvert barrel (see Figure 9.6). Of course, such a configuration will result in additional energy losses due to the curve or bend. Another disadvantage is that debris tends to build up and cause culvert blockages at bend locations. If a culvert operates under inlet control [see Section 2.2 (page 24) and Section 9.5 (page 335)], additional head losses caused by a curve or bend will not affect the headwater elevation. The headwater elevation will be affected if the culvert operates under outlet control.

Changes in the horizontal alignment of a culvert should ideally be accomplished using a curved alignment. In cast-in-place concrete box culvert construction, curved concrete forms may be used. In pipe culvert construction, small deflections at individual pipe joints according to the manufacturer's recommendations can be accomplished to approximate a smooth curve. If a sudden bend is used instead of a curved alignment, the bend deflection angle should be limited to no more than approximately 15 degrees, and such bends should be separated from one another by a distance of at least 50 ft (15 m) (ASCE, 1993).

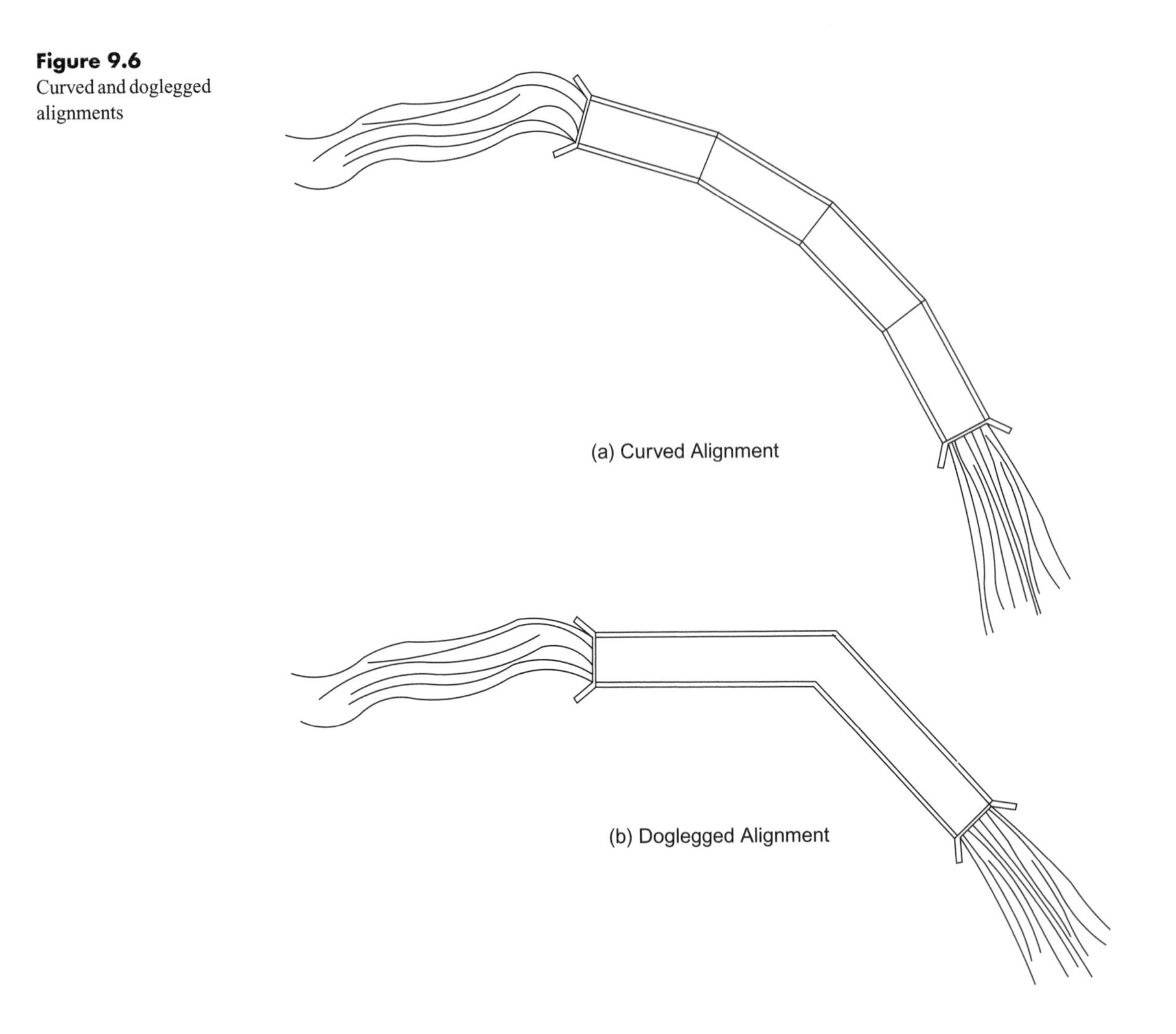

Figure 9.6
Curved and doglegged alignments

Vertical Alignment

In a *sag culvert*, one or more of the straight barrel segments has an adverse slope, as illustrated in Figure 9.7(a), which usually results in part of the culvert flowing full and under pressure. A sag culvert is especially prone to clogging by trapped sediments and debris, but it can be the best option if its use prevents, for instance, a costly utility relocation. Sag culverts are also used frequently at roadway crossings of irrigation canals.

Broken-back culverts [Figure 9.7(b)] have bends in their vertical alignments. The bend in a broken-back culvert may cause the upstream part of the culvert to have a smaller slope than the downstream part, or vice versa. A broken-back culvert may be used to minimize excavation, avoid a rock outcropping, or reduce the exit velocity from the culvert.

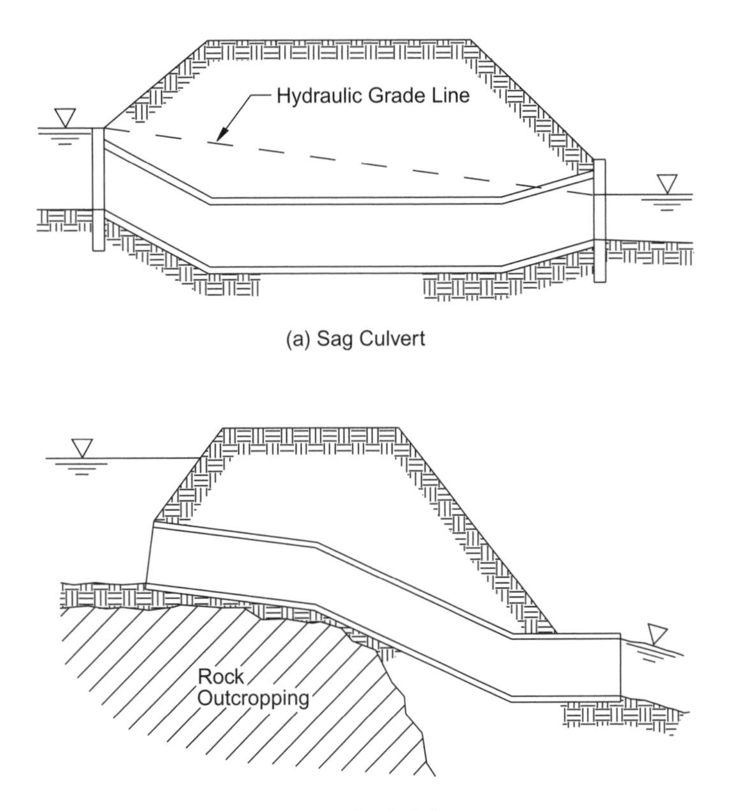

Figure 9.7
Sag and broken-back culverts (Norman, Houghtalen, and Johnston, 2001)

If a broken-back culvert has a steeper slope downstream of the bend, the possibility exists that the hydraulic grade will drop below the crown of the culvert in the vicinity of the bend although it is otherwise above the crown. This phenomenon causes subatmospheric pressure to develop within the culvert at the bend location and can potentially lead to structural and hydraulic problems. If this possibility exists, a vent should be provided to allow atmospheric pressure to exist at the bend location.

Hydraulic analyses of broken-back and sag culverts are complicated by the fact that they usually involve regions of both full (pipe) flow and open-channel flow. Further, open-channel flow may be subcritical in one part of the culvert and supercritical in another. Thus, the potential exists for many flow controls, and each situation must be treated differently. Computer-based modeling tools can quickly evaluate flow regimes for complicated culvert geometries, and hence can be used to optimize designs.

Skewed Culverts

Positioning a culvert alignment to approximate the flow path of the stream channel often involves placing the culvert at an angle that is not perpendicular to the embankment centerline, as shown in Figure 9.8, parts (b) and (c). This culvert alignment is referred to as *skewed* in relation to the normal embankment alignment. In the case of skewed culverts, headwalls may be skewed with respect to the barrel alignment, as illustrated in Figure 9.8(c), to better align with embankment slopes (that is, the culvert is skewed with respect to the embankment, and the headwalls are skewed by an offsetting amount with respect to the culvert so that they will still align with the embankment slopes). If the headwall on a culvert is skewed with respect to the barrel, an additional entrance loss occurs; however, this loss is usually small and may be neglected in many cases. This additional loss can be estimated by using the design charts published in FHWA's Hydraulic Design Series Number 5, *Hydraulic Design of Highway Culverts* (Normann, Houghtalen, and Johnston, 2001), also known as HDS-5.

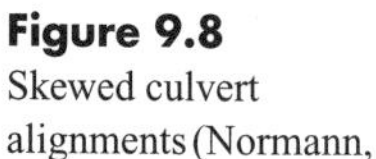

Figure 9.8
Skewed culvert alignments (Normann, Houghtalen, and Johnston, 2001)

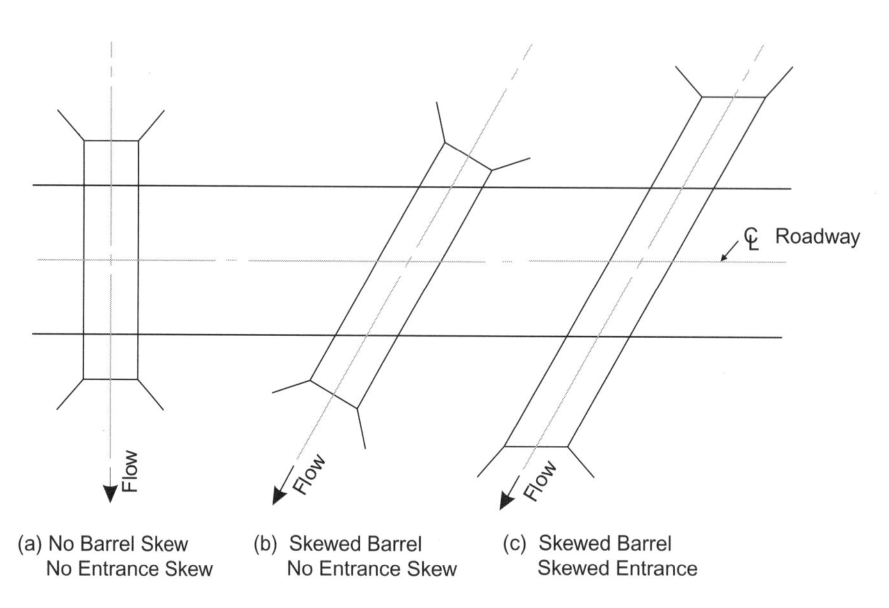

9.5 CULVERT HYDRAULICS

The hydraulics of a culvert are generally complex because of the short conduit length and the several flow control types that may exist. A culvert may flow less than full over part or all of its length, or it may flow full (closed-conduit flow) over its entire length. Culverts are frequently not long enough for uniform flow to be achieved and maintained; therefore, the entire flow profile often consists of gradually and/or rapidly varied flow. The flow control may also change with changes in discharge (that is, the culvert may operate under one hydraulic control for some discharges, and under one or more other controls for other discharges).

In design, the head loss (which approximately equals the difference between the water surface elevations upstream and downstream of the culvert) that can be tolerated in a culvert is often limited. The most widely used design approach is to evaluate the performance of a trial culvert design for several different flow control types. The worst-performing control condition (the condition resulting in the greatest head loss) is then used to evaluate the acceptability of the proposed design. This method is a conservative, worst-case design approach that ignores conditions that may result in better performance. The benefits of this approach are its ease of use and the assurance that the culvert will perform adequately under the most adverse conditions.

Two general methods can be used to predict the hydraulic performance characteristics of a culvert, and both are presented in this section. The graphical method involves the use of nomographs that relate the headwater elevation at the culvert entrance to factors such as the length, size, and roughness of the barrel and the tailwater elevation downstream of the outlet. This method is relatively easy to use and provides results that are adequate for most purposes. It should not be used for culverts containing horizontal curves or bends unless the additional losses are accounted for separately. It should almost never be used for a broken-back or sag culvert.

The second method for predicting culvert performance is to compute the flow profile using the gradually varied flow techniques presented in Section 7.8. This method is much more laborious than the first but can yield much more accurate predictions of culvert performance in many cases. This method is usually required for broken-back and sag culverts. Because of the number of calculations involved, computer programs such as CulvertMaster are especially beneficial when using this method.

Flow Controls

Research by the National Bureau of Standards (NBS) and the FHWA in the United States divides culvert flow controls for a straight, uniformly shaped culvert into two basic classes depending on the location of the control section: inlet control and outlet control (Normann, Houghtalen, and Johnston, 2001). With inlet control, the culvert's entrance characterisics determine its capacity. Conversely, with outlet control, the inlet can accept more flow than the culvert can carry, because of either the friction losses in the barrel or the tailwater elevation. Descriptions of the various flow regimes encountered under both control types follow. Additional flow controls may exist in broken-back or sag culverts, and in those with improved inlet configurations.

Inlet Control. For a culvert operating under *inlet control*, the culvert barrel is capable of conveying a greater discharge than the inlet will accept. The flow control section is just inside the culvert barrel at its entrance, and the flow profile passes through critical depth at this location. The culvert is hydraulically steep, and the flow regime downstream of the control section is supercritical with an S2 profile [see Section 7.8 (page 274) and Figure 7.24]. Conditions downstream of the entrance have no effect on the culvert capacity.

Figure 9.9 illustrates four cases in which a culvert operates under inlet control. Note the drawdown of the S2 water surface profile near the upstream end of the barrel in each case.

Case A in Figure 9.9 represents a situation in which neither the inlet nor the outlet of the culvert are submerged by the headwater or tailwater. The flow approaches normal depth as it moves along the culvert, and, depending on the flow depth in the downstream channel, a hydraulic jump may form downstream of the culvert outlet.

Case B illustrates a situation in which the outlet is submerged but the inlet is not submerged. The culvert is hydraulically steep. An S2 water surface profile forms inside the culvert entrance, and flow approaches normal depth. Submergence of the downstream end of the culvert results in pressure flow near the outlet end of the barrel. A hydraulic jump forms inside the culvert barrel and provides a transition between the supercritical and pressure flow regimes.

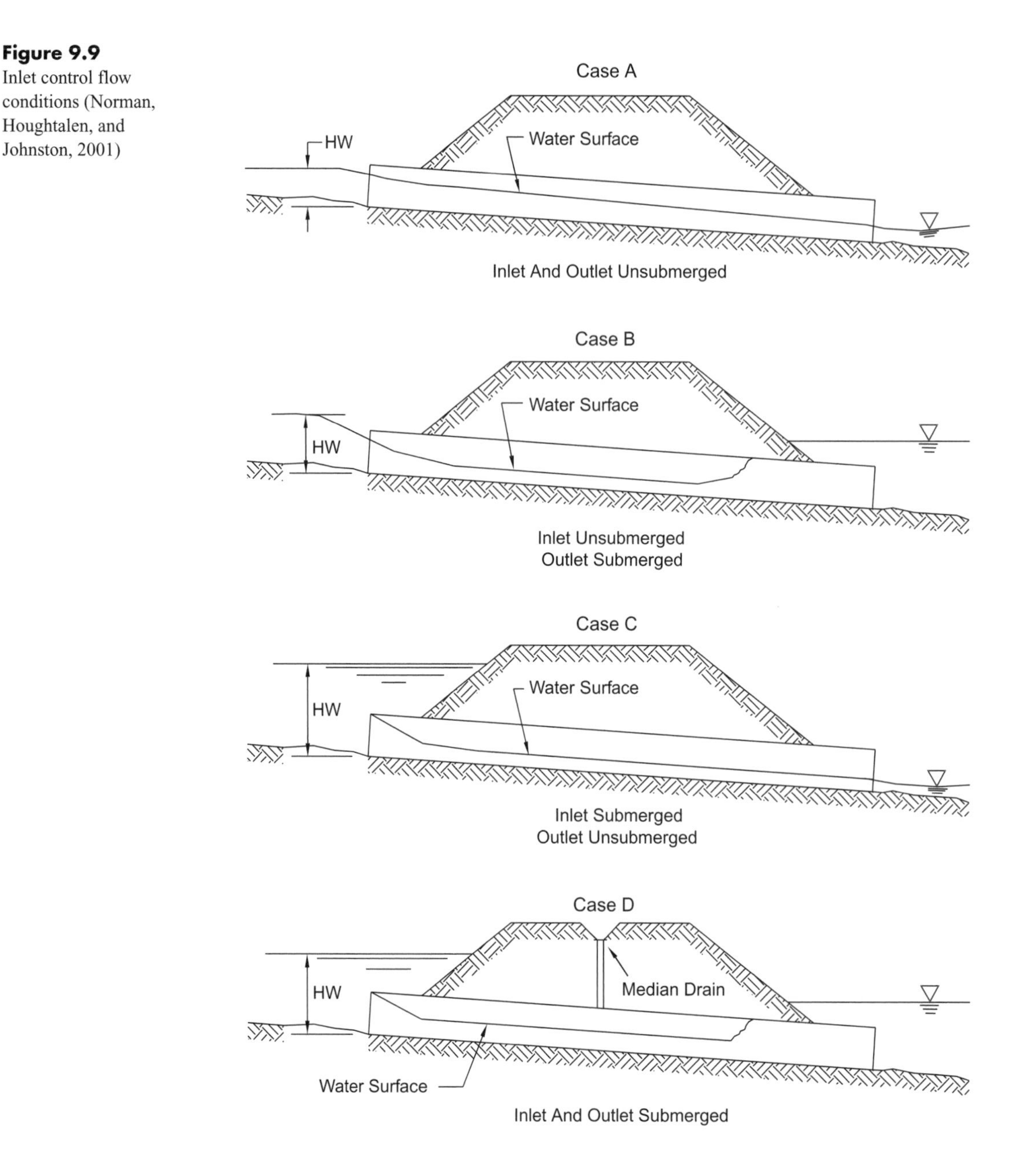

Figure 9.9
Inlet control flow conditions (Norman, Houghtalen, and Johnston, 2001)

In *Case C*, the inlet is submerged by the headwater, but the outlet is not submerged. As in Case A, the flow approaches normal depth as it moves along the culvert. A hydraulic jump may form in the downstream channel.

Case D shows that a culvert may flow partly full even for cases in which both the inlet and the outlet are submerged. The hydraulically steep nature of the culvert promotes the S2 profile near the entrance end of the barrel. A hydraulic jump forms a transition between the upstream supercritical flow and the downstream pressure flow. A vent, such as a median drain, should be provided in a culvert anticipated to operate in this mode to permit atmospheric pressure to exist within the culvert.

Other than the discharge, factors affecting headwater depth for a culvert operating under inlet control include the cross-sectional area and shape of the barrel and the inlet's edge configuration. The barrel slope also influences the headwater depth, but its effect is small. A sharp inlet edge, such as that occurring on a thin-walled metal pipe with a projecting end, causes a large flow contraction at the culvert entrance. A square-edged inlet, or better yet, an inlet with a rounded or pipe socket edge, would reduce the headwater elevation in this case.

For a given allowable headwater elevation, the capacity of a culvert can be enhanced by depressing the invert of the culvert barrel below the channel bed (with the fill retained by wingwalls), thereby increasing the headwater depth. The height of the depression T, as shown in Figure 9.10, is called the *fall*.

Figure 9.10 shows that most of the fall T occurs over the front portion of the culvert approach apron, and this portion of the apron should have a minimum horizontal length of $2T$. The remainder of the fall occurs over the second portion of the approach apron, which consists of a projection of the culvert slope for a distance equal to half the height D of the culvert.

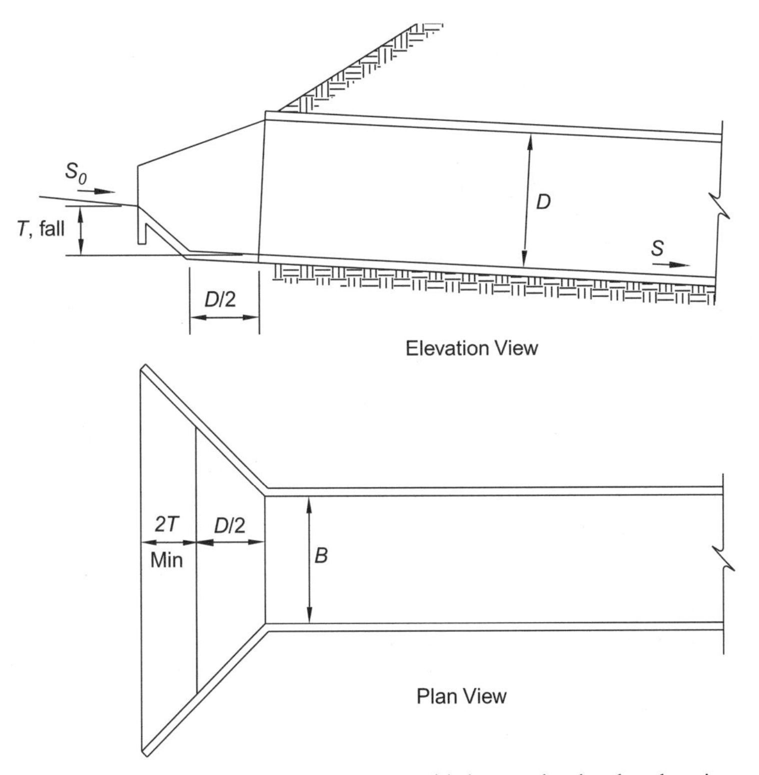

Figure 9.10 Fall at culvert entrance (Norman, Houghtalen, and Johnston, 2001)

If not done properly, provision of a depressed inlet can lead to headcutting and degradation of the upstream channel. Grade control structures in the form of buried riprap or concrete walls placed at various intervals along the channel bottom may become necessary to provide a measure of protection against upstream headcutting and degradation. Also, the approach apron should be paved. When a depressed inlet is used, sufficient elevation change must exist to ensure a positive slope between the upstream and downstream ends of the culvert.

Outlet Control. *Outlet control* occurs when the slope of a culvert is hydraulically mild or when the tailwater elevation is high enough to affect the headwater elevation. In such cases, the inlet is capable of accepting more flow than the barrel and/or downstream flow conditions will permit. The control section is at or near the outlet point, and the flow condition in the barrel near the outlet is critical, subcritical, or full (pressure) flow.

For a culvert operating under outlet control, many factors affect the headwater depth. These factors include discharge; barrel cross-sectional area, shape, length, roughness, and slope; inlet edge configuration; and tailwater depth. Figure 9.11 illustrates seven cases in which outlet control may occur.

Case A, shown in Figure 9.11, represents a situation in which the culvert barrel flows full and under pressure over its entire length. Both the inlet and the outlet are submerged. This condition is often assumed to exist in outlet control calculations, but seldom occurs in actual culvert performance.

Case B is a situation in which the outlet is submerged but the headwater elevation is only slightly higher than the crown of the inlet. In this instance, because of the low submergence of the inlet and the flow contraction that occurs at that location, the crown of the inlet is exposed where the water enters the barrel.

Case C represents a rare situation in which the inlet is submerged, the outlet is not submerged, and the barrel flows full over its entire length. A large headwater depth is required to cause a culvert to flow full over its entire length when the outlet is not submerged. The exit velocity in this case is quite high and can cause severe erosion in the downstream channel.

Case D occurs more commonly than does Case C. Like Case C, it represents a situation in which the inlet is submerged but the outlet is not. If the tailwater elevation is below critical depth, the flow passes through critical depth at the outlet end of the culvert barrel. If the tailwater is between critical depth and normal depth for the barrel, profile calculations begin at the tailwater depth. Because of the hydraulically mild slope of the barrel, an M2 water surface profile forms upstream of the critical depth (control) section. The M2 profile intersects the crown at some distance from the outlet, and upstream of that point the culvert flows full and under pressure.

Case E is similar to Case D. If the tailwater is below critical depth, flow passes through critical depth at the outlet and profile calculations begin at critical depth. If the tailwater depth is between normal depth and critical depth, profile calculating begin at the tailwater elevation. An M2 water surface profile occurs upstream of that point. The normal depth is smaller than the height of the barrel, and the flow profile approaches it asymptotically. Flow in the culvert remains less than full throughout its length.

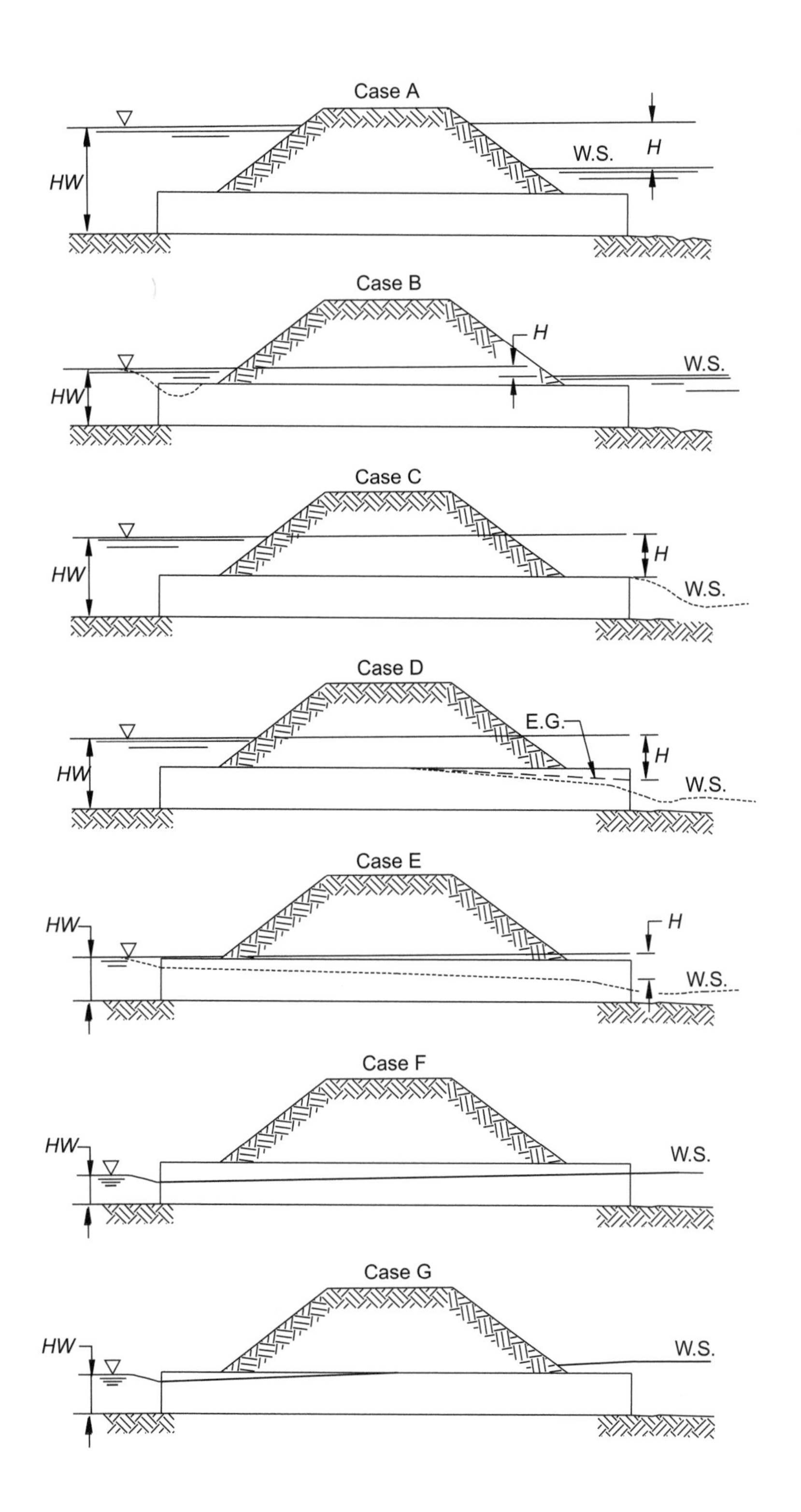

Figure 9.11
Outlet control flow conditions (adapted from Norman, Houghtalen, and Johnston, 2001)

Cases F and *G* can occur when the tailwater at the outlet is above normal depth for a culvert operating with subcritical flow conditions. Depending on the culvert entrance characteristics, the resulting backwater profile may take precedence over inlet control conditions.

Graphical Analysis of Culvert Performance

A traditional approach to culvert performance analysis is graphical in nature and makes use of nomographs published in HDS-5 (Normann, Houghtalen, and Johnston, 2001). The nomographs are published for both inlet and outlet control conditions, for different barrel shapes and materials, and for different entrance edge configurations. Because of the large number of possible combinations of these influential factors, many nomographs are needed.

The following subsections describe and illustrate the use of the charts for both inlet and outlet control conditions. When designing a culvert, the headwater depths computed using inlet and outlet control assumptions are compared, and the larger value is used to determine whether the design is acceptable.

Inlet Control Charts. Figure 9.12 is an *inlet control nomograph* for concrete pipe culverts with circular cross sections. It is only one of the many inlet control charts available for different combinations of pipe material, shape, and entrance edge configuration. This particular chart illustrates the relationship that exists between the culvert diameter, discharge, and headwater depth for a circular concrete culvert with three possible entrance types: (1) square edge of pipe with headwall; (2) groove end of pipe with headwall; and (3) groove end of pipe projecting from fill.

For a specified discharge, culvert diameter, and entrance edge configuration, the nomograph in Figure 9.12 can be used to estimate *HW/D*, which is a dimensionless ratio of the required headwater depth to the diameter. As illustrated with the example given in the figure, a straight line can be drawn from a point on the pipe diameter scale, through the appropriate point on the discharge scale, to a point on *HW/D* scale (1). The value on the *HW/D* scale defines the required depth ratio if the culvert inlet configuration consists of a headwall with a square-edged barrel entrance. If the inlet configuration is instead a groove (socket) pipe end with a headwall or a projecting groove pipe end, a horizontal line should be projected from the *HW/D* scale (1) intersection point to scale (2) or scale (3), respectively.

Example 9.1 – Inlet Control for a Circular Culvert. A circular concrete pipe culvert has a diameter of 36 in., a length of 100 ft, and a slope of 0.50 percent, as illustrated in Figure E9.1.1. The invert of the culvert will match the channel bottom elevation, which is 125.30 ft at the upstream end. Assuming that the culvert operates under inlet control, use Figure 9.12 to estimate the headwater elevations corresponding to discharges of 30 and 60 cfs. The culvert inlet consists of a headwall with a square edge at the barrel entrance.

Solution: A line projected to scale (1) from the points representing 36 in. and 30 cfs on the diameter and discharge scales yields $HW/D = 0.90$. The headwater depth is therefore

$$HW = 0.90D = 0.90(36/12) = 2.70 \text{ ft}$$

The headwater elevation is the sum of the channel bed elevation and the headwater depth, or

$$125.30 + 2.70 = 128.00 \text{ ft}$$

For the discharge of 60 cfs, $HW/D = 1.65$, and the headwater depth is therefore

$$HW = 1.65(3.00) = 4.95 \text{ ft}$$

The headwater elevation is

$$125.30 + 4.95 = 130.25 \text{ ft}$$

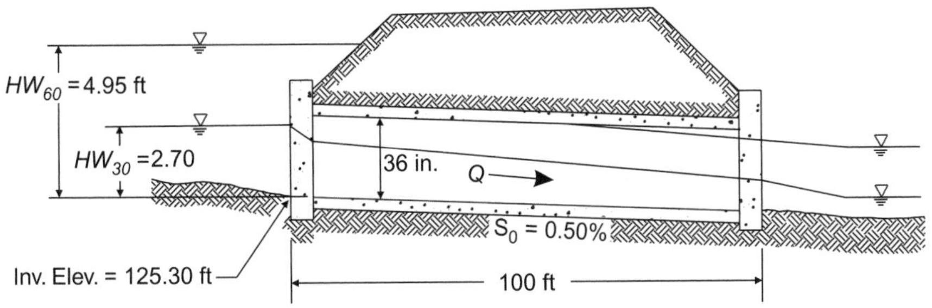

HW60 = 4.95 ft
HW30 = 2.70
36 in.
Q
S0 = 0.50%
Inv. Elev. = 125.30 ft
100 ft

Figure E9.1.1 Culvert for Example 9.1

The next example shows that simply changing the entrance edge configuration can turn an otherwise unacceptable design into an acceptable one, thus illustrating the significant influence that the entrance edge has on the performance of a culvert operating under inlet control.

Example 9.2 – Inlet Control for a Circular Culvert with Different Entrances. Consider the same culvert described in the previous example, but assume that the maximum allowable headwater elevation is 129.65 ft when the discharge is 60 cfs. Clearly, the square edge with headwall design is not acceptable on this basis. Without changing the culvert diameter, determine whether an alternative type of culvert entrance configuration can be used to lower the headwater elevation to below 129.65 ft.

Solution: As described in Example 9.1, the value of HW/D found on scale (1) for the discharge of 60 cfs is 1.65. Extending a horizontal line on the chart from that point, one determines values of HW/D on scales (2) and (3) to be 1.44 and 1.47, respectively. The respective headwater depths are 4.32 and 4.41 ft, and the respective headwater elevations are 129.62 and 129.71 ft. Comparing these values to the allowable headwater depth of 129.65, it is concluded that use of a groove end with a headwall will achieve the desired performance, whereas a projecting groove end (marginally) will not.

Outlet Control Charts. An *outlet control nomograph* such as the one shown in Figure 9.13 is used to predict the culvert *head, H,* for a culvert operating under outlet control. The head is equal to the total head loss through the culvert, but, in practice, it can usually be taken as the difference between the headwater and tailwater elevations. The variables that impact the outlet control head are culvert discharge, cross-section geometry, length, entrance loss coefficient, and roughness. The nomograph presented here is for circular concrete culverts with an n value of 0.012. As with the inlet control nomograph from Figure 9.12, it is only one of many such charts published for different combinations of culvert materials and shapes.

To use the nomograph shown in Figure 9.13, the engineer must first determine which of the two length scales shown should be used based on the entrance loss coefficient k_e. The descriptions and entrance loss coefficient values provided in Table 9.1 (which is a repeat of Table 6.5) can be used to determine which length scale (in this example, $k_e = 0.2$ or $k_e = 0.5$) should be used.

Figure 9.12
Nomograph to compute headwater depth for circular concrete culverts with inlet control (Normann, Houghtalen, and Johnston, 2001)

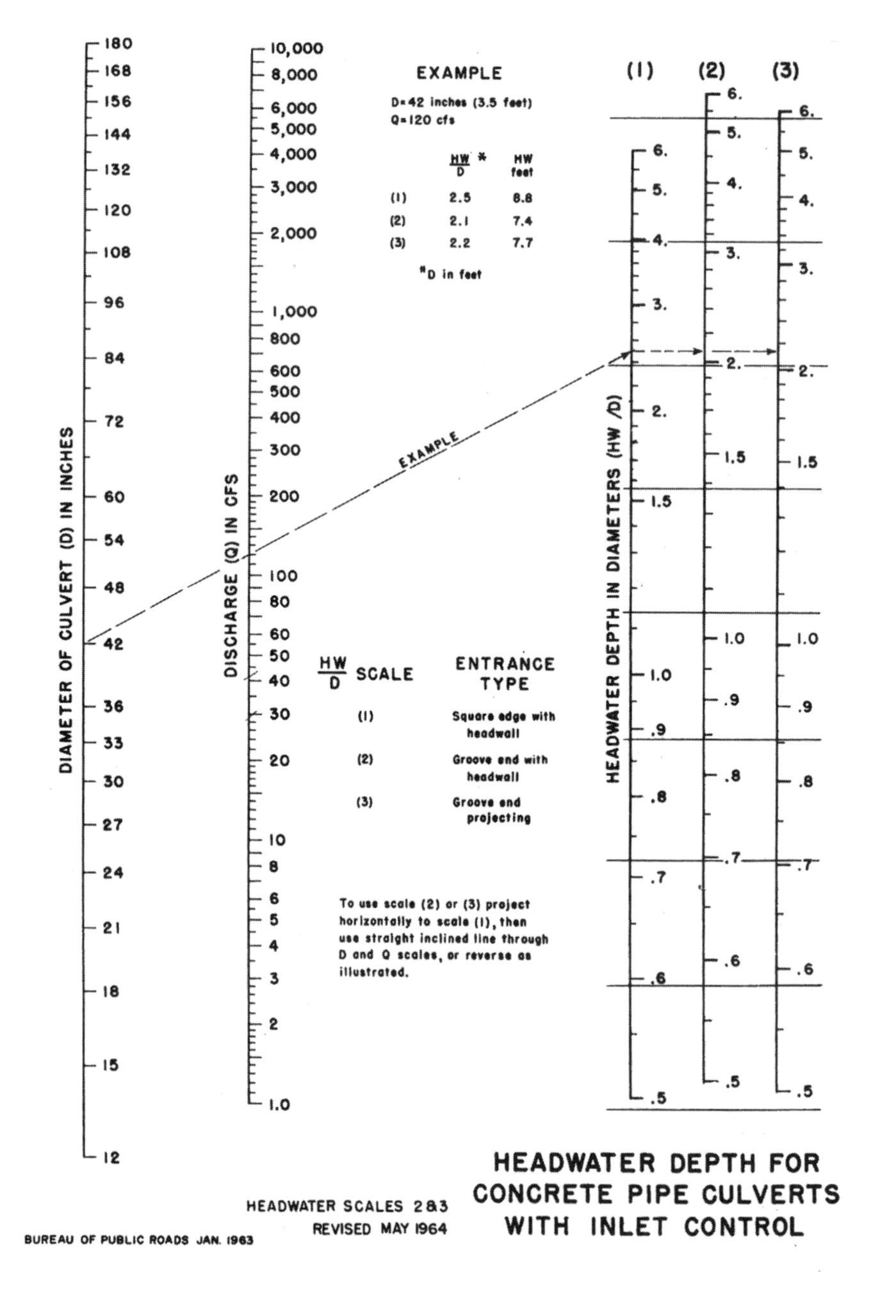

After the appropriate length scale has been determined, a straight line connecting the applicable points on the diameter and length scales is drawn. Then, the head value is determined by drawing a second straight line from the appropriate value on the discharge scale, through the intersection point of the first line and the "turning line," to the head scale. An example of how these lines are constructed is shown on the chart.

Finally, an energy balance for full flow through a culvert with both the inlet and outlet submerged can be written to express the headwater depth as

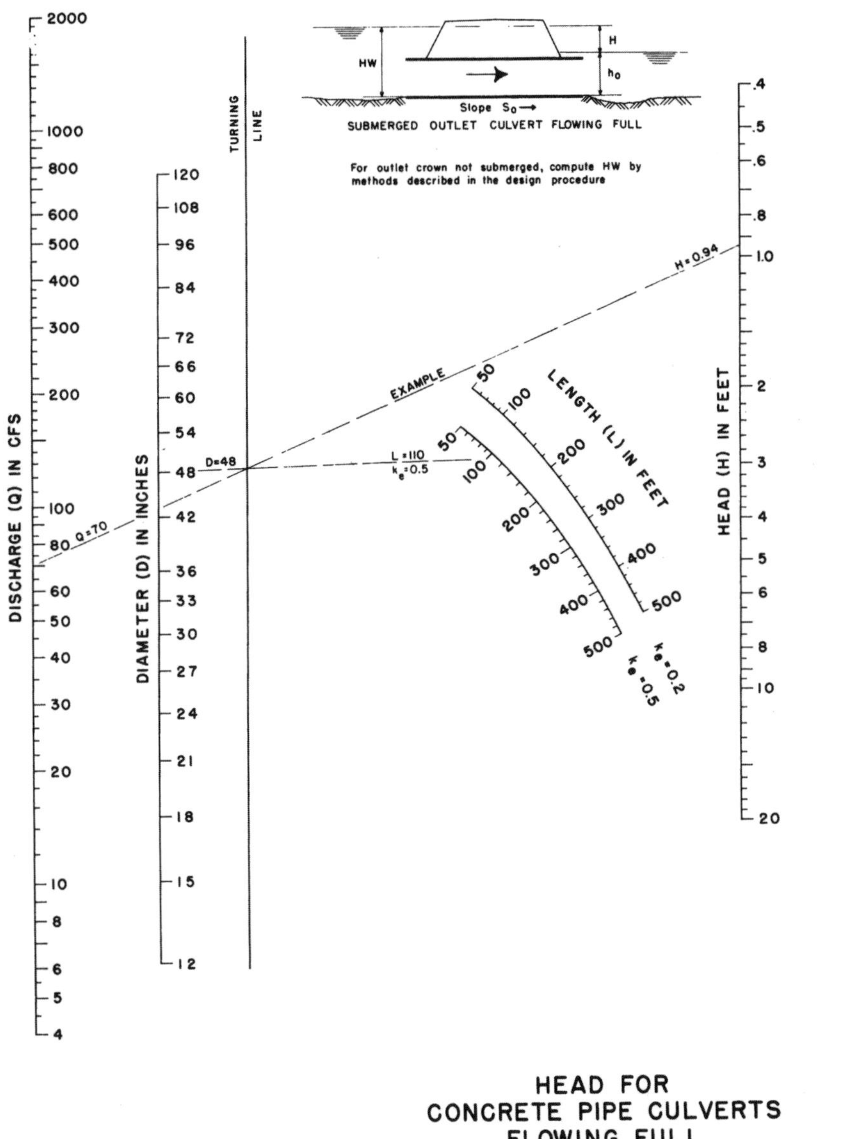

Figure 9.13 Nomograph for computing head on circular concrete pipes with outlet control (Normann, Houghtalen, and Johnston, 2001)

$$HW = h_o + H - S_o L \qquad (9.1)$$

where HW = headwater depth (ft, m)
h_o = tailwater depth (ft, m)
H = total head loss through the culvert (ft, m)
S_o = culvert barrel slope (ft/ft, m/m)
L = culvert length (ft, m)

This expression assumes that the velocity heads in the upstream and downstream channels are equal, which is a reasonable approximation in most instances.

Table 9.1 Entrance loss coefficients for pipes and culverts operating under outlet control (Normann, Houghtalen, and Johnston, 2001)

Structure Type and Entrance Condition	k_e
Concrete pipe	
Projecting from fill, socket or groove end	0.2
Projecting from fill, square edge	0.5
Headwall or headwall and wingwalls	
Socket or groove end	0.2
Square edge	0.5
Rounded (radius = $D/12$)	0.2
Mitered to conform to fill slope	0.7
End section conforming to fill slope	0.5
Beveled edges (33.7° or 45° bevels)	0.2
Side- or slope-tapered inlet	0.2
Corrugated metal pipe or pipe-arch	
Projecting from fill (no headwall)	0.9
Headwall or headwall and wingwalls (square edge)	0.5
Mitered to conform to fill slope	0.7
End section conforming to fill slope	0.5
Beveled edges (33.7° or 45° bevels)	0.2
Side- or slope-tapered inlet	0.2
Reinforced concrete box	
Headwall parallel to embankment (no wingwalls)	
Square-edged on 3 sides	0.5
Rounded or beveled on 3 sides	0.2
Wingwalls at 30° to 75° from barrel	
Square-edged at crown	0.4
Crown edge rounded or beveled	0.2
Wingwalls at 10° to 25° from barrel	
Square-edged at crown	0.5
Wingwalls parallel (extensions of box sides)	
Square-edged at crown	0.7
Side- or slope-tapered inlet	0.2

One method for determining the tailwater depth in Equation 9.1 is to approximate it using a uniform flow formula such as the Manning equation. This approximation requires information on the downstream channel cross-sectional dimensions, slope, and roughness. Also, this approach assumes that the flow in the downstream channel is approximately uniform and that flow controls downstream of the culvert do not cause backwater or similar effects. An alternative (and more accurate) method for determining the tailwater depth is to perform a flow profile analysis for the downstream channel. This method is clearly more laborious than the uniform flow method just described, but is required when backwater or other effects cause the tailwater depth to differ appreciably from that of uniform flow.

Example 9.3 – Outlet Control for a Circular Culvert with Tailwater Effects. Repeat Example 9.1 assuming that the culvert operates under outlet control. The invert elevation at the downstream end of the culvert is 124.80 ft, as illustrated in Figure E9.3.1. Backwater effects due to an obstruction in the downstream channel cause the tailwater depth to be $h_o = 4.20$ ft when $Q = 30$

cfs, and h_o = 4.70 ft when Q = 60 cfs. By comparing the inlet and outlet control solutions, determine whether the culvert operates under inlet or outlet control for discharges of 30 and 60 cfs.

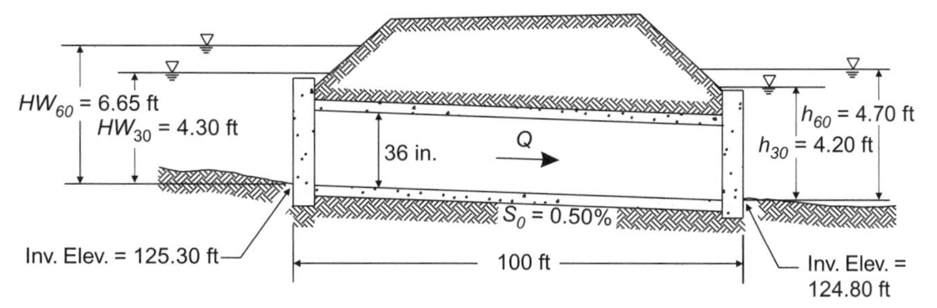

Figure E9.3.1 Culvert for Example 9.3

Solution: From Example 9.1, the circular concrete culvert has a diameter of 36 in., a length of 100 ft, and a slope of 0.50 percent. The channel bed elevation at the upstream end of the culvert is 125.30, and the inlet configuration is a headwall with a square-edged barrel entrance.

From Table 9.1, the entrance loss coefficient for the this type of inlet configuration is 0.5. Using a diameter of 36 in. and a length of 100 ft (on the scale representing k_e = 0.5), a straight-edge can be used on Figure 9.13 to position a point on the turning line. Using that point and the discharge of 30 cfs, a line can then be projected to the head scale to determine H = 0.60 ft. When the discharge is 60 cfs, the head loss is similarly found to be H = 2.45 ft.

Application of Equation 9.1 for the case when Q = 30 cfs yields

$$HW = 4.20 + 0.60 - 0.005(100) = 4.30 \text{ ft}$$

The headwater elevation is therefore

$$125.30 + 4.30 = 129.60 \text{ ft}$$

Because this elevation is higher than the headwater elevation of 128.00 ft computed for inlet control, conservative practice dictates that the outlet control headwater elevation of 129.60 ft governs for a discharge of 30 cfs.

For the case of Q = 60 cfs, Equation 9.1 yields

$$HW = 4.70 + 2.45 - 0.005(100) = 6.65 \text{ ft}$$

The headwater elevation is therefore

$$125.30 + 6.65 = 131.95 \text{ ft}$$

Because this value is again higher than the headwater elevation determined for the assumption of inlet control (130.25 ft), outlet control again governs and the headwater elevation is 131.95 ft for a discharge of 60 cfs.

Figure 9.13 and other similar charts are based on the assumption that the culvert flows full over its entire length and is submerged at both its upstream and downstream ends. Nevertheless, such charts can be used to approximate outlet control behavior when the tailwater elevation drops to or below the crown of the culvert outlet (Cases C through G in Figure 9.11). The procedure yields the best approximation, however, when the culvert flows full over at least a portion of its length. When the culvert flows only partly full over its entire length, as in Case E, the method deteriorates as the headwater elevation drops further below the top of the culvert pipe. Adequate results may be obtained for headwater depths as small as $0.75D$, where D is the height of the culvert barrel. Accurate calculations require flow profile analysis, as described in the next section.

If one chooses the approximate nomograph analysis of outlet control in Case D or E, the tailwater depth used in Equation 9.1 should be the larger of (1) the actual tailwater depth or (2) the arithmetic average of the critical depth in the barrel and the height of the barrel $[(y_c + D)/2]$. The critical depth in the barrel can be determined by using the methods described in Section 7.4, page 246, or it can be estimated for circular pipes using Figure 9.14. As noted in this figure, in no case can the critical depth be greater than the height of the barrel.

Example 9.4 – Outlet Control for a Circular Culvert with Smaller Tailwater Effects. Repeat Example 9.3 (page 344) assuming that the obstruction in the downstream channel has been removed and lowering the tailwater depths to h_o = 1.80 ft when Q = 30 cfs and h_o = 2.70 ft when Q = 60 cfs, as illustrated in Figure E9.4.1. By comparing the inlet and outlet control solutions, determine whether the culvert operates under inlet or outlet control for each of the discharges.

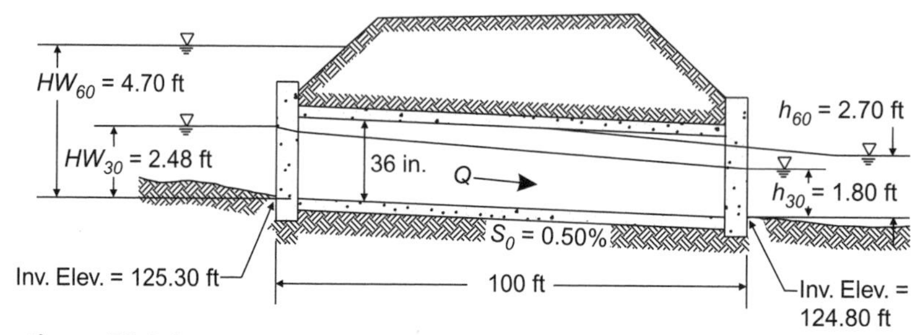

Figure E9.4.1 Culvert for Example 9.4

Solution: From Example 9.3, H = 0.60 ft when Q = 30 cfs. Using Figure 9.14, the critical depth in the culvert barrel for this discharge is about 1.75 ft. When using the outlet control nomographs with Case D or E flow, the tailwater depth should be taken as the larger of the actual tailwater depth or the average of tailwater depth and barrel height. Averaging this depth with the culvert diameter yields 2.38 ft. Comparing this value with the actual tailwater depth of 1.80 ft and taking the larger of the two leads to h_o = 2.38 ft. Substitution into Equation 9.1 yields

$$HW = 2.38 + 0.60 - 0.005(100) = 2.48 \text{ ft}$$

This value is approximately 83 percent of the height of the culvert barrel, and hence the result, while only an approximation, may be adequate. The headwater elevation is

$$125.30 + 2.48 = 127.78 \text{ ft}$$

Comparing this elevation to the inlet control headwater elevation of 128.00 ft found in Example 9.1 (page 340) shows that inlet control governs the performance. The headwater elevation for this discharge would therefore be 128.00 ft.

When Q = 60 cfs, H = 2.45 ft (from Example 9.3). The critical depth for this discharge is about 2.50 ft (from Figure 9.14). Averaging the critical depth with the culvert barrel height yields

$$(3.00 + 2.50)/2 = 2.75 \text{ ft}$$

Because 2.75 ft is larger than the tailwater depth, h_o = 2.75 ft. Substitution into Equation 9.1 yields

$$HW = 2.75 + 2.45 - 0.005(100) = 4.70 \text{ ft}$$

Because this depth is larger than the culvert diameter, the culvert flows full over part of its length; therefore, the quality of the approximation is better than it is for the smaller discharge of 30 cfs. The headwater elevation is

$$125.30 + 4.70 = 130.00 \text{ ft}$$

Because this elevation is again lower than the headwater elevation determined for the assumption of inlet control (130.25 ft), inlet control governs and the headwater elevation is 130.25 ft for a discharge of 60 cfs.

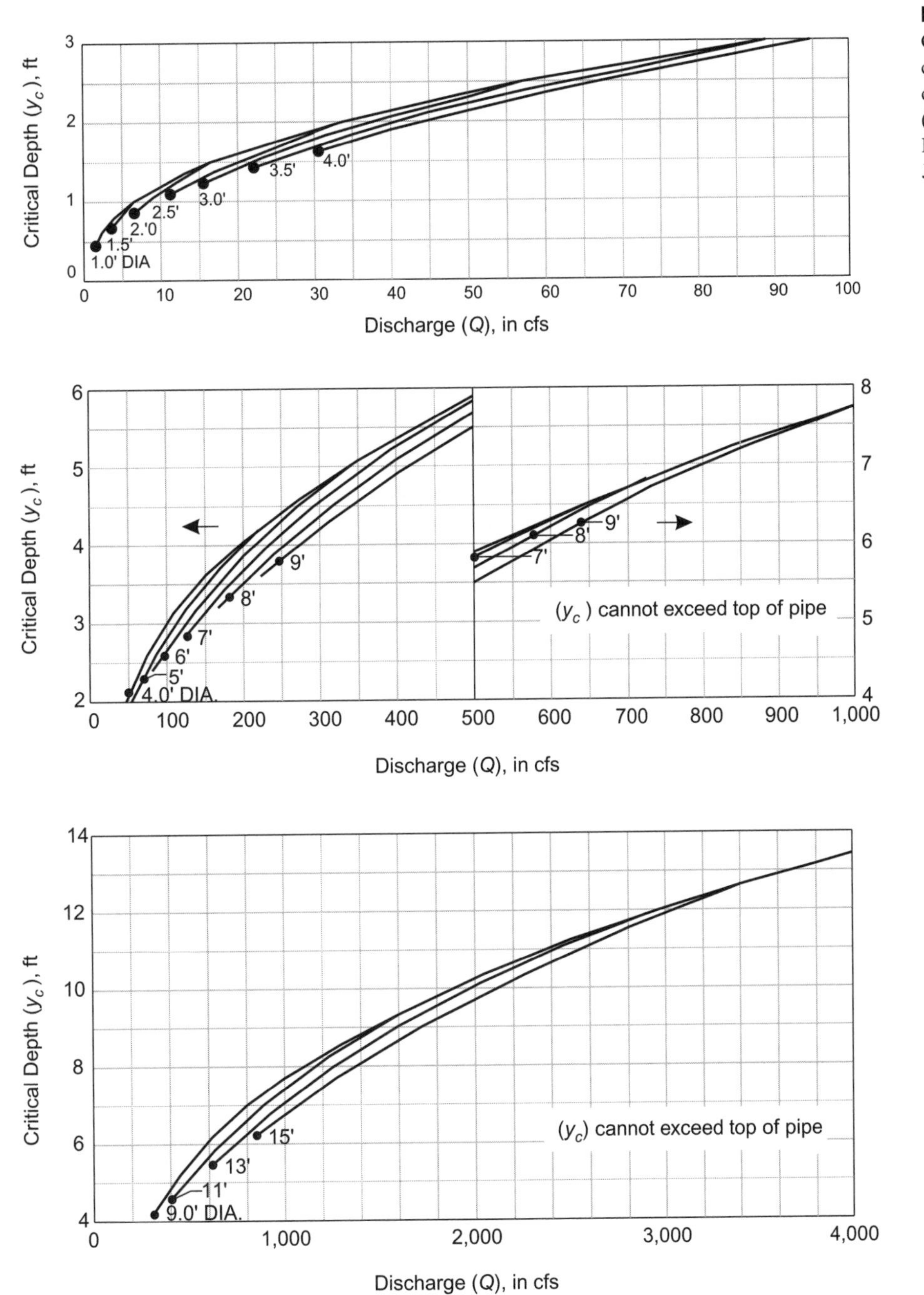

Figure 9.14
Graphs for estimating critical depths for circular pipes (Normann, Houghtalen, and Johnston, 2001)

Flow Profile Analysis of Culvert Performance

As shown in the previous sections and examples, culvert performance analysis using the nomographs published in HDS-5 (Normann, Houghtalen, and Johnston, 2001) is convenient, but doing so yields only approximate solutions in many practical cases. More accurate analysis requires computation of flow profiles and makes use of the closed-conduit and open-channel flow relationships presented in Chapter 6 and Chapter 7. The subsections that follow describe flow profile analysis methods for both inlet and outlet control. The tedious nature of these methods makes the use of computer-based modeling tools attractive. The methods described herein are the basis for the algorithms used in many modern computer tools.

Inlet Control. The entrance to a culvert operating under inlet control behaves as a weir when the headwater depth is smaller than the culvert entrance height. It operates as an orifice if the headwater submerges the entrance. This behavior is shown by the two curves in Figure 9.15. Between the two curves representing weir flow and orifice flow is a transition zone representing an intermediate type of flow. A graph such as Figure 9.15 is a key to inlet control performance evaluation using flow profile analysis.

When the culvert entrance operates as a weir, the *HW/D* ratio can be expressed as (Normann, Houghtalen, and Johnston, 2001)

$$\frac{HW}{D} = \frac{E_c}{D} + K\left(\frac{C_f Q}{AD^{1/2}}\right)^M - 0.5S_o \tag{9.2}$$

An alternative expression is

$$\frac{HW}{D} = K\left(\frac{C_f Q}{AD^{1/2}}\right)^M \tag{9.3}$$

where HW = headwater depth (ft, m)
D = culvert barrel height (ft, m)
E_c = specific energy (ft) corresponding to the barrel critical depth ($E_c = y_c + V_c^2/2g$), (ft, m)
Q = discharge (cfs, m^3/s)
A = full cross-sectional area of the barrel (ft^2, m^2)
S_o = barrel slope (ft/ft, m/m)
C_f = conversion factor (1.0 U.S. Customary, 1.811 SI)
K and M are constants determined from hydraulic model tests

Values of K and M are provided in Table 9.2, along with an indication of whether Equation 9.2 or 9.3 should be used in the calculation. Equation 9.2 is theoretically better than Equation 9.3, but Equation 9.3 is easier to apply in practice. The slope term expressed as ($-0.5S_o$) in Equation 9.2 should be changed to ($+0.7S_o$) if the barrel entrance is mitered. Equations 9.2 and 9.3 are applicable for values of $Q/AD^{1/2}$ less than or equal to about 3.5 for U.S. Customary units, or 1.93 for SI units.

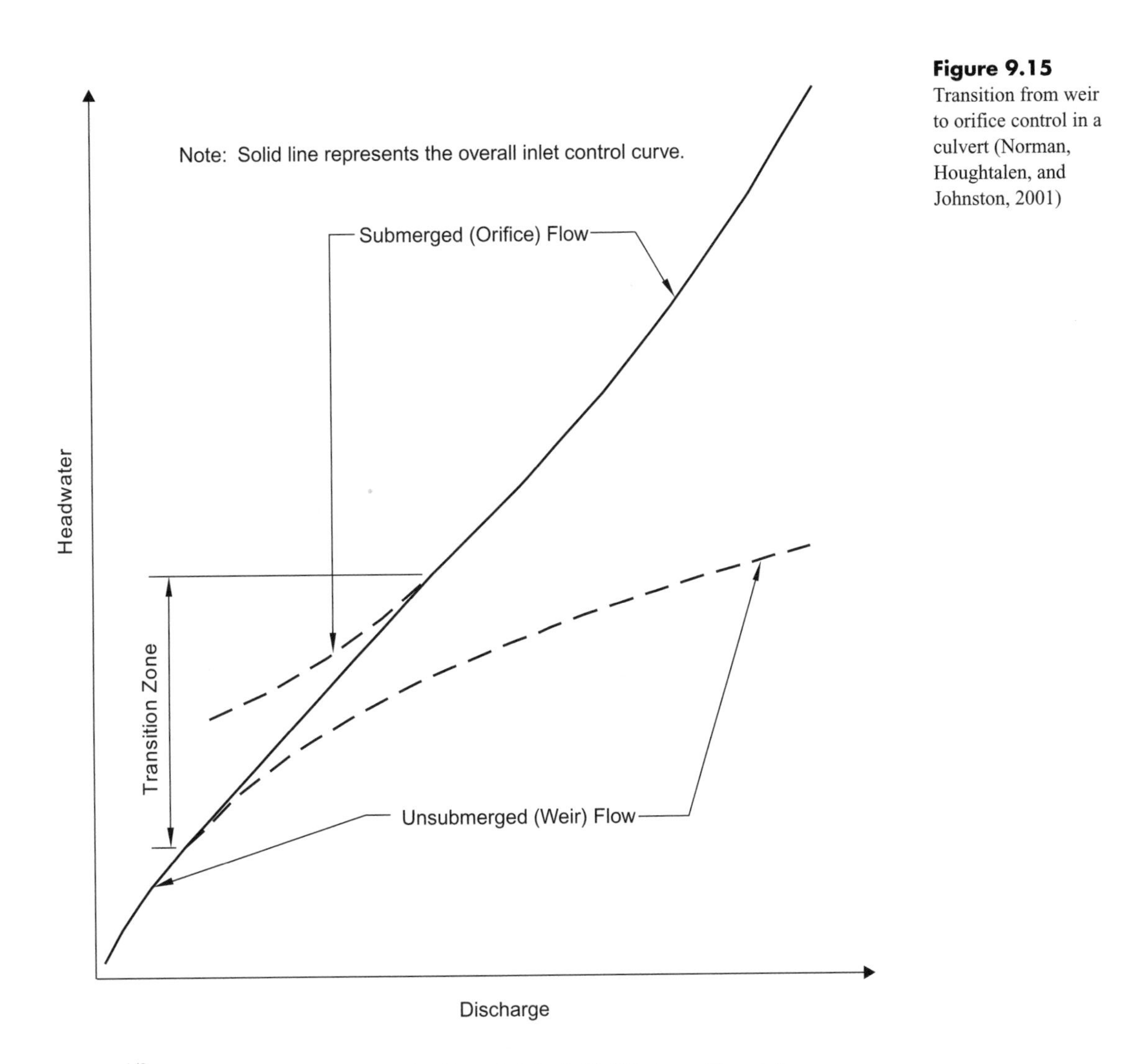

Figure 9.15
Transition from weir to orifice control in a culvert (Norman, Houghtalen, and Johnston, 2001)

If $Q/AD^{1/2}$ is between about 3.5 (1.93 SI) and 4.0 (2.21 SI), transitional flow occurs, and the headwater value must be interpolated as shown in the transition zone in Figure 9.15. Above this range, the culvert entrance acts as an orifice, and the *HW/D* ratio can be expressed as (Normann, Houghtalen, and Johnston, 2001)

$$\frac{HW}{D} = c\left(\frac{C_f Q}{AD^{1/2}}\right)^2 + Y - 0.5S_o \tag{9.4}$$

where *c* and *Y* are constants determined from hydraulic model tests

Values of the constants *c* and *Y* are provided in Table 9.2. Equation 9.4 is valid for values of $Q/AD^{1/2}$ greater than or equal to about 4.0. Below this range, transitional flow occurs. Again, the slope term ($-0.5S_o$) should be replaced with ($+0.7S_o$) if the barrel entrance is mitered.

Table 9.2 Constants for inlet control equations (Normann, Houghtalen, and Johnston, 2001)

Culvert Shape and/or Material	Inlet Edge Description	Unsubmerged (Weir Flow)			Submerged (Orifice Flow)	
		Equation in text	*K*	*M*	*c*	*Y*
Circular, concrete	Square edge with headwall	(9.2)	.0098	2.0	.0398	0.67
	Groove end with headwall		.0018	2.0	.0292	.74
	Groove end projecting		.0045	2.0	.0317	.69
Circular, CMP	Headwall	(9.2)	.0078	2.0	.0379	.69
	Mitered to slope		.0210	1.33	.0463	.75
	Projecting		.0340	1.50	.0553	.54
Circular	Beveled ring, 45× bevels	(9.2)	.0018	2.50	.0300	.74
	Beveled ring, 33.7× bevels		.0018	2.50	.0243	.83
Rectangular box	30× to 75× wingwall flares	(9.2)	.026	1.0	.0347	.81
	90× and 15× wingwall flares		.061	0.75	.0400	.80
	0× wingwall flares		.061	0.75	.0423	.82
Rectangular box	45× wingwall flares, d = 0.043D	(9.3)	.510	.667	.0309	.80
	18× to 33.7× wingwall flares, d =		.486	.667	.0249	.83
Rectangular box	90× headwall with ¾-in. chamfers	(9.3)	.515	.667	.0375	.79
	90× headwall with 45× bevels		.495	.667	.0314	.82
	90× headwall with 33.7× bevels		.486	.667	.0252	.865
Rectangular box	45× skewed headwall; ¾-in. chamfers	(9.3)	.545	.667	.0505	.73
	30× skewed headwall; ¾-in. chamfers		.533	.667	.0425	.705
	15× skewed headwall; ¾-in. chamfers		.522	.667	.0402	.68
	10-45× skewed headwall; 45× bevels		.498	.667	.0327	.75
Rectangular box with ¾-in. chamfers	45× nonoffset wingwall flares	(9.3)	.497	.667	.0339	.803
	18.4× nonoffset wingwall flares		.493	.667	.0361	.806
	18.4× nonoffset wingwall flares; 30×		.495	.667	.0386	.71
Rectangular box w/ top bevels	45× wingwall flares, offset	(9.3)	.497	.667	.0302	.835
	33.7× wingwall flares, offset		.495	.667	.0252	.881
	18.4× wingwall flares, offset		.493	.667	.0227	.887
Corrugated metal boxes	90× headwall	(9.2)	.0083	2.0	.0379	.69
	Thick wall projecting		.0145	1.75	.0419	.64
	Thin wall projecting		.0340	1.5	.0496	.57
Horizontal ellipse, concrete	Square edge w/ headwall	(9.2)	.0100	2.0	.0398	.67
	Groove end w/ headwall		.0018	2.5	.0292	.74
	Groove end projecting		.0045	2.0	.0317	.69
Vertical ellipse, concrete	Square edge w/ headwall	(9.2)	.0100	2.0	.0398	.67
	Groove end w/ headwall		.0018	2.5	.0292	.74
	Groove end projecting		.0095	2.0	.0317	.69
Pipe arch, CM, 18-in. corner radius	90× headwall	(9.2)	.0083	2.0	.0379	.69
	Mitered to slope		.0300	1.0	.0463	.75
	Projecting		.0340	1.5	.0496	.57
Pipe arch, CM, 18-in. corner radius	Projecting	(9.2)	.0300	1.5	.0496	.57
	No bevels		.0088	2.0	.0368	.68
	33.7× bevels		.0030	2.0	.0269	.77
Pipe arch, CM, 31-in. corner radius	Projecting	(9.2)	.0300	1.5	.0496	.57
	No bevels		.0088	2.0	.0368	.68
	33.7× bevels		.0030	2.0	.0269	.77
Arch, CM	90× headwall	(9.2)	.0083	2.0	.0379	.69
	Mitered to slope		.0300	1.0	.0463	.75
	Thin wall projecting		.0340	1.5	.0496	.57

Under inlet control conditions, flow in the barrel just downstream of the culvert entrance is supercritical and forms an S2 water surface profile that asymptotically approaches normal depth. The direct-step method, described in Section 7.8 on page 275, can be used to compute the profile.

If the tailwater level is high and submerges the culvert outlet as shown in Cases B and D of Figure 9.9, one of two possible situations occurs. In the first situation (which is illustrated for Cases B and D in Figure 9.9), pressure flow occurs in the most downstream part of the culvert and a hydraulic jump forms inside the barrel. In the second situation (not illustrated), the S2 water surface profile extends all the way through the culvert to its outlet and a hydraulic jump may form downstream of the outlet. In either case, formation of a hydraulic jump is contingent on a momentum balance being achieved (see Section 7.5, page 260).

Example 9.5 – Inlet Control Equations. Assume that the culvert described in Example 9.1, illustrated in Figure E9.5.1, operates under inlet control. Use Equations 9.2, 9.3, and/or 9.4 to determine the headwater depths for discharges of 30 cfs, 45 cfs, and 60 cfs. Compare the headwater depths for the discharges of 30 cfs and 60 cfs to those determined in Example 9.1.

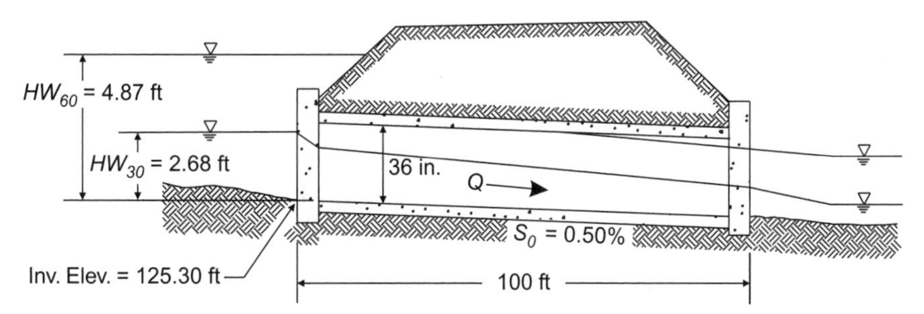

Figure E9.5.1 Culvert for Example 9.5

Solution: From Example 9.1, a circular concrete culvert with a square-edge headwall entrance has $L = 100$ ft, $D = 36$ in., $S_o = 0.5\%$, and an upstream invert elevation of 125.30 ft. For each of the three discharges, the values of $Q/AD^{1/2}$ and the corresponding flow types (based on criteria given in the preceding sections) are as follows:

Q (cfs)	$Q/AD^{1/2}$	Flow Type
30	2.45	Weir flow
45	3.68	Transitional flow
60	4.90	Orifice flow

Consider first the discharge of 30 cfs, for which weir flow occurs. For a circular culvert with headwalls and a square-edged barrel entrance, Table 9.2 indicates that the constants K_1 and M_1 are equal to 0.0098 and 2.0, respectively, and that these values should be used with Equation 9.2. By trial and error, the critical depth in the barrel corresponding to this discharge is $y_c = 1.77$ ft, and the corresponding specific energy is $E_c = 2.51$ ft. Substitution of these values into Equation 9.2 yields

$$HW = 2.51 + 3.00(0.0098)(2.45)^2 - 0.5(0.005)(3.00) = 2.68 \text{ ft}$$

This value compares quite favorably with the depth of 2.70 ft determined in Example 9.1.

Orifice flow occurs for the discharge of 60 cfs; thus, Equation 9.4 applies. From Table 9.2, the values of K_2 and M_2 are 0.0398 and 0.67, respectively. The computed headwater depth is

$$HW = 3.00(0.0398)(4.90)^2 + 3.00(0.67) - 0.5(0.005)(3.00) = 4.87 \text{ ft}$$

Again, this compares favorably with the depth of 4.95 ft determined in Example 9.1.

To determine the headwater depth for a discharge of 45 cfs (for which transitional flow occurs) a linear interpolation procedure will be used. Equation 9.2 applies for $Q/AD^{1/2}$ values up to 3.5, whereas Equation 9.4 applies for $Q/AD^{1/2}$ values greater than 4.0. Thus, the largest discharge that Equation 9.2 applies to for this culvert is 43 cfs, and the smallest discharge that Equation 9.4 applies to is 49 cfs.

Using Equation 9.2, the headwater depth corresponding to a discharge of 43 cfs is found to be 3.48 ft. Similarly, using Equation 9.4, the headwater depth corresponding to a discharge of 49 cfs is found to be 3.91 ft. Linear interpolation between the coordinate pairs (Q = 43 cfs, HW = 3.48 ft) and (Q = 49 cfs, HW = 3.91 ft) produces a headwater depth of HW = 3.62 ft for a discharge of 45 cfs.

Outlet Control. The type of flow profile analysis performed for a culvert operating under outlet control depends on the profile condition from Figure 9.11 that is being analyzed. Equation 9.1 can be applied directly if the culvert outlet is submerged and the flow corresponds to either Case A (a full-flow condition) or Case B (full flow except for the contraction at the culvert entrance) from Figure 9.11. Note, however, that this equation assumes that the velocity heads in the upstream and downstream channels are approximately equal and thus cancel one another out in the energy equation.

Case C is a case of full flow that occurs when the critical depth in the culvert barrel is equal to the height of the barrel (it can be no larger) and the tailwater depth is smaller than the height of the barrel. In this case, the hydraulic grade line elevation at the outlet is equal to the elevation of the crown of the barrel. Assuming that the velocity head in the upstream channel is negligible, the headwater elevation is greater than the outlet crown elevation by an amount consisting of the sum of the velocity head in the barrel, the frictional loss along the length of the barrel, and the entrance loss. The entrance loss can be expressed as the product of the velocity head in the barrel and a minor loss coefficient k_e (see Table 9.1). If the Manning equation is applied for evaluation of frictional losses, the friction slope S_f can be estimated using Equation 6.19. The product of that slope and the barrel length is the total head loss due to friction. Thus, the headwater elevation can be expressed as

$$EL_{hw} = EL_o + D + S_f L + (1 + k_e)\frac{V^2}{2g} \qquad (9.5)$$

where EL_{hw} = headwater elevation (ft, m)
EL_o = invert elevation at the outlet end of the culvert (ft, m)
D = culvert barrel height (ft, m)
S_f = slope of the energy grade line (ft/ft, m/m)
L = culvert barrel length (ft, m)
k_e = culvert entrance loss coefficient
V = full-flow velocity inside the culvert barrel (ft/s, m/s)

Cases D and E in Figure 9.11 occur when both the critical depth and the tailwater depth are smaller than the height of the barrel. A flow profile analysis for either of these cases begins with the determination of whether the tailwater depth is greater than or equal to critical depth or less than critical depth. If the tailwater depth is greater than or equal to critical depth, the depth at the control section at the downstream end of the barrel is equal to the tailwater depth, and computation of the M2 flow profile begins at this depth and proceeds in the upstream direction. If the tailwater depth is smaller than the critical depth, the M2 flow profile calculation starts at critical depth and proceeds upstream.

A Case D condition occurs if the M2 water surface profile reaches the crown of the culvert at some distance upstream of the outlet. Upstream of the point where the water surface intersects the crown, the culvert flows full and the friction slope can evaluated using Equation 6.19. An extension of the friction slope to the inlet of the culvert yields the hydraulic grade line elevation just inside the culvert entrance. The entrance loss and the full-flow velocity head can be added to this value to obtain the headwater elevation.

A Case E condition occurs if the M2 flow profile extends all the way to the culvert inlet without intersecting the crown. In this case, the hydraulic grade line elevation just inside the barrel entrance is equal to the computed water surface elevation at that point. The headwater elevation is determined by adding the velocity head just inside the barrel entrance and the entrance loss to this computed water surface elevation.

Example 9.6 – Determination of Outlet Control with Gradually Varied Flow. A rectangular concrete box culvert has a span of 1.2 m and a rise (height) of 1.0 m. It is 60.0 m in length and has a slope of 0.0015, as illustrated in Figure E9.6.1. Assuming that the culvert operates under outlet control and the tailwater depth is 0.7 m, determine the headwater elevation for a discharge of 2 m^3/s. Use Manning's n = 0.013. The invert elevation at the culvert outlet is 30.5 m, and the culvert entrance has a loss coefficient of 0.2.

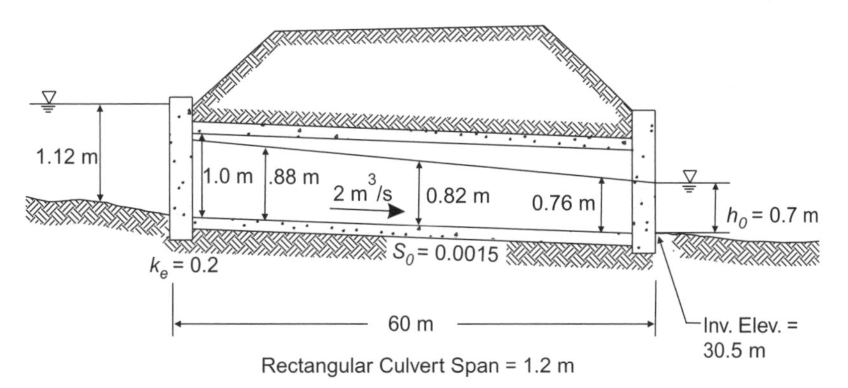

Figure E9.6.1 Culvert for Example 9.6

Solution: Because the tailwater depth is smaller than the barrel height, the outlet is not submerged. The critical depth inside the rectangular barrel is found to be

$$y_c = \{2^2/[9.81(1.2)^2]\}^{1/3} = 0.66 \text{ m}$$

Because this value is smaller than the barrel height, the flow corresponds to either Case D or Case E in Figure 9.7. The normal depth in a rectangular channel with the width, slope, and roughness of the culvert barrel is y_n = 1.06 m. Because this depth is larger than the culvert height, the flow might correspond to Case D. However, profile calculations are necessary to confirm this. Had the normal depth been less than the barrel height, one could say without qualification that the flow would correspond to Case E.

Computation of the M2 flow profile using the direct-step method (see Section 7.8 on page 275) is summarized in the following table. Because the tailwater depth of 0.7 m exceeds critical depth, calculations begin at this elevation.

Depth (m)	Distance from Outlet (m)
0.70	0
0.76	12.18
0.82	28.70
0.88	51.07
0.94	81.71

It can be seen from this summary that the water surface profile does not intersect the crown of the culvert; thus, the flow profile corresponds to Case E.

More detailed M2 water surface profile calculations indicate that the depth of flow at the upstream end of the culvert just inside its entrance is about 0.9 m. The corresponding velocity is

$$V = 2/[0.89(1.20)] = 1.85 \text{ m/s}$$

and the velocity head is 0.18 m. The headwater depth is the sum of the depth inside the entrance, the velocity head, and the entrance loss:

$$HW = 0.9 + (1 + 0.2)(0.18) = 1.116 \text{ m}$$

The culvert invert elevation at its entrance is

$$30.5 + 60(0.0015) = 30.59 \text{ m}$$

and thus the headwater elevation is

$$30.59 + 1.116 = 31.706 \text{ m}$$

Outlet Velocity

It is necessary to evaluate the velocity of flow emerging from a culvert in order to design measures to protect against erosion at the outlet (see Section 9.8). If the outlet is submerged, or if the barrel flows full over its entire length, then the outlet velocity is simply the ratio of the discharge to the cross-sectional area of the barrel. As shown in Figures 9.9 and 9.11, this situation can occur for culverts operating under either inlet or outlet control. When the outlet is not submerged, the procedure for determining the outlet velocity is more involved.

For a culvert operating under inlet control, an accurate estimate of the outlet velocity requires a flow profile analysis. However, the outlet velocity can be approximated using a simpler procedure that is conservative in that it always yields a value greater than or equal to the actual velocity. This procedure assumes that normal depth exists in the culvert at the outlet. Because normal depth may be smaller than the actual depth, the cross-sectional area of the flow is assumed to be smaller than it may be in reality, which in turn results in a higher calculated velocity. Determination of the normal depth and the associated cross-sectional area is a trial-and-error process that may be accomplished using the procedures described in Section 7.3.

When a culvert operates under outlet control and the outlet is not submerged, the depth of flow at the outlet should be taken as the larger of the tailwater depth and the critical depth in the culvert barrel. The cross-sectional area of the flow in the barrel may be computed based on the chosen depth, and the outlet velocity may be determined.

Example 9.7 – Outlet Velocity Determination. Estimate the outlet velocity for the 4-ft by 3-ft culvert from Example 9.6 (page 353).

Solution: As described in Example 9.6, the flow is 2 m^3/s, the outlet is not submerged, and the tailwater depth is 0.70 m. This depth is greater than the critical depth in the barrel, and thus is the depth at which flow exits the barrel. The cross-sectional area of the flow at the barrel outlet is

$$A = 0.70(1.2) = 0.84 \text{ m}^2$$

The outlet velocity is therefore

$$V = Q/A = 2/0.84 = 2.38 \text{ m/s}$$

9.6 ROADWAY OVERTOPPING

If the actual discharge approaching a culvert exceeds the design discharge for the structure, or if the culvert has become clogged by accumulated sediment or debris, the headwater elevation at the upstream end of the culvert will rise and may lead to *overtopping* of the embankment, or flow over the embankment. Overtopping can cause a hazard to the traveling public and, in severe cases, can cause the entire embankment to fail and be washed out.

Flow over a roadway embankment usually occurs at the low point of a sag vertical curve and is similar to the flow over a broad-crested weir. Equation 9.6 relates discharge to the head on the weir:

$$Q = C_d L (HW_r)^{3/2} \quad (9.6)$$

where Q = discharge (cfs, m^3/s)
C_d = discharge coefficient ($ft^{1/2}/s$, $m^{1/2}/s$)
L = crest length normal to the direction of flow (ft, m)
HW_r = head on the weir (ft, m)

HW_r is computed as the difference between the headwater elevation and the roadway embankment elevation. The discharge coefficient is determined as

$$C_d = k_t C_r \quad (9.7)$$

where k_t = correction factor for submergence
C_r = base coefficient value ($ft^{1/2}/s$, $m^{1/2}/s$)

The value of C_r depends on HW_r, the breadth of the embankment in the direction of flow (L_r), and whether the embankment has a paved or gravel surface. The term k_t is a correction factor to account for the effect of potential submergence of the weir and is a function of the ratio of the tailwater height h_t to HW_r. Numerical values of C_r and k_t may be obtained by using Figure 9.16. *It is important to note that the value of C_r obtained from Figure 9.16 will be in $ft^{1/2}/s$. This value must be converted to $m^{1/2}/s$ before solving Equation 9.6 using SI units.*

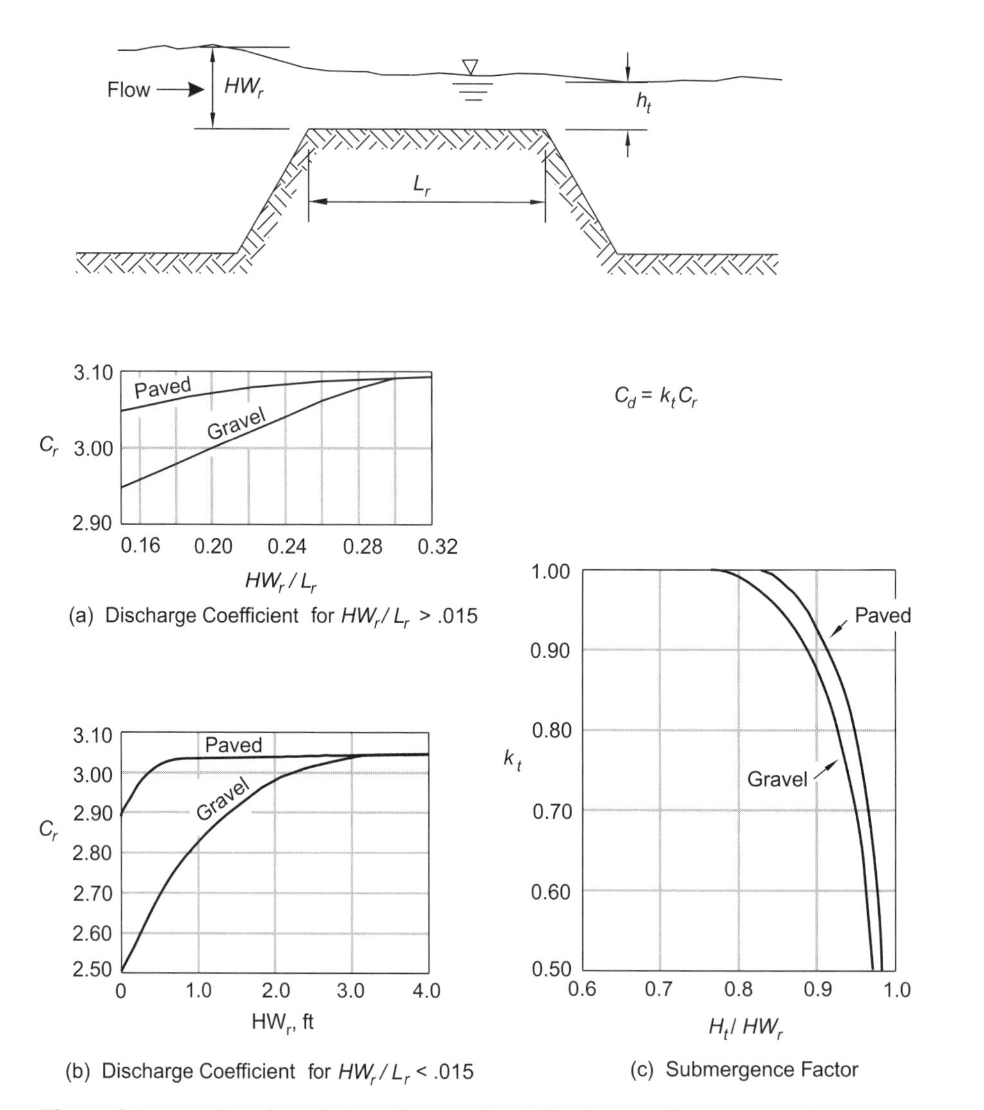

Figure 9.16 Graphs for determining the broad-crested weir discharge coefficient used in computing flow over a roadway (Normann, Houghtalen, and Johnston, 2001)

The weir's crest length and elevation can be difficult to estimate when the crest is defined by a sag vertical curve in a roadway profile. An approximate solution in this situation is to subdivide the total crest length into a series of horizontal segments. For the example shown in Figure 9.17(a), the roadway crest is approximated as three weir segments of lengths X_1, X_2, and X_3. The discharge for each segment can be computed separately, and then the total discharge can be computed as the sum of the individual segment discharges.

Normann, Houghtalen, and Johnston (2001) reported that it is often adequate in culvert design to represent a sag vertical curve by a single horizontal weir segment [see Figure 9.17(b)]. The length of the weir X in this case can be taken as the distance between the points on the sag vertical curve having the same elevation as the headwater. The head on the weir can be approximated as half of the difference between the headwater elevation and the elevation of the lowest point on the vertical curve.

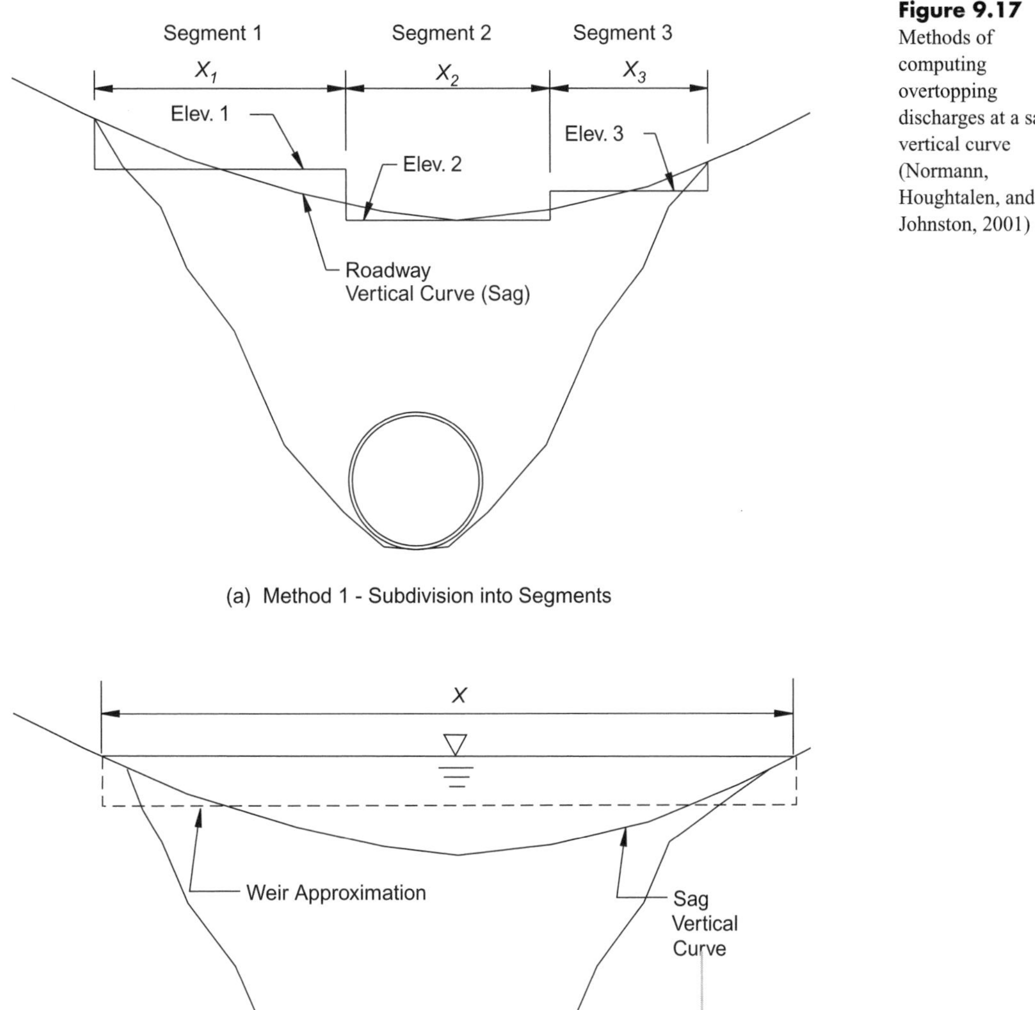

Figure 9.17 Methods of computing overtopping discharges at a sag vertical curve (Normann, Houghtalen, and Johnston, 2001)

If a total discharge exceeding the design capacity of a culvert is specified, a trial-and-error approach is required to determine how much of the flow passes through the culvert barrel and how much overtops the roadway embankment. Clearly, the sum of the two individual discharges must be equal to the specified total. If a performance curve has been developed and graphed for the culvert installation, it can be used to obtain a direct, graphical solution to a problem of this type.

Example 9.8 – Roadway Overtopping. For the culvert described in Example 9.1, assume that the roadway embankment is paved and has a breadth of 70 ft, as illustrated in Figure E9.8.1. The lowest roadway elevation at the sag point of a vertical curve is 130.70 ft. The sag vertical curve crest of the roadway embankment is described by the parabolic equation

$$y = 130.70 + 0.000075x^2$$

where y = crest elevation (ft)
x = distance from the lowest point in the sag (ft)

Estimate the discharge over the roadway embankment when the headwater elevation is 131.20 ft. Assume that a single horizontal segment can be used to approximate the weir and the tailwater elevation downstream of the culvert does not submerge the weir.

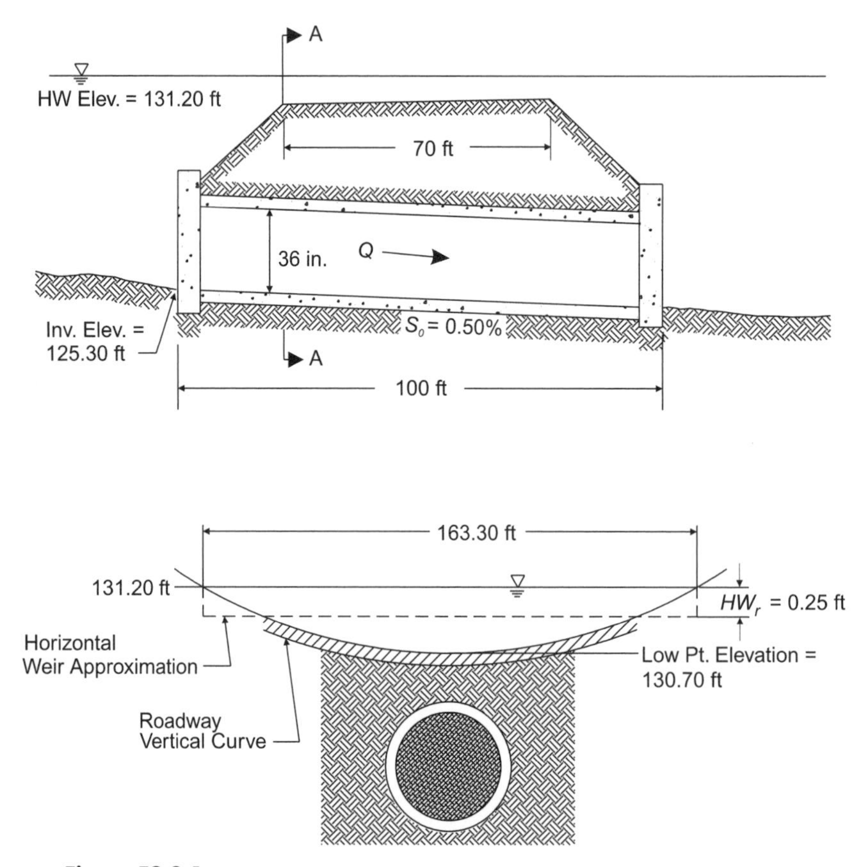

Figure E9.8.1 Culvert and roadway profile for Figure 9.8

Solution: The maximum depth of submergence of the roadway embankment (weir) is the difference between the headwater elevation and the minimum crest elevation, or

$$131.20 - 130.70 = 0.50 \text{ ft}$$

The average head on the weir is HW_r = 0.25 ft (one-half the maximum depth of submergence). The length of the weir crest across which flow occurs is found by computing the values of x in the parabolic equation that satisfy y = 131.20. These values are –81.65 ft and +81.65 ft; thus, the total crest length is L = 163.30 ft.

The ratio of HW_r to L_r is

$$HW_r/L_r = 0.25/70.0 = 0.00357$$

From Figure 9.16, $k_t = 1.00$ and $C_r = 2.97\ \text{ft}^{1/2}/\text{s}$, so

$$C_d = k_t C_r = 2.97\ \text{ft}^{1/2}/\text{s}$$

Substituting this value into Equation 9.6, along with the crest length and average head, yields the discharge over the embankment as

$$Q = 2.97(163.30)(0.25)^{3/2} = 61\ \text{cfs}$$

9.7 PERFORMANCE CURVES

A *performance curve* for a culvert is a graphical relationship between the discharge and the headwater elevation (or depth). As shown in Figure 9.18, the overall performance curve for a culvert is constructed from individual performance curves representing inlet control, outlet control, and roadway overtopping. It therefore illustrates, at a glance, the type of flow control that exists for any discharge of interest, the corresponding headwater elevation, and the depth of flow over the roadway (if any). If detention effects upstream of the culvert are significant and are thus taken into consideration in optimizing the culvert design, then the performance curve is the same as the stage-discharge curve required for the detention basin routing calculations (see Section 12.6, page 505).

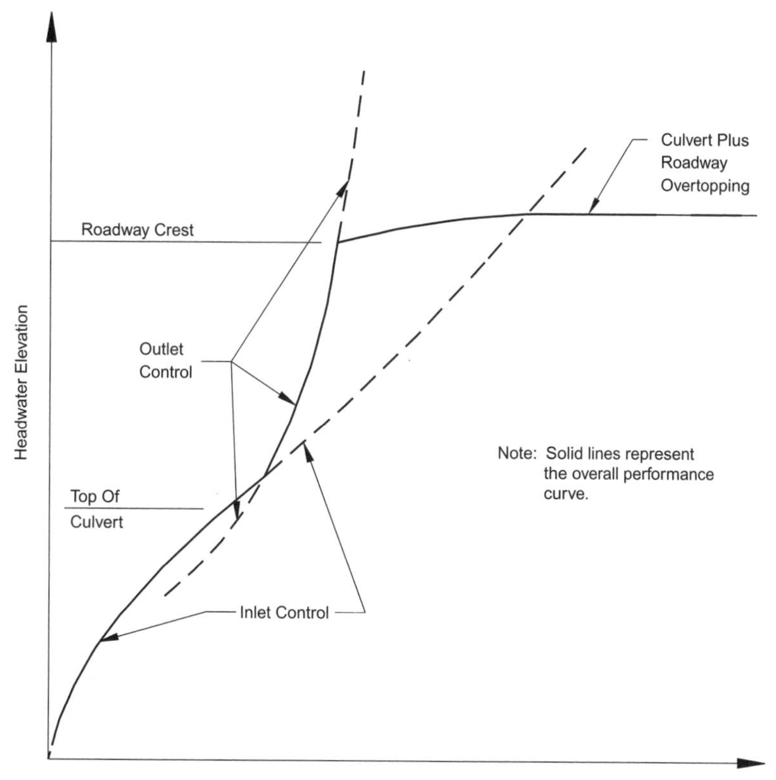

Figure 9.18 Construction of the overall performance curve for a culvert (Normann, Houghtalen, and Johnston, 2001)

An inlet control performance curve such as the one illustrated in Figure 9.18 is constructed by assuming various discharges, determining the corresponding headwater elevations, and plotting the headwater versus discharge data on a graph. Either the nomographs or the inlet control equations presented in Section 9.5 (page 348) can be used for this purpose.

Assumed discharges used in developing the curve should range from zero to a value that is greater than the design discharge. If the inlet control equations are used instead of nomographs, care should be taken to represent the transitional range of flows as well as the weir and orifice flow ranges.

The outlet control performance curve is also graphed by assuming various discharges and determining the corresponding headwater elevation for each. The chosen range of discharges should span the design discharge. The curve can be developed by using nomographs, but it is more accurate to use flow profile analysis as described in Section 9.5 (page 352).

After the inlet and outlet control performance curves have been plotted on the same graph, the overall performance curve is formed from those parts of the individual curves that have the highest headwater for a given discharge. For example, at locations on the graph where the inlet control curve is higher than the outlet control curve, the overall performance curve is the same as the inlet control curve. In a similar vein, the overall performance curve is the same as the outlet control curve where it is higher than the inlet control curve (see Figure 9.18). This procedure for delineation of the overall performance curve applies only to the portion for which the headwater elevation is at or below the roadway crest.

For headwater elevations above the crest, roadway overtopping must also be considered. The head on the weir represented by the sag vertical curve is equal to the difference between the headwater elevation and the roadway crest elevation. Knowing this head, the discharge over the weir can be determined by using Equation 9.6. This discharge, added to the discharge that passes through the culvert for that same headwater elevation, defines the overall performance curve for the region where the headwater elevation is above the roadway crest. As shown in Figure 9.18, this portion of the overall performance curve is generally quite flat compared to the rest of the curve, indicating that the water surface elevation rises much more slowly with increasing discharge once roadway overtopping has begun.

9.8 EROSION PROTECTION

The velocity at which water exits a culvert is usually much higher than the velocity in the channel and can cause objectionable scour and erosion if the excess kinetic energy associated with the flow is not dissipated. Section 9.5 (page 354) describes how to estimate the velocity at which a flow exits a barrel. That exit velocity, along with geometric information related to the culvert design, can be used to predict the size and extent of required riprap erosion protection at the outlet (see Figure 9.19). Section 11.7 describes this process.

Figure 9.19
Riprap outlet protection

For instances in which the kinetic energy caused by the exit velocity of the flow is too great to be safely dissipated using a riprap blanket, several options are available. The exit velocity may be reduced by modifying the culvert barrel near the outlet. For example, one might design a broken-back culvert with a smaller slope at the outlet than at the inlet, thereby reducing the outlet velocity. Another approach is to increase the barrel roughness near the outlet by using a material such as corrugated metal instead of concrete in this area. The rougher corrugated metal increases the depth and reduces the velocity near the barrel outlet.

If barrel modifications such as those just described cannot reduce the outlet velocity to an acceptable limit, one must resort to use of a type of *stilling basin* structure at the outlet, such as an *impact-type energy dissipator* (which uses a baffle in front of the water jet to dissipate energy). Corry et al. (1983) provide further direction on this subject.

In culvert design, it is often desirable to provide erosion protection measures at the entrance as well as the outlet. As flow enters a culvert, it must usually contract into a smaller area than that of the upstream channel. The result is increased velocity, which in turn leads to localized erosion of the channel bed and banks. The erosion protection necessary at an entrance is typically not as extensive as that required at the outlet—a riprap blanket, concrete apron, or slope paving is usually sufficient.

9.9 ENVIRONMENTAL CONSIDERATIONS

Culverts can often be used in lieu of bridges for small streams; however, culverts can pose obstacles to migration pathways if the stream supports fish and similar aquatic organisms. For instances in which fish passage is a concern, it is essential that input and requirements of state and/or federal fish and wildlife protection officials be obtained early in the design process. In some instances, a bridge may be required even though a less expensive culvert could provide the required hydraulic performance.

To ensure that culverts are suitable for fish passage, additional design criteria may be introduced to govern minimum flow depths within the barrel and minimum and/or maximum flow velocities. Seasonalities of fish migration and streamflow are important considerations here. The species of fish present in the stream are also important, as some species are less tolerant of certain flow depths and velocities than are other species. Another consideration is culvert lighting. Fish may not enter dark culverts even when conditions are otherwise favorable. Long culverts may be constructed in sections to allow natural light to enter. Artificial lighting may also be an option in some circumstances.

Perhaps the most significant requirements for fish passage relate to the need to simulate the natural streambed environment within the culvert structure. If an open-bottom culvert is used, such as a field-assembled corrugated arch structure, natural channel materials can be left in place along the culvert bottom. This type of culvert requires adequate foundation support for the culvert arch bearing ends. Localized scour or general channel bed degradation that could undermine the foundations is another concern. If a culvert pipe is used instead of an open-bottom structure, the barrel may be oversized as shown in Figure 9.20. The barrel can then be partially filled with natural streambed materials.

Figure 9.20
Partial filling of a culvert pipe with natural streambed materials

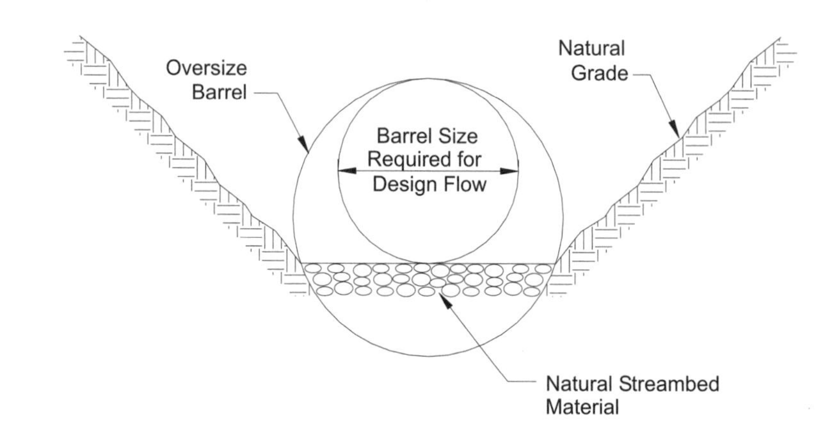

In some cases, special baffles and/or weirs are installed in culvert barrels to simulate natural riffles and pools that may exist in undisturbed areas of the stream. Design procedures for these types of installations, and for fish passage facilities in general, can be found in the *Model Drainage Manual* (AASHTO, 1991) and in Chang and Normann (1976). Fish and wildlife agencies also may provide specific design guidance.

Modeling Focus: Supplemental Culvert Design

An existing roadway stream crossing consists of a 6-ft (span) by 4-ft (high) concrete box culvert with the following characteristics:

- Upstream invert elevation = 124.75 ft
- Downstream invert elevation = 124.55 ft
- Length = 60.0 ft
- Headwall with 30° to 75° wingwall flares, K_e = 0.5

The minimum elevation of the roadway sag vertical curve is 132.55 ft. The culvert outfalls to a downstream trapezoidal channel with the following characteristics:

- Channel invert elevation matches culvert outlet invert
- Bottom width = 6 ft
- Side slopes are = 1.5 H to 1.0 V
- Manning's n = 0.045
- Channel slope = 0.0075 ft/ft

A computer program such as CulvertMaster can be used to analyze the capacity of the culvert and generate a rating curve. For the culvert and channel characteristics described above and a maximum headwater elevation of 132.55 ft (the roadway overtopping elevation), the culvert capacity is found to be 269 cfs. However, allowing for 1 ft of freeboard between the maximum headwater elevation and the minimum roadway elevation (i.e., the maximum headwater elevation is 131.55 ft), the culvert capacity is 236 cfs. The rating curve for the culvert is shown below.

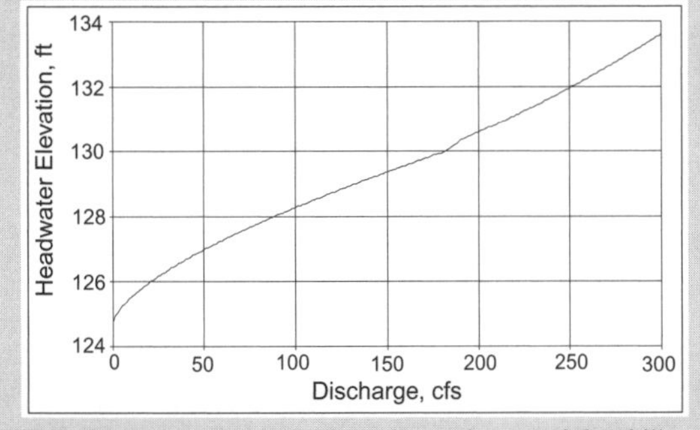

Improvements are being made to the roadway, and the ability of the existing culvert to meet current needs must be evaluated. The culvert was originally designed to convey the 50-year peak flow from the NRCS (SCS) Type II storm with 1 ft of freeboard; however, substantial development has occurred in the area since the original structure was installed.

Assuming that the watershed will be fully developed (subject to zoning regulations) within 5 years, the 50-year design flow is found to be 295 cfs. It is apparent that the capacity of the existing culvert is insufficient to meet the design requirements. Because the existing culvert is in good condition, it will remain in place but will be supplemented by a new culvert in parallel.

The supplemental culvert must be capable of conveying 295 cfs – 236 cfs = 59 cfs at the allowable headwater of 131.55 ft. The engineer decides to evaluate the feasibility of supplementing with a circular, concrete culvert section. The existing headwalls will be extended to accommodate the new culvert.

Assuming that the inverts and length for the new culvert will match the existing culvert and using a "square edge with headwall" entrance type, the engineer enters the needed capacity of 59 cfs and allowable headwater elevation of 131.55 ft and allows the program to solve for the minimum commercially available size that meets the criteria-in this case, a 36-in. pipe.

The 36-in. pipe is inserted into the program's worksheet for the existing system, and the program performs iterative calculations to solve for a headwater elevation of 131.44 ft on both culverts for design flow conditions. This design therefore meets the minimum 1-ft freeboard requirement. The program may also report the exit velocities, which are 9.6 cfs for the box culvert and 9.1 cfs for the 36-in. culvert. Outlet protection such as a riprap apron (see Section 11.7) should be provided.

The engineer is also asked to check the design for the 100-year storm event, which corresponds to a peak discharge of 336 cfs. The headwater on the culverts is found to be 132.46 ft for the 100-year storm, which is slightly below the minimum roadway elevation (i.e., the road will not be overtopped for the 100-year check storm).

The individual and composite rating curves for the proposed culvert system are shown below.

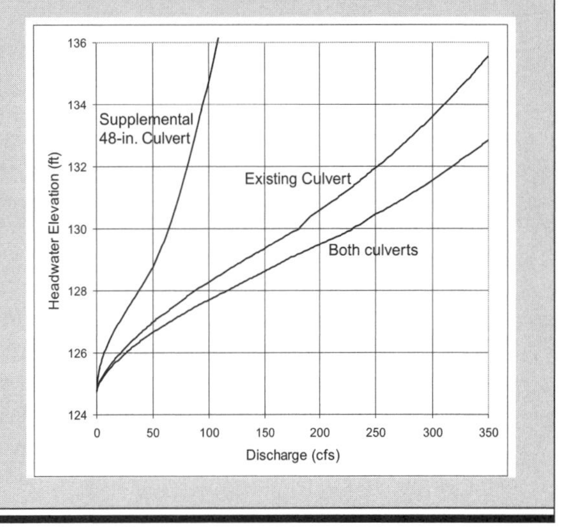

9.10 CHAPTER SUMMARY

A culvert is a relatively short conveyance conduit that passes through an obstruction such as a roadway embankment. Culverts can have a variety of cross-sectional shapes, including circular, elliptical, arched and rectangular geometries, and may consist of multiple barrels. Available materials include concrete, corrugated steel, PVC, corrugated aluminum, and polyethylene.

Culverts are typically designed to convey the peak flow resulting from a design storm of a particular recurrence interval (for instance, the peak flow resulting from a 24-hour, 25-year storm event) with a headwater elevation that does not exceed an allowable level. If significant upstream ponding occurs, the detention effects will result in a reduction of the design storm peak flow that should be considered in the evaluation. As an alternative to the design-storm approach, culverts may be designed using economic analysis. The goal of this approach is to minimize total cost if both flood damages and the capitalized cost of the culvert are considered.

Culvert performance may also be checked for a storm event of greater magnitude than the design storm (often, the 100-year event). This evaluation typically involves an analysis of roadway overtopping in addition to culvert hydrualics.

Various end treatments are used at culvert entrances and exits to provide erosion control and improve hydraulic efficiency. The culvert barrel may simply project from the embankment, or it may be mitered to conform to the embankment slope. Precast end sections are often available from pipe manufacturers. Alternatively, headwalls—with or without wingwalls—may be used at one or both ends. Finally, specialized inlet types, such as side-tapered and slope-tapered inlets, may be necessary if certain geometric constraints are present.

Although culvert barrels are typically straight and roughly perpendicular to the roadway or obstruction they cross, there are many exceptions. A skewed culvert is a culvert that is not perpendicular to its obstruction, typically in order to better align with the natural direction of the stream. A horizontal alignment may be curved or doglegged to avoid obstacles such as rock outcroppings. A broken-back culvert has bends in its vertical alignment, and may be used to reduce exit velocities, for instance. With a sag culvert, a portion of the culvert barrel is lower than the culvert entrance and exit. Such culverts tend to clog, but they may be required in, for example, certain drainage canal crossings.

Culvert hydraulics are complex, and several flow control types may exist for a single culvert and discharge. The basic approach to hydraulic evaluation of a culvert is to analyze its performance for each of the possible flow control types and allow the assumption that results in the worst performance (that is, the highest headwater) to govern the design. Two general methods of evaluation exist: (1) the graphical method, which uses nomographs to find the headwater elevation for a particular set of conditions; and (2) the method of calculating the complete flow profile, as presented in Section 7.8. The second method is used by most culvert-analysis computer programs. It is the more computationally intensive of the two approaches, but it is more accurate in many situations.

Flow in a culvert can be classified as functioning under either inlet control or outlet control. Under inlet control, the barrel of the culvert is capable of conveying a greater discharge than the inlet will accept. For an outlet-control situation, the inlet is capable of accepting more flow than the barrel characteristics or downstream tailwater condition will permit.

A performance curve graphically shows the relationship between discharge and headwater elevation. For a culvert, the performance curve is constructed by plotting the rating curves that would result for both inlet- and outlet-control assumptions. Whichever curve has the higher headwater elevation for a given discharge is assumed to govern the design and is used to define the performance curve. If the discharge is increased to the point where the corresponding headwater is higher than the culvert embankment, the resulting flow over the roadway (for example) should be reflected in the performance curve. For hydraulic analysis purposes, the roadway is typically assumed to be a broad-crested weir.

Culvert outlet velocities should be evaluated to determine what energy dissipation and erosion protection measures are necessary, if any. Riprap can be used to both dissipate energy and provide erosion protection. Another option is to modify the culvert barrel to reduce the outlet velocity. Other protective measures include impact-type dissipators and stilling basins.

Other factors that should be considered in culvert design relate to possible environmental impacts. For instance, a culvert may be an obstacle to fish passage. To minimize these impacts, the culvert may be designed to maintain a certain minimum water level and/or minimum or maximum velocity, and to simulate the natural streambed environment.

REFERENCES

American Association of State Highway and Transportation Officials (AASHTO). 1991. *Model Drainage Manual*. Washington, D.C.: AASHTO.

American Society of Civil Engineers (ASCE). 1993. *Design and Construction of Urban Storm Water Management Systems*. ASCE Manuals and Reports on Engineering Practice, No. 77. New York: ASCE.

Brice, J. C. 1981. *Stability of Relocated Stream Channels*. Report No. FHWA/RD-80/158. U.S. Geological Survey.

Brown, S. A., S. M. Stein, and J. C. Warner. 2001. *Urban Drainage Design Manual*. Hydraulic Engineering Circular No. 22, 2d ed., FHWA-SA-96-078. Washington, D.C.: U.S. Department of Transportation.

Chang, F. M., and J. M. Normann. September, 1976. *Design Considerations and Calculations for Fishways Through Box Culverts*, Unpublished Report, Federal Highway Administration, U.S. Department of Transportation, Washington, D.C.

Corry, M. L, P. L. Thompson, F. J. Watts, J. S. Jones, and D. L. Richards. 1983. *Hydraulic Design of Energy Dissipators for Culverts and Channels*. Hydraulic Engineering Circular No. 14. Washington, D.C.: Federal Highway Administration, U.S. Department of Transportation.

Dyhouse, Gary, J. A. Benn, David Ford Consulting, J. Hatchett, and H. Rhee. 2003. *Floodplain Modeling Using HEC-RAS*. Waterbury, Connecticut: Haestad Methods.

Harrison, L. J., J. L. Morris, J. M. Normann, and F. L. Johnson. 1972. *Hydraulic Design of Improved Inlets for Culverts*. Hydraulic Engineering Circular No. 13. Washington, D.C.: Federal Highway Administration, U.S. Department of Transportation.

Normann, J. M., R. J. Houghtalen, and W. J. Johnston. 2001. *Hydraulic Design of Highway Culverts*, 2d ed. Washington, D.C.: Hydraulic Design Series No. 5, Federal Highway Administration (FHWA).

PROBLEMS

Note: Problems 1, 3, 4, 5, and 11 may be solved with the nomographs provided in this book. Problems 2, 8, and 10 may be solved using the nomographs in HDS 5. Alternatively, the CulvertMaster software provided with this book can be used.

9.1 Considering only inlet control, determine the flow in a 24-in. concrete pipe culvert with a square-edge headwall. The headwater elevation is 6 ft. Assume Manning's $n = 0.013$.

9.2 Considering only inlet control, what is the headwater elevation required to convey 5.7 m^3/s in a 1200-mm corrugated metal pipe with a mitered entrance? Assume Manning's $n = 0.02$.

9.3 A 60-in. concrete pipe is to be used as a culvert to convey 400 cfs under a road. Calculate the reduction in headwater elevation resulting from replacing a square-edge inlet ($K = 0.5$) at the headwall with a groove-end inlet ($K = 0.02$).

9.4 A culvert is to be designed to carry 250 cfs under a roadway. The culvert is 75 ft long and the upstream invert elevation is 100 ft. A 60-in., groove-end concrete pipe is to be installed at the natural streambed slope of 0.01. Assume that there is a free fall at the exit (no tailwater). Determine if the culvert is operating under inlet or outlet control and report the headwater elevation.

9.5 Repeat Problem 9.4 for a culvert with a tailwater at an elevation of 4 ft above the crown of the downstream end.

9.6 A 1200-mm CMP culvert is operating under inlet control. The inlet type is a headwall and the slope is 0.008. Determine the flow type and the headwater depth for flows of 1.70 m^3/s and 3.4 m^3/s. Assume Manning's $n = 0.02$.

9.7 The 50-year flow at a design site is 11.3 m^3/s. The maximum allowable headwater for a proposed culvert is 4.57 m. The culvert will have a 90° headwall with chamfers. Design the smallest rectangular concrete barrel that will pass the peak flow without exceeding the allowable maximum headwater. Use $n = 0.012$. Because of structural considerations, the box culvert must be 1520 mm high. Other pertinent design data are

- Stream bed slope = 0.01
- Culvert length = 61 m
- Tailwater elevation = 2.74 m above the culvert outlet invert

9.8 Determine the outlet velocity of a 4-ft–diameter culvert conveying 150 cfs and operating under inlet control. Use the conservative approximation method.

9.9 A performance chart for a roadway culvert is shown below. The roadway at the culvert location has a low point of 37.2 m. The vertical alignment of the road is described by the parabolic equation $y = 37.2 + 1.52 \times 10^{-3} x^2$ (x and y in meters), where y is the crest elevation and x is the distance from the location of the low point. Assume a 10-m-wide paved road surface. Calculate the total flow through the culvert and over the roadway when the headwater elevation is 38.1 m.

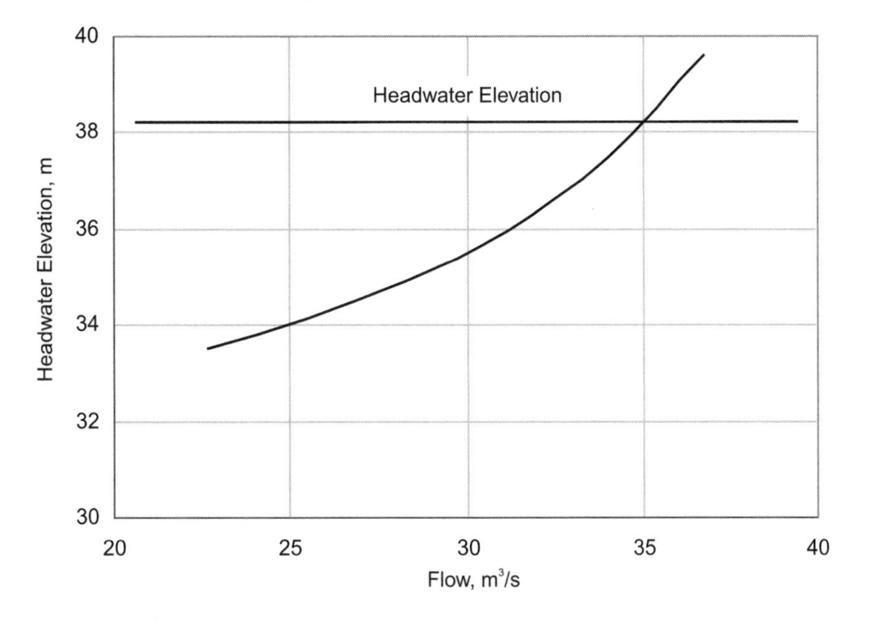

9.10 Construct a performance curve for a CMP culvert with a diameter of 4 ft and a length of 150 ft. The pipe entrance is projecting from the fill and the pipe slope is 0.01. Assume that Manning's $n = 0.024$ and that there is free flow away from the exit under all flow conditions. Use a flow range from 0 to 200 cfs.

9.11 Construct a performance curve for a 60-in. circular concrete pipe culvert. The pipe is 100 ft long and the slope is 0.008. The entrance is a groove end with a headwall and Manning's $n = 0.013$. The tailwater is 8 ft above the invert of the exit end for all flow conditions. Use a flow range of 0 to 400 cfs.

This combination inlet in a local depression captures runoff in a parking lot.

CHAPTER

10

Gutter Flow and Inlet Design

Storm sewer systems, which are most commonly designed and constructed for service in urbanized areas, are an integral part of the infrastructure. They are typically designed as conveyance systems for smaller, more-frequent storms in order to prevent the nuisances and flood damages that would otherwise accompany them. Runoff from larger, less-frequent events is generally not conveyed entirely by storm sewers; rather, it flows over the land surface in roadways and in natural and constructed open channels.

Stormwater enters the subsurface sewer conveyance system through inlets in roadway gutters, parking lots, depressions, ditches, and other locations. These surface components are critical, and they must be properly designed to ensure that the overall system functions as intended. For example, an engineer may design a pipe system capable of handling the 10-year storm event, but the level of protection provided will be lower (often much lower) if the system's inlet capacities are not sufficient to intercept the 10-year flows. Inadequate attention to roadway drainage and inlets can cause undue hazards to the traveling public and can compromise the degree of protection afforded to private and public properties located along the sides of a roadway.

This chapter is the first of two chapters on the design and hydraulic analysis of storm sewer systems. It focuses on surface drainage—specifically, the physical and hydraulic characteristics of roadway gutters and grate, curb, combination, and slotted-drain inlets. The various types of gutter sections, including detailed information on the hydraulic analysis of uniform and composite gutters, are presented in Section 10.1. Section 10.2 covers the various inlet types and methods for evaluating the hydraulic performance of grate, curb, combination, and slotted-drain inlets. Determination of inlet design flows and proper placement of inlets are discussed in Section 10.3.

Chapter 11 continues the presentation on storm sewer systems with information on the analysis and design of the subsurface portion of the system.

10.1 ROADWAY GUTTERS

During design storm events, the *depth* of open-channel flow in roadway gutters must be limited to prevent overflowing of curbs and consequent flooding of adjacent properties. The *spread* (that is, the top width) of the gutter flow should also be limited to keep water from extending too far into roadway travel lanes. Excessive spread can cause safety hazards such as vehicular hydroplaning and limited visibility due to splash and spray. At locations where the depth and/or spread of the flow into the travel lane begins to exceed the allowed value, stormwater inlets or catch basins are required to intercept some or all of the water and convey it into the subsurface storm sewer system.

The relationship between the depth of flow in a gutter and its spread depends on the gutter type and the cross slope of the pavement and/or gutter. As illustrated in Figure 10.1, gutters fall into two basic categories: *conventional curb and gutter sections* and *shallow swale sections*. With conventional curb and gutter, the flow cross-section extends from the curb toward the roadway centerline. The basic types of curb and gutter illustrated in the figure are *uniform, composite,* and *curved sections*. For streets where curbs are not required, shallow swale sections along the side of the road or in the median can be used instead. Swale sections include *V-shape gutters, V-shape median swales,* and *circular swales.* In urbanized areas, composite gutters and V-shape gutters are the most common types.

Figure 10.1
Typical gutter cross sections

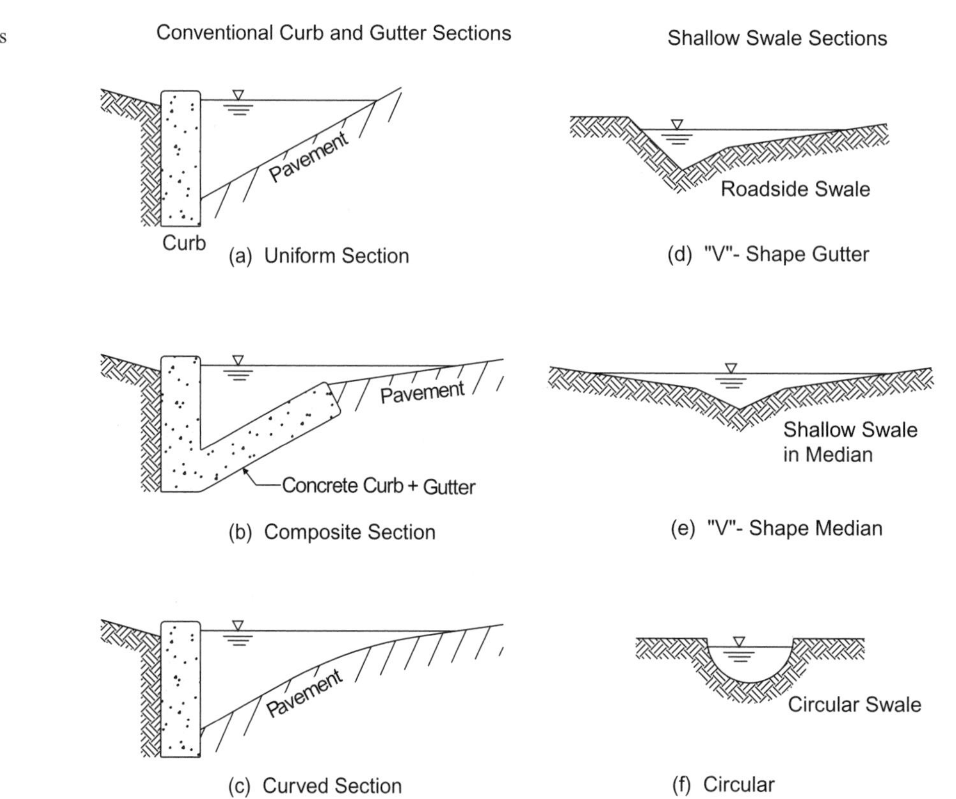

To maintain proper drainage, roadway *longitudinal slopes* (slopes along the length of the road) are usually specified to be no smaller than about 0.4 percent. It is difficult to maintain positive drainage with smaller slopes, resulting in localized puddling. Pavement *cross slopes* (measured perpendicular to the road direction) necessary to induce flow off of the pavement surface and into the gutter range from 1.5 percent to as high as about 4 percent, but are typically specified at around 2 percent (ASCE, 1992).

The allowable spread of open-channel flow in a gutter section is generally a function of the roadway classification, the design traffic speed, and whether that portion of the gutter is in a *sag* location (that is, at the low point of a roadway vertical curve). Design criteria suggested by the FHWA (Brown, Stein, and Warner, 2001) are presented in Table 10.1; other criteria may be specified by local jurisdictions.

Table 10.1 Suggested maximum flow spread in roadway sections (Brown, Stein, and Warner, 2001)

Roadway Classification	Design Speed/Location	Maximum Spread
High volume, divided, or bidirectional	< 45 mph (70 km/h)	Shoulder + 3 ft (0.9 m)
	> 45 mph (70 km/h)	Shoulder
	Sag point	Shoulder + 3 ft (0.9 m)
Collector streets	< 45 mph (70 km/h)	½ driving lane
	> 45 mph (70 km/h)	Shoulder
	Sag point	½ driving lane
Local streets	Low traffic	½ driving lane
	High traffic	½ driving lane
	Sag point	½ driving lane

Flow in Uniform Gutter Sections

In a *uniform gutter section* (Figure 10.2), the depth of flow is related to the spread and the pavement cross slope as

$$d = TS_x \tag{10.1}$$

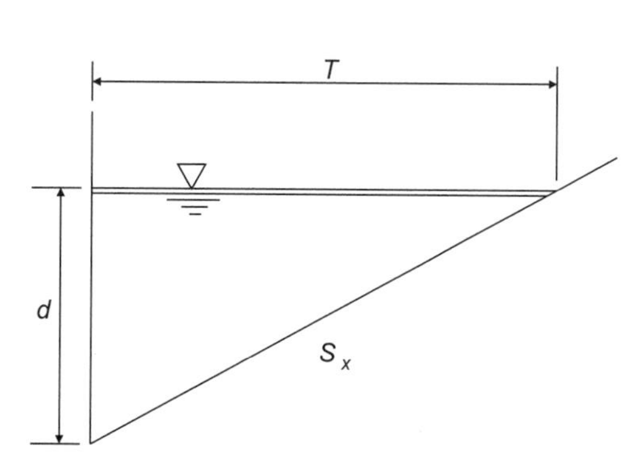

Figure 10.2 Uniform gutter cross section

The cross-sectional area of the flow is

$$A = Td/2 = T^2S_x/2 \quad (10.2)$$

and the wetted perimeter is

$$P = T(1 + S_x) \approx T \quad (10.3)$$

where d = depth of flow (ft, m)
T = spread (ft, m)
S_x = pavement cross slope (ft/ft, m/m)
A = cross-sectional flow area (ft^2, m^2)
P = wetted perimeter (ft, m)

Substitution of these expressions into the Manning equation yields

$$Q = \frac{C_f}{n} S_x^{5/3} T^{8/3} S_o^{1/2} \quad (10.4)$$

where Q = flow in gutter section (cfs, m^3/s)
C_f = 0.56 for U.S. customary units or 0.376 for SI units
n = Manning's roughness coefficient
S_o = gutter longitudinal slope (ft/ft, m/m)

Manning's n values for use in Equation 10.4 are presented in Table 10.2. The values in this table may need to be modified to account for obstructions in the gutter, such as parked cars. Note that Equation 10.4 can be used to determine the discharge if the spread is specified, and vice versa.

Table 10.2 Manning's n for streets and gutters (Brown, Stein, and Warner, 2001)

Type of Gutter[a] or Pavement	Manning's n
Concrete gutter, troweled finish	0.012
Asphalt pavement	
Smooth texture	0.013
Rough texture	0.016
Concrete gutter with asphalt pavement	
Smooth	0.013
Rough	0.015
Concrete pavement	
Float finish	0.014
Broom finish	0.016

a. For gutters with small slopes, where sediment may accumulate, increase above n values by 0.002.

Equation 10.4 is a modified form of the Manning equation presented in Section 6.2 (page 198) and Section 7.2 (page 226). The modification is necessary because the hydraulic radius does not adequately describe the gutter section, particularly when the spread may be more that 40 times the depth at the curb face (ASCE, 1992).

Example 10.1 – Computing Gutter Capacity. An asphalt-paved street has a cross slope $S_x = 0.02$ and a longitudinal slope $S_o = 0.03$. As shown in Figure E10.1.1, the gutter cross section is uniform with a 6-in.-high concrete curb. The street is in a residential area, and the distance from the street centerline to the face of the curb is 18 ft. Assuming that the street width allows for a 10-ft-wide driving lane and 8 ft of parking space on either side of the centerline, estimate the conveyance capacity of each side of the street such that maximum spread criteria are not violated.

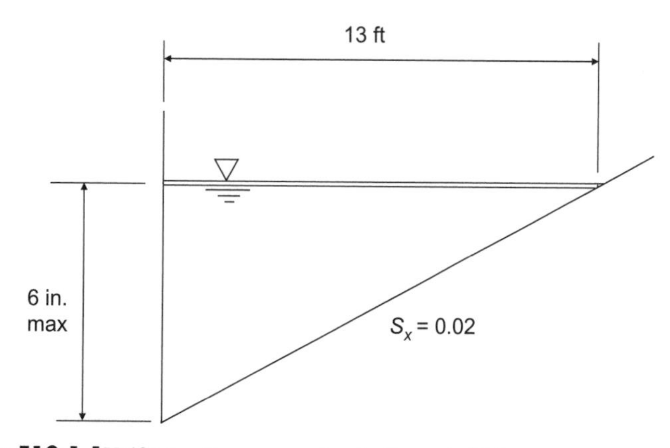

Figure E10.1.1 Uniform gutter cross section for Example 10.1

Solution: From Table 10.1, the maximum recommended spread is one-half of a driving lane, or 5 ft in this instance. Adding to that the 8 ft of parking space along the curb gives an allowable spread of T = 13 ft. Note that the depth of flow at the curb face is

$$d = TS_x = 13(0.02) = 0.26 \text{ ft}$$

which is smaller than the height of the curb. Therefore, curb overtopping will not occur, and the spread is the controlling criterion for the allowable depth of flow.

For Manning's $n = 0.015$ (Table 10.2), the conveyance capacity of each side of the street is estimated from Equation 10.4 to be

$$Q = \frac{0.56}{0.015} 0.02^{5/3} 13^{8/3} 0.03^{0.5} = 8.9 \text{ cfs}$$

Thus, the total capacity of the street (both gutters) is 2(8.9) = 17.8 cfs.

Flow in Composite Gutter Sections

As shown in Figure 10.3, the total discharge in a *composite gutter section* is partitioned into the discharge within the depressed (gutter) section and the discharge in the roadway above the depressed section, as expressed in Equation 10.5:

$$Q = Q_w + Q_s \tag{10.5}$$

where Q = total discharge (cfs, m^3/s)
Q_w = discharge within the depressed (gutter) section (cfs, m^3/s)
Q_s = discharge within the roadway above the depressed section (cfs, m^3/s)

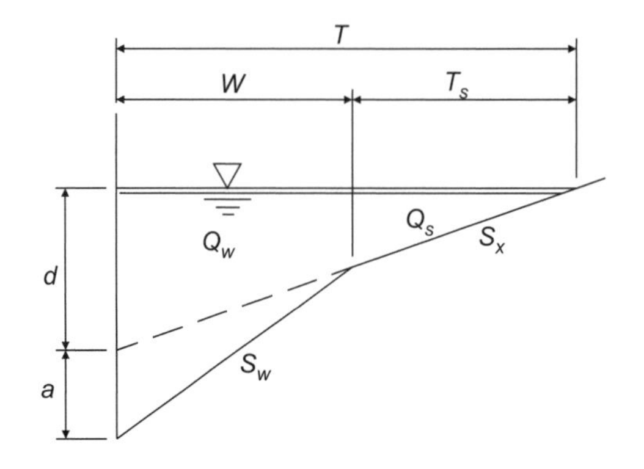

Figure 10.3
Composite gutter cross section

If the spread, T, of the flow is also partitioned, it can be written as

$$T = W + T_s \tag{10.6}$$

where W = width of the gutter depression (ft, m)
T_s = width of flow in the roadway above depressed section (ft, m)

Note that for this type of gutter section, the actual depth of flow at the curb face is

$$y = d + a = TS_x + a \tag{10.7}$$

where y = depth of flow at the curb face (ft, m)
d and a are as shown for the composite section in Figure 10.3 (ft, m)

Note further that the dimension a is *not* the difference in elevation between the lowest point of the cross section and the top of the gutter depression; rather, it is smaller than that difference by the amount WS_x. Thus, the gutter cross slope is

$$S_w = S_x + a/W \tag{10.8}$$

where S_w = gutter depression cross slope (ft/ft, m/m)

To determine the total discharge in a composite gutter section, the discharge Q_s can be determined using Equation 10.4 if the spread T is replaced by T_s, or

$$Q_s = \frac{C_f}{n} S_x^{5/3} T_s^{8/3} S_o^{1/2} \tag{10.9}$$

If the discharge in the depressed gutter section is expressed as

$$Q_w = E_o Q \tag{10.10}$$

where E_o = fraction of the total discharge that flows in the depressed gutter section ($0 \le E_o \le 1$)

then, substituting for Q_w from Equation 10.5 and solving for Q gives

$$Q = \frac{Q_s}{1 - E_o} \tag{10.11}$$

The fraction E_o of the total discharge that flows in the depressed gutter section is given by Brown, Stein, and Warner (2001).

$$E_o = \left\{ 1 + \frac{S_w / S_x}{\left(1 + \frac{S_w / S_x}{(T/W) - 1} \right)^{8/3} - 1} \right\}^{-1} \tag{10.12}$$

Equations 10.5 through 10.12 permit an easy determination of the discharge in a composite gutter section when the spread is specified. Unfortunately, a trial-and-error solution of these equations is necessary to determine the spread when the discharge is known. This solution is accomplished by assuming different spread values until the computed discharge agrees with the given value. A computer program can also be used to perform this calculation.

It should be noted that when a roadway is rehabilitated and existing pavement is overlaid with new asphalt, the existing concrete gutter may be paved over as well. In this way, a composite gutter section may be transformed into a uniform section.

Example 10.2 – Composite Gutter Capacity. Repeat Example 10.1 to find the capacity of the gutter if the gutter has a composite cross section with a gutter depression width of W = 2 ft. As shown in Figure E10.2.1, the curb is again 6 in. in height, and the vertical distance from the lowest point in the cross section to the top of the gutter depression is 1.5 in. (0.125 ft).

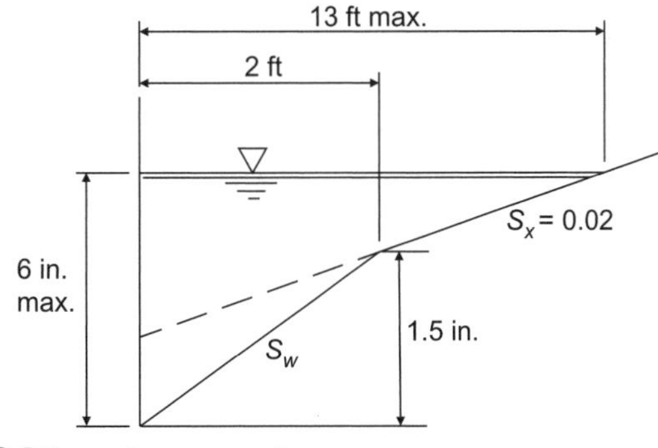

Figure E10.2.1 Composite gutter section

Solution: With the substitutions $T = 13$ ft and $W = 2$ ft, Equation 10.6 yields $T_s = 11$ ft.

The gutter cross slope is

$$S_w = 0.125/2 = 0.0625$$

The gutter depression, from Equation 10.8, is

$$a = W(S_w - S_x) = 2(0.0625 - 0.02) = 0.085 \text{ ft}$$

The depth of flow at the curb face is

$$y = TS_x + a = 13(0.02) + 0.085 = 0.345 \text{ ft}$$

Again, the spread, rather than curb overtopping, is the limiting criterion. Substituting these values into Equation 10.9 yields the discharge Q_s as

$$Q_s = \frac{0.56}{0.015} 0.02^{5/3} 11^{8/3} 0.03^{1/2} = 5.7 \text{ cfs}$$

The fraction E_o of the total discharge that flows in the depressed gutter section is found from Equation 10.12 as

$$E_o = \left[1 + \frac{0.0625/0.02}{\left(1 + \frac{0.0625/0.02}{(13/2) - 1}\right)^{8/3} - 1} \right]^{-1} = 0.43$$

The total conveyance capacity from Equation 10.11 is

$$Q = 5.7/(1 - 0.43) = 10 \text{ cfs}$$

Considering both sides of the street, the conveyance capacity is 20 cfs.

"That's the best you can do—'Time to call the gutter man'?"

Other Gutter Sections

For gutter-section shapes other than uniform and composite, such as shallow triangular swale sections, the Manning equation presented in Section 7.2 (page 226; Equation 7.9) can be applied. The FHWA (Brown, Stein, and Warner, 2001) also provides guidance for gutter sections not detailed here.

10.2 INLETS

A stormwater *inlet* is a structure for intercepting stormwater on the ground surface or in a roadway gutter and conveying it into the storm sewer piping system. It consists of an *inlet opening* at the surface and an *inlet box*, which is a subsurface box, manhole, or other structure that supports the inlet opening and connects to the subsurface piping system. Additional information on inlet boxes and manholes is provided in Section 10.1.

All basic inlet types can be classified further as either *continuous-grade* inlets (also referred to as inlets on grade) or *sag* inlets. A sag inlet is constructed in the low point of a road or other surface where water drains to the inlet from all directions. With a continuous-grade inlet, surface water can flow away from the inlet opening in one or more directions. The cutaway storm sewer system profile in Figure 10.4 shows inlets in a roadway gutter for both continuous-grade and sag installations.

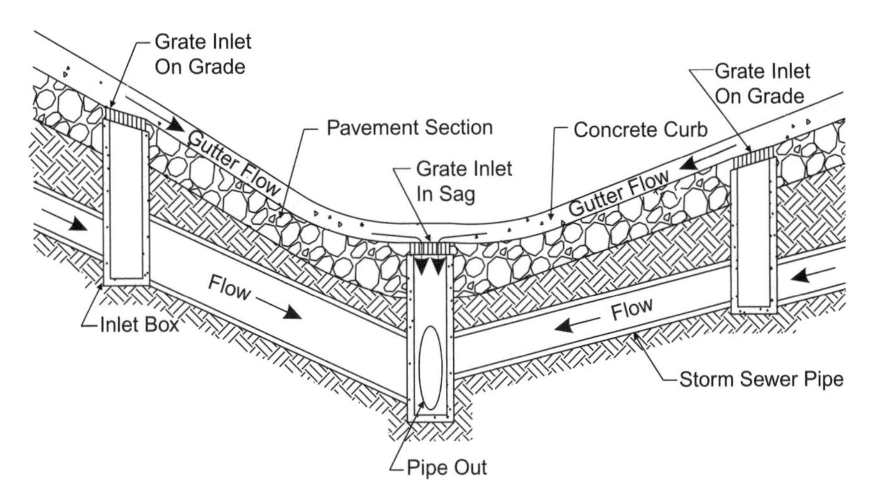

Figure 10.4
Cutaway view of a storm sewer system profile (vertical scale exaggerated)

Inlet opening geometries vary considerably from one locale to another, but all can be classified as one of four general types, as illustrated in Figure 10.5. *Grate inlets* consist of an opening in a gutter or swale, with the opening covered by a grate. *Curb-opening inlets* consist of an opening in the face of a curb. *Combination inlets* have both grate and curb openings and combine the advantages of each of the individual inlet types. Finally, *slotted drain inlets* consist of a grate opening oriented along the longitudinal axis of a pipe and manufactured integrally with the pipe.

In most cases, the tops of stormwater inlets and inlet openings are constructed of cast-in-place reinforced concrete, and grates, if present, are cast or fabricated metal. In many locales, precast concrete inlet structures are also available.

Figure 10.5
Common inlet types

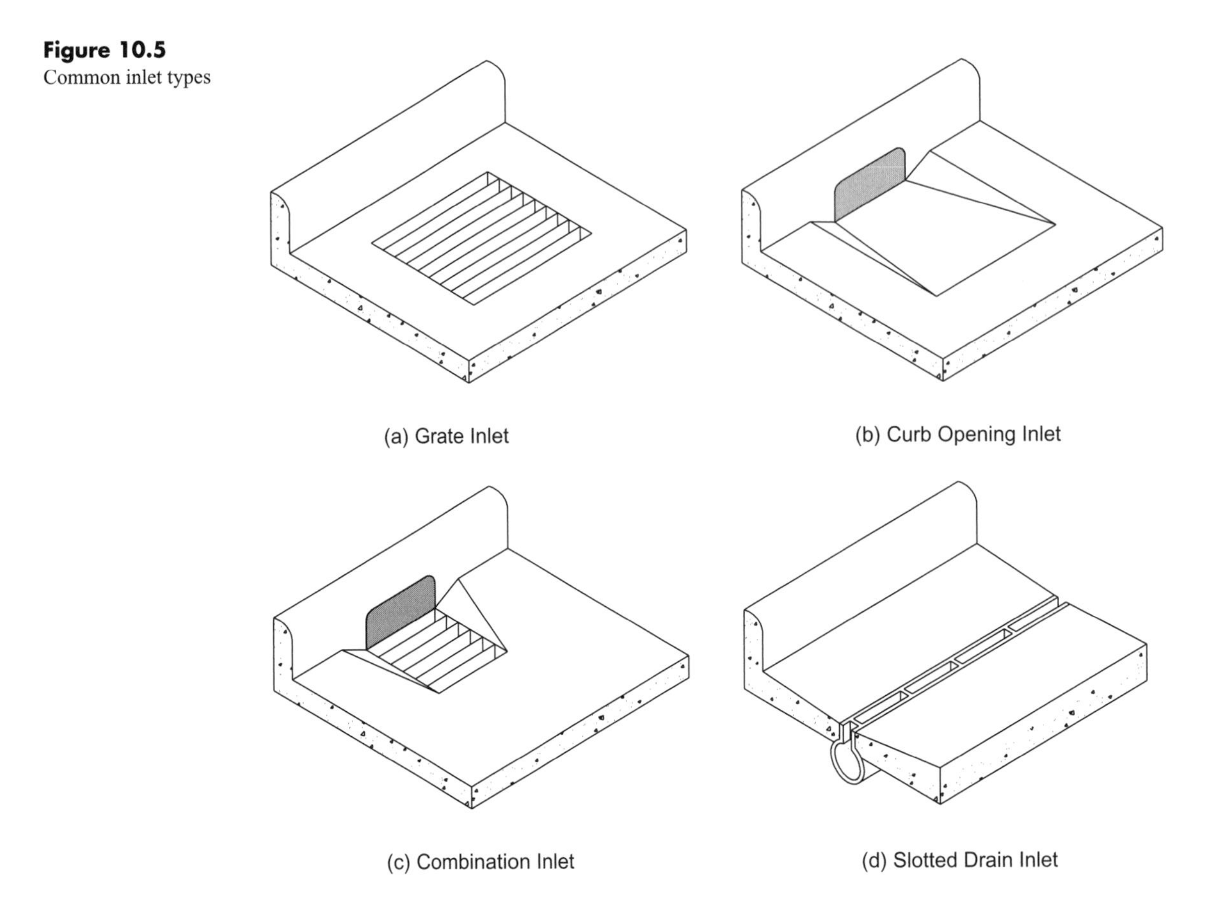

Hydraulic Performance Concepts

From a hydraulic performance standpoint, a particular inlet type may be more suitable than another for use with a given installation. Grate inlets have been found to perform well over a wide range of longitudinal gutter slopes. Their capacity decreases as the gutter slope increases, but this loss of capacity is not as great as it is for curb-opening inlets. If necessary, a *curved vane* or *tilt-bar grate,* which has its bars oriented perpendicular to the direction of flow and tilted in the upstream direction, can be used to increase the interception capacity of a grate inlet on a steep slope. Grate inlets on grade are able to intercept more water than combination inlets on grade because the grate lies in the path of the water flowing in the gutter, but they suffer from a propensity to be easily clogged by trash and debris. They can also present hazards to bicyclists and persons wearing heeled footwear, although bicycle-safe grate designs have been developed and are widely used. Because grate inlets clog easily, they are not generally recommended for use in sag locations.

Curb-opening inlets are less apt to clog than are grate inlets and are particularly attractive for use in sags and other locations where the gutter slope is relatively flat. Because flowing water must change direction to enter a curb-opening inlet, its interception capacity decreases as the gutter slope and approach velocity increase. For public-safety reasons, the height of the opening should be no greater than about 6 in. (15 cm).

Combination inlets, which have both grate and curb openings, offer the advantages of both inlet types. The curb opening is usually the same length as the grate, but it may extend beyond the grate in either or both directions. When the curb opening is longer than the grate, it becomes effective for intercepting trash and debris before it has a chance to clog the grate opening. For this reason, so-called *sweeper inlets* with this configuration are often recommended for use in sags where excessive ponding due to debris accumulation could pose a safety hazard.

Slotted-drain inlets, because of their longer lengths, can be very effective in intercepting sheet flows at locations such as loading docks and garage entrances. They are also effective for use in locations, such as parking lot entrances, where they can intercept flow before it enters the roadway. Slotted drains are very susceptible to clogging and are therefore not recommended for use in sags or other locations where trash and debris loadings may be considerable. Slotted drains are also susceptible to crushing loads because of their shallow burial depths, and thus require great care in pipe bedding and backfilling operations.

Given a specific set of inlet dimensions, the performance of a particular type of inlet depends on a number of factors. Among them are the depth and velocity of the flow in the approach gutter(s), the longitudinal and cross slopes of the roadway, and the Manning roughness coefficient for the gutter. The performance also depends significantly on whether the inlet is installed in a depression, such as the sag point of a vertical curve, or on a continuous grade. Inlets installed in sag points should be sized to capture the entire design discharge approaching them. Inlets on continuous grades may be designed to intercept either all or part of the design flow in the approach gutter. The *efficiency* of an inlet is defined as the ratio of the discharge intercepted by the inlet to the total discharge approaching the inlet:

$$E = Q_{int}/Q \qquad (10.13)$$

where E = efficiency of the inlet
Q_{int} = discharge intercepted by the inlet (cfs, m^3/s)
Q = total discharge approaching the inlet (cfs, m^3/s)

Bypass flow, or *carryover,* is the term given to any discharge not intercepted by an inlet and is defined as

$$Q_b = Q - Q_{int} \qquad (10.14)$$

where Q_b = bypass flow (cfs, m^3/s)

The following subsections present procedures for estimating inlet performance characteristics for each of the four inlet types. These procedures are based on research conducted by the Bureau of Reclamation and Colorado State University on behalf of the Federal Highway Administration, and are also presented by Brown, Stein, and Warner (2001). Some locales have developed their own standard inlet designs, and their design procedures may vary from those described here.

Grate Inlets

The methods for evaluating the capacity of a grate inlet (see Figure 10.6) differ depending on whether the inlet is located on a continuous grade or in a sag. The subsections that follow present information on determining grate capacity for both conditions.

Capacity of a Grate Inlet on a Continuous Grade. A grate inlet's interception capacity on a continuous grade depends on the depth of flow, the size and configuration of the grate, and the velocity of the approaching flow. The FHWA has developed a number of standardized grate designs, which are described in Table 10.3 and illustrated in Figures 10.7 through 10.12 (Brown, Stein, and Warner, 2001; FHWA, 1977). These grates may be used with uniform or composite gutter sections, or in the bottom of a channel such as a median swale or roadside ditch.

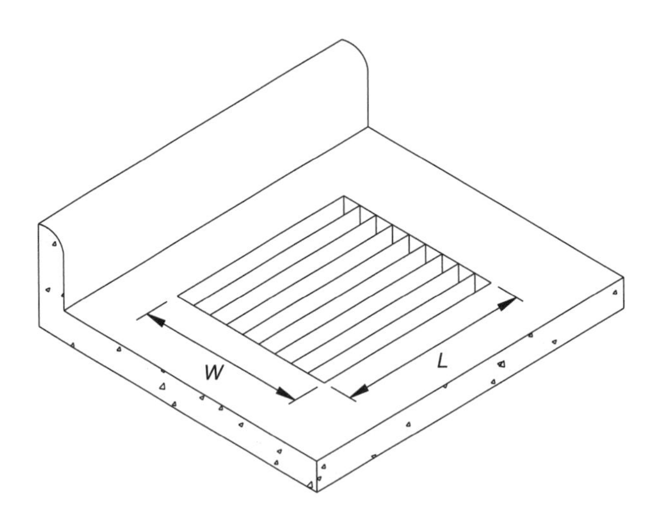

Figure 10.6
Grate inlet

Table 10.3 FHWA standard grate types (Brown, Stein, and Warner, 2001; FHWA, 1977)

Grate type	Description	Figure Reference
Reticuline	"Honeycomb" pattern of lateral and longitudinal bearing bars	10.7
P-50	Parallel bar grate with bar spacing 48 mm (1.875 in.) on center	10.8
P-50 × 100	Parallel bar grate with bar spacing 48 mm (1.875 in.) on center, and with 10-mm (0.375-in.) diameter lateral rods spaced at 102 mm (4 in.) on center	10.8
P-30	Parallel bar grate with bar spacing 29 mm (1.125 in.) on center	10.9
Curved vane	Curved-vane grate with longitudinal bar spacing 82 mm (3.25 in.) on center and transverse bar spacing 114 mm (4.5 in.) on center	10.10
45°–60 tilt bar	45° tilt-bar grate with longitudinal bar spacing 57 mm (2.25 in.) on center and transverse bar spacing 102 mm (4 in.) on center	10.11
45°–85 tilt bar	45° tilt-bar grate with longitudinal bar spacing 83 mm (3.25 in.) on center and transverse bar spacing 102 mm (4 in.) on center	10.11
30°–85 tilt bar	30° tilt-bar grate with longitudinal bar spacing 82 mm (3.25 in.) on center and transverse bar spacing 102 mm (4 in.) on center	10.12

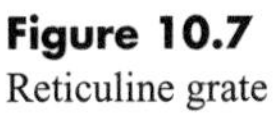

Figure 10.7
Reticuline grate

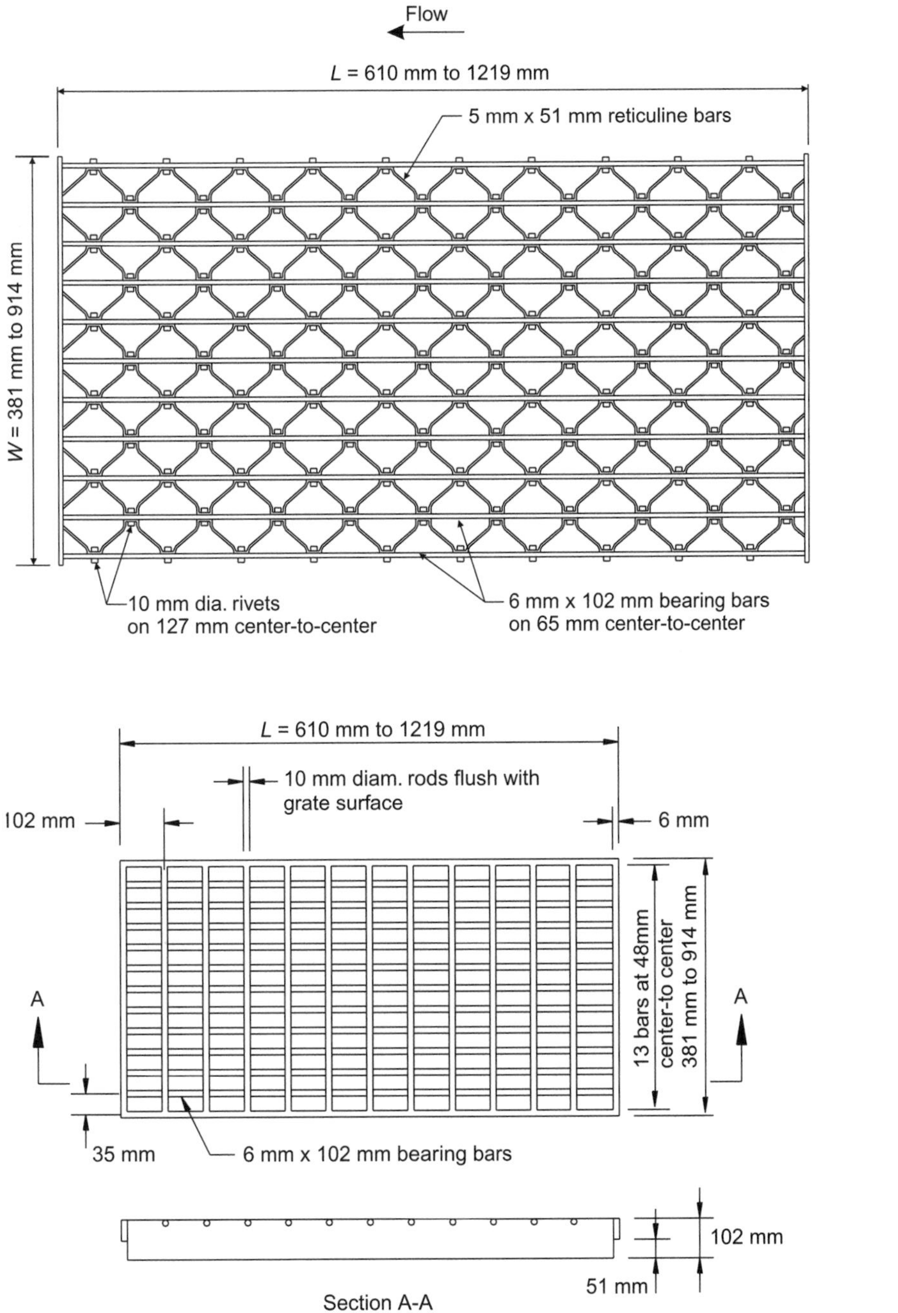

Figure 10.8
P-50 and P-50 × 100 grates

Figure 10.9
P-30 grate

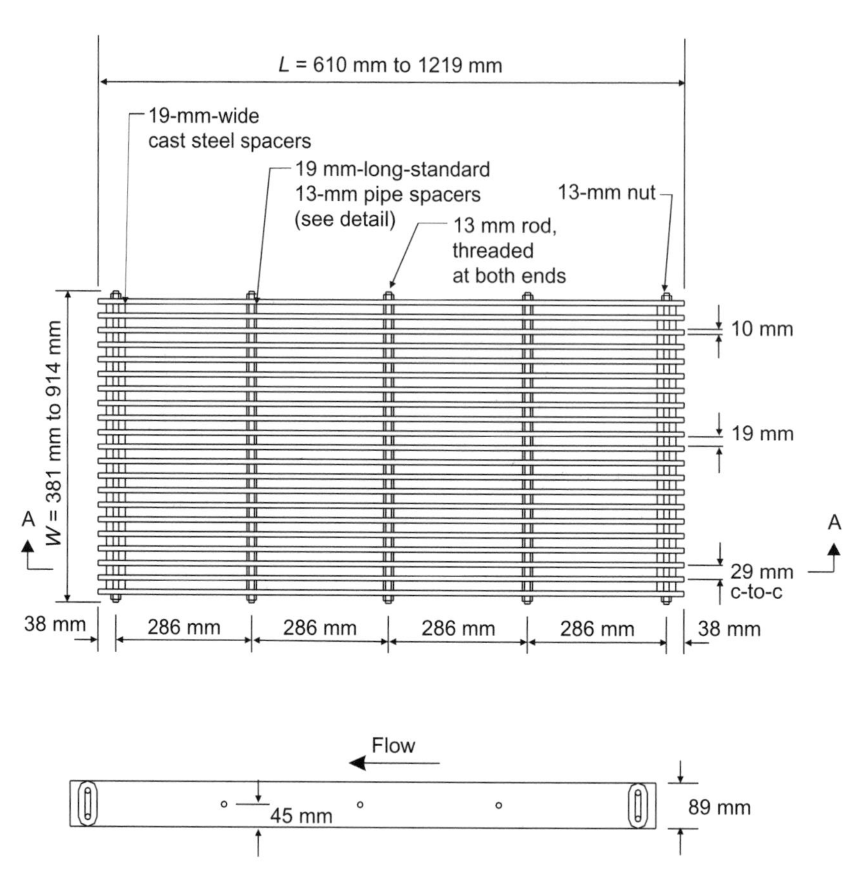

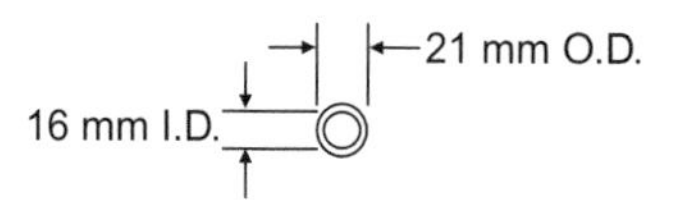

13 mm Pipe Spacer Detail

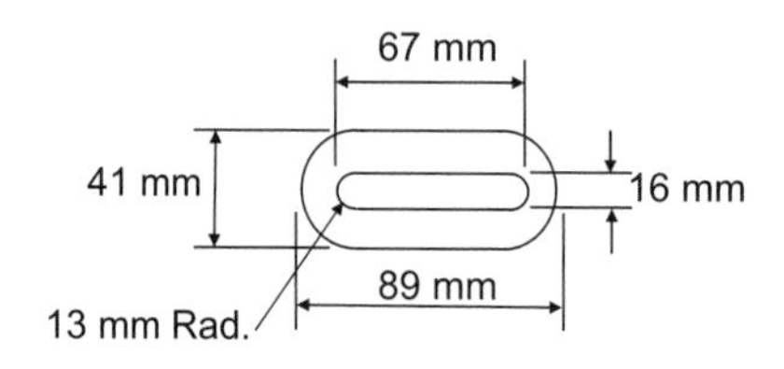

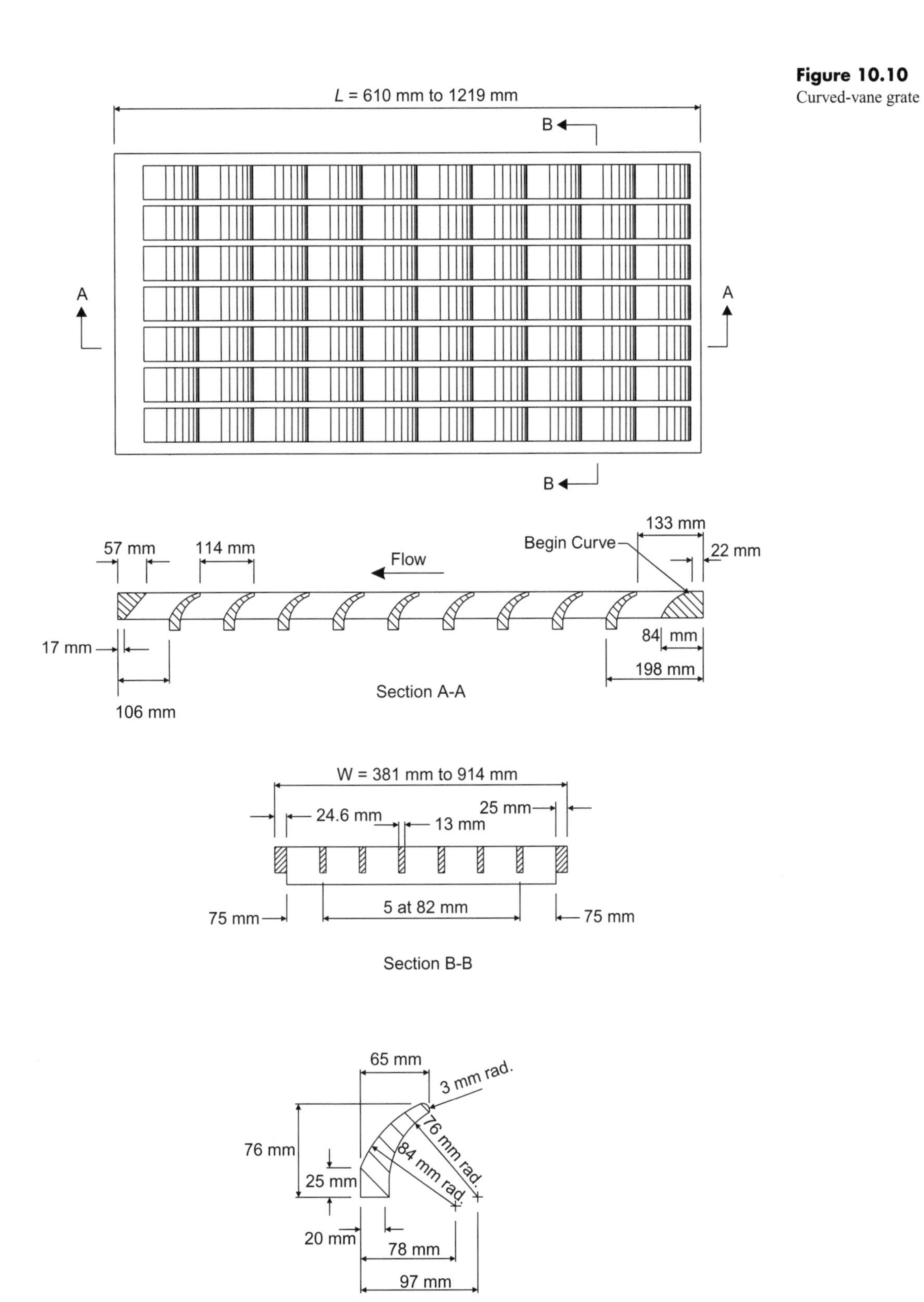

Figure 10.10
Curved-vane grate

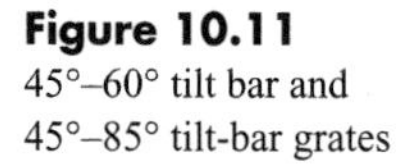

Figure 10.11
45°–60° tilt bar and
45°–85° tilt-bar grates

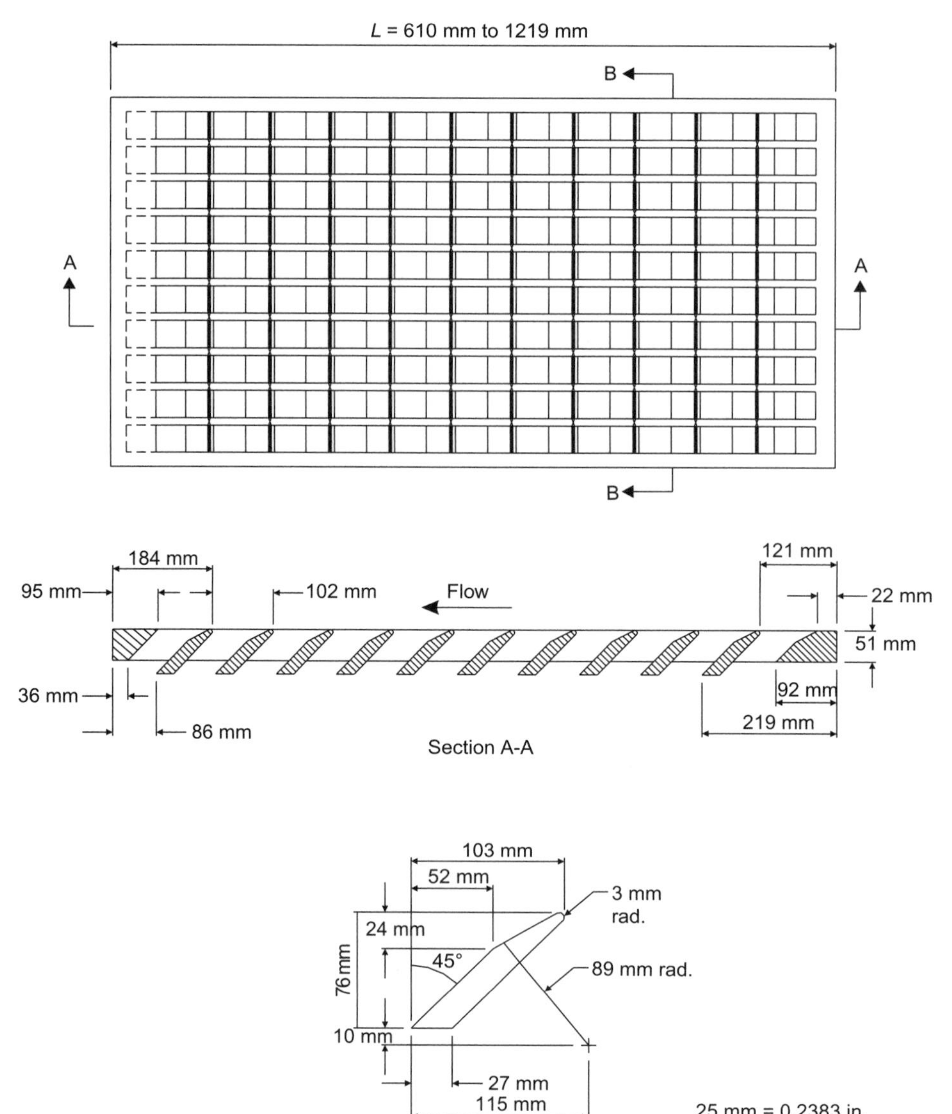

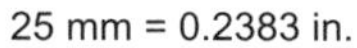

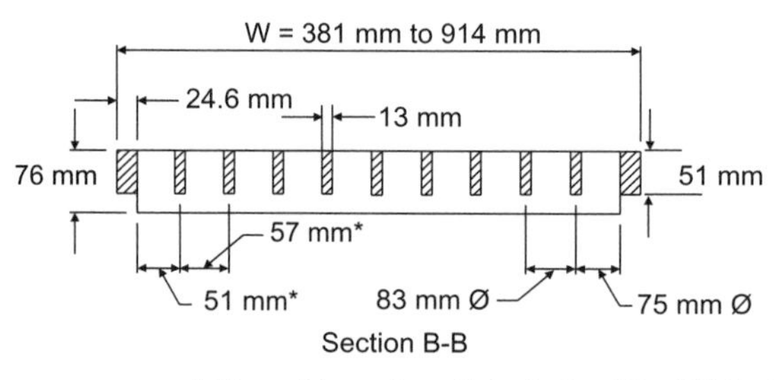

* These Dimensions Refer to the 45° - 60° Grate
Ø These Dimensions Refer to the 45° - 85° Grate

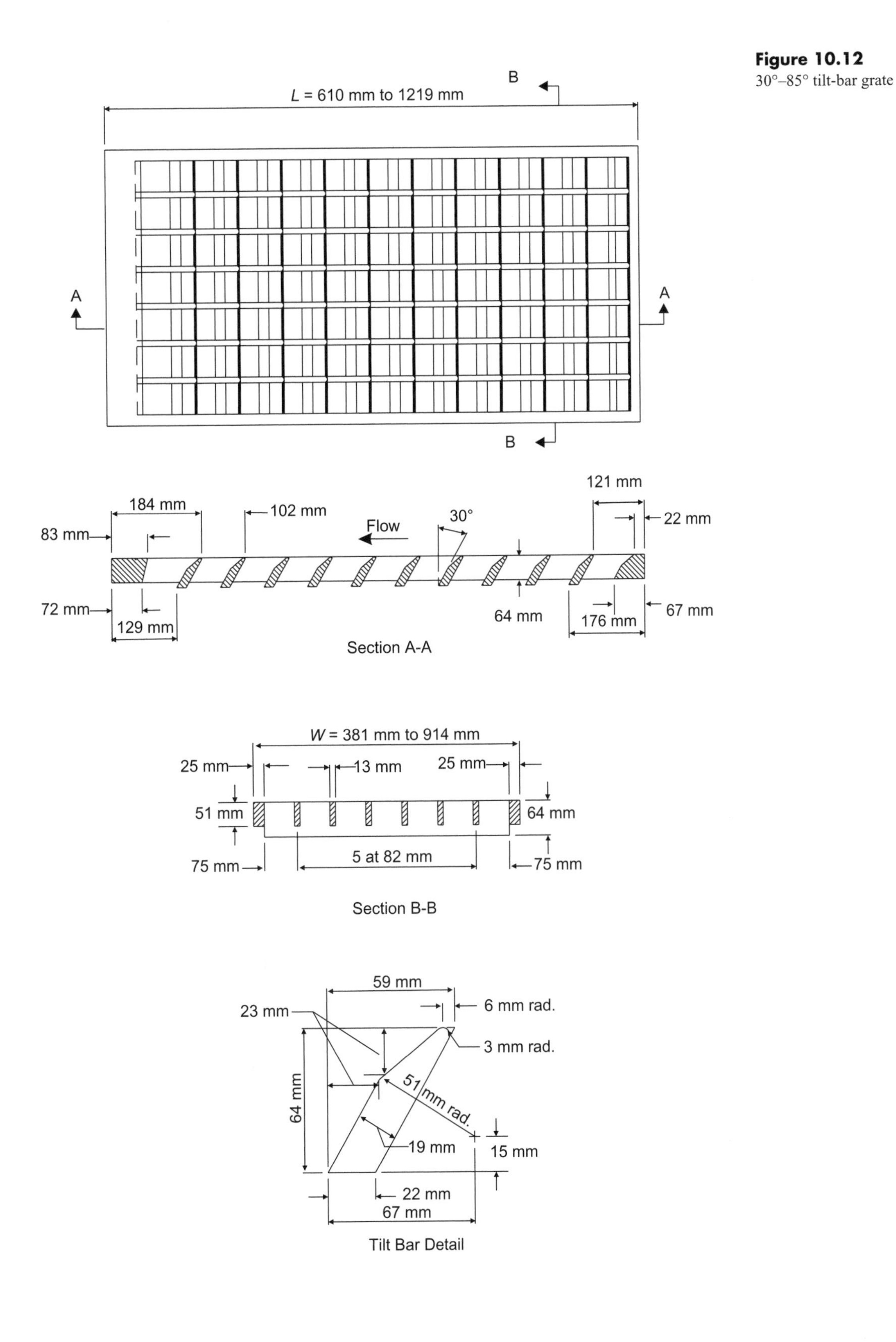

Figure 10.12
30°–85° tilt-bar grate

Analysis of the interception capacity of a grate inlet on a continuous grade first requires the determination of the *frontal flow,* Q_w, which is the discharge in the gutter over the width W of the grate, and *side flow,* Q_s, which is the discharge in the gutter above the width of the grate. If the gutter section is uniform, the ratio E_o of frontal flow to total flow is given by

$$E_o = 1 - \left(1 - \frac{W}{T}\right)^{2.67} \tag{10.15}$$

With a composite gutter section, the grate width is often equal to the width of the gutter depression. In this instance, the definitions of frontal flow and side flow are the same as those presented in the discussion of composite gutters in Section 10.1. Therefore, the ratio E_o for the composite gutter when the spread is greater than the grate width is determined using Equation 10.16:

$$E_o = \left\{ 1 + \frac{S_w / S_x}{\left(1 + \frac{S_w / S_x}{(T/W) - 1}\right)^{8/3} - 1} \right\}^{-1} \tag{10.16}$$

where E_o = ratio of frontal flow to total gutter flow (Q_w/Q)
S_w = cross slope for depressed portion of gutter
S_x = cross slope for portion of gutter above gutter depression
T = spread (ft, m)
W = width of grate (ft, m)

For a composite gutter section that has a grate width smaller than the width of the gutter depression, E_o is first computed using Equation 10.12, assuming that W is the full width of the depressed portion of the gutter (that is, in the same way that it was computed in Section 10.1). An adjusted value for the ratio, E_o', is then computed using Equation 10.17. This adjusted value should be used in place of E_o for Equations 10.19 through 10.26.

$$E_o' = E_o \left(\frac{A_w'}{A_w} \right) \tag{10.17}$$

where E_o' = adjusted frontal-flow area ratio for grates in composite cross-sections
A_w' = cross-sectional gutter flow area for the grate width (ft^2, m^2)
A_w = cross-sectional gutter flow area in the depressed gutter width (ft^2, m^2)

For a grate inlet in the bottom of a trapezoidal channel, as shown in Figure 10.13, the Manning equation may be used to solve for the flow depth in the channel (see Section 7.2, page 226). E_o is then given by

$$E_o = \frac{W}{B + dz} \tag{10.18}$$

where B = bottom width of the trapezoidal channel (ft, m)
d = depth of flow (ft, m)
z = measure of the steepness of the channel sides

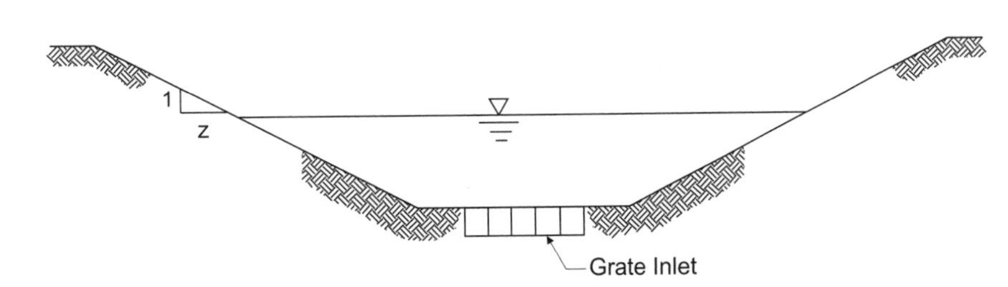

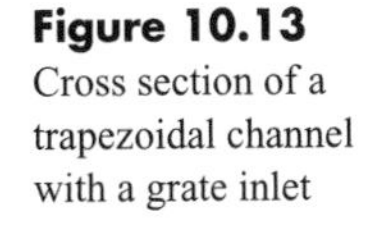

Figure 10.13
Cross section of a trapezoidal channel with a grate inlet

Once E_o (or E'_o) is known, the gutter flow and side flow approaching the grate are given by the equations

$$Q_w = E_o Q \tag{10.19}$$

and

$$Q_s = (1 - E_o)Q \tag{10.20}$$

The frontal flow intercepted by the inlet, denoted by $(Q_w)_{int}$, is

$$(Q_w)_{int} = R_f Q_w \tag{10.21}$$

where

$$R_f = \begin{cases} 1 - K_c(V - V_o); & V \geq V_o \\ 1; & V \leq V_o \end{cases} \tag{10.22}$$

K_c = 0.09 for U.S. customary units or 0.295 for SI units
V = Q/A, the velocity of flow in the gutter (ft/s, m/s)
V_o = the velocity at which splash-over occurs (ft/s, m/s)

Note that R_f cannot be greater than one nor less than zero. V_o, the *splash-over velocity*, is the minimum velocity at which some of the frontal flow passes over the top of the grate without being intercepted. This value depends on the grate type and its length in the direction of flow. V_o (and R_f) can be determined from Figure 10.14.

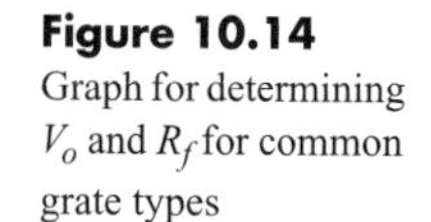

Figure 10.14
Graph for determining V_o and R_f for common grate types

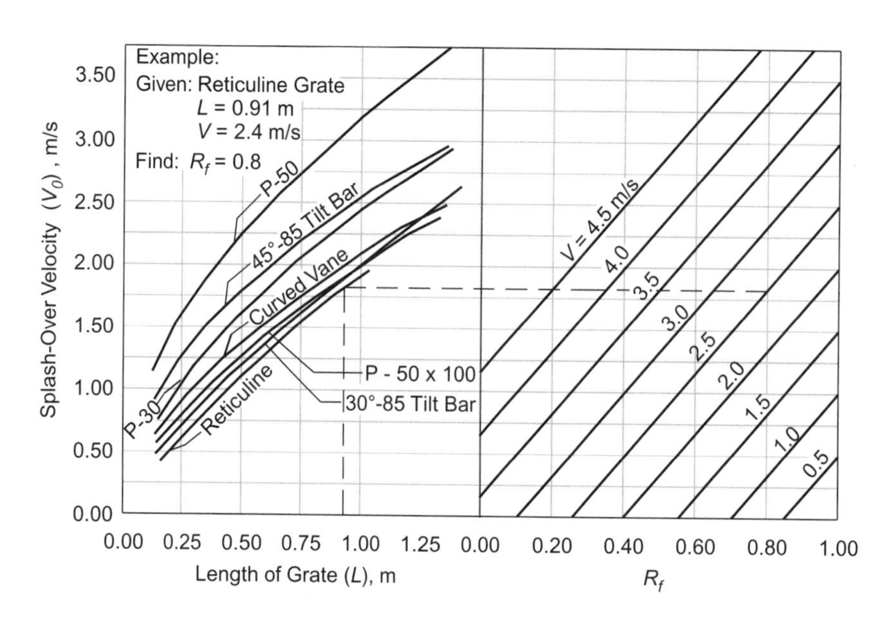

The side flow intercepted by a grate inlet on a continuous grade, denoted by $(Q_s)_{int}$, is

$$(Q_s)_{int} = R_s Q_s \tag{10.23}$$

where

$$R_s = \left(1 + \frac{K_s V^{1.8}}{S_x L^{2.3}}\right)^{-1} \tag{10.24}$$

K_s = 0.15 for U.S. customary units or 0.0828 for SI units
S_x = roadway cross slope (ft/ft, m/m)
L = grate length (ft, m)

Note that R_s as computed by Equation 10.24 will always be less than one, and hence some of the side flow will always bypass a grate inlet on a continuous grade regardless of the flow velocity or inlet length. This is a deficiency of the equation, but it is not of great concern for practical purposes.

Combining the above equations gives the total discharge intercepted by a grate inlet on a continuous grade as

$$Q_{int} = (Q_w)_{int} + (Q_s)_{int} = R_f Q_w + R_s Q_s = Q[R_f E_o + R_s(1 - E_o)] \tag{10.25}$$

The efficiency of the inlet is, therefore,

$$E = R_f E_o + R_s(1 - E_o) \tag{10.26}$$

The orientation of an inlet grate can have a significant influence on the interception capacity of the structure. Analyzing the capacity of an inlet grate requires that the relationships used are appropriate for the intended orientation in which the grate will be installed.

Example 10.3 – Interception Capacity of a Grate Inlet on a Grade. Estimate the interception capacity of a P-50 inlet grate on a continuous grade if the grate width is $W = 2$ ft and the grate length is $L = 3$ ft. The gutter has the composite cross section from Example 10.2, which is shown in Figure E10.3.1. Assume that the inlet is placed at a location with the maximum permissible spread, so that the total discharge approaching the inlet is $Q = 10$ cfs.

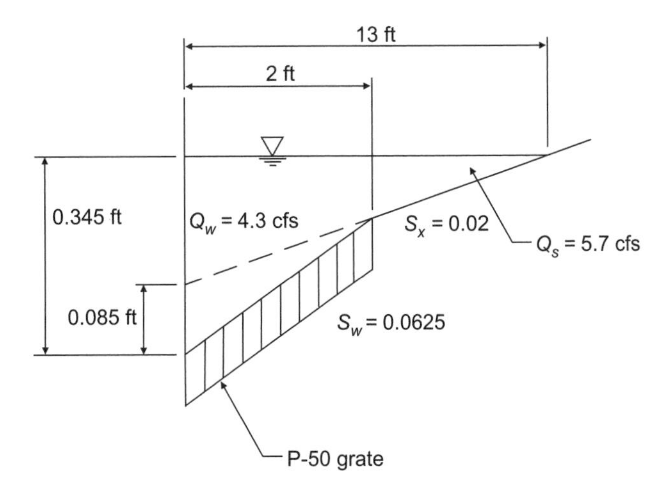

Figure E10.3.1 Composite gutter section

Solution: From Example 10.2, the ratio of frontal flow to the total gutter flow is $E_o = 0.43$. Therefore, $Q_w = 4.3$ cfs and $Q_s = 5.7$ cfs. The total cross-sectional area of the flow is

$$A = \left(\frac{2(11)(0.02)+0.125}{2}\right)2 + 0.5(11)^2(0.02) = 1.27 \text{ ft}^2$$

The gutter flow velocity is therefore

$$V = 10/1.27 = 7.9 \text{ ft/s}$$

Figure 10.14 shows that the splash-over velocity for a P-50 grate with a length of 3 ft = 0.9 m is approximately $V_o = 3$ m/s = 10 ft/s. Since $V \leq V_o$, according to Equation 10.22, $R_f = 1$ and all of the frontal flow will be intercepted by the inlet.

The fraction of the side flow that will be intercepted is estimated using Equation 10.24 as

$$R_S = \left(1 + \frac{0.15(7.9)^{1.8}}{0.02(3)^{2.3}}\right)^{-1} = 0.039$$

The total flow intercepted by the inlet is, from Equation 10.25,

$$Q_{int} = 10[1(0.43) + 0.039(0.57)] = 4.5 \text{ cfs}$$

and its efficiency is, therefore,

$$E = 4.5/10 = 0.45$$

The bypass flow is $Q_b = 5.5$ cfs.

Capacity of a Grate Inlet in a Sag. A grate inlet in a sag location behaves hydraulically like a weir at small ponding depths and like an orifice at large depths. The depth at which the transition from weir to orifice behavior occurs depends on the grate's bar spacing and overall dimensions. The transition depth increases as the overall size of the grate increases.

The length of the perimeter of the grate opening is the critical performance variable when the inlet behaves as a weir. When the inlet behaves as an orifice, the cross-sectional area of the grate opening is the critical variable. Clogging of a grate can reduce the value of either of these variables, and thus grate inlets often are not recommended for use in sag locations.

The capacity of a grate operating as a weir can be estimated as

$$Q = C_w P d^{3/2} \tag{10.27}$$

where Q = capacity of the grate operating as a weir (cfs, m^3/s)
C_w = weir coefficient, use 3.0 for U.S. customary units or 1.66 for SI units
P = length of the grate's effective perimeter (ft, m)
d = average depth of flow across the grate (ft, m) (see Figure 10.15)

When one edge of a grate inlet is placed against a curb face, the length of that edge should be disregarded when computing P.

The capacity of a grate operating as an orifice can be estimated as

$$Q = 0.67 A_g (2gd)^{1/2} \tag{10.28}$$

where Q = capacity of the grate operating as an orifice (cfs, m^3/s)
A_g = clear opening area of the grate (ft^2, m^2)
g = acceleration due to gravity (ft/s^2, m/s^2)

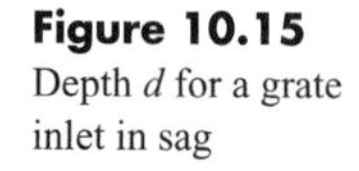

Figure 10.15
Depth d for a grate inlet in sag

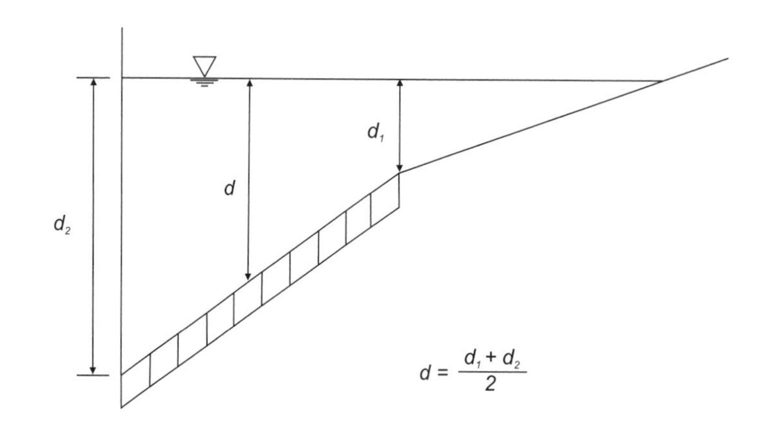

The clear-opening area of a grate used in Equation 10.28 is equal to the total grate area less the area occupied by the longitudinal and transverse bars when projected onto a horizontal plane. When the clear opening area is computed in this manner, it can be increased by about 10 percent for curved-vane and 30° tilt-bar grates (Brown, Stein, and Warner, 2001). However, curved-vane and tilt-bar grates are not recommended for use in sag locations where there is a chance that a grate inlet would operate as an orifice.

In an analysis of the capacity of a grate inlet in a sag, a conservative practice is to estimate the capacity using both Equations 10.27 and 10.28 and to then adopt the smaller of the two estimates. For design problems in which the required inlet dimensions are being determined, the larger of the sizes obtained by solving the two equations should be used.

Example 10.4 – Capacity of a Grate Inlet in Sag. Determine the required size of a grate inlet in a sag if the roadway has a uniform cross slope and dimensions as specified in Example 10.1, as shown again in Figure E10.4.1. Assume that the design discharge for the inlet is $Q = 3.5$ cfs and that the width of the grate is $W = 2$ ft.

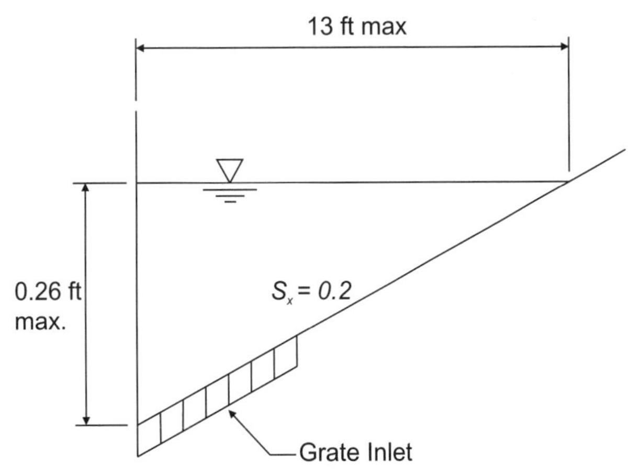

Figure E10.4.1 Uniform gutter cross section for Example 10.1

Solution: As shown in Example 10.1, the maximum permissible spread is $T = 13$ ft. The depth d_1 is

$$d_1 = (13 - 2)0.02 = 0.22 \text{ ft}$$

and depth d_2 at the curb face is

$$d_2 = 13(0.02) = 0.26 \text{ ft}$$

Therefore, the average depth is computed as

$$d = (d_1 + d_2)/2 = (0.26 + 0.22)/2 = 0.24 \text{ ft}$$

Assuming that the inlet behaves hydraulically as a weir, Equation 10.27 yields the required perimeter of the grate as

$$P = \frac{Q}{C_w d^{3/2}} = \frac{3.5}{3.0(0.26)^{1/2}} = 9.9 \text{ft}^2$$

Because one edge of the grate is adjacent to the curb face and is therefore ineffective, the required length of the grate is

$$L = P - 2W = 9.9 - 2(2) = 5.9 \text{ ft}$$

If the inlet is assumed to behave hydraulically as an orifice, its required clear-opening area is given by Equation 10.28:

$$A_g = \frac{Q}{0.67(2gd)^{1/2}} = \frac{3.5}{0.67[2(32.2)(0.24)]^{1/2}} = 1.3 \text{ ft}^2$$

If a P-50 grate has a length of 5.9 ft and a width of 2 ft, then the clear-opening area is well in excess of 1.3 ft^2 (shown in Figure 10.8); therefore, weir behavior controls the design.

To provide a factor of safety against clogging of the grate, one might assume (for example) that its perimeter should be increased by 25 percent. Applying this assumption gives the required perimeter as

$$P = 1.25(9.9) = 12.4 \text{ ft}$$

Therefore, the inlet length should be

$$L = 12.4 - 2(2) = 8.4 \text{ ft}$$

Curb-Opening Inlets

Like those of grate inlets, the methods used to compute the capacity of a curb-opening inlet (see Figure 10.16) are different if the inlet is on a continuous grade than if it is in a sag location. The subsections that follow present methods for evaluating both conditions.

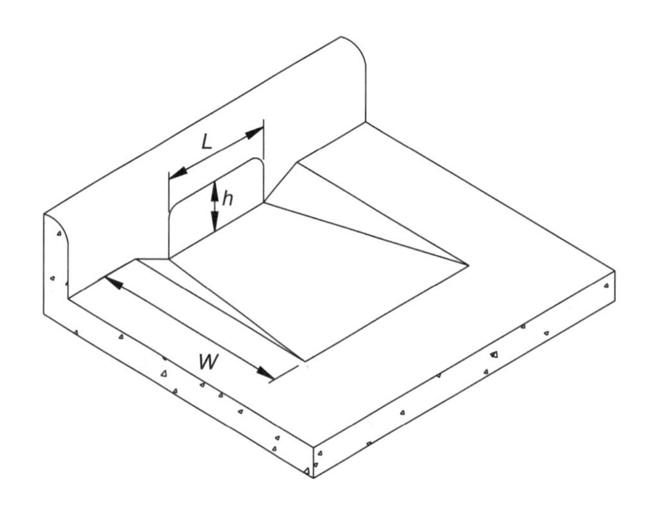

Figure 10.16
Curb-opening inlet

Capacity of a Curb-Opening Inlet on a Continuous Grade. The depth of flow at the curb face and the length of the inlet opening are the primary variables governing the interception capacity of a curb-opening inlet on a continuous grade. If the gutter cross-slope is increased in the vicinity of the inlet so as to correspondingly increase the depth of flow at the curb face, then the capacity of the inlet can be greatly improved.

When curb-opening inlets must be provided with top-slab supports for structural reasons (see Figure 10.17), the supports should be rounded and set back several inches from the curb face. If the supports are placed flush with the curb face, tests have shown that they can reduce the interception capacity of the downstream opening by as much as 50 percent. The capacity of the downstream opening may be reduced to near zero if trash and debris are collected by a structural support.

Figure 10.17
Curb-opening inlet with supports

As noted earlier in Section 10.2, the height of the opening in a curb-opening inlet should be no more than about 6 in. (150 mm). The length of the opening required for total interception of the gutter flow when the gutter has a uniform cross-section is

$$L_T = C_f Q^{0.42} S_o^{0.3} \left(\frac{1}{nS_x} \right)^{0.6} \tag{10.29}$$

where L_T = length of curb opening necessary for interception of the total gutter discharge (ft, m)
Q = gutter discharge (cfs, m^3/s)
S_o = longitudinal slope of the gutter (ft/ft, m/m)
n = Manning roughness coefficient for the gutter
S_x = cross slope (ft/ft, m/m)
C_f = 0.6 for U.S. customary units or 0.817 for SI units

The efficiency of a curb-opening inlet of length L less than L_T is

$$E = 1 - \left(1 - \frac{L}{L_T} \right)^{1.8} \tag{10.30}$$

where E = efficiency of the curb-opening inlet
L = length of the curb opening (ft, m)

If the gutter cross slope is increased in the vicinity of a curb-opening inlet to create a gutter depression (see Figure 10.18), either along the entire gutter or locally at the inlet, then the interception capacity of the inlet improves significantly. In such cases, the length required for total interception can be computed using Equation 10.29 if an equivalent cross slope is used in place of S_x.

The equivalent cross slope is

$$S_e = S_x + S_{wp} E_o \qquad (10.31)$$

where S_e = equivalent cross slope (ft/ft, m/m)
S_x = pavement cross slope (ft/ft, m/m)
S_{wp} = a/W, the slope of the depression relative to the pavement cross slope S_x (see Figure 10.18)
E_o = fraction of the total gutter flow within the gutter width W

The value of E_o can be determined using Equation 10.12.

Equation 10.30 can again be used to determine the inlet efficiency if its actual length is shorter than the length required for total interception.

Figure 10.18
Depressed curb-opening inlet

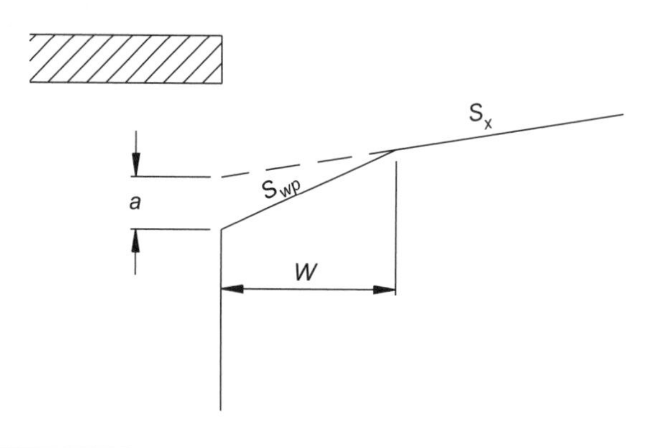

Example 10.5 – Determining the Capacity of a Curb-Opening Inlet on a Grade. Determine the required length of a curb-opening inlet on a continuous grade ($S_o = 0.01$) if it is to intercept a total approaching flow of $Q = 0.14$ m^3/s. Assume that the gutter is depressed in the inlet area so that the opening height is 13 cm and the top slab is 8 cm thick. (Thus, the total curb height at the inlet opening is 21 cm.) The gutter width is $W = 0.6$ m, the depression dimension is $a = 0.08$ m (Figure 10.18), and the pavement cross slope is $S_x = 0.02$. Assume that the ratio $E_o = 0.40$ and Manning's $n = 0.015$.

Solution: The equivalent cross slope of the roadway is determined from Equation 10.31 as

$$S_e = 0.02 + (0.08/0.6)0.40 = 0.07$$

Substitution of this result into Equation 10.29 yields the required inlet length as

$$L_T = 0.817(0.14)^{0.42}(0.01)^{0.3}\left(\frac{1}{0.015(0.07)}\right)^{0.6} = 5.5 \text{ m}$$

As a factor of safety, this inlet length should be increased by an assumed amount to allow for clogging.

Capacity of a Curb-Opening Inlet in a Sag. A curb-opening inlet in a sag behaves hydraulically as a weir when the depth of water ponding at the curb face is less than or equal to the height h of the curb opening. For ponding depths greater than $1.4h$, the inlet opening behaves as an orifice. For intermediate depths, a transition between weir and orifice behavior occurs. In inlet capacity analysis, a conservative approach is to estimate the capacity assuming both weir and orifice behaviors, and to then adopt the smaller of the two capacities. Similarly, a conservative approach to design is to adopt the larger of the two required inlet sizes.

The weir equation for a curb-opening inlet without a gutter depression yields an estimate of the inlet capacity as

$$Q = C_w L d^{3/2} \quad (10.32)$$

where Q = inlet capacity (cfs, m^3/s)
C_w = weir coefficient equal to 3.0 for U.S. customary units or 1.66 for SI units

If the gutter in the vicinity of a curb-opening inlet in a sag is depressed to increase the depth at the curb face, then the location of the weir behavior moves from the curb face to the lip of the gutter and its length is effectively increased. In such cases, the weir equation for the inlet capacity takes the form

$$Q = C_w(L + 1.8W)d^{3/2} \quad (10.33)$$

where W = width of the gutter (ft, m)
d = depth at curb measured from the normal cross slope (ft, m)

In this expression, the depth d at the curb face is measured relative to the normal roadway cross slope (that is, $d = TS_x$).

For curb-opening inlets with lengths in excess of 12 ft (3.6 m), Equation 10.32 produces a higher estimate of the capacity than does Equation 10.33. This is a deficiency of the equations, as an inlet with a depressed gutter will perform at least as well as one without the depression. It is therefore recommended that Equation 10.32 be used in lieu of Equation 10.33 for inlets with lengths greater than 12 ft (3.6 m), regardless of whether there is a gutter depression. In this case, the value of d in Equation 10.32 is equal to the total depth at the curb face, including the depression.

When a curb-opening inlet behaves as an orifice, the *effective depth of flow at the curb face, d_o*, is considered in the capacity calculation. The determination of d_o depends on the type of *throat* opening at the inlet. *Horizontal, inclined, and vertical throat configurations,* and the method used to determine d_o for each, are illustrated in Figure 10.19. Note that d_i is the depth at the lip of the curb opening.

The capacity of a curb-opening inlet behaving as an orifice can be estimated as

$$Q = 0.67hL(2gd_o)^{1/2} \quad (10.34)$$

where h = curb opening height (ft, m)
d_o = effective depth of flow at the curb face (see Figure 10.19)

Figure 10.19
Curb-opening inlet throat types

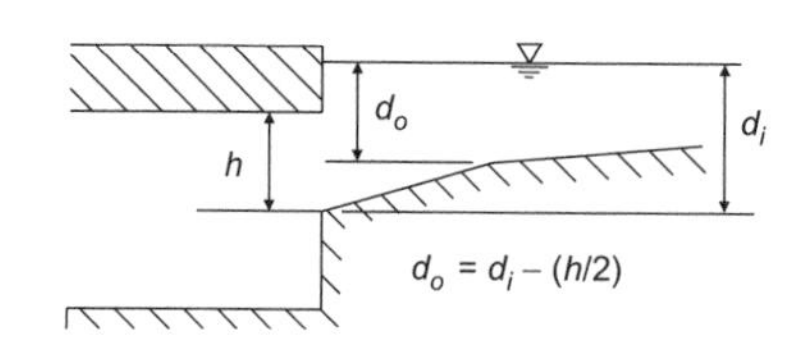

(a) Horizontal Throat

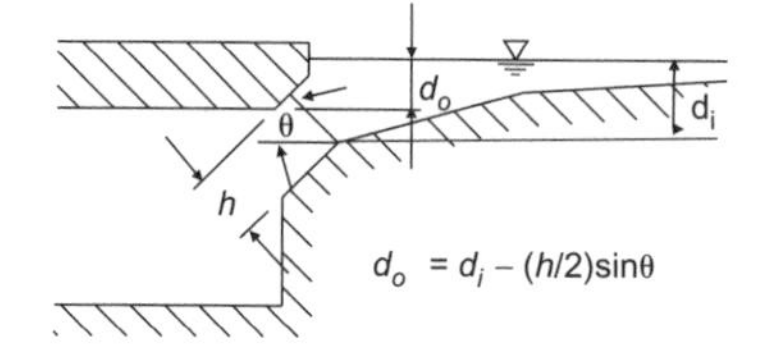

(b) Inclined Throat

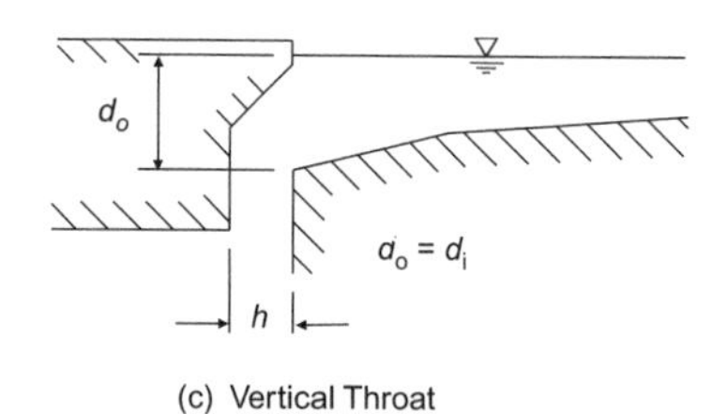

(c) Vertical Throat

Example 10.6 – Capacity of a Curb Inlet in a Sag. A roadway has a composite gutter section with a pavement cross slope $S_x = 0.02$, a normal gutter cross slope $S_w = 0.0625$, and a curb-face height of 15 cm. The gutter width is $W = 0.61$ m. Thus, the top of the curb is normally 11 cm higher than the elevation of the lip of the gutter. A curb-opening inlet with a horizontal throat is to be designed at a sag point in the roadway with a design discharge of $Q = 0.3$ m^3/s. The gutter is depressed at the inlet location so that an opening height of 12 cm and a top slab thickness of 8 cm yield a total curb height of 20 cm. The allowable spread at the inlet is $T = 4$ m. Determine the required curb-opening length.

Solution: Because the gutter is depressed at the inlet, Equation 10.33 can be used to estimate the required inlet length under the assumption that it behaves as a weir. Setting

$$d = TS_x = 4(0.02) = 0.08 \text{ m}$$

gives the required length as

$$L = \frac{Q}{C_w d^{3/2}} - 1.8W = \frac{0.3}{1.66(0.08)^{3/2}} - 1.8(0.61) = 6.89 \text{ m}$$

Because this length is greater than 3.6 m, Equation 10.32 should be applied instead. The depth d used in Equation 10.32 is the depth at the curb face. This value is equal to the depth at the curb face at a normal gutter cross-section, plus an additional 5 cm for the depressed gutter. [The 5-cm depression is computed as the difference between the normal curb height (15 cm) and the height at the inlet (20 cm).] At a normal gutter section, the curb-face depth is

$$y = (T - W)S_x + W_{SW} = (4 - 0.61)(0.02) + 0.61(0.0625) = 0.11 \text{ m}$$

and thus the depth for use in Equation 10.32 is

$$d = 0.11 + 0.05 = 0.16 \text{ m}$$

Therefore, if weir behavior controls the design, the required inlet length is

$$L = \frac{Q}{C_w d^{3/2}} = \frac{0.3}{1.66(0.16)^{3/2}} = 2.82 \text{ m}$$

If orifice behavior controls the design, then the required inlet length is determined using Equation 10.34 with h = 13 cm = 0.13 m, which gives

$$d_o = 0.16 - (0.12/2) = 0.10 \text{ m}$$

Substituting this result into Equation 10.34 and solving for L gives

$$L = \frac{Q}{0.67h(2gd_o)^{1/2}} = \frac{0.3}{0.67(0.12)[2(9.81)(0.10)]^{1/2}} = 2.66 \text{ m}$$

Comparing the lengths shows that weir behavior controls the design, and the required length is L = 2.58 m. This length should be increased by an assumed amount in order to account for clogging of the inlet opening.

Combination Inlets

As previously discussed, a combination inlet is a configuration in which a grate inlet and curb inlet are used together (see Figure 10.20). Like both calculations for grate and combination inlets, the method used to compute combination inlet capacity depends on whether the inlet is on grade or in a sag.

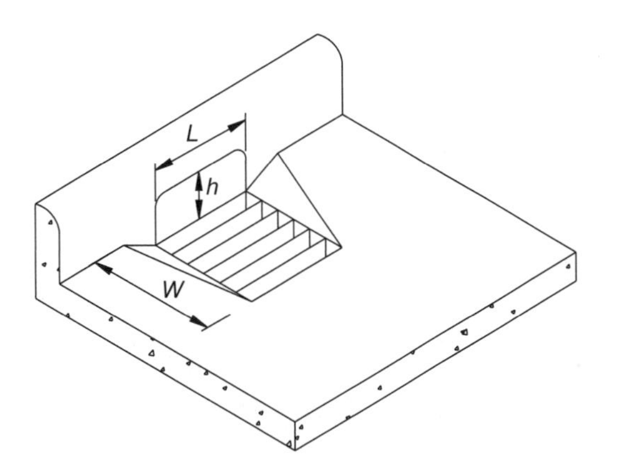

Figure 10.20
Combination inlet

Capacity of a Combination Inlet on a Continuous Grade. When the length of the curb opening is the same as the length of the grate for a combination inlet on a continuous grade, the capacity of the inlet is assumed to be no greater than the capacity of the grate alone. Thus, the interception capacity of the inlet should be determined using the procedure for grate inlets on grade.

If a combination inlet on a continuous grade has a sweeper configuration, which means that the curb opening extends upstream of the grate, then the interception capacity is the sum of the capacity of the curb opening upstream of the grate and the grate capacity. When performing the calculations, the frontal flow Q_W that reaches the grate should be reduced by the amount that is intercepted by the curb opening.

Capacity of a Combination Inlet in a Sag. The procedure for determining the capacity of a combination inlet in a sag depends on the length of the curb opening relative to the length of the grate. If the inlet has a sweeper configuration in which the curb opening is longer than the grate, then the curb opening can be very effective in capturing trash and debris before it has a chance to clog the grate.

Under weir-flow conditions, the capacity of a combination inlet with equal grate- and curb-opening lengths is essentially that of the grate alone. However, if the inlet has a sweeper configuration, it is reasonable to assume that the inlet's capacity is equal to the sum of the weir capacities of the grate and the portion of the curb opening that extends beyond the grate. When the capacity of a sweeper inlet is computed in this manner, it is suggested that the weir length applicable to the grate be computed as

$$P = L + W \tag{10.35}$$

where P = applicable weir length for the grate (ft, m)
L = length of the grate (ft, m)
W = width of the grate (ft, m)

Note that the edge of the grate adjacent to the extended curb opening is neglected in this determination of P, as is the edge of the grate adjacent to the curb face.

Under orifice-flow conditions, the capacity of a combination inlet in a sag is the sum of the capacities of the grate and the curb opening. The capacity of the grate is given by Equation 10.28 and the capacity of the curb opening by Equation 10.34.

Slotted-Drain Inlets

Slotted-drain inlets may be installed either along the length of the gutter, as shown in Figure 10.21, or perpendicular to the flow direction to intercept sheet flows. The calculation of hydraulic capacity for both installation types is presented in this section.

Capacity of a Slotted-Drain Inlet on a Continuous Grade. The interception capacity of a slotted-drain inlet on a continuous grade depends significantly on the orientation of the longitudinal axis of the slotted drain with respect to the surface-flow direction. When slotted drains are installed in the normal fashion (along the length of the gutter), as illustrated in Figure 10.21, and when the slot width is at least 1.75 in. (45 mm), FHWA tests show that the required length of a slotted drain for total interception of the gutter flow is given by Equation 10.29. In other words, the slotted drain performs similarly to a curb-opening inlet in this instance. If the actual length of the slotted drain is shorter than the length required for total interception, then its efficiency is given by Equation 10.30.

If a slotted drain is installed in a configuration for sheet-flow interception, its capacity can be approximated by treating it as a grate inlet with a grate length equal to the width of the slotted opening. Thus, its capacity can be approximated using Equations 10.21 and 10.22, where the frontal flow is the total sheet-flow discharge approaching the slotted drain opening. Use of this approximation requires some judgment about the appropriate value of the splash-over velocity. Although no specific recommendation is made here, intuition suggests that because of the generally shallow nature of sheet flow, the factor R_f can often be taken as unity.

Figure 10.21
Slotted-drain inlet

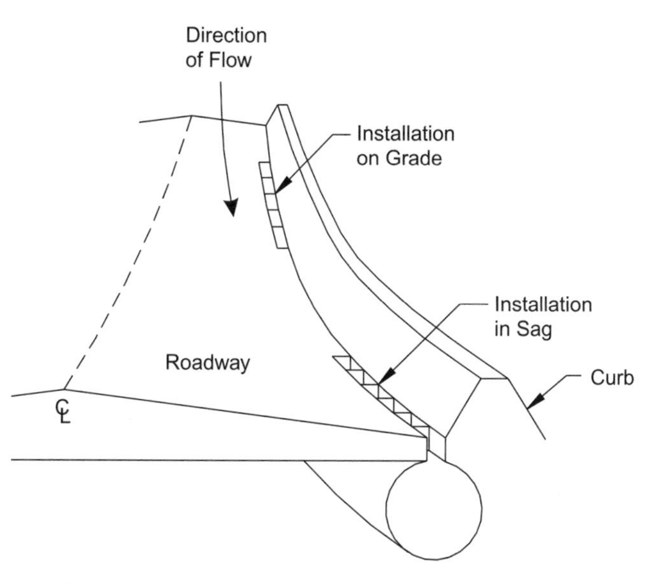

Credit: McCuen, 1998. Reprinted by permission of Pearson Education, Inc., Upper Saddle River, NJ.

Capacity of a Slotted-Drain Inlet in a Sag. When installed in a sag condition, a slotted drain will perform as a weir for flow depths up to about 0.2 ft (0.06 m), although this depth depends on the slot width. For such operating conditions, the capacity of the slotted drain can be estimated as

$$Q = C_w L d^{3/2} \tag{10.36}$$

where C_w =weir coefficient
L =length of the slot (ft, m)
d =depth at curb measured relative to the normal cross slope (i.e. $d = TS_x$) (ft, m)

The weir coefficient C_w in this expression depends on the flow depth and the slot length, but a typical value is 2.48 for U.S. customary units or 1.4 for SI units.

Equation 10.36 is based on the assumption that the slotted drain is installed next to a curb face, so that only one edge of the slot behaves as a weir. The capacity of a slotted drain when both edges of the slot serve as weirs should be twice that predicted by Equation 10.36, though it is expected that the weir coefficient would require modification in such cases. Manufacturers of slotted-drain products can typically provide guidance on the required length of drain for various types of installations.

For flow depths greater than about 0.4 ft (0.12 m), a slotted drain behaves hydraulically as an orifice, with its capacity given by

$$Q = 0.8LW(2gd)^{1/2} \tag{10.37}$$

where W = slot width (ft, m)
d = depth of water at the slot opening (ft, m)

"Looks like it's clearing up. Maybe we should get started again."

10.3 INLET LOCATIONS AND DESIGN FLOWS

The surface portion of a stormwater conveyance system cannot be viewed as simply a collection of individual inlets. Rather, it is a network of interdependent inlets connected by gutters and other conveyances. The gutter downstream of a particular inlet on grade must have sufficient capacity to carry both that inlet's bypass flow and any additional flow entering the gutter. The next downstream inlet on grade will intercept only a portion of the total gutter flow—again, some flow is carried over to the next downstream inlet. This interdependency continues in a downstream direction until the flow reaches a sag inlet, which captures 100 percent of the inflow.

Determining the locations where inlets are required and their corresponding design discharges is at least in part a trial-and-error process. Nevertheless, required locations of many inlets can be determined simply by referring to plans and profiles for the roadway. According to Brown, Stein, and Warner (2001), inlets should be provided at the following locations, at a minimum:

- At all low points (sag points) in gutters
- Immediately up-grade of median breaks, entrance/exit ramp gores, and street intersections (that is, at any location where water could flow onto the traveled surface of a roadway)
- Immediately up-grade of bridges, to prevent drainage onto bridge decks, and immediately down-grade of bridges, to intercept bridge-deck runoff [see Young, Walker, and Chang (1993) for information on drainage of bridge decks]
- Immediately up-grade of roadway cross-slope reversals
- Immediately up-grade of pedestrian cross-walks
- Behind curbs, shoulders, or sidewalks as necessary to drain low areas

Inlet locations based on the above criteria should be plotted on a plan view of the roadway system. Topographic information and roadway profile drawings can then be used to delineate the boundaries of the drainage basin contributing runoff to each inlet. Estimates of design flows to each inlet are then computed for a selected design-storm (see Table 2.2, page 37). The design flow in the approach gutter to each inlet can be compared with the allowable gutter conveyance capacity to check that flow depth and spread criteria are not violated. In all cases, one should strive to place inlets using projections of land-parcel or lot lines so the inlets will not interfere with potential driveway locations. In sag vertical curves, such placement often requires adjustment of roadway grades and/or vertical curve lengths.

Provided that the gutter capacity is not exceeded at an inlet, the selection of an inlet type and the inlet design can proceed as outlined in the previous subsections. Designs of sag inlets should be such that the entire design flow approaching the inlet is intercepted. Conversely, inlets on continuous grades may be designed to intercept only a fraction of the approaching flow, leaving the bypass flow to proceed to the next downstream inlet.

For any inlet, the total design flow consists of the runoff from its local drainage basin plus any bypass flow from upstream inlets. However, if the bypass flow is large, or if the times of concentration are significantly different for the bypass flow and the local runoff to an inlet, then the total design discharge may be computed using the procedure outlined by Guo (1997).

In the event that the discharge approaching an inlet is in excess of the conveyance capacity of the gutter section, a trial-and-error approach is required to determine locations for additional inlets. These inlets are virtually always continuous grade, but are usually necessary only on extended lengths of roadway where inlets are not already required on the basis of the geometric criteria previously listed.

The trial-and-error approach entails selection of a trial inlet location. Based on the trial location, drainage basins are again delineated and design discharges computed. A comparison of the design discharges with allowable gutter capacities is then made to check the adequacy of the selected inlet locations. Where possible, inlet positions determined in this manner should coincide with locations where projections of lot or parcel lines intersect the gutter. As noted above, this practice avoids potential conflicts with driveways or access roads.

In roadway sag locations, especially where there is no outlet for stormwater other than through the storm sewer system, *flanking inlets* should be provided on either side of the sag inlets. These inlets provide relief if the sag inlet becomes clogged. This practice should not be necessary in newly developed areas if proper attention has been given to the major storm-drainage system, but it may be required in retrofit or rehabilitation projects. Flanking inlets should be designed to intercept all of the design flow without violating allowable spread criteria if the sag inlets are completely clogged.

Modeling Focus: Location of Inlets on Grade

Computer programs can be useful tools when determining where inlets should be placed to meet design constraints such as allowable gutter spread and depth. A number of programs have been written to perform these types of calculations, including the FlowMaster software packaged with this book. Like the methods presented in this chapter, FlowMaster's computational methods are based on FHWA's HEC-22 publication (Brown, Stein, and Warner, 2001).

Consider that an engineer is designing a subdivision and plans to use curved-vane grate inlets measuring 2 ft long by 2 ft wide as the standard inlet size. He or she begins by proposing inlets at the minimum number of locations using the guidelines provided in Section 10.3. The next step is to evaluate the performance of the gutters and inlets, given these inlet locations.

The first inlet the engineer examines, inlet A, is the most-upstream inlet on a section of roadway. The engineer delineates the contributing drainage area for this inlet and calculates that the peak runoff flow rate to the inlet for the design storm is 4.0 cfs.

This section of roadway has a 3-percent longitudinal slope and a 2-percent cross slope. A composite gutter section with a a 2-ft–wide depressed gutter having a cross slope of 3-percent will be used on this project. According to local regulations, the maximum amount of gutter spread permitted for this roadway is 8 ft.

The engineer uses a computer program or hand calculations to determine whether the gutter can handle a flow of 4.0 cfs without violating the maximum-spread constraint. Assuming a spread of 8 ft and solving for flow, the engineer determines that the maximum capacity of this gutter is 3.0 cfs.

Based on this calculation, the engineer knows that an additional inlet must intercept, at a minimum, (4.0 cfs – 3.0 cfs) = 1.0 cfs.

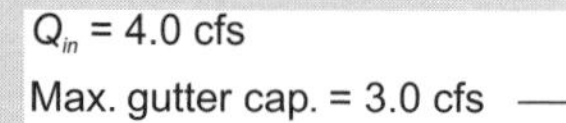

Anticipating that some of the flow from the additional inlet (inlet B) will be bypassed to inlet A, the engineer decides to locate inlet B so that its drainage area is roughly half that of the original area. The flow approaching inlet B is calculated as 2.1 cfs.

Because 2.1 cfs is less than 3.0 cfs (the maximum allowable gutter capacity), the location of inlet B appears acceptable from the standpoint of upstream gutter capacity. The computer program or hand calculations can now be used to evaluate the performance of inlet B. It is determined that inlet B intercepts 1.4 cfs and bypasses 0.7 cfs to inlet A. (The gutter spread upstream of inlet B is found to be 7.0 ft.)

Summing the bypass flow from inlet B and the flow from inlet A's additional contributing drainage area, the engineer determines that the flow approaching inlet A will be 0.7 cfs + (4.0 cfs – 2.1 cfs) = 2.6 cfs. Because this flow is less than 3.0 cfs, the gutter spread constraint is met. Inlet A is computed to intercept 1.6 cfs and bypass 1.0 cfs. (The gutter spread immediately upstream of the inlet is 7.6 ft.)

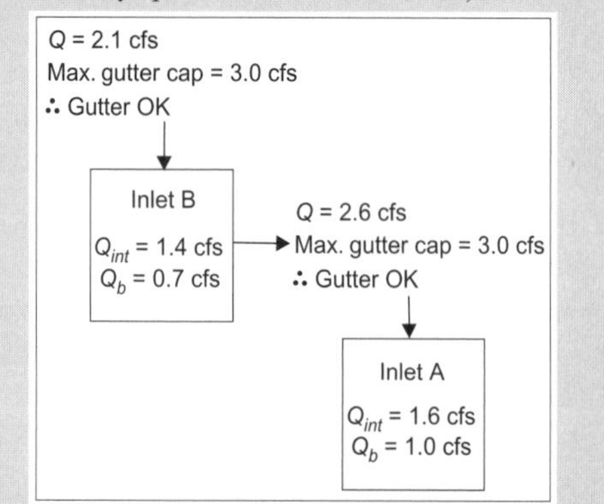

The design of the remaining inlets in the subdivision proceeds in a similar fashion, working downstream from the most upstream inlets, transferring bypass flows downstream, checking gutter spread, and adding or enlarging inlets as needed. Note that in this simple example, the peak bypass flow and the peak additional flow from the drainage area are assumed to be additive. If the times of concentration for these flows differed significantly, this would not be the case.

10.4 CHAPTER SUMMARY

Inlets provide a means for stormwater to enter the subsurface sewer system. Most inlets are located in roadway, driveway, or parking lot gutters. These gutters collect runoff and convey it to the inlet. The two basic types of gutters are the conventional curb and gutter section (including uniform, composite, and curved gutters) and shallow swale sections (including V-shape, V-shape median, and circular gutters).

Gutter-capacity calculations are performed to ensure that allowable gutter spread and depth criteria are not violated for the system design. These calculations are typically performed using the modified versions of the Manning equation presented in this chapter.

Common inlet types are grate, curb-opening, combination, and slotted-drain. These inlets are found on both continuous grades and in sag locations. For an inlet on a continuous grade, the inlet may only intercept a portion of the flow and bypasses remaining flow to the next downstream inlet.

Common grate inlet types include parallel bar, vaned, tilt-bar, and reticuline of various shapes and sizes. For a grate inlet on grade, the flow captured by the inlet is computed as the sum of the intercepted frontal flow and side flow. The frontal flow intercepted is a function of the cross-sectional geometry of the gutter, the splash-over velocity for the inlet (the velocity at which frontal flow begins to bypass the inlet), and the flow velocity. The side flow intercepted is a function of flow velocity, roadway cross slope, and grate length. A grate inlet in a sag functions hydraulically as a weir for shallow flows and as an orifice for deeper flows.

The determination of capture efficiency for a curb inlet on grade is based on the inlet length required to capture 100 percent of the flow. Inlet performance can be significantly increased by increasing the gutter cross slope. Similar to a grate inlet, a curb inlet in sag operates as a weir for shallow flows and as an orifice for deeper flows. The orifice behavior of a curb-opening inlet is influenced by the type of throat opening (horizontal, inclined, or vertical).

Because a combination inlet consists of a grate inlet and curb inlet together, the calculation methods presented for both of these inlet types apply. However, for a combination inlet on a continuous grade or for a combination inlet in sag behaving as a weir, the inlet capacity is assumed to be the same as that of the grate acting alone if the curb opening and the grate are the same length. If the curb opening is longer than the grate, the capacity for the length of curb opening extending beyond the grate should be considered. For a combination inlet in sag operating as an orifice, the capacity is computed as the sum of the grate and curb opening capacities.

The capacity of a slotted-drain inlet on a continuous grade depends on the orientation of the drain relative to the direction of surface flow. When a slotted drain is installed along the length of a gutter, the slotted drain performs similarly to a curb-opening inlet on a continuous grade. If the direction of the drain is normal to the flow direction, it is treated as a grate inlet in capacity calculations. If the slotted-drain inlet is installed in a sag, it is analyzed as a weir or an orifice, depending on the flow depth.

Various geometric considerations on a site control the initial placement of inlets. For instance, inlets should be placed in low points and immediately upstream of intersections. Inlets are then added or resized as necessary to meet gutter-flow constraints. A computer program that performs gutter and inlet calculations can be of great assistance in this trial-and-error process.

REFERENCES

American Society of Civil Engineers (ASCE). 1992. *Design and Construction of Urban Storm Water Management Systems.* New York: ASCE.

Brown, S. A., S. M. Stein, and J. C. Warner. 2001. *Urban Drainage Design Manual.* Hydraulic Engineering Circular No. 22, Second Edition, FHWA-SA-96-078. Washington, D.C.: U.S. Department of Transportation.

Federal Highway Administration (FHWA). 1977. *Bicycle Safe Grate Inlets Study: Vol. 1 and 2—Hydraulic and Safety Characteristics of Selected Grate Inlets on Continuous Grade.* FHWA-RD-77-44. Washington, D.C.: U.S. Department of Transportation.

Guo, J. C. Y. 1997. *Street Hydraulics and Inlet Sizing Using the Computer Model UDINLET.* Highlands Ranch, Colorado: Water Resources Publications.

McCuen. 1998. *Hydrologic Analysis and Design.* Upper Saddle River, New Jersey: Pearson Education.

Young, G. K., S. E. Walker, and F. Chang. 1993. *Design of Basic Bridge Deck Drainage.* Hydraulic Engineering Circular No. 21, GHWA-SA-92-010. Washington, D.C.: U.S. Department of Transportation.

PROBLEMS

10.1 A roadway gutter has a longitudinal slope of 0.01 and a cross slope of 0.02. Use Manning's n = 0.016.

a) Find the spread at a flow of 1.8 cfs.

b) Find the gutter flow at a spread of 8.2 ft.

10.2 A composite curb and gutter section is located on a roadway with a longitudinal slope of 0.01 and a cross slope of 0.02. The gutter width is 2 ft and the depression is 2 in. Assume Manning's n = 0.016.

a) What is the gutter flow if the spread is 8.2 feet?

b) What is the spread if the flow is 4.2 cfs?

10.3 Find the interception capacity of a 610-mm–wide by 914-mm–long P-50 grate located in a composite gutter section having a longitudinal slope of 0.01 and a roadway cross-slope of 0.02. The width W of the depressed portion of the gutter is 0.6 m, the depth of the depression a is 50 mm, and the gutter cross slope is 0.103. The flow in the gutter is 0.065 m^3/s and Manning's n = 0.016.

10.4 A uniform roadway gutter has a grated inlet in a depressed sag configuration. The flow is 0.20 m^3/s and the depth of flow at the curb is 122 mm. Determine the length of 610-mm–wide, P-50-type grate needed to intercept the flow in a sag configuration.

10.5 A gutter has the following characteristics:

- $S_0 = 0.01$
- $S_x = 0.03$
- $Q = 2$ cfs
- $n = 0.016$

Find Q_{int} for a 10-ft–long curb opening.

10.6 A 3.05-m curb-opening inlet is located on a grade with a slope of 0.01 and $n = 0.012$. The gutter section is depressed, and its dimensions are as follows:

- $S_x = 0.03$
- $S_w = 0.05$
- $a = 51$ mm
- $W = 0.91$ m

Determine the flow intercepted by the inlet if its efficiency if 50 percent. (*Hint:* use the included FlowMaster software to graph efficiency versus discharge.)

Workers install a section of concrete pipe for a storm sewer.

CHAPTER

11

Storm Sewer Pipe System and Outlet Design

This chapter is the second of two chapters related to the design and analysis of storm sewer systems. Chapter 10 focused on the gutter and inlet components that make up the surface portion of a system, whereas this chapter covers the subsurface portion of the system.

Storm sewers are typically *dendritic* (treelike) subsurface systems that collect and direct flows from inlets into larger and larger conveyance structures (typically pipes) or systems as one moves downstream in the watershed. The downstream ends of storm sewers, called *outfalls,* are generally located at treatment and/or detention or retention facilities, receiving waters, or at locations where alternate conveyance facilities, such as open channels, are more practical.

Chapter 2 briefly introduced many of the basic physical components and materials found in storm sewer systems. Section 11.1 of this chapter expands on that discussion by providing more detailed information on storm sewer pipe materials and sizes, and manhole construction.

The horizontal and vertical alignments of storm sewers are governed by a number of factors, as will be discussed in Section 11.2. These factors include topography, standardization of horizontal locations within roadway rights of way, subterranean conditions, such as the presence of bedrock and other utilities, and minimum and maximum flow velocities. Storm sewer alignments, both horizontal and vertical, are often shown in construction documents on "plan and profile" sheets such as the one shown in Figure 11.1

Estimation of design flows in storm sewers and determination of the pipe sizes required for conveyance of these flows are usually accomplished using the rational method and the Manning equation. This procedure was introduced in Example 5.8 and is elaborated on in Section 11.3.

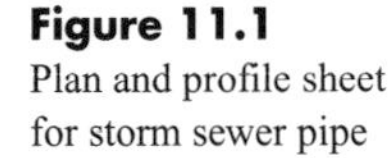

Figure 11.1
Plan and profile sheet for storm sewer pipe runs

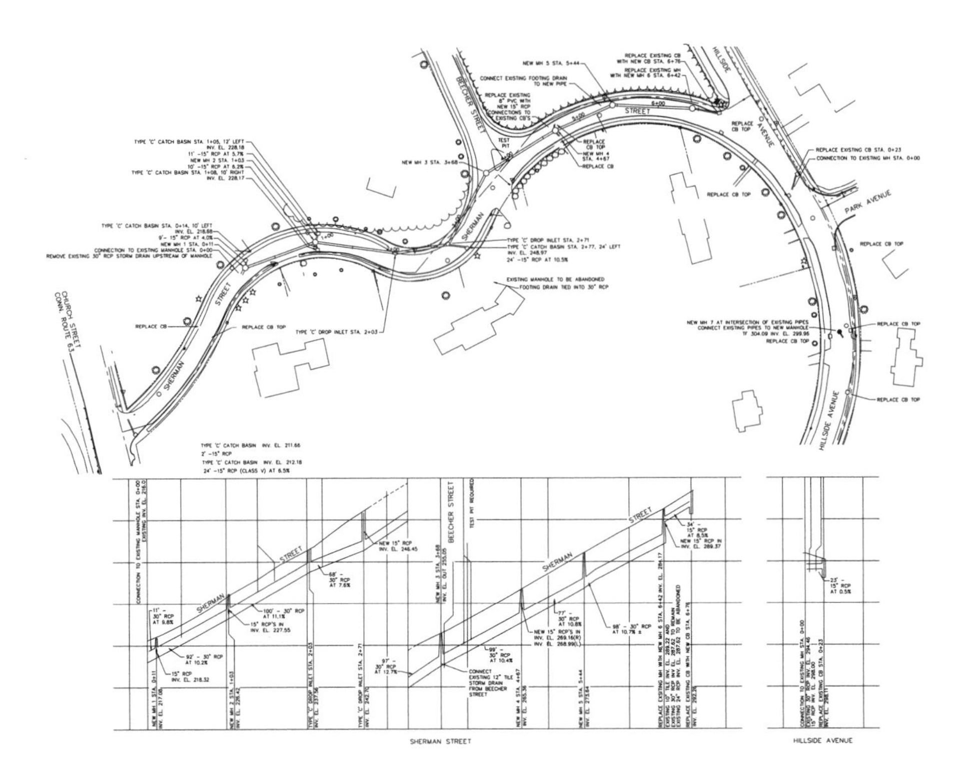

In many cases, plotting the hydraulic and energy grade lines for design flows on a profile of a storm sewer is desirable in order to ensure that the design is adequate. This verification is particularly important when one or more parts of the system are surcharged. To compute the hydraulic and energy grades, it is necessary to evaluate not only the hydraulic profile and friction losses in the pipes, but also minor losses in inlet boxes, manholes, and junction structures. The classical approach to computing minor losses through various appurtenances was presented in Section 6.3, and this method can be applied to sewer systems. However, other techniques have been developed specifically for common sewer conditions and configurations. Two of these methods, the energy-loss method and composite energy-loss method, are described in Section 11.4.

Section 11.5 covers final design, and presents procedures for computing and plotting the hydraulic and energy grade lines through a storm sewer system for both surcharged and open-channel flow conditions. The latter part of this section provides discussions on additional design considerations, including structural loads on sewer pipes and flow routing of storm sewers. Because of the number of calculations required to design or analyze storm sewer hydraulics, computer programs are typically used for this purpose. Section 11.6 outlines a typical procedure for designing a storm sewer system with the aid of a computer.

The energy associated with an excessive flow velocity at a storm sewer outfall point should be dissipated to prevent potentially destructive and unsightly erosion from occurring at that location. As described in Section 11.7, erosion protection may consist of a simple riprap blanket, a concrete energy dissipator, or a stilling basin.

11.1 STORM SEWER TYPES, COMPONENTS, AND MATERIALS

This section provides information on the components and materials commonly used in the construction of storm sewer systems. Most of these components were first presented in Chapter 2, but are they are covered in more detail here.

It is important to note that storm sewers, as considered in this book, are separate from systems intended to convey sanitary wastes. *Combined sewers*, which are not addressed in this book, convey both storm runoff and sanitary flows. The choice of cross-sectional shape and material for a sewer conduit can be influenced by whether the system is combined or separate. In a combined sewer, for example, it may be desirable to provide a *cunette* (smaller channel) in the sewer invert to handle low flows. Also, materials such as concrete or steel may be undesirable in combined sewers because of the potential for corrosion caused by hydrogen sulfide gas released from raw sewage.

The subsections that follow first present information on pipe materials frequently used in storm sewer system design and then describe common appurtenances such as inlets, catch basins, manholes, junctions, and outlets.

Storm Sewer Pipes

A variety of pipe materials is used in the construction of storm sewers. Concrete and steel are by far the most popular and are discussed in detail in this section. Another material that may be used as an alternative to corrugated steel is corrugated aluminum, which is resistant to corrosion. Because of its lower stiffness, however, aluminum pipe requires more care during construction. Other pipe materials include high-density polyethylene (HDPE) and polyvinyl chloride (PVC), as well as various forms of cast-in-place concrete sections.

Concrete Pipe. Concrete is the most-used pipe material for storm sewer construction. It is available both with and without reinforcing steel and comes in a range of wall thicknesses and strength classes. In addition to pipe, precast concrete manholes, inlets, outlet sections, and special structures, such as tees and bends, are available. In the United States, these materials are manufactured in accordance with applicable standard specifications of the American Society for Testing and Materials (ASTM). The American Concrete Pipe Association (ACPA) publishes considerable technical data, as well as software that can be downloaded from the Internet to assist in specifying appropriate concrete-pipe products. The NSSN web site (www.nssn.org) provides links to standards information from more than 600 international, national, and regional standards organizations.

As illustrated in Figure 11.2, concrete pipe is manufactured in a variety of cross-sectional shapes: circular, elliptical, arch, and rectangular (box). *Plain* or nonreinforced circular concrete pipe is available in inside diameters ranging from 4 to 36 in. (100 to 900 mm). Typically, plain concrete pipes are available in 2-in. (50-mm) increments for diameters of 4 to 12 in. (100 to 300 mm), and in 3-in. (75-mm) increments for diameters of 12 to 36 in. (300 to 900 mm). *Reinforced-concrete pipe* is more commonly used and is available in diameters ranging from 12 to 180 in. (300 to

4500 mm). Generally, the minimum recommended diameter for storm sewer applications is 12 in. (300 mm), due to the potential for clogging by debris and for maintenance requirements, but some jurisdictions may specify a larger diameter as the minimum acceptable size.

Table 11.1 provides a listing of the commonly available sizes of circular reinforced-concrete pipe and corresponding available pipe-wall thickness classifications A, B, and C. The thinnest wall is designated as A and the thickest wall is designated as C. Pipe manufacturers typically provide additional information on the permissible loads for the various classfications.

Figure 11.2 Cross-sectional shapes available from concrete pipe manufacturers

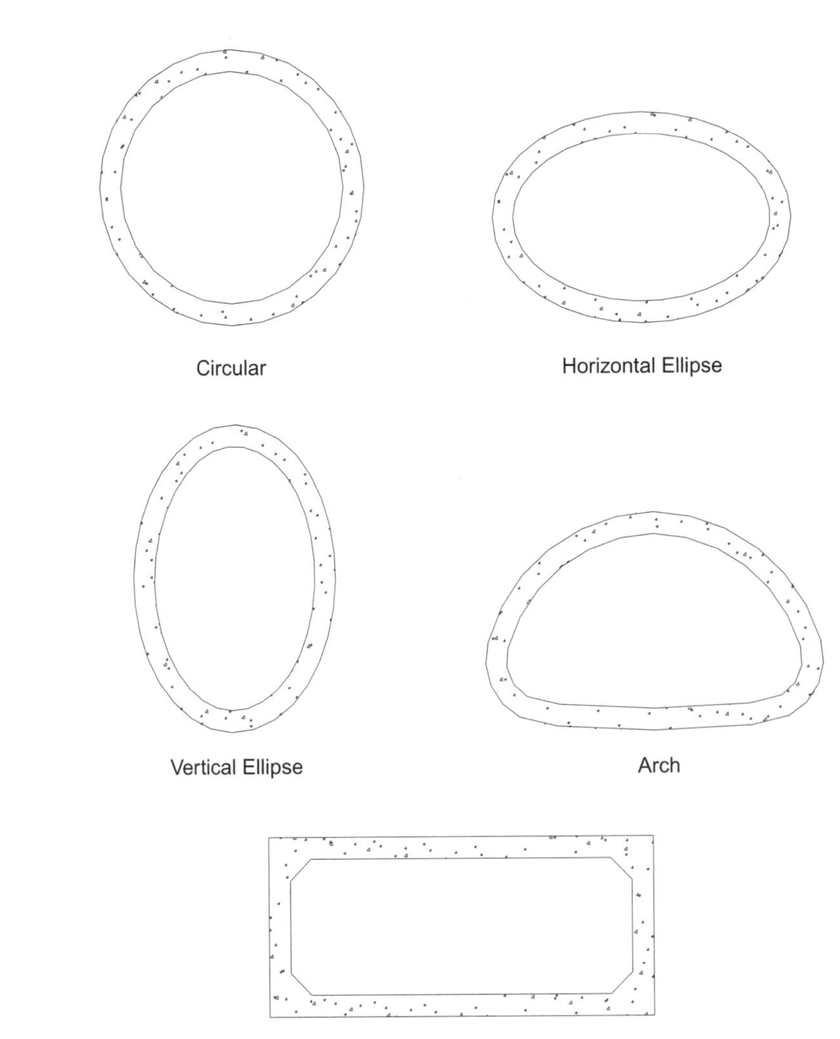

Reproduced by permission of the American Concrete Pipe Association.

Table 11.1 Standard sizes of circular reinforced-concrete pipe with tongue and groove joints (adapted from ACPA, 2000)

Internal Diameter		Min. Wall Thickness, in. (mm)					
		Wall A		Wall B		Wall C	
in.	mm	in.	mm	in.	mm	in.	mm
12	300	$1\frac{3}{4}$	44	2	50		
15	375	$1\frac{7}{8}$	48	$2\frac{1}{4}$	56		
18	450	2	50	$2\frac{1}{2}$	63		
21	525	$2\frac{1}{4}$	56	$2\frac{3}{4}$	69		
24	600	$2\frac{1}{2}$	63	3	75	$3\frac{3}{4}$	94
27	675	$2\frac{5}{8}$	66	$3\frac{1}{4}$	82	4	100
30	750	$2\frac{3}{4}$	69	$3\frac{1}{2}$	88	$4\frac{1}{4}$	106
33	825	$2\frac{7}{8}$	72	$3\frac{3}{4}$	94	$4\frac{1}{2}$	113
36	900	3	75	4	100	$4\frac{3}{4}$	119
42	1050	$3\frac{1}{2}$	88	$4\frac{1}{2}$	113	$5\frac{1}{4}$	131
48	1200	4	100	5	125	$5\frac{3}{4}$	144
54	1350	$4\frac{1}{2}$	113	$5\frac{1}{2}$	138	$6\frac{1}{4}$	156
60	1500	5	125	6	150	$6\frac{3}{4}$	169
66	1650	$5\frac{1}{2}$	138	$6\frac{1}{2}$	163	$7\frac{1}{4}$	182
72	1800	6	150	7	175	$7\frac{3}{4}$	194
78	1950	$6\frac{1}{2}$	163	$7\frac{1}{2}$	188	$8\frac{1}{4}$	207
84	2100	7	175	8	200	$8\frac{3}{4}$	219
90	2250	$7\frac{1}{2}$	188	$8\frac{1}{2}$	213	$9\frac{1}{4}$	232
96	2400	8	200	9	225	$9\frac{3}{4}$	244
102	2550	$8\frac{1}{2}$	213	$9\frac{1}{2}$	238	$10\frac{1}{4}$	257
108	2700	9	225	10	250	$10\frac{3}{4}$	269
114	2850	$9\frac{1}{2}$	238				
120	3000	10	250				
126	3150	$10\frac{1}{2}$	263				
132	3300	11	275				
138	3450	$11\frac{1}{2}$	288				
144	3600	12	300				
150	3750	$12\frac{1}{2}$	313				
156	3900	13	327				
162	4050	$13\frac{1}{2}$	339				
168	4200	14	350				
174	4350	14 1/2	363				
180	4500	15	375				

Concrete pipe is also manufactured in elliptical (horizontal or vertical) and arch-shaped cross sections for use in applications where vertical or horizontal clearances are limited. Rectangular concrete box sections are also available. For noncircular sections, the pipe or box size is specified by both its internal horizontal dimension (*span*) and its internal vertical dimension (*rise*). Standard product sizes are provided in various ACPA publications.

Two basic types of joints are used to connect segments of concrete pipe. *Bell-and-spigot joints* are common for smaller pipe diameters (4 to 36 in. or 100 to 900 mm), and *tongue-and-groove joints* are used with larger pipe sizes (12 to 180 in. or 300 to 4500 mm). Joints are usually sealed with a bituminous mastic material supplied in rope form by the pipe manufacturer, but mortar, compression rubber gaskets, or O-ring rubber gaskets can also be used. These latter types of joints are useful in situations where water-tightness is a concern. In high-pressure applications, such as force mains, joints can be manufactured with steel end rings and O-ring gaskets.

Corrugated-Steel Pipe. *Corrugated-steel pipe* is another frequently used material in storm-sewer construction. Like concrete pipe, it is manufactured in a number of cross-sectional shapes and sizes and in strict accordance with applicable standard specifications. The American Iron and Steel Institute (AISI) publishes a number of design documents, some of which are available for download from their website.

Circular corrugated-steel pipe has either *annular* (ringed) or *helical* (spiral) corrugations, and is available in diameters ranging from 4 to 144 in. (100 to 3600 mm). Standard sizes and thicknesses for steel pipes of various corrugation sizes are shown in Table 11.2. Like concrete pipe, 12-inch-diameter (300-mm) steel pipe is generally the minimum recommended size for storm-sewer applications. For corrosion protection, corrugated-steel pipe may be specified as galvanized, bituminous-coated, or coated and asphalt-paved. Steel thicknesses should be specified to withstand the loads expected to occur in a particular application.

In addition to circular pipe sections, corrugated-steel products include arch pipes, structural-plate pipe arches, and structural-plate arches. Structural-plate products consist of galvanized plates that require bolted assembly in the field, as illustrated in Figure 11.3. The shape of steel arch pipe is similar to that of concrete arch pipe. A structural-plate arch is an arch span similar to a typical bridge, in that it has no bottom. Other available steel drainage-system products include bends, tees, wyes, transitions, manhole risers, and end sections.

Table 11.2 Standard sizes of corrugated-steel pipe (adapted from AISI, 1999)

Internal Diameter		Available Steel Thicknesses											
in.	mm	in.	mm	in.	mm	in.	mm	in.	mm	in.	mm	in.	mm
1½ × ¼-in. (38 × 6.5-mm) Corrugations													
6	150	0.052	1.32	0.064	1.63								
8	200	0.052	1.32	0.064	1.63								
10	250	0.052	1.32	0.064	1.63								
12	300	0.052	1.32	0.064	1.63								
15	375	0.052	1.32	0.064	1.63								
18	450	0.052	1.32	0.064	1.63								
2 × ½-in. (68 × 13-mm) Corrugations													
12	300	0.052	1.32	0.064	1.63	0.079	2.01						
15	375	0.052	1.32	0.064	1.63	0.079	2.01						
18	450	0.052	1.32	0.064	1.63	0.079	2.01						
21	525	0.052	1.32	0.064	1.63	0.079	2.01						
24	600	0.052	1.32	0.064	1.63	0.079	2.01						
30	750	0.052	1.32	0.064	1.63	0.079	2.01						
36	900	0.052	1.32	0.064	1.63	0.079	2.01	0.109	2.77	0.138	3.51		
42	1050	0.052	1.32	0.064	1.63	0.079	2.01	0.109	2.77	0.138	3.51		
48	1200	0.052	1.32	0.064	1.63	0.079	2.01	0.109	2.77	0.138	3.51	0.168	4.27
54	1350	0.064	1.63	0.079	2.01	0.109	2.77	0.138	3.51	0.168	4.27		
60	1500	0.079	2.01	0.109	2.77	0.138	3.51	0.168	4.27				
66	1650	0.079	2.01	0.109	2.77	0.138	3.51	0.168	4.27				
72	1800	0.109	2.77	0.138	3.51	0.168	4.27						
78	1950	0.109	2.77	0.138	3.51	0.168	4.27						
84	2100	0.109	2.77	0.138	3.51	0.168	4.27						
90	2250	0.109	2.77	0.138	3.51	0.168	4.27						
96	2400	0.138	3.51	0.168	4.27								
3 × 1-in. or 5 × 1-in. (75 × 25-mm or 125 × 25-mm) Corrugations													
54	1350	0.064	1.63	0.079	2.01	0.109	2.77	0.138	3.51	0.168	4.27		
60	1500	0.064	1.63	0.079	2.01	0.109	2.77	0.138	3.51	0.168	4.27		
66	1650	0.064	1.63	0.079	2.01	0.109	2.77	0.138	3.51	0.168	4.27		
72	1800	0.064	1.63	0.079	2.01	0.109	2.77	0.138	3.51	0.168	4.27		
78	1950	0.109	2.77	0.138	3.51	0.168	4.27						
84	2100	0.064	1.63	0.079	2.01	0.109	2.77	0.138	3.51	0.168	4.27		
90	2250	0.064	1.63	0.079	2.01	0.109	2.77	0.138	3.51	0.168	4.27		
96	2400	0.064	1.63	0.079	2.01	0.109	2.77	0.138	3.51	0.168	4.27		
102	2550	0.064	1.63	0.079	2.01	0.109	2.77	0.138	3.51	0.168	4.27		
108	2700	0.064	1.63	0.079	2.01	0.109	2.77	0.138	3.51	0.168	4.27		
114	2850	0.064	1.63	0.079	2.01	0.109	2.77	0.138	3.51	0.168	4.27		
120	3000	0.064	1.63	0.079	2.01	0.109	2.77	0.138	3.51	0.168	4.27		
126	3150	0.079	2.01	0.109	2.77	0.138	3.51						
132	3300	0.079	2.01	0.109	2.77	0.138	3.51	0.168	4.27				
138	3450	0.079	2.01	0.109	2.77	0.138	3.51	0.168	4.27				
144	3600	0.109	2.77	0.138	3.51	0.168	4.27						

Figure 11.3
Assembly of structural-plate pipe arches for an underground stormwater-detention facility

Manning's roughness values for corrugated-steel pipe vary considerably depending on the type of corrugation, pipe size, how full the pipe is, and whether or not it has asphalt lining. Table 11.3 provides data for selecting n-values for corrugated-steel pipe.

Joints for corrugated-steel pipes usually consist of bands that overlap at the ends of adjoining pipe sections. Mastic, sleeve-type, or O-ring gaskets are used with the bands to seal against leakage. For unusual conditions such as high pressures or in locations with high disjointing forces, alternate types of joints can be specified.

Table 11.3 Values of Manning's roughness coefficient (n) for standard corrugated-steel pipe (AISI, 1999)

	Condition		
Corrugation Type and Pipe Size (if applicable)	**Full Unpaved**	**25% Paved**	**Part Full Unpaved**
Annular 2⅔in. (68 mm), circular	0.024	0.021	0.027
Helical 1½ × ¼ in. (38 × 6.5 mm), circular			
8 in. (200 mm)	0.012	—	—
10 in. (250 mm)	0.014	—	—
Helical 2⅔ × ½ in. (68 × 13 mm), circular			
12 in. (300 mm)	0.011	—	0.012
15 in. (375 mm)	0.012	—	0.013
18 in. (450 mm)	0.013	—	0.015
24 in. (600 mm)	0.015	0.014	0.017
30 in. (750 mm)	0.017	0.015	0.019
36 in. (900 mm)	0.018	0.017	0.020
42 in. (1050 mm)	0.019	0.018	0.021
48 in. (1200 mm)	0.020	0.020	0.022
54 in. (1350 mm) and larger	0.021	0.018	0.023
Annular 2⅔ in. (68 mm), arch	0.026	—	0.029
Helical 2⅔ ×½ in. (68 ×13 mm), arch			
17 × 13 in. (430 × 330 mm)	0.013	—	0.018
21 × 15 in. (530 × 380 mm)	0.014	—	0.019
28 × 20 in. (710 × 510 mm)	0.016	—	0.021
35 × 24 in. (885 × 610 in.)	0.018	—	0.023
42 × 29 in. (1060 × 740 mm)	0.019	—	0.024
49 × 33 in. (1240 × 840 mm)	0.020	—	0.025
57 × 38 in. (1440×970 mm)	0.021	—	0.025
64 × 43 in. (1620×1100 mm) and larger	0.022	—	0.026
Annular 3 × 1 in. (75 × 25 mm), circular	0.027	0.023	—
Helical 3 × 1 in (75 × 25 mm), circular			
36 in. (900 mm)	0.022	0.019	—
42 in. (1050 mm)	0.022	0.019	—
48 in. (1200 mm)	0.023	0.020	—
54 in. (1350 mm)	0.023	0.020	—
60 in. (1500 mm)	0.024	0.021	—
66 in. (1650 mm)	0.025	0.022	—
72 in. (1800 mm)	0.026	0.022	—
75 in. (1950 mm) and larger	0.027	0.023	—
Annular 5 × 1 in. (125 × 25 mm), circular	0.025	0.022	—
Helical 5 × 1 in. (125 × 25 mm), circular			
48 in. (1200 mm)	0.022	0.019	—
54 in. (1350 mm)	0.022	0.019	—
60 in. (1500 mm)	0.023	0.020	—
66 in. (1650 mm)	0.024	0.021	—
72 in. (1800 mm)	0.024	0.021	—
75 in. (1950 mm) and larger	0.025	0.022	—
All pipe with smooth interior[a]		All diameters 0.012	

a.Includes full-paved, concrete-lined, spiral-rib, and double-wall pipe.

Storm Sewer Appurtenances

Storm sewer systems consist of a number of appurtenances other than the sewer conduits (pipes). Among these appurtenances are inlet boxes and catch basins; manholes (access holes) and junction structures; and outlet structures and associated energy-dissipation devices. Increasingly, additional types of appurtenances are finding their way into storm sewer systems, primarily in response to water-quality concerns.

Many of the elements described in this section were introduced in Section 2.2, which included several photographs. This section provides more detailed descriptions.

Inlet Boxes. An inlet box is a structure that supports an inlet opening and connects it to the subsurface piping system. Inlet boxes also allow access to the subsurface system for cleaning and inspection. The walls of the inlet box may be precast or cast-in-place concrete, brick, masonry block, or vertically mounted corrugated pipe. Three typical inlet box configurations are shown in Figure 11.4.

Ideally, the base of an inlet box will be shaped and formed to more smoothly channel flows from the inlet pipe(s) to the outlet pipe. As described in Section 11.4 (page 437), shaping of the base minimizes minor energy losses due to flow expansion and contraction within the structure. In new construction, where concrete is used almost exclusively, experienced construction personnel can pour and finish a shaped base very quickly and without concrete forms. When a precast-concrete structure is used, ready-mix concrete can be poured into the base and shaped after the precast section has been set in place.

Catch Basins. Although the two terms are often used synonymously, stormwater inlets and catch basins are not the same. Both are intended to convey surface water into the storm sewer pipe system, and from the surface they appear to be identical. However, a catch basin differs from an inlet in the important respect that it has a *sump,* which is a portion of the subsurface structure that is lower than the invert of the outlet pipe. Also, a catch basin sometimes has one or more baffles (such as the hinged cast iron hood in Figure 11.5) to enhance capture of suspended sediment and other debris in the stormwater before it can enter the outlet pipe. In some cases, outlet pipes from catch basins may be equipped with traps to prevent objectionable sewer odors, but this is normally required only for combined sewers. Catch basins offer water-quality benefits not offered by regular inlets, but they require more regular maintenance to keep them free of trapped debris.

Manholes. Like inlets and catch basins, *manholes* (or *access holes*) can provide access to the sewer system for routine inspection and maintenance. Manholes are usually installed where there is a change in horizontal (plan-view) pipe direction or pipe slope, where several pipes join, or where the pipe size changes. However, manholes are typically used for these purposes only where use of an inlet box would be inappropriate. Manholes or inlet boxes should also be installed to provide regular access intervals along straight sections of sewer. Recommended maximum distances between access locations depend on pipe diameter and are indicated in Table 11.4.

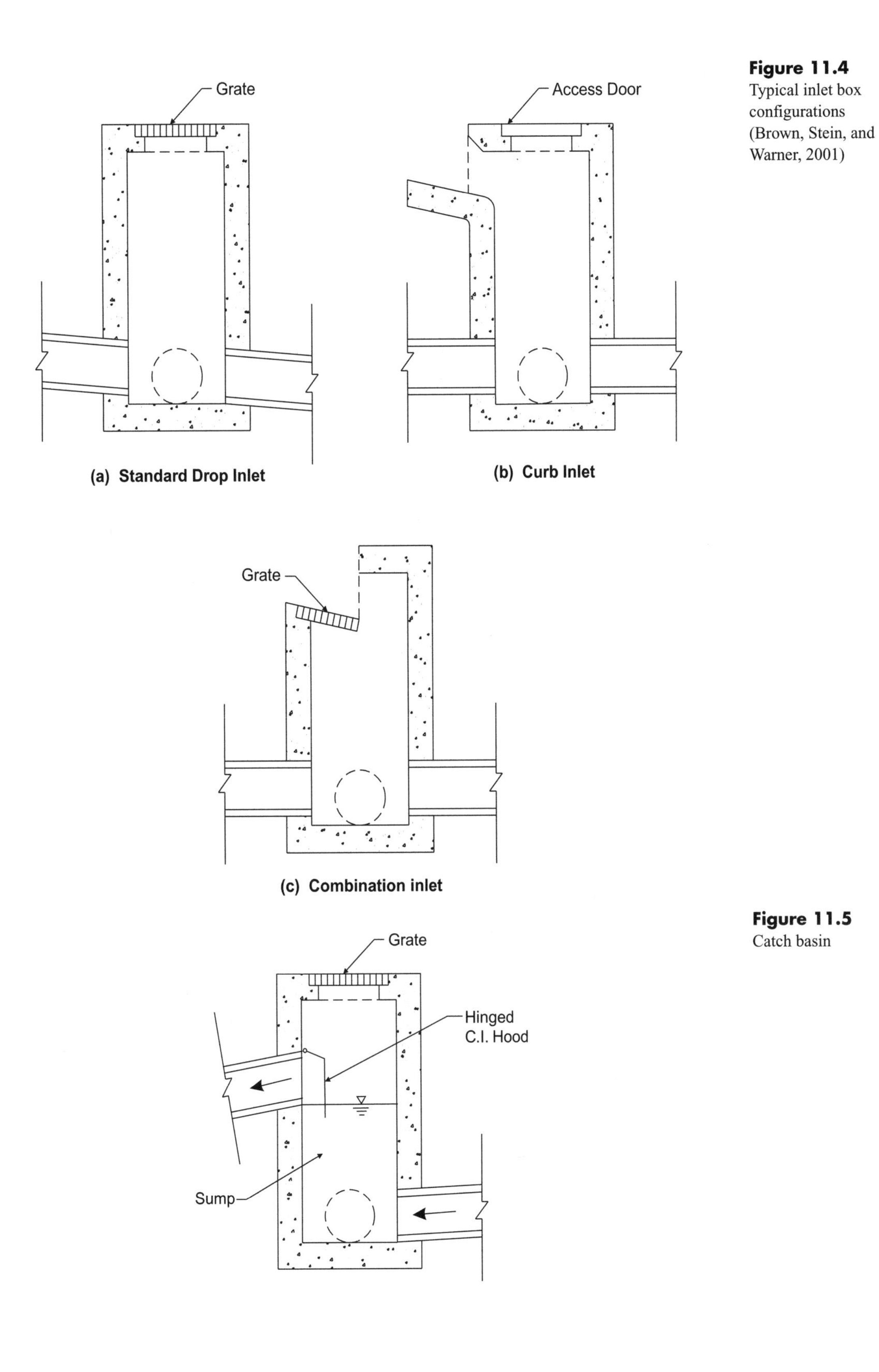

Figure 11.4 Typical inlet box configurations (Brown, Stein, and Warner, 2001)

Figure 11.5 Catch basin

Table 11.4 Manhole (or inlet box) spacing criteria (adapted from Brown, Stein, and Warner, 2001)

Pipe Diameter		Maximum Spacing	
in.	mm	ft	m
12–24	300–600	300	91
27–36	675–900	400	122
42–54	1050–1350	500	152
60 and larger	1500 and larger	1000	305

Illustrations of precast-concrete manhole assemblies are provided in ACPA (2000) (see Figure 11.6). The *riser* or barrel of a manhole typically has a circular cross section, though not necessarily so. In old construction, manhole barrels were usually constructed of brick and mortar. The risers of new manholes are almost exclusively precast concrete with bituminous (mastic) joint fillers, though some are constructed of corrugated pipe. Internal diameters of manhole risers are usually 4 ft (1.2 m), but precast sections are available in other diameters. Manhole steps (usually aluminum) should be provided to facilitate access.

In new construction, the upper portion of the subsurface structure usually consists of a precast *eccentric cone* to transition between the diameter of the riser and the diameter of the opening. For applications where the depth of cover over the sewer is limited, precast flat-slab tops may be used instead. Precast-concrete *grade rings* or courses of brick and mortar may be placed on top of either the cone or flat slab to serve as final grade adjustments for the cast-iron manhole *frame and cover*. Where applicable, frame and cover should be rated for heavy-duty service and should be able to withstand heavy vehicular traffic loadings.

Typically, a manhole *base* is made of precast concrete, and it has holes that have been appropriately sized and located for connection of the storm sewer piping. As with inlet boxes, it is preferable for the base of the manhole to be formed to minimize energy losses through the structure. A *transition section* connects the base to the riser. For large-diameter sewers, manholes may be formed using special *tee sections* in the sewer pipe (see Figure 11.7). In such cases, minor energy losses through the manhole structure are negligible, particularly when the sewer flow is less than full.

Junction Structures. *Junction structures*, like manholes, are used where two or more pipes join one another. However, junction structures tend to be associated with large-pipe installations for which more-conventional precast manhole sections are not large enough, or with installations having complicated connecting-pipe alignments and geometries. Junction structures are usually constructed of cast-in-place reinforced concrete, although specialized corrugated-steel junctions can also be fabricated. Many junction structures do not provide direct access to the surface; however, manhole structures may be integrated for this purpose.

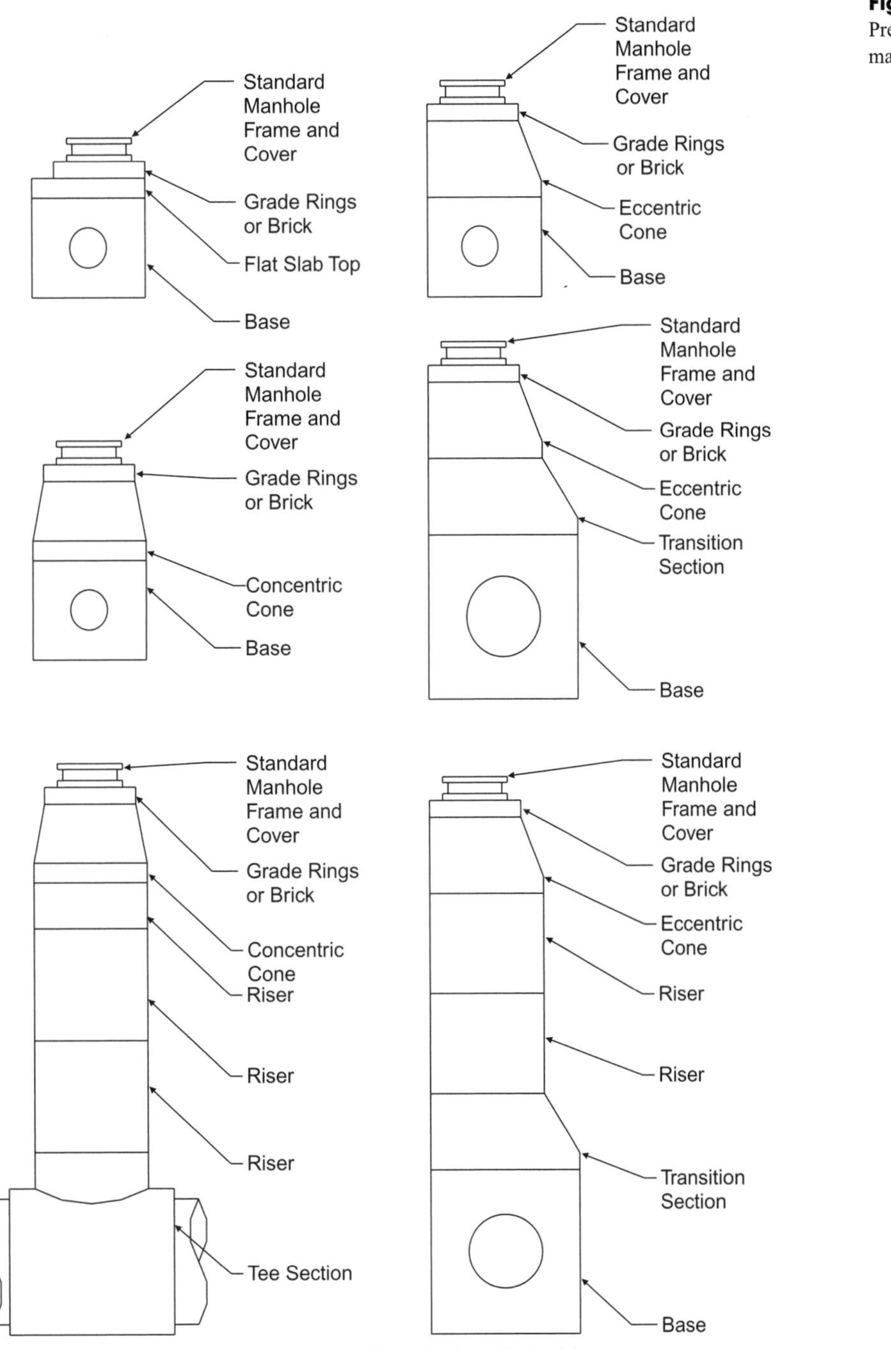

Credit: Reproduced by permission of the American Concrete Pipe Association.

Figure 11.6
Precast-concrete manhole assemblies

Figure 11.7
Large-diameter manhole construction

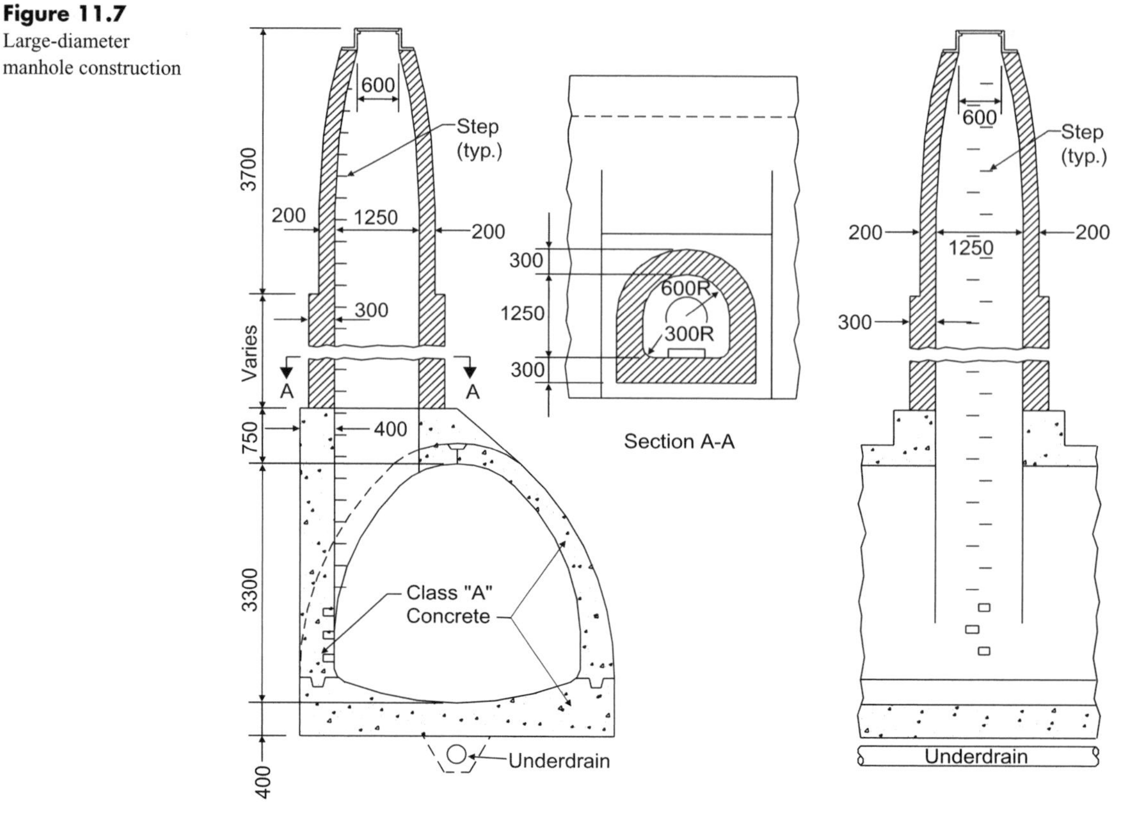

Credit: Steel and McGhee, 1979. Reproduced with permission of the McGraw-Hill Companies.

Outlet Structures. Exit velocities at storm sewer outfalls are often high enough to cause erosion problems and even undermining of the outfall pipe. *Outlet structures* can be designed to limit or prevent undermining by dissipating flow energy. As described in Section 11.7 (page 464), an outlet structure may consist of a simple concrete endwall with riprap and a filter blanket downstream of the outfall point, or a more complex *energy dissipator* or *stilling basin* configuration may be used. Prefabricated, flared end-sections are also available in both concrete and corrugated-pipe materials. In any case, it is recommended that a concrete *toe-wall* (also called a *cradle* or *cut-off wall*) similar to that in Figure 9.5 (page 330) and Figure 11.11 (page 465) be placed at the outfall point to prevent undermining of the pipe and/or outlet structure. Sometimes, a storm sewer may outfall to a water body whose water-surface elevation may, at times, be high enough to back up and flood the system. This backup can be prevented by a valve at the outlet, such as a flap gate or duckbill check valve.

Other Structures. Many types of structures and facilities other than those described in the previous subsections can be included in storm sewer systems. Underground detention basins, for example, may be fitted with manifold-like distribution chambers to distribute incoming flows to multiple storage chambers or to collect outflows from multiple storage chambers. Increasing environmental and water-quality concerns have given rise to the use of water-quality-enhancement structures, including filters and vortex-type oil and sediment separators.

11.2 ALIGNMENT OF STORM SEWERS

In an urbanized setting where much of the utility infrastructure is buried in public rights of way, a degree of standardization for utility locations within the right of way can minimize potential conflicts between the alignments of the various utility lines and can make maintenance operations more efficient. Locations of utilities such as sanitary sewers, potable-water lines, electrical lines, and gas lines are more easily standardized because they are present in nearly every public street right-of-way and, except for sanitary sewers, are not gravity-driven. Nevertheless, consideration should also be given to standardizing storm sewer alignments.

Horizontal Alignment

The horizontal (plan-view) alignment of a storm sewer normally should be straight between manholes, inlets, and other structures. This restriction may necessitate very close manhole spacing in roadway curves with small radii.

Curved storm sewers may be acceptable in some instances if local jurisdictions permit them, but should not be used with small-diameter pipes because they tend to cause maintenance problems. Curved sewer alignments are usually constructed by deflecting pipe joints when they are assembled. Associations of pipe manufacturers provide design guidance on maximum deflection angles.

Storm sewers may be located behind the curb and outside of paved roadway areas, or they may be located under the pavement at a standard distance from the curb line. Installation behind the curb means less disruption when excavating and repairing the sewer; however, this option often requires that temporary or permanent easements be obtained from private properties adjacent to the public right of way.

Some review agencies suggest that a minimum horizontal distance of about 10 ft (3 m) be maintained between the centerlines of potable water and storm sewer pipes. In cases of large-diameter pipes, this centerline-to-centerline distance should be increased to maintain a minimum separation distance of 5–6 ft between the pipes.

Vertical Alignment

The vertical alignment of a storm sewer should maintain a depth of cover sufficient to prevent the crushing of pipes by traffic loads. Methods for developing detailed estimates of loads and pipe strengths are published by associations of pipe manufacturers, although a depth of cover of about 3 ft (0.9 m) is common. The vertical alignment of a storm sewer should be shown on a profile drawing and should be laid out to minimize conflicts with other buried utilities. Vertical separation distances between storm sewers and other buried utility lines should be at least 1 ft (0.3 m) to allow for differential settlement. Where groundwater tables are high, or where bedrock is encountered at shallow depths, the alignment should also strive to minimize potential buoyancy problems and/or construction difficulties caused by these factors.

There are a few approaches for setting the relative invert elevations of pipes entering and leaving a manhole or junction structure. For many engineers, the traditional rule of thumb has been to match the crown elevations of incoming and outgoing pipes.

Benefits of this approach are said to be that the drop in elevation helps to offset energy losses through the structure, and the likelihood of having pressurized flow in a smaller upstream pipe is reduced. An alternative is to match pipe invert elevations, or provide a small minimum drop in elevation across the structure. When inverts are matched, the bottom of the structure can be formed to convey some flows with less energy loss. Finally, others assert that the most hydraulically efficient configuration is to match pipe springline (centerline) elevations.

In reality, the hydraulic benefit of any of these strategies over another is probably negligible. From a least-cost perspective, if a minimum cover constraint is driving the elevations of pipe inverts, then matching pipe crowns will typically result in the least amount of excavation. However, if the ground surface is fairly flat, or if the storm sewer slopes in the direction opposite that of the ground slope, then minimum pipe slope requirements will typically drive the invert elevations. In this situation, matching pipe inverts will minimize excavation. In the absence of regulations to the contrary or extenuating circumstances, the engineer may choose to match inverts to minimize construction costs in this case.

Sewer Slopes

The slope of a storm sewer pipe should be sufficient to maintain a self-cleansing flow velocity, thus preventing sediment deposition that can lead to maintenance problems. According to ASCE (1992), the minimum flow velocity should be about 2 to 3 ft/s (0.6 to 0.9 m/s) when the sewer is flowing full. To prevent erosion of the pipe interior by suspended sediment and debris, the full-flow velocity should be less than approximately 15 ft/s (4.5 m/s).

For any given pipe diameter and material, a constitutive relationship such as the Manning equation can be applied to determine the minimum pipe slope required to maintain a self-cleansing, full-flow velocity. Storm sewers flow full only occasionally, but it is typically only during storm events that they convey a flow at all. Therefore, checking minimum velocities for non-full-flow conditions is not as important as it is, for example, in sanitary sewer design. Table 11.5 is a listing of minimum pipe slopes required to maintain full-flow velocities of 2 ft/s (0.6 m/s) and 3 ft/s (0.9 m/s) for concrete and corrugated-metal pipes. Manning's n values used to construct the table are $n = 0.013$ for concrete and $n = 0.024$ for corrugated metal. Note that, with few exceptions, all of the slopes in the table are quite small—less than minimum slopes normally specified for construction quality control. Thus, in most cases of practical interest, self-cleansing velocity criteria will be satisfied regardless of the pipe slope specified by the designer.

11.3 INITIAL PIPE SYSTEM DESIGN

Following the selection of a design storm return period, which will frequently be determined by the local regulating authority, and the computation of flows intercepted by each inlet in the storm sewer system (see Section 10.3), one can proceed with an *initial design* of the pipes in the system. The purpose of the initial design is to estimate the design flow for each pipe in the system and make an initial selection of the size required for each pipe.

Table 11.5 Minimum slope required to maintain pipe cleansing velocities for pipes flowing full

Internal Diameter		Pipe Slope Required to Maintain V_{min}			
		Concrete Pipe (n = 0.013)		Corrugated-Metal Pipe (n = 0.024)	
in.	mm	V_{min} = 2 ft/s (0.6 m/s)	V_{min} = 3 ft/s (0.9 m/s)	V_{min} = 2 ft/s (0.6 m/s)	V_{min} = 3 ft/s (0.9 m/s)
12	300	0.0019	0.0044	0.0066	0.0149
15	375	0.0014	0.0032	0.0049	0.0111
18	450	0.0011	0.0025	0.0039	0.0087
21	525	0.0009	0.0021	0.0031	0.0071
24	600	0.0008	0.0017	0.0026	0.0059
30	750	0.0006	0.0013	0.0020	0.0044
36	900	0.0004	0.0010	0.0015	0.0034
42	1050	0.0004	0.0008	0.0012	0.0028
48	1200	0.0003	0.0007	0.0010	0.0023
54	1350	0.0003	0.0006	0.0009	0.0020
60	1500	0.0002	0.0005	0.0008	0.0017
66	1650	0.0002	0.0005	0.0007	0.0015
72	1800	0.0002	0.0004	0.0006	0.0014
78	1950	0.0002	0.0004	0.0005	0.0012
84	2100	0.0001	0.0003	0.0005	0.0011
90	2250	0.0001	0.0003	0.0005	0.0010
96	2400	0.0001	0.0003	0.0004	0.0009

The initial design begins with the most upstream pipe(s) and progresses in a downstream direction. The design discharge for each pipe includes any flow from the inlet at the pipe's upstream end and the flow from the upstream piping system. Typically, initial pipe sizes are determined by assuming that the pipe slope will be approximately the same as that of the ground surface, subject to some minimum slope constraint. The Manning equation or another method is then used to solve for the required pipe size given the pipe roughness, design discharge, and slope. The calculated size is then rounded up to the next commercially available size that is at least as large as the minimum allowable size.

The required pipe sizes are determined as each pipe is considered in turn. When sizing pipes for a storm sewer, the diameter of a pipe exiting a structure should not be smaller than the largest incoming pipe, regardless of its conveyance capacity. In other words, pipe diameters should never decrease downstream, even when pipe slopes may be such that a smaller downstream pipe would have sufficient capacity.

Design discharges are usually determined using the rational method (see Section 5.4, page 140), although other methods, such as NRCS (SCS) methods, may be used instead. Example 11.1 presents a simplified but realistic calculation of storm sewer design flows and initial pipe sizing for a typical case using the rational method. In this example, the travel time calculations are simplified by assuming the pipes flow full.

The next two subsections clarify the initial design process for special cases involving additional considerations in the computation of design flows. Example 11.2 provides a detailed initial design, including inlet flow calculations, inlet design, pipe sizing, and more accurate travel time calculations that consider velocities for partially full pipes.

Based on hydraulic and energy grade line computations and possible constraints on pipe slopes and vertical alignments, the initially selected pipe sizes may require changes during the *final design* process, which is presented in Section 11.5.

Example 11.1 – Computing Flows for Storm Sewer Pipes with the Rational Method. Figure E11.1.1 is a plan view of a storm sewer system draining three subbasins. Use the rational method to determine the peak discharge in each pipe given that the inlets do not bypass any flow, and size each pipe assuming that the pipes flow full. The tables that follow the figure list the IDF data and subbasin and pipe characteristics. Assume also that the pipes will be concrete with $n = 0.013$.

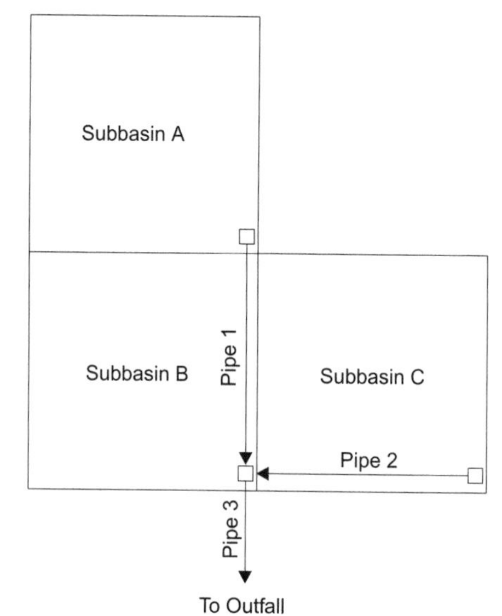

Figure E11.1.1 System for Example 11.1

Table 11.6 IDF data for Example 11.1

Duration (min)	Intensity (mm/hr)
5	213.0
10	178.0
15	151.0
30	112.0
60	75.0

Table 11.7 Subbasin characteristics for Example 11.1

Subbasin	A (ha)	C	t_c (min)
A	2.5	0.6	20
B	1.5	0.8	10
C	2.0	0.8	15

Table 11.8 Pipe characteristics for Example 11.1

Pipe	Length (m)	Slope (%)
1	150	1.0
2	120	1.2
3	150	0.9

Flow into Pipe 1 occurs from Subbasin A only. Using the time of concentration as the storm duration, the 25-year rainfall intensity is obtained from Table 11.6 as $i = 133.6$ mm/hr. The peak discharge used in sizing Pipe 1 is therefore

$$Q = 0.6(2.5)(133.6) = 200.4 \text{ ha-mm/hr} = 0.56 \text{ m}^3\text{/s}$$

Assuming that Pipe 1 is flowing full, its required diameter D may be found using Manning's equation as follows:

$$D = \left(\frac{Qn}{0.312S^{1/2}}\right)^{3/8} = \left(\frac{.56(0.013)}{0.312(0.01)^{1/2}}\right)^{3/8} = 0.58 \text{ m}$$

Rounding up to the next commercially available size, Pipe 1 should have a diameter of 600 mm.

The cross-sectional area of Pipe 1 is therefore 0.28 m^2, and the average velocity in Pipe 1 is

$$V = Q/A = 0.55/0.28 = 2.0 \text{ m/s}$$

The travel time in Pipe 1 is

$$t = L/V = 150/2.0 = 75 \text{ s} = 1.3 \text{ min}$$

Note that the travel time is roughly estimated by assuming that the pipe flows full.

Pipe 2 is treated the same way as Pipe 1, recognizing that runoff from Subbasin C only enters Pipe 2. The peak discharge from Subbasin C is $Q = 0.62$ m^3/s, and the required diameter of Pipe 2 is $D = 600$ mm. The travel time in Pipe 2 is $t = 54$ s = 0.9 min. The cross-sectional area of Pipe 2 is A = 0.28 m^2.

Pipe 3 must be sized to handle the runoff from all three of the subbasins, which have a total area of 6.0 ha. The runoff coefficient for the combined areas is computed as a composite value, and is

$$C = \frac{2.5(0.6) + 1.5(0.8) + 2(0.8)}{6} = 0.72$$

The time of concentration is computed as the longest of the travel times to the upstream end of Pipe 3. These travel times are:

The time of concentration of Subbasin B (10 minutes).

The time of concentration of Subbasin A plus the travel time in Pipe 1 (20 + 1.3 = 21.3 min).

The time of concentration of Subbasin C plus the travel time in Pipe 2 (15 + 0.9 = 15.9 min).

Thus, the time of concentration for Pipe 3 is 21.3 min, and the corresponding rainfall intensity is 134.6 mm/hr. The peak discharge for Pipe 3 is

$$Q = 0.72(134.6)(6) = 581.5 \text{ ha-mm/hr} = 1.62 \text{ m}^3\text{/s}$$

The required diameter of Pipe 3 is 880 mm. Rounding up to the next available size yields a diameter of 900 mm.

Multiple Design-Storm Return Periods

Frequently, a storm sewer system serves a drainage basin in which multiple land uses are present. For example, one part of a storm sewer system may serve a residential area, while another branch of the same system may serve a commercial area such as a local shopping center. For such a case, regulations might state that the residential part of the system should be designed for a 2-year return period, and the commercial portion for a 10-year period.

Because high-return-period flows are larger than short-return-period flows, the portion of the overall system that must convey runoff from both types of land uses should be designed using the higher flows. Portions of the system that serve only single land uses should be designed according to their respective return periods.

In the final design phase when hydraulic and energy grade lines are computed, calculations for each design storm should be performed. First, the entire system should be analyzed for lower-return-period discharges, even though some parts of the system were initially designed for higher discharges. The final set of computations then analyzes the entire system for high-return-period discharges. In this calculation run, it is expected that the portion of the system designed for low-return-period discharges will be inadequate to convey the higher flows, and therefore the major drainage system will be activated in that portion of the system. This multi-step procedure is necessary to ensure that the portion of the system serving multiple land uses is capable of conveying overall system inflows corresponding to the highest applicable design storm return period.

Directly and Indirectly Connected Areas

The rational method for design flow estimation, as presented in Example 11.1, is applicable in instances where the drainage basin contributing runoff to an inlet is homogeneous in terms of its land use characteristics and small enough that the rational method is considered appropriate (local practices vary regarding acceptable maximum basin size). However, many urban drainage areas are non-homogeneous in the sense that the impervious areas of the drainage basin contributing to each inlet can be subdivided into *directly connected areas* and *indirectly connected areas*. Runoff from directly connected impervious areas such as streets and driveways proceeds directly to the nearest conveyance (storm sewer, drainage ditch, etc.) without traversing a pervious area such as a lawn. Runoff from indirectly connected impervious areas; however, flows across impervious areas where it has the chance to at least partially infiltrate before reaching an inlet. An example of an indirectly connected impervious area is a roof drained by a downspout that discharges onto a lawn.

Although the directly connected impervious area of a drainage basin is smaller than the total basin area, the peak rate of runoff from the directly connected area alone may be larger than the peak rate of runoff from the basin as a whole. This discrepancy occurs because the time of concentration associated with the directly connected area is usually smaller than the time of concentration for the basin as a whole, which translates to a higher design storm intensity. In such an instance, design flows to each inlet and pipe should be computed twice, once assuming that only the directly connected area contributes runoff, and a second time assuming that the entire drainage basin contributes runoff. The larger of the runoff estimates may then be taken as a conservative estimate of the design discharge.

The initial design process, including consideration of directly connected impervious areas, is illustrated in Example 11.2.

Example 11.2 – Inlet Design and the Initial Piping Design Process. The layout of a proposed storm sewer system to serve three city blocks of single-family residential lots is illustrated in Figure E11.2.1. Design the system for a 2-year event using the IDF data in the table below.

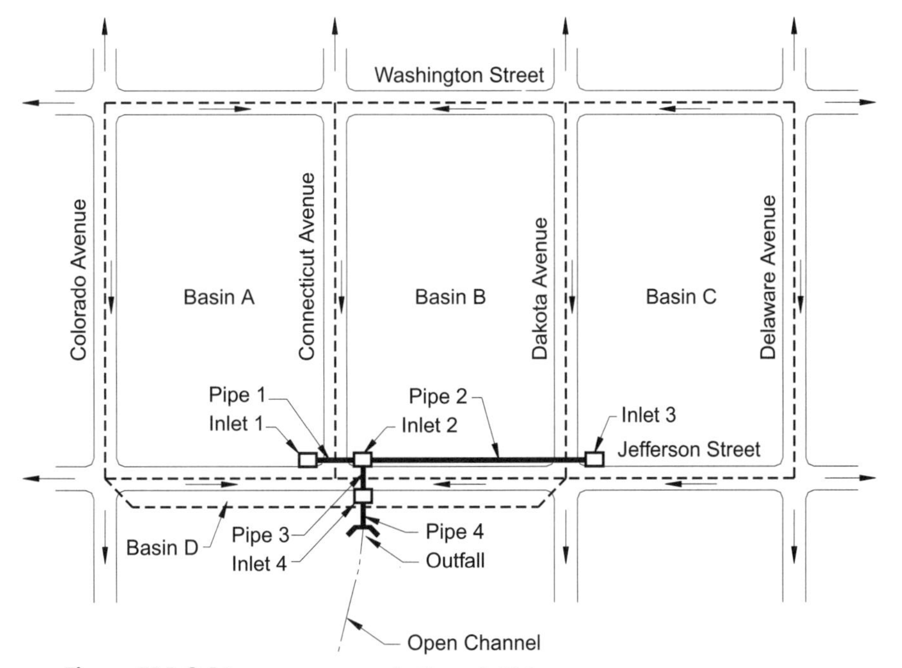

Figure E11.2.1 Storm sewer system for Example 11.2

Duration (min)	Intensity (in/hr)
5	5.88
10	4.80
15	4.04
30	2.78
60	1.79

Drainage area data

Each of the city blocks (that is, Basins A through C) is 330 ft by 660 ft, which corresponds to a total drainage area of 5.0 ac for each, of which impervious areas ($C = 0.95$) total 1.7 ac. The remaining area of each block consists of lawns and landscaped areas with C = 0.15. Of the 1.7 ac of impervious area, 1.0 ac is directly connected. The time of concentration is 20 minutes for each 5.0 ac area, or 10 minutes if only the directly connected impervious area is considered.

Basin D consists of one-half of the right-of-way of Jefferson St. between Colorado Ave. and Dakota Ave. It has an area of 0.45 ac, a runoff coefficient of 0.95, and a time of concentration of 5 minutes.

Gutter and inlet design criteria

Street slopes are $S_o = 0.50$ percent in the directions shown in Figure E11.2.1. The composite gutters have a width $W = 2$ ft, a curb height of 6 in., and a gutter cross slope of $S_w = 6.25$ percent (see Figure E11.2.2). Street cross slopes are $S_x = 2$ percent. The maximum allowable spread T is 15 ft. Manning's $n = 0.015$ for the gutters.

All of the inlets shown in Figure E11.2.1 are curb-opening inlets in sag locations. Additional inlets, if required on the basis of gutter flow calculations, will be continuous-grade inlets. The inlets have horizontal throats with h = 5 in., and the gutters are depressed an additional 2 in. at inlet locations (see Figure E11.2.3). Required inlet lengths should be increased by 25 percent to account for potential clogging. The minimum permissible inlet length is 3 ft. The inlet throat bottom elevations are given in the following table.

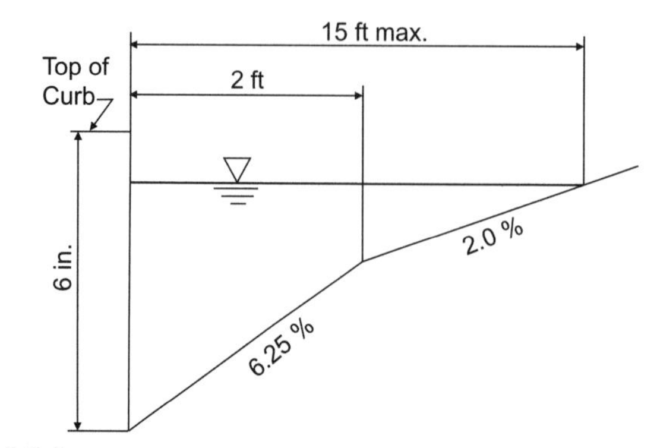

Figure E11.2.2 Gutter cross-section for Example 11.2

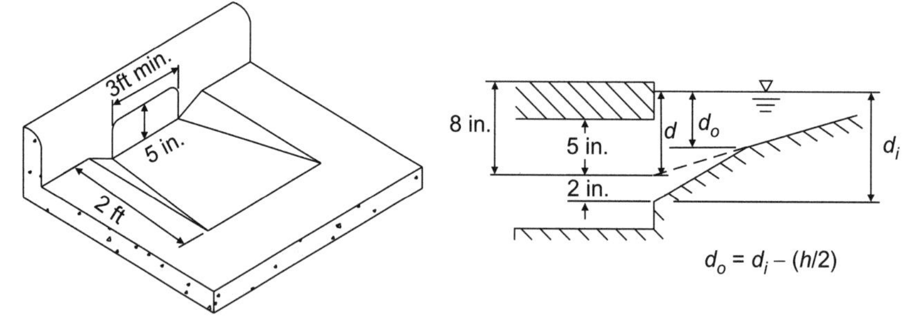

Figure E11.2.3 Curb inlet for Example 11.2

Inlet Label	Throat Elevation (ft)
Inlet 1	212.20
Inlet 2	211.73
Inlet 3	213.50
Inlet 4	211.85

Pipe design criteria

Pipes are to be reinforced concrete with n = 0.013. The minimum acceptable pipe diameter, as dictated by local regulations, is 15 in. Pipe crowns should be matched at junctions where incoming and outgoing pipes are of different diameters. Based on the flatness of the local topography, pipe slopes should be no greater than the minimum allowable slope of 0.5 percent specified by the local review agency. Pipe 4 outfalls to an open channel with an invert elevation of 205.73 ft. This controlling outfall elevation is used with the pipe slopes and crown matching criteria to establish pipe invert elevations throughout the sewer system. Pipe lengths are provided in the following table.

Pipe Label	Length (ft)
Pipe 1	70
Pipe 2	315
Pipe 3	40
Pipe 4	40

Solution: The design proceeds by determining, first, the design flow rates and street conveyance capacities in order to evaluate whether the placement of inlets shown in Figure E11.2.1 is adequate. Inlets are then designed to intercept the design flows, and pipes are designed to convey flows to the system outfall point.

Step 1: Compute design flows for inlets

Because the drainage basins A, B, and C are essentially identical in terms of their hydrologic characteristics, the peak runoff rate computed for one is also applicable to the other two. Noting that impervious areas in each basin consist of both directly-connected and indirectly-connected areas, the peak runoff calculations must be performed twice.

The first set of peak runoff calculations for basins A, B and C assumes that only the directly-connected impervious areas are contributing runoff to the basin outlet. Thus, the area and time of concentration for use in the rational method are A = 1.0 ac and t_c = 10 min. The rainfall intensity corresponding to a t_c of 10 minutes is i = 4.80 in/hr, and the peak runoff rate is

$$Q = CiA = 0.95(4.80 \text{ in/hr})(1.0 \text{ ac}) = 4.6 \text{ cfs}$$

The second set of peak runoff calculations for Basins A through C computes the runoff from the entire basin area of A = 5.0 ac. The composite runoff coefficient for the entire basin is

$$C = [(1.7 \text{ ac})(0.95) + (5.0 \text{ ac} - 1.7 \text{ ac})(0.15)]/5.0 = 0.42$$

The time of concentration of the entire basin is 20 minutes; thus, i = 3.62 in/hr. The peak runoff rate is

$$Q = CiA = 0.42(3.62 \text{ in/hr})(5.0 \text{ ac}) = 7.6 \text{ cfs}$$

The larger of the two peak discharges, 7.6 cfs, controls the design of the gutters, inlets, and pipes.

For Basin D, which consists only of the right-of-way of Jefferson St., A = 0.45 ac, C = 0.95, and t_c = 5 minutes. The rainfall intensity corresponding to a 5-minute duration is i = 5.88 in/hr. The peak discharge from Basin D is therefore

$$Q = CiA = 0.95(5.88 \text{ in/hr})(0.45 \text{ ac}) = 2.5 \text{ cfs}$$

Step 2: Perform gutter capacity calculations

The total discharge of 7.6 cfs approaching each of the Inlets 1 through 3 arrives at each inlet by two routes. For example, in Basin A, flows approach Inlet 1 via both Jefferson St. and Connecticut Ave. Assuming that the flows in each approaching curb are more or less equal (a reasonable assumption for this example because of the basin geometries), the flow in each approach gutter is about 3.8 cfs. Based on similar reasoning, the flow approaching Inlet 4 from each direction in Jefferson St. is about 1.25 cfs. The gutter capacities must be checked to ensure that these discharges do not exceed permissible flow depth or spread limits.

Design criteria provide for a maximum permissible spread of T = 15 ft. With a gutter width of W = 2 ft, the allowable side flow width from Equation 10.6 is

$$T_s = T - W = 15 - 2 = 13 \text{ ft}$$

The discharge capacity Q_s in the roadway area can be determined using Equation 10.9 as

$$Q_s = \frac{0.56}{n} S_x^{5/3} T_s^{8/3} S_o^{1/2} = \frac{0.56}{0.015}(0.02)^{5/3}(13)^{8/3}(0.005)^{1/2} = 3.6 \text{ cfs}$$

The total discharge capacity Q of the gutter can be determined using Equations 10.11 and 10.12. The fraction E_o is determined from Equation 10.12 to be

$$E_o = \left\{ 1 + \frac{S_w / S_x}{\left(1 + \frac{S_w / S_x}{(T/W) - 1}\right)^{8/3} - 1} \right\}^{-1} = \left\{ 1 + \frac{0.0625/0.02}{\left(1 + \frac{0.0625/0.02}{(15/2) - 1}\right)^{8/3} - 1} \right\}^{-1} = 0.372$$

Then, from Equation 10.11, the total discharge capacity is found as

$$Q = \frac{Q_s}{1-E_o} = \frac{3.6 \text{ cfs}}{1-0.372} = 5.7 \text{ cfs}$$

Because the gutter capacity of 5.7 cfs is greater than the peak runoff rate (3.8 cfs or 1.25 cfs, depending on location), the adequacy of the gutters is established and no additional inlets are required.

Step 3: Perform Inlet Design

Because the gutter is depressed at the inlet location, Equation 10.33 should be used to estimate a required inlet length under the assumption that it behaves as a weir. The weir coefficient is $C_w = 3.0$, and the gutter width is $W = 2$ ft. The depth d required for the expression (see Figure E11.2.3) is

$$d = TS_x = 15(0.02) = 0.30 \text{ ft}$$

Therefore, rearranging Equation 10.32 to solve for L, the required length for Inlets 1 through 3 (Q = 7.6 cfs) is

$$L = \frac{Q}{C_w d^{3/2}} - 1.8W = \frac{7.6}{3.0(0.30)^{3/2}} - 1.8 \times 2 = 11.8 \text{ ft}$$

The required length of Inlet 4 (Q = 2.5 cfs) is similarly calculated as

$$L = \frac{Q}{C_w d^{3/2}} - 1.8W = \frac{2.5}{3.0(0.30)^{3/2}} - 1.8 \times 2 = 1.5 \text{ ft}$$

To find the required inlet length assuming orifice flow, it is necessary to first compute the flow depth at the curb face d_i, which can be used to find the depth of the centroid of the inlet throat opening, d_o (see Figure E11.2.3):

$$d_i = T_s S_x + WS_w + (\text{2-in. inlet depression}) = 13(0.02) + 2(0.0625) + 2/12 = 0.55 \text{ ft}$$

$$d_o = d_i - h/2 = 0.55 - (5/2)/12 = 0.34 \text{ ft}$$

Then, the required length of Inlets 1 through 3 can be determined by solving Equation 10.34 for length:

$$L = \frac{Q}{0.67h(2gd_o)^{1/2}} = \frac{7.6}{0.67(5/12)[2 \times 32.2 \times 0.34]^{1/2}} = 5.8 \text{ ft}$$

The required length of Inlet 4 is

$$L = \frac{Q}{0.67h(2gd_o)^{1/2}} = \frac{2.5}{0.67(5/12)[2 \times 32.2 \times 0.34]^{1/2}} = 1.9 \text{ ft}$$

For each inlet, the longer of the two calculated inlet lengths should be used. Thus, for Inlets 1 through 3, weir behavior controls the design, and the required length of each inlet is 11.8 ft. Orifice behavior controls the length of Inlet 4; its required length is 1.9 ft. Increasing required inlet lengths by 25 percent to account for clogging, the required length (rounded to the nearest foot) for Inlets 1 through 3 is

$$L = 1.25(11.8) = 15 \text{ ft}$$

and for Inlet 4 is

$$L = 1.25(1.9) = 2 \text{ ft}$$

Invoking the design criterion regarding minimum permissible inlet length, an inlet of length $L = 3$ ft will be used for Inlet 4.

Step 4: Design Pipes

Pipes 1 and 2 are terminal pipes in the system in the sense that each receives inflow from only one inlet. The design discharge for each of these pipes is therefore $Q = 7.6$ cfs, and the pipes can be sized to flow full under design flows. Using concrete pipes with $n = 0.013$ and $S_o = 0.005$, the required pipe diameters can be found by rearranging the form of the Manning equation given in Equation 6.18:

$$V = \frac{C_f}{n} R^{2/3} S_o^{1/2}$$

The velocity V can be rewritten as

$$V = \frac{Q}{A} = \frac{Q}{\pi D^2 / 4}$$

where D is the pipe diameter, and the hydraulic radius R is

$R = D/4$

Rearranging to solve for diameter D and using substitution, the Manning equation becomes:

$$D = \left(\frac{nQ}{0.464 S_o^{1/2}} \right)^{3/8} = \left(\frac{0.013(7.6)}{0.464(0.005)^{1/2}} \right)^{3/8} = 1.511$$

Rounding up to the next commercially available size, Pipes 1 and 2 should have diameters of 21 in.

Pipe 3 carries flows from both pipes 1 and 2 (that is, Basins A and C), plus additional flow entering locally at that inlet from Basin B. Its design discharge can be found by applying the rational method to the total area of Basins A, B, and C (15 ac) with a composite runoff coefficient and a controlling time of concentration. Because Basins A, B, and C are nearly identical, the C value of 0.42 computed previously applies.

The controlling time of concentration is the longest travel time to the upstream end of Pipe 3, which in this case would be the longest of the following:

a) The time of concentration to Inlet 1 plus the travel time of flow in Pipe 1

b) The time of concentration to Inlet 3 plus the travel time of flow in Pipe 2

c) The time of concentration to Inlet 2

The time of concentration to Inlet 2 is not larger than the others, the velocities in Pipes 1 and 2 can be assumed to be equal in this case, and Pipe 2 is longer than Pipe 1. Therefore, the longest travel time approaching Inlet 2 must be from Basin C/Inlet 3, and the velocity in Pipe 2 must be determined.

The velocity in a circular pipe flowing partially full can be determined with the aid of Figure 7.8, which is shown again here as Figure E11.2.4. Using the Manning equation, the full-flow velocity V_{full} for a 21-in. pipe is

$$V_{full} = \frac{1.49}{n} R^{2/3} S_o^{1/2} = \frac{1.49}{0.013} \left(\frac{21/12}{4} \right)^{2/3} 0.005^{1/2} = 4.67 \text{ ft/s}$$

and the full-flow capacity Q_{full} is

$Q_{full} = AV_{full} = 4.67[\pi(21/12)^2/4] = 11.2$ cfs

Thus, for pipes 1 and 2,

$Q/Q_{full} = 7.6/11.2 = 0.68$

From Figure E11.2.4, $Q/Q_{full} = 0.68$ corresponds to $y/D = 0.60$ and $V/V_{full} = 1.08$. Assuming open channel flow, the actual flow velocity is therefore

$V = 1.08(4.66) = 5.03$ ft/s

Pipe 2, which has a length of 315 ft, has a travel time of

$t_t = (315 \text{ ft})/(5.03 \text{ ft/s}) = 63 \text{ s} = 1.0 \text{ min}$

Thus, the travel time to Pipe 3 is

(Basin C t_c) + (Pipe 2 t_t) = 20 min + 1 min = 21 min

Interpolating from the 2-year IDF table, the rainfall intensity for a duration of 21 min. is $i = 3.54$ in/hr. The design discharge for Pipe 3 is

$Q = CiA = 0.42(3.54)(15.0) = 22$ cfs

Applying Manning's equation with $n = 0.013$ and $S_o = 0.005$, the required pipe diameter (after rounding to the next commercially available size) is $D = 30$ in.

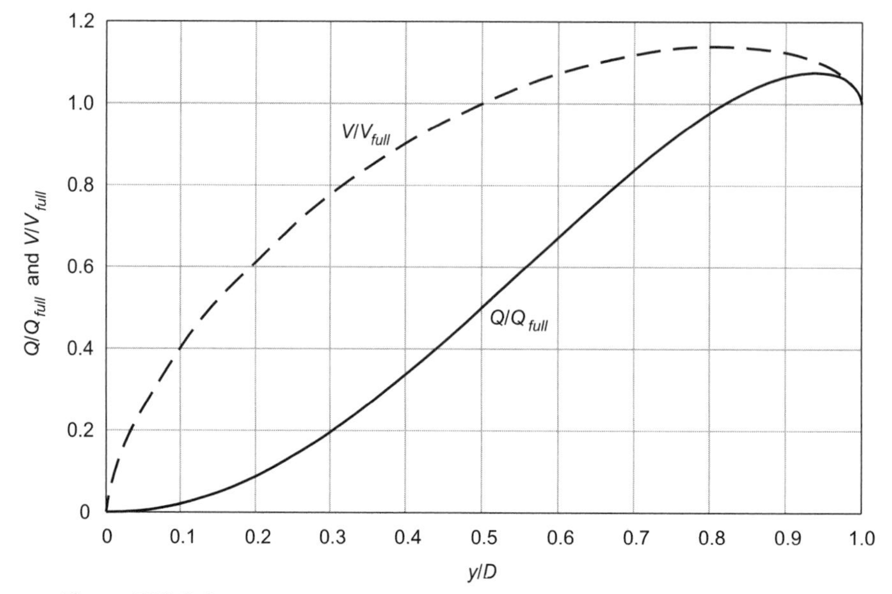

Figure E11.2.4 Graph for determining velocity in a circular pipe flowing partially full

Pipe 4 must convey the total discharge from all four basins. Its design discharge can be found using a composite runoff coefficient for the entire contributing area and the longest flow time. Each of the Basins A through C has a composite $C = 0.42$ and $A = 5.0$ ac, and Basin D has $C = 0.95$ and $A = 0.45$ ac. Therefore, the composite C for all four basins is computed as

$$C = \frac{3(5)(0.42) + 0.45(0.95)}{15.45} = 0.44$$

The longest travel time to the upstream end of Pipe 4 can be seen to be equal to the time of concentration for Basin C plus the travel times in Pipes 2 and 3; thus, it is necessary to find the flow velocity in Pipe 3. Using Manning's equation, the full-flow capacity of a 30-in. pipe is computed as $Q_{full} = 29.1$ cfs, and the full-flow velocity is $V_f = 5.93$ ft/s.

$$Q/Q_{full} = 22/29.1 = 0.76$$

Therefore, from Figure E11.2.4, $V/V_f = 1.10$. Assuming open-channel flow, the actual flow velocity is

$$V = 1.10(5.93) = 6.52 \text{ ft/s}$$

Because Pipe 3 has a length of 40 ft, this translates into a travel time of 6 sec = 0.1 min. The travel time in Pipe 2 was computed as 1 min. Thus, the total travel time to Pipe 4 is

$$20 \text{ min} + 1 \text{ min} + 0.1 \text{ min} = 21.1 \text{ min}$$

Interpolating from the IDF data, the design rainfall intensity is $i = 3.53$ in/hr. The design discharge for Pipe 4 is therefore

$$Q = CiA = 0.44(3.53)(15.45) = 24 \text{ cfs}$$

Applying the Manning equation with $n = 0.013$ and $S_o = 0.005$, the required pipe diameter (after rounding to the next commercially available size) is $D = 30$ in.

Step 5: Set Initial Invert Elevations

If pipe crowns are matched at Inlet 2, then the invert elevation of Pipe 3 at that structure should be 9 in. lower than the invert elevations of Pipes 1 and 2 where they enter the structure. At Inlet 4, where the incoming and outgoing pipes are of the same diameter, an additional invert elevation drop of 0.1 ft is provided through the structure per standard practice in the project locale. Figure E11.2.5 is a profile drawing showing the completed initial design.

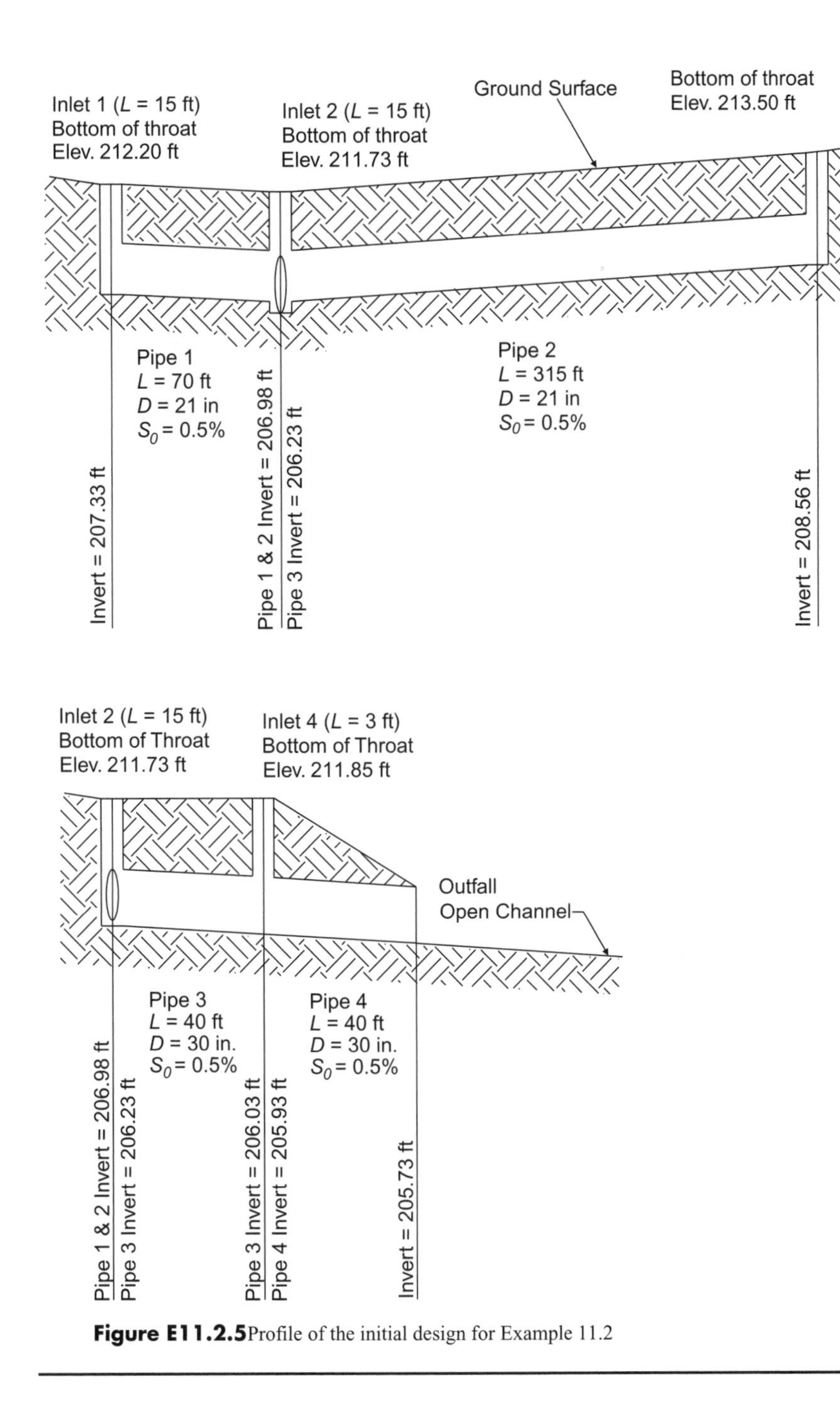

Figure E11.2.5 Profile of the initial design for Example 11.2

"Like so many New Yorkers, I've lived here all my life and never been to the top of the Empire State Building."

11.4 MINOR LOSSES IN JUNCTION STRUCTURES

The completion of a storm sewer design, which will be presented in Section 11.5, requires the evaluation of the hydraulic grade line through the system. To do this, the head losses throughout the system, including both frictional and minor losses, must be computed.

In Section 6.3 (page 202), the classical method for computing minor head losses due to pipe entrances, exits, transitions, and various appurtenance was introduced. The classical approach computes head loss as the product of a minor loss coefficient and the absolute difference between the velocity heads upstream and downstream of an appurtenance, and it can be applied to computing minor losses through junction structures in storm sewer systems such as manholes and inlet boxes. However, at stormwater inlets, manholes, and junction structures, where flow geometries are often complex, additional methods have been developed for minor loss prediction. Two of these methods—the energy-loss and composite energy-loss methods—are presented by Brown, Stein, and Warner (2001).

Energy-Loss Method

The incoming pipe(s) to an inlet, manhole, or junction structure in a stormwater conveyance system can occur in a number of different configurations with respect to their invert elevations. The various configurations that can occur are

a) All incoming pipe invert elevations lie below the elevation of the predicted depth of water in the structure.

b) All incoming pipe invert elevations lie above the elevation of the predicted depth of water in the structure.

c) One or more of the incoming pipe invert elevations lie above the elevation of the predicted depth of water in the structure, and one or more incoming pipe invert elevations lie below the elevation of the predicted depth of water in the structure.

Note that a stormwater inlet capturing water on the ground surface and discharging into an outlet pipe corresponds to configuration (b), provided that the free water depth inside the structure lies below the elevation of the stormwater inlet opening. If a stormwater inlet not only captures water at the ground surface for discharge to the outlet pipe but also acts as a junction between the outlet pipe and another incoming pipe, it corresponds to configuration (c).

For structures in which all of the incoming pipe invert elevations lie above the predicted free water surface elevation within the structure (that is, *plunging flow* occurs from all of the incoming pipes), the outlet pipe behaves hydraulically as a culvert. In that case, the water surface elevation within the structure can be predicted using the methods presented in Chapter 9, and the water surface elevations in each of the upstream pipes can be determined independently of one another as free outfalls.

The *energy-loss method* for estimation of head losses at inlets, manholes, and junctions is applicable only to configuration (a) and to the incoming pipes in configuration (c) whose invert elevations lie below the predicted free water surface elevation within the structure. When one or more incoming pipes meet these criteria, the method may be applied to each to determine the corresponding head loss.

For incoming pipes to which the energy-loss method is applicable, the head loss through the structure can be computed using Equation 6.26, repeated here as Equation 11.1

$$h_L = K\frac{V_o^2}{2g} \tag{11.1}$$

where h_L = minor loss (ft, m)
V_o = velocity in the outlet pipe (ft/s, m/s)
K = adjusted minor loss coefficient
g = gravitational acceleration (32.2 ft^2/s, 9.81 m^2/s)

Laboratory research has shown that K can be expressed as

$$K = K_o C_{D1} C_{D2} C_Q C_P C_B \tag{11.2}$$

where K_o=initial head loss coefficient based on relative size of structure
C_{D1}=correction factor for pipe diameter
C_{D2}=correction factor for flow depth
C_Q=correction factor for relative flow
C_P=correction factor for plunging flow
C_B=correction factor for benching

Equations 11.1 and 11.2 can be applied to each of several incoming pipes at a structure, provided their invert elevations lie below the predicted water surface elevation within the structure.

The initial head loss coefficient, K_o, depends on the size of the structure relative to the outlet pipe diameter, and on the angle θ between the inlet and outlet pipes (see Figure 11.8):

$$K_o = 0.1\left(\frac{B}{D_o}\right)(1-\sin\theta)+1.4\left(\frac{B}{D_o}\right)^{0.15}\sin\theta \qquad (11.3)$$

where B = structure diameter (ft, m)
D_o = outlet pipe diameter (ft, m)
θ = angle between inlet and outlet pipes

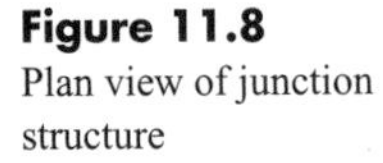
Figure 11.8
Plan view of junction structure

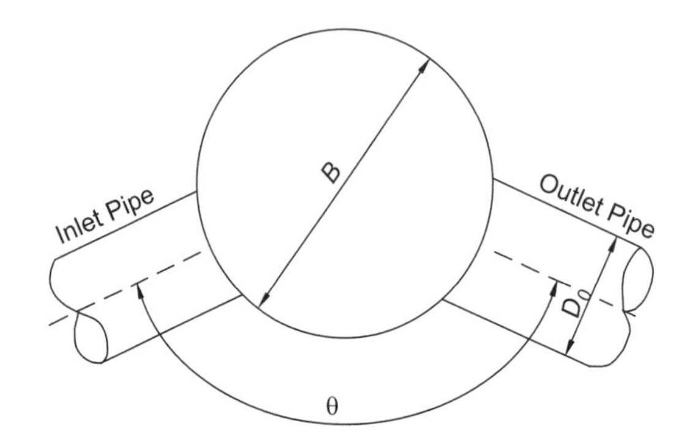

If the structure is not circular, an *equivalent structure diameter* should be used. The equivalent diameter is defined as the diameter of a circular structure having the same area as the actual non-circular one. For example, if the inside dimensions (plan view) of a stormwater inlet were 3 ft by 6 ft (A = 18 ft^2), the equivalent diameter used in Equation 11.3 would be $B = (4A/\pi)^{1/2} = [4(18)/3.14]^{1/2} = 4.79$ ft.

Computing C_{D1} (Correction for Relative Pipe Diameter). The correction factor for pipe diameter, C_{D1}, needs to be considered only in cases where the predicted water depth d in the structure is at least 3.2 times as great as the outlet pipe diameter. When applying this and other correction factors, the depth of water d should be taken as the difference between the hydraulic grade line elevation at the upstream end of the outlet pipe and the invert elevation of the outlet pipe.

In such cases, C_{D1} is determined as

$$C_{D1} = \left(\frac{D_o}{D_i}\right)^3 \qquad (11.4)$$

where D_i = inlet pipe diameter (ft, m)

When the predicted depth in the structure is not at least 3.2 times the outlet pipe diameter, $C_{D1} = 1$.

Computing C_{D2} (Correction for Flow Depth). The correction factor for flow depth, C_{D2}, needs to be considered only in cases where the predicted depth in the structure is less than 3.2 times the outlet pipe diameter; otherwise, $C_{D2} = 1$. The correction factor is determined as

$$C_{D2} = 0.5\left(\frac{d}{D_o}\right)^{0.6} \tag{11.5}$$

where d = water depth in the structure computed as the difference in HGL and invert at the upstream end of the outlet pipe (ft, m)

Computing C_Q (Correction for Relative Flow). The correction factor for relative flow, C_Q, is required in applications where there are two or more incoming pipes at a structure. When there is only one incoming pipe, $C_Q = 1$. The correction factor depends on the angle θ between the outlet pipe and the incoming pipe of interest, and on the ratio of the discharge in the incoming pipe of interest to the outlet pipe:

$$C_Q = 1 + (1 - 2\sin\theta)\left(1 - \frac{Q_i}{Q_o}\right)^{0.75} \tag{11.6}$$

where Q_i = discharge in the incoming pipe of interest (cfs, m^3/s)
Q_o = discharge in the outflow pipe (cfs, m^3/s)

Computing C_P (Correction for Plunging). When one or more of the incoming pipes to a structure have invert elevations higher than the elevation of the free water surface in the structure d, they are said to have *free outfalls,* and the water plunges into the structure. Plunging flow can also enter a structure from the inlet opening. The resulting turbulence and energy dissipation within the structure has an effect on the head loss for other incoming pipes whose flow is not plunging (that is, their invert elevations lie below the free water surface). The coefficient C_P is computed and applied in the head loss calculations only for the pipe(s) whose flow is not plunging. The coefficient is determined as

$$C_P = 1 + 0.2\left(\frac{h}{D_o}\right)\left(\frac{h-d}{D_o}\right) \tag{11.7}$$

where h = difference in elevation between the highest incoming pipe invert and the centerline of the outlet pipe (ft, m)

For cases in which no plunging incoming pipes are present, or in which $h \le d$, the value of the coefficient is $C_P = 1$.

Selecting C_B (Correction for Benching of Structure Invert). *Benching* of the invert of a structure (see Figure 11.9) can reduce head losses by effectively directing the path of the flow through the structure. In practice, the base of a manhole can be benched by pouring a cast-in-place base that is shaped to the pipes connecting to the structure. Benching can also be achieved by pouring fresh concrete into the inside base of a precast manhole section and shaping it to the required form. The correction factor for benching, C_B, is presented in Table 11.9 and depends on the depth of water in the structure relative to the outlet pipe diameter. For the use of that table, submerged flow is considered to occur if $d/D_o \geq 3.2$, and unsubmerged flow is considered to occur if $d/D_o \leq 1.0$. For values of d/D_o between 1.0 and 3.2, linear interpolation between the tabulated values may be performed.

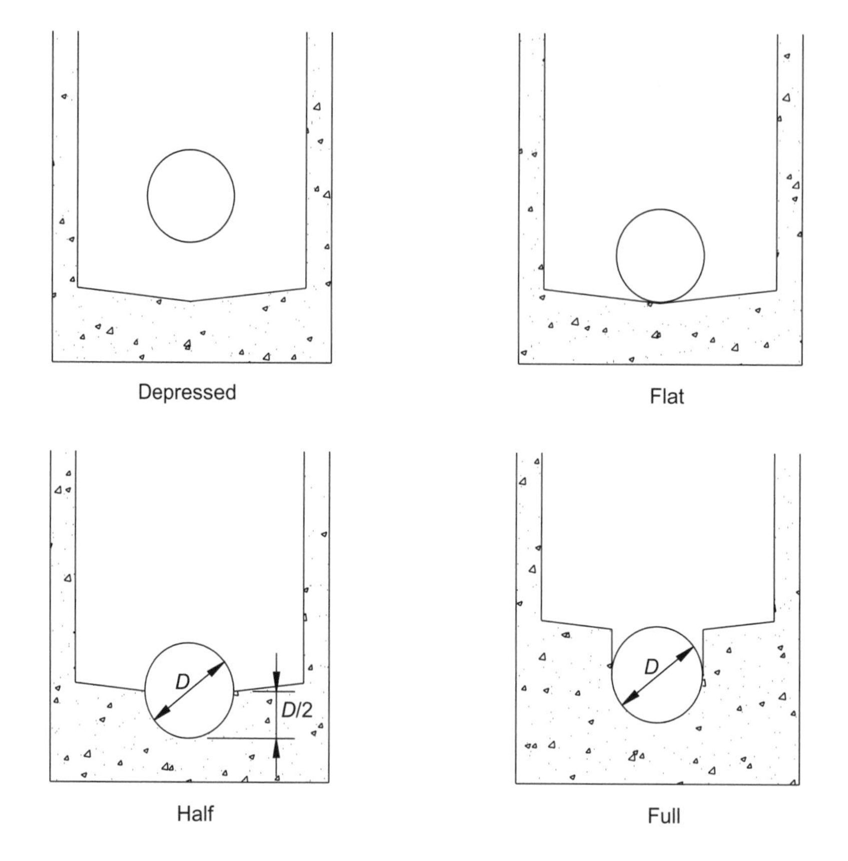

Figure 11.9 Benching the invert of a structure

Table 11.9 Correction factors for benching

Bench Type	Submerged	Unsubmerged
Flat or depressed floor	1.00	1.00
Half bench	0.95	0.15
Full bench	0.75	0.07

Example 11.3 – Applying Minor Losses. Consider a manhole with two inlet pipes, as shown in Figure E11.3.1. The inlet pipe invert elevations have been set so that their crown elevations match the crown of the outlet pipe. If the HGL elevation at the upstream end of the outlet pipe is 32.0 m and is controlled by downstream conditions, determine the EGL and HGL elevations for all three pipes. Assume that no benching is provided in the manhole.

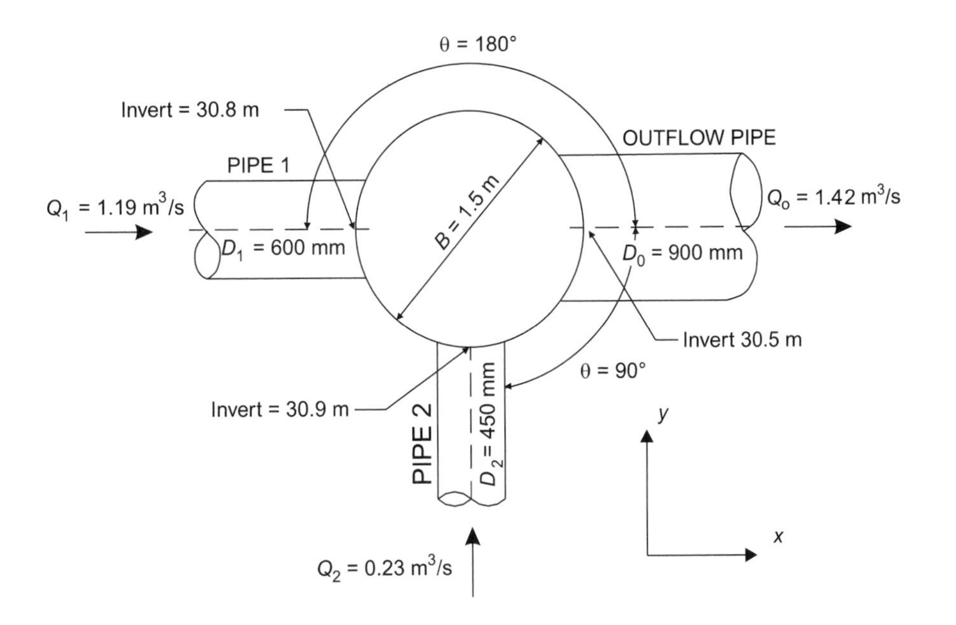

Figure E11.3.1 Manhole configuration for Example 11.3

Solution: The velocity in the outlet pipe is computed as

$V_o = Q_o/A_o$ = 1.42/0.64 = 2.23 m/s.

The EGL at the upstream end of the outlet pipe is therefore

$$EGL_o = HGL_o + V_o^2/2g = 32.0 + 2.32^2/2(9.81) = 32.27 \text{ m}$$

For Pipe 1, θ = 180° so, using Equation 11.3,

$$K_{o1} = 0.1\left(\frac{B}{D_O}\right)(1-\sin\theta) + 1.4\left(\frac{B}{D_O}\right)^{0.15}\sin\theta = 0.1\left(\frac{1.5}{0.90}\right)(1-\sin 180°) + 1.4\left(\frac{1.5}{0.90}\right)^{0.15}\sin 180° = 0.17$$

Similarly, for Pipe 2, where θ = 90°,

$$K_{o2} = 0.1\left(\frac{1.5}{0.90}\right)(1-\sin 90°) + 1.4\left(\frac{1.5}{0.90}\right)^{0.15}\sin 90° = 1.51$$

The depth, d, of flow in the manhole structure is the difference in elevation between the outlet pipe HGL and its invert, or d = 1.5 m. Also, d/D_o = 1.64. Because $d/D_o < 3.2$, the correction factor for pipe diameter is equal to 1 for both incoming pipes:

$$(C_{D1})_1 = (C_{D1})_2 = 1$$

The correction factor for flow depth is determined using Equation 11.5 for each of the incoming pipes:

$$(C_{D2})_1 = (C_{D2})_2 = 0.5\left(\frac{d}{D_o}\right)^{0.6} = 0.5\left(\frac{32.0-30.5}{0.90}\right)^{0.6} = 0.67$$

For application of Equation 11.6, the angle θ_1 = 180° and sin θ_1 = 0. Also, θ_2 = 90° and sin θ_2 = 1. The coefficients C_Q for each of the two incoming pipes are therefore

$$C_{Q1} = 1 + (1-2\sin\theta)\left(1-\frac{Q_1}{Q_o}\right)^{0.75} = 1 + \left(1-\frac{1.19}{1.42}\right)^{0.75} = 1.26$$

$$C_{Q2} = 1 + (1 - 2\sin\theta)\left(1 - \frac{Q_2}{Q_o}\right)^{0.75} = 1 + (1 - 2)\left(1 - \frac{0.23}{1.42}\right)^{0.75} = 0.12$$

Because no plunging flow is in this example, $C_{P1} = C_{P2} = 1$. Further, there is no benching in the structure so $C_{B1} = C_{B2} = 1$.

The adjusted loss coefficient for Pipe 1 is computed using Equation 11.2 as

$$K_1 = K_{o1}(C_{D1})_1(C_{D2})_1 C_{Q1} C_{P1} C_{B1} = 0.17(1)(0.67)(1.26)(1)(1) = 0.14$$

Similarly,

$$K_2 = K_{o2}(C_{D1})_2(C_{D2})_2 C_{Q2} C_{P2} C_{B2} = 1.51(1)(0.67)(0.12)(1)(1) = 0.12$$

The head loss for Pipe 1 is computed using Equation 11.1 as

$$h_{L1} = K_1 \frac{V_o^2}{2g} = 0.14 \frac{2.23^2}{2(9.81)} = 0.04 \text{ m}$$

and the head loss for Pipe 2 is

$$h_{L2} = K_2 \frac{V_o^2}{2g} = 0.12 \frac{2.23^2}{2(9.81)} = 0.03 \text{ m}$$

The EGL at the downstream end of Pipe 1 is the sum of the EGL at the upstream end of the outlet pipe plus the head loss, or

$$EGL_1 = EGL_o + h_{L1} = 32.27 + 0.04 = 32.31 \text{ m}$$

Similarly, for Pipe 2,

$$EGL_2 = EGL_o + h_{L2} = 32.27 + 0.02 = 32.28 \text{ m}$$

The HGL in Pipe 1 is computed as

$$\text{HGL}_1 = \text{EGL}_1 - \frac{V_1^2}{2g} = 32.31 - \frac{[1.19/(\pi \times 0.3^2)]^2}{2 \times 9.81} = 31.41 \text{ m}$$

For Pipe 2,

$$\text{HGL}_2 = \text{EGL}_2 - \frac{V_2^2}{2g} = 32.29 - \frac{[0.23/(\pi \times (0.225)^2)]^2}{2 \times 9.81} = 32.22 \text{ m}$$

Composite Energy-Loss Method

The composite energy-loss method can be used in situations similar to those for which the energy-loss method, discussed in the previous section, applies. However, the composite method is better suited to analyzing losses in structures with many inflow pipes. It is only applicable to subcritical flows in pipes.

The method is used to compute a unique head loss through the junction structure for each of the incoming pipes. This head loss is then added to the energy grade at the upstream end of the outlet pipe (EGL_o) to obtain the EGL at the downstream end of the incoming pipe (EGL_i). The velocity head for the incoming pipe can then be subtracted from EGL_i to obtain the hydraulic grade, HGL_i. The head loss through the structure for a particular inflow pipe is again given by Equation 6.26, repeated here as Equation 11.8:

$$h_L = K\frac{V_o^2}{2g} \qquad (11.8)$$

where K = the adjusted minor loss coefficient for the composite energy-loss method

The adjusted minor loss coefficient K is defined as

$$K = (C_1 C_2 C_3 + C_4)C_B \qquad (11.9)$$

where C_1 = coefficient for the relative access-hole diameter
C_2 = coefficient for the water depth in the access hole
C_3 = coefficient for lateral flow, the lateral angle, and plunging flow
C_4 = coefficient for the relative pipe diameters
C_B = coefficient for benching (given in Table 11.9)

Empirical equations describing these coefficients were developed from laboratory studies and analyses. These equations are presented in the subsections that follow.

Computing C_1 (Relative Access-Hole Diameter Coefficient). The energy-loss coefficient relative to the access-hole diameter is related to the ratio of the access-hole diameter, B, to the outlet-pipe diameter, D_0, and is given by

$$C_1 = \begin{cases} \dfrac{0.9\left(\dfrac{B}{D_o}\right)}{6+\dfrac{B}{D_o}}, & \dfrac{B}{D_o} < 4 \\[2ex] 0.36, & \dfrac{B}{D_o} \geq 4 \end{cases} \qquad (11.10)$$

Thus, the value of this coefficient increases with the ratio B/D_0 until that ratio is equal to 4. For $B/D_0 \geq 4$, C_1 has the constant value $C_1 = 0.36$. Note that the value of C_1 is the same for all incoming pipes.

Computing C_2 (Water Depth in Access Hole Coefficient) and C_3 (Lateral Flow, Lateral Angle, and Plunging Flow Coefficient). The coefficients C_2 and C_3 represent the composite effect of all the inflow pipes, the outflow pipe, and the access hole. Their calculation is affected by and also affects the calculation of the access-hole water depth, d. Because of this interdependence, an iterative method is used to calculate these coefficients. The first step is to compute an initial estimate of d with the equation

$$d = HGL_o + C_1 C_B \frac{V_o^2}{2g} - Z_o \qquad (11.11)$$

where HGL_0 = hydraulic grade elevation at the upstream end of the outlet pipe (ft, m)
Z_o = elevation of the outlet pipe invert, (ft, m)

This value is used to calculate an estimate of C_2 as

$$C_2 = \begin{cases} 0.24\left(\dfrac{d}{D_o}\right)^2 - 0.05\left(\dfrac{d}{D_o}\right)^3, & \dfrac{d}{D_0} \leq 3 \\ 0.82, & \dfrac{d}{D_o} > 3 \end{cases} \qquad (11.12)$$

The analysis of the factors affecting energy losses for lateral flows resulted in an equation for C_3 that is the most complex of any of the coefficients. For a simple, two-pipe system with no plunging flow, $C_3 = 1.0$. (A pipe has plunging flow if the critical flow depth elevation in the pipe, $y_c + Z_i$, is higher than the access-hole depth elevation, $d + Z_o$.) Otherwise, C_3 is expressed in the form

$$C_3 = 1 + C_{3A} + C_{3B} + C_{3C} + C_{3D} \qquad (11.13)$$

where the individual terms in Equation 11.13 are given by Equations 11.14 through 11.18.

The calculations for C_3 consider the angle θ_i between the inlet and outlet pipes. As this angle deviates from 180 (straight-line flow), the energy loss increases because the flow cannot smoothly transition to the outlet pipe. All inflow pipe angles are measured clockwise from the outlet pipe. The calculation of C_3 accounts for inlet-flow plunging by considering the inlet as a fourth, synthetic inflow pipe with the corresponding angle set to 0 .

C_3 can have a value ranging from 1 for no lateral flow to potentially very high values for greater plunge heights. Because empirical studies do not support this result, a value of 10 is set as a realistic upper limit on C_3.

The term C_{3A} represents the energy loss from plunging flows and is valid for three inlet pipes plus the plunging flow from the inlet:

$$C_{3A} = \sum_{i=1}^{4}\left(\frac{Q_i}{Q_o}\right)^{0.75}\left[1 + 2\left(\frac{Z_i}{D_o} - \frac{d}{D_o}\right)^{0.3}\left(\frac{Z_i}{D_o}\right)^{0.3}\right] \qquad (11.14)$$

whereQ_1, Q_2, Q_3=discharge from inflow pipes 1, 2, and 3 (cfs, m^3/s)
Q_4 = discharge into the access hole from the inlet (cfs, m^3/s)
Z_1, Z_2, Z_3 = invert elevations of the inflow pipes relative to the outlet pipe invert (ft, m)

The term C_{3B} represents the energy loss due to change in direction between the inlet and outlet pipes. If the horizontal momentum check HMC_i as computed using Equation 11.15 is less than 0, then the flow is assumed to be falling from such a height that horizontal momentum can be neglected and $C_{3B} = 0$. Otherwise, HMC_i is given by Equation 11.15. Note that the inlet flow is not considered in this calculation.

$$HMC_i = 0.85 - \left(\frac{Z_i}{D_o}\right)\left(\frac{Q_i}{D_o}\right)^{0.75} \tag{11.15}$$

If $HMC_i \geq 0$, C_{3B} is computed as

$$C_{3B} = 4\sum_{i=1}^{3} \frac{\cos\theta_i \times HMC_i}{\left(\frac{d}{D_o}\right)^{0.3}} \tag{11.16}$$

where θ_1, θ_2, θ_3 = angle between the outlet main and inflow pipes 1, 2, and 3, degrees
HMC_i = horizontal momentum check for pipe i computed using Equation 11.15

If there is more than one inflow pipe, C_{3C} is calculated for all combinations of inflow pipes with $HMC_i > 0$. The pair that produces the highest value is then used for the calculations of C_{3C} and C_{3D}.

$$C_{3C} = 0.8\left(\frac{Z_A}{D_o} - \frac{Z_B}{D_o}\right) \tag{11.17}$$

where Z_A, Z_B = invert elevation, relative to the outlet pipe invert, for the inflow pipes that produce the largest value of C_{3D}

$$C_{3D} = \left|\left(\frac{Q_A}{Q_o}\right)^{0.75} \sin\theta_A + \left(\frac{Q_B}{Q_o}\right)^{0.75} \sin\theta_B\right| \tag{11.18}$$

where Q_A, Q_B = discharges for the pair of inflow pipes producing the largest value of C_{3D}
θ_A, θ_B = angle between the outlet main and inflow pipes for the pair of inflow pipes producing the largest value of C_{3D}

With the initial estimates of C_2 and C_3, the access-hole depth is recalculated with

$$d = \text{HGL}_o + \frac{V_0^2}{2g} + C_1C_2C_3C_B\frac{V_0^2}{2g} - Z_o \tag{11.19}$$

where HGL_o = hydraulic gradeline at the upstream end of the inlet pipe
Z_o = invert elevation of the outlet pipe at the upstream end

This new value of d is compared to the previous estimate, and the entire procedure for calculating C_2, C_3, and d continues until the method converges on the value of d.

Computing C_4 (Relative Pipe Diameter Coefficient). The last coefficient to be determined in Equation 11.9, the coefficient related to relative pipe diameters, C_4, is computed for each inflow pipe using

$$C_{4i} = 1 + \left(\frac{Q_i}{Q_o} + 2\frac{A_i}{A_o}\cos\theta_i \right)\frac{V_i^2}{V_o^2} \qquad (11.20)$$

where A_i, A_o = cross-sectional areas of the inlet and outflow pipes (ft^2, m^2)
θ_i = angle between the outlet pipe and inflow pipe i
V_i, V_o = velocity of flow in the inlet pipe and outflow pipes (ft/s, m/s)

Because this term represents an exit loss from each inflow pipe, C_{4i} is set to 0 for each pipe with plunging flow. In addition, if θ_i is less than 90° or greater than 270° for any inlet pipe, the corresponding term cos θ_i is set to 0 in Equation 11.20. The upper limit for the value of C_4 is 9.0.

Because the complex, iterative nature of the composite energy-loss method, its application is not well suited to manual calculations and thus it is not illustrated here through an example problem. However, it is readily adaptable to computer solutions.

11.5 FINAL DESIGN OF THE PIPE SYSTEM

Following completion of the initial design process as illustrated in Example 11.2, one proceeds to computation of the hydraulic and energy grade lines (the HGL and EGL) in order to ensure that the system performs adequately. Computation of the HGL and EGL elevations is especially important in cases where the storm sewer system outlet may be submerged, or where the subsurface drainage structures are shallow in depth.

When computing hydraulic and energy grade lines in a system, the direction (that is, upstream or downstream) in which the calculations should proceed depends on whether a pipe has surcharging or open-channel flow conditions. This is discussed further in the subsections and examples that follow.

Surcharged Pipes

In a surcharged pipe, the HGL is above the crown of the pipe (there is pressure flow) and calculations proceed upstream from a known tailwater condition at the outfall point. Checks must be made throughout the calculations to ensure that the HGL remains above the crown of the pipe. If it drops below the pipe crown at any point, the condition should be treated as open-channel flow.

Energy losses, whether they refer to frictional losses along pipes or to minor losses at structures, apply to the EGL rather than the HGL. Therefore, computation of the HGL elevations in a storm sewer system is actually accomplished by first computing the EGL elevations. Once these elevations are known, HGL elevations may be found by subtracting the velocity head $V^2/2g$. EGL elevations must always decrease in a downstream direction, and must always increase in an upstream direction. However, HGL elevations, depending on relative velocities in the pipes connected to one another at a structure, may increase or decrease at structure locations regardless of the direction considered. For example, an HGL elevation may increase in a downstream direction within a pipe if there is a hydraulic jump.

In design of a surcharged storm sewer system, the HGL must be maintained at elevations below the ground surface, especially at locations of manholes and inlets. If the HGL elevation is higher than the ground surface at a manhole or inlet location, it indicates that, for the design storm event, stormwater would exit the system through the top of the structure and flow onto the land surface. In extreme cases, these high HGL elevations can cause popping of manhole covers due to excessive pressures in the storm sewer. At inlet locations, it is recommended that the HGL elevation be no higher than about 6 in. (15 cm) below the grate elevation or the elevation of the bottom of the throat. [Local regulations may require more than 6 in. 15 cm).] Otherwise, the hydraulic behavior of the grate or throat may be impacted. When making HGL checks at inlet locations, conservative design practice is to use the highest HGL elevation computed for the structure.

Example 11.4 illustrates final design calculations for the storm sewer system originally presented in Example 11.2, where discharges and pipe sizes were computed. Because of the high tailwater on the system, all of the pipes will operate under pressure-flow conditions. The basic steps involved in the final design calculations for the pipe system are:

1. Compute the EGL in the outfall channel and the EGL at the downstream end of the outfall pipe. Consider minor losses at the pipe exit. Typically, exit losses are computed using Equation 6.25 with $K = 1.0$:

$$h_L = 1.0\left|\frac{V_2^2 - V_1^2}{2g}\right|$$

2. Compute the velocity head in the most downstream pipe and subtract this value from the EGL to compute the HGL at the downstream end of the pipe.
3. Compute the head loss in the most downstream pipe, and add this value to the EGL at the downstream end of the pipe to compute the EGL at the upstream end of the pipe. Subtract the velocity head from step 2 to find the HGL at this locaton.
4. Use a method such as the energy-loss or composite energy-loss method to compute the head loss through the structure for each of the incoming pipes. Add these values to the EGL of the outgoing pipe to determine the EGL at the downstream end of each of the incoming pipes.
5. Compute the velocity head for each of the incoming pipes, and subtract this value from the EGL at the downstream end of each pipe to determine the HGL at these locations.
6. If an HGL elevation is higher than the allowable maximum elevation necessary to maintain minimum freeboard at a structure, increase the size of the outgoing pipe, redo the affected calculations, and recheck the HGL elevations.
7. If the structure has more than a single incoming pipe, select one of the incoming branches and continue the calculations by repeating steps 3 through 7 until the EGL and HGL elevations for the entire branch have been determined.
8. Use Equation 6.25 with $K = 0.5$ to compute the pipe entrance loss at terminal upstream structures, and compute the EGL and HGL at this location. Compare

this value for the HGL (which was computed assuming outlet-control conditions) with the inlet-control HGL computed using methods presented in Section 9.5. The higher of these elevations controls the value of the HGL at this location.

9. Repeat steps 3 through 8 until the entire system has been evaluated.
10. Check to ensure that all velocities are within acceptable limits and adjust pipe sizes if necessary.

Example 11.4 – Final Design Process for a System with Pressure Flow. Compute the HGL and EGL elevations for the storm sewer system from Example 11.2 (page 427) and shown again in Figure E11.4.1. Backwater from a culvert downstream of the outfall causes the tailwater at the storm sewer outfall point to be at an elevation of 209.70 ft. The tailwater velocity in the channel downstream of the outfall is 2 ft/s. Use the energy-loss method to evaluate minor losses at structures, and assume that the inlets are constructed with no benching of their inverts.

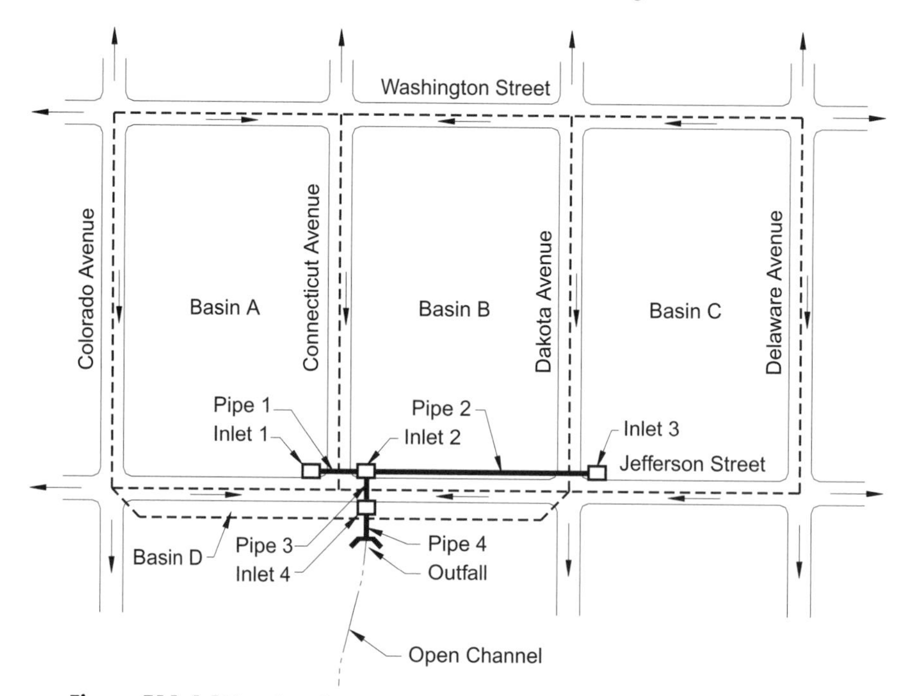

Figure E11.4.1 Plan view of storm sewer system for and Example 11.4

Pipe data

The pipe lengths, diameters, discharges, and invert elevations given or computed in Example 11.2 will be used in the calculations. This information is presented again in the following table.

Pipe Label	Length L (ft)	Diameter (in.)	Upstream Invert Elevation (ft)	Downstream Invert Elevation (ft)	Discharge Q (cfs)
Pipe 1	70	21	207.33	206.98	7.6
Pipe 2	315	21	208.56	206.98	7.6
Pipe 3	40	30	206.23	206.03	22.0
Pipe 4	40	30	205.93	205.73	24.0

Inlet data

The curb inlets designed in Example 11.2 will be used for the calculations. The inlet boxes are rectangular with a length equal to the inlet throat opening length and a width (perpendicular to the gutter flow line) of 3 ft. Information for individual inlets is given in the following table.

Inlet Label	Bottom of Throat Elev. (ft)	Length L (ft)	Equivalent Diameter B^a (ft)
Inlet 1	212.20	15	7.6
Inlet 2	211.73	15	7.6
Inlet 3	213.50	15	7.6
Inlet 4	211.85	3	3.4

a. Equivalent diameter B for inlet boxes is computed as $B = (4A/\pi)^{1/2}$, where $A = 3L$ for 3-ft-wide boxes.

Solution: Because the sewer outfall is submerged by the tailwater, all or part of the storm sewer system will be surcharged and calculations should proceed upstream through the system from the outfall point. Following are the calculations for HGL and EGL through the system, and the results are tabulated in the table following the calculations.

EGL and HGL at outfall

The HGL immediately downstream of the outfall point is the same as the tailwater elevation in the channel, or HGL = 209.70 ft. Thus, the EGL immediately downstream of the outfall point is

$$EGL = HGL + V^2/(2g) = 209.70 + 2^2/[2(32.2)] = 209.76 \text{ ft}$$

The exit loss at the outfall point can be determined using Equation 6.25, where V_1 is the velocity in Pipe 4 and $V_2 = 2$ ft/s is the tailwater velocity. The minor loss coefficient $K = 1.0$. Because Pipe 4 will be flowing full due to the tailwater elevation, its velocity is

$$V = \frac{Q}{A} = \frac{4Q}{\pi D^2} = \frac{4(24)}{2.5^2 \pi} = 4.89 \text{ ft/s}$$

Therefore, the exit loss is

$$h_L = K\left|\frac{V_2^2 - V_1^2}{2g}\right| = 1.0\frac{4.89^2 - 2^2}{2(32.2)} = 0.31 \text{ ft}$$

The EGL immediately upstream of the outfall point (at the downstream end of Pipe 4) is equal to the EGL downstream of the outfall point plus the head loss, or

$$EGL = 209.76 + 0.31 = 210.07 \text{ ft}$$

The velocity head for Pipe 4 is

$$V^2/(2g) = 4.89^2/[2(32.2)] = 0.37 \text{ ft}$$

The HGL at the downstream end of Pipe 4 is therefore

$$HGL = EGL - V^2/(2g) = 210.07 - 0.37 = 209.70 \text{ ft}$$

EGL and HGL at Inlet 4

The EGL at the upstream end of Pipe 4 (at the outlet of Inlet 4) is computed as the sum of the EGL elevation at the downstream end of Pipe 4, plus frictional losses that occur in Pipe 4. The slope of the EGL for Pipe 4 can be evaluated by rearranging the Manning equation (Equation 7.9), and the frictional loss can be determined from Equation 7.8 as the product of the EGL slope and the pipe length:

$$S_f = \frac{(nV)^2}{1.49^2 R^{4/3}} = \frac{(nV)^2}{1.49^2 (D/4)^{4/3}} = \frac{[0.013(4.89)]^2}{2.22(2.5/4)^{4/3}} = 0.00341$$

$$h_L = S_f L = 0.00341(40) = 0.14 \text{ ft}$$

The EGL elevation at the upstream end of Pipe 4 is

$$EGL = 210.07 + 0.14 = 210.21 \text{ ft}$$

Because the velocity head in Pipe 4 is 0.37 ft, the HGL elevation at the upstream end of Pipe 4 is

$$HGL = EGL - V^2/(2g) = 210.21 - 0.37 = 209.84 \text{ ft}$$

This value is higher than the pipe crown elevation, confirming that Pipe 4 flows full for its entire length.

The minor loss at Inlet 4 is evaluated using Equations 11.1 and 11.2. The initial loss coefficient K_o is found using Equation 11.3 with $D_o = 30$ in. = 2.5 ft and $\theta = 180°$. The equivalent diameter of the inlet box is 3.4 ft. Therefore,

$$K_o = 0.1\left(\frac{B}{D_O}\right)(1-\sin\theta)+1.4\left(\frac{B}{D_O}\right)^{0.15}\sin\theta$$

$$= 0.1\left(\frac{3.4}{2.5}\right)(1-\sin 180°)+1.4\left(\frac{3.4}{2.5}\right)^{0.15}\sin 180° = 0.14$$

The pipe diameter correction factor $C_{D1} = 1$ because the predicted depth in the inlet will not be greater than 3.2 times the outlet pipe diameter. The correction factor for flow depth, C_{D2}, is evaluated from Equation 11.5 with d being the difference in elevation between the HGL and pipe invert elevations at the upstream end of Pipe 4:

$$C_{D2} = 0.5(d/D_o)^{0.6} = 0.5[(209.84 - 205.93)/2.5]^{0.6} = 0.65$$

The correction factor C_Q is equal to 1 because there is only one inflow pipe to Inlet 4.

Because flow plunges into Inlet 4 from its throat opening, the factor C_P must be determined using Equation 11.7. In that expression, d is 3.91 ft (as above) and h is the elevation difference between the bottom of the inlet throat and the centerline of the outlet pipe, or

$$h = 211.85 - 207.18 = 4.67 \text{ ft}$$

Therefore,

$$C_P = 1+0.2\left(\frac{h}{D_o}\right)\left(\frac{h-d}{D_o}\right) = 1+0.2\left(\frac{4.67}{2.5}\right)\left(\frac{4.67-3.91}{2.5}\right) = 1.11$$

The correction factor for benching is $C_B = 1$ because no benching is provided in the inlet.

Applying Equations 11.1 and 11.2, the minor loss in Inlet 4 is

$$h_L = K_o C_{D1} C_{D2} C_Q C_P C_B \frac{V_o^2}{2g} = 0.14(1)(0.65)(1)(1.11)(0.37) = 0.04 \text{ ft}$$

The EGL elevation at the downstream end of Pipe 3 is equal to the EGL at the upstream end of Pipe 4, plus the minor loss, and is therefore

$$EGL = 210.21 + 0.04 = 210.25 \text{ ft}$$

The diameter of Pipe 3 is 30 in. = 2.5 ft. The velocity for this pipe is

$$V = \frac{Q}{A} = \frac{4Q}{\pi D^2} = \frac{4(22)}{2.5^2\pi} = 4.48 \text{ ft/s}$$

The HGL at the downstream end of Pipe 3 is then computed as

$$HGL = EGL - V^2/2g = 210.25 - 4.48^2/[2(32.2)] = 209.94 \text{ ft}$$

The highest of the HGL elevations computed at Inlet 4 is HGL = 209.94 ft. This elevation is 1.91 ft below the bottom of the throat elevation; therefore, the hydraulic behavior of the inlet opening will not be affected.

HGL and EGL at Inlet 2

The HGL and EGL elevations at the upstream end of Pipe 3 (at the outlet of Inlet 2) are higher than the corresponding values for the downstream end of Pipe 3 by the amount $h_L = S_f L$. The pipe length is $L = 40$ ft and, from the Manning equation,

$$S_f = \frac{(nV)^2}{1.49^2 (D/4)^{4/3}} = \frac{[0.013(4.48)]^2}{1.49^2 (2.5/4)^{4/3}} = 0.00286$$

Therefore, $h_L = 0.11$ ft and the EGL and HGL elevations are

$$EGL = 210.25 + 0.11 = 210.36 \text{ ft}$$

$$HGL = 210.36 - 4.48^2/[2(32.2)] = 210.05 \text{ ft}$$

Minor losses in Inlet 2 are evaluated using Equations 11.1 through 11.7 with

$$d = 210.05 - 206.23 = 3.82 \text{ ft}$$

and

$$h = 211.73 - 207.48 = 4.25 \text{ ft}$$

The minor losses must be evaluated separately for each of the two inlet pipes.

For inlet Pipe 1, $\theta = 90°$, $K_o = 1.65$,$C_{D1} = 1$, $C_{D2} = 0.64$, $C_P = 1.06$, and $C_B = 1$.

The factor C_Q is evaluated using $Q_I = 7.6$ cfs and $Q_o = 22$ cfs:

$$C_Q = 1 + (1 - 2\sin\theta)\left(1 - \frac{Q_I}{Q_o}\right)^{0.75} = 1 + (1 - 2\sin 90°)\left(1 - \frac{7.6}{22}\right)^{0.75} = 0.27$$

Therefore, the minor loss to be used for Pipe 1 is

$$h_L = 1.65(1)(0.64)(0.27)(1.06)(1)[4.48^2/2(32.2)] = 0.09 \text{ ft}$$

and the EGL elevation at the downstream end of Pipe 1 is

$$EGL = 210.36 + 0.09 = 210.29 \text{ ft}$$

The diameter of Pipe 1 is 21 in. = 1.75 ft. The velocity for this pipe is

$$V = \frac{Q}{A} = \frac{4Q}{\pi D^2} = \frac{4(7.6)}{1.75^2 \pi} = 3.16 \text{ ft/s}$$

The HGL elevation at the downstream end of Pipe 1 is therefore

$$HGL = EGL - V^2/2g = 210.45 - 3.16^2/[2(32.2)] = 210.29 \text{ ft}$$

For Pipe 2, $\theta = 90°$, $K_o = 1.65$, $C_{D1} = 1$, and $C_{D2} = 0.64$. The factor C_Q is evaluated using $Q_I = 7.6$ cfs and $Q_o = 22$ cfs, and is 0.27. The factor $C_P = 1.06$ and $C_B = 1$. Therefore, the minor loss to be used for Pipe 2 is

$$h_L = 1.65(1)(0.64)(0.27)(1.06)(1)[4.48^2/2(32.2)] = 0.09 \text{ ft}$$

and the EGL elevation at the downstream end of Pipe 2 is

$$EGL = 210.36 + 0.09 = 210.45 \text{ ft}$$

The diameter, flow, and therefore velocity are the same for Pipe 2 as for Pipe 1. The HGL elevation at the downstream end of Pipe 2 is

$$HGL = EGL - V^2/2g = 210.45 - 3.16^2/2(32.2) = 210.29 \text{ ft}$$

Note that, in this example, the minor loss applicable to each of the incoming pipes (Pipes 1 and 2) was identical. In general, this will not be true (see Example 11.3), but, in this case, the inlet pipes were of the same size, with the same flow rates and identical geometric orientations relative to the outflow pipe.

The highest of the HGL elevations in Inlet 2 is HGL = 210.29 ft. This elevation is 1.44 ft below the inlet throat elevation, and thus the hydraulic behavior of the throat will not be affected.

HGL and EGL at Inlet 1

The EGL and HGL elevations immediately downstream of Inlet 1 (at the upstream end of Pipe 1) are determined by computing frictional losses in Pipe 1, $h_L = S_fL$. First, the friction slope is computed as

$$S_f = \frac{(nV)^2}{1.49^2(D/4)^{4/3}} = \frac{[0.013(3.16)]^2}{1.49^2(1.75/4)^{4/3}} = 0.00227$$

The head loss in Pipe 1 is therefore

$$h_L = S_fL = 0.00227(70) = 0.16 \text{ ft}$$

Thus, the HGL and EGL elevations are

$$EGL = 210.45 + 0.16 = 210.61 \text{ ft}$$

$$HGL = 210.61 - 3.16^2/[2(32.2)] = 210.45 \text{ ft}$$

The minor loss due to the pipe entrance within the inlet can be evaluated using Equation 6.26. Assuming that the pipe entrance has a square edge, Table 6.5 on page 203 yields $K = 0.5$. Thus, the minor loss is

$$h_L = KV^2/(2g) = 0.5(3.15)^2/[2(32.2)] = 0.08 \text{ ft}$$

The EGL elevation inside Inlet 1 is therefore

$$EGL = 210.61 + 0.08 = 210.69 \text{ ft}$$

Assuming, conservatively, that the velocity of the flow inside Inlet 1 is negligible, the HGL elevation is the same as the EGL, and is HGL = 210.69 ft.

To check that inlet control does not control the HGL elevation within Inlet 1, the inlet control chart (Figure 9.12) is employed with $D = 21$ in = 1.75 ft and $Q = 7.6$ cfs to obtain HW/D = 0.85 from the scale representing a squared-edged entrance. Thus, under inlet control, the water depth in the inlet would be

$$HW = 0.85D = 0.85(1.75) = 1.49 \text{ ft}$$

Adding this to the invert elevation of the inlet yields a HGL elevation of HGL = 208.82 ft. This is lower than the value of HGL = 210.69 ft determined above, and thus the actual HGL is not governed by inlet control.

The controlling HGL in Inlet 1 is governed by downstream conditions, and HGL = 210.69 ft. This is 1.51 ft below the throat elevation, and hence the hydraulic behavior of the throat will not be affected.

HGL and EGL at Inlet 3

The EGL and HGL elevations immediately downstream of Inlet 3 (at the upstream end of Pipe 2) are determined by computing frictional losses in Pipe 2. The friction slope for Pipe 2 is the same as for Pipe 1, so $S_f = 0.00227$. The head loss in Pipe 2 is therefore

$$h_L = S_fL = 0.00227(315) = 0.72 \text{ ft}$$

and thus the HGL and EGL elevations are

$$EGL = 210.45 + 0.72 = 211.17 \text{ ft}$$

$$HGL = 211.17 - 3.16^2/[2(32.2)] = 211.01 \text{ ft}$$

The minor loss due to the pipe entrance within the inlet can be evaluated using Equation 6.26. Assuming that the pipe entrance has a square edge, Table 6.5 yields $K = 0.5$. Thus, the minor loss is

$$h_L = 0.5(3.16)^2/[2(32.2)] = 0.08 \text{ ft}$$

The EGL elevation inside Inlet 1 is therefore

$$EGL = 211.17 + 0.08 = 211.25 \text{ ft}$$

Assuming, conservatively, that the velocity of the flow inside Inlet 3 is negligible, the HGL elevation is the same as the EGL and is HGL = 211.25 ft.

A check of inlet control yields HW = 1.49 ft and HGL = 210.05 ft. This is lower than the controlling HGL = 211.25 ft, and thus inlet control does not govern the HGL elevation. The governing HGL =

211.25 ft is 2.25 ft below the throat elevation, and hence the hydraulic behavior of the throat will not be affected. The following table summarizes the results of this example. Figure E11.4.2 is a profile of the storm sewer system illustrating the computed HGL and EGL elevations.

Junction Structure	Downstream Structure	EGL (ft)	Velocity (ft/s)	HGL (ft)	Upstream Structure	EGL (ft)	Velocity (ft/s)	HGL (ft)
Outfall	Outfall Channel	209.76	2.00	209.70	Pipe 4	210.07	4.89	209.70
Inlet 4	Pipe 4	210.21	4.89	209.84	Pipe 3	210.25	4.48	209.94
Inlet 2	Pipe 3	210.36	4.48	210.05	Pipe 1 Pipe 2	210.45 210.45	3.15 3.15	210.29 210.29
Inlet 1	Pipe 1	210.61	3.15	210.45	None	210.69	0.00	210.69
Inlet 3	Pipe 2	211.17	3.15	211.01	None	211.25	0.00	211.25

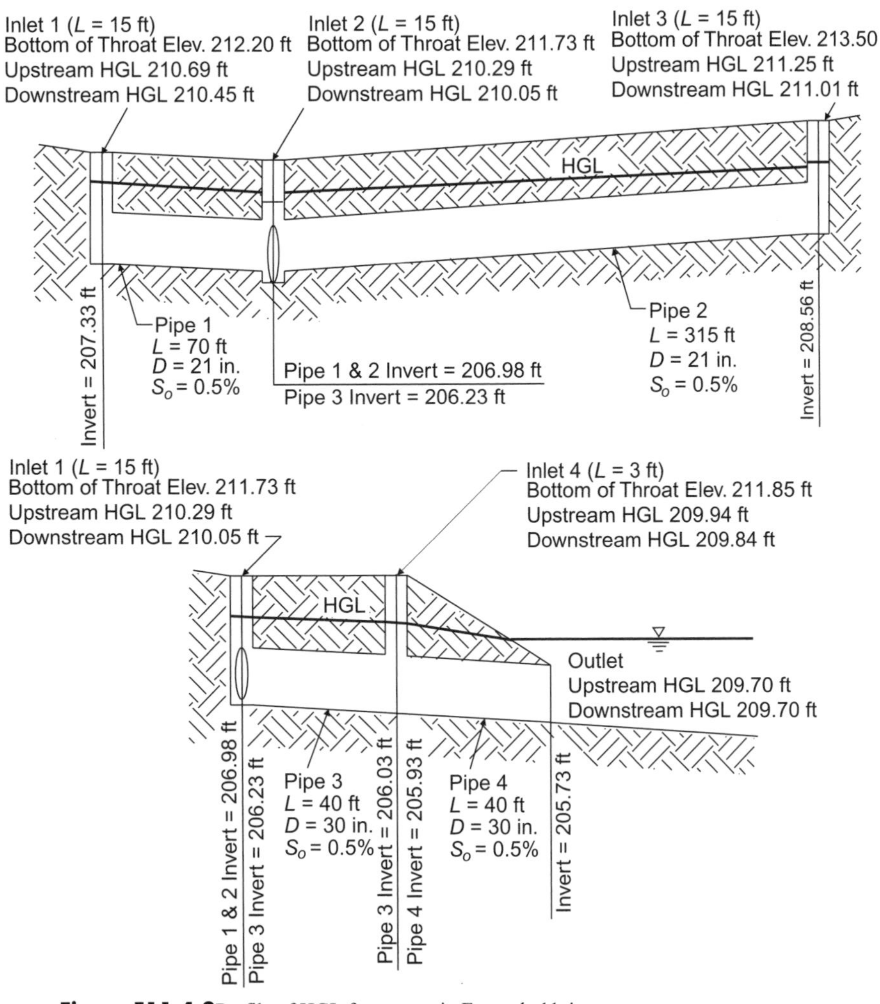

Figure E11.4.2 Profile of HGL for system in Example 11.4

Open-Channel Flow

If a pipe conveys open-channel flow, calculations proceed in an upstream direction if the flow is subcritical or in a downstream direction if the flow is supercritical. Hydraulic jumps, if they occur in the system, may be analyzed using the concept of specific force discussed in Section 7.5 on page 258. Calculations are complicated by the need to compute gradually-varied flow profiles, which require substantial amounts of information on cross-sectional areas, hydraulic radii, and so forth for various depths of flow. These relationships are complex for pipe cross-sections, and make the HGL and EGL calculations quite laborious. Computer modeling tools capable of performing the tedious calculations are obviously desirable in such instances.

Example 11.5 demonstrates the evaluation of the EGL and HGL for the final design of a storm-sewer system with open-channel flows. The basic steps are as follows:

1. Compute normal depth and critical depth for every pipe in the system for use in classifying the profile type for each pipe.
2. Determine whether the flow profiles for the system will be subcritical, supercritical, or mixed. Subcritical flow profiles will be evaluated from the downstream end to the upstream end; the reverse is true for supercritical profiles.
3. If the flow in the outfall pipe is subcritical, start at the downstream end of the system and compute the EGL in the outfall channel and the EGL at the downstream end of the outfall pipe. Consider minor losses at the pipe exit. Typically, exit losses are computed using Equation 6.25 with $K = 1.0$:

$$h_L = 1.0\left|\frac{V_2^2 - V_1^2}{2g}\right|$$

   If the flow is supercritical, assume that the depth at the upstream end of the pipe is equal to critical depth. Determine the entrance losses at the upstream end using the appropriate entrance loss coefficient.

4. Compute the HGL and EGL at the downstream (upstream for supercritical) end of the pipe.
5. Use the direct-step or standard-step method (from Section 7.8) to evaluate the flow profile through the pipe.
6. Compute the HGL and EGL at the upstream (downstream for supercritical) end of the pipe.
7. Use a method such as the energy-loss or composite energy-loss (subcritical only) method to compute the head loss through the structure for each of the incoming pipes. Use these values to determine the EGL and HGL at the downstream end of the incoming pipe (upstream end of outgoing pipe for supercritical).
8. If an HGL elevation is higher than the allowable maximum elevation necessary to maintain minimum freeboard at a structure, increase the size of the outgoing pipe, redo the affected calculations, and recheck the HGL elevations.

9. For subcritical flow, if the structure has more than a single incoming pipe, select one of the incoming branches. Continue the calculations by repeating steps 4 through 9 until the EGL and HGL elevations for the entire branch have been determined.
10. Use Equation 6.25 with $K = 0.5$ to compute the pipe entrance loss at terminal upstream structures, and compute the EGL and HGL at this location. Compare this value for the HGL (which was computed assuming outlet-control conditions) with the inlet-control HGL computed using methods presented in Section 9.5. The higher of these elevations controls the value of the HGL at this location.
11. Repeat steps 3 through 11 until the entire system has been evaluated.
12. Check to ensure that all velocities are within acceptable limits and adjust pipe sizes if necessary.

Example 11.5 – Final Design Process for a System with Open-Channel Flow. Compute the EGL and HGL elevations for the system from Examples 11.2 on page 427 and 11.4 on page 446. However, assume this time that the tailwater depth in the downstream channel is 2 ft. The tailwater velocity is 4.5 ft/s.

Solution: In this example, the tailwater elevation is lower than it was in Example 11.3, where all pipes operated under pressure flow conditions. For this example, the flow profile types must be classified and the HGL determined using procedures for evaluating open-channel flow profiles with the Manning equation and minor losses with the energy-loss method. Profiles showing HGL through the system are given in Figure E11.5.3.

Computing normal and critical depths

The solution procedure begins with the determination of normal and critical depths for each pipe. This determination permits the pipes to be classified as to whether they are hydraulically steep, mild, etc., and hence provides guidance as to the types of flow profiles that will occur in the pipes.

Pipe 1 (D = 1.75 ft) conveys a discharge of Q = 7.6 cfs with n = 0.013 and S_o = 0.005. Geometric properties of the circular pipe were provided in Table 7.1 and are repeated here in Figure E11.5.1 and in Table 11.10.

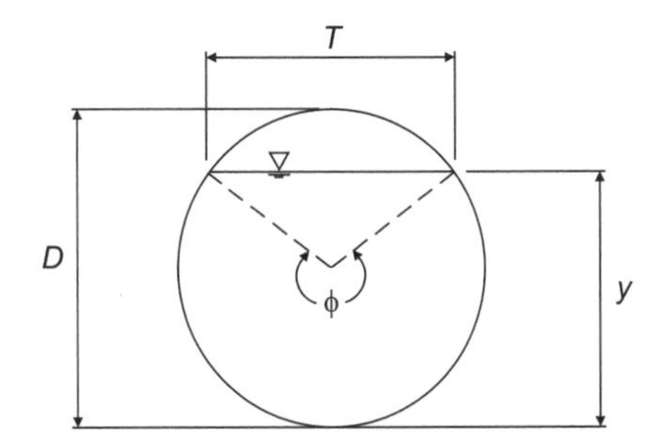

Figure E11.5.1 Definition figure for open-channel flow in a circular channel

Table 11.10 Geometric properties of circular open channels

Area *(A)*	Wetted Perimeter *(P)*	Hydraulic Radius *(R)*	Top Width *(T)*	Hydraulic Depth *(D_h)*
$\frac{D^2(\phi - \sin\phi)}{8}$	$\frac{D\phi}{2}$	$\frac{D}{4}\left(1 - \frac{\sin\phi}{\phi}\right)$	$D\sin\left(\frac{\phi}{2}\right)$	$\frac{D}{8}\left(\frac{\phi - \sin\phi}{\sin(\phi/2)}\right)$

Applying Equation 7.23 from page 235 to first determine ϕ (trial and error required), the normal depth y_n can be determined using Equation 7.20. Equation 7.23 is

$$Q = \frac{1.49}{n}\left[\frac{D^2}{8}(\phi - \sin\phi)\right]^{5/3}\left(\frac{D\phi}{2}\right)^{-2/3} S_o^{1/2}$$

and Equation 7.20 is

$$y = \frac{D}{2}\left[1 - \cos\left(\frac{\phi}{2}\right)\right]$$

Alternatively, Figure 7.8 (repeated here as Figure E11.5.2) can be used to find normal depth from the full-flow capacity and velocity for the pipe. For instance, the diameter of Pipe 1 is 21 in., its discharge (from Example 11.4) is 7.6 cfs, and its slope is 0.005. Applying the Manning equation, the full-flow capacity of the pipe is

$$Q_{full} = \frac{1.49}{n} AR^{2/3} S_o^{1/2} = \frac{1.49}{0.013} \frac{\pi(1.75^2)}{4}\left(\frac{1.75}{4}\right)^{2/3} 0.005^{1/2} = 11.23 \text{ cfs}$$

The ratio Q/Q_{full} is 7.6/11.23 = 0.68. Reading from Figure E11.5.2, y/D = 0.605. Therefore,

$y = 0.605(1.75) = 1.06$ ft

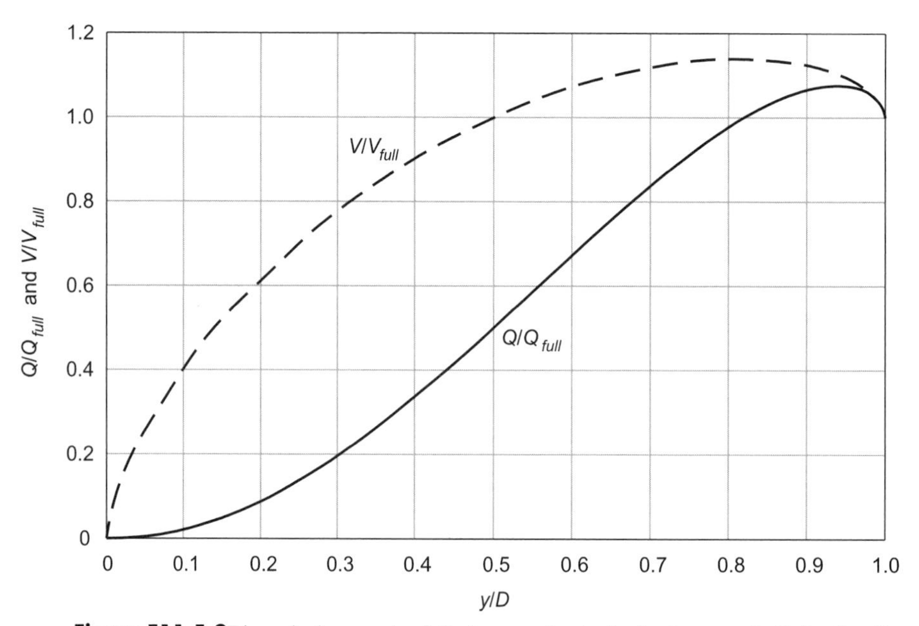

Figure E11.5.2 Dimensionless graph of discharge and velocity fractions vs. depth fraction for circular pipe

Equation 7.20 and Equation 7.23 can be used to verify that 1.06 ft is the normal depth. The same procedure is followed to obtain the normal depths for the other three pipes. These depths are given in the table that follows.

The critical depth in each pipe is found using Equation 7.38 (page 247):

$$\frac{Q^2 T}{gA^3} = 1$$

By assuming a trial depth, one can use Equation 7.20 to determine ϕ, followed by use of table to determine the top width and cross-sectional area. These values are then inserted into Equation 7.38, and the solution proceeds in a trial-and-error fashion until Equation 7.38 converges. For example, for Pipe 1, critical depth is assumed to be 1.0. Equation 7.20 is used to find the corresponding value of ϕ:

$$1 = \frac{1.75}{2}\left[1 - \cos\left(\frac{\phi}{2}\right)\right]$$

The value of ϕ is found to be 196.43° = 3.428 rad. A and T can then be computed as

$$A = \frac{D^2(\phi - \sin\phi)}{8} = \frac{1.75^2(3.428 - \sin 3.428)}{8} = 1.42\ \text{ft}^2$$

$$T = D\sin\left(\frac{\phi}{2}\right) = 1.75\sin\left(\frac{3.428}{2}\right) = 1.73\ \text{ft}$$

Checking this assumption,

$$\frac{Q^2T}{gA^3} = \frac{7.6^2(1.73)}{32.2(1.42)} = 1.08$$

Because 1.08 is not equal to 1.0, another depth should be assumed and the process repeated. In this manner, y_c for Pipe 1 is found to be 1.02 ft. The same procedure can be used to compute the critical depths for the remaining pipes, which are listed in the following table.

Pipe	y_n (ft)	y_c (ft)
1	1.06	1.02
2	1.06	1.02
3	1.63	1.60
4	1.73	1.67

Note that the normal depth is larger than the critical depth in all four pipes; thus, all four pipes have hydraulically mild slopes. Gradually varied flow profiles in the pipes will be of the M types (M1, M2, or M3). M1 and M2 profiles are subcritical flows, and hence are controlled by downstream conditions (calculations proceed upstream). An M3 profile has supercritical flow, and is controlled by upstream conditions (calculations proceed downstream). Detailed water surface profile calculations follow for each pipe and structure.

EGL and HGL at the Outfall

The tailwater depth is 2 ft and the tailwater velocity is 4.5 ft/s. Its velocity head is

$$V^2/2g = 4.5^2/2(32.2) = 0.31\ \text{ft}$$

The HGL (water surface) elevation just downstream of the outfall point is

$$HGL = 205.73 + 2.00 = 207.73\ \text{ft}$$

and the EGL elevation is

$$EGL = 207.73 + 0.31 = 208.04\ \text{ft}$$

The minor energy loss that occurs at the outfall can be evaluated using Equation 7.16 (page 230) with K = 1.0. Writing Bernoulli's equation (Equation 7.3, page 223) for points just upstream (section 1) and just downstream (section 2) of the outfall, and assuming α = 1 and $V_1 > V_2$, one obtains

$$y_1 + \frac{V_1^2}{2g} = y_2 + \frac{V_2^2}{2g} + K\left|\frac{V_2^2 - V_1^2}{2g}\right|$$

$$y_1 + \frac{V_1^2}{2g} = 2.00 + \frac{4.5^2}{2(32.2)} + 1.0\left|\frac{4.5^2}{2(32.2)} - \frac{V_1^2}{2g}\right|$$

$$y_1 + \frac{V_1^2}{2g} = 2.00 + 0.31 + \frac{V_1^2}{2g} - 0.31$$

$$y_1 = 2.00\ \text{ft}$$

Cancelling the velocity head for section 1 from both sides, this expression yields y_1 = 2.00 ft. Therefore, A_1 = 4.21 ft^2 and V_1 = 24/4.21 = 5.70 ft/s, which is greater than the downstream channel velocity of 4.5 cfs, so the assumption that $V_1 > V_2$ is correct.

Immediately upstream of the outfall point in Pipe 4, the water surface (HGL) elevation is HGL = 207.73 ft and the EGL elevation is

$$EGL = 207.73 + 5.70^2/2(32.2) = 208.23 \text{ ft}$$

EGL and HGL through Pipe 4

Because Pipe 4 has a mild (M) slope and the flow depth of y = 2.00 ft at its downstream end is larger than both normal and critical depths, the water surface profile within Pipe 4 is of the M1 type (see Section 7.8). Calculation of the M1 profile can be accomplished via the direct-step method, and is summarized in the following table (see Example 7.20 for an illustration of the detailed calculations required).

y (ft)	*V* (ft/s)	*E* (ft)	Distance Upstream (ft)
2.00	5.70	2.50	0
1.98	5.76	2.49	7
1.96	5.81	2.48	15
1.94	5.87	2.48	22
1.92	5.93	2.47	30
1.90	6.00	2.46	37
1.88	6.06	2.45	45

Pipe 4 has a length of 40 ft, and thus the depth and specific energy at its upstream end are y = 1.89 ft and E = 2.46 ft. The water surface (HGL) elevation at its upstream end is

$$HGL = 205.93 + 1.89 = 207.82 \text{ ft}$$

and the EGL elevation is

$$EGL = 205.93 + 2.46 = 208.39 \text{ ft}$$

EGL and HGL at Inlet 4

The minor energy loss in Inlet 4 can be evaluated using Equations 11.1 through 11.7 with θ= 180°, d = 1.89 ft, h = 211.85 – 207.18 = 4.67 ft, and B = 3.4 ft. The factor K_o = 0.14 and C_{D1} = 1. The factor C_{D2} = 0.42 and C_Q = 1. Because there is a plunging flow from the inlet throat, the factor C_P = 1.42. With no benching of the invert, C_B = 1. The velocity for a depth of 1.89 ft is 6.02 ft/s. Therefore, the minor loss is

$$h_L = 0.14(1)(0.42)(1)(1.42)(1)(6.02^2/64.4) = 0.05 \text{ ft}$$

The EGL elevation at the outlet of Pipe 3 is

$$EGL = 208.39 + 0.05 = 208.44 \text{ ft}$$

Subtracting the invert elevation of Pipe 3, the specific energy at the downstream end of Pipe 3 is

$$E = 208.44 - 206.03 = 2.41 \text{ ft}$$

By trial and error, the depth at the downstream end of Pipe 3 corresponding to this specific energy level is y = 1.97 ft. The HGL elevation at the downstream end of Pipe 3 is therefore

$$HGL = 206.03 + 1.96 = 208.00 \text{ ft}$$

EGL and HGL through Pipe 3

Because the depth of 1.90 ft at the downstream end of the pipe is larger than both the normal and critical depths, the profile is of the M1 type. Profile calculations are summarized in the following table:

y (ft)	V (ft/s)	E (ft)	Distance Upstream (ft)
1.97	5.30	2.41	0
1.95	5.36	2.40	6
1.93	5.41	2.38	12
1.91	5.47	2.37	18
1.89	5.53	2.36	24
1.87	5.59	2.35	30
1.85	5.65	2.35	36
1.83	5.71	2.34	42

Because Pipe 3 is 40 ft long, the depth at its upstream end is y = 1.84 ft and the specific energy at its upstream end is E = 2.34 ft. Therefore, the HGL and EGL elevations at its upstream end are

$$HGL = 206.23 + 1.84 = 208.07 \text{ ft}$$

$$EGL = 206.23 + 2.34 = 208.57 \text{ ft}$$

EGL and HGL at Inlet 2

At Inlet 2, there are two incoming pipes (Pipes 1 and 2). The minor energy loss needs to be evaluated separately for each.

For Pipe 1, $\theta = 90°$, d = 1.84 ft, B = 7.6 ft, and

$$h = 211.73 - 207.48 = 4.25 \text{ ft}$$

The factor K_o = 1.65 and C_{D1} = 1. The factor C_{D2} = 0.42 and C_Q = 0.27. Because there is plunging flow from the inlet throat, the factor C_P = 1.36. With no benching of the invert, C_B = 1. Therefore, the minor loss is

$$h_L = 1.65(1)(0.42)(0.27)(1.36)(1)(5.68^2/64.4) = 0.13 \text{ ft}$$

In this example, Pipes 1 and 2 have identical discharges relative to the outlet discharge and identical geometries. Therefore, the minor energy losses for these pipes are the same.

The EGL elevation at the outlets of Pipes 1 and 2 is therefore

$$EGL = 208.57 + 0.13 = 208.70 \text{ ft}$$

Subtracting from this the invert elevations of those pipes, the specific energy at the downstream end of each pipe is E = 1.72 ft. By trial and error, the depth at the downstream ends of Pipes 1 and 2 corresponding to this specific energy level is y = 1.54 ft. The HGL elevation at the downstream ends of Pipes 1 and 2 is therefore

$$HGL = 206.98 + 1.54 = 208.52 \text{ ft}$$

EGL and HGL through Pipe 2

The depth y = 1.54 ft at the downstream end of Pipe 2 is larger than both the normal and critical depths, so the profile is of the M1 type. The calculations are summarized in the following table.

Because the length of Pipe 2 is 315 ft, which is longer than the 150-ft M1 profile, the actual depth of flow at points further than 150 ft from the downstream end of the pipe is $y = y_n$ = 1.06 ft. Therefore, the HGL and EGL elevations at the upstream end of Pipe 2 are

$$HGL = 208.56 + 1.06 = 209.62 \text{ ft}$$

$$EGL = 208.56 + 1.45 = 210.01 \text{ ft}$$

y (ft)	V (ft/s)	E (ft)	Distance Upstream (ft)
1.54	3.39	1.72	0
1.49	3.48	1.68	14
1.44	3.59	1.64	28
1.39	3.71	1.60	41
1.34	3.85	1.57	55
1.29	4.00	1.54	69
1.24	4.17	1.51	83
1.19	4.36	1.49	98
1.14	4.58	1.47	114
1.09	4.83	1.45	133
1.06 (= y_n)	4.99	1.45	150

EGL and HGL through Pipe 1

Because the HGL and EGL elevations at the downstream end of Pipe 1 are the same as those for Pipe 2, and because the pipes and their discharges are identical, the water surface profile computed for Pipe 2 is also applicable to Pipe 1.

Pipe 1 has a length of 70 ft, and thus the depth at its upstream end is 1.28 ft. The specific energy at its upstream end is 1.53 ft. The HGL and EGL elevations at its upstream end are

$$HGL = 207.33 + 1.29 = 208.62 \text{ ft}$$

$$EGL = 207.33 + 1.54 = 208.87 \text{ ft}$$

EGL and HGL at Inlet 1

The minor loss at the pipe entrance in Inlet 1 can be evaluated using Equation 7.16 (page 230) with K = 0.50 (square-edged pipe entrance). The velocity in Pipe 1 immediately downstream of the inlet is 4.03 ft/s. Therefore, the minor loss is

$$h_L = KV^2/2g = 0.50(4.00)^2/2(32.2) = 0.12 \text{ ft}$$

Adding this value to the EGL at the upstream end of Pipe 1, and assuming (conservatively) that the velocity inside the inlet is negligible, the HGL and EGL elevations inside Inlet 1 are

$$HGL = EGL = 208.87 + 0.12 = 208.99 \text{ ft}$$

From Example 11.4, the HGL and EGL elevations in Inlet 1, if inlet control were governing, would be HGL = EGL = 208.82 ft. Because this elevation is lower than the computed value of 208.99 ft, inlet control does not govern, and the final HGL and EGL for Inlet 1 are 208.99 ft.

EGL and HGL at Inlet 3

The minor loss at the pipe entrance in Inlet 3 can be evaluated using Equation 7.16 with K = 0.50 (square-edged pipe entrance). The velocity in Pipe 3 immediately downstream of the inlet is 4.99 ft/s. Therefore, the minor loss is

$$h_L = 0.50(4.99)^2/2(32.2) = 0.19 \text{ ft}$$

Adding this value to the EGL at the upstream end of Pipe 2, and assuming (conservatively) that the velocity inside the inlet is negligible, the HGL and EGL elevations inside Inlet 3 are

$$HGL = EGL = 210.01 + 0.19 = 210.20 \text{ ft}$$

From Example 11.4 (page 446), the HGL and EGL in Inlet 1, if inlet control were governing, would be HGL = EGL = 210.05 ft. Because this elevation is lower than the computed value of 210.20 ft, inlet control does not govern, and the final HGL/EGL elevation for Inlet 1 is 210.20 ft.

The example results are summarized in the following table. Profiles showing the HGL through the system are given in Equation 11.5.3.

Junction Structure	Downstream				Upstream			
	Downstream Structure	EGL (ft)	Velocity (ft/s)	HGL (ft)	Upstream Structure	EGL (ft)	Velocity (ft/s)	HGL (ft)
Outfall	Outfall Channel	208.04	4.50	207.73	Pipe 4	208.23	5.70	207.73
Inlet 4	Pipe 4	208.39	6.02	207.82	Pipe 3	208.44	5.30	208.00
Inlet 2	Pipe 3	208.57	5.68	208.07	Pipe 1 Pipe 2	208.70 208.70	3.39 3.39	208.52 208.52
Inlet 1	Pipe 1	208.87	4.00	208.62	None	208.99	0.00	208.99
Inlet 3	Pipe 2	210.01	4.99	209.62	None	210.20	0.00	210.20

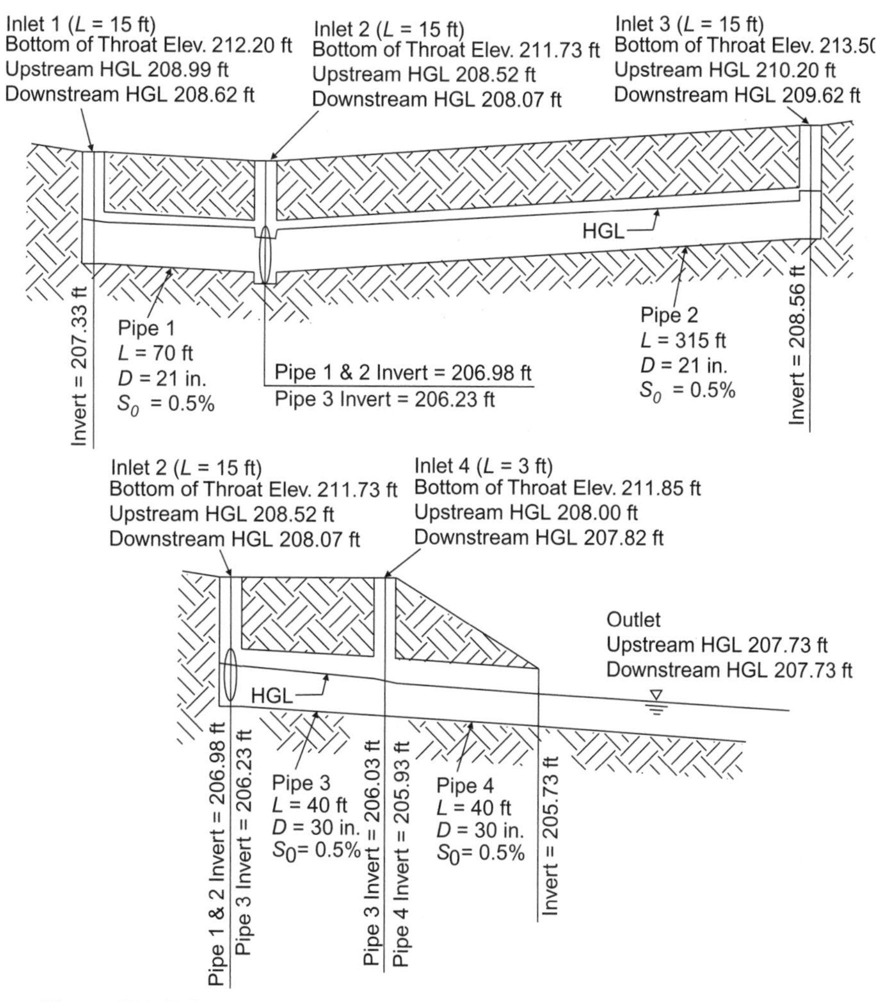

Figure E11.5.3 Profiles showing HGL for Example 11.5

External Forces

External forces on a buried pipe are caused by a combination of dead loads due to the weight of the soil and other materials overlying the pipe, and live loads caused by traffic loading on the ground surface. These external forces may cause both compressive and bending stresses to develop in pipe walls, and can cause a pipe to collapse if its strength is insufficient. Because of these adverse effects, some review agencies specify criteria pertaining to minimum and/or maximum pipe burial depths.

The external forces and corresponding stresses on a buried pipe depend on many things, including the pipe diameter and/or shape, the pipe material, the width of the trench, the burial depth, the specific weight of the backfill material and its degree of compaction, the way in which the pipe is bedded in the trench, and the types of live loads that might be applied at the ground surface. The external forces and stresses also depend on whether the pipe is installed in a trench excavated into a natural ground surface, or whether it is installed in a trench excavated into a constructed embankment.

Because of the variety of pipe materials and installation methods encountered in practice, estimation of the required pipe strength is generally an empirical process. For additional information on this subject, readers are referred to literature and design guidelines published by associations of pipe manufacturers such as the American Concrete Pipe Association (ACPA) and the American Iron and Steel Institute (AISI). Manufacturers and suppliers of plastic pipe materials may also be consulted for guidance.

Structural Loads and Stresses in Pipes

In addition to determining the required diameters and layout of the pipes and/or culverts in a project, the engineer must consider loads and stresses on the conduits so that they will be structurally sound. In particular, the engineer needs to consider issues such as pipe materials and pipe wall thicknesses. In the case of concrete pipe, for example, the engineer needs to specify the pipe class (a measure of its strength) and whether it should be reinforced.

Fundamental to the problem of specifying pipe materials and strength classes is an analysis of the loads on a pipe and the resulting stresses on the pipe walls. *Loads* on pipes may be caused by *internal forces* (fluid pressure or other momentum-related hydraulic forces; see “Internal Pressure Forces” in Chapter 13 on page 557), which tend to cause tensile stresses in pipe walls, or by *external forces,* which tend to cause compressive and/or bending stresses to develop in pipe walls. External forces may arise from *dead loads* due to overlying soil used to backfill pipe trenches, or from *live loads* due to traffic on the ground surface. Stresses in pipes may also develop as a consequence of thermal effects, wherein temperature changes may cause *axial*, that is, longitudinal, forces to develop in pipe walls. When pipes are installed above the ground surface, or are exposed within a pumping station, gravitational forces can cause the pipe to behave much like a beam. Differential settlement of soil and pipe bedding materials can also cause beam loads to develop in buried pipes.

Flow Routing in Storm Sewers

The storm sewer design process, as described in this chapter, is based on estimates of peak discharges for each element (inlet, pipe, etc.) in the system, with hydraulic sizing of elements to accommodate the peak discharge estimates. In most instances, storm sewer systems are small enough that there is so little in-pipe storage compared with the flow through the pipe, so minimal flow attenuation occurs in the system. Flow routing is not necessary in these cases.

An alternative—but much more complex—approach is to route runoff hydrographs through an existing or proposed sewer system to check its performance. Observations of unacceptable system performance attributes, based on the results of flow routing simulations, provide guidance for design modifications and/or needed retrofits. Flow routing in storm sewers is necessary in continuous simulation where inflow hydrographs, as opposed to just peak flow rates, are computed or specified. This may be necessary for large drainage basins where the rational method is not appropriate, in instances where temporary water storage is a concern, or in instances where one desires to simulate water quality attributes as well as flows.

A number of computer algorithms are available for dynamic simulation of storm sewer systems. This is accomplished this by solving the complete Saint-Venant equations expressing conservation of mass and momentum (see Appendix C). There are two practical approaches for solving these equations: explicit and implicit solutions. The widely distributed SWMM is an example of an explicit solution. However, most commercial solutions rely on implicit schemes to solve these equations since explicit schemes exhibit problems with robustness and stability for large time steps. Implicit solutions being more robust have proven reliable in complex situations (Price, 1994). ASCE (1992) provides descriptions of several additional models with flow routing capabilities, and also provides direction to a number of published reviews of available models.

11.6 USING COMPUTER MODELS FOR STORM SEWER DESIGN AND ANALYSIS

The calculations required to analyze the hydraulics of storm system design can be complicated and tedious. Computer programs have greatly simplified the design process, allowing the engineer to focus on making design decisions instead of becoming mired in repetitive computations. In addition, computer programs allow the engineer to quickly view and report results in a variety of useful formats, such as plan and profile drawings, graphs, and tables.

The following steps outline the basic process of using a model for design.

1. Obtain a computer program capable of performing the necessary hydraulic and hydrologic calculations.
2. Obtain topographic data for the project area from surveys, construction drawings, topographic maps, or digital elevation models (see Section 2.5).
3. Identify the roadway and gutter cross-section geometry, preferred inlet types, preferred pipe material, minimum pipe size and slope, and design storm characteristics. (Design storms are discussed in Chapter 4.)
4. Lay out the storm sewer system, including preliminary locations of inlets, manholes, outfalls, and pipes. Identify inlet opening elevations.

5. For each inlet, delineate the drainage area and compute the time of concentration and other parameters necessary for runoff calculations (for example, runoff coefficient or curve number; see Chapter 5).
6. Use the computer program to compute the peak design flow to each inlet and check the capacities of the surface structures (see Chapter 10). Resize the inlets, add inlets, or alter roadway and gutter cross-sectional geometry if necessary. In some instances (such as with StormCAD), the computer program may be capable of automatically adjusting inlet sizes.
7. Establish the horizontal layout for the sewer pipes based on the inlet layout.
8. If you are using a computer program capable of automatically sizing pipes and setting invert elevations, the following steps apply:
    a. Establish pipe design constraints (for example, cover, slope, size, and flow velocity requirements)
    b. Assume all pipes are the minimum allowable size. Using the design storm, perform a design run on the network to set pipe inverts, increase pipe sizes as necessary, and perform hydraulic calculations.
9. If you are using a program that does not have automated design capabilities, do the following:
    a. Approximate pipe slopes based on the slope of the ground surface or minimum slope requirement, whichever is greater. Estimate the design flow for each pipe based on cumulative upstream area characteristics, controlling time of concentration, and system flow time if significant.
    b. For each pipe, use the Manning equation or a similar method (see Sections 6.2 and 7.2) to solve for the pipe size required to convey the peak flow. Assign initial pipe sizes based on next larger commercially available size in the desired material, subject to the minimum size constraint.
    c. Set initial pipe invert elevations based on design constraints such as minimum slope and cover.
    d. Run the program to calculate hydraulic grades, velocities, and so forth.
10. Evaluate the acceptability of the initial design, considering factors such as hydraulic grade, too much or too little capacity, whether design constraints are met, and vertical alignment issues (for instance, minimum distances at utility crossings and cover at intermediate locations along the pipe length). Problems with pipes having excessive velocity and/or slope may be alleviated by using drop structures.
11. Make adjustments to the design and run the calculations again. Re-examine the acceptability of the design, make more adjustments if necessary, and recalculate. Repeat this process until an acceptable design is obtained. Some programs have features such as color coding, annotation, and profiling that can help in identifying problem areas.
12. If multiple design alternatives exist (such as a choice in pipe material or storm sewer route), compare these alternatives by considering factors such as performance, cost, and durability, and develop a design recommendation.

Figure 11.10 is a flowchart showing the typical process for the design of a storm sewer system.

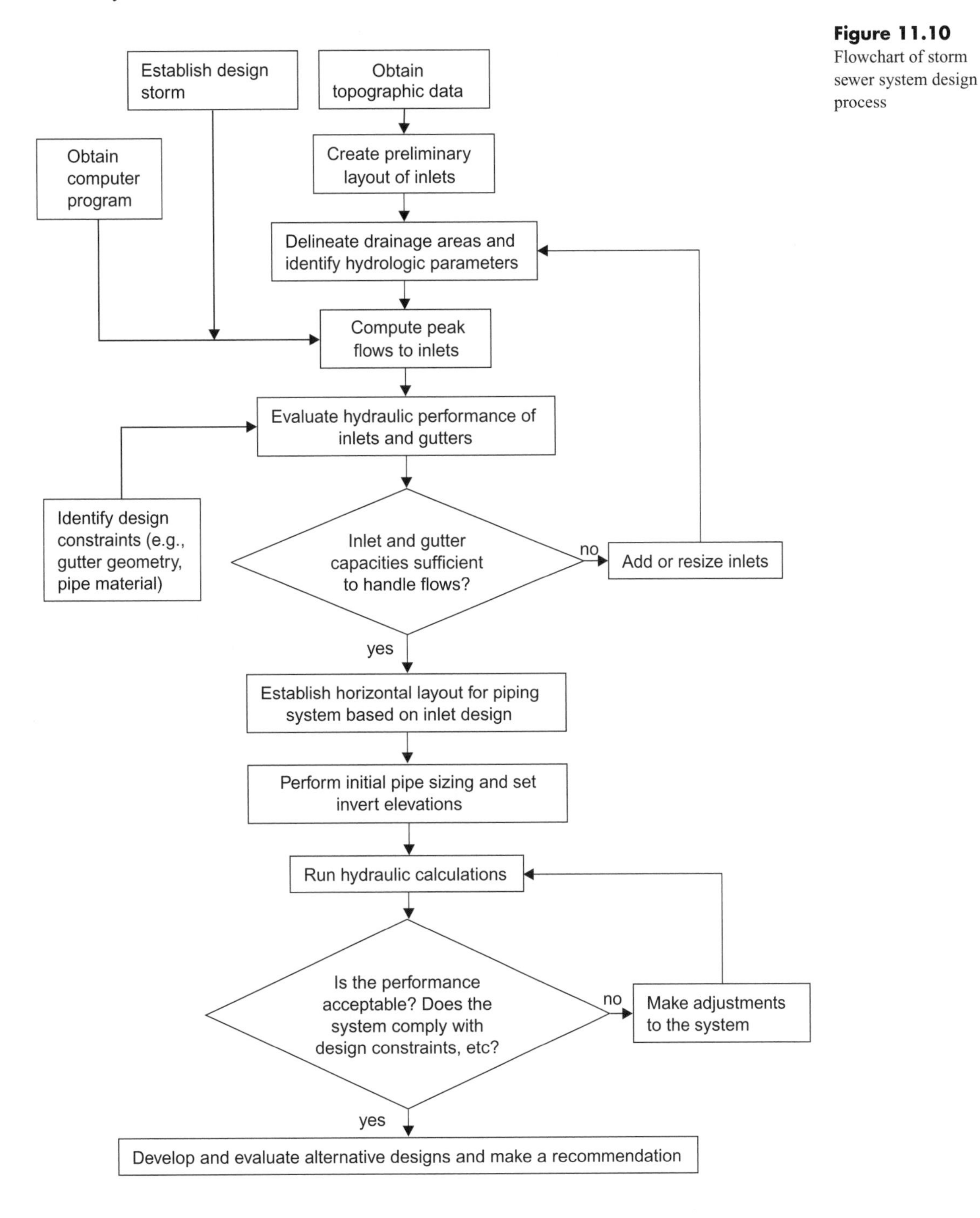

Figure 11.10
Flowchart of storm sewer system design process

11.7 OUTFALL DESIGN AND ENERGY DISSIPATION

The discharge exiting a storm sewer system at its outfall point often has a velocity high enough to cause destructive and unsightly erosion. Protection against erosion is therefore necessary at storm sewer outfalls. Methods of erosion protection vary, but riprap aprons are frequently used. *Riprap* is angular stone of a size and gradation sufficient to resist the mobilizing forces of flowing water. For extreme instances in which exit velocities are very high, a special type of energy dissipator structure may be required, such as an impact-type dissipator or stilling basin. However, because of the potentially high cost of special energy dissipators, designers should give consideration to system design modifications that reduce exit velocities and thus eliminate the need for them. Guidance for design of energy dissipators can be obtained from the U.S. Bureau of Reclamation (USBR, 1987; USBR, 1983).

This section provides guidance on design of riprap erosion protection and is based on a procedure used in the Denver, Colorado Urban Drainage and Flood Control District (UDFCD, 2001). This procedure has been suggested for use in applications where all of the following are true.

1. The outlet Froude number is less than or equal to 2.5 [that is, the values of parameters $Q/D^{2.5}$ or $Q/WH^{1.5}$ are less than or equal to 14 ft$^{0.5}$/s (7.7 m$^{0.5}$/s)].
2. The outlet pipe slope is approximately the same as that of the downstream riprap apron.
3. The invert of the outlet pipe is flush with the level of the riprap surface.

In number (1) above,

Q = discharge (cfs, m^3/s)
D = diameter of a circular outlet pipe (ft, m)
W = width of rectangular conduit (ft, m)
H = height of rectangular conduit (ft, m)

Plan and profile views showing the general configuration of a riprap apron for a trapezoidal channel are illustrated in Figure 11.11. The recommended maximum side slope for the channel is 4:1 (horizontal:vertical), and a slope that should not exceed 3:1. The height of the riprap should extend to the height of the culvert or normal depth in the channel, whichever is smaller. The thickness of the riprap in the channel sides and bottom should be 1.5 times the median size of the riprap, d_{50}. Note the extra thickness of the riprap ($2d_{50}$) in the area adjacent to the pipe outlet, where the hydraulic forces tending to mobilize the riprap are generally the largest. At the pipe outlet, either a concrete cradle (see Figure 9.6) or a headwall should be provided to prevent undermining and possible settlement or failure of the pipe. At the downstream end of the riprap apron, an end slope is provided on the riprap to accommodate potential degradation of the downstream channel.

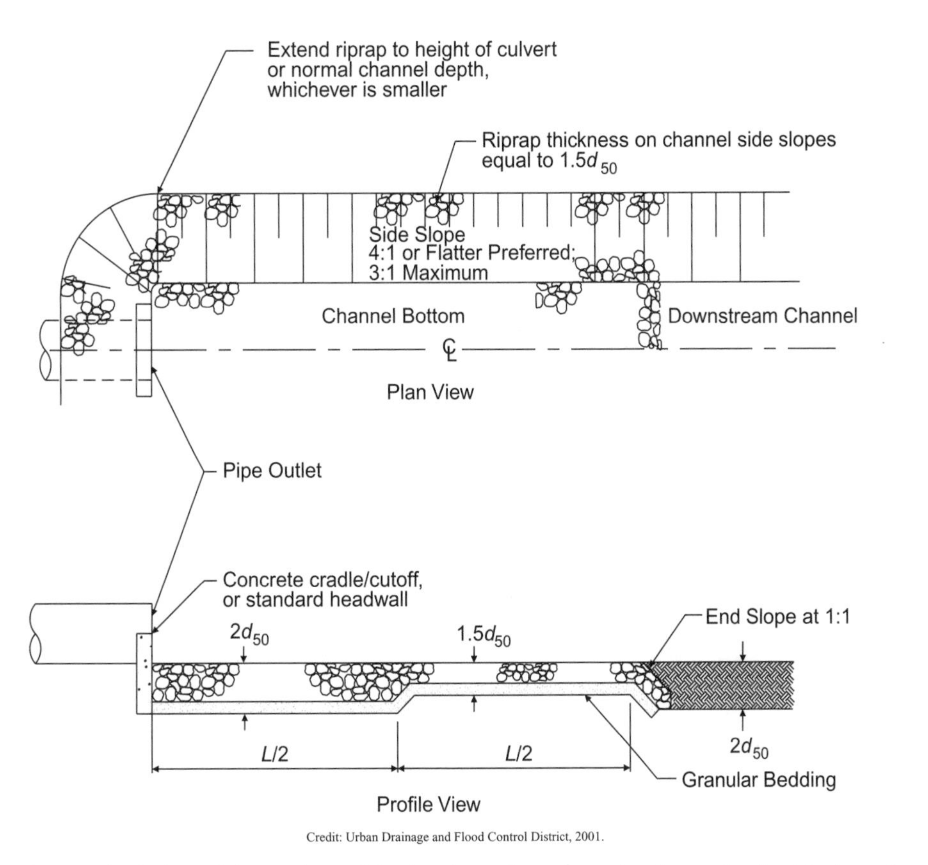

Credit: Urban Drainage and Flood Control District, 2001.

Figure 11.11
A riprap apron

The required size of the rock used in the riprap layer may be determined with the aid of Figures 11.12 and 11.13, with the former used for circular outlet pipes and the latter for rectangular conduits. The variable y_t used in the figures denotes the tailwater depth in the channel at the downstream end of the riprap apron. If y_t is unknown, or if it is suspected that a hydraulic jump occurs at the outlet, it is recommended that one use $y_t/D = y_t/H = 0.4$ with these figures. The gradations for the indicated rock types (L, M, and so forth) are provided in Table 11.11.

Figures 11.12 and 11.13 are based on an assumption that the flow in the outlet pipe is subcritical. Nevertheless, these figures can be used for supercritical pipe flows if the values of D or H are modified and replaced, respectively, by D_a or H_a as given in Equations 11.21 and 11.22.

$$D_a = \frac{D + y_n}{2} \tag{11.21}$$

$$H_a = \frac{H + y_n}{2} \tag{11.22}$$

where y_n = normal depth of the supercritical flow in the conduit (ft, m)

Maximum values of D_a and H_a, respectively, are D and H.

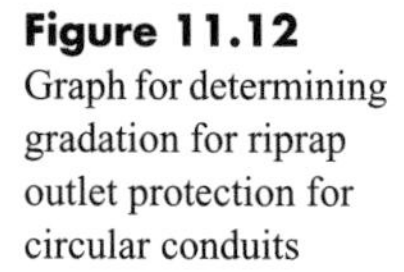

Figure 11.12
Graph for determining gradation for riprap outlet protection for circular conduits

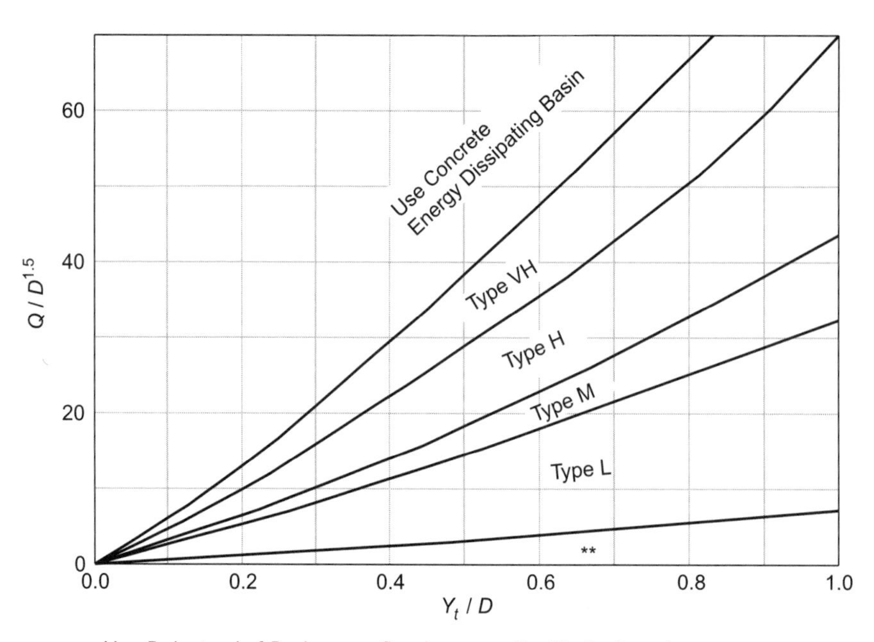

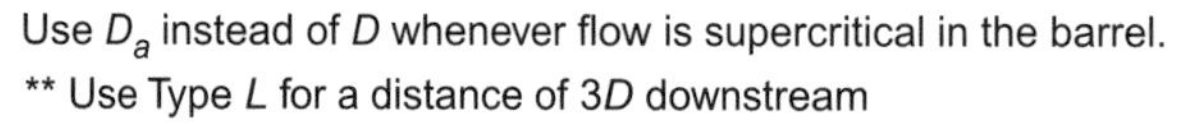

Credit: Urban Drainage and Flood Control District, 2001.

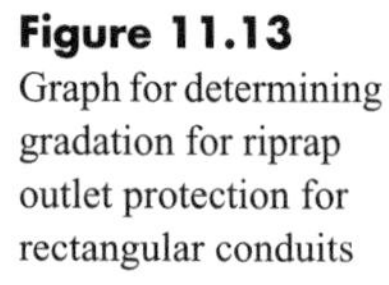

Figure 11.13
Graph for determining gradation for riprap outlet protection for rectangular conduits

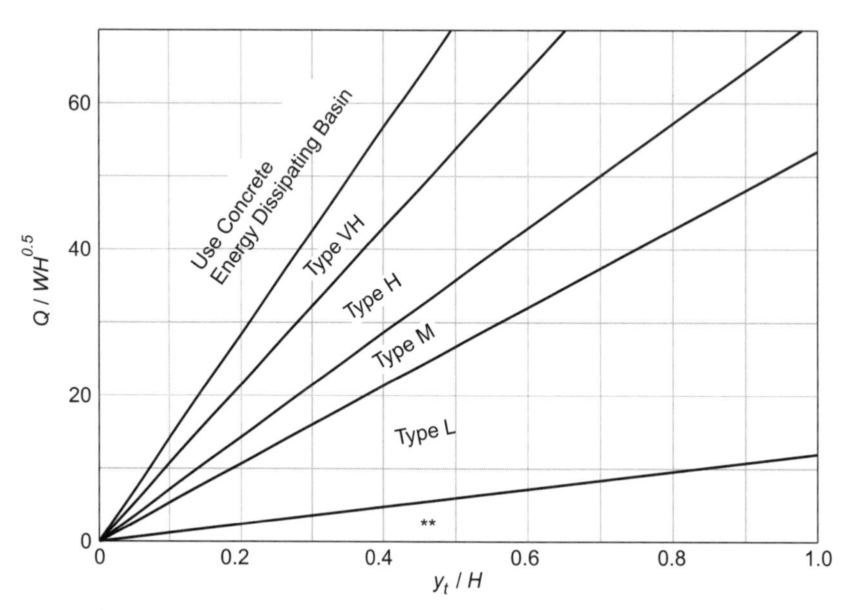

Use H_a instead of H whenever culvert has supercritical flow in the barrel.
** Use Type L for a distance of $3H$ downstream

Credit: Urban Drainage and Flood Control District, 2001.

Table 11.11 Classification and gradation of ordinary riprap (adapted from UDFCD, 2001)

Type[a]	% Smaller by Weight	Rock Size		Median Size, d_{50}	
		in.	mm	in.	mm
VL	70–100	12	300	6	150
	50–70	9	230		
	35–50	6	150		
	2–10	2	50		
L	70–100	15	380	9	230
	50–70	12	300		
	35–50	9	230		
	2–10	3	80		
M	70–100	21	530	12	300
	50–70	18	460		
	35–50	12	300		
	2–10	4	100		
H	100	30	760	18	460
	50–70	24	610		
	35–50	18	460		
	2–10	6	150		
VH	100	42	1070	24	610
	50–70	33	840		
	35–50	24	610		
	2–10	9	230		

a. Bury types VL and L with native topsoil and revegetate to protect from vandalism.

The length L of the riprap apron (see Figure 11.11) depends on the degree of desired protection. The greater the difference between the pipe outlet velocity and the downstream tailwater velocity (that is, the required velocity reduction), the longer the required length of the apron. The procedure recommended by the UDFCD (2001) assumes that the velocity at any cross section downstream from the pipe outlet point is related to the angle of the lateral expansion, θ, of the jet. Assuming that the expanding jet has a rectangular shape, the required apron length can be determined as

$$L = \frac{1}{2\tan\theta}\left(\frac{A_t}{y_t} - W\right) \tag{11.23}$$

where L = length of apron (ft, m)
W = width of the outlet pipe (use $W = D$ for circular conduits) (ft, m)
y_t = tailwater depth (ft, m)
θ = expansion angle of the jet
A_t = the tailwater cross-sectional area, Q/V_t (ft^2, m^2)
V_t = allowable non-eroding tailwater velocity (ft/s, m/s)

The expansion factor, $1/(2\tan\theta)$, in Equation 11.23 can be obtained from Figure 11.14 or 11.15, depending on whether the outlet conduit is circular or rectangular. For example, Figure 11.14 is read by finding the value of the expansion factor corresponding to the intersection of y_t/D with the line for $Q/D^{2.5}$.

Figure 11.14
Graph for obtaining the expansion factor for circular conduit for use with Equation 11.23

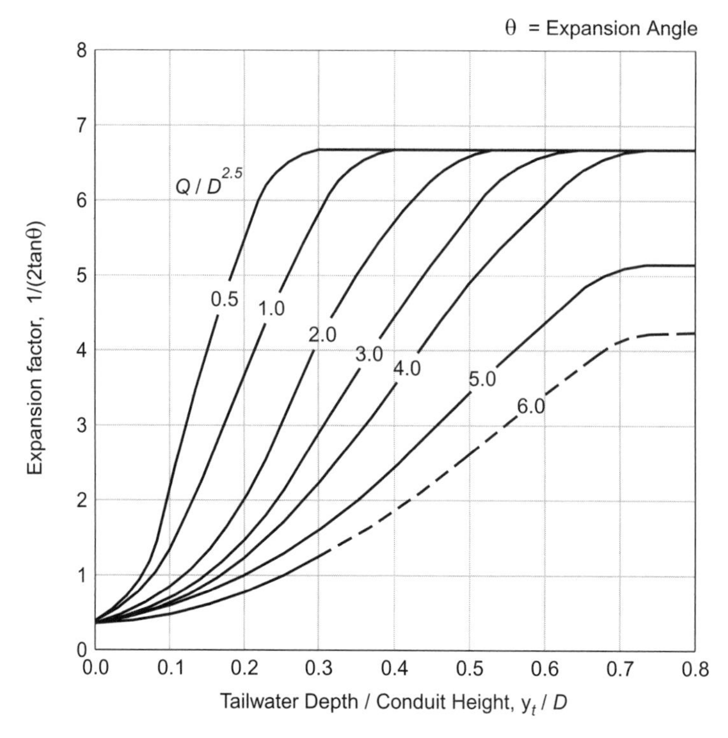

Credit: Urban Drainage and Flood Control District, 2001.

Figure 11.15
Graph for obtaining the expansion factor for rectangular conduit for use with Equation 11.23

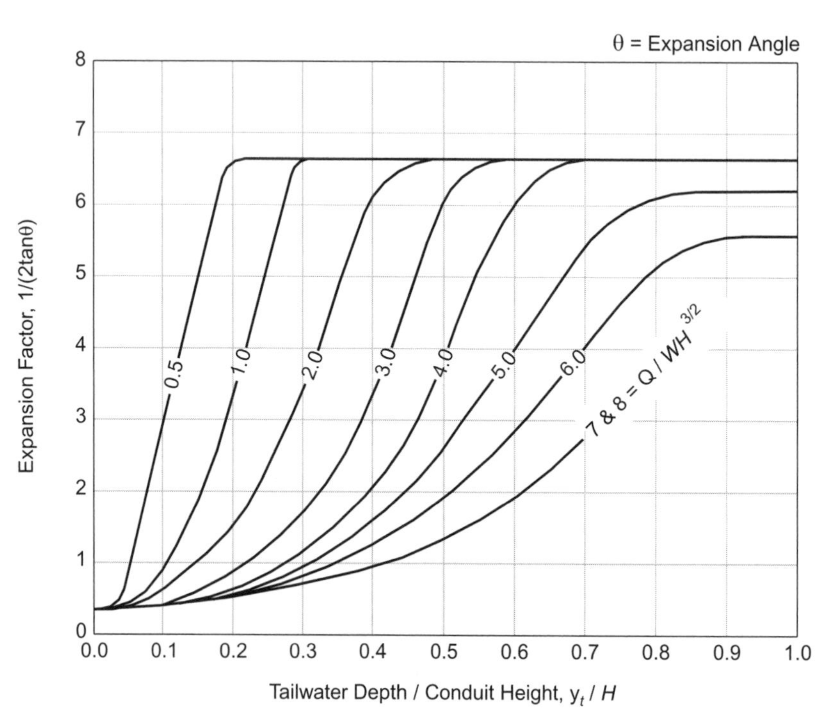

Credit: Urban Drainage and Flood Control District, 2001.

In some cases, Equation 11.23 will yield unreasonable results. For circular conduits, the required length L should not be less than $3D$, nor more than $10D$ if $Q/D^{2.5}$ is less than 6 $ft^{0.5}/s$ (3.3 $m^{0.5}/s$). If $Q/D^{2.5}$ is greater than 6 $ft^{0.5}/s$ (3.3 $m^{0.5}/s$), then the apron length is computed as

$$L = 10D + (Q/D^{2.5}C_f - 6)(D/4) \tag{11.24}$$

where C_f = 1 (U.S. Customary) or 1.811 (SI)

For rectangular channels, the required length L should not be less than $3H$, nor more than $10H$ if $Q/WH^{1.5}$ is less than 8 $ft^{0.5}/s$ (4.4 $m^{0.5}/s$). If $Q/WH^{1.5}$ is greater than 8 $ft^{0.5}/s$ (4.4 $m^{0.5}/s$), then the apron length is computed as

$$L = 10H + (Q/WH^{1.5}C_f - 8)(H/4) \tag{11.25}$$

To minimize piping of natural soil materials into the riprap layer, a filter fabric or granular filter layer, or both, should be placed under the riprap. Filter fabric is prone to tear during the dumping of rock to form the actual riprap layer, and also provides less resistance to the mobilizing forces of the flowing water than does a granular filter material. Nevertheless, filter fabrics have proven to be an adequate replacement for granular filter materials, particularly in mildly sloping channels. If a granular filter is used, its gradation should be determined using an appropriate geotechnical filter design method. The UDFCD (2001) provides guidance on this, as do many geotechnical engineering texts (such as Holtz and Kovacs, 1981).

Example 11.6 – Riprap Blanket Design. Determine the required length and gradation of a riprap apron downstream of the storm sewer system designed in Example 11.5. Recall that the downstream channel velocity is V_t = 4.5 ft/s, the downstream depth is y_t = 2.00 ft, the pipe exit velocity is V = 5.70 ft/s, the discharge is Q = 24 cfs, and the pipe diameter is D = 30 in. = 2.5 ft.

Solution: Because the depth of flow in the exit pipe is larger than critical depth, the flow is subcritical and the Froude number is less than unity. The design procedure described is therefore applicable.

The required riprap gradation can be determined using Figure 11.12 with

$Q/D^{1.5} = 24/2.5^{1.5} = 6.1$ and $y_t/D = 2/2.5 = 0.80$

From the figure, Type L riprap (Table 11.11) should be selected. Because this stone is small enough to be subject to vandalism, it is recommended that it be buried using natural channel materials after it has been placed.

The expansion factor, $1/(2\tan\theta)$, in Equation 11.23 can be obtained from Figure 11.14 with

$Q/D^{2.5} = 24/2.5^{2.5} = 2.4$ and $y_t/D = 2/2.5 = 0.80$

The expansion factor is therefore

$1/(2\tan\theta) = 6.7$

The required length of the apron is found using Equation 11.23 as

$$L = 6.7\left(\frac{24/4.5}{2.00} - 2.5\right) = 1\text{ ft}$$

Because this length is shorter than the minimum length of $3D$, the apron length is set equal to

$L = 3(2.5) = 7.5$ ft

The geometric configuration of the riprap layer should be as shown in Figure 11.11.

Modeling Focus: Storm Sewer Design

A storm sewer system is being designed for a commercial site, and the proposed horizontal layout for the system is shown in the figure below. However, pipe sizes and invert elevations must still be determined, and the performance of the inlets and gutters must be checked.

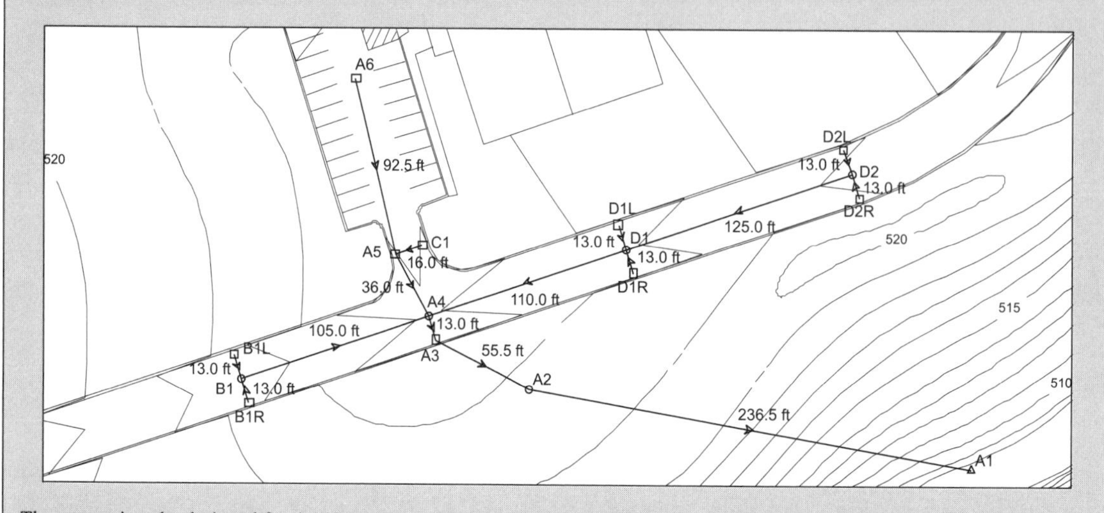

The system is to be designed for the 10-year storm using the rational method. Ten-year IDF data is given in the table below.

Duration (min)	5	10	15	20	25	30	35	45	60
Intensity (in./hr)	4.00	3.00	2.65	2.30	2.05	1.80	1.65	1.40	1.10

Pipes should be designed to flow full and pipe crown elevations should be matched at inlet locations. Other design constraints are provided in the table on the left.

All gutters have a uniform cross-section with a 2-percent cross slope, and all inlets are combination-type with the characteristics shown on the right (note that these characteristics correspond to a DI-10A inlet type in StormCAD's default inlet library). All pipes will be concrete (Manning's $n = 0.013$) and circular with the lengths shown in the figure above.

Criterion	Value
Pipe cover	2 ft min., 10 ft max.
Flow velocity	2 ft/s min., 15 ft/s max.
Slope	1% min, 10% max.
Gutter spread	6 ft max.
Gutter depth	6 in. max.
Inlet capture efficiency	50% min.

Parameter	Value
Inlet box dimensions	3 ft x 3 ft
Curb opening characteristics	Vertical throat
Height	4.5 in.
Length	3 ft
Local depression depth	1 in.
Local depression width	28 in.
Grate characteristics	45° tilt bar grate
Width	20 in.
Length	3 ft

Information on the node structures (inlets, manholes, and outlet) is provided in the next table, including hydrologic characteristics, whether inlets are on grade or in sag locations, "bypass targets" that define the next downstream inlet for inlets on grade, roadway slopes, and head loss calculation data.

(continued on facing page)

Label	Structure Type	Ground Elev, (ft)	Area (ac)	Inlet C	Time of Conc. (min)	Inlet Location	Bypass Target	Long Slope (ft/ft)	Head Loss Method	HEC-22 Benching Method	Head Loss Coeff.
A1	Outlet	510.60	N/A	N/A	N/A	N/A	N/A	N/A	N/A	N/A	N/A
A2	Manhole	518.20	N/A	N/A	N/A	N/A	N/A	N/A	HEC-22 Energy	Flat	N/A
A3	Inlet	516.80	0.60	0.45	5.00	In Sag	N/A	N/A	HEC-22 Energy	Flat	N/A
A4	Manhole	516.90	N/A	N/A	N/A	N/A	N/A	N/A	HEC-22 Energy	Flat	N/A
A5	Inlet	517.00	0.39	0.45	10.00	In Sag	N/A	N/A	HEC-22 Energy	Flat	N/A
A6	Inlet	517.80	0.75	0.41	10.00	In Sag	N/A	N/A	Standard	N/A	0.50
B1	Manhole	518.40	N/A	N/A	N/A	N/A	N/A	N/A	HEC-22 Energy	Flat	N/A
B1L	Inlet	518.20	0.20	0.42	7.00	On Grade	A5	0.020	Standard	N/A	0.50
B1R	Inlet	518.20	0.20	0.90	5.00	On Grade	A3	0.020	Standard	N/A	0.50
C1	Inlet	517.00	0.37	0.63	8.00	In Sag	N/A	N/A	HEC-22 Energy	Flat	N/A
D1	Manhole	517.90	N/A	N/A	N/A	N/A	N/A	N/A	HEC-22 Energy	Flat	N/A
D1L	Inlet	517.70	0.46	0.35	10.00	On Grade	C1	0.010	Standard	N/A	0.50
D1R	Inlet	517.70	0.08	0.90	5.00	On Grade	A3	0.010	Standard	N/A	0.50
D2	Manhole	519.20	N/A	N/A	N/A	N/A	N/A	N/A	HEC-22 Energy	Flat	N/A
D2L	Inlet	518.80	0.25	0.42	5.00	On Grade	D1L	0.010	Standard	N/A	0.50
D2R	Inlet	518.80	0.14	0.56	5.00	On Grade	D1R	0.010	Standard	N/A	0.50

All of the input data, including design constraints, are entered into a design program capable of performing automatic design, such as StormCAD. The results are summarized in the following tables.

Label	Intercepted Flow (cfs)	Bypassed Flow (cfs)	Capture Efficiency (%)	Gutter Spread (ft)	Gutter Depth (ft)	Upstream HGL (ft)
A3	1.34	0.00	100.0	3.87	0.08	514.19
A5	0.57	0.00	100.0	1.75	0.01	514.68
A6	0.92	0.00	100.0	2.24	0.04	515.51
B1L	0.27	0.04	86.3	3.66	0.07	515.45
B1R	0.53	0.20	73.1	5.04	0.10	515.56
C1	0.93	0.00	100.0	2.25	0.04	514.67
D1L	0.41	0.13	76.1	5.15	0.10	515.01
D1R	0.34	0.08	80.0	4.32	0.08	514.96
D2L	0.34	0.08	80.0	4.69	0.09	516.08
D2R	0.27	0.05	84.3	4.19	0.08	516.05

Label	Sec. Size (in.)	U.S. Invert Elev. (ft)	D.S. Invert Elev. (ft)	Const. Slope (%)	Total Flow (cfs)	Ave. Vel. (ft/s)	U.S. Cover (ft)	D.S. Cover (ft)
A2-A1	15	512.39	508.60	1.60	5.84	6.45	4.57	0.75
A3-A2	15	512.94	512.39	1.00	5.86	5.37	2.61	4.57
A4-A3	15	513.07	512.94	1.00	4.88	4.03	2.58	2.61
A5-A4	12	513.84	513.32	1.44	3.25	4.57	2.16	2.58
A6-A5	12	514.80	513.84	1.04	1.90	3.23	2.00	2.16
B1-A4	12	515.07	513.32	1.67	0.74	1.94	2.33	2.58
B1L-B1	12	515.20	515.07	1.00	0.27	1.59	2.00	2.33
B1R-B1	12	515.20	515.07	1.00	0.53	2.26	2.00	2.33
C1-A5	12	514.00	513.84	1.00	0.93	1.50	2.00	2.16
D1-A4	12	514.42	513.32	1.00	1.08	2.33	2.48	2.58
D1L-D1	12	514.70	514.42	2.15	0.41	1.82	2.00	2.48
D1R-D1	12	514.70	514.42	2.15	0.28	1.52	2.00	2.48
D2-D1	12	515.67	514.42	1.00	0.60	2.22	2.53	2.48
D2L-D2	12	515.80	515.67	1.00	0.34	1.87	2.00	2.53
D2R-D2	12	515.80	515.67	1.00	0.27	1.66	2.00	2.53

11.8 CHAPTER SUMMARY

The basic components of storm sewers are pipes, inlet boxes, manholes, junction structures, catch basins, and outlet structures. Some storm systems may include other facilities such as detention structures and pumping stations, which are covered in Chapters 12 and 13, respectively.

The most common materials used for storm sewer piping are concrete and corrugated steel. The most common cross-sectional shape is circular, but shapes such as boxes, arches, and ellipses are also available. Inlet boxes support the inlet opening, connect the surface drainage system to the subsurface piping system, and can serve as junctions where pipe alignment or size changes. Catch basins are modified versions of inlet boxes for capturing suspended sediment and other debris in the bottom of the structure. Manholes also serve as pipe junctions, and they provide access to the sewer system at locations where no additional surface flow enters. Manholes and inlets should be located at regular intervals to provide access for cleaning and maintenance of the system. Other types of junction structures may be used instead of manholes in large or more complex installations. Storm sewer outlet structures commonly consist of a concrete endwall and some type of erosion protection, such as riprap. Frequently, energy dissipation devices are also required.

Storm sewer pipes typically run in straight lines to connect inlets, manholes, and other structures. A certain amount of pipe curvature can be accomplished by deflecting pipe joints. Pipes must be placed to avoid conflicts with other utilities and obstructions. Minimum cover depths on pipes must be maintained to prevent crushing, and minimum and maximum slope criteria should be met. It is advisable to check flow velocities to help ensure that pipes can be self-cleansing.

Once the design flows for each pipe have been determined by hydrologic analysis and inlet design calculations, and a horizontal layout of the system is complete, the engineer can perform the initial design of pipe sizes and invert elevations. Typically, pipe slopes are assumed to be roughly the same as the slope of the ground surface, and an equation for computing pipe capacity, such as the Manning equation, is used to compute minimum pipe size requirements assuming a full-flow condition. These sizes are then rounded to the next available size, and pipe inverts are assigned based on minimum/maximum cover and slope criteria.

The final design calculations for a storm sewer system involve determining the hydraulic grade through the system. In addition to considering the head losses due to pipe friction, these calculations must also account for minor losses through structures such as manholes and inlet boxes. Two methods for computing these minor losses presented in FHWA's HEC-22 publication are the energy-loss method and the composite energy-loss method. These methods consider factors such as the junction geometry, shaping of the structure to minimize losses, the relative elevations of incoming pipes, and so forth.

For systems or portions of systems operating under pressure-flow conditions, the hydraulic grade can be computed using the equations presented in Chapter 6. In the case of systems with open-channel flow conditions, the water surface profile evaluation techniques described in Chapter 7 should be used. In either case, the computed hydraulic grade elevations must be compared with the elevations of the tops of inlet structures and manholes to ensure that no flooding will occur for the design event. Final design calculations can be quite cumbersome to do by hand, so computerized modeling techniques are generally used.

Other considerations when designing a storm sewer system include structural loads and stresses in pipes and proper outfall design. Riprap is a common measure for preventing erosion and dissipating flow energy at storm sewer outlets. Various procedures are available for estimating riprap apron dimensions and the gradation of stone required, such as the Denver, Colorado Urban Drainage and Flood Control District approach described in this chapter.

REFERENCES

American Concrete Pipe Association (ACPA). 2000. *Concrete Pipe Design Manual.* 13th Printing (revised). Irving, Texas: ACPA.

American Iron and Steel Institute (AISI). 1999. *Modern Sewer Design.*, 4th ed. Washington, D.C.: AISI.

American Society of Civil Engineers (ASCE). 1992. *Design and Construction of Urban Storm Water Management Systems.* New York: ASCE.

Brown, S. A., S. M. Stein, and J. C. Warner. 2001. *Urban Drainage Design Manual.* Hydraulic Engineering Circular No. 22, FHWA-SA-96-078. Washington, D.C.: U.S. Department of Transportation.

Chang, F. M., R. T. Kilgore, D. C. Woo, and M. P. Mistichelli. 1994. *Energy Losses through Junction Manholes, Vol. 1, Research Report and Design Guide.* McLean, Virginia: Federal Highway Administration.

Holtz, R. D. and W. D. Kovacs. 1981. *An Introduction to Geotechnical Engineering.* Englewood Cliffs, New Jersey: Prentice Hall.

Steel, E. W. and T. J. McGhee. 1979. *Water Supply and Sewerage.* 5th ed. New York: McGraw-Hill.

U.S. Bureau of Reclamation (USBR). 1983. *Design of Small Canal Structures.* Lakewood, Colorado: USBR.

U.S. Bureau of Reclamation (USBR). 1987. *Design of Small Dams.* 3d ed. Lakewood, Colorado: USBR.

Urban Drainage and Flood Control District (UDFCD). 2001. *Urban Storm Drainage Criteria Manual, Vol. 2, Major Drainage.* Denver, Colorado: UDFCD.

Yen, B. "Hydraulics of Storm Sewer Systems." In *Stormwater Collection System Design Handbook,* ed. Mays, L. New York: McGraw Hill.

PROBLEMS

11.1 A stormwater collection system is being designed for a small subdivision. The layout of the system is shown schematically in the figure and the table presents hydrologic and ground elevation data. (Catchment and ground elevation information for each row apply to the "From" inlet.)

The design is based on a time of concentration of 20 min. to the most upstream inlets (A-6, C-2, and B-1) and then governing system flow times (from estimated flow velocities in the system) as the design proceeds downstream. Using the rational method and the 10-year rainfall data provided, calculate the design flow in each segment of the system. Next, appropriately size concrete pipes ($n = 0.013$) for the system and set pipe invert elevations. Assume that the pipes can flow full, the minimum allowable pipe size is 12 in., and the minimum allowable slope is 0.005. Pipe crown elevations should match at each structure. The ground elevation at the outfall is 88.7 ft. Record your answers in the results table.

Duration (min)	10-Year Intensity (in./hr)
5	7.44
10	6.18
15	5.20
30	3.80
60	2.53

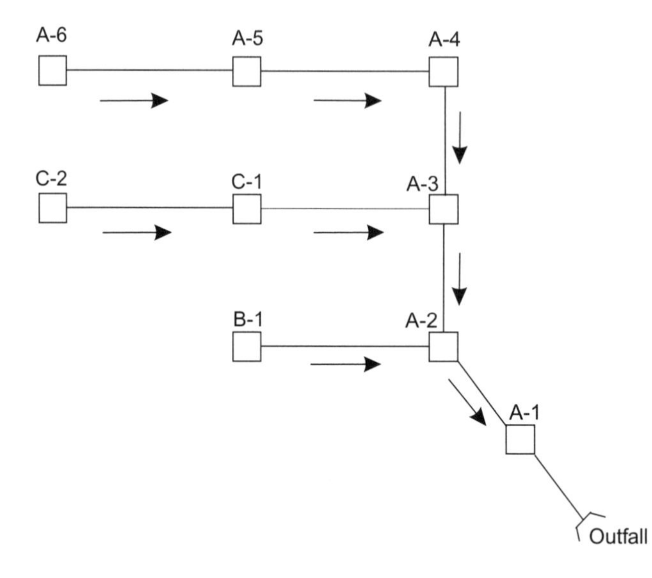

Inlet Numbers From	Inlet Numbers To	Section Length (ft)	Area To Inlet (ac)	Area Accumulated (ac)	System Flow Time (min)	Accumulated Average Runoff Coeff.	Ground Elevation (ft)
A6	A5	500	2.69	2.69	20.0	0.30	98.6
A5	A4	500	3.61	6.30	21.4	0.30	94.9
A4	A3	400	3.88	10.18	22.6	0.41	91.8
C2	C1	500	5.59	5.59	20.0	0.30	96.5
C1	A3	500	6.44	12.03	21.1	0.30	92.3
A3	A2	400	3.92	26.13	23.8	0.39	89.7
B1	A2	400	2.51	2.51	20.0	0.37	92.9
A2	A1	400	32.5	32.5	24.9	0.43	89.5
A1	Outfall	150	38.04	38.04	26.0	0.45	88.7

Results Table				
Pipe	Discharge (cfs)	Pipe Size (in.)	Upstream Inv. Elevation (ft)	Downstream Inv. Elevation (ft)
A6-A5				
A5-A4				
A4-A3				
C2-C1				
C1-A3				

Results Table				
Pipe	**Discharge (cfs)**	**Pipe Size (in.)**	**Upstream Inv. Elevation (ft)**	**Downstream Inv. Elevation (ft)**
A3-A2				
B1-A2				
A2-A1				
A1-Outfall				

11.2 Use the Energy-Loss Method to compute the adjusted minor loss coefficient K for computing the head loss through a manhole with the characteristics given in the table. The structure has one incoming pipe, matching pipe crown elevations, and half benching.

Structure Diameter (B)	3 ft
Outlet Pipe Diam. (D_o)	24 in.
Incoming Pipe Diam.	18 in.
Angle between pipes (θ)	2.08 radians
Water depth in structure (d)	1.2 ft

11.3 Develop the hydraulic and energy grade lines for the three segments of the conveyance system described in the table and figure below. The water level in the receiving body is 22.9 m. The total flow through the system is 5.7 m^3/s. All pipes are 1500 mm concrete pipe ($n = 0.013$). Assume that the adjusted loss coefficient through the access holes is $K = 0.5$. What is the elevation of the water surface in Access Holes 1 and 2?

Nodes				
Upstream	**Downstream**	**Length (m)**	**Upstream Inv. Elev. (m)**	**Downstream Inv. Elev. (m)**
N/A	Access Hole 2	152	22.9	22.2
Access Hole 2	Access Hole 1	122	22.2	21.0
Access Hole 1	Outfall	91	21.0	20.4

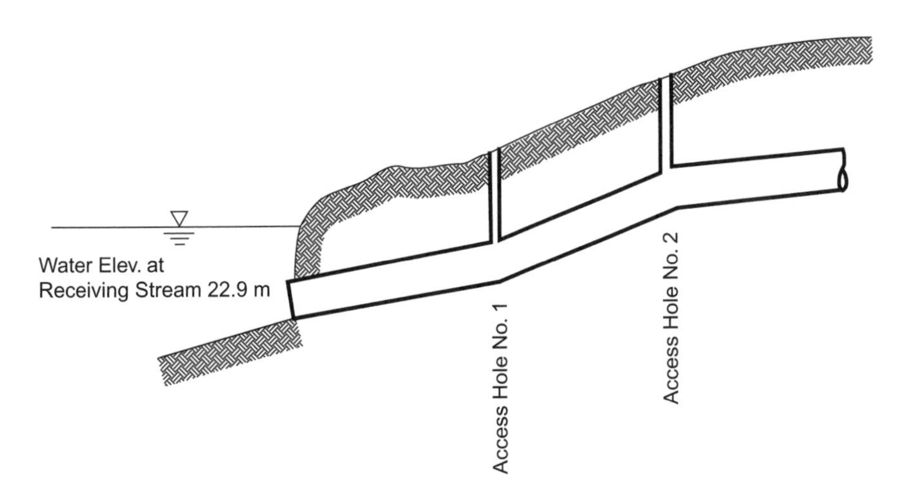

Water exits a stormwater detention pond through a weir box outlet structure. Vertical bars extending from the top of the weir keep debris from entering the outlet, but require frequent cleaning.

CHAPTER

12

Stormwater Detention

Stormwater detention is the temporary storage of runoff in underground containers, ponds, basins, or depressions. Stormwater detention is typically necessary in new developments because of the higher runoff rates caused by the increased impervious area (streets, driveways, roofs, sidewalks, etc.). Higher runoff rates from a site typically increase the severity and frequency of flooding for downstream properties and conveyance structures. If runoff is collected in a detention facility, it can be released to downstream properties in a controlled manner such that downstream flooding and other adverse impacts are prevented or alleviated. Stormwater detention is a widely practiced approach to stormwater management in urbanized areas of the United States.

This chapter provides basic information on stormwater detention design and analysis. Section 12.1 presents an overview of stormwater detention concepts, and Section 12.2 discusses the basic classifications of detention facilities. Section 12.3 provides a conceptual review of stormwater runoff hydrographs and how they are shaped and altered by urbanization and detention facilities.

Section 12.4 covers design criteria and other considerations, and Section 12.5 describes the various physical components and geometries of detention facilities. Included are discussions of pond outlet structures and overflow spillways. The section also addresses other design considerations, such as benching of pond side slopes and facility maintenance.

Section 12.6 discusses hydraulic relationships for detention facilities. These relationships include detention pond stage versus surface area, stage versus storage, and stage versus discharge relationships. Detention *routing* is a process that uses these relationships to transform a detention inflow hydrograph into an outflow hydrograph. Section 12.7 presents the storage-indication method of routing, a numerical integration routing method, and modeling examples incorporating these techniques. An overview of the iterative process of designing detention facilities is provided in Section 12.8.

12.1 OVERVIEW OF STORMWATER DETENTION

Figure 12.1 illustrates the basic concepts involved in stormwater detention analysis. An inflow runoff hydrograph from one or more contributing drainage areas is directed to a storage facility—in this case, a detention pond. (Section 5.5 covers the development of runoff hydrographs.) Runoff is then released from the facility at a controlled, lower rate through a properly sized outlet structure, such as a culvert. The result is a pond outflow hydrograph that is substantially flatter (that is, has a lower peak discharge) than the inflow hydrograph.

Figure 12.1
Conceptual drawing of a detention pond

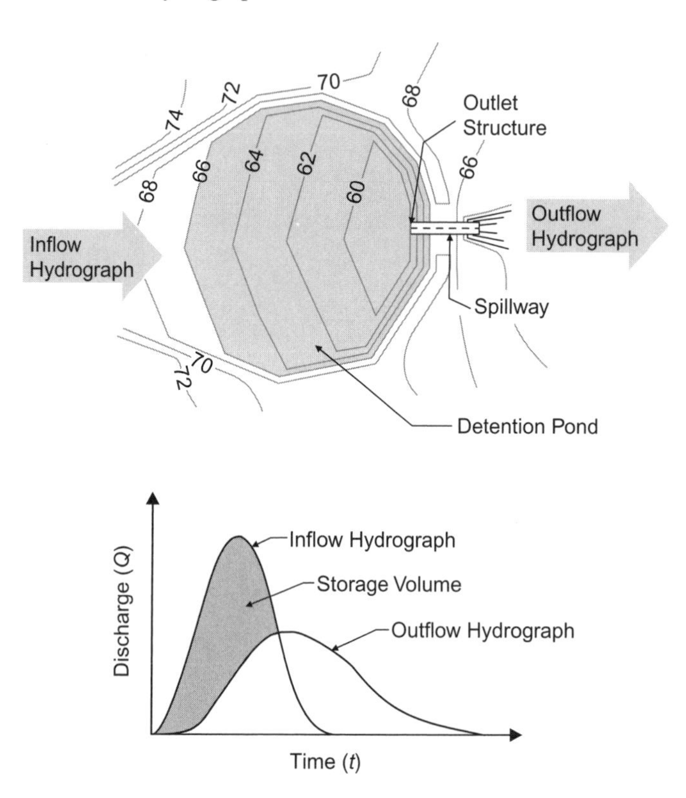

The Advantages of Implementing Detention Ponds

Stormwater detention is mandated by law for new land development in many jurisdictions within the United States. The intent of required detention is to reduce flood damages by limiting post-development peak discharges to be equal to or less than those that occurred prior to development, or to a rate based on other criteria specified by the applicable stormwater authorities. However, even if these peak discharge objectives are met, most detention ponds do not mitigate the increase in runoff *volume* caused by land development. Thus, with detention, the peak flow from a site will not be higher than it was prior to development, but the flow will occur over a longer period.

In addition to peak discharge reduction, stormwater detention improves the water quality of stormwater runoff. The residence time resulting from the low velocity of flow through a detention pond may provide sufficient time for suspended particulate matter and adsorbed contaminants to settle out of the water column. As a *best management practice (BMP)*, detention ponds can help reduce pollutant loads to receiving waters (see Section 15.3). Achievement of significant loading reductions often requires that increased attention be paid to the selection of pond shape, pond depth, maintenance plans, and vegetation.

Stormwater Detention versus Retention

The use of the terms stormwater *detention* and stormwater *retention* has generated much confusion, as many engineers use the terms synonymously. For the purposes of this text, retention facilities are distinguished from detention facilities by the absence of outlet structures. Stormwater retention involves capturing all runoff from a drainage basin, with no release of water onto downstream lands. Retention ponds are sometimes used on mining sites or other areas where no discharge is allowed for water quality reasons. They may also be used to avoid increasing the runoff volume from a site. The mechanisms of infiltration and evaporation serve as the only outflows from a retention facility.

Unlike retention ponds, detention ponds have outlet structures. Stormwater detention involves a temporary storage of runoff, with a peak release rate to downstream properties that is usually substantially lower than the detention facility peak inflow rate. The mechanisms of infiltration and evaporation play a role in stormwater detention, but they are often negligible compared to the inflow and outflow rates and are usually ignored. This chapter addresses only detention facility design, which is more widely used in urban land development than retention, and focuses on the peak discharge reduction aspect of detention facility design.

Additional References on Stormwater Detention

The hydraulics of detention pond sizing have been the subject of considerable research, and the reader can turn to numerous publications for more details, including Akan (1989); Aron and Kibler (1990); Glazner (1998); Kamaduski and McCuen (1979); Kessler and Diskin (1991); and McEnroe (1992). In addition to these papers, extensive guidance on the design and construction of the impoundments can be found in Task Committee (1985); ASCE (1982); ASCE (1992); Guo (2001); Loganathan, Kibler and Grizzard (1996); Paine and Akan (1993); Urbonas and Roesner (1993); and Urbonas and Stahre (1993).

12.2 TYPES OF STORMWATER DETENTION FACILITIES

Several types of stormwater detention facilities exist. They may be classified on the basis of whether they are dry or wet during dry-weather periods, and whether they are on-line or off-line. Each of these types of ponds has strengths and weaknesses with respect to a number of issues. A review of various pond types and their inherent strengths and weaknesses is presented in this section.

Dry versus Wet Ponds

A *dry pond* has an outlet that is positioned at or below the lowest elevation in the pond, such that the pond drains completely between storm events. A *wet pond*, however, has its lowest outlet at an elevation above the pond bottom. Water remains in the pond during dry-weather periods between storm events and can be depleted only by infiltration into the soils and/or by evaporation from the water surface. Wet ponds can be provided with a gated low-level outlet that can be opened to permit drainage of the pond for maintenance purposes, but that outlet is kept closed during normal operation. Figure 12.2 demonstrates the distinction between wet ponds and dry ponds.

Considerations in pond-type selection should include, at minimum, hydrologic, aesthetic, and water quality issues. A wet pond is typically suitable if inflows into the pond maintain a minimum water level by exceeding the rate of loss caused by infiltration and evaporation. This inflow can come from either a continuous base flow through the wet pond or from frequent rainfall events. However, it should be recognized that a wet pond can decline to low or completely empty levels during extended

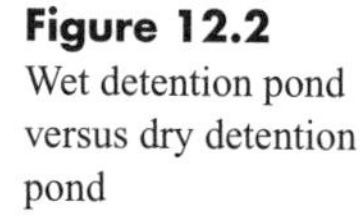

Figure 12.2
Wet detention pond versus dry detention pond

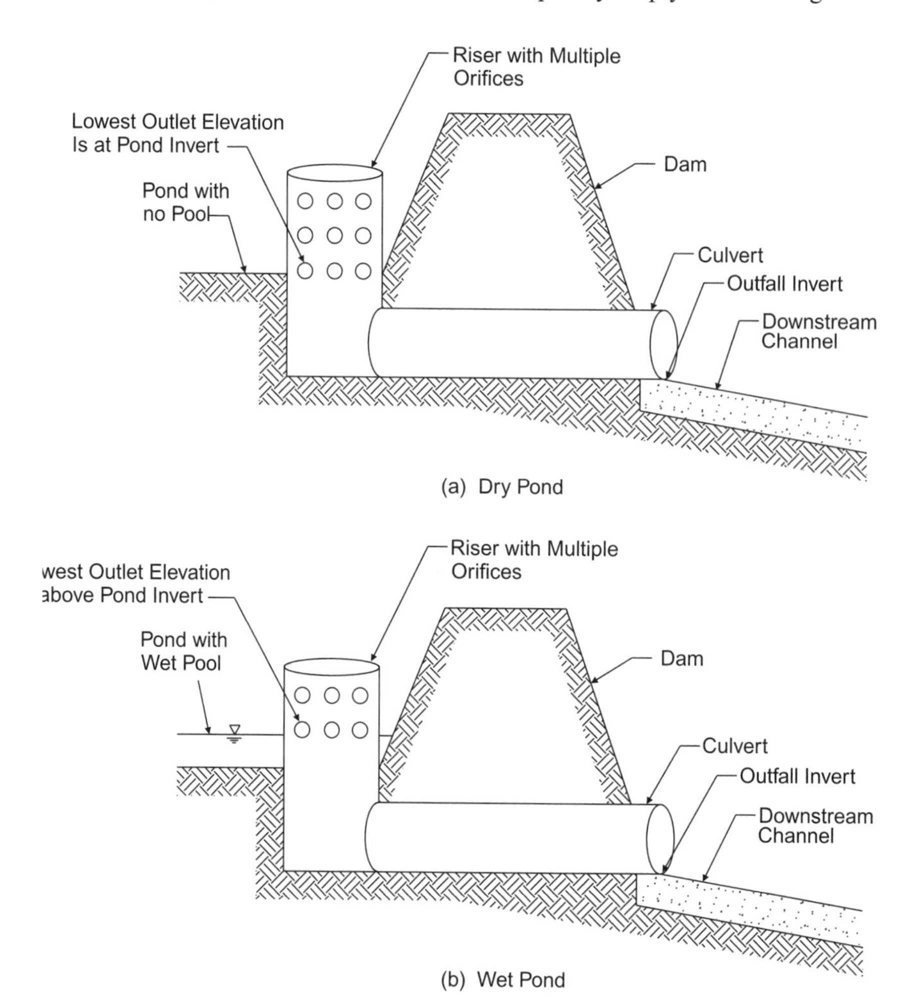

dry-weather periods. Dry ponds should be considered in areas where base flow and/or abundant rainfall are not present.

Aesthetically, a wet pond is usually more desirable than a dry one. A wet pond can provide a focal point for a land development project and usually can be integrated easily into required open-space areas. Nevertheless, a wet pond can be a haven for mosquitoes and a source of odors and may have unsightly floating debris and scum. Also, wet ponds are more subject to filling with sediment because they allow more time for particles to settle.

Dry ponds tend to trap transported debris and require frequent maintenance to keep from becoming eyesores. The bottoms of dry ponds are often soggy, waterlogged areas where weeds and other vegetation thrive, creating a hospitable environment for stench and vermin.

From a water quality perspective, wet ponds are more desirable than dry ponds. During storm events, the water in a properly designed wet pond will be displaced in whole or in part by incoming runoff. The new stormwater resides in the pond until the next runoff event. This extended residence time provides an opportunity for sediments and adsorbed pollutants to settle out of the water column. Vegetation in the pond and along its banks can also provide water quality benefits during dry-weather periods.

Aboveground versus Underground Detention

Detention ponds can be designed in an almost unlimited number of shapes and sizes. An *aboveground* facility usually consists of a depressed or excavated area in the land surface. An earthen dam or berm may or may not be required. In cases where there is not ample surface area available to meet storage requirements, *underground detention* may be necessary. Underground detention can be as simple as a series of appropriately sized pipes or pre-fabricated custom chambers manufactured specifically for underground detention.

"It's been a wonderfully buggy summer, hasn't it?"

Aboveground and underground detention ponds function in the same manner; they detain water and attenuate the outflow rate. The same routing fundamentals apply for each type of pond, as well. The main difference in the two analyses is the way in which pond volume data are computed. For aboveground earthen ponds, contour data are typically used to determine the volume characteristics of the pond. For sloped underground pipes or custom structures, the calculations depend on the particular geometry of the storage facility.

On-Line versus Off-Line Detention

On-line and off-line detention ponds are distinguished from one another on the basis of whether they are positioned along the alignment of a main stormwater drainage pathway. An *on-line pond*, as its name suggests, is positioned along the pathway and all runoff from the upstream drainage basin must pass through it. An *off-line pond* is located away from the main drainage pathway, and only a portion of the total contributing stormwater runoff is diverted through it. The advantages of off-line detention are that the outlet structure can be smaller and, in some cases, less storage volume is required (see Section 12.5).

Figure 12.3 illustrates the configuration of on-line and off-line detention facilities. Only a fraction of the main channel's flow is diverted into the pond with the off-line configuration.

Off-line ponds are not as effective as on-line ponds (whether wet or dry) for water quality enhancement because only a fraction of the total runoff is diverted into them.

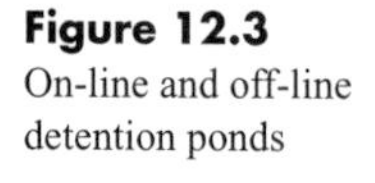

Figure 12.3
On-line and off-line detention ponds

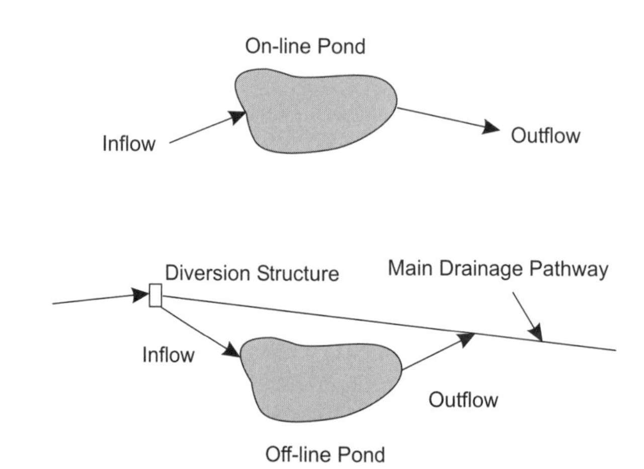

Regional versus On-Site Detention

In some instances, the runoff from several land development projects may be directed into a *regional detention pond* rather than individual *on-site detention* ponds for each development project. In the case of regional detention, the objective is to mitigate the peak flow rate for a large combined watershed area, whereas the objective for on-site detention is to mitigate the peak outflow from each specific development site.

Figure 12.4 illustrates the concepts of regional and on-site detention facilities for a watershed. The regional ponds are designed to handle flows from relatively large contributing areas, and on-site ponds are designed on a site-by-site basis while typically ignoring the overall watershed characteristics. Because development increases runoff *volumes* as well as discharges, and because detention ponds affect the timing characteristics of runoff hydrographs (see Section 12.3), the combined hydrographs from all the on-site detention ponds will not necessarily be effective in mitigating the flow rate for the overall watershed. In fact, on-site detention near the watershed outfall may in some cases increase the peak flow rate from the watershed as a whole because it can delay the hydrograph peak from the site to coincide more closely with that of the larger watershed.

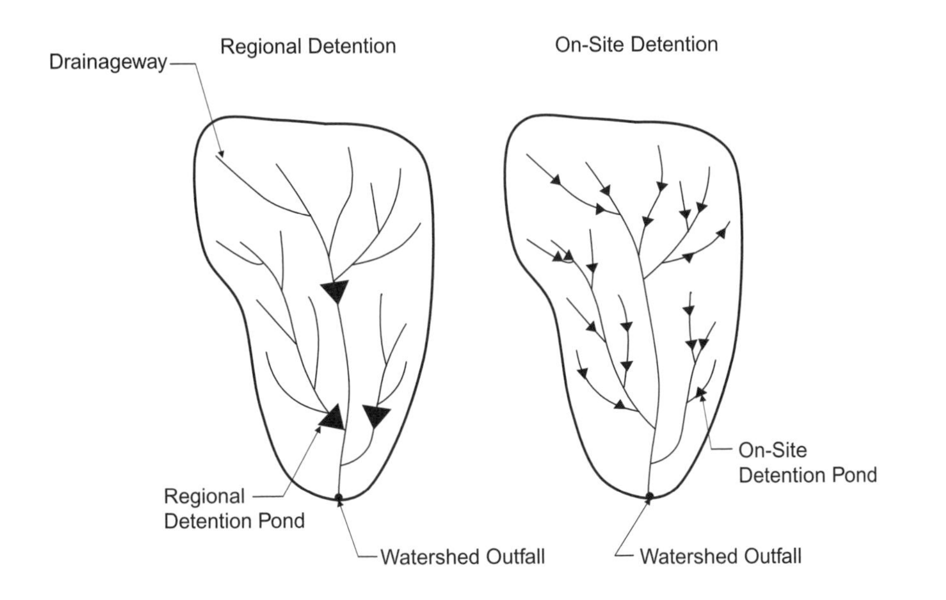

Figure 12.4
Regional versus on-site detention

Both regional and on-site detention philosophies have advantages and disadvantages. If only regional detention is used, then the overall watershed peak discharge for that region can be attenuated, but the conveyance systems upstream of the ponds usually must be larger to handle the increased, unattenuated flows. Regional ponds are often more difficult to implement for political and economic reasons. The design, planning, and construction of regional ponds usually involve multiple property owners, developers, and possibly even multiple drainage jurisdictions. Conversely, if only on-site detention is used, better water quality will be achieved on an on-site basis, and many intermediate conveyance systems will be smaller because of the on-site attenuation from the ponds. However, the attenuation of the peak flow from the overall watershed will be in question because the watershed's pond system was developed in a piecemeal fashion. Further complicating the question of whether to implement regional or on-site detention is the fact that the designated "regional" watershed is actually a subwatershed of an even larger region. When one considers this larger picture, what were considered regional detention facilities begin to look like the on-site ponds of the original view, with similar advantages and disadvantages. The issues are simply transferred to a larger scale.

A hybrid system of regional and on-site ponds offers the strengths of both regional and on-site detention. This approach makes it possible to design for each individual site, while also considering the larger watershed picture. Figure 12.5 demonstrates how both philosophies can be combined within the same watershed.

Figure 12.5
Combination of regional and on-site detention

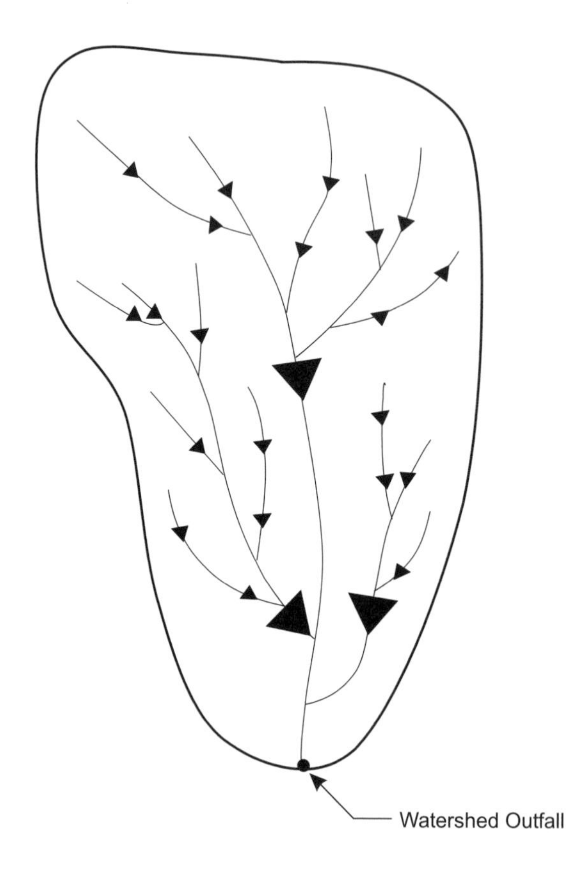

On-site retention is another approach to mitigating some of the adverse effects of development. It involves retention of at least a portion of the runoff in depressions and artificial storage devices, allowing the water to slowly percolate into the soil over an extended period of time. In theory, if the infiltration capacity of the soils is large enough, the detention storage requirements for an overall development project can be greatly reduced or even eliminated. However, more research and pilot programs on retention are needed to address potential water quality issues surrounding the recharge of water into the groundwater system, facility life expectancy, maintenance requirements, and other issues.

Independent versus Interconnected Ponds

Ponds are considered to be *interconnected* if they are directly connected and interact hydraulically. However, ponds may be connected in series without being considered "interconnected" from a hydraulic modeling standpoint. For instance, Figure 12.6 illustrates a situation in which each pond can be routed separately using a procedure such as the *storage indication method*, which is discussed in Section 12.7. The ponds are connected, but the downstream pond does not influence the hydraulic behavior of the outlet structure for the upstream pond because the downstream water surface never submerges it. Storage indication routing is applicable to each pond because their outlets function *independently* of one another.

Figure 12.7 illustrates an interconnected pond scenario in which a tidal water surface rises high enough to affect the water surface elevation in the downstream pond, which in turn affects the water surface elevation in the upstream pond as the culvert outfall is submerged. In this case, the hydraulic behaviors of the two ponds are *interdependent*. With interconnected conditions, reverse flow through the outlet structure can occur if the downstream water surface elevation rises above the elevation in the connected upstream pond. If desired, flap gates can be installed to prevent reverse flow in these situations.

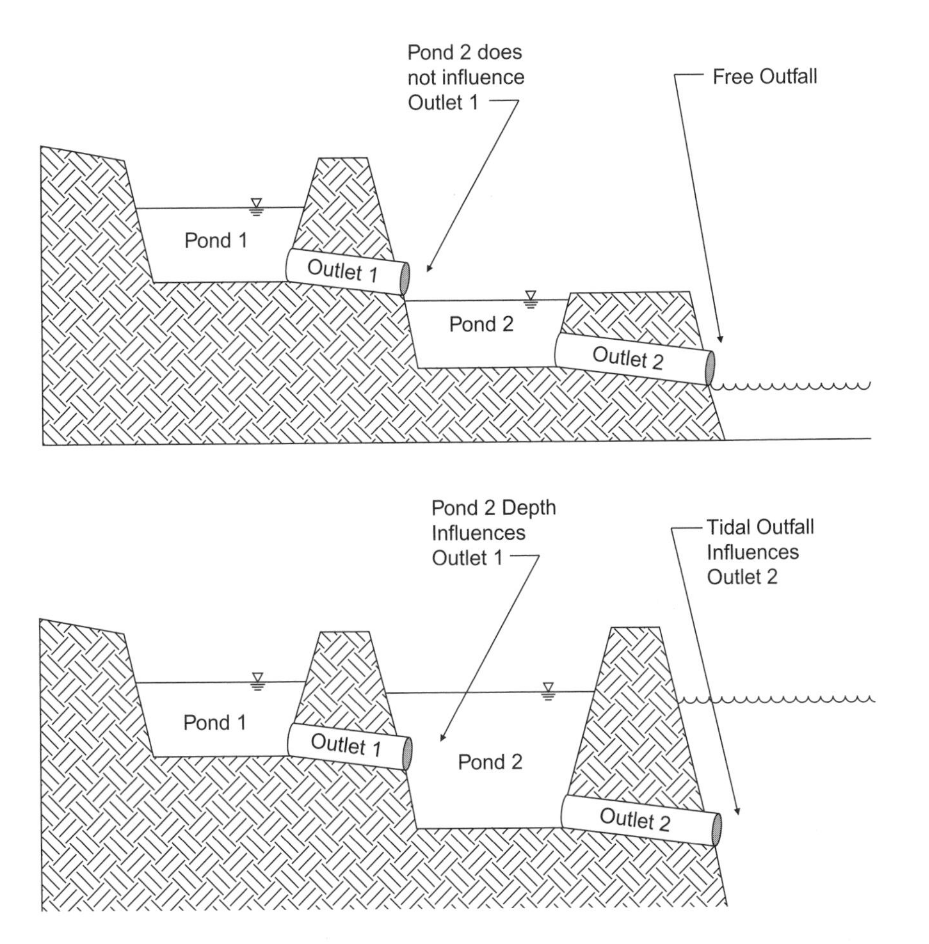

Figure 12.6
Ponds in series

Figure 12.7
Interconnected ponds

12.3 STORMWATER DETENTION MODELING CONCEPTS

Stormwater detention design requires the estimation of a complete runoff hydrograph, as opposed to the single peak discharge typically used in storm sewer design. Section

5.5 presents procedures for inflow hydrograph estimation. This section provides a conceptual review of stormwater hydrographs and discusses how they are shaped and altered by the introduction of detention facilities.

Figure 12.8 illustrates the basic hydrograph concepts associated with detention. The inflow hydrograph represents the runoff from a watershed. Its principal attributes are the peak discharge rate and the time from the beginning of the storm to the peak discharge. The area under the hydrograph represents the total volume of runoff resulting from this storm event.

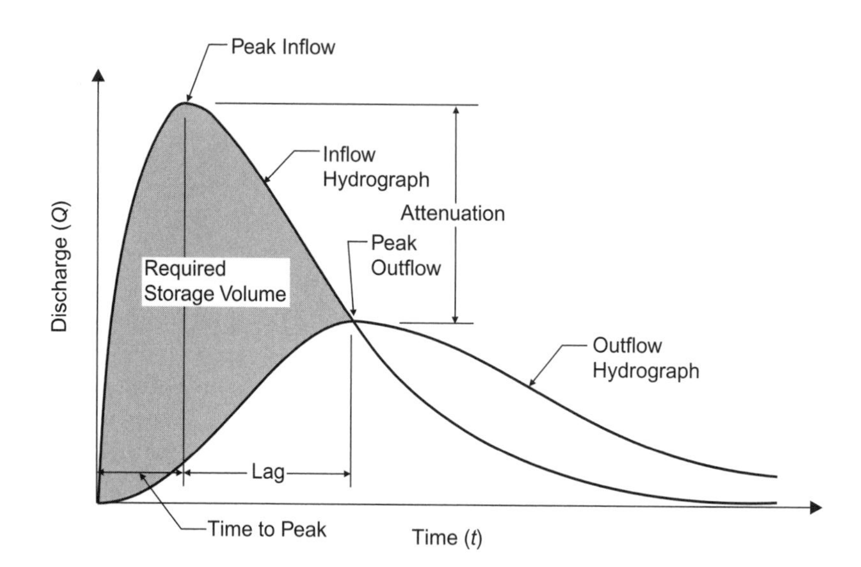

Figure 12.8
Attenuation of an inflow hydrograph and lag in hydrograph peak due to detention routing

The outflow hydrograph represents the discharge from a detention facility. The outflow hydrograph is obtained by performing a *routing analysis* (see Section 12.6 and Section 12.7) on a given detention pond, outlet configuration, and inflow hydrograph.

Attenuation (reduction) of the peak discharge resulting from a storm event is accomplished by a detention facility or any type of temporary storage. With the reduction of the peak comes a *lag* in the time at which the peak occurs. Figure 12.8 illustrates attenuation and lag. Note that the peak discharge for the detention pond inflow hydrograph is larger and occurs earlier than peak discharge for the outflow hydrograph.

If pond infiltration and evaporation losses occurring in the detention facility during the storm event are negligible, the areas under the two hydrographs (the runoff volumes) must be equal for ponds that do not provide retention. The shaded area in Figure 12.8 under the inflow hydrograph represents the maximum volume of water present in the pond during the storm event. If this volume is greater than the storage capacity of the pond, the pond will overflow. (A detention pond will typically be equipped with an emergency spillway to handle the overflow.)

The following three scenarios summarize the relationship of a pond's inflow rate, outflow rate, storage, and water surface elevation for any given time interval within a stormwater runoff event:

- If the average inflow rate is greater than the average outflow rate during a time interval, the volume of stored water in the pond increases during the time interval, and the water surface elevation in the pond increases.
- If the average inflow rate is equal to the average outflow rate during a time interval, the volume of stored water does not change during the interval, and the water surface elevation remains constant.
- If the average outflow rate is greater than the average inflow rate during a time interval, the volume of stored water decreases during the time interval, and the water surface elevation in the pond decreases.

Most detention ponds have *uncontrolled outlets* (their outlets have no valves or gates that can be opened or closed). For a pond with an uncontrolled outlet into a free outfall, the peak of the outflow hydrograph from the pond will occur at the point where the outflow hydrograph intersects the receding limb of the inflow hydrograph (see Figure 12.8).

This concept can be used as a check on the validity of a set of routing calculations, and can be reasoned as follows. Prior to the time of the intersection point, the inflow rate is larger than the outflow rate, and the volume of water in the pond is increasing. After that intersection time, the volume is decreasing because the outflow rate is greater than the inflow rate (see Figure 12.8). Thus, the maximum stored volume of water in the pond occurs at the time where the two hydrographs intersect. Because the outflow rate from a pond increases as the headwater depth in the pond increases, it follows that the maximum outflow rate must occur at the maximum depth, which also corresponds to the maximum storage during the event. Finally, at the point of intersection, the outflow is equal to the inflow rate, resulting in a zero rate of change of stored volume at the moment in time when the water surface crests in the pond.

12.4 DETENTION DESIGN CONSIDERATIONS

When designing a detention facility, the engineer must meet certain design criteria. These design criteria are typically provided by one or more governmental regulating bodies, such as the local public works department and the state environmental department. Before beginning the design of a detention facility, the engineer should contact the relevant authorities to obtain their requirements.

A common criterion for detention design is that the peak discharge from a site after development cannot exceed that of the undeveloped site, but the recurrence interval or intervals for which the criterion must be met vary among jurisdictions. Another frequent requirement is that the discharge from the site should not exceed the capacity of downstream conveyances for a particular storm event.

An additional factor that the engineer may need to consider is the effect of detention on the shape of the discharge hydrograph. For instance, detention at a site that is near the outlet of a watershed may actually delay the hydrograph peak for the site such that it coincides more closely with the peak of the overall basin. A possible result is that the peak discharge from the larger basin will increase even though pre- versus post-development criteria for the particular site are met. The effects of the increased volume discharging from a site due to development should also be considered.

Pre-Development versus Post-Development Criteria

As a watershed is developed, the peak discharge and the amount of runoff will tend to increase, and the time to peak will typically decrease. The intent of stormwater detention is typically to limit post-development peak discharges from a site to discharges no greater than those that occurred prior to development. For instance, the post-development peak discharges from the site for the 2-, 5-, 10-, and 25-year storms should be equal to or less than the peak discharges for those same storms under pre-developed site conditions. Thus, the conveyance capacities of existing downstream structures will still be sufficient if these structures were adequately sized prior to development.

A variation of the pre- versus post-development criteria is to attenuate post-development peak discharges for a given set of storms to be significantly less than pre-development peak flows for those same storms. For instance, regulations might require that the 25-year post-development peak not exceed the 5-year pre-development peak. Pond outflows are significantly attenuated, but the discharge will occur over a longer time.

Downstream Systems

Sometimes, allowable detention facility discharges may be based on criteria other than pre-development flows. For instance, the pond may need to release flows at rates significantly lower than pre-development values to avoid exceeding the capacity of one or more existing downstream conveyance structures. It is important to note, however, that even if peak discharge objectives are met, detention ponds typically do not mitigate the increase in runoff volume caused by land development. Although the peak flow rate from a site for a given storm event may be the same as it was prior to development, the duration of the discharge can be substantially longer (see page 491).

Mandated Release Rates

Some drainage jurisdictions mandate a specific allowable peak outflow rate based on site area. For example, if the mandated release rate for a locale is 0.1 cfs/ac and the development site is 100 ac, then the allowable release rate from the site is 10 cfs.

Design Storms

The pre- versus post-development criteria described in the preceding sections must be met for one or more specified design storm events. For most localities, these design events will be specified in the local regulations. For example, the regulations may specify that post-development peak flows be less than pre-development peaks for 24-hour synthetic storms having return frequencies of 2, 5, and 10 years. It is advisable, however, to check the proposed design for additional storms of various durations. It is also good practice to provide an emergency overflow structure capable of handling flows from the 100-year storm (for instance). The magnitude of the largest storm that the facility must be designed to accommodate may depend on risk factors such as site area and land use.

Comparing Storm-Event Magnitudes

The 2-year return event is often used as the smallest storm evaluated when designing stormwater facilities. However, with respect to 24-hour total rainfall depths, the 2-year event is not really as "small" as it may sound; it often represents more than 40 percent of the 100-year event. The table shown below compares northwest Illinois' (United States) rainfall depths for various return frequencies to its 100-year rainfall event to demonstrate the significant size of events with "lower" return frequencies.

A second comparison in this table applies the NRCS (SCS) Curve Number method (see Section 5.2, page 119) for computing runoff volume for a developed area with $CN = 85$ to demonstrate the resulting runoff depths for these events. In this example, the runoff depth for the 2-year event is about 30 percent of the runoff depth for the 100-year event. Thus, if a detention pond were designed for the 2-, 5-, 10-, 25-, 50-, and 100-year events, the volume of water impounded for the 2-year storm could be expected to be about 30 percent of the 100-year impounded volume (that is, the pond would be about 30 percent full, not including freeboard).

These comparisons demonstrate that design engineers need additional rainfall data (less than the 2-year event) to understand and evaluate rainfall events that are similar in magnitude to those that stormwater facilities will often experience (that is, multiple times in a typical year).

The following table compares rainfall depths and runoff volumes for 24-hour storm events in northwest Illinois (Huff and Angel, 1992).

Return Event	100-Year, 24-Hour Rainfall Depth*	Percent of 100-Year Rainfall Depth	Runoff Depth for CN=85 and I_a=0.35	Percent of 100-Year Runoff Depth
2 month	1.40 in. (35.6 mm)	19%	0.39 in. (9.9 mm)	7%
6 month	2.08 in. (52.8 mm)	28%	0.85 in. (21.6 mm)	15%
1 year	2.57 in. (65.3 mm)	35%	1.24 in. (31.5 mm)	22%
2 year	3.11 in. (79.0 mm)	42%	1.68 in. (42.7 mm)	30%
5 year	3.95 in. (100.3 mm)	54%	2.41 in. (61.2 mm)	43%
10 year	4.63 in. (117.6 mm)	63%	3.03 in. (77.0 mm)	54%
25 year	5.60 in. (142.2 mm)	76%	3.93 in. (99.8 mm)	70%
50 year	6.53 in. (165.9 mm)	89%	4.81 in. (122.2 mm)	86%
100 year	7.36 in. (186.9 mm)	100%	5.60 in. (142.2 mm)	100%

Peak Flow Timing Issues

Figure 12.9 illustrates problems that can arise from changes in the times at which peak runoff rates occur and demonstrates the need to consider regional as well as local effects in detention facility design and implementation. The provision of detention can actually lead to increases in downstream discharges if timing effects are not carefully considered. To counter this effect, some jurisdictions will suspend the detention requirement in certain cases, and may ask that other conveyance system improvements be made in lieu of detention.

The figure shows hydrographs for two tributaries (A and B) to the main stem of a stream. A detention facility is to be provided for a development site on tributary B. Two scenarios are shown for the total combined flow downstream of the confluence of the two tributaries. One scenario represents the total combined flow for post-development conditions if the detention pond is provided. The other scenario represents the total combined flow for post-development conditions assuming that no detention is provided. Both scenarios are compared to the pre-developed hydrograph for the overall watershed (areas A and B).

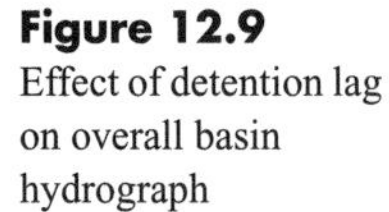

Figure 12.9
Effect of detention lag on overall basin hydrograph

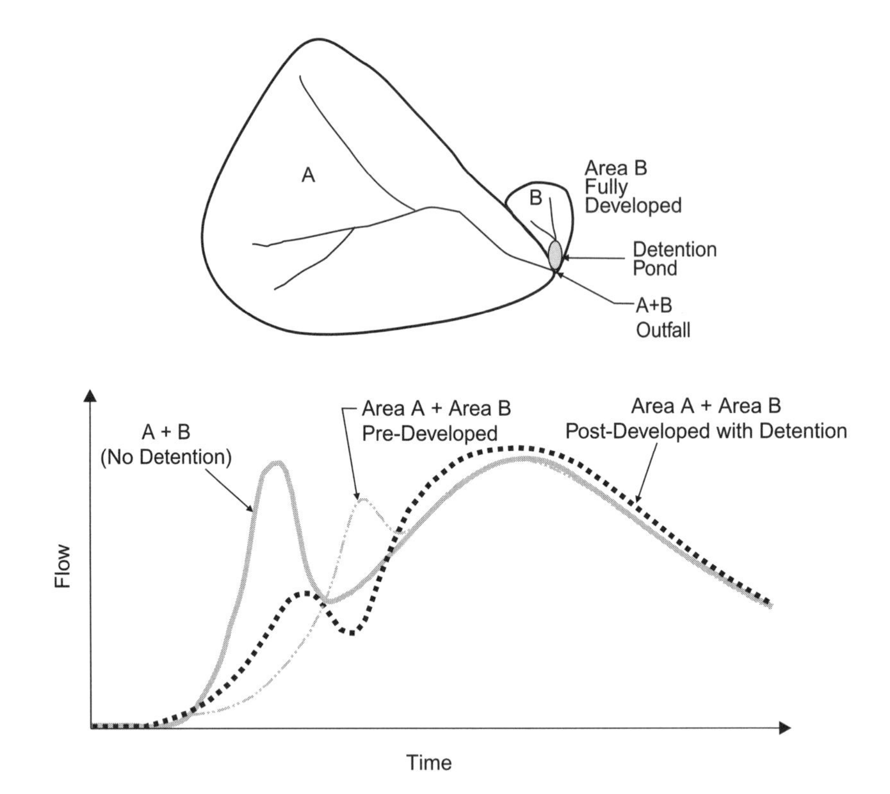

Although the scenario with the detention pond successfully attenuates the post-developed flow from tributary B to a peak discharge lower than the pre-developed condition for that tributary, it actually *increases* the peak discharge of the total combined flow downstream of the confluence. Conversely, the scenario without detention has little impact on the combined flow downstream of the confluence during the period when the runoff hydrograph from the overall watershed is peaking, but it does greatly increase the discharges on the rising limb of the overall hydrograph until most of the water from tributary B has drained from the system. Removal of detention from tributary B also negates any possible water quality improvement that might be associated with detention.

Bypassing Flows from Off-Site

In most cases, some discharge from off-site drainage areas will flow onto the project site. The drainage system on the site can be designed either to convey and detain off-site discharges with the on-site flow, or to keep the flow separated, diverting it around or through the site. This situation is illustrated in Figure 12.10.

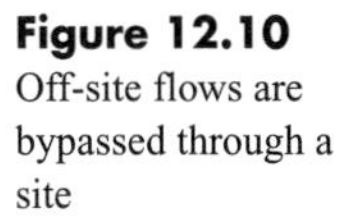

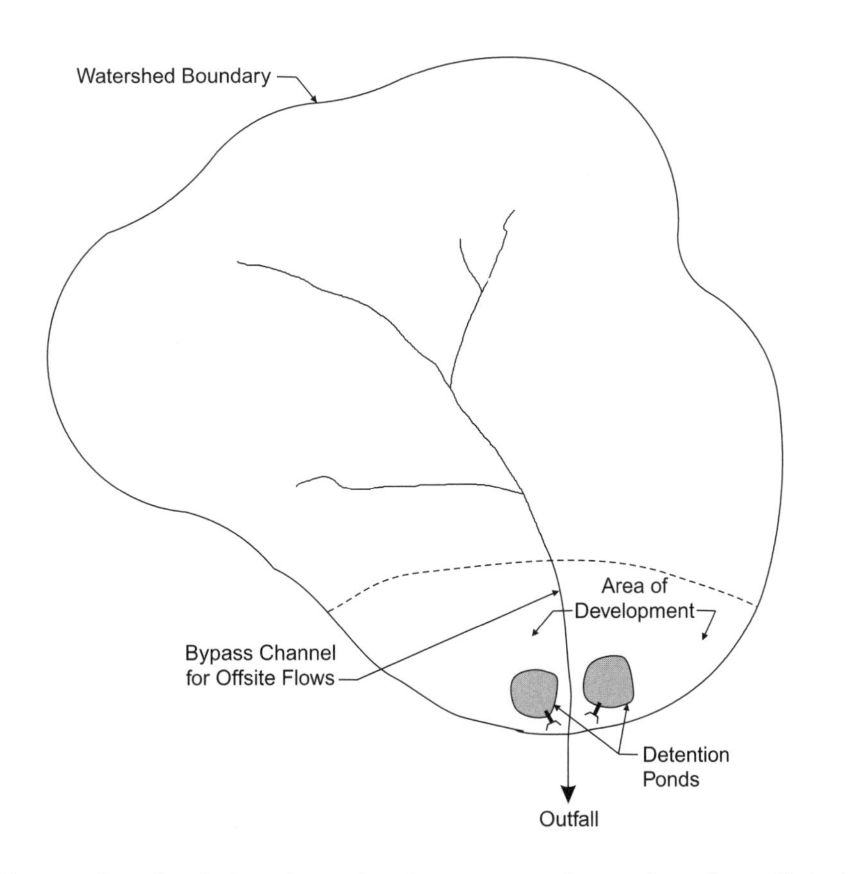

Figure 12.10
Off-site flows are bypassed through a site

With some detention designs, it may be advantageous to bypass flows from off-site if they would greatly increase the required size of the detention pond outlet structure. However, due to the timing issues previously discussed, the resulting peak discharge from the site (computed by summing the bypass and detention outflow hydrographs) can actually be higher in some instances than if there were no detention at all.

Volume Mitigation Issues

When a drainage basin is urbanized, the addition of impervious surfaces, such as roofs and pavement, and the compaction of soils result in an increase in the volume of runoff produced for a given storm event. If evaporation and pond infiltration losses are neglected, and if the amount of water in a pond is the same after a storm event as before the event, then the total pond inflow volume must be equal to the total outflow volume. That is, the areas under the two hydrographs in Figure 12.8 are equal. For this reason, the duration for which the outflow hydrograph is nonzero must be longer than the duration of the inflow hydrograph. This increased flow duration can lead to erosion and/or sedimentation problems downstream of the pond.

Figure 12.11 illustrates a comparison of a hydrograph for pre-developed conditions with post-development inflow and outflow hydrographs for a detention pond. Although the peak discharge on the outflow hydrograph from the pond is slightly smaller than the peak discharge on the pre-development hydrograph, a relatively high

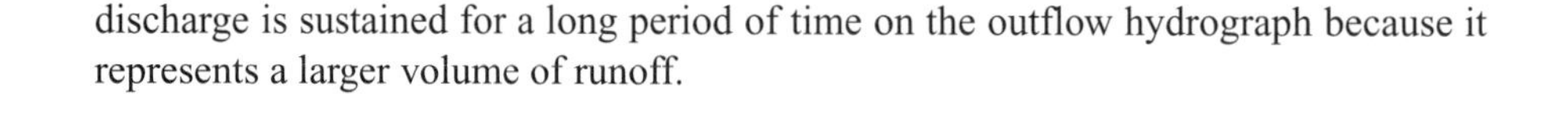
discharge is sustained for a long period of time on the outflow hydrograph because it represents a larger volume of runoff.

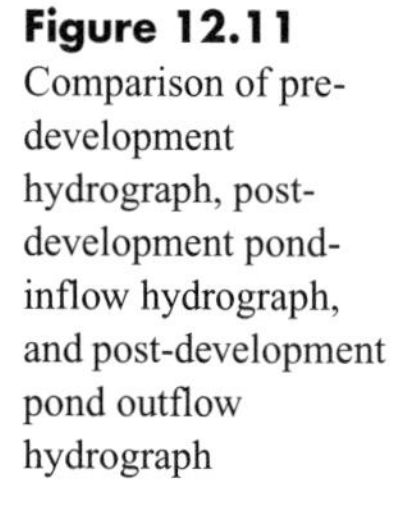
Figure 12.11
Comparison of pre-development hydrograph, post-development pond-inflow hydrograph, and post-development pond outflow hydrograph

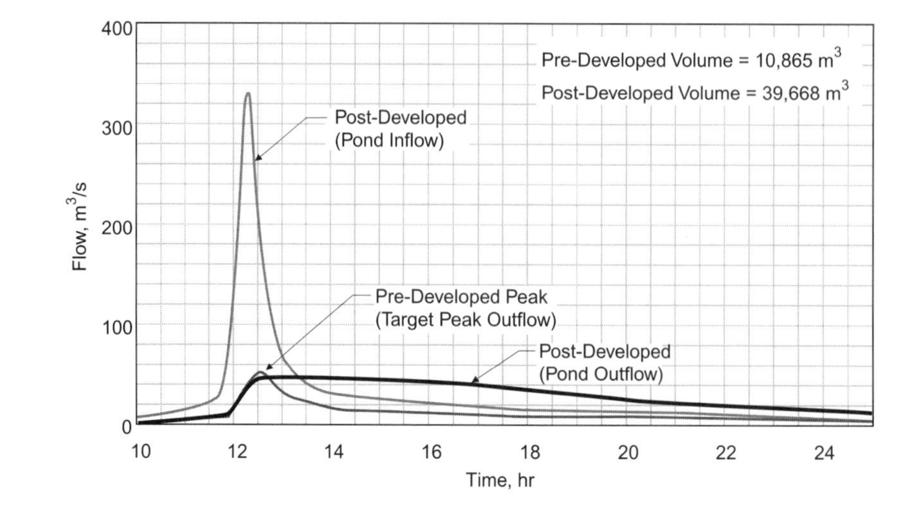

Figure 12.12 shows the increase in runoff volume that resulted from the development. The detention pond has been designed to prevent the peak of the outflow hydrograph from exceeding that of pre-developed conditions, but the pond still must pass the entire volume of post-development runoff downstream. The shaded area represents the increase in the volume of runoff and can be viewed as the additional volume of water that was lost to rainfall abstractions under pre-development conditions.

Figure 12.12
Net increase in runoff volume resulting from development

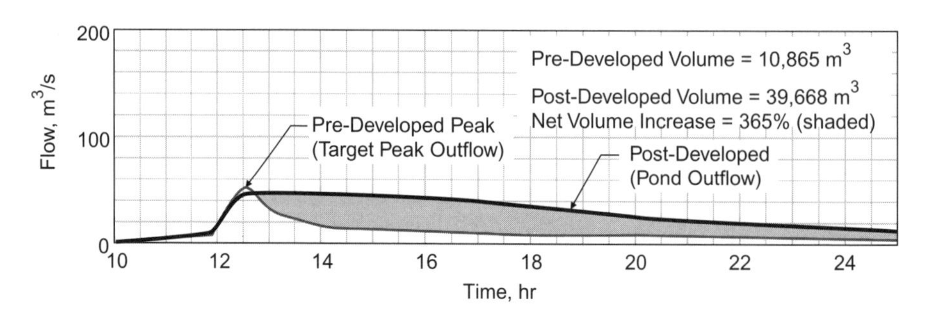

The classical focus of detention facility design has been to reduce the post-development peak discharge of runoff to no greater than the peak discharge that occurred prior to development. Recognizing the increase in runoff volume caused by urbanization, review agencies may in some cases require that the post-development peak discharge be limited to some fraction of the pre-development peak discharge to reduce the likelihood of pond failure.

12.5 COMPONENTS OF DETENTION FACILITIES

Detention facilities, whether dry or wet, on-line or off-line, usually consist of an aboveground pond that is excavated and graded into the land surface. In some instances, they may consist of shallow ponding areas in parking lots or on rooftops. In areas where available space is at a premium, it may be economically advantageous to use underground storage consisting of pipes and/or vaults. Thus, the physical components of individual detention facilities can be quite variable, but some general observations can be made. The following subsections provide discussions of the most common elements of detention facilities.

Dam Embankments

In most cases, a stream or localized depression is enlarged and/or dammed to create a detention pond. The design of the dam should conform to accepted geotechnical engineering practices. Governmental dam safety and permitting requirements such as the Dam Safety Act in the United States may apply if the embankment height or storage volume exceeds a regulatory limit.

Freeboard

Freeboard is defined as the vertical distance from the maximum water surface elevation (stage) for a storm event to the top of the dam or to the lowest pond bank elevation. The storage available within the freeboard elevation range adds a "safety" volume to the required detention storage volumes. Some jurisdictions and reviewing agencies mandate freeboard requirements for the maximum design event.

Pond Bottoms and Side Slopes

Side slopes of ponds, especially for wet ponds that become attractions for people, should be gentle. Side slopes should also be mowable for maintenance of dry ponds and dry elevations of wet ponds. Barrier vegetation, such as cattails, can be installed along the perimeter of a wet pond to limit access to the water's edge. Provision of a shelf in wet pond side slopes limits water depths in the vicinity of the banks, providing an additional safety measure. An extended shelf width near the inlet of a wet pond can effectively act as a forebay in which incoming velocities are reduced and much of the large particulate matter settles out.

Detention pond bottoms should have a slope that is sufficient to ensure complete drainage between storm events in dry ponds, and to facilitate pond drainage for maintenance in the case of wet ponds. For dry ponds, the engineer should consider providing a low-flow channel along the bottom of the pond from its inlet to its outlet structure to assist in draining the pond bottom.

A common mistake made in erosion control design around ponds is to assume that the impounded water within the pond will serve as a natural energy dissipation device for incoming flows from conveyance elements such as channels, pipes, and culverts. This may in fact be true during the course of an event when the pond is storing a significant volume of water. However, in the early part of a high-intensity storm, conveyance elements may discharge significant flows into a pond before it has impounded much water. This situation results in exposed outfalls with the potential to cause significant erosion. For both wet and dry ponds, riprap or similar protective measures should be provided near the inlet to prevent inflows from eroding the side slopes of the pond.

Outlet Structures

A detention facility's *outlet structure* allows flow to discharge from the pond at a controlled rate. Early detention ponds were usually designed to control only a single runoff event having a specified recurrence interval; thus, these ponds had very simple outlet structures. This practice provided little to no benefit during smaller or larger storm events.

Modern detention facility design typically focuses on control of multiple storm events and often requires several outlet openings at different pond stages. These multiple openings can consist of multiple pipe outlets of differing diameters or invert elevations, *orifices* at various levels, *overflow weirs*, or any combination of these elements. Figure 12.13 depicts a profile view of an outlet structure configuration consisting of several structural components.

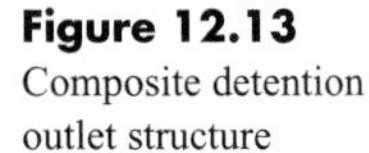

Figure 12.13
Composite detention outlet structure

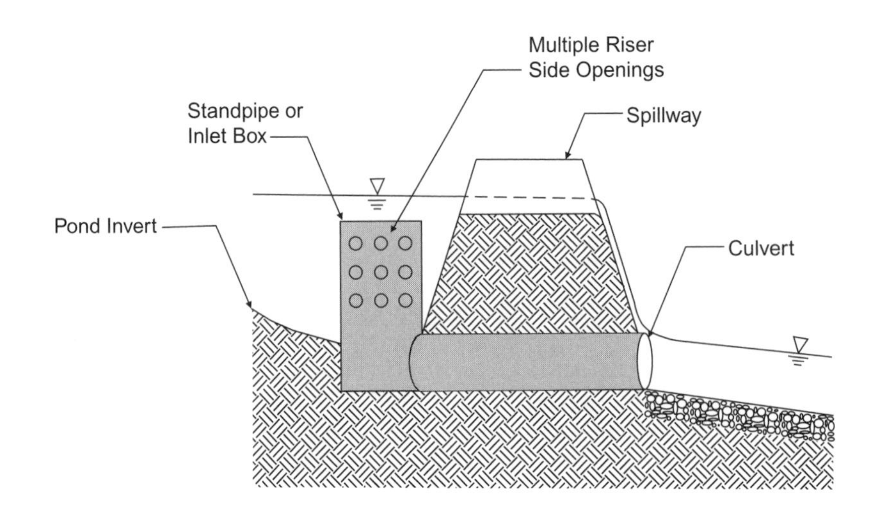

If pipe outlets, such as culverts, are used, they can be integrated into the side slopes of a pond perimeter by using headwalls or special end sections. For safety reasons, large-diameter pipes should be equipped with trash racks or screens to prevent entry by children (see Section 9.3). In dry ponds, it is recommended that the invert of the lowest pipe be set at an elevation below the bottom of the pond, and that a *fall* similar to that illustrated in Figure 9.4 be employed at the pipe inlet. In wet climates, the fall will help drain the soils in the pond bottom and prevent it from becoming waterlogged.

Orifice openings, of various sizes and/or at various elevations, are often provided in the walls of *riser pipes,* also called *standpipes* or *inlet box* structures (see Figure 12.14) in detention facilities. Orifices are an effective means of controlling small flow rates. However, orifices can become clogged by floating or suspended debris, especially if the openings are small, and should be checked regularly for blockages.

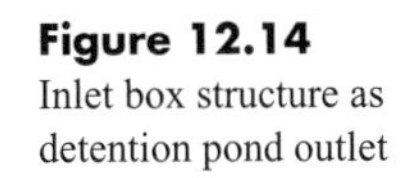

Figure 12.14
Inlet box structure as detention pond outlet

Weirs of various shapes are also effective outflow control devices and are almost universally used in wet detention facilities. Weirs can be provided in the bank of a pond at its outlet, or they can consist of the top edge of a riser pipe or inlet box structure. If riser structures are used, their tops should be fitted with *anti-vortex devices* to prevent objectionable flow patterns and trash racks or screens to prevent debris from entering the structure. Ladder rungs or similar safety devices should also be provided inside of risers.

Regardless of the type of flow control device used in a detention facility, consideration should be given to providing an *anti-seepage collar* around any outlet pipe conveying flow through a dam embankment. The collar will help minimize piping of the surrounding soil particles caused by water leakage, which could ultimately lead to embankment failure.

In the case of a wet pond, a *low-level outlet* might be provided so that the pond can be completely drained for maintenance purposes. The low-level outlet should be fitted with a valve or gate to prevent the water in the pond from draining through it. The valve or gate should be periodically opened for short time periods to prevent it from seizing so that it will be operable when maintenance is needed. An alternative is to pump the pond dry when maintenance is required. Requirements for pump size, pump quantity, and pumping time should be computed at the time of the pond design in order to estimate the cost and level of effort required for the pumping procedure.

Oveflow Spillways

All detention ponds, regardless of type, should be fitted with an *overflow spillway*. The purpose of an overflow spillway is to safely convey discharge when the design storage capacity is exceeded during a large storm event. Overflow spillways are commonly referred to as *emergency spillways* because they are designed to operate under extreme rainfall conditions to prevent dam overtopping.

Many types of outlet devices are lumped into the general definition of spillways, including broad-crested weirs, side overflow weir spillways, and ogee spillways. Overflow spillways for detention ponds range from little more than localized depressions in the pond embankment to elaborate structural facilities. Erosion protection such as riprap or paving helps to prevent excessive damage to the spillway due to high velocity flows. For ponds with earthen dam embankments, erosion during overtopping can cause the embankment to fail. Therefore, adequate erosion protection for a spillway located in a dam is critical and may take the form of riprap or concrete paving.

The storm recurrence interval for which an overflow spillway should be designed is an issue that requires an exercise of professional judgment, as well as consideration of the adverse effects of downstream flooding or damages. In areas where the major drainage system is evaluated for a designated storm recurrence interval (frequently the 100-year storm), the same recurrence interval might be used for overflow spillway design. If failure of a dam embankment could cause considerable damage or loss of life downstream, regulatory requirements pertaining to dams may be appropriate. For detailed coverage of spillway hydraulics, the reader is referred to USBR (1987).

Outfall Sheet Flow Conditions

In some predevelopment situations, such as a property that resides within an open meadow on the side of a hill, the overall flow pattern of runoff exiting the property is sheet flow in nature. If the property is developed and detention is implemented, discharges from the property become concentrated at a single outfall point. Directing this concentrated flow onto a natural sheet flow area with no channel can lead to significant erosion.

In these cases, an appropriate outfall solution must be designed. Two alternative solutions are:

- Acquire an easement across the downstream property and design and construct a surface channel or underground pipe system to convey the water safely to the next viable downstream conveyance system.
- Design a flow diffusor that will emulate a sheet flow discharge onto the downstream property. Care should be taken to ensure that prolonged discharges do not erode the downstream property, and that slope stability on the downstream property is not compromised by saturation of the soils.

Diversion Structures and Conveyances

Designs of off-line ponds (those constructed adjacent to the main drainage pathway) must provide for diversion of flows to the pond facility. A *diversion structure* is placed along the main drainage pathway upstream of the pond, and pipes or open channels convey diverted water from this structure to the pond, and from the pond outlet back to the main drainage pathway.

The diversion structure for an off-line detention facility can be one of two types. The first type diverts a fixed percentage of the total discharge to the pond, regardless of whether the total discharge is small or large. This type of diversion can be accomplished practically using a *flow splitter*. The fraction of the total flow that is diverted to the pond is a decision that is made during the design process.

The second type of diversion structure diverts all or part of the total discharge above some base flow level. Flows below that base level are not diverted to the detention pond. During the design process, the base level and the percentage of flow that is diverted once the base flow level is exceeded are determined.

Nix and Durrans (1996) found that diversion of discharges above a base level is more effective at reducing the required size of a detention facility than is diverting a specified fraction of the total discharge. This makes sense as the threshold diversion method delays the use of pond storage capacity until the threshold discharge is reached, rather than using some storage beginning with the onset of the event.

12.6 ROUTING DATA: STORAGE AND HYDRAULIC RELATIONSHIPS

Hydrograph routing through a detention facility is the process of computing an outflow hydrograph and water surface elevations in the pond for a specified inflow hydrograph. In addition to the inflow hydrograph, pond routing requires two key types of data to describe the pond's storage and hydraulic characteristics:

1. *Stage versus storage data* to describe the pond's shape and size (may be developed from *stage versus surface area data*)
2. *Stage versus discharge data* to describe the hydraulic behavior of the pond's outlet structure(s)

The following subsections describe these relationships.

Stage versus Surface Area

A *stage versus surface area* relationship for a detention facility is represented by a graph or table showing the water surface area of the pond or other storage facility as a function of the water surface elevation (stage) or water depth. This relationship is used as an intermediate step in calculating the stage versus storage relationship. For a storage facility having a regular shape (such as storage in underground pipes), geometric equations may be applied to determine the surface area corresponding to any stage of interest. For ponds that are excavated and graded, contour areas may be acquired by using a planimeter or CAD drawings. If several ponds will be routed, a separate stage versus surface area relationship must be developed for each pond.

The choice of whether to use stage (water surface elevation) or water depth to develop the relationship is largely a matter of personal preference. Either approach will yield the same routed outflow results. However, stage is normally used because it can be directly related to site elevations and construction drawings. If ponds are connected to other conveyance elements, it is important to know their relative vertical positions.

If possible, stage versus water surface area information should begin at the dry invert for a pond, even if it is designed to be a wet pond. By following this rule, any water surface elevation within the pond can be modeled during the routing process. For example, it would permit a wet pond to be modeled for a scenario in which a prolonged drought has occurred such that the water surface elevation in the pond at the beginning of the runoff event is below the lowest pond outlet. In cases of existing wet pond facilities where contour data is not available below the water surface, the normal pool elevation can be taken as the "invert" of the pond. Stage versus water surface area information should extend from the pond invert to elevations at least as high as the top of the dam embankment or to the tops of the pond banks.

In this chapter, the water surface elevation (or alternatively, the depth of water) in a pond is denoted by h. The horizontal area occupied by the water surface is denoted by A. For most ponds, A is a function of h. An exception is a pond with vertical sides, for which A is constant.

Example 12.1 – Stage versus Surface Area Relationship. An underground storage vault is to be constructed to serve as a detention pond for a heavily urbanized area. The walls of the vault will be vertical, but its floor will be sloped along its length to promote drainage. In plan view, the vault is rectangular with a length of 23 m and a width of 12 m. The slope of the floor is 2 percent along the 23-m dimension. The invert of the lower end of the vault is at an elevation of 37 m. The elevation of the top of the storage vault is 39 m. Determine the stage versus surface area relationship for the vault.

Solution: The difference in the elevation of the floor along the length of the vault is 23(0.02) = 0.46 m. Because the floor elevation at its lower end is at an elevation of 37 m, the elevation at its high end is 37.46 m. For any stage between these elevations, the rectangular water surface area is equal to the width of the vault (W = 12 m) times the length of the water surface area. That length depends on the depth of water in the vault and can be expressed in terms of the stage, h, as

$$L = (h - 37)/0.02$$

The corresponding water surface area is

$$A = WL = 12(h - 37)/0.02$$

These expressions apply to values of h ranging from 37 m to 37.46 m. For stages above 37.46 m, the surface area is constant because of the vertical walls and is equal to

$$A = 12(23) = 276 \text{ m}^2$$

The following table provides tabular data on the stage (elevation) versus surface area relationship between elevations 37 m and 39 m. Figure E12.1.1 is a graph of the relationship.

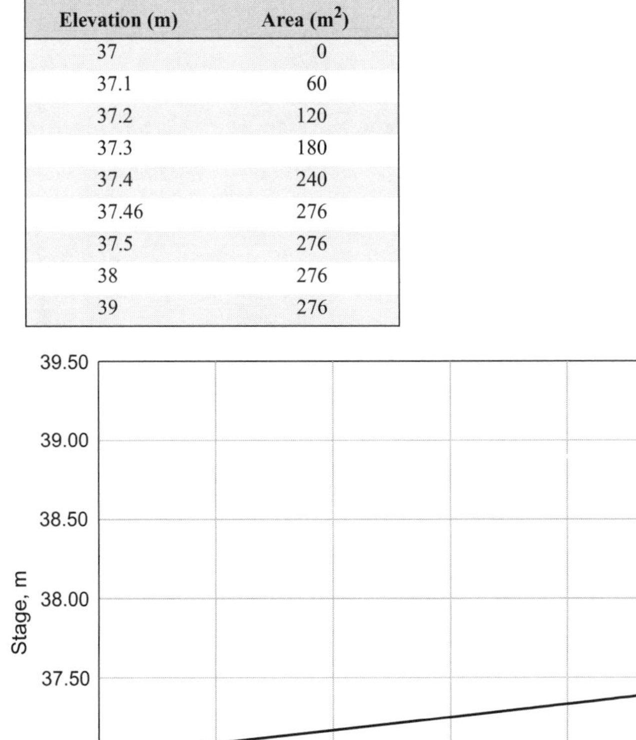

Elevation (m)	Area (m^2)
37	0
37.1	60
37.2	120
37.3	180
37.4	240
37.46	276
37.5	276
38	276
39	276

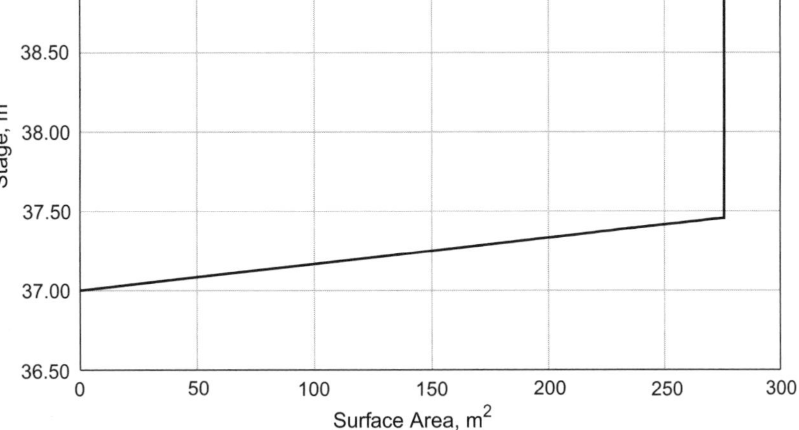

Figure E12.1.1 Elevation versus surface area relationship for underground vault in Example 12.1

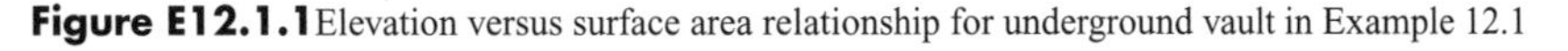

Stage versus Storage Volume

The *stage versus storage relationship* for a detention facility relates pond water surface elevation (the stage) or water depth (h) to the volume of water stored (S). It provides a geometric description of the pond that is used during routing to determine the rise or fall of the water surface elevation given a change in the volume of stored water.

Storage volumes are determined incrementally over a range of pond stages using the stage versus surface area relationship described previously. The precision of the stage versus storage curve increases as the incremental elevation change between stages decreases. The designer must judge the appropriate degree of approximation when setting the stage increment size to be used.

Figure 12.15 demonstrates how stage-storage data are used in routing to determine the relative rise or fall in water surface for a given change in volume. In this example, h_1 and S_1 are the elevation and volume, respectively, at the beginning of the time interval, and ΔS is the change in volume during the time interval calculated from the difference between the integrated inflow and outflow hydrographs. At the end of the time interval, the stored volume is $S_2 = S_1 + \Delta S$. The stage h_2 at the end of the time interval can be found from the curve based on S_2. In Figure 12.15, the net change in storage is positive, so the net change in storage elevation is also positive (that is, the water is rising).

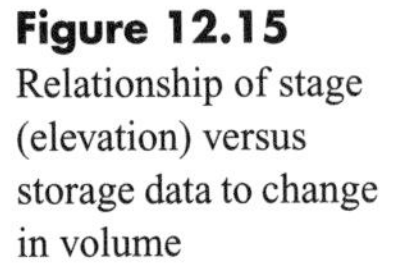

Figure 12.15
Relationship of stage (elevation) versus storage data to change in volume

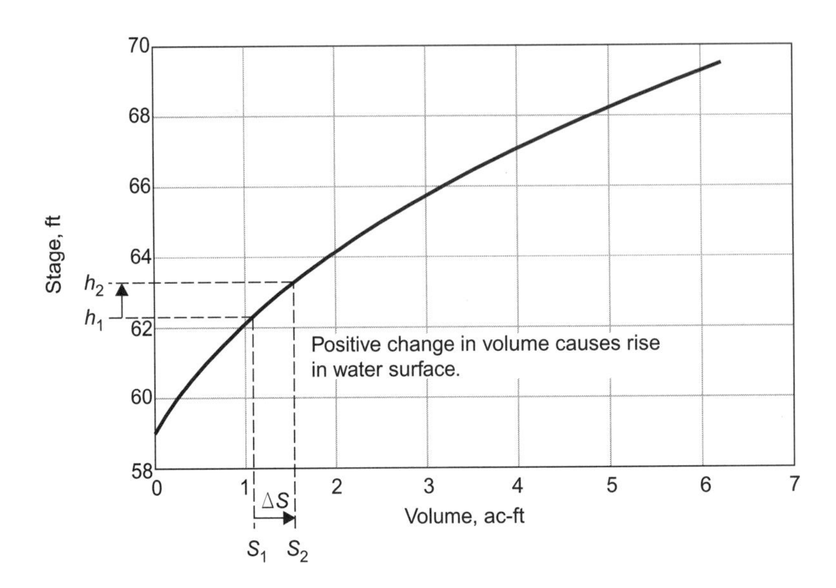

Several approaches to developing stage-storage relationships exist. If a detention facility has a regular geometric shape (for example, it is a cylindrical tank), a geometric equation can be applied to describe the stage-storage relationship. In most cases, however, the engineer must use a stage-surface area relationship to develop stage-storage data. Three methods for determining the stage-storage relationship—the *average-end-area method*, the *conic method*, and *numerical integration*—are presented in this section.

Average-End-Area Method. The average-end-area method, also known as the *trapezoidal method,* uses the same concept frequently used in computations for estimating earthwork volumes. In this method, the incremental storage volume ΔS available between two water surface elevations is computed as:

$$\Delta S = \frac{A_1 + A_2}{2}(h_2 - h_1) \qquad (12.1)$$

where ΔS = change in volume during the time interval
h_1 = water surface elevation at 1 (ft, m)
h_2 = water surface elevation at 2 ($h_2 > h_1$) (ft, m)
A_1 = water surface area corresponding to stage h_1 (ft^2, m^2)
A_2 = water surface area corresponding to stage h_2 (ft^2, m^2)

By Strategic Use of Postcards, Thoreau Manages to Keep Walden Pond Unspoiled.

Computation of incremental storage amounts for a series of known stages and surface areas, followed by summing of the incremental storages, yields the desired relationship between h and S.

A limitation of the average-end-area method is that it applies a linear averaging technique to describe the surface area of a pond, whereas, for a pond with sloped sides, the surface area is actually a second-order (or higher) function. However, the error associated with this approximation is well within the error bounds of other key design assumptions, such as design storm determination and watershed homogeneity. This is especially true if the selected stage increment size ($h_2 - h_1$) is small. If more accuracy is needed, the conic method described in the next section may be applied to represent second-order effects.

Example 12.2 – Computing Storage Volume Using the Average-End-Area Method. Use the average-end-area method to calculate and graph the stage versus storage relationship for the elevation-area data obtained in Example 12.1.

Solution: The first two columns of the following table contain the stage versus surface area data from Example 12.1. To compute the stage-storage relationship using the average-end-area method, adjacent area values in column 2 are added and then divided by the stage increment for those areas.

The steps used in computing the stage-storage relationship are described as follows, and the results are reported in the table. The relationship is plotted in Figure E12.2.1.

Column 3 = [Column 2 (current row)] + [Column 2 (previous row)]

Column 4 = [Column 1 (current row)] – [Column 1 (previous row)]

Column 5 = [Column 3 (current row) ÷ 2] × [Column 4 (current row)]

Column 6 = [Column 5 (current row)] + [Column 6 (previous row)]

(1)	(2)	(3)	(4)	(5)	(6)
Stage (m)	Area (m^2)	$A_n + A_{n-1}$ (m^2)	$h_n - h_{n-1}$ (m)	ΔS (m^3)	Cumulative S (m^3)
37.00	0	0	0.0	0	0
37.10	60	60	0.1	3	3
37.20	120	180	0.1	9	12
37.30	180	300	0.1	15	27
37.40	240	420	0.1	21	48
37.46	276	516	0.06	15	63
37.50	276	552	0.04	11	75
38.00	276	552	0.5	138	213
39.00	276	552	1.0	276	489

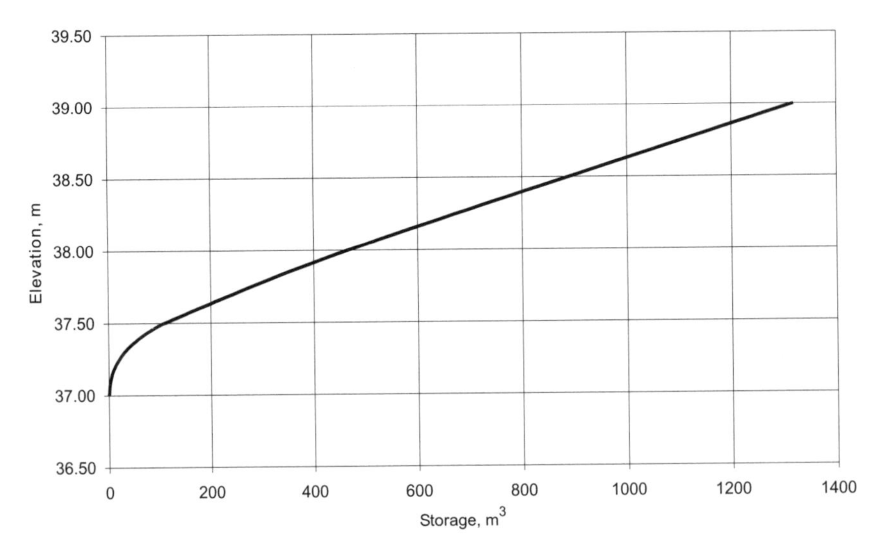

Figure E12.2.1 Plot of stage versus storage for Example 12.2

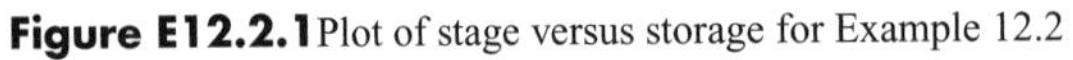

Conic Method. This method applies a conic geometry to each stage increment to approximate the nonlinear relationship between stage and water surface area, as illustrated in Figure 12.16. The incremental storage volume between two stages h_1 and h_2 ($h_2 > h_1$) is defined as

$$\Delta S = \left(\frac{h_2 - h_1}{3}\right)(A_1 + A_2 + \sqrt{A_1 A_2}) \qquad (12.2)$$

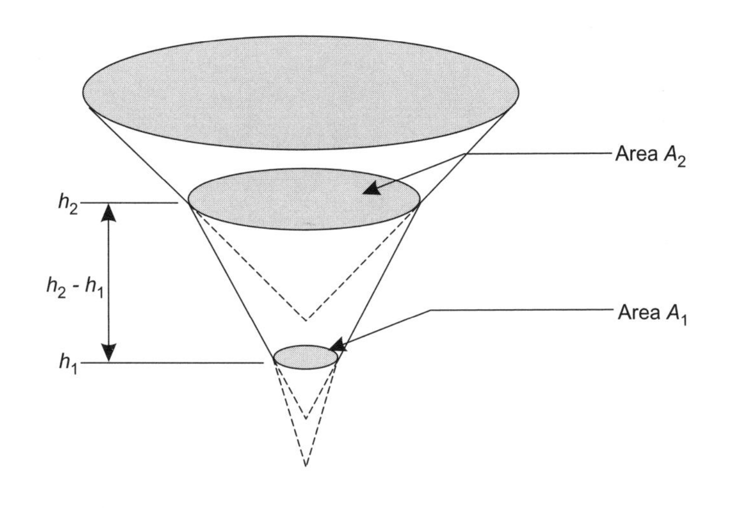

Figure 12.16
Illustration of conic method for computing pond volume

Example 12.3 – Computing the Storage Volume Using the Conic Method. Use the conic method to calculate and graph the stage (elevation) versus storage relationship for a pond with the elevation versus surface area relationship given in the following table.

Stage (ft)	Area (acre)
100.00	0.158
100.50	0.170
101.00	0.182
101.50	0.194
102.00	0.207
102.50	0.221
103.00	0.234
103.50	0.248
104.00	0.263
104.50	0.278
105.00	0.293
105.50	0.309
106.00	0.325
106.50	0.342
107.00	0.359

Solution: The following table is used to determine the cumulative volume at each elevation. Each column represents a step in the calculations used for the conic method. A plot of the elevation-storage relationship is given in Figure E12.3.1.

Columns 1 and 2 are the elevation vs. surface area relationship, and are taken from the previous table.

Column 3 = {[Column 1 (current row)] – [Column 1 (previous row)]} / 3

Column 4 = [(Column 2 (current row)] + [Column 2 (previous row)] + {[Column 2 (current row)] × [Column 2 (previous row)]}$^{0.5}$

Column 5 = (Column 3) × (Column 4)

Column 6 = [Column 6 (previous row)] + [Column 5 (current row)]

(1)	(2)	(3)	(4)	(5)	(6)
Stage (ft)	Area (ac)	$(h_2 - h_1)/3$ (ft)	$A_1 + A_2 + \sqrt{A_1 A_2}$ (ac)	ΔS (ac-ft)	Cumulative S (ac-ft)
100.00	0.158	0.000	0.000	0.000	0.000
100.50	0.170	0.167	0.492	0.082	0.082
101.00	0.182	0.167	0.528	0.088	0.170
101.50	0.194	0.167	0.564	0.094	0.264
102.00	0.207	0.167	0.602	0.100	0.364
102.50	0.221	0.167	0.641	0.107	0.471
103.00	0.234	0.167	0.682	0.114	0.585
103.50	0.248	0.167	0.723	0.121	0.705
104.00	0.263	0.167	0.766	0.128	0.833
104.50	0.278	0.167	0.811	0.135	0.968
105.00	0.293	0.167	0.856	0.143	1.111
105.50	0.309	0.167	0.903	0.150	1.262
106.00	0.325	0.167	0.951	0.158	1.420
106.50	0.342	0.167	1.000	0.167	1.587
107.00	0.359	0.167	1.051	0.175	1.762

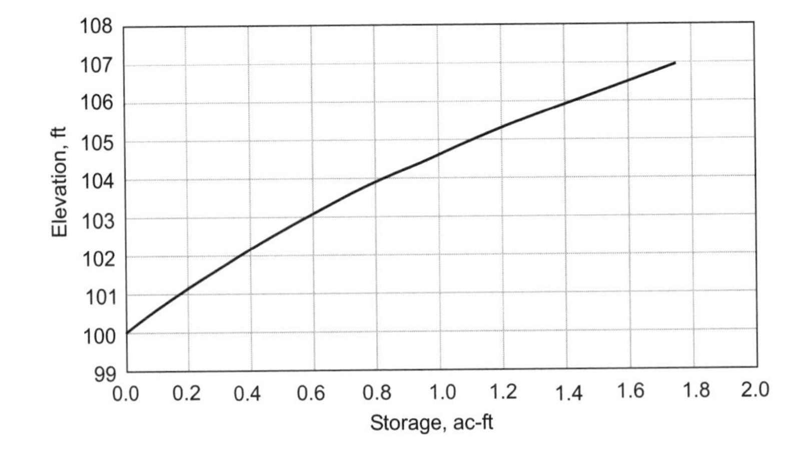

Figure E12.3.1 Stage versus storage relationship for Example 12.3

Numerical Integration. Another approach to development of a stage-storage relationship is similar to the average-end-area method, but involves numerical or graphical *integration* of a stage versus surface area relationship. If the water surface elevation of a pond changes by a small increment dh, the corresponding increment of storage change is dS, where $dS = A(h)dh$. Integrating, the volume of storage between two stages h_1 and h_2 ($h_2 > h_1$) is

$$\Delta S = \int_{h_1}^{h_2} A(h)dh \qquad (12.3)$$

That is, the incremental amount of storage between two stages is equal to the corresponding area under a stage versus surface area curve. That area can be determined either by applying a trapezoidal rule (which reduces to Equation 12.1), by using a more sophisticated numerical integration algorithm (see Chapra and Canale, 1988), by counting squares, or by digitizing.

Stage versus Discharge

The discharge (Q) from a detention facility depends on the stage (h) of the water surface and the geometric and hydraulic characteristics of the outlet(s), and possibly on tailwater effects that may influence outlet hydraulics. In cases where interconnected ponds are routed, the discharge from one pond into another depends on the stages in the ponds and the hydraulic characteristics of the channel or conduit connecting them.

The hydraulic characteristics of a pond's outlet structure are described by a *stage versus discharge relationship,* which is applied during routing to determine the increase or decrease in outflow given a change in the stage (water surface elevation). Figure 12.17 demonstrates how stage-discharge information is related to routing. The relationship is determined by computing the discharge for various water surface stages. The precision of the curve increases as the incremental change in stage used to compute the relationship decreases. The designer must judge the appropriate degree of approximation when selecting the stage increment size for development of the *rating curve*.

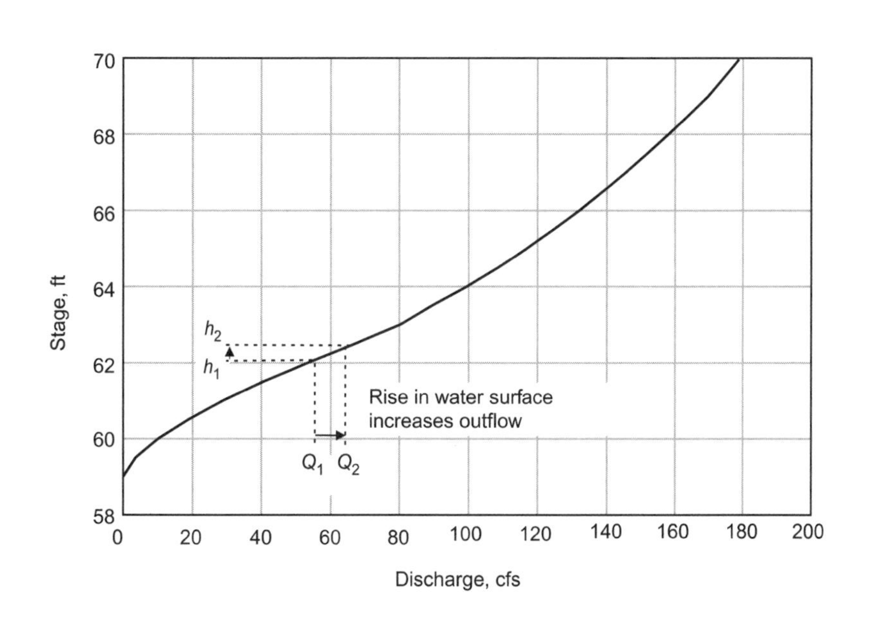

Figure 12.17
Stage versus discharge relationship

In Figure 12.17, h_1 and Q_1 are the stage and discharge, respectively, at the beginning of a time interval. The stage at the end of the time interval, h_2, is determined from the stage-storage curve based on the change in storage volume. Using h_2, the discharge Q_2 at the end of the time interval can be obtained from the stage versus discharge relationship. In this illustration, the change in stage was positive (rising), so the change in discharge is also positive (increasing).

Most dry detention facilities have both low- and high-level outlets. The *low-level outlet* may consist of a pipe or culvert positioned with its invert either at or near the bottom of the pond, one or more orifices, or a weir. The *high-level outlet* is usually an overflow spillway, and is expected to function only during extreme runoff events with recurrence intervals that exceed the pond's design storm recurrence interval. When ponds are designed to control runoff rates for several different recurrence intervals, intermediate outlets are usually provided as well. They are similar to low-level outlets in design but are installed at higher elevations.

The stage versus discharge relationship for a pond is developed using hydraulic relationships applicable to the number and types of individual outlets. The relationships for several common outlet structure types have already been presented in this book, including weirs (Section 7.6), orifices (Section 7.7), and culverts (Section 9.5). This section describes overflow spillway structures, which are commonly found in detention ponds and are a specific application of the weir equation. The development of composite rating curves for situations in which multiple structures are present is also described.

Composite Stage Versus Discharge Relationship for Structures in Parallel. The *composite stage versus discharge relationship* for a detention pond represents the total discharge from a pond (possibly occurring through several outlets) as a function of the water surface elevation (or depth). Many detention ponds have several outlet structures that work independently of one another. In this situation, the discharges through the various outlets are summed to obtain the total discharge for each water surface elevation (stage) selected to compute the composite stage versus discharge relationship. Figure 12.18 illustrates a set of outlet structures working in parallel, and Figure 12.19 shows the stage versus discharge relationship for this outlet configuration. Example 12.4 illustrates how the composite relationship for a composite structure consisting of a weir and an orifice is developed.

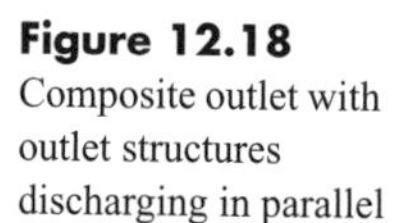

Figure 12.18
Composite outlet with outlet structures discharging in parallel

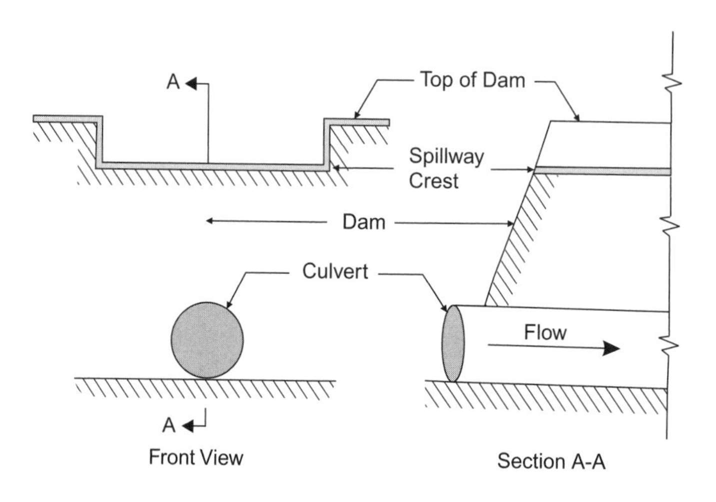

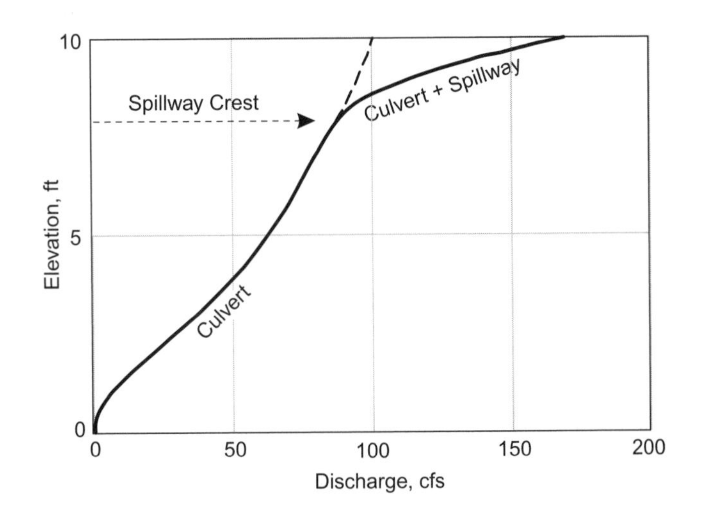

Figure 12.19 Rating curve for composite outlet with outlet structures discharging in parallel

Example 12.4 – Stage versus Discharge for a Composite Outlet Structure. A simple composite outlet structure is being designed with an orifice and a weir operating in parallel. The orifice is a 6-in. diameter orifice plate attached to the entrance of a culvert with an invert elevation of 100.00 ft and an orifice coefficient of 0.60. (The culvert is large enough that it does not affect the hydraulic capacity of the orifice; the orifice can be assumed to have a free outfall.) The concrete rectangular contracted weir has a weir coefficient approximated as 0.49 for the entire range of headwaters, a weir length of 15 ft, and a crest elevation at 105.00 ft. It operates under free outfall conditions.

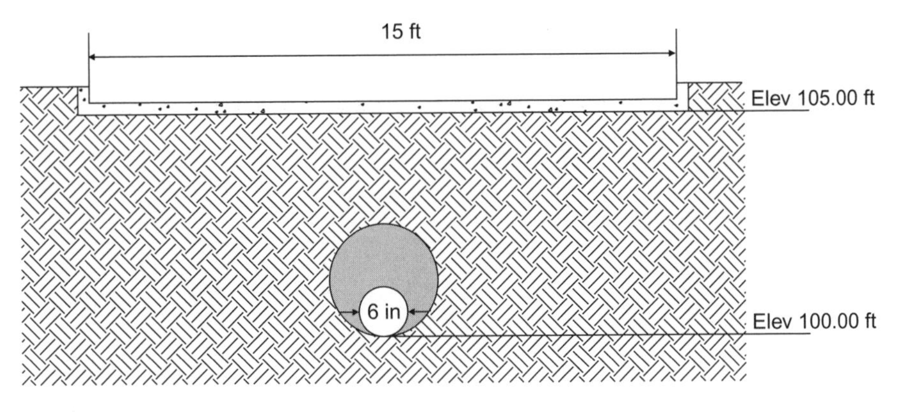

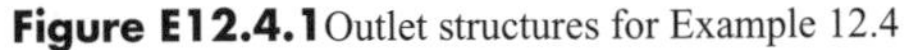

Figure E12.4.1 Outlet structures for Example 12.4

Calculate and graph the elevation vs. discharge rating curve from elevations 100.00 ft to 107.00 ft for each 0.50 ft increment. (Note that a smaller increment would yield a smoother, more precise rating curve.)

Solution: From Chapter 7, the rectangular weir equation for a weir (Equation 7.64) with two contractions is

$$Q = \frac{2}{3} C_d \sqrt{2g} (L - 0.2H) H^{3/2}$$

The orifice equation (Equation 7.68) is

$$Q = C_d A_o (2gh_o)^{1/2}$$

The following table shows the stage-storage calculation results described by the following steps. Figure E12.4.1 is a plot of the composite structure rating curve.

Orifice Centroid Elevation = 100.00 ft + (1/2 × 0.5 ft) = 100.25 ft

Orifice Area = $\pi D^2/4 = 0.196$ ft^2

Column 1 is the range of rating table elevations (100.0 ft to 107.0 ft, in 0.5-ft increments)

Column 2 is the head on the orifice = (Column 1) – (Orifice Centroid Elevation)

Column 3 is the orifice discharge = Solution to orifice equation for $C_d = 0.60$, $A_o = 0.196$, h_o = Column 2.

Column 4 is the head on the weir = (Column 1) – (Weir Crest Elevation)

Column 5 is the weir discharge = Solution to rectangular weir equation for $C_d = 2.6$, $L = 15.0$, H = Column 4

Column 6 is the composite structure discharge = (Column 3) + (Column 5)

(1)	(2)	(3)	(4)	(5)	(6)
	Centroid = 100.25 ft		Crest = 105.00 ft		Total Discharge
Elevation (ft)	Head (ft)	Discharge (cfs)	Head (ft)	Discharge (cfs)	(cfs)
100.00	0.00	0.00	0.00	0.00	0.00
100.50	0.25	0.47	0.00	0.00	0.47
101.00	0.75	0.82	0.00	0.00	0.82
101.50	1.25	1.06	0.00	0.00	1.06
102.00	1.75	1.25	0.00	0.00	1.25
102.50	2.25	1.42	0.00	0.00	1.42
103.00	2.75	1.57	0.00	0.00	1.57
103.50	3.25	1.70	0.00	0.00	1.70
104.00	3.75	1.83	0.00	0.00	1.83
104.50	4.25	1.95	0.00	0.00	1.95
105.00	4.75	2.06	0.00	0.00	2.06
105.50	5.25	2.16	0.50	13.79	15.95
106.00	5.75	2.26	1.00	39.00	41.27
106.50	6.25	2.36	1.50	71.65	74.01
107.00	6.75	2.45	2.00	110.31	112.76

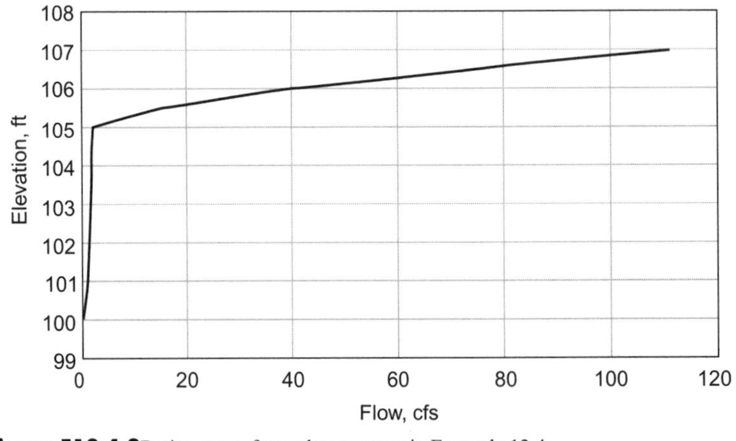

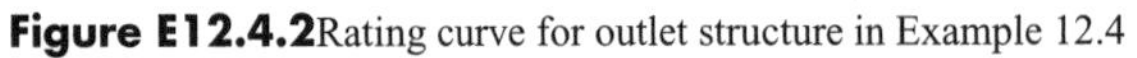

Figure E12.4.2 Rating curve for outlet structure in Example 12.4

Composite Stage Versus Discharge Relationship for Structures in Series. When outlet structures are in series, it is possible that backwater effects caused by hydraulic controls at downstream structures will affect the discharges through connecting upstream structures. In these cases, hydraulic grade line calculations are required to determine the interactions between individual structure components.

Figure 12.20 illustrates a pond outlet with structures in parallel and in series. In this example, at lower stages, the water surface elevation in the culvert remains low enough that it does not influence the hydraulic performance of the v-notch weir, and the v-notch weir controls the discharge. As the water surface rises above 4 ft, water flows over the crest of the inlet box, and the culvert still has enough capacity for the discharge to be simply the sum of the flows through the v-notch weir and over the top of the box. When the water surface reaches an elevation of approximately 6 ft, however, the v-notch weir/inlet box combination becomes capable of passing more flow than the culvert. The inlet box riser becomes submerged, and the culvert controls the discharge. For elevations of more than 8 ft, water flows over the spillway, and the spillway and culvert together control the discharge.

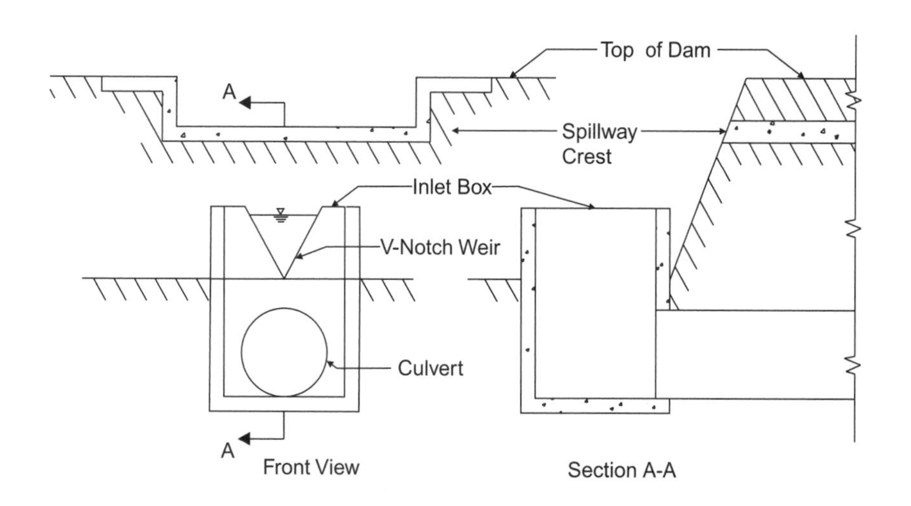

Figure 12.20
Composite outlet with outlet structures discharging in parallel and in series

12.7 STORAGE ROUTING CALCULATIONS

Routing is the process of analyzing the difference between flow entering and leaving the pond for a series of time increments in order to determine the change in volume and water surface elevation. Figure 12.22 graphically depicts an inflow hydrograph and the routed outflow hydrograph for a specific pond geometry (stage versus storage relationship), outlet structure (stage versus discharge relationship), and initial water surface elevation.

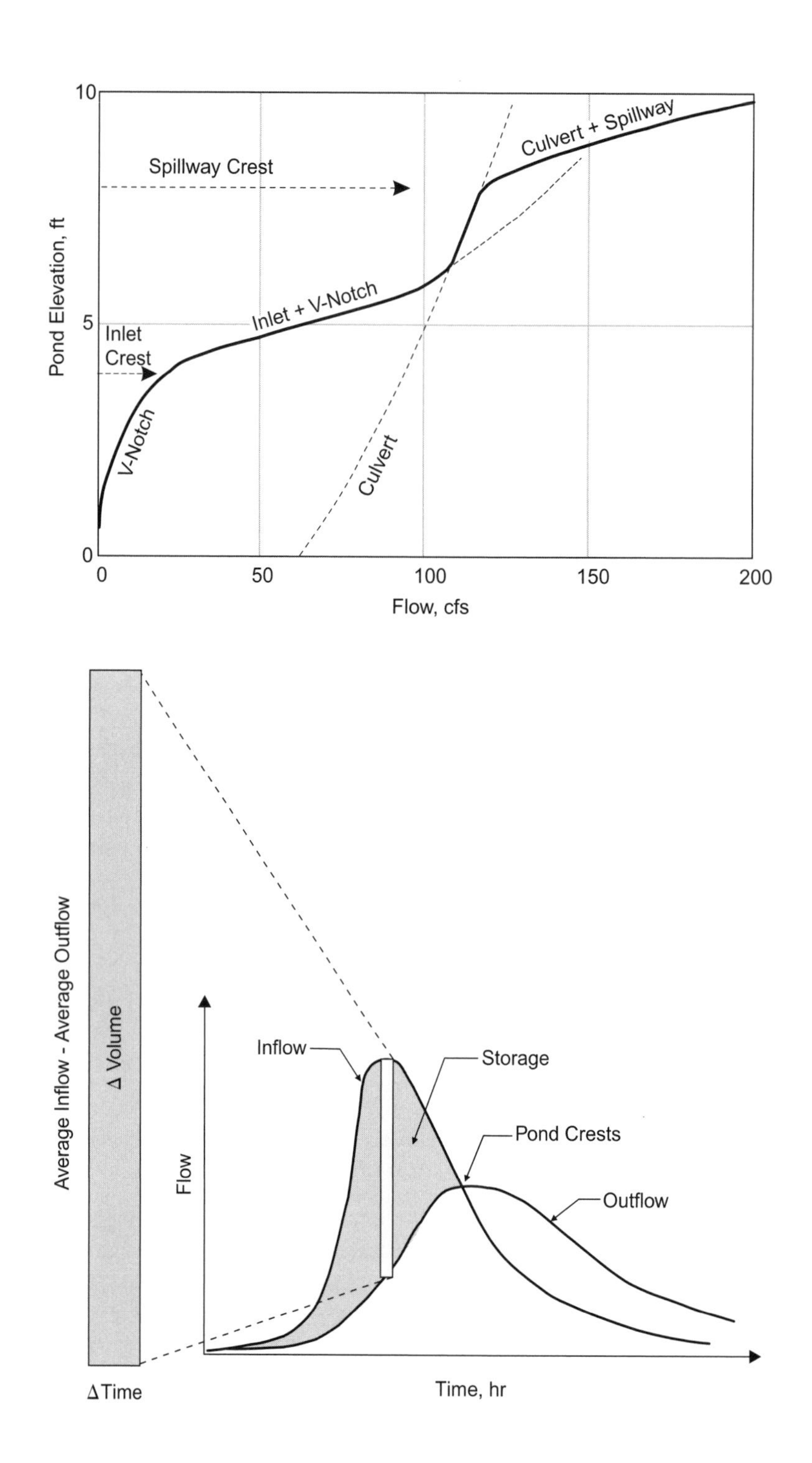

Figure 12.21
Rating curve for composite outlet with outlet structures discharging in parallel and in series

Figure 12.22
Change in pond volume for a routing time step

Fundamentally, routing of a pond amounts to a solution of the conservation of mass equation

$$\frac{dS}{dt} = I(t) - O(t) \tag{12.4}$$

where S = volume of water in the pond at time t (ft^3, m^3)
$I(t)$ = specified inflow rate at time t (cfs, m^3/s)
$O(t)$ = outflow rate at time t (cfs, m^3/s)

Equation 12.4 states that the change in storage during a time period is equal to the difference between the inflow and the outflow. A number of routing methods are available for evaluating detention facilities. The most common of these is the *storage-indication method*, which is described in the next section. This method is based on a finite-difference approximation of Equation 12.4. A second approach to storage routing is based on the numerical integration of a modified form of Equation 12.4. This integration can be performed using any one of a number of alternative numerical integration algorithms; in this book, a fourth-order Runge-Kutta algorithm is used.

Storage-Indication Method

Figure 12.22 shows the inflow and outflow hydrographs for a pond. The graph magnifies a time increment (step) for which the inflow is greater than the outflow. The net change in pond volume during the time increment is computed as the area between the hydrographs for that time increment (that is, the difference in flow multiplied by the time step). Because the inflow is greater than the outflow for this time increment, the change in volume is positive, which means that the pond water surface (stage) is rising.

At the point where the inflow equals the outflow (that is, where the inflow and outflow hydrographs intersect), the change in storage is zero, the water surface elevation (stage) is constant, and the pond is cresting. After the pond has crested, the outflow rate is greater than the inflow rate, resulting in a net loss in volume that causes the water surface (stage) to fall during a time step. The shaded area of Figure 12.22 is the bounded area for which inflow is greater than outflow (that is, for the period when the pond is filling); this area is equal to the maximum volume stored.

Figure 12.23 shows a portion of the inflow and outflow hydrographs for a pond for a period in which the inflow discharge is greater than the outflow discharge. This figure is useful for understanding the relationships given in Equations 12.5 through 12.7, which are applicable for any time period, regardless of the relative magnitudes of the inflow and outflow discharges.

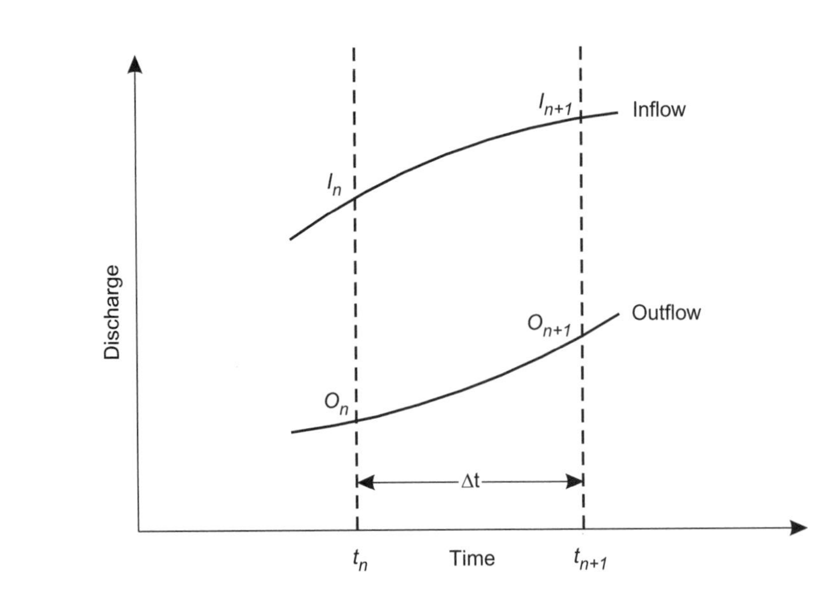

Figure 12.23
Definition sketch for terms in Equation 12.5

Over a time interval $\Delta t = t_{n+1} - t_n$, the inflow varies from I_n to I_{n+1}, and therefore the average inflow discharge over the time interval can be approximated as $(I_n + I_{n+1})/2$. Over the same time interval, the average outflow discharge can be approximated as $(O_n + O_{n+1})/2$. The change in storage over the time interval is $\Delta S = S_{n+1} - S_n$, and the time rate of change of storage may be approximated as $\Delta S/\Delta t = (S_{n+1} - S_n)/\Delta t$. Substitution of the average inflow and outflow rates, and of the approximate rate of change of storage, into Equation 12.4 yields:

$$\frac{S_{n+1} - S_n}{\Delta t} = \frac{I_n + I_{n+1}}{2} - \frac{O_n + O_{n+1}}{2} \tag{12.5}$$

where S_{n+1} = storage at time t_{n+1} (ft³, m³)
S_n = storage at time t_n (ft³, m³)
Δt = change in time, $t_{n+1} - t_n$ (s)
I_n = inflow at time t_n (cfs, m³/s)
I_{n+1} = inflow at time t_{n+1} (cfs, m³/s)
O_n = outflow at time t_n (cfs, m³/s)
O_{n+1} = outflow at time t_{n+1} (cfs, m³/s)

Because the inflow hydrograph is known in detention pond routing, the terms I_n and I_{n+1} are known for every time interval in the routing process. For the first time interval, at the beginning of a pond inflow event, the terms O_n and S_n are also initially known. Note that O_n and S_n are related, as for each pond stage h_n a corresponding S_n and a corresponding O_n exist. Rearranging Equation 12.5 with unknown quantities on the left and known quantities on the right yields the storage indication equation:

$$\left(\frac{2S_{n+1}}{\Delta t} + O_{n+1}\right) = \left(\frac{2S_n}{\Delta t} - O_n\right) + (I_n + I_{n+1}) \tag{12.6}$$

Thus, knowledge of the quantities S_n, O_n, and I_n at the beginning of a time interval (at time t_n), as well as knowledge of the inflow discharge I_{n+1} at the end of the time interval, can be used to compute the value of the left side of Equation 12.6. However the left side of the equation has two unknown variables (O_{n+1} and S_{n+1}). Additional information must be established to expose the relationship between O_{n+1} and S_{n+1} so that these two values can be determined.

A *storage indication curve* is a graph that permits the two unknowns to be determined by illustrating the relationship between O and $(2S/\Delta t + O)$. Such a graph can be constructed because for any stage h in a pond, corresponding values of O and S exist, and thus corresponding values of O and $(2S/\Delta t + O)$ also exist.

In performing storage indication calculations for a specified time interval, Equation 12.6 is applied to compute the numerical value of the quantity $(2S_{n+1}/\Delta t + O_{n+1})$. Next, the value of $(2S_{n+1}/\Delta t - O_{n+1})$ for the time interval must be found. Because the value of the left side of Equation 12.6 $(2S_{n+1}/\Delta t + O_{n+1})$ is known, and because this value can be used with the storage-indication curve to find O_{n+1}, $(2S_{n+1}/\Delta t - O_{n+1})$ can be computed as

$$\left(\frac{2S_{n+1}}{\Delta t} - O_{n+1}\right) = \left(\frac{2S_{n+1}}{\Delta t} + O_{n+1}\right) - 2O_{n+1} \tag{12.7}$$

The term on the left side of the equation can now be used in the next time step to compute $(2S_{n+2}/\Delta t + O_{n+2})$, and the process repeats through every time step.

The following section describes the full procedure used in storage indication routing and presents an example problem. It should be noted that the storage indication routing procedure discussed is not applicable to interconnected ponds. For interconnected pond analysis, a more complex algorithm is required because the tailwater conditions for the outlet are time-variable. For information on routing interconnected ponds, the reader is referred to Glazner (1998).

Stormwater Detention Analysis Procedure. This chapter has introduced each of the basic components required for modeling how a pond functions during a runoff event. A determination of the detailed behavior of a pond during a runoff event can be accomplished through the following steps if the storage indication method is used.

1. Calculate the inflow hydrograph to the pond. In most design cases, this inflow hydrograph will be the post-development runoff hydrograph from a development site. Methods for calculating hydrographs are presented in Section 5.5.

2. Calculate the stage versus storage relationship for the selected pond design. Stages used for development of the relationship should range from the dry invert of the pond to the top of the dam or pond sides.

3. Calculate the stage versus discharge relationship for the pond's outlet structure(s). Again, stages used for development of the relationship should range from the dry invert of the pond to the top of the dam or pond sides, and should be the same stages used for development of the stage vs. storage relationship in step 2. This holds true even if the pond discharges corresponding to low stages are zero.

4. Establish the relationship between stage, storage, and discharge by creating a storage indication curve for $2S/t + O$ versus O.

5. Use the inflow hydrograph from step 1 and the storage indication curve from step 4 to route the inflow hydrograph through the pond.

Example 12.5 – Storage Indication Routing. Route the inflow hydrograph given in the following table through the pond described by the stage versus storage data computed in Example 12.3 and the stage versus discharge data computed in Example 12.4. Determine the maximum water surface elevation for the detention basin, and the maximum storage volume, maximum outflow rate, and peak lag resulting from detention.

Time (minutes)	Inflow (cfs)
0	0.00
10	0.58
20	5.94
30	15.44
40	20.75
50	20.27
60	17.44
70	14.19
80	11.42
90	9.44
100	8.12
110	7.27
120	6.61
130	5.49
140	3.32
150	1.52
160	0.66
170	0.29
180	0.12
190	0.05
200	0.02
210	0.00

Solution: The results from Examples 12.3 and 12.4 are combined in the following table. Note the importance of using the same elevations for storage and discharge in order to directly correlate flow and storage at each elevation.

Elevation (ft)	Total Volume (ac-ft)	Total Flow (cfs)
100.00	0.000	0.00
100.50	0.082	0.47
101.00	0.170	0.82
101.50	0.264	1.06
102.00	0.364	1.25
102.50	0.471	1.42
103.00	0.585	1.57

Elevation (ft)	Total Volume (ac-ft)	Total Flow (cfs)
103.50	0.705	1.70
104.00	0.833	1.83
104.50	0.968	1.95
105.00	1.111	2.06
105.50	1.262	15.95
106.00	1.420	41.27
106.50	1.587	74.01
107.00	1.762	112.76

Perform the five steps outlined for storage indication pond routing, as follows:

a) Compute the post-developed pond inflow hydrograph. This can be accomplished using the concept of rainfall hyetographs discussed in Section 4.5 (page 91) in combination with infiltration, runoff, and hydrograph procedures discussed in Chapter 5. In this example, the final inflow hydrograph is given.

b) Calculate the stage versus storage curve from the pond grading plan or storage structure geometry. This information was calculated in Example 12.3 on page 503 (no calculations necessary for this example).

c) Calculate the stage versus outflow curve for the pond outlet structure. This information was calculated in Example 12.4 on page 507 (no calculations necessary for this example).

d) Calculate and graph the storage indication curve that correlates $(2S/t + O)$ and O. To calculate this curve, the rating elevations for the stage versus storage curve must be identical to those used for the stage versus outflow curve. Use a 10-minute time step in this example. The results are shown in the following table, where Column 4 values were computed using:

Column 4 = {2.0 × [(Column 2) × 43,560 ft^2/ac] ÷ (10 min × 60 s/min)} + (Column 3)

(1)	(2)	(3)	(4)
Elevation (ft)	**Volume (ac-ft)**	**Outflow (cfs)**	**$2S/\Delta t + O$ (cfs)**
100.00	0.000	0.00	0.00
100.50	0.082	0.47	12.38
101.00	0.170	0.82	25.50
101.50	0.264	1.06	39.39
102.00	0.364	1.25	54.10
102.50	0.471	1.42	69.81
103.00	0.585	1.57	86.51
103.50	0.705	1.70	104.07
104.00	0.833	1.83	122.78
104.50	0.968	1.95	142.50
105.00	1.111	2.06	163.38
105.50	1.262	15.95	199.19
106.00	1.420	41.27	247.45
106.50	1.587	74.01	304.44
107.00	1.762	112.76	368.60

Because the maximum inflow is about 20 cfs, we already know that the maximum outflow value should be equal to or less than 20 cfs. The storage indication curve shown in Figure E12.5.1 is an expanded view for the outflow range below 20 cfs, which will make it easier to read the smaller flow values during the routing tabulation.

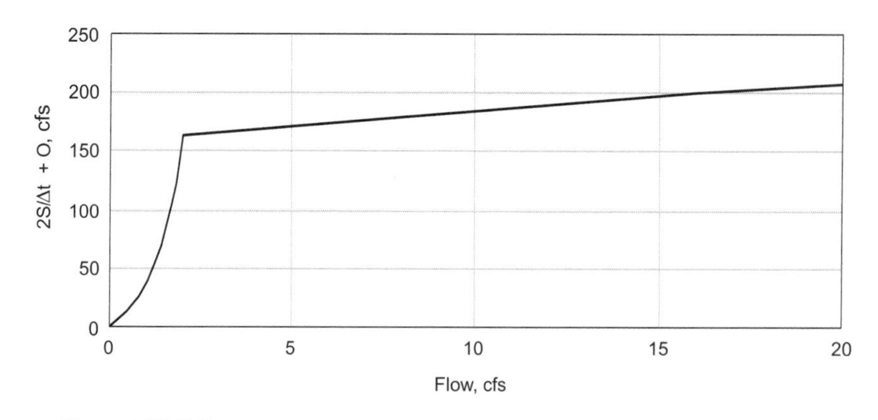

Figure E12.5.1 Storage indication curve for Example 12.5

e) Use the inflow hydrograph from Step 1 and the storage indication curve from Step 4 to route the hydrograph using the systematic worksheet approach as shown in the table following the column descriptions

Columns 1 and 2 represent the given inflow hydrograph.

Under initial conditions, columns 3 through 6 have values of 0. Because the pond is initially empty, the initial elevation (column 6) is 100.00 ft.

For subsequent time steps,

[Column 3 (current row)]= [Column 2 (previous row)]+ [Column 2 (current row)]

[Column 5 (current row)] = [Column 3 (current row)] + [Column 4 (previous row)]

Determine column 6 values by using the column 5 values and solving for the associated outflow from the storage indication curve calculated in Step 4.

Solve column 7 (current row) using stage vs. outflow curve created in Step 3.

Finally, to complete the current row, solve

[Column 4 (current row)] = [Column 5 (current row)] – 2 × [Column 6 (current row)].

The tabulated results shown stop at t = 300 minutes into the routing. To complete the routing, the calculations need to be continued until the pond is fully drained (the routed outflow in column 6 recedes to zero). The graph for the inflow and outflow hydrographs shown in Figure E12.5.2 displays the entire routing time.

(1) Time (min)	(2) Inflow (cfs)	(3) $I_1 + I_2$ (cfs)	(4) $2S/\Delta t - O$ (cfs)	(5) $2S/\Delta t + O$ (cfs)	(6) Outflow (cfs)	(7) Elevation (ft)
0	0	0	0.00	0.00	0.00	100.00
10	0.58	0.58	0.54	0.58	0.02	100.02
20	5.94	6.52	6.52	7.06	0.27	100.28
30	15.44	21.38	26.18	27.90	0.86	101.09
40	20.75	36.19	59.69	62.37	1.34	102.26
50	20.27	41.02	97.36	100.71	1.68	103.40
60	17.44	37.71	131.26	135.07	1.90	104.31
70	14.19	31.63	158.77	162.89	2.06	104.99
80	11.42	25.61	163.97	184.38	10.21	105.29
90	9.44	20.86	164.07	184.83	10.38	105.30
100	8.12	17.56	163.35	181.63	9.14	105.25
110	7.27	15.39	162.70	178.74	8.02	105.21

(1)	(2)	(3)	(4)	(5)	(6)	(7)
Time (min)	**Inflow (cfs)**	$I_1 + I_2$ **(cfs)**	$2S/\Delta t - O$ **(cfs)**	$2S/\Delta t + O$ **(cfs)**	**Outflow (cfs)**	**Elevation (ft)**
120	6.61	13.88	162.22	176.58	7.18	105.18
130	5.49	12.1	161.71	174.32	6.30	105.15
140	3.32	8.81	160.86	170.52	4.83	105.10
150	1.52	4.84	159.78	165.70	2.96	105.03
160	0.66	2.18	157.86	161.96	2.05	104.97
170	0.29	0.95	154.73	158.81	2.04	104.89
180	0.12	0.41	151.11	155.14	2.02	104.80
190	0.05	0.17	147.29	151.28	2.00	104.71
200	0.02	0.07	143.41	147.36	1.98	104.62
210	0	0.02	139.52	143.43	1.95	104.52
220	0	0	135.65	139.52	1.93	104.42
230	0	0	131.84	135.65	1.91	104.33
240	0	0	128.07	131.84	1.89	104.23
250	0	0	124.34	128.07	1.86	104.13
260	0	0	120.66	124.34	1.84	104.04
270	0	0	117.03	120.66	1.82	103.94
280	0	0	113.45	117.03	1.79	103.85
290	0	0	109.92	113.45	1.77	103.75
300	0	0	106.44	109.92	1.74	103.66

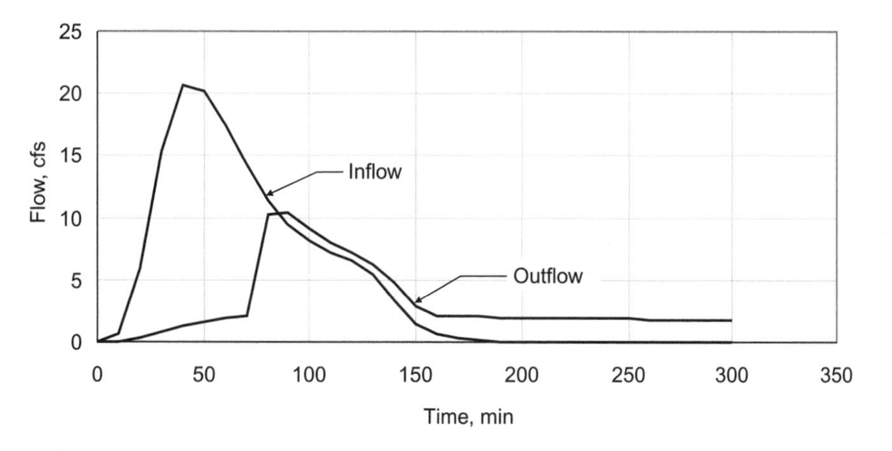

Figure E12.5.2 Pond inflow and outflow hydrographs

From the routing table:

Maximum stage = 105.30 ft, which is above spillway crest

Maximum storage = 1.2 ac-ft (determined from the stage versus storage data for an elevation of 105.30 ft)

Maximum outflow = 10.38 cfs

Outflow peak time = 90 min

Peak flow was delayed 50 min from inflow peak time of 40 min

Modeling Focus: Detention Pond Analysis

A proposed pond is to attenuate peak flows from a new 100-ac residential development with 1-ac lots. According to local criteria, the peak outflows from the pond may not exceed pre-development peak flows from the site for the 2-, 25-, and 100-year events. In addition, the local reviewing agency is requiring that the 100-year peak flow not exceed the maximum capacity of the downstream conveyance system of 32 cfs, and that a minimum freeboard of 1 ft be maintained. The analysis should be performed using an NRCS (SCS) 24-hour, Type II rainfall distribution and the hydrologic parameters given in the tables below.

Storm	24-Hr Rainfall Depth, in.
2 year	3.11
25 year	5.30
100 year	7.36

Watershed Condition	Area (ac)	*CN*	t_c (hr)
Predevelopment	50	61	1.0
Developed	50	68	0.5

A variety of computer programs are available for performing hydrologic calculations. The design storms are created by applying the synthetic storm distribution to 24-hr rainfall depths for the 2-, 25-, and 100-year events in local area. [Note that, in the case of the PondPack software that accompanies this text, the NRCS (SCS) Type II rainfall distribution has already been entered.] The times of concentration, drainage areas, and curve numbers for pre-developed and developed conditions are then entered, and the program computes the hydrographs.

The proposed pond outlet structure is a single, 24-in. circular concrete culvert with a free outfall. Data describing the culvert and the pond volume (in the form of contour areas) are provided in the tables that follow.

Contour Data

Elevation (ft)	Area (ac)
450.0	1.607
450.5	1.645
451.0	1.684
451.5	1.723
452.0	1.762
452.5	1.801
453.0	1.842
453.5	1.882
454.0	1.923
454.5	1.965
455.0	2.006
455.5	2.049

Outlet Structure Data

Culvert entrance type	Groove end projecting
Diameter	24 in.
Upstream invert elevation	450.0 ft
Downstream invert elevation	449.5
Length	38 ft
Manning's n	0.013
Entrance loss coefficient, K_e	0.20

PondPack is an example of a computer program capable of performing not only hydrograph calculations, but also the hydraulic calculations necessary to describe the outlet structure rating curve and the detention pond routing calculations. HEC-HMS is an example of a program capable of computing hydrographs and performing routing; however, the stage-storage relationship for the outlet structure must be developed externally. Other programs may compute only the hydrographs, making it necessary to perform hydraulic and routing calculations externally, for instance with the aid of a spreadsheet.

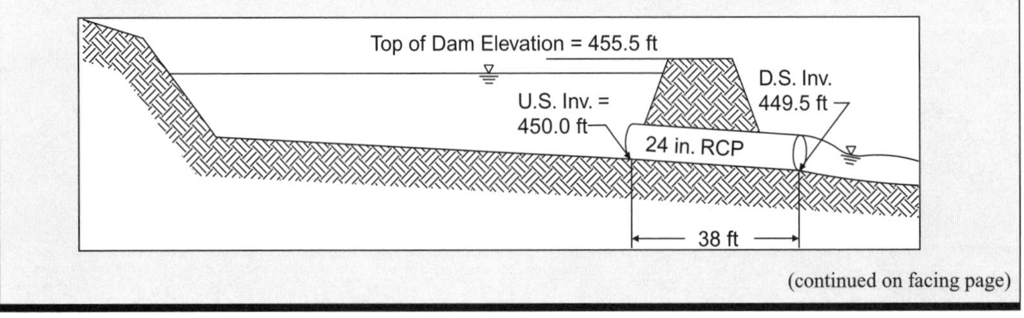

(continued on facing page)

For this example, all calculations are performed using PondPack. Therefore, the next step is to input the elevation-area data describing the pond volume and the physical data describing the culvert outlet. The analysis is run for the entire system, with the storage-indication method being used for the routing calculations. The computed pond volume curve and outlet structure rating curve are shown, as well as the pre-development hydrograph and pond inflow and outflow hydrographs.

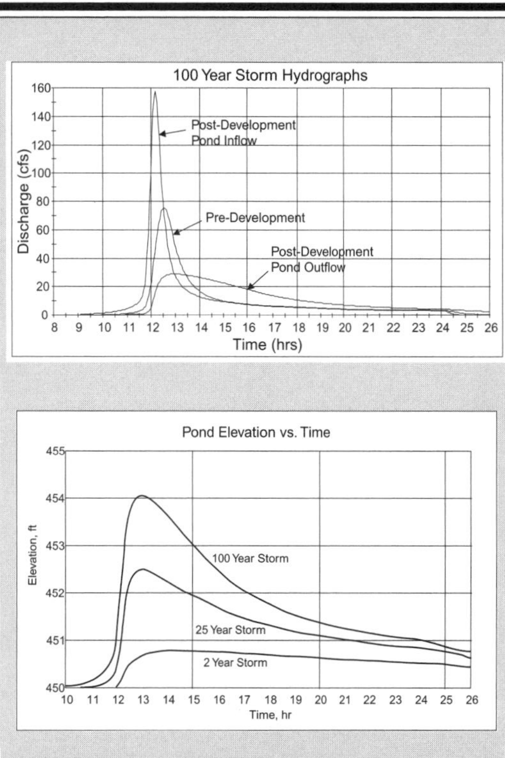

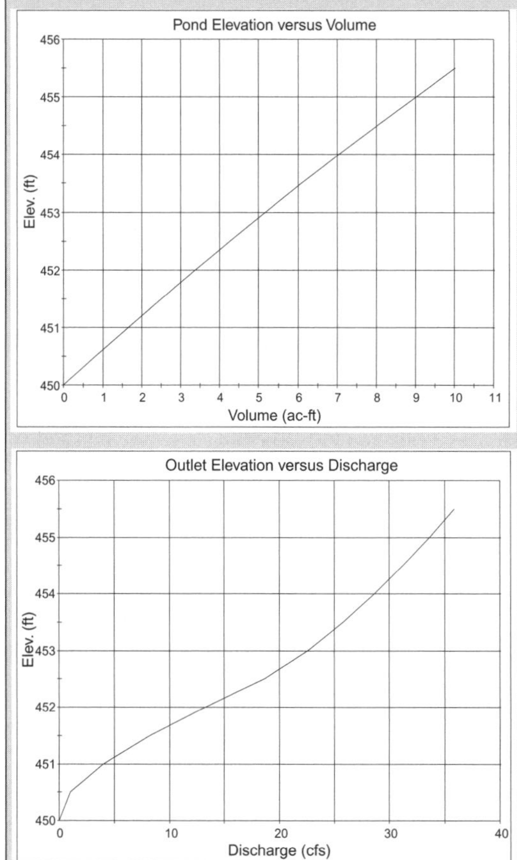

The results of the analysis are summarized in the following table. The allowable outflow is taken as the smaller of the pre-development peak flow and the downstream capacity of 32.0 cfs, and the freeboard is the difference between maximum water surface elevation and the top of the pond (455.50 ft). The pond meets all design criteria. Note that the engineer could add an emergency spillway with an invert above elevation 454.05 ft (100-year maximum pond elevation), which will provide additional emergency outflow capacity but not affect pond performance for events equal to or smaller than the 100-year design storm.

Return Freq.	Predevelopment Conditions			Allow. Outflow (cfs)	Developed Conditions (Pond Inflow)			Developed Conditions (Pond Outflow)		
	T_p (hr)	Q_{Peak} (cfs)	Volume (ac-ft)		T_p (hr)	Q_{Peak} (cfs)	Volume (ac-ft)	Q_{Peak} (cfs)	Max. Elev. (ft)	Freeboard (ft)
2 years	12.65	6.73	1.70	6.73	12.20	24.03	2.85	2.73	450.79	4.71
25 years	12.55	41.86	7.26	32.00	12.20	96.87	9.65	18.73	452.51	2.99
100 years	12.55	75.22	12.35	32.00	12.20	157.55	15.43	28.88	454.05	1.45

Pond Routing by Numerical Integration

Use of an accurate numerical integration method for pond routing requires that the governing equation of conservation of mass be expressed in a slightly different form. Note that a small change in storage dS is accompanied by a small change in stage dh:

$$dS = A(h)dh \tag{12.8}$$

where S = storage volume (ft^3, m^3)
h = stage (ft, m)
$A(h)$ = water surface area at stage h (ft^2, m^2)

Note also that the outflow rate O depends on h. Thus, Equation 12.4 can be rewritten as

$$\frac{dh}{dt} = \frac{I(t) - O(h)}{A(h)} \tag{12.9}$$

where $I(t)$ = inflow rate at time t (cfs, m^3/s)
$O(h)$ = outflow rate at stage h (cfs, m^3/s)

Recalling that a derivative is the slope of a line tangent to a curve, the right side of this expression is simply the slope of a graph of h as a function of t. Thus, if the stage h_n at the beginning of a time interval (at time t_n) is known, and if the slope m of the relationship between h and t can be found, then the value of the stage h_{n+1} at the end of the time interval is

$$h_{n+1} = h_n + m\Delta t \tag{12.10}$$

where h_n = stage at beginning of time interval n (ft, m)
h_{n+1} = stage at end of time interval n (ft, m)
m = slope of the relationship between h and t
Δt = length of time interval (s)

Noting that, from Equation 12.9,

$$m = \frac{I(t) - O(t)}{A(h)} \tag{12.11}$$

one can write

$$h_{n+1} = h_n + \left(\frac{I(t) - O(h)}{A(h)}\right)\Delta t \tag{12.12}$$

This last expression permits values of h to be computed at the end of each of a number of sequential time intervals. From those computed stages, corresponding outflow rates from the pond may be computed from a stage versus discharge relationship. Note that this approach to routing does not require development of a storage versus discharge relationship; it requires only a stage versus surface area relationship and a stage versus discharge relationship.

Applying Equation 12.12 requires an approximation of the slope term (the term in parentheses) on the right side, and alternative numerical integration methods are available for computing this approximation. One such method, called the 4th-order (or classical) Runge-Kutta method, was described by Chapra and Canale (1988) and is outlined below. It approximates the slope m as

$$m = \frac{1}{6}\left(k_1 + 2k_2 + 2k_3 + k_4\right) \tag{12.13}$$

where

$$k_1 = \frac{I(t_n) - O(h_n)}{A(h_n)} \tag{12.14}$$

$$k_2 = \frac{I\left(t_n + \frac{\Delta t}{2}\right) - O\left(h_n + \frac{k_1 \Delta t}{2}\right)}{A\left(h_n + \frac{k_1 \Delta t}{2}\right)} \tag{12.15}$$

$$k_3 = \frac{I\left(t_n + \frac{\Delta t}{2}\right) - O\left(h_n + \frac{k_2 \Delta t}{2}\right)}{A\left(h_n + \frac{k_2 \Delta t}{2}\right)} \tag{12.16}$$

$$k_4 = \frac{I\left(t_n + \Delta t\right) - O\left(h_n + k_3 \Delta t\right)}{A(h_n + k_3 \Delta t)} \tag{12.17}$$

Note that the value of k_2 depends on k_1, k_3 depends on k_2, and so forth. Thus, determination of the stage h of the pond as a function of time involves application of Equation 12.12 and Equations 12.14 through 12.17 for each time interval. Once h has been determined as a function of time, the outflow rate O is found using the stage versus discharge relationship. The procedure is illustrated in the following example.

Example 12.6 – Detention Routing with Numerical Integration Using 4th-order Runge-Kutta approximation. The hydrograph given in the following table is to be routed through a reservoir. The reservoir has a water surface area of 50 ac, and has sides that are nearly vertical rock walls. The base flow is assumed to be 25 cfs, and the reservoir is initially under steady-state conditions where the inflow and outflow rates are equal. Outflow occurs over a spillway on a dam at the downstream end of the reservoir. The spillway is a 100-ft-long contracted weir with the outflow rate given by

$$O(h) = 0.5(2g)^{1/2}(100 - 0.2h)h^{3/2}$$

where h is the height (ft) of the reservoir water surface above the weir crest.

Time (minutes)	Inflow (cfs)
0	25
10	25
20	25
30	125
40	400
50	866
60	1304
70	1540
80	1529
90	1338
100	1024
110	750
120	541
130	388
140	283
150	207
160	153
170	115
180	89
190	70
200	57
210	48
220	41
230	37
240	33
250	29
260	28
270	27
280	26
290	25

Solution: Details of the solution are presented here and tabulated in the table of results following the calculations. In the table, Column 1 is simply the time index *n,* and Columns 2 and 3 are the values of t_n and I_n (the inflow hydrograph).

For Column 4, an initial value of h_n is computed for the assumed steady-state weir flow of 25.0 cfs:

$$25.0 = 0.5[2(32.2)]^{1/2}(100 - 0.2h_0)h_0^{3/2}$$

By trial and error, h_0 is found to be 0.157 ft, and this value is placed in the first row of Column 4 (note that the initial assumed outflow of 25.0 cfs was placed in Column 14).

The water surface area of the pond A is constant at 50 ac (2,178,000 ft^2). Thus, the value of k_1 for the first row of Column 5 is computed using Equation 12.14 as

$$k_1 = \frac{I(t_n) - O(h_n)}{A} = \frac{25.0 - 25.0}{2{,}178{,}000} = 0$$

The value of k_2 (Column 8) is given by Equation 12.15 as

$$k_2 = \frac{I\left(t_n + \frac{\Delta t}{2}\right) - O\left(h_n + \frac{k_1 \Delta t}{2}\right)}{A}$$

The two terms in the numerator of this equation are given in columns 6 and 7. The values in column 6 are computed by averaging the inflow for the current time step, n, and the next time step, $n + 1$. The value in Column 7 is computed using the discharge equation for the weir:

$$O(h) = 0.5(2g)^{1/2}(100 - 0.2h)h^{3/2}$$

assuming that the head h on the weir is equal to the quantity

$$h_n + \frac{k_1 \Delta t}{2}$$

For the first time increment, this quantity is

$$0.157 + (0)(600)/2 = 0.157 \text{ ft}$$

noting that Δt is in seconds. The value in Column 7 is therefore 25.0 cfs.

With the values for Columns 6 and 7 computed, the value of k_2 for the first row is found to be

$$k_2 = (25.0 - 25.0)/2{,}178{,}000 = 0$$

The value of k_3 (Column 10) is given by Equation 12.16:

$$k_3 = \frac{I\left(t_n + \frac{\Delta t}{2}\right) - O\left(h_n + \frac{k_2 \Delta t}{2}\right)}{A}$$

The first term in the numerator is already given in Column 6. The value of the second term is given in Column 9. This value is computed using the discharge equation for the weir, assuming that the head h on the weir is equal to the quantity

$$h_n + \frac{k_2 \Delta t}{2}$$

For the first row, this quantity is $0.157 + 0 = 0.157$ ft. The value in column 9 is therefore 25.0 cfs.

Column 10 for the first row can then be computed as

$$k_3 = (25.0 - 25.0)/2{,}178{,}000 = 0$$

The value of k_4 (Column 12) is given by Equation 12.17:

$$k_4 = \frac{I(t_n + \Delta t) - O(h_n + k_3 \Delta t)}{A}$$

The value of the first term in the numerator is simply the flow from the inflow hydrograph in Column 3 for the next time step ($t_{n+1} = t_n + \Delta t$). The value of the second term is given in Column 11. This value is computed using the discharge equation for the weir, assuming that the head h on the weir is equal to the quantity

$$h_n + k_3 \Delta t$$

For the first row, this quantity is $0.157 + 0 = 0.157$ ft. The value in Column 11 is therefore 25.0 cfs.

Column 12 for the first row can then be computed as

$$k_4 = (25.0 - 25.0)/2{,}178{,}000 = 0$$

The slope m (Column 13) is then found using Equation 12.13:

$$m = \frac{1}{6}(k_1 + 2k_2 + 2k_3 + k_4)$$

Therefore, $m = 0$ for the first row. The head on the weir for the next row, h_{n+1}, is then computed as

$$h_{n+1} = h_n + m\Delta t$$

For the first row, the value in Column 13 is computed as

$$h_{n+1} = 0.157 + (0)(600) = 0.157 \text{ ft}$$

The outflow from the reservoir (Column 14) for the second row, $O(h_{n+1})$, is therefore given by

$$O(h_{n+1}) = 0.5(2g)^{1/2}(100 - 0.2h_{n+1})h_{n+1}^{3/2} = 25.0 \text{ cfs}$$

The remainder of the columns in the second row are completed by repeating the procedure used for the first row, and these steps are repeated until the entire outflow hydrograph is computed. Note that this solution was made simpler by the fact that the water surface area for the reservoir is constant. In the case of a reservoir with a changing water surface area, the denominators for k_1, k_2, k_3, and k_4, as given in Equations 12.14 through 12.17, would have to be calculated for each row.

(1)	(2)	(3)	(4)	(5)	(6)	(7)	(8)	(9)	(10)	(11)	(12)	(13)	(14)
n	t_n (min)	I_n (cfs)	h_n (ft)	k_1	$I(t_n+\Delta t/2)$ (cfs)	$O(k_2)$ (cfs)	k_2	$O(k_3)$ (cfs)	k_3	$O(k_4)$ (cfs)	k_4	m	O_n (cfs)
0	0	25.0	0.157	0.000E+00	25.0	25.009	-4.251E-09	25.009	-4.111E-09	25.009	-3.981E-09	-3.451E-09	25.0
1	10	25.0	0.157	-4.024E-09	25.0	25.008	-3.892E-09	25.008	-3.896E-09	25.008	-3.768E-09	-3.895E-09	25.0
2	20	25.0	0.157	-3.768E-09	75.0	25.008	2.295E-05	26.668	2.219E-05	28.249	4.442E-05	2.245E-05	25.0
3	30	125.0	0.171	4.440E-05	262.5	31.661	1.060E-04	36.543	1.037E-04	45.083	1.630E-04	1.045E-04	28.3
4	40	400.0	0.233	1.629E-04	633.0	60.118	2.630E-04	69.960	2.585E-04	97.048	3.531E-04	2.598E-04	45.2
5	50	866.0	0.389	3.529E-04	1085.0	139.612	4.341E-04	150.024	4.293E-04	208.382	5.030E-04	4.304E-04	97.3
6	60	1304.0	0.648	5.029E-04	1422.0	285.659	5.217E-04	288.696	5.203E-04	376.353	5.343E-04	5.202E-04	208.7
7	70	1540.0	0.960	5.343E-04	1534.5	474.262	4.868E-04	465.254	4.909E-04	561.900	4.440E-04	4.890E-04	376.3
8	80	1529.0	1.253	4.444E-04	1433.5	652.818	3.584E-04	634.723	3.667E-04	714.898	2.861E-04	3.635E-04	561.1
9	90	1338.0	1.471	2.867E-04	1181.0	776.821	1.856E-04	754.267	1.959E-04	800.485	1.026E-04	1.921E-04	713.5
10	100	1024.0	1.586	1.034E-04	887.0	822.232	2.974E-05	805.469	3.743E-05	815.722	-3.018E-05	3.460E-05	798.7
11	110	750.0	1.607	-2.958E-05	645.5	807.709	-7.448E-05	797.543	-6.981E-05	782.868	-1.111E-04	-7.153E-05	814.4
12	120	541.0	1.564	-1.107E-04	464.5	757.372	-1.345E-04	752.097	-1.320E-04	723.550	-1.541E-04	-1.330E-04	782.1
13	130	388.0	1.485	-1.539E-04	335.5	689.744	-1.626E-04	687.857	-1.618E-04	653.521	-1.701E-04	-1.621E-04	723.1
14	140	283.0	1.387	-1.700E-04	245.0	617.723	-1.711E-04	617.499	-1.710E-04	582.348	-1.723E-04	-1.711E-04	653.4
15	150	207.0	1.285	-1.723E-04	180.0	547.573	-1.688E-04	548.282	-1.691E-04	514.810	-1.661E-04	-1.690E-04	582.3
16	160	153.0	1.183	-1.661E-04	134.0	482.698	-1.601E-04	483.853	-1.606E-04	453.315	-1.553E-04	-1.605E-04	514.8
17	170	115.0	1.087	-1.554E-04	102.0	424.560	-1.481E-04	425.892	-1.487E-04	398.772	-1.422E-04	-1.485E-04	453.4
18	180	89.0	0.998	-1.423E-04	79.5	373.553	-1.350E-04	374.827	-1.356E-04	351.117	-1.291E-04	-1.354E-04	398.8
19	190	70.0	0.916	-1.291E-04	63.5	329.178	-1.220E-04	330.379	-1.225E-04	309.824	-1.161E-04	-1.224E-04	351.2
20	200	57.0	0.843	-1.161E-04	52.5	290.894	-1.095E-04	291.971	-1.099E-04	274.263	-1.039E-04	-1.098E-04	309.9
21	210	48.0	0.777	-1.039E-04	44.5	257.989	-9.802E-05	258.905	-9.844E-05	243.670	-9.305E-05	-9.831E-05	274.3
22	220	41.0	0.718	-9.307E-05	39.0	229.648	-8.753E-05	230.477	-8.791E-05	217.381	-8.282E-05	-8.780E-05	243.7
23	230	37.0	0.666	-8.284E-05	35.0	205.362	-7.822E-05	206.028	-7.853E-05	194.759	-7.427E-05	-7.843E-05	217.4
24	240	33.0	0.618	-7.428E-05	31.0	184.361	-7.041E-05	184.899	-7.066E-05	175.118	-6.709E-05	-7.059E-05	194.8
25	250	29.0	0.576	-6.710E-05	28.5	166.046	-6.315E-05	166.576	-6.340E-05	158.094	-5.973E-05	-6.332E-05	175.1
26	260	28.0	0.538	-5.974E-05	27.5	150.286	-5.638E-05	150.723	-5.658E-05	143.400	-5.344E-05	-5.651E-05	158.1
27	270	27.0	0.504	-5.345E-05	26.5	136.633	-5.057E-05	136.996	-5.073E-05	130.634	-4.804E-05	-5.068E-05	143.4
28	280	26.0	0.474	-4.805E-05	25.5	124.734	-4.556E-05	125.038	-4.570E-05	119.477	-4.338E-05	-4.566E-05	130.6
29	290	25.0	0.446										119.5

Inspection of the table shows that the peak outflow discharge is about 814 cfs, and that it occurs at time $t = 110$ min.

12.8 ITERATIVE POND DESIGN

The previous sections of this chapter explained the fundamental elements of stormwater detention modeling. The examples presented in the chapter up to this point have been detention facility *analyses* (that is, the final design configurations have been specified for the facility shape, size, and outlet structure components).

Design of a new detention pond involves an iterative (trial and error) design procedure. Figure 12.24 provides a flowchart for the basic steps in design of a detention facility. First, the inflow hydrograph is developed, and can be used with data on allowable outflows to estimate the required detention volumes for the design storm event(s). From this estimate, the engineer can develop a grading plan for the pond area, and then develop stage-storage data for the proposed pond. Based on the stage-storage data, the engineer can project what stage the pond will reach for each design storm volume estimate.

Next, a proposed outlet structure is selected and its stage-discharge relationship developed. Typically, the engineer can tell if the outlet structure configuration is feasible by checking to see if the discharge values corresponding to the predicted pond stages for the various design storms roughly match allowable outflow values. If not, the engineer should resize the outlet as needed to increase or decrease capacity.

Once a likely outlet structure configuration is found, the final routing calculations for the pond can occur. If the design criteria are met, the design can be finalized. Otherwise, the engineer must go back and adjust the pond volume and/or outlet configuration as necessary and perform the routing again. Many iterations of the procedure may be required to develop an acceptable design. Although the example in this text shows how detention pond routing can be performed manually, engineers typically use computer programs to perform these calculations.

"Kiss me. I'm Claude Monet."

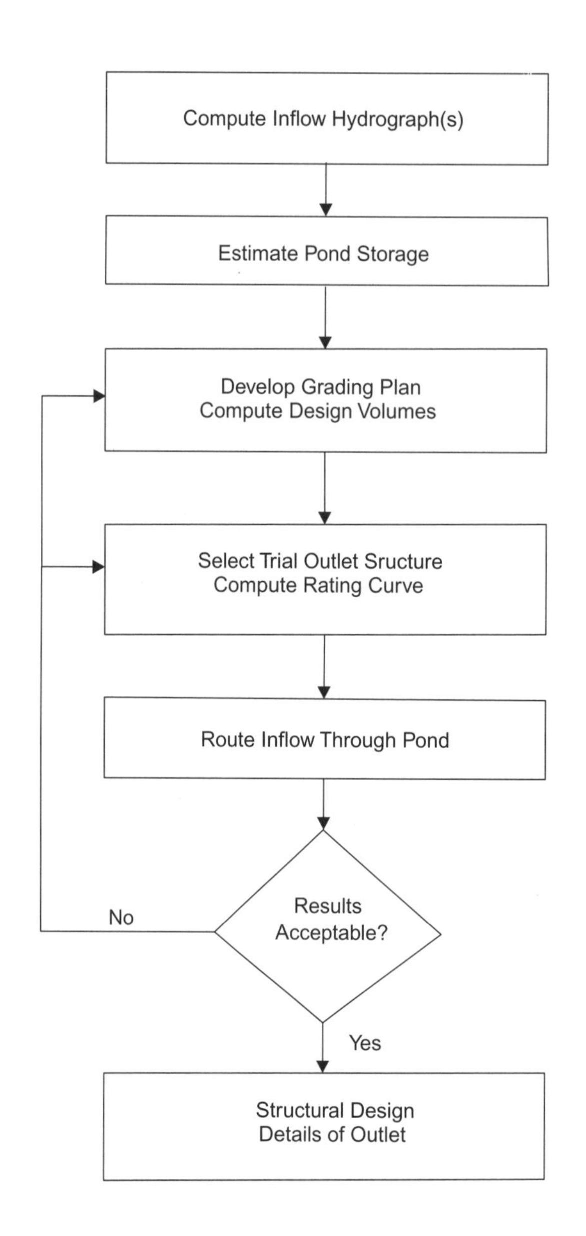

Figure 12.24
Flowchart illustrating steps in designing stormwater detention

12.9 CHAPTER SUMMARY

Stormwater detention is used to mitigate the effects of increased runoff rates caused by development. Several facility types or classifications exist, including:

- Dry ponds versus wet ponds
- Aboveground versus underground detention
- On-line versus off-line detention
- Regional versus on-site detention
- Individual ponds versus interconnected ponds

A detention facility has the effect of attenuating an inflow hydrograph, meaning that the peak flow is delayed and reduced. A common design requirement in facility design is that the peak outflow from a site under developed conditions be equal to or less than the peak outflow under undeveloped conditions. The design storms to which this requirement applies vary. However, even with detention, the total runoff volume increases when a site is developed.

The basic parts of a typical detention pond are the dam, pond bottom and sides, outlet structure(s), and emergency spillway. Also, a freeboard requirement (for example, 1 ft) may exist to ensure some minimum distance between the maximum water surface elevation and the top of the dam for a particular storm event (often the 100-year storm).

The stage versus storage volume and stage versus outlet discharge relationships must be known before an inflow hydrograph can be routed through a detention facility using the storage indication method. The basic steps of detention facility analysis using the storage indication method are:

1. Calculate the inflow hydrograph to the pond.
2. Calculate the stage versus storage relationship for the selected pond design.
3. Calculate the stage versus discharge relationship for the pond's outlet structure(s).
4. Establish the relationship between stage, storage, and discharge by creating a storage indication curve for $(2S/t + O)$ versus O.
5. Use the inflow hydrograph from step 1 and the storage indication curve from step 4 to route the inflow hydrograph through the pond.

A fourth-order Runge-Kutta approximation can also be used to perform rating calculations.

Detention facility design is a trial-and-error process that typically involves multiple iterations of the preceding steps until an acceptable combination of storage facility geometry and outlet structure is found.

REFERENCES

Akan, A. O. 1989. "Detention Pond Sizing for Multiple Return Periods." *Journal of Hydraulic Engineering* 115(5): 650.

ASCE. 1982. *Stormwater Detention Facilities.* Reston, Va.: ASCE.

ASCE Task Committee. 1985. *Detention Outlet Control Structures.* Reston, Va: ASCE.

American Society of Civil Engineers (ASCE). 1992. *Design and Construction of Urban Storm Water Management Systems.* New York: ASCE.

Aron, G. and Kibler, D. F. 1990. "Pond Sizing for Rational Formula Hydrographs." *Water Resources Bulletin,*26(2): 255.

Brown, S. A., S. M. Stein, and J. C. Warner. 1996. *Urban Drainage Design Manual.* Hydraulic Engineering Circular No. 22, FHWA-SA-96-078. Washington, D.C.: U.S. Department of Transportation.

Chapra, S. C., and R. P. Canale. 1988. *Numerical Methods for Engineers.* 2d ed. New York: McGraw-Hill.

Glazner, Michael K. 1998 "Pond Routing Techniques: Standard vs. Interconnected." *Engineering Hydraulics and Hydrology.* ed. by Strafaci, A. M. Waterbury, Connecticut: Haestad Methods.

Guo, J. C. Y. 2001. "Design of Off-Line Detention Systems." In *Stormwater Collection Systems Design Handbook,* ed. L. W. Mays. New York: McGraw-Hill.

Kamaduski, G. E. and R. H. McCuen. 1979. "Evaluation of Alternative Stormwater Detention Policies." *Journal of Water Resources Planning and Management* 105(WR2): 171.

Kessler, A. and M. H. Diskin. 1991. "The Efficiency Function of Detention Reservoirs." *Water Resources Research* 27(3): 259.

Loganathan, G. V., D. F. Kibler, and T. J. Grizzard. 1996. "Urban Stormwater Management." In *Water Resources Handbook,* ed. L. W. Mays. New York: McGraw-Hill.

McEnroe, B. M. 1992. "Preliminary Sizing of Detention Reservoirs to Reduce Peak Discharge." *Journal of Hydraulic Engineering* 118(11): 1540.

Nix, S. J. and S. R. Durrans. 1996. "Off-Line Stormwater Detention Systems." *Journal of the American Water Resources Association* 32, no. 6: 1329–1340.

Paine, J. N. and A. O. Akan. "Design of Detention Systems." In *Stormwater Collection Systems Design Handbook,* ed. L. W. Mays. New York: McGraw-Hill.

Urbonas, B. R. and L. A. Roesner. 1993. "Urban Design and urban Drainage for Flood Control." In *Handbook of Hydrology,* ed. D. Maidment. New York: McGraw-Hill.

Urbonas, B. and P. Stahre. 1993. *Stormwater Best Management Practices including Detention.* Englewood Cliffs, New Jersey: Prentice-Hall.

U.S. Bureau of Reclamation (USBR). 1987. *Design of Small Dams.* Washington, D.C.: U.S. Department of the Interior, USBR.

PROBLEMS

12.1 Use the average-end-area method to develop and graph an elevation versus storage relationship for a reservoir having the elevation versus surface area relationship given in the following table.

Elevation (m)	Area (ha)	Volume (m^3)
37.6	0.00	
37.8	0.08	
38.0	0.20	
38.2	0.31	
38.4	0.42	
38.6	0.62	
38.8	0.68	
39.0	0.75	
39.2	0.90	
39.4	1.12	
39.6	1.42	

12.2 Develop and graph an elevation versus discharge relationship at 0.5-ft increments for a dry detention pond with a dam elevation of 248 ft. The outlet structure consists of the following:

- Two 6-in. orifices ($C_d = 0.6$), both having invert elevations of 234.3 ft, which is the elevation of the bottom of the pond
- Three 6-in. diameter orifices ($C_d = 0.6$) with invert elevations of 240.1 ft
- A rectangular, sharp-crested, contracted overflow weir with a crest elevation of 244.0 ft and a length of 10 ft ($C_d = 0.62$)

12.3 A trapezoidal detention pond has a base with dimensions W =15.2 m and L = 30.5 m, side slopes of 3 horizontal to 1 vertical (that is, $Z = 3$), and a maximum depth D = 4.5 m. Tabulate the stage versus storage volume (V) relationship at 0.1-m increments.

Hint: According to HEC-22 (Brown, Stein, and Warner, 2001), the volume of a trapezoidal pond with equal side slopes can be computed as

$$V = LWD + (L + W)ZD^2 + (4/3)z^2D^3$$

12.4 The outlet structure of a wet detention pond consists of a vertical riser pipe with an open top. The diameter of the riser is 600 mm, the top of the riser is at an elevation of 7.5 m, and the elevation of the top of the pond is 10.5 m. Develop an elevation-discharge relationship at 0.1-m increments for the outlet structure of the elevation range of 7.5 to 10.5 m. Evaluate the structure as both a weir (C_d = 0.62) and an orifice ($C_d = 0.6$), and assume that the structure type yielding the lower discharge for a particular elevation controls.

12.5 The detention pond described in Problem 12.3 is used to reduce peak runoff. The bottom of the pond is at an elevation of 6.0 m. The outlet structure described in Problem 12.4 is used to control discharge. Assume that, at the start of the storm, the water surface elevation in the basin is equal to the elevation of the top of the riser (7.5 m). The pond inflow hydrograph is given in the following table.

Use the storage indication method to route the hydrograph through the detention facility and complete the hydrograph table provided. What is the peak pond outflow? What is the highest water surface elevation in the pond during the storm?

Time (Harmon)	Pond Inflow (m^3/s)	Pond Outflow (m^3/s)
12:00	0.00	
12:05	0.01	
12:10	0.21	
12:15	0.36	
12:20	0.54	
12:25	0.99	
12:30	1.47	
12:35	2.43	
12:40	2.64	
12:45	2.58	
12:50	2.37	
12:55	1.71	
13:00	1.44	
13:05	1.17	
13:10	0.96	
13:15	0.78	
13:20	0.66	
13:25	0.45	
13:30	0.33	
13:35	0.27	
13:40	0.09	
13:45	0.00	
13:50	0.00	
13:55	0.00	
14:00	0.00	

12.6 Repeat Problem 12.5 using a 4th-order Runge-Kutta method to perform the routing calculations.

12.7 A detention pond has the elevation-area relationship provided for computing its volume. The pond's outlet structure consists of an orifice plate with two 6-in. diameter orifices ($C_d = 0.6$), both with invert elevations of 150.00 ft, and a 10-ft-long, rectangular, contracted weir ($C_d = 0.62$) with a crest elevation of 155.00 ft. All outlet structures have a free outfall tailwater condition. The pond inflow hydrograph is provided.

a) Use the conic method to graph and tabulate the elevation-volume relationship for an elevation increment of 0.5 ft.

b) Graph and tabulate the elevation-discharge relationship for the composite outlet structure from elevation 150.00 ft to 156.00 ft at 0.5-ft increments.

c) Route the inflow hydrograph through the pond using the storage-indication method and tabulate the outflow hydrograph. What are the peak pond outflow, water surface elevation, and storage volume? Approximately how long is the pond outflow peak delayed compared to the inflow hydrograph time to peak? How much freeboard does the pond have for this event?

Elevation (ft)	Area (ac)	Volume (ft^3)
150.00	0.119	
150.50	0.133	
151.00	0.147	
151.50	0.162	
152.00	0.178	
152.50	0.194	
153.00	0.212	
153.50	0.230	
154.00	0.248	
154.50	0.268	
155.00	0.288	
155.50	0.309	
156.00	0.331	

Time (min)	Pond Inflow (cfs)	Pond Outflow (cfs)
0	0.00	
10	0.59	
20	4.64	
30	10.40	
40	12.85	
50	11.98	
60	10.00	
70	7.98	
80	6.33	
90	5.18	
100	4.42	
110	3.93	
120	3.56	
130	2.95	
140	1.78	
150	0.81	
160	0.35	
170	0.15	
180	0.06	
190	0.03	
200	0.01	
210	0.00	
220	0.00	

Stormwater normally flows under the levee through the culvert on the far left, and additionally through the remaining culverts for larger flows. When the water behind the levee is high, the culverts are closed, and flow is pumped to the other side.

CHAPTER

13

Stormwater Pumping

In most cases, storm sewers are designed as gravity systems that roughly parallel natural drainage pathways. They accept water at inlets located throughout a drainage basin and transport it to an outfall point at a lower elevation in the basin. There is no need to add energy to the flow to move the water in the desired direction. However, sometimes stormwater discharge must be pumped to a higher elevation in a drainage basin. Examples of such cases are:

- Natural or constructed levees that block drainage from reaching stream or river channels
- Highways or roadways that are depressed with respect to surrounding ground elevations
- Obstructions that necessitate flow diversion
- Detention ponds with bottom elevations that are lower than an available outfall discharge point

The total capitalized cost of a stormwater conveyance system with pumping can be high compared to an otherwise similar system without pumping. Elements that contribute to the higher cost include pump maintenance, debris-removal requirements, added energy costs, and the need to periodically retire and replace pumps. Design issues affecting these additional costs are discussed in Section 13.1.

Section 13.2 provides a description of the various types of pumps and installation methods commonly used for stormwater pumping applications, including discussions on pump wet wells and dry wells, sizing of pumping wells, and debris handling and removal.

The operating characteristics of a pump, when installed in a piping system, can be ascertained on the basis of pump and system curves. These characteristics include the energy head produced by the pump, the corresponding pump discharge, the brake horsepower requirements for the pump motor(s), and the pump installation requirements for avoiding cavitation damage to the impeller and other internal parts. Section 13.3 presents methods for determining these operating characteristics for single- and multiple-pump installations.

Section 13.4 discusses piping-system requirements for pumping. Considerations in the selection of piping materials include estimates of internal fluid pressures, backflow prevention, and requirements related to maintenance and vibration isolation. The startup and stopping of a pump (or pumps) at a pumping station are usually controlled by float or pressure-sensing devices. Section 13.5 describes typical control devices and methods.

Peak discharges of stormwater runoff from a drainage basin are often larger than flow rates that can be economically pumped. Because of this disparity, temporary water-storage facilities (essentially, detention basins) are often required at pumping stations. Section 13.6 presents methods for developing an estimate of the required storage volume. This estimate of the storage volume, along with other trial-design selections involving pump models and piping-system components, must then be refined through a trial-and-error process that evaluates alternative designs. Section 13.7 describes this iterative design process.

In critical locations where temporary pump failures due to power outages cannot be tolerated, it may be necessary to provide standby power facilities. Pump stations should also be aesthetically pleasing and have appropriate safety and monitoring devices. Brief summaries of these miscellaneous aspects of pump-station design are presented in Section 13.8.

13.1 NEED FOR STORMWATER PUMPING

As noted, stormwater pumping is generally undesirable, but is unavoidable in some instances. Pumps require maintenance and repair, periodic replacement, provisions for an electrical or fuel power supply, and regular removal of trapped debris. Over a long period of service, the operation and maintenance costs associated with a stormwater pumping installation can be significant. Thus, at the planning stage of a design, one should carefully consider alternative system routes and designs that eliminate the need for pumping. The long-term costs associated with pumping can often justify a considerably higher initial capital cost for a nonpumped system.

The most obvious alternative to a stormwater pumping installation is to use a gravity system that outfalls at the nearest available location. This outfall may be some distance away, requiring more pipe and associated storm sewer appurtenances like manholes. In such instances, downhill pipe slopes will generally be adverse to the ground surface slope, with the net effect that pipes must be deeply buried and excavation and backfilling costs may be high. Other alternatives to stormwater pumping include recharge basins, siphons, or tunnels.

Pumps are often used to transport tributary streams across levees or flood walls when the main channel is at flood stage (or high tide). The figure before the start of this chapter is an example of this type of application. When the main stream is at normal elevation, drainage can occur under gravity-flow conditions, as shown in Figure 13.1(a). When the main stream is at flood stage, the flood gates through which the tributary stream flows must be closed to prevent backflow, resulting in flooding due to interior drainage behind the levee. Through a combination of ponding and pumping, this flooding can be minimized, as shown in Figure 13.1(b). Figure 13.2 shows pumped flow exiting through a tidal gate in a levee.

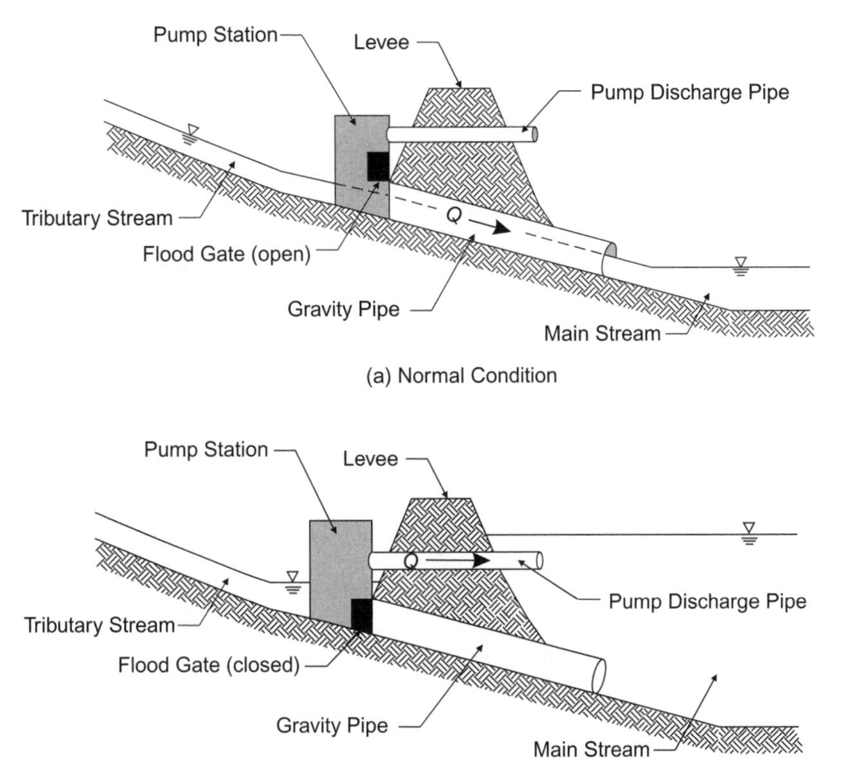

Figure 13.1
Stormwater pumping for interior drainage

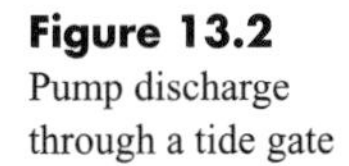

Figure 13.2
Pump discharge through a tide gate

Credit: Jim Brozena

Another common location for a stormwater pump station is at a low point within a highway interchange that has grade separation if the surrounding terrain is flat. Because these stations are usually built in conjunction with new highway work, the designer has the opportunity to develop a grading plan that minimizes the drainage area contributing to the station. Therefore, the drainage areas for highway pump stations are usually small, and peak inflow can be estimated using the rational method.

If stormwater detention is required, it may be necessary to limit the peak flow from the station the pre-development peak flow or other specified flow rate. In such cases, the station may be equipped with a "storage unit" in addition to the wet well (Smith, 2001). Providing detention storage also reduces the required pumping capacity. These storage units can also be used to contain hazardous material spills that may occur on the highway. Because design flows are often associated with large storms that can knock out power, standby generators, either permanent or portable, are usually required.

Pumping station installation is often an inexpensive alternative in terms of construction costs, but it may not be advantageous when future operation and maintenance costs are considered. Further, highway pumping station sites are often congested and difficult to access. U.S. Federal Highway Administration (FHWA) guidance states, "...the use of stormwater pumping stations is recommended only where no other practicable alternative is available. Alternatives to pumping stations include siphons, recharge basins, deep and long storm drain systems, and tunnels" (Smith, 2001).

When there is no option but to pump stormwater, a great amount of care and thought is required from the designer. Each situation is unique and requires careful pump selection and station design to maximize operating efficiencies and minimize costs. Standardized designs and equipment are certainly desirable to the extent that they can be implemented, but the engineer must take care to ensure that a design is optimal for the site in question. The U.S. Army Corps of Engineers publishes a series of manuals on flood control pumping station design (USACE, 1989, 1995, 1999).

13.2 PUMP TYPES AND INSTALLATION METHODS

Pumps are one of the most common machine types in existence. At the most fundamental level, a pump and its motor are machines for conversion of electrical or fuel energy into mechanical energy (in this case, rotation of a drive shaft), and subsequently into fluid energy (in this case, increased energy head). Turbine generators are the opposites of pumps; they convert fluid energy into mechanical and then electrical energy.

Pump stations can be configured in a number of ways. For most stormwater pumping applications, pumps are submerged in a *wet well,* which is an underground chamber or vault for storing stormwater. If the wet well is relatively shallow, it may be possible to locate the pumps above ground, connecting them to the wet well through suction piping. With larger pump stations, the pumps may be located in a separate dry well adjacent to the wet well.

Pump Types

Pumps come in a wide variety of types and sizes, with each type best suited to a particular range of applications. The two broad classifications of pumps are centrifugal and positive displacement. *Centrifugal pumps* have rotating impellers and are further classified by impeller type. *Axial-flow, radial-flow,* and *mixed-flow* pumps are the pump types most commonly used in stormwater applications. In the U.S., these three types are considered centrifugal pumps, although, strictly speaking, the term applies to radial-flow pumps only.

An *axial-flow pump* has an impeller that resembles the propeller on a boat. Flow moves axially along the drive shaft, which is usually encased within a protective shield. Axial-flow pumps do not handle debris particularly well, as it can damage the impeller. Fibrous materials also tend to wind around the drive shaft and clog the pump. These pumps are best suited to applications involving high discharges and low energy heads.

A *radial-flow pump* has an impeller that is shrouded within a volute (spiral) casing. Water typically enters the pump near the center of the impeller, and vanes on the impeller cast it radially outwards towards the casing wall. Depending on the impeller configuration, radial-flow pumps can handle debris quite well and are available for a wide range of operating conditions. They are, however, best suited to low-discharge, high-energy-head applications. Figure 13.3 a shows a small end suction radial flow pump located in a dry well.

Figure 13.3
Radial-flow pump

Mixed-flow pumps, as the name implies, are hybrids with aspects of both axial- and radial-flow pumps. The flow direction caused by the impeller has both axial and radial components. Mixed-flow pumps handle debris better than axial-flow pumps but worse than radial-flow pumps. They are best suited for intermediate discharge and energy-head applications.

Positive-displacement pumps may also be used for stormwater pumping. These pumps work by using a moving element to displace water from the pump casing and simultaneously raise its pressure. For stormwater applications, *screw pumps* are the most commonly used type of positive-displacement machinery, and may be used to lift flow over a levee or floodwall. A screw pump has a helical impeller reminiscent of a corkscrew that causes water to be displaced axially along the impeller as it rotates. Other types of positive-displacement pumps include reciprocating-piston, rotary, and pneumatic pumps (Sanks, 1998).

One way to approach the problem of selecting the pump type is to use the pump characteristic known as *specific speed.* In the United States, specific speed is commonly defined as

$$N_s = \frac{\omega\sqrt{Q}}{h_P^{3/4}} \tag{13.1}$$

where N_s = specific speed (gal$^{1/2}$/min$^{3/2}$ft$^{3/4}$, but units are usually unstated)
ω = rotational speed of the pump (rpm)
Q = pumping rate (gpm)
h_P = energy head produced by the pump (ft)

Outside of the United States, specific speed is typically defined as

$$N_s = \frac{\omega\sqrt{Q}}{(gh_P)^{3/4}} \tag{13.2}$$

where N_s = specific speed (dimensionless)
ω = angular speed of the pump (rad/s)
Q = pumping rate (m^3/s)
g = 9.81 m/s^2 = acceleration of gravity
h_P = energy head produced by the pump (m)

In these equations, both Q and h_P are taken at the pump's best efficiency point. These definitions imply that high specific speeds tend to be associated with pumps delivering high discharges at low heads and, conversely, low specific speeds tend to be associated with pumps delivering small discharges at high heads. Figure 13.4 relates pump efficiency to specific speed and discharge. It also shows the specific speed range for various impeller shapes, indicating that axial-flow pumps tend to have relatively high specific speeds, radial-flow pumps tend to have relatively low specific speeds, and mixed-flow pumps have specific speeds somewhere in between. In stormwater pumping applications, discharges are often high while head requirements are low, indicating a common requirement for high-specific-speed pumps.

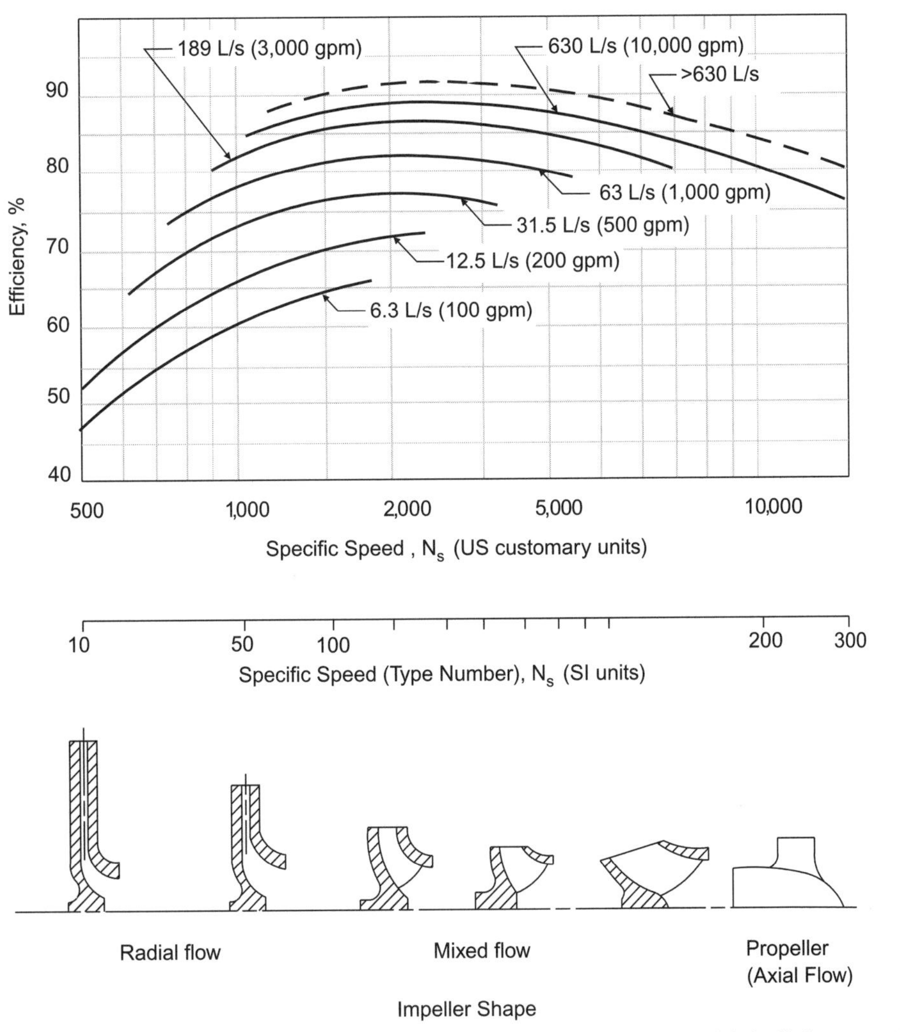

Reprinted from Pumping Station Design, 2nd ed., ISBN 0750694831, Robert L Sanks, Ph.D., P.E., Copyright 1998, with permission from Elsevier.

Figure 13.4
Pump efficiency as related to specific speed and discharge

Example 13.1 – Selecting Pump Type Use Figure 13.4 to determine the type of pump (axial-, radial-, or mixed-flow) best suited to an application where the pumping rate is to be Q = 10 cfs, the pumping head is h_P = 20 ft, and the motor speed is ω = 1,800 rpm.

Solution: The discharge of 10 cfs is equivalent to 4,500 gpm. Using Equation 13.1, the specific speed

$$N_s = \frac{\omega\sqrt{Q}}{h_P{}^{3/4}} = \frac{1800\sqrt{4500}}{20^{3/4}} = 13{,}000$$

From Figure 13.4, an axial-flow pump appears to be the best choice for this installation.

Pump Installations

There are two general types of pump installation: wet-well and wet/dry-well. Pumping-station installations include the pumps, motors, and other associated mechanical and electrical equipment. They frequently also include water-storage facilities necessary to temporarily store accumulated stormwater runoff until it can be pumped to a discharge location.

In *wet-well installations*, the pumps are submerged within a pool of water in the sump, with the pump motors and other controls usually installed above the pump and connected via a vertical drive shaft. Vertically installed axial-flow pumps, as shown in Figure 13.5, are well suited to this type of installation. Figure 13.6 shows a typical motor room for a vertical pump installation with the motors on the right and the motor control center on the left. *Submersible pumps*, which are available in large sizes for stormwater pumping applications, are also installed in wet wells. These types of pumps are designed for easy removal and often include rail systems to facilitate this task.

In *wet/dry-well installations*, which are most common with larger pumping stations, the pumps are usually mounted on the floor of a dry chamber. Horizontal piping supplies the pumps from a separate wet well or temporary water-storage facility (such as a detention pond or outdoor forebay).

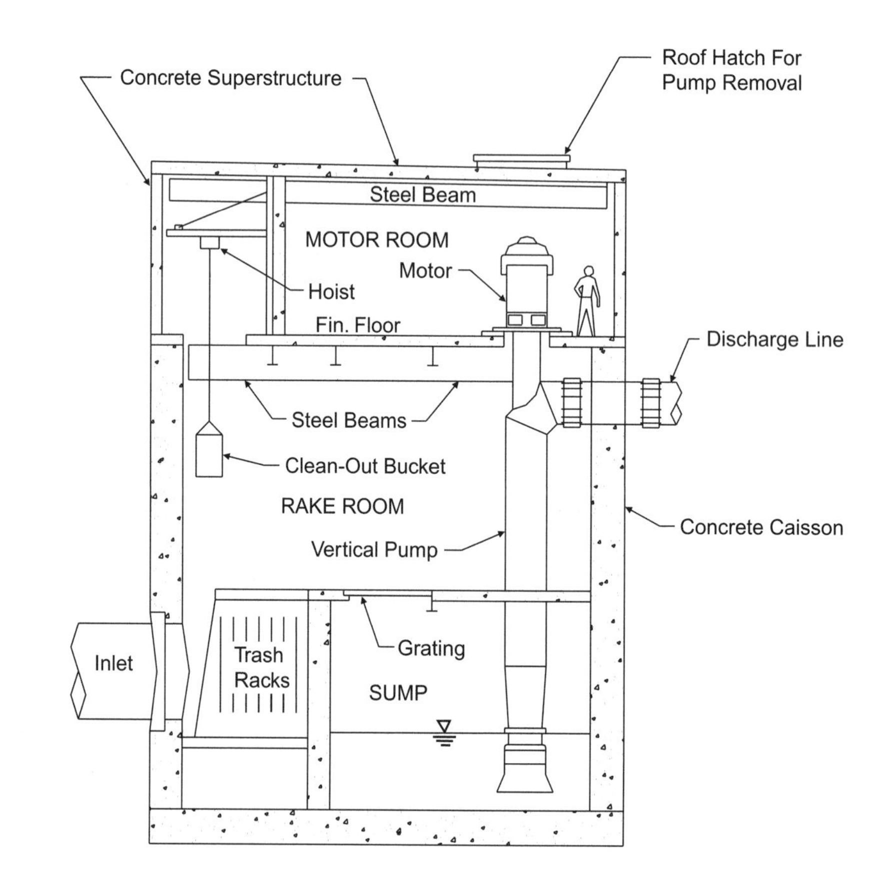

Figure 13.5
Vertically installed axial-flow pumps in a large wet-well installation (Brown, Stein, and Warner, 1996)

Credit: Jim Brozena

Figure 13.6
Motor control center and motors for vertical pumps

Centrifugal pumps are often used in dry-well installations, although axial-flow pumps can be mounted with horizontal or vertical drive shafts. Motors are generally located adjacent to the pumps on the floor of the dry well. A typical layout of a dry-well installation is shown in Figure 13.7. Note that a small sump pump, in addition to the stormwater pumps, should be provided in a wet/dry-well installation. The sump pump discharges any water that accumulates in the bottom of the dry well, either from leakage or from groundwater infiltration. Wet/dry-well installations may have higher construction costs, but offer several advantages over wet-well installations. In particular, pump access is easier, which reduces operation and maintenance costs.

In some instances, pumps may be mounted in the open at ground level in wet/dry-well installations. This option can be particularly attractive when pumping from an open detention pond. Pump enclosures or security fencing are recommended with installations of this type to preclude unauthorized access to pumps and appurtenances. In such applications, care must be taken to ensure that the available net positive suction head (NPSH) on the pumps is adequate (see Section 13.3).

Regardless of the type of installation, the spacing of pumps and the positioning of their suction inlets should be consistent with the manufacturer's recommendations and with the standards published by the Hydraulic Institute (1998). Of particular importance are minimum sump dimensions near the pump intake points, which should be adhered to in order to prevent vortexing and/or cavitation damage. Also, manufacturers can provide minimum submergence criteria for submersible pumps. The distribution of flow to the pumps should be as uniform as possible; water should not have to flow past one pump to reach another. Figure 13.8 illustrates one type of sump design, and Figure 13.9 provides corresponding guidance on the dimensions shown.

The design of the intakes is often complex in that it must account for high and low flow conditions. For example, Figure 13.10 shows a series of culverts on the left that pass flow to the river through a levee. The intake screens for the pump station are on the right. When the river (behind the levee) is high, the culverts are closed and all flow diverted is to the pump station.

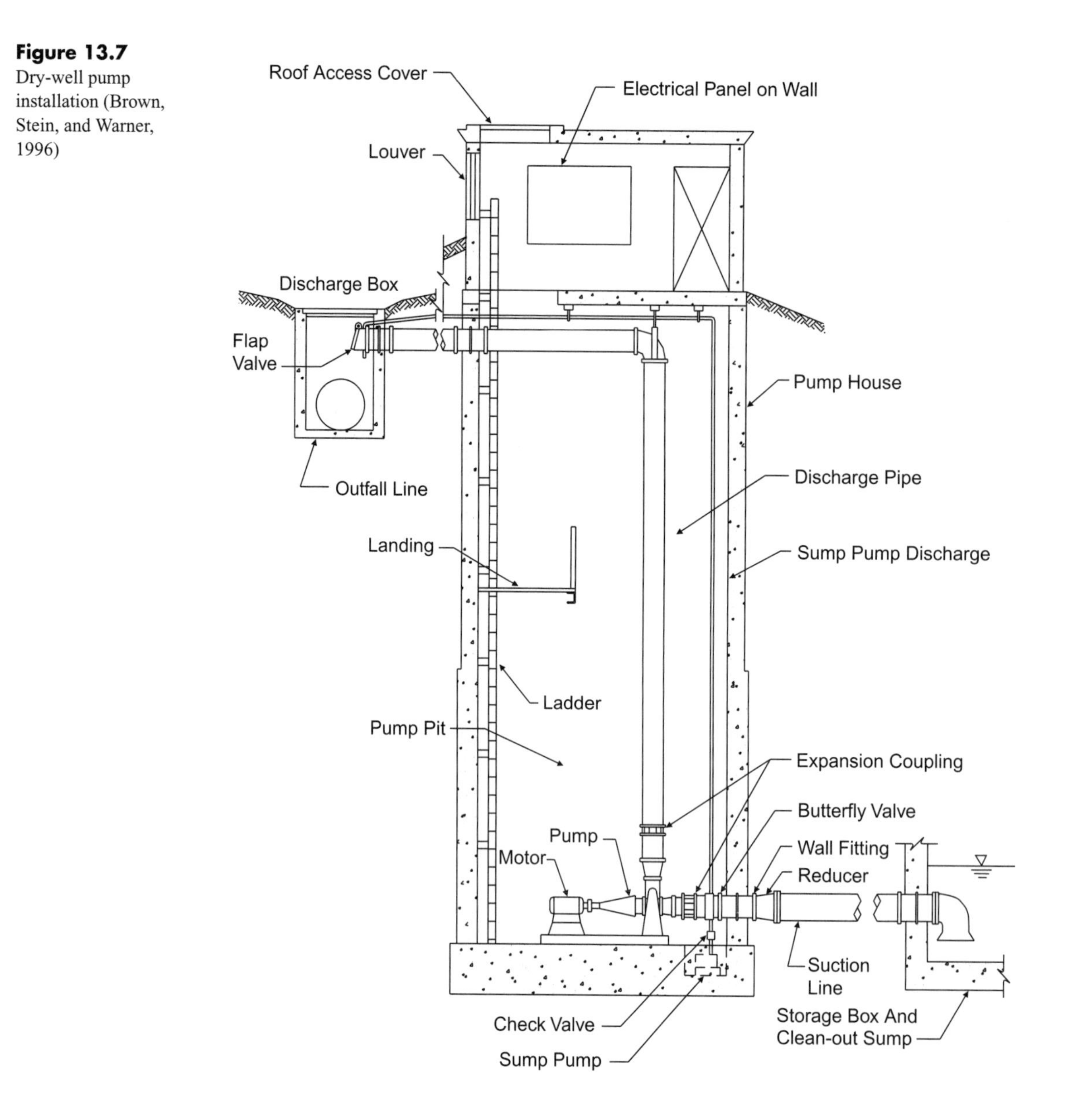

Figure 13.7
Dry-well pump installation (Brown, Stein, and Warner, 1996)

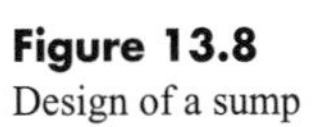
Figure 13.8
Design of a sump

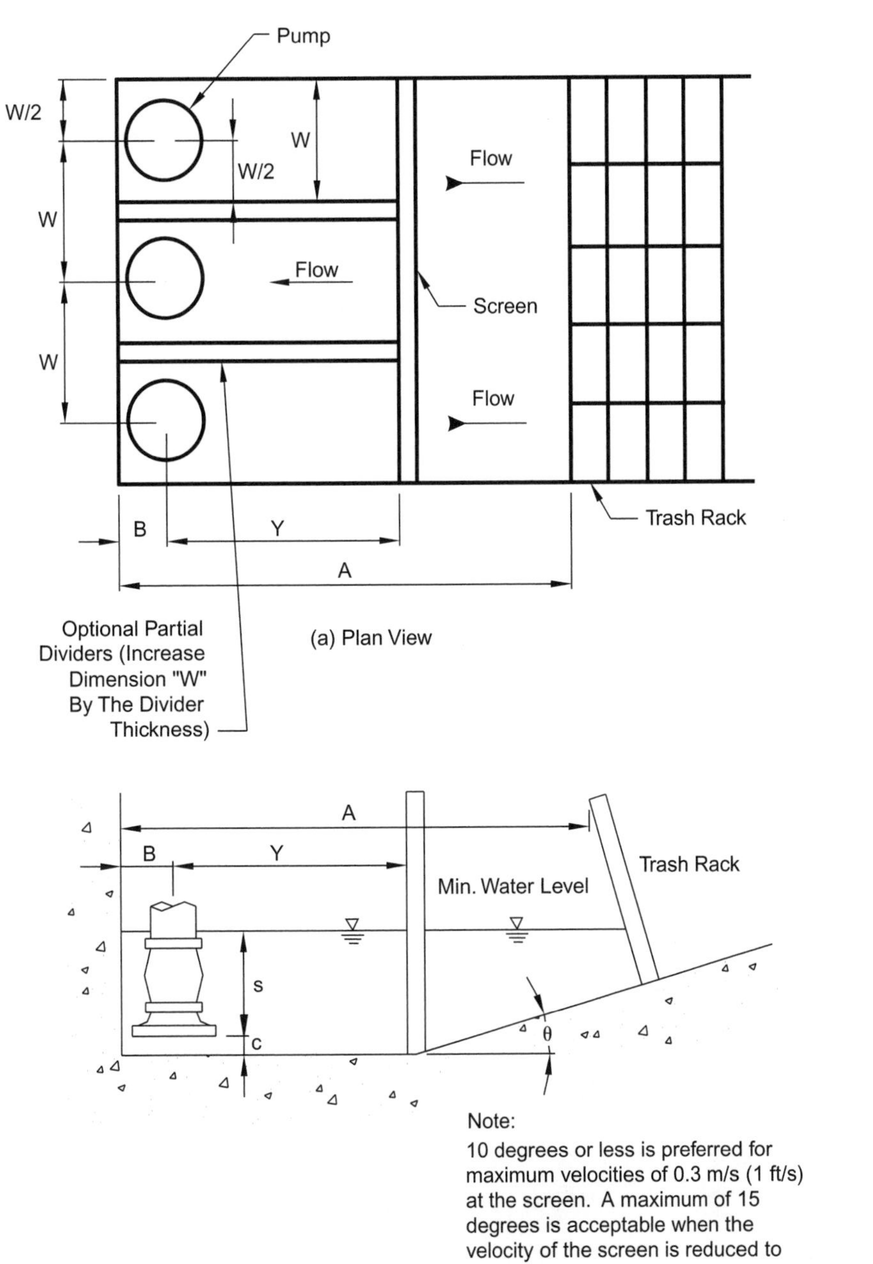

Reproduced by permission of the Hydraulic Institute.

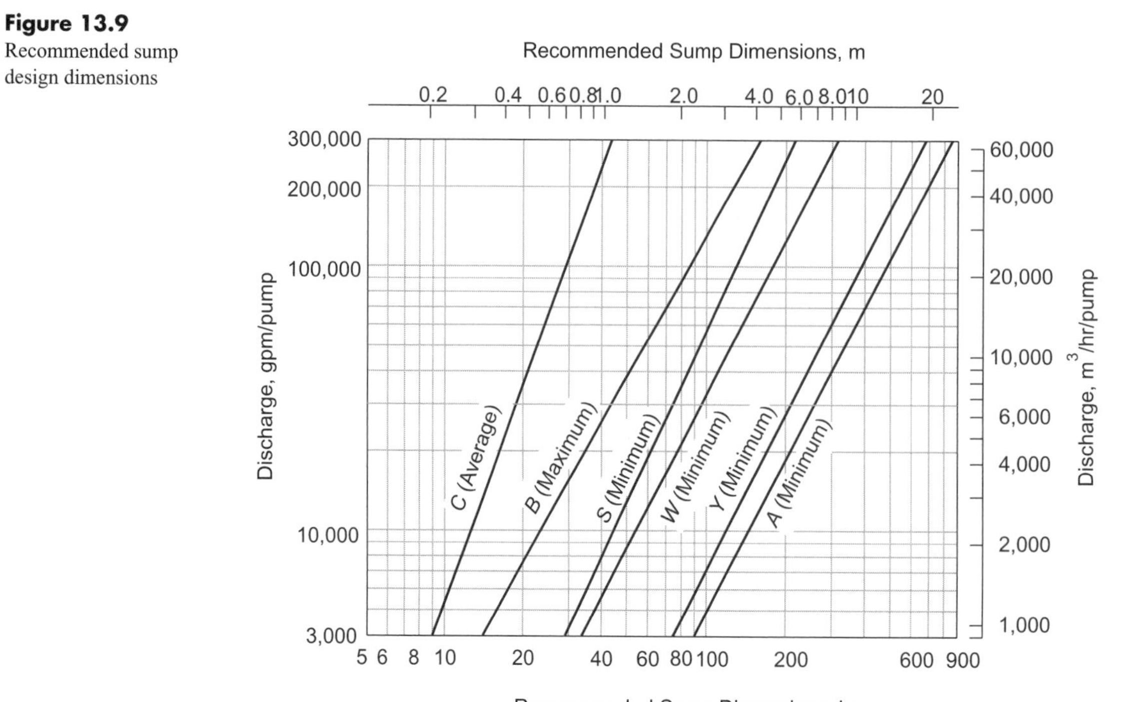

Reproduced by permission of the Hydraulic Institute.

Figure 13.9 Recommended sump design dimensions

Figure 13.10 Intake for flood control pump station

13.3 PUMP PERFORMANCE AND SELECTION

The operating characteristics of a pump installed in a piping system can be determined from *pump performance curves* and piping *system curves*. These characteristics include the energy head produced by the pump for a corresponding pump discharge, the efficiency and brake horsepower requirements for the pump motor(s), and the pump installation requirements for avoiding cavitation damage to the impeller and other internal parts. The following subsections describe various ways of quantifying these pump performance characteristics for single- and multiple-pump installations.

Pump Performance Curves

The operating characteristics of a pump, in isolation from a piping system, are typically summarized in a set of *performance curves*. The pump manufacturer prepares these curves based on hydraulic bench testing and publishes them in pump-selection manuals. Figure 13.11 illustrates a typical set of pump performance curves. Note that the actual appearances of pump performance curves vary from one manufacturer to another.

The horizontal axis of the graph in Figure 13.11 is the discharge Q through the pump and the left-hand vertical axis is the energy head h_P produced by the pump. The energy head h_P is also called the *total dynamic head (TDH),* and it corresponds to the difference in head across the pump. For a given pump and impeller diameter, Q and h_P are related: as Q increases, h_P decreases, and as h_P increases, Q decreases. The concave-downward curves in Figure 13.11 are called *head characteristic curves,* and they indicate the relationship between Q and h_P. Usually, as in Figure 13.11, several such curves are shown with each pertaining to a different impeller diameter. Note that relatively large impellers yield higher energy heads for a given discharge than do small impellers.

Superimposed over the Q versus h_P curves are pump-efficiency contours, which, for this type of graph, have circular or elliptical shapes that form an approximate bulls-eye with a center corresponding to maximum efficiency. For a given point on a Q versus h_P curve, the *efficiency* ε of the pump can be estimated by interpolating between the contours. Pumps should be selected for maximum efficiency, as low efficiencies typically result in increased power requirements. The pump efficiency is defined as

$$\varepsilon = \frac{P_{out}}{P_{in}} \qquad (13.3)$$

where ε = pump efficiency
P_{out} = *water power* produced by the pump (ft-lb/s, HP, W, etc.)
P_{in} = power supplied to the pump by its drive shaft (ft-lb/s, HP, W, etc.)

The efficiency is always less than unity (or 100 percent) because power is lost in bearing friction, turbulence, and other energy-dissipating mechanisms. In some cases, such as with close-coupled, submersible pumps, the efficiency is not simply the pump efficiency, but rather the *overall efficiency* of the pump and motor combined (also known as the *wire-to-water efficiency*).

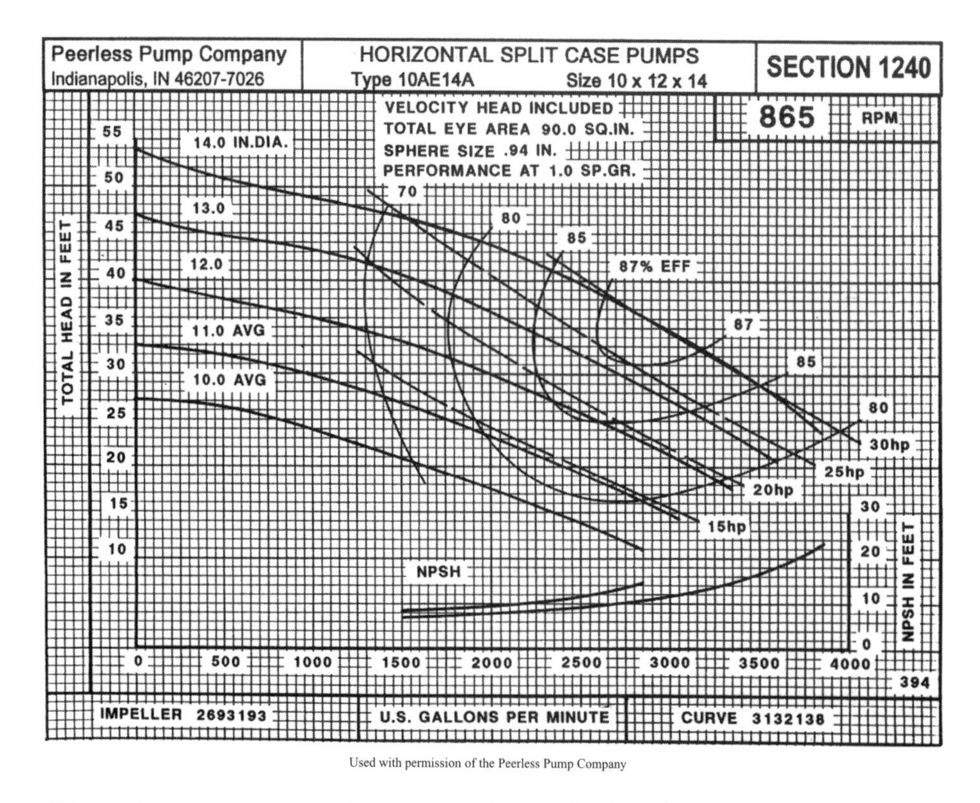

Figure 13.11 Sample set of pump performance curves

Used with permission of the Peerless Pump Company

The *water power* produced by a pump, denoted by P_{out} in Equation 13.3, can be expressed in terms of the discharge, energy head, and specific weight of the fluid as

$$P_{out} = Q\gamma h_P \quad (13.4)$$

where P_{out} = water power produced by the pump (ft-lb/s, W)
Q = discharge (cfs, m^3/s)
γ = specific weight of the fluid (lb/ft^3, N/m^3)
h_P = head supplied by the pump (ft, m)

To convert from units of ft-lb/s to horsepower, divide the result by 550.

The power provided to a pump by its drive shaft is known as the *brake power* (often the *brake horsepower*) and is represented in Equation 13.3 by P_{in}. Rearranging that expression and substituting Equation 13.4 yields the brake power requirement for a pump motor as

$$P_{in} = \frac{Q\gamma h_P}{\varepsilon} \quad (13.5)$$

where P_{in} = brake power (ft-lb/s, W)

This brake power requirement is sometimes shown on a set of pump curves as a function of the discharge. If pump curves do not contain a discharge-versus-power curve or curves (see Figure 13.11), the power requirement can be computed using Equation 13.5.

Another type of information usually shown on a set of pump curves is the required net positive suction head (NPSH). This parameter is related to potential cavitation problems that could ultimately damage the pump. The required NPSH for a pump is usually shown, as in Figure 13.11, as a function of the discharge. Cavitation and NPSH are discussed in more detail later in this section.

The performance curves for a pump, summarized in the preceding paragraphs, depend on fluid properties and on the rotational speed of the drive shaft. Pump performance curves are nearly always published for water at a standard temperature of 20° C (68° F), with the rotational speed usually noted on the chart. Pump performance curves should be approximately correct if actual water temperatures do not significantly differ from standard temperature, but will be incorrect for large temperature departures or for motor speeds other than that shown on the chart. If the motor speed is different, pump *affinity laws* may be applied to develop an accurate set of performance curves.

The pump curves shown in Figure 13.11 are typical of those from pump catalogs where a number of curves are presented for each pump for a range of impeller sizes. The engineer must select the correct impeller size for the system head curve. Another type of pump curve, shown in Figure 13.12, depicts the results of a pump test conducted by the manufacturer before a specific pump is being shipped out for installation.

The reader is referred to Sanks (1998), Karassik et al. (2000), and U.S. Army Corps of Engineers (1989, 1995, 1999) for more information on pumps, pump selection, and the design of pumping stations.

Example 13.2 – Determining Water Power and Brake Horsepower Required. A pump with an operating efficiency of 80 percent delivers a discharge of 19 m^3/min at a total head of 6 m. Determine the water power produced by the pump and the brake horsepower required for the motor to drive the pump.

Solution: The discharge of 19 m^3/min is equivalent to 0.32 m^3/s. From Equation 13.4, the water power produced by the pump is

$$P_{out} = Q\gamma h_P = 0.32(9810)(6) = 18{,}835.2 \text{ W} = 25.2 \text{ HP}$$

The brake horsepower requirement for the motor is equal to the water horsepower produced by the pump divided by the pump's operating efficiency:

$$P_{in} = Q\gamma h_P/\varepsilon = 25.2/0.80 = 31.5 \text{ HP}$$

In practice, one would select a commercially available motor providing at least this much brake horsepower.

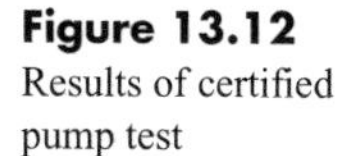

Figure 13.12
Results of certified pump test

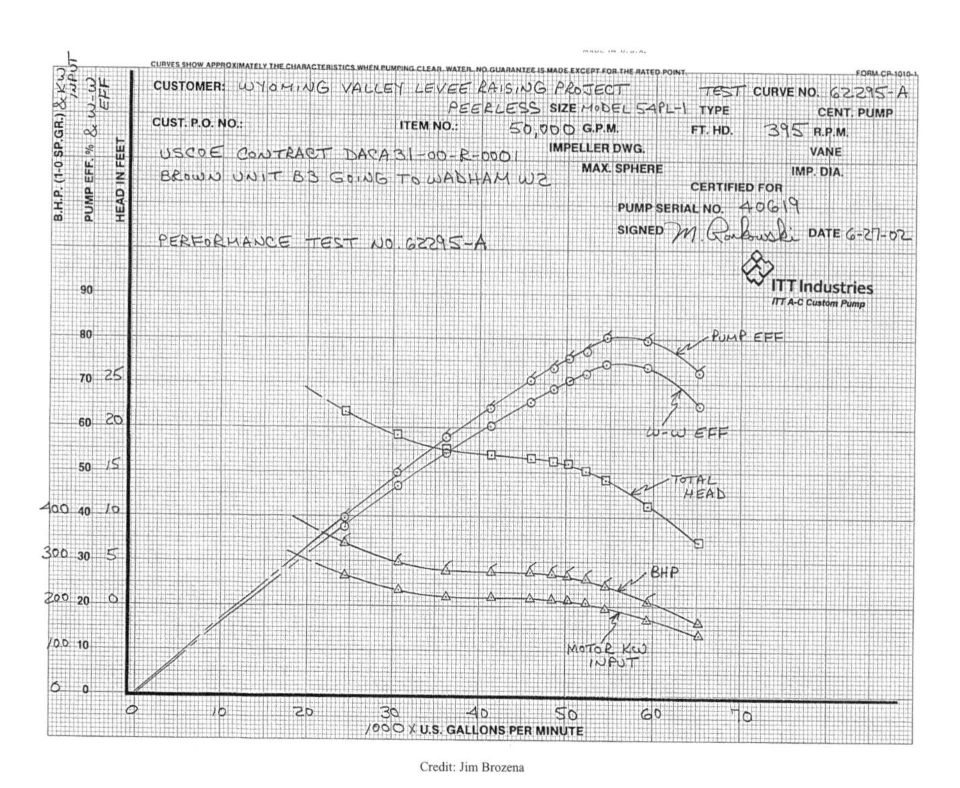

Credit: Jim Brozena

System Curves

System curves for a pump installation depict the relationship between the discharge and the energy head that must be supplied by the pump(s) to maintain that discharge. System curves reflect the effects of head losses caused by friction and minor losses in the piping system, as well as the effect of the difference between the total energy heads at the system intake and discharge points.

The energy equation is applied to develop and plot a system curve for a given piping configuration. The following description refers to the piping layout in Figure 13.13. The upstream point 1 for the energy equation is at the free water surface of the lower (suction) reservoir, and the downstream point 2 is at the free water surface at the upper (discharge) reservoir. Because $p_1 = p_2 = 0$ and $V_1 = V_2 = 0$, the energy equation can be written as

$$h_P = \Delta Z + h_L \tag{13.6}$$

where ΔZ = the difference between the two free water surface elevations, (ft, m)
h_L = energy losses in the piping system due to frictional and minor losses (ft, m)

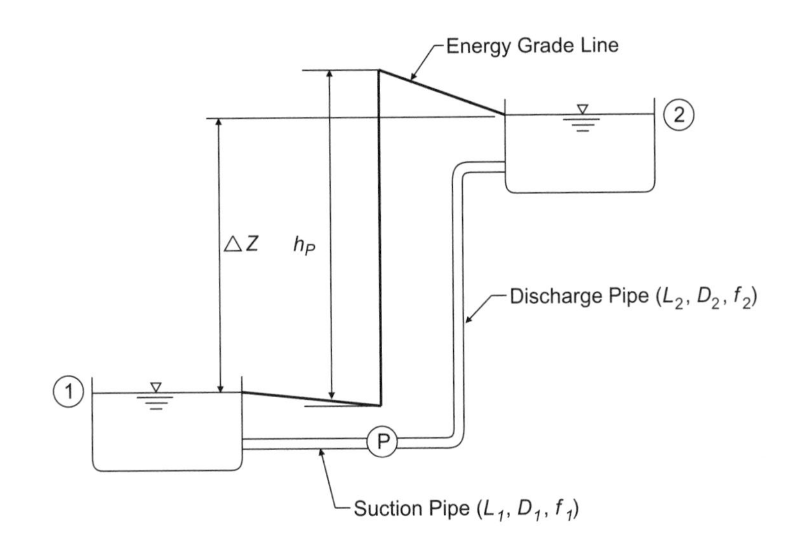

Figure 13.13 Piping system for illustration of system curve development

If, for example, frictional energy losses are expressed using the Darcy-Weisbach equation, Equation 13.6 can be expressed as

$$h_P = \Delta Z + f_1 \frac{L_1}{D_1} \frac{Q^2}{2gA_1^2} + f_2 \frac{L_2}{D_2} \frac{Q^2}{2gA_2^2} + \sum K_i \frac{Q^2}{2gA_i^2} \tag{13.7}$$

where f_1 and f_2 = Darcy-Weisbach friction factors for pipes 1 and 2, respectively
L_1 and L_2 = length of pipes 1 and 2, respectively (ft, m)
D_1 and D_2 = diameter of pipes 1 and 2, respectively (ft, m)
g = gravitational acceleration constant (ft/s^2, m/s^2)
A_1 and A_2 = cross-sectional area of pipes 1 and 2, respectively (ft^2, m^2)
K_i = minor loss coefficient for fitting i
A_i = cross-sectional area of fitting i

The second and third terms on the right-hand side of this expression are the frictional losses in the pipes upstream and downstream of the pump. The last term represents the sum of all the minor losses that occur in the two pipes.

It can be seen from Equation 13.7 that the energy head h_P that must be supplied by the pump increases with the square of the discharge. For a given piping system, a range of values of Q may be substituted into Equation 13.7 to compute corresponding values of h_P. The resulting graph of h_P versus Q is the system curve.

In most stormwater pumping, the lift provided by the pump (that is, the z term in Equation 13.7) is usually larger and more variable than the other terms. Therefore, rather than a single system head curve, there is actually a family of system head curves corresponding to various water levels in the wet well and receiving water. For example, a pump may come on when the water level in the receiving stream is 2 ft above the wet well, but may also need to pump against 20 ft of head difference during the height of a flood. The entire range of system head curves must be developed.

Example 13.3 – Developing System Head Curves. A centrifugal pump is installed in a dry well. Suction piping leading to the pump is ductile iron (absolute roughness = 0.00085 ft) with a length of 10 ft and a diameter of 12 in. The entrance to the suction pipe is rounded and its length includes a gate valve. The discharge pipe has a free discharge to the atmosphere at its downstream end and is ductile iron with a diameter of 10 in. and a length of 100 ft. It includes one 90-degree bend, a gate valve, and a check valve. Minor loss coefficients for the various valves and fittings are provided in the following table (Mays, 2001).

Fitting Type	Minor Loss Coefficient, *K*
Pipe entrance (rounded)	0.20
Gate valves (fully open)	0.10
Check valve	1.00
90° bend	0.30
Exit loss	1.00

Develop the system curve for a 2-ft elevation difference between the water level in the suction reservoir and the pipe centerline at the discharge point, which will be the elevation when the pump first comes on. Repeat the calculation for an elevation difference of 20 ft, which corresponds to the pump's "off" elevation. Assume a water temperature of 60°F ($\nu = 1.217 \times 10^{-5}$ ft^2/s).

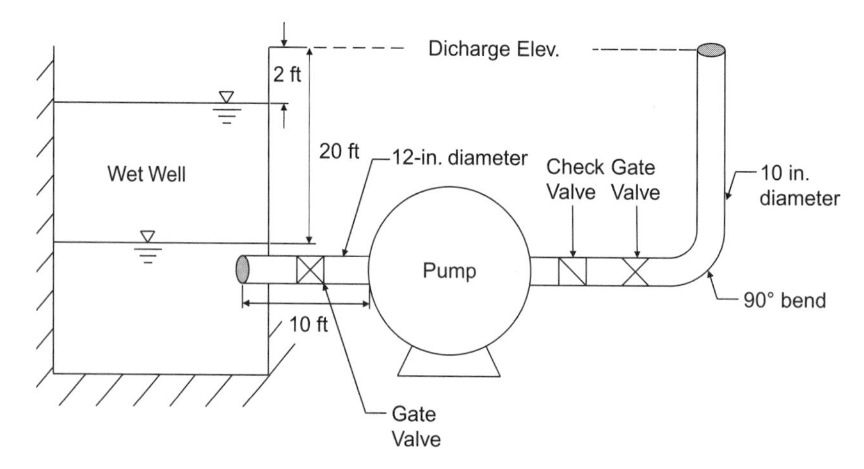

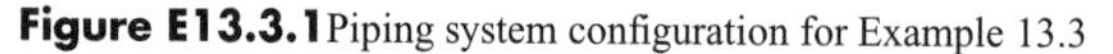
Figure E13.3.1 Piping system configuration for Example 13.3

Solution: The cross-sectional area of the 12-in. suction pipe (Pipe 1) is

$$A_1 = \pi D^2/4 = \pi/4 = 0.7854 \text{ ft}^2$$

The cross-sectional area of the 10-in. discharge pipe (Pipe 2) is

$$A_2 = \pi D^2/4 = \pi(10/12)^2/4 = 0.5454 \text{ ft}^2$$

A tabulation of the calculations of the two required system curves is provided in the table following the calculations. The first column contains assumed discharges for development of the curve. The second and third columns contain the Reynolds number of the flow in each of the two pipes, which are computed as

$$Re_1 = VD/\nu = [QD/Ai/\nu = (Q/0.7854)/(1.217 \times 10^{-5}) = 104{,}620Q$$

$$Re_1 = VD/\nu = [QD/A]/\nu = (0.8333Q/0.5454)/(1.217 \times 10^{-5}) = 125{,}544Q$$

The fourth and fifth columns contain the friction factors computed using Equation 6.11:

$$\frac{1}{\sqrt{f}} = -2\log\left(\frac{\varepsilon / D}{3.7} + \frac{5.74}{\mathrm{Re}^{0.9}}\right)$$

$$f_1 = \left[-2\log\left(\frac{\varepsilon / D_1}{3.7} + \frac{5.74}{\mathrm{Re}_1^{0.9}}\right)\right]^{-2} = \left[-2\log\left(0.0002297 + \frac{5.74}{\mathrm{Re}_1^{0.9}}\right)\right]^{-2}$$

$$f_2 = \left[-2\log\left(\frac{\varepsilon / D_2}{3.7} + \frac{5.74}{\mathrm{Re}_2^{0.9}}\right)\right]^{-2} = \left[-2\log\left(0.0002757 + \frac{5.74}{\mathrm{Re}_2^{0.9}}\right)\right]^{-2}$$

Columns 6 and 7 contain the frictional losses in each of the two pipes, which are computed as

$$h_{Lf1} = f_1 \frac{L_1}{D_1}\frac{Q^2}{2gA_1^2} = f_1 \frac{10Q^2}{2(32.2)(0.7854)^2} = f_1\left(0.2517Q^2\right)$$

$$h_{Lf2} = f_2 \frac{L_2}{D_2}\frac{Q^2}{2gA_2^2} = f_2 \frac{100Q^2}{2(32.2)(0.5454)^2} = f_2\left(5.220Q^2\right)$$

Columns 8 and 9 contain the sums of the minor losses in each of the pipes:

$$h_{Lm1} = \sum K_1 \frac{V_1^2}{2g} = (0.20 + 0.10)\frac{(Q/0.7854)^2}{2(32.2)} = 0.007552Q^2$$

$$h_{Lm2} = \sum K_2 \frac{V_2^2}{2g} = (1.00 + 0.30 + 0.10 + 1.00)\frac{(Q/0.7854)^2}{2(32.2)} = 0.01812Q^2$$

Column 10 is the total energy-head requirement for an elevation difference $Z = 2$ ft; this was computed by adding 2 ft to the sum of the entries in Columns 6 through 9. Column 11 contains the total energy-head requirements for an elevation difference of 20 ft; these were computed by adding 20 ft to the sums of the entries in columns 6 through 9. Note that the entries in Column 11 are equal to those in Column 10 plus the 18 ft of additional elevation difference. Head requirements for other ΔZ values could be obtained in a similar fashion.

(1)	(2)	(3)	(4)	(5)	(6)	(7)	(8)	(9)	(10)	(11)
Q (cfs)	**Re_1**	**Re_2**	**f_1**	**f_2**	**h_{Lf1} (ft)**	**h_{Lf2} (ft)**	**h_{Lm1} (ft)**	**h_{Lm2} (ft)**	**h_P (ft) ($\Delta Z = 2$ ft)**	**h_P (ft) ($\Delta Z = 20$ ft)**
0	0	0	0.000	0.000	0.00	0.00	0.00	0.00	2.00	20.00
1	104,674	125,641	0.022	0.022	0.01	0.14	0.01	0.13	2.28	20.28
2	209,348	251,282	0.021	0.021	0.02	0.53	0.03	0.50	3.08	21.08
3	314,023	376,923	0.020	0.021	0.05	1.17	0.07	1.13	4.40	22.40
4	418,697	502,564	0.020	0.020	0.08	2.05	0.12	2.01	6.25	24.25
5	523,371	628,205	0.020	0.020	0.12	3.19	0.19	3.14	8.63	26.63
6	628,045	753,847	0.020	0.020	0.18	4.57	0.27	4.52	11.52	29.52
7	732,720	879,488	0.019	0.020	0.24	6.20	0.37	6.15	14.94	32.94
8	837,394	1,005,129	0.019	0.020	0.31	8.08	0.48	8.03	18.88	36.88
9	942068	1,130,770	0.019	0.020	0.39	10.20	0.61	10.16	23.34	41.34
10	1,046,742	1,256,411	0.019	0.020	0.49	12.58	0.76	12.55	28.33	46.33
11	1,151,417	1,382,052	0.019	0.020	0.59	15.20	0.91	15.18	33.84	51.84
12	1,256,091	1,507,693	0.019	0.020	0.70	18.07	1.09	18.07	39.87	57.87
13	1,360,765	1,633,334	0.019	0.020	0.82	21.19	1.28	21.20	46.42	64.42
14	1,465,439	1,758,975	0.019	0.020	0.95	24.55	1.48	24.59	53.50	71.50
15	1,570,113	1,884,616	0.019	0.020	1.09	28.16	1.70	28.23	61.09	79.09

Figure E13.3.2 illustrates the two required system curves and is a graph of the energy-head requirements in columns 10 and 11 of the preceding table versus the corresponding discharges in column 1 of that table.

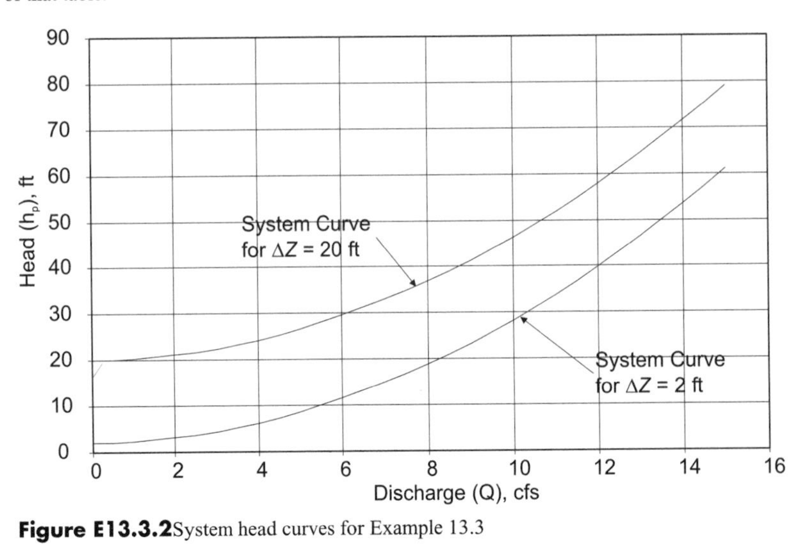

Figure E13.3.2 System head curves for Example 13.3

Operating Point and Pump Selection

The reader should observe from the two preceding subsections that a pump performance curve and a piping-system curve both show h_P as a function of Q. Both curves can be plotted on a single graph, as illustrated in Figure 13.14. The point where the two curves intersect, the *operating point*, is the actual combination of discharge and energy head that will be produced by the pump when installed in that piping system.

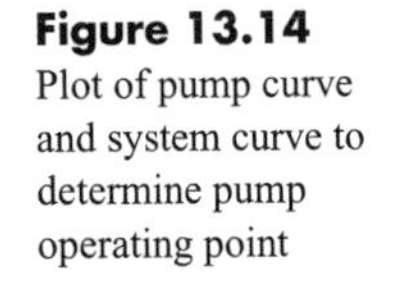

Figure 13.14
Plot of pump curve and system curve to determine pump operating point

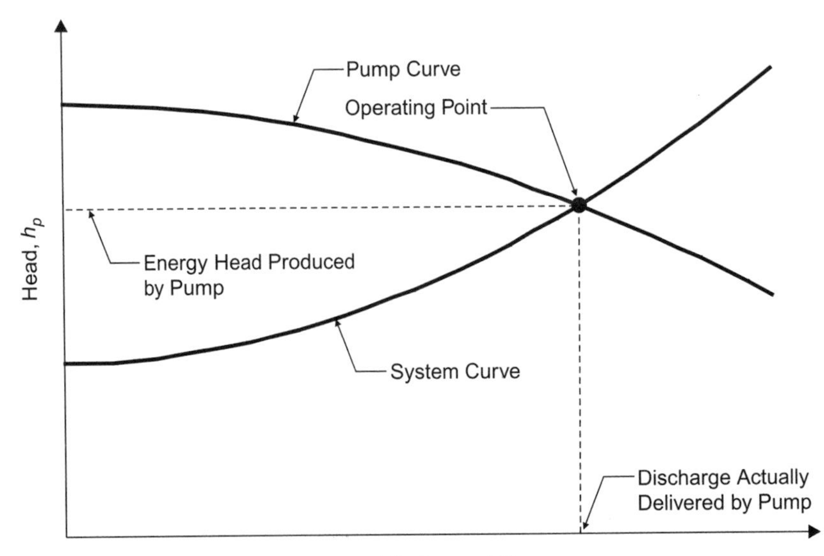

After the h_P and Q of the operating point have been determined, the pump's efficiency, brake power requirement, and required NPSH at that operating point can be determined from their respective curves, as described earlier. All of these operating characteristics, as well as other issues addressed in this chapter, should be considered when judging the acceptability of a pump's performance. Generally, one seeks to maximize the efficiency and minimize the power requirement, while meeting a desired discharge goal.

The operating point will move as the suction and discharge water levels (and hence system head curves) change. A pump can therefore operate at a very high flow rate when the discharge side head is low but a much lower rate as the water level increases. The engineer must consider operation over this full range of conditions. A pump will run efficiently at one operating point but will lose efficiency as it moves away from that point.

Cavitation and Limits on Pump Location

The physical properties of water, such as density and viscosity, depend on both temperature and pressure. For most civil and environmental engineering problems, fluid pressures are not expected to depart significantly from atmospheric pressure, and one can neglect pressure variations and treat the fluid properties as functions of temperature only.

Pumps represent one instance in which neglect of pressure considerations can cause problems with a design. As water enters a pump and nears the impeller, its velocity increases. To compensate for this increase in kinetic energy, the fluid pressure (potential energy) must drop. If the absolute fluid pressure falls below the *vapor pressure* corresponding to the fluid's temperature, the fluid vaporizes [even though its temperature is typically well below the "normal" boiling temperature of 212°F (100°C)]. Water-vapor pockets form in the fluid as it passes through the low-pressure area. When these vapor pockets subsequently reach a high-pressure area within the pump, they implode. This process is called *cavitation*. The vapor-pocket implosions, combined with any particulate matter in the fluid, erode the impeller and other internal parts of the pump. Left unchecked, the pump will perform poorly and eventually fail.

During bench testing of a pump by its manufacturer for preparation of its performance curves, tests are conducted to define its cavitation characteristics. These are usually quantified through a parameter known as the required net positive suction head (NPSH), defined as

$$\text{NPSH} = \frac{p_s}{\gamma} + \frac{V_s^2}{2g} - \frac{p_v}{\gamma} \qquad (13.8)$$

where NPSH=required net positive suction head (ft, m)

p_s =absolute fluid pressure at the entrance to the suction side of the pump (lb/ft^2, Pa)

V_s =flow velocity at the suction side of the pump (ft/s, m/s)

p_v =vapor pressure of the fluid (lb/ft^2, Pa)

To prevent cavitation damage in a pump installation, the available net positive suction head must be at least as great as the required net positive suction head.

This can be accomplished by rearranging Equation 13.8 to yield the required minimum absolute fluid pressure at the suction side of the pump:

$$p_s = p_v + \gamma\left(NPSH - \frac{V_s^2}{2g}\right) \quad (13.9)$$

If the actual absolute fluid pressure is lower than the amount given by this expression, cavitation can be expected and the design should be revised. Often, the pump can be installed at a lower elevation (relative to the water level in the suction reservoir), which will increase p_s. If that is not possible, the suction piping should be redesigned to reduce the energy losses, or a different pump should be selected, or both.

Example 13.4 – Determining Cavitation Limits. A pump installed in a dry-well draws water from a suction reservoir having a water-surface elevation of 61 m. At its operating point, the pump delivers a discharge of 0.12 m^3/s. Energy losses in the suction piping for this discharge, including frictional losses and all minor losses, total 1.7 m. If the pump installation is near sea level with atmospheric pressure of about 101.3 kPa and if the NPSH requirement for the pump is 4.6 m, determine the maximum elevation at which the pump can be installed to avoid possible cavitation. The diameter of the pump entrance is 300 mm. The water temperature is 15.5° C.

Solution: The cross-sectional area of the pump entrance is 0.07 m^2, so the velocity at that point is

$$V_s = 0.12/0.07 = 1.71 \text{ m/s}$$

From Appendix A, the vapor pressure of the water is 1.79 kPa. Using Equation 13.9 gives the fluid pressure at the suction side of the pump to be at least

$$p_s = 1790 + 9810\left(4.6 - \frac{1.71^2}{2(9.81)}\right) = 45.5 \text{ kPa}$$

The required gage pressure is equal to this amount less the atmospheric pressure of 101.3 kPa or

$$p_s = 45.5 - 101.3 = -55.85 \text{ kPa}$$

The minimum required pressure head at the pump entrance is therefore

$$p_s/\gamma = -55.84/9810 = -5.70 \text{ m.}$$

The maximum permissible pump mounting elevation can be found by subtracting the suction-pipe energy losses and the minimum pressure-head requirement from the suction-reservoir water-surface elevation:

$$\text{Max. permissible elevation} = 61 - (-5.70) - 1.7 = 65 \text{ m}$$

Cavitation should not occur if the pump is installed at an elevation lower than 65 m. Of course, one would probably design the pump in this installation to be below the suction-reservoir water-surface elevation of 61 m, thus minimizing the likelihood of cavitation.

Installations with Multiple Pumps

Multiple-pump installation facilities are often desirable because of the potential for individual pump failure and because a pumping station may be required to handle a wide range of flow rates. For cases in which the length of discharge piping is short and the required pump head is small, it is economical to provide a separate piping system for each pump. In this case, the pumps and their associated piping systems are constructed in parallel and can be designed and analyzed independently, as previously described.

For instances in which the length of discharge piping downstream of the pumps is long, it may be economical to install a *manifold* such that two or more pumps installed in *parallel* discharge to a common pipe. In either of these situations, an *equivalent pump performance curve* must be derived from the individual pump performance curves to determine the operating point for simultaneous pump operation. The manner in which the equivalent pump curve is derived depends on the number of pumps that operate simultaneously and whether the pumps are identical.

When two pumps are installed in parallel, energy considerations dictate that the energy heads produced by the pumps must be equal. If for a given pump head h_P one pump delivers a discharge of Q_1 and the second pump delivers a discharge of Q_2, then the total discharge delivered by the pair of pumps is $Q = Q_1 + Q_2$. The equivalent pump curve for a system of two pumps in parallel is illustrated in Figure 13.15. Again, this result is easily generalized to the case of three or more pumps.

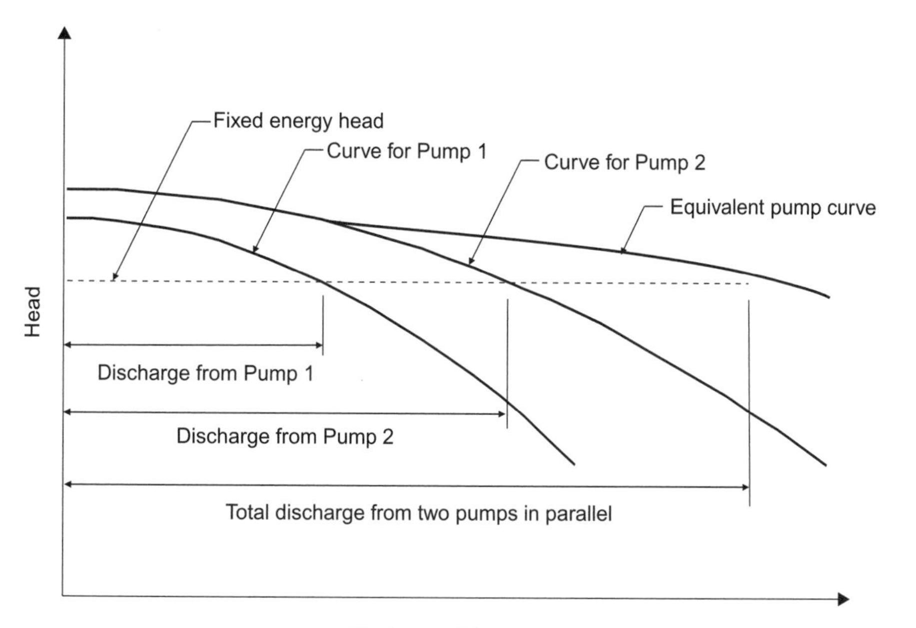

Figure 13.15 Equivalent pump curve for a system of two pumps in parallel

Figure 13.16 shows a system curve, a pump curve for a single pump, and an equivalent pump curve for two pumps in parallel. For this example, it is assumed that the parallel installation is constructed with identical pumps and that there is a single discharge piping system. From inspection of the operating points in the figure, it can be seen that the parallel installation does not double the discharge delivered to the system. In other words, the marginal benefit of the second pump, in terms of the increase in discharge yielded, is less than the discharge produced by the operation of one pump alone. If a third pump were added, its marginal benefit would be less than that of the second pump, and so on. The cause of this behavior is the increased frictional and minor losses in the piping system resulting from the increase in discharge. This is evident in Figure 13.16 from the increasing slope of the system curve.

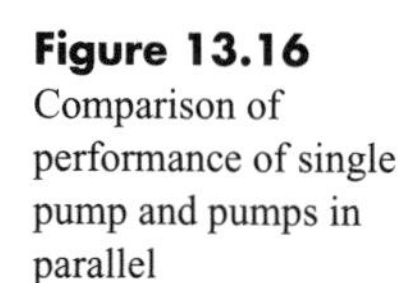

Figure 13.16
Comparison of performance of single pump and pumps in parallel

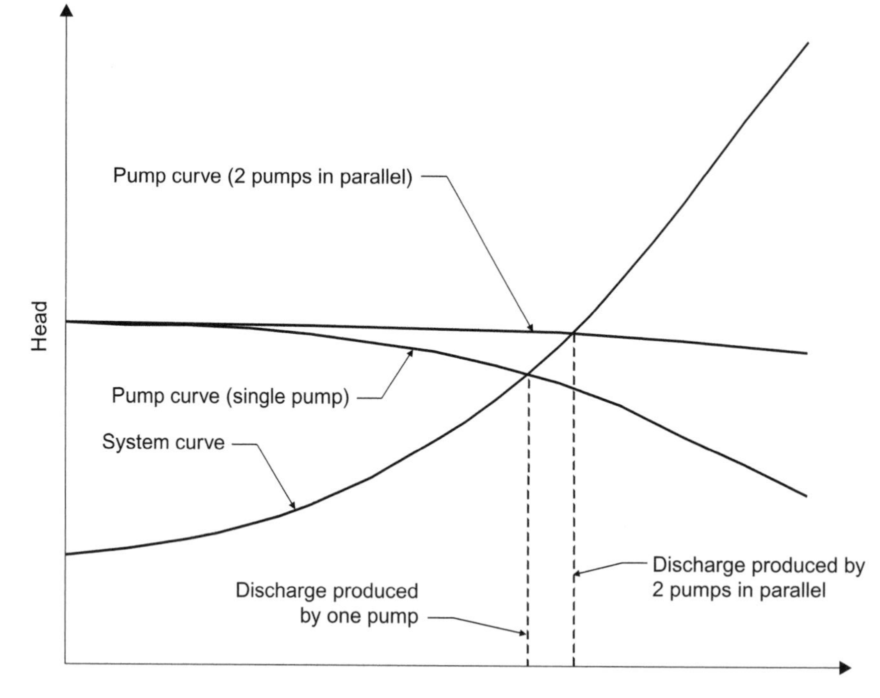

13.4 PIPING SYSTEM REQUIREMENTS

Because the pump discharge piping conveys stormwater under pressure, there are stresses on the pipes that can, if not accounted for, result in leaks or even system failure. Therefore, pipe material, joint, and installation requirements must be given careful consideration. Additional design issues center on valves and backflow prevention and on pipe restraints and vibration isolation.

Pipe Requirements

As described in Section 11.1, pipe materials used in storm sewer design and construction are typically concrete or corrugated metal. Either of these materials can be used for those portions of a pumping system where the pipe is buried below ground level. The pipe, however, must be specified so that it will have the strength necessary to resist the external and internal pressures expected to develop during pump operation (see "External Forces" on page 460, "Internal Pressure Forces" on this page, and Section 6.4, page 210). An appropriate pipe strength class and/or wall thickness must be selected. Pipe-jointing methods should also be given consideration, as leakage could cause erosion of soil materials surrounding the pipe and, ultimately, system failure.

Alternatives to concrete and corrugated metal pipes that are more desirable for use within a pumping station are steel, ductile iron, and PVC pipes. Economy, consistency, and ease of construction might even dictate that they be used for the entire pumping system. Major considerations in pipe-material selection include size availability, jointing methods, and the need for special linings or coatings. Pressure pipe standards published by the American Water Works Association (AWWA) may be consulted for further guidance.

Within a pumping station, exposed pipes should be jointed, supported, and braced to form a rigid structure capable of withstanding the hydrodynamic forces developed by momentum changes in bends and fittings. In buried portions of the piping system, concrete thrust blocks or similar restraints may be necessary at pipe bends. Sections 3.3 and 7.5 present methods for determining the magnitudes and directions of hydrodynamic forces.

Internal Pressure Forces

The figure is a free-body diagram of the left half of a circular pipe with diameter D and wall thickness t. A force balance in the horizontal direction in the figure requires that the force caused by the internal pressure, p, must be equal to the force induced by the tensile stress, σ, in the pipe walls. Considering a unit length of pipe (into the page), this may be expressed as $pD = 2t\sigma$ or as

$$\sigma = \frac{pD}{2t}$$

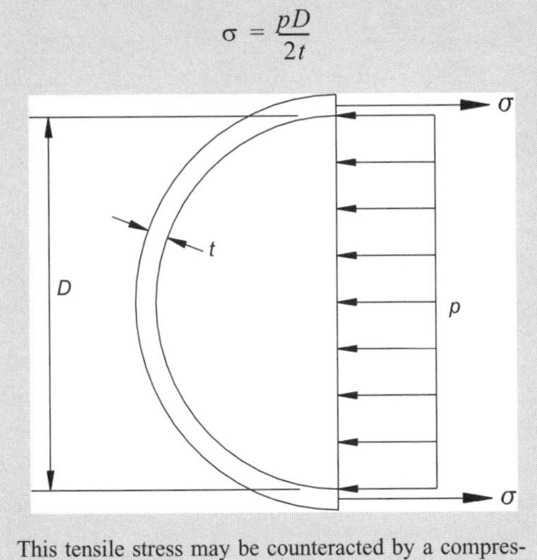

This tensile stress may be counteracted by a compressive stress due to external loading on the pipe.

The derivation of the previous equation assumes that the pipe walls are thin, or $t < D/10$. For thick-walled pipes ($t > D/10$), the tensile stress can be computed as (Seely and Smith, 1959)

$$\sigma = \frac{pr_i^2}{r_o^2 - r_i^2}\left(1 + \frac{r_o^2}{r^2}\right)$$

where r_i = inside radius of the pipe

r_o = outside radius of the pipe

r = radius associated with the location where the stress is desired ($r_i \le r \le r_o$)

It may be seen from this expression that the stress is largest at the inside face of the pipe wall, and that it is smallest at the outside face of the pipe wall.

In storm sewers and culverts, where internal fluid pressures are usually small, the tensile stress predicted by either of the two equations is usually negligible. However, internal fluid pressures in storm sewers and culverts can, in some instances, give rise to a need for special pipe joints to resist the internal pressures and prevent leakage into the surrounding soils. Non-negligible internal pressures and corresponding tensile stresses in pipe walls can occur when pumps are placed in a system.

Valves and Fittings

The following subsections describe standard fittings used in pump installations, with descriptions of their functions. Figure 13.17 shows a possible arrangement of these elements in a pump station.

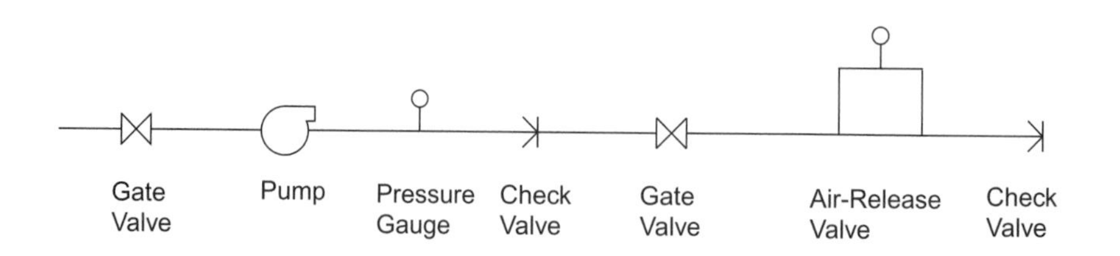

Figure 13.17 Example of a possible arrangement of equipment in a pump station

Shutoff Valves. *Shutoff valves,* such as gate or butterfly valves, serve to isolate a pump from the main piping network. They are often installed immediately upstream and downstream of the pump so that it can be easily removed from service for maintenance and eventual replacement. These valves should remain either completely open or completely closed; they should not be used to throttle the flow.

Reducers and Flexible Connections. *Reducers* are required if the pump inlet and discharge diameters are different from the piping system diameters and should be eccentric (flat on one edge) to minimize air trapping. Reducers consisting of *flexible connections* minimize the transfer of vibration energy from the motor and pump to the piping system and may eliminate the need for expansion and contraction joints if large temperature fluctuations will be encountered.

Flap Gates. *Flap gates* and/or *check valves,* which restrict flow to one direction, are required for backflow prevention. Flap gates may be installed on the downstream end of the pump's discharge pipe and are desirable if tailwater conditions downstream of that point could cause backups into the system. Check valves, installed between the pump and the shutoff valve in the downstream discharge pipe, may be desirable if the pipe is long and/or large in diameter and if its geometric configuration is such that water would flow backwards during nonpumping times. Check valves are also required to prevent backflows when two or more pumps are connected via a manifold to a common discharge pipe. Check valves should be installed in horizontal portions of the piping system and should be designed to prevent valve slamming, which can sudden velocity changes and, consequently, hydraulic transients (water hammer).

Air-Release Valves. *Air-release* or *vacuum valves* permit trapped air to escape from pipes during pump startup. Similarly, when pumps shut off, vacuum valves enable air to enter the pipe to prevent its possible collapse by external forces.

Pressure Gauges and Clean-Outs. For purposes of operation and maintenance, consideration might be given to installation of a *pressure gauge* on the pump discharge pipe. The gauge can be used to periodically check pump performance and can provide warnings of impending problems or failures. *Clean-outs,* which serve as access points for easy removal of debris and blockages, can also be useful in stormwater applications.

13.5 PUMP CONTROLS

Pump controls start and stop pumps based on water-level sensors in the wet-well or temporary-storage facility. Sensors include floats, pressure transducers, and other devices, and they have the capability to control several pumps simultaneously. Careful consideration must be given to the design of pump controls, as an inadequate design can lead to excessive pump cycling and premature motor failure. The pump-control scheme also has implications for required temporary water-storage volumes; a poor scheme may result in a larger storage volume than necessary. It is recommended that designers work closely with pump manufacturers to ensure that a design does not violate any of the pump's mechanical constraints.

The American Association of State Highway and Transportation Officials *Model Drainage Manual* (AASHTO, 1991) describes two types of pump-control methods commonly used in stormwater pumping applications. In the *common off-elevation scheme*, individual pumps start at successively higher water-level elevations, but they all stop simultaneously at the same elevation (see Figure 13.18). This scheme is advantageous if large amounts of suspended matter (such as sediment) are present in the stormwater, as it limits the amount of time that solids have to settle.

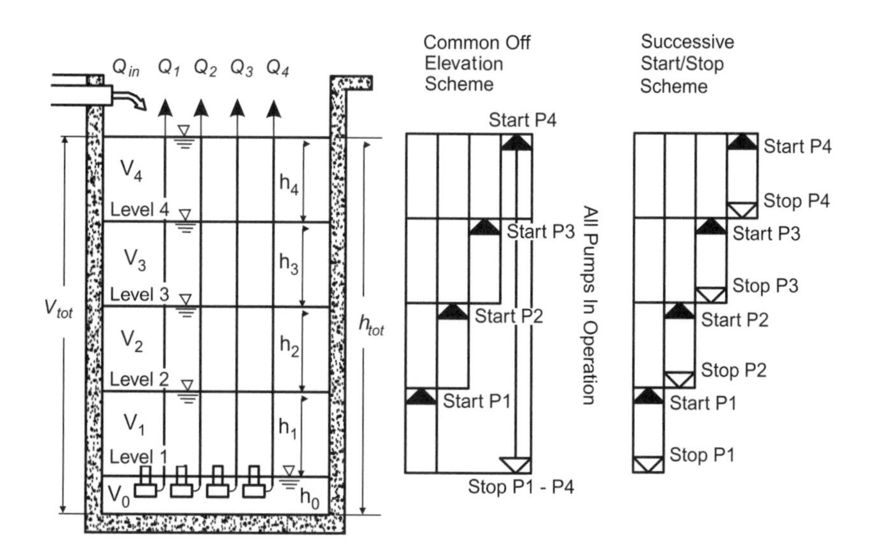

Figure 13.18 Multiple-pump control schemes

In the *successive start/stop scheme*, the start elevation for one pump is the stop elevation for the subsequent pump. In other words, the start elevation for pump 1 is the stop elevation for pump 2, the stop elevation for pump 2 is the start elevation for pump 3, and so on, as depicted in Figure 13.18. A benefit to this scheme is that pumps are successively shut off as the water surface in the suction reservoir drops; thus, no more pumps are operated than is necessary.

In addition to these schemes and their infinite variations, control schemes are employed to equalize wear on the individual pumps in a system. Some of these schemes simply alternate the first pump to start, while schemes that are more complex continuously alternate all pumps during operation.

As indicated previously, start/stop elevations for individual pumps should be devised to avoid short pump-cycling times. Because of the high power requirement and consequent heat generation during pump startup, sufficient time must be provided between starts to prevent problems with pump-motor overheating and failure. Temporary water-storage volumes must be large enough to allow minimum cycling times; however, overly large storage volumes promote settling of suspended solids in the stormwater, with consequent maintenance requirements. Although the pump manufacturer should be consulted first, some guidance on minimum cycling times is provided in Table 13.1.

Table 13.1 Guidelines for minimum pump cycling times (Brown, Stein, and Warner, 1996)

Motor HP	Motor kW	Min. Cycling Time (min)
0–15	0–11	5
20–30	15–22	6.5
35–60	26–45	8
65–100	49–75	10
150–200	112–149	13

13.6 STORAGE REQUIREMENTS

Because stormwater runoff rates and volumes tend to be large and quickly generated, temporary storage (that is, *detention*) of the runoff is often required. The discharge that can be economically pumped may be much lower than the peak rate of runoff; thus, temporary storage offers a means to reduce the required pumping capacity. The storage of stormwater can also provide water quality benefits because a portion of the particles suspended in the water has time to settle out. When temporary storage is needed (as it usually is), the designer must estimate not only the peak discharge from the drainage basin, but also the time distribution of runoff in the form of a hydrograph. The runoff hydrograph may be estimated using synthetic unit hydrograph procedures (see Section 5.5), or, in the case of small drainage basins, using a hydrograph-based application of the rational method.

As described in more detail in Section 13.7, determination of the required volume of a temporary-storage facility is accomplished through a trial-and-error process that also includes pump selection, piping diameters, and control strategies. However, to begin evaluating the performance of alternative designs, Baumgardner (1981) presented a technique that can be used to make a first estimate of the required temporary-storage volume. As shown in Figure 13.19, the method is graphical and simply involves drawing a horizontal line through the incoming runoff hydrograph at a discharge equal to the average pumping rate. The volume of water under the hydrograph and above the horizontal line is an estimate of the required storage capacity. If the inflow hydrograph can be reasonably approximated as triangular (see Figure 13.20), similar triangles can be used to estimate the required storage volume as

$$S = \frac{1}{2}\left(\frac{t_b}{Q_p}\right)(\Delta Q)^2 \qquad (13.10)$$

where S = required storage volume (ft^3, m^3)
t_b = base time of the inflow hydrograph (s)
Q_p = peak inflow rate (cfs, m^3/s)
ΔQ = difference between peak inflow and outflow rates (ft^3/s, m^3/s)

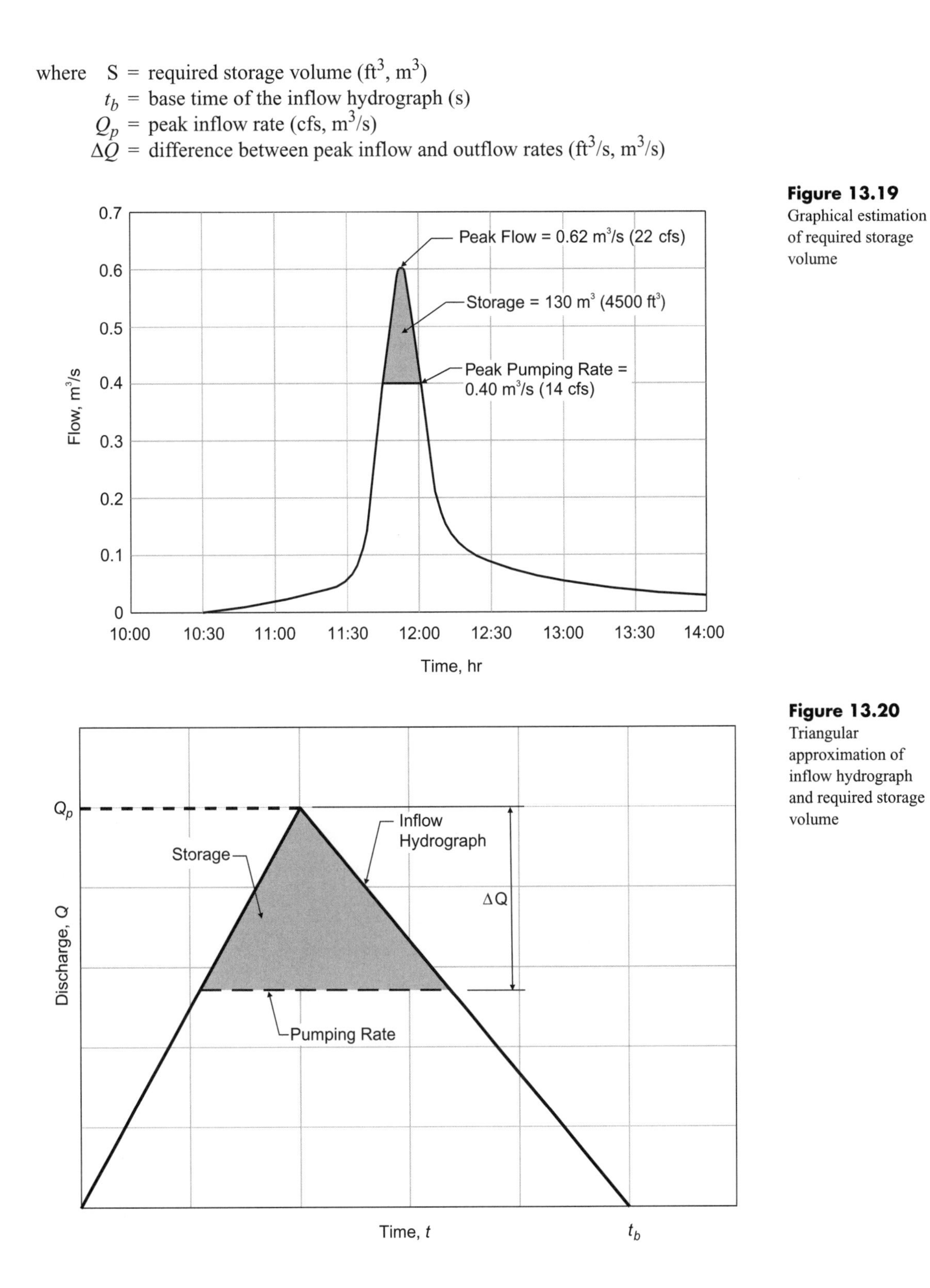

Figure 13.19 Graphical estimation of required storage volume

Figure 13.20 Triangular approximation of inflow hydrograph and required storage volume

Example 13.5 - Determining Required Storage with Baumgardner's Technique. An inflow hydrograph is shown in columns 1 and 2 of the following table. Determine the required storage volume if the maximum pumping rate is 0.071 m^3/s.

Solution: The solution to this problem could be obtained by graphing the inflow hydrograph and outflow rate, as shown in Figure 13.19, and determining the volume between the two curves. However, it will be solved instead using the numerical procedure illustrated in the following table.

Entries in column 3 in the table are the outflow hydrograph (i.e., the pumping rate); they are equal to the inflow rate when the inflow rate is less than 0.071 m^3/s and are equal to 0.071 m^3/s when the inflow rate is larger than that amount. Entries in column 4 are equal to the difference between entries in columns 2 and 3; note that these values are zero when the inflow rate is less than or equal to 0.071 m^3/s. Averages of two successive entries in column 4 are listed in column 5. Those averages, multiplied by the time-step interval (5 min, or 300 sec), yield the incremental storage volumes in column 6. Finally, column 7 contains running sums of the incremental volumes in column 6. The last entry in column 7, 682.8 m^3, is an estimate of the required temporary-storage volume.

(1)	(2)	(3)	(4)	(5)	(6)	(7)
t (min)	I (m^3/s)	Q (m^3/s)	$I - Q$ (m^3/s)	Avg $I - Q$ (m^3/s)	Incr. Vol. (m^3)	Cum. Vol. (m^3)
0	0	0	0			0
5	0.003	0.003	0	0	0	0
10	0.006	0.006	0	0	0	0
15	0.009	0.009	0	0	0	0
20	0.011	0.011	0	0	0	0
25	0.014	0.014	0	0	0	0
30	0.017	0.017	0	0	0	0
35	0.020	0.020	0	0	0	0
40	0.023	0.023	0	0	0	0
45	0.025	0.025	0	0	0	0
50	0.028	0.028	0	0	0	0
55	0.031	0.031	0	0	0	0
60	0.034	0.034	0	0	0	0
65	0.071	0.071	0	0	0	0
70	0.127	0.071	0.056	0.028	8.4	8.4
75	0.326	0.071	0.255	0.156	46.7	55.1
80	0.538	0.071	0.467	0.361	108.3	163.4
85	0.609	0.071	0.538	0.503	150.8	314.1
90	0.481	0.071	0.410	0.474	142.2	456.3
95	0.340	0.071	0.269	0.340	101.9	558.2
100	0.184	0.071	0.113	0.191	57.3	615.5
105	0.142	0.071	0.071	0.092	27.6	643.1
110	0.113	0.071	0.042	0.057	17.0	660.0
115	0.099	0.071	0.028	0.035	10.5	670.5
120	0.093	0.071	0.022	0.025	7.5	678.0
125	0.076	0.071	0.005	0.014	4.1	682.1
130	0.071	0.071	0.000	0.003	0.8	682.8
135	0.065	0.065	0	0	0	682.8
140	0.059	0.059	0	0	0	682.8
145	0.057	0.057	0	0	0	682.8
150	0.054	0.054	0	0	0	682.8

Baumgardner's technique is quite simple and does not take into account the actual variation of pumping rate caused by system-head changes and pump controls, during a storm event. The FHWA (1982) and Brown, Stein, and Warner (1996) have presented a mass curve routing procedure for estimation of the required storage capacity. This routing procedure is a vast improvement over the simple graphical method, but it still suffers from a lack of consideration of system head change effects on discharges. The discharge delivered by a pump is typically quite sensitive to changes in head (see Figure 13.11), and thus the mass curve routing method may yield inaccurate storage-volume estimates. Use of an analysis method such as a fourth-order Runge-Kutta approximation method as discussed in Section 13.7 will yield more accurate results.

13.7 PUMPING STATION DESIGN

Design of a pumping facility is a trial-and-error process involving evaluations of multiple combinations of design variables, such as storage volume, pump model, piping diameters, and control system. None of these design variables should be determined independently, as they are all related. Further, because of the large number of combinations of design variables, experience and professional judgment play significant roles in developing the alternative designs to be evaluated. The following sections provide a general overview of the necessary design tasks.

Hydrology

One of the first tasks in pumping-station design is to estimate the design inflow hydrograph to the pumping facility. Hydrologic methods discussed in Chapter 5 can be used for this purpose. In locations such as highway or roadway depressions, stormwater pumping installations are typically designed to convey 50-year design storm flowrates (FHWA, 1982; Brown, Stein, and Warner, 1996), but it is also recommended that checks of station performance and flooding limits be made for the 100-year design storm event. Designs for more frequent return periods may be acceptable in some cases, but should be agreed upon between the designer, owner, and appropriate review agencies. In areas where storms occur quite frequently, water-storage facilities may not drain completely between storms, and continuous simulation may be necessary to assess the adequacy of a design.

Development of a Trial Design

The initial trial-design variables can be selected after the design inflow hydrograph has been established. As noted earlier, the trial-design variables that must be selected for any evaluation include:

- Piping diameters and layouts
- Pump models
- Temporary water-storage volume
- Pump-control strategies

All of these variables are interrelated and should not be chosen independently. For example, choosing excessively small pipe diameters or a small storage volume will necessitate the use of large pumps; poor design of the control strategy may result in excessive pump cycling times; and so on.

The first design variable that should be selected is the size and layout of the piping system. Piping-system layout is usually governed by legal right-of-way and discharge-point considerations. Piping diameters should be chosen to maintain flow velocities within reasonable bounds. Reasonable velocities are important not only for hydrodynamic reasons (such as the hydraulic forces that develop at bends and fittings), but also to minimize the effects of hydraulic transients (such as water hammer). The designer should also consider at this stage whether to use a dry/wet- or dry-well installation, as this choice will affect the economics of the trial design.

Subsequently, system curves can be developed from the piping-system design. Based on these system curves and an estimate of the desired maximum pumping rate, manufacturer guidance and pump performance curves are used to select an initial trial pump model. Maximum pumping rates are determined by downstream system capacities, legal requirements limiting downstream peak discharges to predevelopment levels, or economics.

Pumping stations are usually equipped with, at a recommended minimum, two or three pumps, although guidelines vary. Consideration should be given to oversizing the pumps to partially compensate for pump failure. For example, in a two-pump system, the pumps may be sized so that each is capable of handling 70 percent of the total design flow. In a three-pump system, each pump might be capable of handling 50 percent of the design flow. The *firm capacity* of a pumping installation is generally taken as the maximum discharge that could be delivered by the facility if one pump were out of service. When designing pumping stations with multiple pumps, it is advantageous if all the pumps are identical to facilitate ease of maintenance.

An initial estimate of the storage-volume requirement can be obtained graphically, as described in Section 13.6. If storage is to be provided underground, this estimate can be used to select trial storm-sewer pipe and/or storage-tank dimensions. If the storage is to be provided in a surface-water pond, a topographic map can be used to contour the pond to provide the necessary volume.

Development of a trial control strategy is based on the design inflow hydrograph, the selected pump make and model, and the storage-volume characteristics. As already noted, the objective here should be to maintain minimum pump-cycling times.

Trial Design Evaluation

After a set of trial-design variables has been selected, as described in the preceding subsection, the task of routing the design inflow hydrograph through the facility can proceed. This routing process, carefully performed, will provide information on actual pumping rates and efficiencies, pump-cycling times, storage-volume adequacy, and other characteristics. These results can then be reviewed to assess the acceptability of a design and develop cost estimates that can be compared to those of other alternatives.

Because of the relative complexity of flow routing in a pumping facility (as compared to a simpler storage facility without pumping, such as a detention pond), flow routing should be accomplished using a numerical integration approximation method such as the Runge-Kutta method presented in Section 12.7 rather than the storage indication method. Runge-Kutta algorithms are also described by Chow, Maidment, and Mays (1988) and by Chapra and Canale (1988). Application of the Runge-Kutta method requires the development of stage versus discharge and stage versus water-surface area relationships for the trial design.

As with detention analysis, the stage-discharge relationship, $O(h)$, describes the discharge O from a storage facility as a function of the water-surface elevation (or *stage*) h of the suction reservoir or wet-well. In the context of pumping-station design, this relationship is dictated by the pump-control strategy and by the loci of operating points defined by the pump and system curves. Note here that multiple system curves must be considered, corresponding to the different suction-reservoir elevations expected to occur during passage of a stormwater runoff event through a facility. It is important to consider the effects of the suction reservoir or wet-well stage on the actual pumping rate, because the pumping rate is typically quite sensitive to energy-head changes. It is also important to recognize that the $O(h)$ relationship will exhibit sudden changes (discontinuities) at stages where pumps start or shut off. The relationship may also be different during periods when the reservoir level is receding than when it is rising (called a *hysteresis* phenomenon), depending on the pump-control strategy.

A stage versus water-surface-area relationship, $A_s(h)$, describes the suction-reservoir water-surface area, A_s, as a function of the water-surface elevation h. This relationship can be developed from geometric relationships for underground storage facilities, or from a topographic map depicting contours of a surface-water facility.

Given the design inflow hydrograph $I(t)$, and the $O(h)$ and $A_s(h)$ relationships, routing of flows through a pumping facility is accomplished using the following form of the continuity equation:

$$\frac{dh}{dt} = \frac{I(t) - O(h)}{A_s(h)} \qquad (13.11)$$

where h = stage (ft, m)
t = time (s)
$I(t)$ = inflow at time t (cfs, m^3/s)
$O(h)$ = outflow at stage h (cfs, m^3/s)
$A_s(h)$ = water surface area at stage h (ft^2, m^2)

Integration of this equation (that is, determination of the relationship between h and t) can be accomplished using the Runge-Kutta method. Once h is known as a function of t, the discharge Q and other variables of interest can also be found as a function of t.

13.8 MISCELLANEOUS DESIGN CONSIDERATIONS

When designing a pump station, the design engineer must consider a number of factors beyond selection of the pump size and arrangement and the sizing of the wet well. Among these are the station's architecture, power requirements, debris removal facilities, monitoring devices, safety, and accessibility.

Architectural Considerations

In some cases, such as when submersible pumps are employed, stormwater pumping stations are completely below ground level. Because the ground around the pump station may become saturated during storm events, the designer should consider the possibility that the station may become buoyant. Adding weight to the station or providing valves that drain the groundwater into the wet-well can keep the station from floating. The structural design of pump stations can vary from completely underground stations to large, attractive structures as shown in Figure 13.21.

Figure 13.21
Flood control pump station structure

If the pump station is above ground, it is often designed as an integral, nonobtrusive part of the surrounding neighborhood. Pump stations can be designed to be aesthetically pleasing without significantly increasing their cost through the use of clean lines and textured finishes (such as textured concrete). Strategically placed screening and/or landscaping features can further improve the station's appearance and can help reduce noise propagation during periods when the station is operating. Because of the need for frequent maintenance of pump stations, adequate access roadways and parking space must be provided for maintenance personnel.

"I don't know anything about architecture, but I know what I don't like."

Power and Control Requirements

Pump stations are usually powered by electricity, but they can also be powered by diesel, gasoline, or natural or LP gas engines. Electrical power, however, is usually the most cost-effective and reliable option if electrical service is readily available. Stormwater pumps usually are equipped with constant-speed, three-phase induction motors. Motor voltages depend on the size of the pump, but are generally 440 to 575 V, with a maximum recommended motor capacity of 300 HP (225 kW).

Power-system failures during storms can cause pumps to stop, with consequent flooding of adjacent lands and improved areas. In locations where temporary flooding cannot be tolerated, a standby power supply must be provided. In some instances, two independent electric feed lines may be supplied to the station, with an automatic transfer switch controlling which feed is used. In most cases, a standby generator is required, which can be powered by a diesel, gasoline, or LP or natural gas engine.

In addition to high-voltage power wiring, pump stations can contain a great deal of low-voltage control wiring connecting various sensors to control equipment and the *SCADA* (supervisory control and data acquisition) system. Control wiring should be shielded from power wiring per the manufacturer requirements. The wiring behind a typical control panel is shown in Figure 13.22.

Trash Racks and Grit Chambers

Stormwater often conveys considerable amounts of trash and debris, which, if not removed from the flow, could damage or clog the pumps. *Trash racks* to filter out debris can be manufactured from steel bars. Bar spacing should be about 1.5 inches, but may depend on the maximum size of a sphere that can be passed by the pumps. Trash racks should be installed in modular units to facilitate easy removal and should be inclined from the vertical to facilitate self-cleaning as water levels rise and fall. Clear-opening areas of trash racks should be large enough to not cause excessive energy-head losses in the flow. In some cases, screening may be accomplished at inlets to the stormwater system, negating the need for additional screens at the pump intakes. A typical trash rack for a large pump station is shown in Figure 13.23.

Figure 13.22
Wiring inside a typical control panel

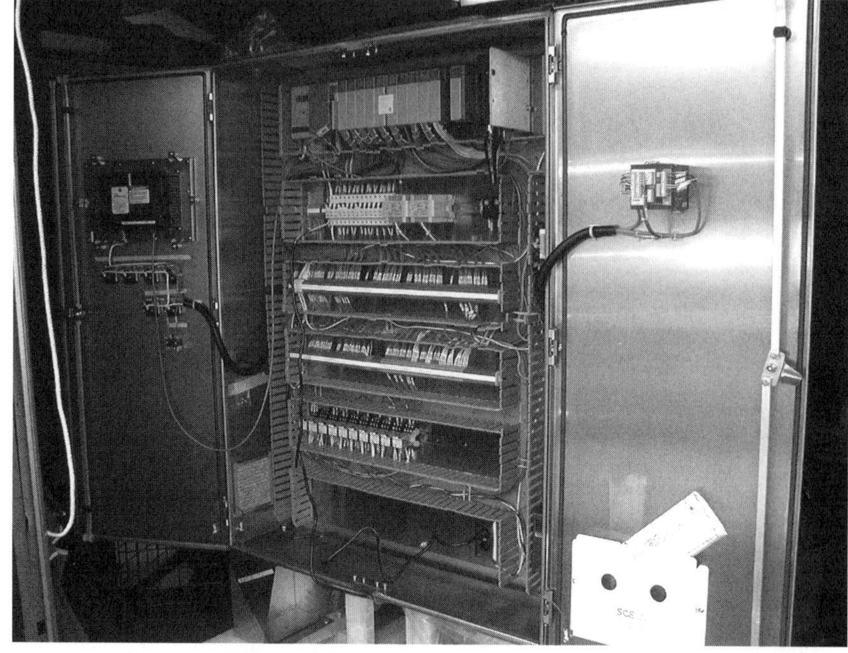

Credit: Jim Brozena

Figure 13.23
Trash rack for pump station

In areas where roadways are routinely sanded during winter storms or in other locations where stormwater may contain significant accumulations of sediment, it may be desirable to provide a *grit chamber* where solids can settle out of the flow. The grit chamber may be designed to be either within or upstream of the wet-well, depending on its ease of accessibility for required sediment removal. An efficient means for sediment removal should be provided.

Monitoring Devices

Power outages and mechanical problems are routinely encountered at pumping facilities. Warning lights and remote alarms can be employed to notify appropriate maintenance personnel, who can then take necessary actions. Alarms are provided for power supplies, pump operation, high water levels, presence of explosive gases, and/or station security.

Safety and Ventilation Requirements

A safety engineer should perform a thorough review of a pumping-station design. Safety shrouds should be provided around moving parts, and adequate space must be provided for maintenance personnel to access motors, pumps, and electrical panels. Ladders and stairwells should be provided, as necessary, and lighting, heating, and ventilation should enable personnel to perform their duties in a safe and efficient manner. In some cases, a pumping station may be classified as a confined space for which special access and safety requirements are defined by code.

Wet- and dry-wells should always be provided with ventilation equipment such as exhaust fans to ensure a safe working environment for maintenance personnel. Exhaust fans should be switched on at least 10 minutes prior to entry to a wet- or dry-well. In some cases, it may be desirable to also provide automatic switching equipment for fans. For example, a sensor could be provided to start the fans if explosive gases were detected.

Doorways, Hatches, and Monorails

Because of the need to periodically remove pumps, motors, or other equipment for maintenance or replacement, all doorways and hatches in a pumping station should be large enough and appropriately positioned to enable these operations.

In some cases, especially in small stations, portable hoisting units can be used for equipment removal. Often, however, it is desirable to provide monorails and lifting tackle as apart of the station infrastructure.

13.9 CHAPTER SUMMARY

In some locations, it may be necessary to use a pump station to remove stormwater. For instance, stormwater may have to be pumped through a levee or floodwall. Pump types used in stormwater applications include axial-, radial-, and mixed-flow pumps, and positive-displacement pumps such as screw pumps.

Initial pump selection is typically based on the specific speed characteristic. High specific speeds correspond to pumps delivering high discharges at low heads, and low specific speeds correspond to pumps that produce lower discharges at high heads. Axial-flow pumps tend to have high specific speeds, and radial-flow pumps tend to have lower ones.

Pump stations can have either a wet-well or wet/dry-well configuration. With wet-well installations, the pump is submerged in a pool of water in the sump. In a wet/dry-well installation, the pump is not submerged in the wet well; rather, it is located in an adjacent dry well and is connected to the wet well with piping. Wet/dry-well installations are usually associated with larger pump stations.

Pump performance curves are used to determine how the pump will operate under a range of conditions. A pump characteristic curve shows the relationship between flow rate and pump head. Other curves show pump efficiency at various operating points and the minimum NPSH required by the pump to prevent cavitation. A system curve that reflects the head loss in the pumping system for a range of discharges can be plotted with the pump characteristic curve; the intersection of these curves is the operating point of the pump for the assumed upstream and downstream boundary conditions and flow rate.

Because of the possibility of pump failure and the wide range of flow rates that a pump station must be able to handle, multiple pumps may be installed. An equivalent pump performance curve for both pumps operating at the same time should be developed to see how the system will perform under this condtion.

In addition to the pumps themselves, stormwater pumping systems include system piping and several types of valves and fittings. The engineer must ensure that the specified pipes have sufficient strength to resist the internal and external forces that they will be subjected to. Typical pumping systems include shutoff valves, reducers, flap gates, air-release valves, gauges, cleanouts, and pump on/off controls.

Several methods exist for estimating the amount of storage needed in the wet-well. A simple approach for preliminary volume estimation is Baumgardner's method, in which the required volume is taken as the area under the inflow hydrograph in excess of the average pumping rate.

Prior to development of a trial pump station design, the inflow hydrograph for one or more design storms must be developed. Methods such as those presented in Chapter 5 may be used for this purpose. Other elements of the trial design are pipe sizing and layout, pump selection, wet-well volume, and a pump control scenario. Once a trial design has been developed, it is evaluated by considering stage-discharge and stage-surface area (or volume) relationships and applying a routing technique to solve the continuity equation. The principles involved are similar to those used in routing detention ponds, which were presented in Chapter 12.

Additional considerations in developing the final design of the pumping station station architecture and power requirements, facilities for removing grit and debris, monitoring devices to alert personnel to malfunctions, safety measures, adequate ventilation, and pump accessibility.

REFERENCES

American Association of State Highway and Transportation Officials (AASHTO). 1991. *Model Drainage Manual.* Washington, D.C.: AASHTO.

Baumgardner, R. H. 1981. "Estimating Required Storage to Reduce Peak Pumping Rates at Storm Water Pumping Stations." Washington, D.C.: Federal Highway Administration Workshop Paper, Office of Engineering.

Brown, S. A., S. M. Stein, and J. C. Warner. 1996. *Urban Drainage Design Manual.* Hydraulic Engineering Circular No. 22, FHWA-SA-96-078. Washington, D.C.: U.S. Department of Transportation.

Chapra, S. C. and R. P. Canale. 1988. *Numerical Methods for Engineers,* 2d ed. New York: McGraw-Hill.

Chow, V. T., D. R. Maidment, and L. W. Mays. 1988. *Applied Hydrology.* New York: McGraw-Hill.

Federal Highway Administration (FHWA). 1982. *Highway Storm Water Pumping Stations,* Vol. 1 and Vol. 2. FHWA–IP-82-17. Washington, D.C.: U.S. Department of Transportation.

Hydraulic Institute. 1998. *Centrifugal/Vertical Pump Intake Design.* Parsippany, New Jersey: Hydraulic Institute.

Karassik, I. J., J. P. Messina, P. Cooper, and C. C. Heald. 2000. *Pump Handbook,* 3d ed. New York: McGraw-Hill.

Mays, L. W. 2001. *Water Resources Engineering.* New York: John Wiley and Sons.

Sanks, R. L. 1998. *Pumping Station Design,* 2d ed. Boston: Butterworth-Heinemann.

Seely, F. B., and J. O. Smith. 1959. *Advanced Mechanics of Materials.* 2nd ed. New York: John Wiley and Sons.

US Army Corps of Engineers. 1995. *General Principles of Pumping Station Design and Layout.* EM 1110-2-3102. Washington, D.C.: USACE.

US Army Corps of Engineers. 1989. *Structural and Architectural Design of Pumping Stations.* EM 1110-2-3104. Washington, D.C.: USACE.

US Army Corps of Engineers. 1999. *Mechanical and Electrical Design of Pumping Stations.* EM 1110-2-3105. Washington, D.C.: USACE.

PROBLEMS

13.1 A stormwater storage system consists of a 158-m–long, 1220 mm pipe and a 6.1-m–diameter wet well, as shown in the following figure. The storage volumes of the two components as a function of elevation are given in the table. Construct a stage-storage curve showing the storage in the system as a function of the elevation.

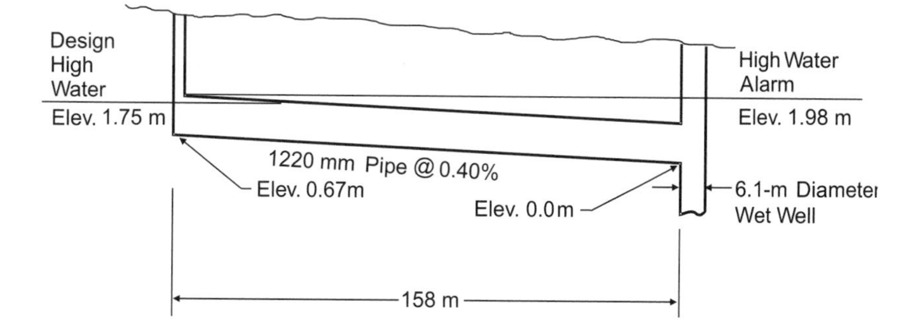

Elevation (m)	Volume (m³)	
	Pipe	Tank
0.00	0.0	0.0
0.15	1.3	4.9
0.30	7.1	9.8
0.46	19.0	14.7
0.61	37.7	19.6
0.76	62.6	24.5
0.91	90.2	29.4
1.07	118.0	34.3
1.22	143.5	39.2
1.37	163.4	44.1
1.52	176.3	49.0
1.68	183.0	53.9
1.83	184.9	58.8
1.98	184.9	63.7
2.13	184.9	68.6

13.2 A pump at a small stormwater pump station is designed to pump from a sump with a water level that can vary from 510 to 515 ft to the other side of a levee with water level during a flood that can vary from 518 ft to 525 ft. The design flow rate is approximately 2,500 gpm (5.58 cfs). The suction piping has a diameter of 12 in. and is 20 ft long. The discharge piping also has a diameter of 12 in. and a length of 150 ft. All pipes have a relative roughness of $\varepsilon = 0.0015$. The sum of all minor loss coefficients in the station is $K = 15$.

Compute and plot the system head curves for the minimum (3 ft) and maximum (15 ft) on the pump curves in Figure 13.11. Select the impeller diameter that will yield close to 2,500 gpm for the maximum lift.

What are the flow, head, horsepower, and NPSH required for the maximum and minimum lift? Record your answers in the table provided.

Property	Max Lift	Min Lift
Flow, gpm		
Head, ft		
NPSH required, ft		
Power, hp		

13.3 A pump delivers runoff from a detention pond to a spillway through an 800-ft–long, 24-in.–diameter concrete pipe. The performance curve for the pump is shown. The pump starts when the water surface elevation in the detention pond is 70 ft and continues to operate until the pond is empty with a bottom elevation of 60 ft. What power must be supplied to the pump when the water surface elevation at the spillway is 100 ft and the elevation in the detention pond is 60 ft? Use a relative roughness $\varepsilon = 0.005$ ft.

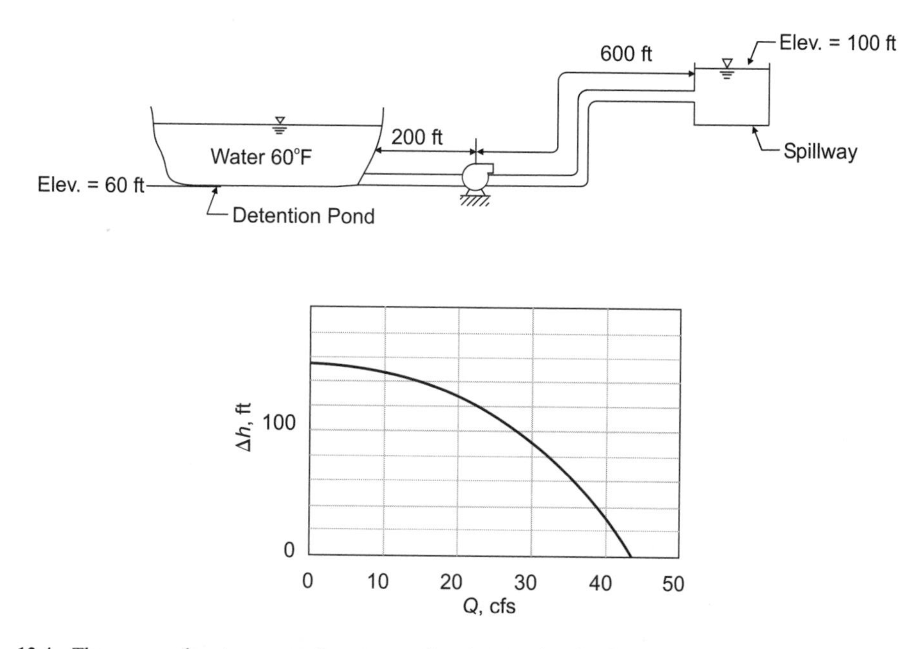

13.4 The pump and system curves for a proposed station are given by the coordinates in the table. What is the operating point for a single pump in this system? For two identical pumps in parallel?

Q (L/s)	Pump Head (m)	System Head (m)
0.0	40.2	15.2
31.5	39.0	20.7
63.1	33.8	37.8
94.6	21.3	70.1

13.5 Figure 13.18 illustrates two types of pump control schemes described in this chapter. For this problem, four pumps are located at the bottom of a 20-by-20–ft wet well. The bottom of the wet well is at an elevation of 0.0 and, for the distances shown in Figure 13.18, $h_1 = h_2 = h_3 = h_4 = 6$ ft. Assume that the pumps are identical and that each has its own piping system. A constant flow of 10 cfs enters the wet well for a period of 60 minutes.

If the discharge produced by any single pump depends on the wet-well water surface elevation h as $Q = 400 + 25h$ (where Q is in gpm and h is in ft), compute and plot stage-discharge curves for each of the two control strategies. In other words, graph the discharge produced by the combination of four pumps as a function of the water surface elevation in the wet well. Be careful to note that the relationships may exhibit hysteresis.

Stormwater regulations are concerned with protecting both natural and constructed environments from damage by flooding and pollution.

CHAPTER

14

Regulatory and Environmental Issues

Stormwater conveyance systems must be designed to meet minimum requirements specified by local, regional, state, or national review agencies. These requirements relate to both stormwater quality and quantity. This chapter provides a broad overview of regulatory and environmental issues, but cannot accommodate all of the differences from one political jurisdiction to another.

Section 14.1 provides an international overview of water resources guidelines and regulations. World Bank water resources policies for international development and aid, policies of the European Union, and water-related environmental regulation in Canada are discussed. A summary of the U.S. federal, state, and local laws and regulations applicable to urban stormwater management is presented in Section 14.2.

14.1 INTERNATIONAL OVERVIEW OF REGULATORY GUIDELINES AND POLICIES

Stormwater management practices throughout the world vary widely. Industrialized countries typically have policies concerning flood control and water quality in place, but many developing countries do not. Further, the ways in which stormwater can be effectively managed differ with the level of urbanization, local climate and topography, and a host of other factors.

Because of the enormous range of differences in stormwater management needs and practices around the globe, it is difficult to present even a summary in the space of a chapter. The goals of this section are to provide the reader with an overview of international stormwater-related policies, some insight into stormwater management in developing countries, illustrative examples of stormwater practices in the European Union, and an overview of stormwater regulations in Canada.

Milestones in International Water Resources Policies

The major milestones in the development of current international water resources policies are summarized in Table 14.1 and briefly reviewed here. The 1977 United Nations Mar del Plata Conference (United Nations, 1977) was the first to identify the need for Integrated Water Resources Management (IWRM). The assessments presented at this conference showed that the lack of adequate water supply and sanitation caused many more premature deaths and much more illness than previously believed. Consequently, a call was made to designate the 1980s as the "International Drinking Water Supply and Sanitation Decade." Furthermore, it was concluded that lack of international coordination was a major reason for the poor state of the world's water affairs. The conference report is known as the Mar del Plata Action Plan (United Nations, 1977).

Table 14.1 Milestones in international water resources policies

Year	Agreement	Milestone
1977	Mar del Plata Action Plan	• Identified need for integrated water resources management
1987	Brundtland Commission – "Our Common Future"	• Shifted focus to sustainable development
1992	Agenda 21	• Promotion of a dynamic, interactive, iterative, and multi-sectoral approach to water resources management • Projects and programs that are both economically efficient and socially appropriate
1996	Brown Agenda	• Defined problems in urban areas to include air, soil, and water pollution, and noise and traffic congestion
1997	Local Agenda 21	• Calls for dedicated local actions to address environmental problems
1997	Charter of European Cities and Towns Towards Sustainability (Aalborg Charter)	• Consensus among cities and towns on sustainability issues

In 1987, the World Commission on Environment and Development (popularly known as the Brundtland Commission) presented its report Our Common Future (WCED, 1987). The creation and general acceptance of this report was a watershed event in that it marked the first time the world community agreed on the need to reverse current development trends and instead focus on sustainable development. The report called for an international conference on environment and development—a conference that eventually took place in Rio de Janeiro in 1992. The major outcome of the 1992 United Nations Conference on Environment and Development was *Agenda 21* (United Nations, 1992), a framework for the implementation of sustainable development throughout the world during the twenty-first century.

Agenda 21 and Water Management

Agenda 21 states that the general objective of water management is to make certain that adequate supplies of good-quality water are maintained for the entire human population, while preserving the hydrological, biological, and chemical functions of ecosystems; adapting human activities to remain within the capacity limits of the natural environment; and combating vectors of water-related diseases. In order to fully utilize the earth's limited water resources and to safeguard those resources against pollution, innovative new technologies, as well as existing indigenous technologies, are needed.

Among the principal objectives of Agenda 21 (United Nations, 1992) are:

- The promotion of a dynamic, interactive, iterative, and multisectoral approach to water resources management that integrates technological, socioeconomic, environmental, and human health conditions; and
- The design, implementation and evaluation of projects and programs that are both economically efficient and socially appropriate within clearly defined strategies, and that calls upon the full participation of indigenous people and local communities (including women and children) in water management policy-making and decision-making.

The promotion of integrated water, sanitation, drainage, and solid waste management (environmental infrastructure) is cited as central to the efforts for sustainable urban development. Agenda 21 also stresses the importance of adequate environmental and social impact assessments. It furthermore advocates the application of precautionary measures in regard to water resources management with a focus on waste minimization and prevention.

Chapter 18 of Agenda 21 lays down the fundamental principles for sustainable water resources management. The chapter builds on the Mar del Plata Action Plan, but adds relevant lessons from the Water Supply and Sanitation Decade, which suffered from poorly targeted intervention projects and a lack of coordination and integration. It has been widely accepted and forms the basis of guidelines and policies for the UN family of agencies and for most international development aid agencies. Many countries are also in the process of adopting these principles in their environmental and water resources legislation.

Local Agenda 21. In 1997, a worldwide *Local Agenda 21* (United Nations, 1997) was formulated. This document makes a plea for dedicated local actions, in both industrialized and developing countries, that combine the reduction of environmental degradation with improvements in local socioeconomic conditions. It was adopted by the UN General Assembly Special Session on the Environment on June 23, 1997.

The European version of Local Agenda 21 is called the Charter of European Cities and Towns Towards Sustainability (ECSCT, 1994), or Aalborg Charter. The Aalborg Charter aimed at building consensus among its members' cities and towns on 14 key sustainability issues. Many local authorities in Europe are now developing Local Agenda 21s, and ponds for stormwater treatment and daylighting of stormwater conveyance systems are frequent components of these plans.

Additional International Guidelines

The Brown Agenda. The 1996 United Nations Conference on Human Settlements, in Istanbul, commonly referred to as Habitat II, stated that health problems related to adverse environmental conditions and ineffective and inadequate health services exact a heavy toll on the quality of life and the overall contribution to society of millions of people (United Nations, 1996). Particular adverse environmental conditions identified at the conference were lack of access to safe water and sanitation, inadequate waste management, poor drainage, air pollution, and exposure to excessive noise levels. The unique environmental problems of urban areas identified in the Brown Agenda include air, soil, water, and noise pollution, and traffic congestion. These have a direct and immediate impact on human health and safety, especially for the poor, and on business productivity. Habitat II concluded that implementing the Brown Agenda (improving water supply, drainage, sanitation, solid waste management, etc.) remains a priority for most cities in developing countries.

Policies and Guidelines in International Development Aid. The World Bank (WB) is the largest funding institution for international development projects, and its policies and guidelines are adopted by many other international development agencies. WB policies are widely criticized by NGOs and others, and since the 1990s, public pressure brought to bear against the WB has led it to adopt and/or strengthen a number of environmental policies that govern its lending. The WB is in the process of converting its old Operational Directives into Operational Policies, commonly known as the 4.0 series.

The Operational Policies differentiate between, and outline requirements for, sectorial approaches and integrated approaches to development projects. In the sectorial approach to a water project, for example, only water-resource-related issues need be considered. The integrated approach requires an environmental assessment (EA) to review a project's effect on the natural environment, human health, and societal/cultural establishments. WB policies require an EA to be conducted by independent experts for those projects likely to have potentially significant adverse environmental, human and/or societal/cultural impacts.

The WB developed a water resources policy paper entitled Water Resources Management (World Bank, 1993). This paper promotes a major shift in water resources management, from a fully segregated approach to a comprehensive and integrated approach. It specifically calls for linking land-use management to sustainable water-resource management and for addressing quantity and quality concerns in an integrated manner.

The current WB philosophy is that the world cannot afford to view the environment as a technical issue to be addressed independently from overall municipal and industrial strategic decision-making. This thinking is best summed up as "environmental management, not just pollution control" (World Bank, 1998). This means that WB policies are meant to affect, among other things, inter alia stormwater management. However, it is debatable whether the WB has been able to implement this new approach. Consequently, in February 2003, the WB's executive board approved and released The New Water Resources Strategy (World Bank, 2003). This strategy is strongly based on the 1993 Water Resources Management policy paper and intends to vigorously implement its policies.

Much of the WB's urban sanitation work is done in cooperation with the United Nations Development Programme's (UNDP) Water and Sanitation Program (WSP). The WSP promotes the Strategic Sanitation Approach (Wright, 1997), a key principle of which is that development should be demand-led. This principle provides that projects should develop those types of systems that beneficiaries are asking for and are prepared to pay for. This approach demands extensive and informed dialogue between the beneficiaries and the project developers in order to achieve success.

The UNDP states that some of the most important results that could presently be produced in the field of international development involve water. The organization is in the process of shifting the direction of its efforts to be in line with Agenda 21. To this end, the UNDP has produced a Strategy Framework Document (UNDP, 1998).

The United Nations Environment Programme (UNEP) has found that technology and technology selection for improving water quality, although very important, cannot be detached from other equally important factors for successful project implementation. A framework is required that considers integrated waste management, transsectorial issues, and the effect on all stakeholders, while granting each a representative voice. Simply solving the problem of wastewater drainage without taking into account solid waste and stormwater drainage will not provide improvements sufficient to protect either public health or the environment. To promote sustainable urban development, the International Environmental Technology Centre (IETC) of the UNEP is producing an *International Source Book on Environmentally Sound Technologies for Wastewater and Stormwater Management* (IETC, 2002), to be published by the International Water Association. The IETC has also developed a Framework for Wastewater and Stormwater Management, which can be found on UNEP's web site (www.unep.org).

The European Commission Directorate-General for Development is another major source of development aid for projects that implement the Agenda 21 principles. The protection and enhancement of natural resources are recognized as important dimensions to the development support provided by the European Union (EU). *Towards Sustainable Water Resources Management* (European Commission, 1998) states that attention must be paid to the improvement of urban drainage, the prevention of floods, and the extension of capacity to withstand periodic droughts.

These directives and documents complement the guiding principles for international development projects. As such they represent a conglomeration of "soft themes and policies" coupled to "hard technical realities" in developing water resources. This is a direct consequence of the complexity and subtle nuances in addressing combined human and environmental issues rather than just technical implementation issues. Naturally, the technical implementation of water-related development projects is recognized as important, and it is the central necessity of integrated management systems to be able to interlink the "soft themes" with the "hard technical realities."

The Urban Stormwater Initiative (Environment Australia, 2002) is an example of a combined "hard" and "soft" effort aimed at enhancing water quality in the waterways of major coastal cities by improving stormwater management. It promotes the construction of infrastructure that is innovative and exemplifies best-practice techniques in managing the quality of urban stormwater. Particular attention is given to projects that turn a problem into a resource, such as reduction of water shortages through the capture, treatment, and reuse of urban stormwater. Strong industry source control, community education, and water-sensitive urban design are also an integral part to this effort.

Effective interagency and intersectorial planning is essential. Therefore, intersectorial policy and planning weaknesses need to be identified, and water should form a part of overall urban planning. Water resources management in the twenty-first century will become increasingly influenced by the principles outlined in Agenda 21, especially in developing countries. The complete integration of technical realities with the limitations on the environmental and human capacity for change in short periods of time will become a central aspect of water resource management and development.

Drainage Management in Developing Countries

In developing countries, urban drainage channels often serve as trash dumps during dry weather conditions. This has two significant consequences. First, when precipitation events occur, there is a great risk of channels clogging, leading to flooding and the dispersal of human refuse. Second, the runoff entrains the garbage, contributing to pollution in receiving waters. In addition, many storm conveyance systems function as combined sewers transporting untreated sanitary wastewater. These factors have severe effects on public health that can be prevented through urban drainage infrastructure improvements. In order to develop efficiently functioning urban drainage systems, both an urban development plan and a master drainage plan are necessary. Most cities in developing countries grow quickly and with little coherent planning. A lack of adequate urban land-use planning and enforcement, coupled with very limited available funds for the provision of basic water services, make the development of efficient drainage systems a difficult task. To overcome these difficulties and successfully implement improvements, the technical tools and management principles developed in industrialized countries must be adapted to a developing country's present conditions and needs. The following examples from Thailand and India illustrate these points.

Thailand. A 2002 WB report on environmental degradation in Thailand (World Bank, 2002) illustrates commonly encountered legal and institutional problems associated with improving environmental conditions in developing countries. Although Thailand has made significant progress in establishing environmental laws and institutions, polluters have had little incentive to control and abate pollution. Implementation of regulatory measures has suffered from weak monitoring and enforcement. The report notes that Thailand's existing command and control measures have not been particularly effective. Implementation of existing regulations has been further hampered by duplication and overlap of responsibilities between the numerous agencies and departments involved in pollution control.

This situation makes a joint approach to urban drainage and solid waste management both difficult and vitally important. Because existing drains are frequently open channels misused for garbage disposal, the regulatory, enforcement, and technical approaches toward both water resources management and solid waste management must be fully integrated in order to be effective.

Indore, India. The World Resources 1996-97 Report (World Resources Institute, 1997) underlined the importance of a community's perception of health risks, such as polluted water, inadequate drainage, or lack of garbage collection, to the successful development of an efficient water resources infrastructure. The report cites the experience and behavior of the inhabitants of Indore's slums as an example. The majority of the slums are located along the banks of the many rivers and canals that crisscross the city and on the city's floodplains. Monsoon waters regularly flood these communities. Although the residents are acutely aware of the fact that the diseases they continuously suffer are tied to the flooding, social and economic factors prevent them from leaving these high-risk areas. City planners designed a project to improve health conditions in the slums by reducing flooding in these areas and the city proceeded to construct closed drainage channels and upgrade sanitary conditions. Despite these improvements, many community members now perceive flooding problems to be worse than before.

An important source of dissatisfaction with the new drainage system is the reduced ability of the community to apply its own risk-reduction strategies. Because the drains are closed, residents can no longer predict the severity of an approaching flood. With the open drains, community members would adjust their flood response behavior according to both rainfall intensity and the volume of water in the open drains. Further, unlike the wider open channels, the closed channels and their inlets are easily blocked by plastic bags and trash.

Surat, India. Surat, a city located in the Indian state of Gujarat, suffered from an epidemic outbreak of plague (a disease caused by the bacterium Yersinai pestis) in 1994. The faulty drainage system could not cope with an enduring rain period, and the result was large-scale waterlogging. Hundreds of cattle and other animals died because of the floods, and the municipal authorities were not prompt enough in cleaning the city, which led to massive sanitation problems.

Despite being one of the richest civic bodies in the country, the Surat Municipal Corporation had failed to provide basic infrastructure to a majority of the city's population. Over 70 percent lacked sewers, and in the northern part of the city there were no sewers at all. The incident led to a complete overhaul of the city administration, and a massive cleaning operation involving government, community, and private organizations was undertaken. Two years later, the concerted efforts of the partners had successfully transformed the city from one of the filthiest to one of the cleanest in India.

"I adore the beauty and tranquillity of these raw-sewage days."

Regulation of International Watercourses

The 1966 Helsinki Rules on the Uses of the Waters of International Rivers (SOURCE) have until recently been the primary legal basis for transboundary issues related to international rivers. The Helsinki Rules encompass two principles: (1) the prohibition of appreciable harm to any party through the deprivation of water rights, pollution, or other means; and (2) the right of each basin state along an international waterway to a reasonable and equitable utilization of the international waterway.

In 1974, the Helsinki Convention on the Protection of the Marine Environment of the Baltic Sea was signed by all of the then seven Baltic Sea States. It has been in force since 1980 and is the first international instrument to deal with all sources of pollution to the Baltic Sea. The major aim is to minimize the load of pollution entering the Baltic Sea from land-based sources.

The UN Economic Commission for Europe established the Convention on the Protection and Use of Transboundary Watercourses and International Lakes (SOURCE), which came into force in Europe in 1996. The convention is based on the principles of Agenda 21 and seeks to protect the quality and supply of fresh water throughout Europe.

In 1997, the UN General Assembly adopted the Convention on the Law of the Non-navigational Uses of International Watercourses (United Nations, 1997). This convention drew from the earlier Helsinki Rules and the Convention on the Protection and Use of Transboundary Watercourses and International Lakes. In particular, it defines the process by which watercourse states come to an agreement, the principle of equitable and reasonable utilization and participation, and the obligation not to cause significant harm. Numerous regional agreements for specific river basins, lakes, or inland seas, have developed from or moved under the auspices of the Convention of the Non-Navigational Uses of International Watercourses. Among these are regional agreements for the Danube, Indus, Niger, Rhine, and Zambezi rivers; the Baltic Sea; and Lake Victoria.

In 1816, one of Europe's oldest treaties established the Rhine as a navigable waterway (Weber, 2000). The necessity for transboundary management of the river Rhine was recognized for shipping as early as the end of the eighteenth century. The transboundary pollution aspect was given attention in 1932, when the Dutch government protested against emissions of residual salt into the French part of the river. After World War II, the pollution of the river increased, and, in 1950, the International Rhine Commission started to study wastewater and water quality problems. The international cooperation to abate pollution was strengthened by the draft resolution International Commission for the Protection of the Rhine against Pollution, released in Berne in 1963. This was followed by the European Council Decision to incorporate the Convention for the Protection of the Rhine Against Chemical Pollution (77/586/EEC) into the Berne resolution of 1963. In order to accelerate ecological improvements in the Rhine, ministers from riparian countries (countries that abut a natural stream, river, or lake) decided in 1986 to establish the Rhine Action Programme, with the following aim:

> The ecosystem of the Rhine must become a suitable habitat to allow the return to this great European river of the higher species which were once present here and have since disappeared (such as salmon). (ICPR, 1992)

The International Commission for the Protection of the Danube River was modeled after the Rhine Commission. Its task is to implement the Danube River Protection Convention (SOURCE), which was signed in 1994 and came into force in 1998. Currently, action is directed toward

- Improvement of the ecological and chemical status of the river;
- Prevention of accidental pollution events; and
- Minimization of the effects of floods.

One of the major areas of effort is in the reduction of pollution from municipal discharges of both stormwater and sanitary wastewater.

Additional international agreements of importance to water resources management include those on climate change, biological diversity, wetlands, and desertification. Climate change is related to stormwater management as climate change is expected to effect precipitation patterns and water levels. This issue was addressed at the Conference on Environment and Development in Rio de Janeiro (UNCED, 1992). The Convention on Biological Diversity was also signed at UNCED, Rio de Janeiro in 1992. The Ramsar Convention, which was signed in Ramsar, Iran in 1971, defines the concept of wetland and calls for their protection (Carp, 1971). It is relevant to stormwater management because poor stormwater management is a threat to biological diversity in receiving water bodies. The United Nations Conference on Drylands was held in Nairobi, Kenya in 1977. Stormwater management as it affects soil erosion was identified as an important process in desertification. In June 1994, the United Nations Convention to Combat Desertification was adopted in Paris, France (United Nations, 1992).

Water and Environment in the European Union

The EU aims to integrate environmental aspects into all development activities. This integration became a legal obligation under the Treaty on European Union (Maastricht Treaty).

When the European Community was created in 1956, the water management systems in different member states could be classified broadly into three categories:

- Decentralized local action
- Coordinated regional action
- Uniform national standards systems

Due to the prevailing sectorial approach, most countries have numerous laws related to stormwater management. For example, a country might have an environmental act, a water act, a water utility act, a town and country planning act, and/or building acts and regulations.

The rules and regulations at the national and regional levels tend to be legally binding, while the local rules are more subject to political decision-making. In most countries, the environmental agency or equivalent ministry provides guidelines on stormwater management. In addition, local stormwater and wastewater utilities have their own rules concerning pre-treatment of water that is discharged into their networks. In Germany and Sweden, some water utilities have developed separate stormwater fees to encourage developers to install on-site facilities for stormwater treatment.

Environmental management standards like the European Union's Eco-Management and Audit Scheme (EMAS) (Hillary, 1994) and the International Organization for Standardization's ISO 14000 have become fashionable among water utilities in several European countries. Utilities that have adopted these standards are becoming increasingly wary about what they allow into their networks. Information on EMAS, ISO 14000, and other environmental management standards can be found on the International Network for Environmental Management's web site (www.inem.org).

For a long time, stormwater management focused on safeguarding land or property from damages resulting from excessive moisture and flooding. Since the mid-nineteenth century, the dominant idea has been that stormwater and other wastewater should be collected in urban areas and disposed of outside of the urban environment as quickly and efficiently as possible. However, stormwater-induced overflows from combined sewer systems have since become an issue, as wastewater treatment plants typically have a design capacity to treat only three to four times the dry-weather wastewater flow. Therefore, the dominant European strategy has turned to developing separate drainage systems for stormwater that discharge the stormwater directly to a surface water body. However, both the quality and the quantity of unmitigated stormwater discharges can have significant environmental impacts that must be considered.

Among the problems encountered in the development of separate drainage systems for stormwater are the impracticality of installing them in old, crowded city centers and ensuring the complete disassociation of all sanitary sewer connections from the old conduit, which generally becomes the stormwater conduit in the separate system. Only a few missed sanitary sewer connections could make stormwater runoff hygienically unsafe.

Although stormwater detention ponds are popular in the United States and Canada, stormwater infiltration/percolation and on-site detention/retention methods have generally been preferred in Europe. For instance, Switzerland's federal water pollution control law now requires that stormwater be infiltrated whenever possible. A permit must be obtained for cases in which stormwater is discharged into surface water.

In many German states, infiltration of stormwater runoff has become the state of the art for sustainable urban storm drainage. In Northrhine-Westphalia, for example, legal regulation demands that newly developed properties infiltrate rainwater into the groundwater or discharge it to receiving waters. The technical and economic advantages are the use of smaller diameter storm sewers and more efficient operation of wastewater treatment plants because the inflow during wet-weather conditions is uniform. However, local geology and housing density have to be duly considered. Impervious soils or high groundwater tables make infiltration inappropriate (Coldewey et al., 2001). In densely populated areas, the capability for full disconnection from the sanitary sewer system may be as low as 2 to 3 percent of all stormwater connections, while the capability may be 50 percent at larger commercial sites and housing associations.

More recently, a significant European interest in vegetated ponds and open conveyance systems for stormwater treatment has emerged. The aesthetic and biodiversity advantages especially have promoted this interest. Vegetated roofs are also seen to have beneficial impacts on stormwater runoff quantity and quality.

Each of these options has significant development potential, but the major challenge is combining them with the existing infrastructure in a creative and productive way so that development moves in a sustainable direction. As the practical understanding of integrated urban drainage is limited, a learn-by-doing approach is recommended in which small, innovative interventions are followed by evaluation and feedback.

The European Union Regulatory Approach. Until very recently, the EU has applied an approach whereby standards are uniformly applied in a nation. This approach has caused concerns about high costs, and questions have been raised about the efficiency of the overall system in meeting water quality goals.

The uniform standards approach is, in fact, a departure from the river basin approach that was widely used in national systems in Western Europe. Germany, France, Spain, and the United Kingdom have river basin authorities of one kind or another, and those systems are now changing to come into compliance with EU requirements. Nevertheless, they still have some flexibility within their own areas of authority. Such flexibility allows for the setting of priorities and realistic targets consistent with available resources.

The 1976 Dangerous Substances Directive (76/464/EEC) was one of the first water-related directives to be adopted, and it had the ambitious objective of regulating potential aquatic pollution for thousands of chemicals. Among other directives of importance to stormwater management are the Bathing Water Directive (76/160/EEC), the Fish Water Directive (78/659/EEC), and the Shellfish Water Directive (79/923/EEC).

In December 2000, the EU implemented new regulations for water management through the EU Water Framework Directive (2000/60/EC). The major innovations were a call for a river basin approach encompassing both point and nonpoint pollution, and replacement of the uniform standards approach with a combined approach. Under this combined approach, pollution is controlled at the source through emission limits, and environmental quality standards are applied to the chemical and physical properties and the ecologic status of receiving waters.

The directive also requires holistic and integrated management of groundwater, surface water, and coastal waters. River basins are lumped into districts, for which there should be a "competent authority" to provide regulation. These authorities are required to be operational by December 2003.

The EU will decide the minimum standards for good water quality. These standards will be based on eco-regions and water categories: rivers, lakes, transitional waters, coastal waters, artificial surface water bodies, or heavily modified surface water bodies. The standards also state that, in addition to existing EU directives, other international conventions should be honored.

The competent authorities (defined by member states) are to classify the status of their waters and to report this information to the EU. For waters not meeting the requirements for "good" status, management plans for achieving "good" status within 15 years must be developed.

The directive calls for two kinds of monitoring: surveillance monitoring to assess water status and operational monitoring to monitor the impacts of rehabilitation measures.

The major objectives of the directive are to

- Prevent further deterioration of, protect, and enhance the status of aquatic ecosystems and, with regard to their water needs, terrestrial ecosystems and wetlands directly dependent on the aquatic ecosystems.
- Promote sustainable water-use based on long-term protection of available resources.
- Achieve enhanced protection and improvement of the aquatic environment.
- Ensure a progressive reduction of groundwater pollution.
- Contribute to mitigating the effects of floods and droughts.

European Experiences. Stormwater management in Europe is affected by factors such as local climate, regulations, and institutional settings. In this section, the stormwater management practices of selected countries are reviewed.

England and Wales. In 1974, the British Parliament decided to transfer the responsibility for water and sewage services from local authorities to regional water authorities. These regional authorities were responsible for the supply of water as well as sanitary and stormwater drainage and treatment. The boundaries of these authorities were set in accordance with the watershed areas. This approach formed a nationalized industry, organizationally and constitutionally.

The regional water authorities were also responsible for environmental protection and regulated water use in agriculture and industry. However, in doing so the regional water authorities became both service providers and regulatory bodies.

In 1989 the regional water authorities were privatized. The regulatory authority remained in public hands through a complicated departmental system. The principal attributes of this system are as follows:

- The setting of policies and standards is the responsibility of the politically elected government, and these policies and standards are executed through the Department of Environment, Transport and Regions.
- Protection of the water environment is the responsibility of the Environment Agency (formerly the National Rivers Authority), which took over the regulatory roles of the regional water authorities.

Economic regulation is carried out by the Director General of Water Services, who is supported by the Office of Water Services (OFWAT).

On a local level, many municipalities have adopted Local Agenda 21s to promote sustainable urban development. A Local Agenda 21 for the Thames River Basin, Thames 21, has been developed for catchment and land-use planning.

The first edition of the Urban Pollution Management Manual (UPM Manual) (Foundation for Water Research, 1994) was published in November 1994 following a major research and development program sponsored by the entire UK water industry. Many of the planning concepts and enabling tools in the UPM Manual were substantially new at that time and addressed issues that were, and continue to be, of great importance to the industry. The intervening period has seen widespread adoption and application of the manual's procedures throughout the UK, particularly in areas with acute combined sewer overflow problems.

The UPM Manual points out the need to consider the wastewater system, which consists of the drainage system, the treatment plant, and the receiving water, as a single entity. A change to one part of the wastewater system will have implications for other parts of the system that must be taken into account in the planning process.

A common misconception that has developed is that an Urban Pollution Management (UPM) study means costly, detailed simulation modeling. In reality, many UPM studies can be undertaken using relatively simple tools. The more complex and costly tools are only employed when the economic case for using them has been proven.

Some three years after publication of the original UPM Manual, the sponsors agreed that it was appropriate to update the original guidance in light of the experience gained over the intervening period. Thus, the second edition of the UPM Manual (UPM 2) was published (Foundation for Water Research, 1998).

UPM 2 identifies two approaches for the protection of river-based aquatic life from wet-weather episodes, as described below.

- Fundamental Intermittent Standards (FIS) directly relate to the characteristics of events that cause stress in river ecosystems. These standards are expressed in terms of concentration-duration thresholds with an allowable return period or frequency.
- High-percentile standards (such as the 99th percentile where water quality standards are exceeded in one case in a hundred) are based on the extrapolation of the 90/95 percentile thresholds from the Environment Agency's Ecosystem (River Ecosystem) classes used to plan and assess river quality.

Significant advances in UPM 2 include:

- The document is no longer focused on the specific circumstances pertaining to England and Wales. The guidance presented is equally applicable elsewhere. References to specific planning documents which have a finite life span have been removed to avoid rapid dating of the document.
- A more comprehensive range of environmental standards appropriate to the management of urban discharges is identified. More emphasis is placed on

the potential role of high-percentile standards, while the range of Fundamental Intermittent standards is expanded and the previous Derived Intermittent Standards are abandoned.

The Construction Industry Research and Information Association (CIRIA) is a UK-based research association concerned with improving performance in the area of construction and the environment. CIRIA has published a best practices manual for sustainable urban drainage, as well as sustainable urban drainage manuals for Scotland, Northern Ireland, England, and Wales.

The Chartered Institution of Water and Environmental Management (CIWEM) is a multidisciplinary professional and examining body for engineers, scientists, and other qualified personnel engaged in water and environmental management. CIWEM has produced a number of manuals of interest to stormwater management.

Scotland. In Scotland, local authorities own the water utilities. So far it has not been possible to privatize the water industry. Instead, the water and sewer services were transferred to three independent, government-established organizations in 1996. The 1996 Scottish reorganization corresponds to the 1974 reorganization in England and Wales in that it created three multi-purpose water authorities to take over the duties of local authorities and water boards.

Under the current system, the ultimate responsibility for the regulation of urban drainage rests with the Scottish Environment Protection Agency (SEPA). At the outset, SEPA policy was designed to protect water quality from pollution by surface water runoff through active regulation by SEPA. However, current policy is to use cooperative approaches as much as possible.

SEPA promotes Sustainable Urban Drainage Systems (SUDS) that allow water to be treated prior to release to Scottish waters. To promote and develop SUDS, SEPA and a range of partner organizations formed the Sustainable Urban Drainage Scottish Working Party (SUDSWP) in 1997. This group has been instrumental in changing attitudes toward sustainable urban drainage systems in Scotland.

The establishment of SUDSWP was based on the recognition that a partnership approach is essential to resolving the problems of urban drainage, and that no single organization can act alone to secure improvements in water quality, prevent flooding, or improve local amenity value. An important achievement, partly funded by SUDSWP, was the publication of the Sustainable Urban Drainage Systems Design Manual for Scotland and Northern Ireland (Martin, 2000).

Current SEPA policy does not necessitate a discharge permit for water from a SUD system if the system has been designed in accordance with the *Sustainable Urban Drainage Systems Design Manual* for Scotland and Northern Ireland. However, it is recommended that conditional prohibition notices be served to the developer prior to construction to ensure that the SUD system is constructed in accordance with the agreed design.

France. In France, municipalities and their territorial services are under the control of the prefects who represent the State. With the development of government environmental regulations and economic policies, the prefects have ever-increasing responsibilities regarding water policy issues. This system fosters serious problems if individual prefects give more weight to economic development than to environmental issues.

The French rely extensively on separate sanitary and stormwater drainage systems. However, these sewers have not always been carefully designed. Further, no specific policy for the control of stormwater quality or quantity exists. The French government has decided to fund some general Agenda 21 activities by levying a tax on all polluting activities. However, a ramification of this tax is that the Agences lose some of their income—income that could be used to develop badly needed intervention capacity in the area of diffuse pollution.

Germany. Germany is a federal republic with laws at both the federal and state levels. In the past, Germany had developed fairly autonomous river basin regulatory organizations. However, because of Germany's adoption of many of the top-down approaches prescribed by the EU, much of the flexibility at the local level has been lost.

An interesting feature of the German system is a law regarding "balanced development" that was enacted in 1987. Balanced development can be interpreted as "no impact" development. Any negative impact expected to endure longer than five years must be mitigated. The consequence of this law with regard to stormwater regulation is that stormwater from newly developed areas must be controlled in terms of both quality and quantity so that no negative impact compared to the undeveloped situation occurs.

The Netherlands. Because the Netherlands has significant areas lying below sea level, drainage has been very important to the country. The Netherlands is at the forefront of cooperative water management. The famous Waterschappen act as water boards and are responsible for local flood protection and water quality and quantity management, including wastewater treatment. The central government formulates the main lines for strategic policies on water issues at the national level, and the provincial governments define their strategic policies for non-state-controlled waters and the regional framework for flood protection.

The Ministry of Housing, Spatial Planning, and the Environment is responsible for general environmental policy; setting of water quality targets and emission standards; and executing laws concerning drinking water, sewerage, and spatial planning.

The water boards and the municipalities are responsible for the operational management and the actual implementation of policy issues. The water boards are also responsible for local flood protection and water quantity and quality management, including wastewater treatment. The water management task at the municipal level is limited to the management of sanitary drainage systems, which is usually conducted by local public works departments or joint agencies.

Canadian Stormwater Regulations

In Canada, development proponents are typically subject to legislation at the provincial level. Federal-provincial/territorial agreements exist to permit cursory review and approval of development applications with regard to federal legislation. This subsection presents a framework for Canadian stormwater-related regulations to demonstrate how the different levels of government interact and uses Ontario as an example of how stormwater management operates on a provincial level.

Canada Water Conservation Assistance Act. Canadian stormwater regulations developed with a focus on flood control. The first federal legislation dealing with water resources management was the Canada Water Conservation Assistance Act, which was passed in 1953. Prior to this law, the provinces were charged with the responsibility of managing their own natural resources, and the federal government was responsible for fisheries, navigable waterways, and irrigation, among other duties. Flood control was not dealt with specifically, and provinces encountered difficulty when entering areas of federal control.

The Canada Water Conservation Assistance Act enabled the federal government to assist the provinces by providing up to 37.5 percent of the cost of water conservation or water control works that would normally exceed the financial means of the province and municipalities. This assistance was provided through federal-provincial/territorial cost-sharing arrangements.

Canada Water Act of 1970. Eventually, it was felt that the Canada Water Conservation Assistance Act was too restrictive as it focused on works rather than the non-structural solutions that later became an integral part of the water resources management strategy. As a result, this Act was superseded by the Canada Water Act of 1970, which reflected a change in the attitudes of both professionals and the public. This Act provided for more public consultation, comprehensive planning that included all water uses, and comprehensive analysis that considered larger geographical areas for a more holistic approach. Most important, the consideration of non-structural alternatives became a critical component in the development of the federal regulations.

Flood Damage Reduction Program. The traditional model for flood management in place in the 1970s employed structural works for prevention and financial disaster assistance in times of need. Payouts continued to escalate as assistance was provided to new developments that were encouraged by the false sense of security engendered by structural works (Environment Canada, 1993). In 1975, the federal government established the Flood Damage Reduction Program. The program represented a shift from how to deal with problems after they occurred to how to prevent flooding damage in the first place. The main premise of this legislation was to keep floodplains free of development. Similar programs already in place in the provinces of British Columbia, Ontario, and Alberta aided the national program by building public support through the information already available. In 1999, the federal government discontinued the Flood Damage Reduction Program. Currently, the mapping of floodplains is conducted at the provincial and sometimes local level.

Other Federal Environmental Legislation. Other major federal environmental legislation includes the Canadian Environmental Protection Act, the Fisheries Act, and the Navigable Waters Protection Act. The Canadian Environmental Protection Act can be viewed as "umbrella" legislation that deals with a broad range of environmental issues (Research Advisory Panel, 2001). The Fisheries Act is administered by the Department of Fisheries and Oceans (DFO) and protects against "harmful alteration, disruption, or destruction of fish habitat (R.S., c. F-14. s-1)." The Navigable Waters Act is administered by Transport Canada and is concerned with the protection of waterways for ship passage.

Stormwater Regulation in Practice: Ontario. In practice, most interaction with government regulation is at the provincial level; each province or territory has differing legislation. The regulatory framework and application of the policies described in this subsection provide an introductory overview for Ontario. Development proponents should fully investigate or retain professionals with specific knowledge of the pertinent approval framework and procedures, as no two projects will follow the same road to final approval.

Examples of provincial legislation in Ontario are as follows (Research Advisory Panel, 2001).

- The Ontario Water Resources Act governs all aspects of water and wastewater activities.
- The Environmental Protection Act provides guidelines for ensuring a thorough review of proposed infrastructure, municipal servicing, and stormwater management facilities prior to granting Certificates of Approval.
- The Environmental Bill of Rights deals with the public's access to information.
- The Environmental Assessment Act works in conjunction with the Bill of Rights.

Under the Conservation Authorities Act (1946), regional authorities were given the task of establishing and undertaking the management of natural resources for lands deemed to be under their jurisdiction. The Conservation Authorities (CAs) were given the responsibility of carrying out flood control measures and enforcing the floodplain-free development principles described previously. During the review process for a development proposal, municipalities work in conjunction with the CAs. Local-level governments such as municipalities and regional municipalities issue design standards and criteria that deal primarily with the detailed design of stormwater management facilities and conveyance systems.

CAs and provincial government agencies may, in many cases, be empowered to review and comment on behalf of federal authorities. As an example, the Ontario Ministry of Natural Resources administers and enforces some of the sections of the Fisheries Act (federal legislation) regarding habitat on behalf of the Department of Fisheries and Oceans, where the CA does preliminary review of proposed stormwater management schemes. For scenarios in which stormwater is discharged directly into a watercourse, the CA will recommend the level of review that must be satisfied depending on the assessed level of disturbance to fish habitat (including both physical disturbances and thermal impacts on receiving waters). A "fill construction" permit may be requested if material is proposed for addition or removal in fill-regulated areas on the subject lands.

Permits are required for construction of sewers, water mains, and appurtenances under the Ontario Water Resources Act, which is administered by the Ontario Ministry of the Environment (MOE). Permits for operation and maintenance are not typically required, as municipalities will assume responsibility for infrastructure management. However, the current trend in Ontario is toward private operators for water and wastewater infrastructure management.

Municipalities will review and provide comments on development proposals on sewer allocation, capacity, and availability through their sewer use bylaws. Quantity controls may be imposed at this stage or through the CA policies. As discussed earlier, the notion of stormwater management in Canada was founded on flood control; thus, CA policies may reflect this by imposing quantity control restrictions for stormwater discharges. The regulating authority for the development will be influenced by locations of the site and storm sewer outfall.

The municipality may have quality control objectives implemented as part of an Official Plan or sewer use by-laws that are more stringent than CA quality control goals. Quality control levels are governed by the more stringent requirements of the CA or municipality. Quality criteria may be met through on-site controls such as infiltration trenches, grassed swales, oil-grit separators, or some combination of these or other best management practices. (For information on water quality best management practices, see Chapter 15.) Permits are not specifically required for stormwater discharge pollutant control, though the MOE has outlined provincial water quality objectives that are targets for the protection of aquatic life.

Projects located in the Great Lakes drainage basin may be subject to additional regulation. The International Joint Commission has designated 42 Areas of Concern (AOC), 16 of which are in Canada. Remedial Action Plans (RAPs) have been developed to restore beneficial uses to the AOCs. Implementation of the RAPs may require preparation of Pollution Prevention and Control Plans for control of contamination from urban runoff and sewage.

14.2 REGULATORY REQUIREMENTS IN THE UNITED STATES

Historically, stormwater management programs in almost all U.S. communities addressed drainage issues only. These reluctant communities were only able and willing to participate in stormwater management at the most basic and minimal level, with the frequent argument that scarce local resources were needed for more pressing issues.

However, with the development of new and increasingly complex laws and regulations concerning water quality, stormwater quality management is no longer a luxury seen only in rapidly growing affluent communities that have the resources and motives. Cities and municipalities around the country are now being mandated to assume roles as water resources managers and enforcers.

A host of laws, regulations, and ordinances exist related to the planning, design, and construction of stormwater management systems. They stem from all branches of government, from federal to local levels. Further, they are in a continual state of change and place a corresponding obligation on designers to keep abreast of their developments. Summaries of some of the more important U.S. laws and regulations are presented in the following subsections. The focus in this text is federal laws and regulations. Less emphasis is given to state and local laws and regulations because of the many differences that exist from one political jurisdiction to another. Nevertheless, state and local laws and regulations must be adhered to in any particular project, and the onus falls on the designer to ensure that their requirements are met.

Regulations pertaining to stormwater generally make a distinction between point and nonpoint sources of pollution. Point sources of water pollution are defined in Section 502 of the 1972 Amendments to the Water Pollution Control Act as "any discernible, confined, and discrete conveyance, including but not limited to any pipe, ditch, channel, tunnel, well, discrete fissure, container, rolling stock, concentrated animal feeding operation, or vessel or other floating craft, from which pollutants are or may be discharged." Nonpoint sources are the remaining pollutant sources not included in this definition. This definition of point source appears to include almost all water discharges, but important court actions have been necessary to clarify it further. The chief contributors to nonpoint source pollution are runoff from agricultural, mining, and forestry operations, and runoff from urban areas.

Federal Laws and Regulations

Federal laws and regulations concerning stormwater drainage systems, pollutants, and receiving waters are extensive. The following subsections present the development of these regulations, and include discussions of recent and emerging regulations. Table 14.2 summarizes the major federal legislation addressing clean water.

Table 14.2 Clean Water Act and major amendments (Copeland, 1999)

Year	Act	Public Law
1948	Federal Water Pollution Control Act	P.L. 80-845
1956	Water Pollution Control Act of 1956	P.L. 84-660
1961	Federal Water Pollution Control Act Amendment	P.L. 87-88
1965	Water Quality Act of 1965	P.L. 89-234
1966	Clean Water Restoration Act	P.L. 89-753
1970	Water Quality Improvement Act of 1970	P.L. 91-224, Part I
1972	Federal Water Pollution Control Car Amendments	P.L. 92-500
1977	Clean Water Act of 1977	P.L. 95-217
1981	Municipal Wastewater Treatment Construction Grants Amendment	P.L. 97-117
1987	Water Quality Act of 1987	P.L. 100-4

The Water Pollution Control Act of 1948. Before 1948, almost all water pollution control authority was vested in state and local governments. The legal powers of different state agencies varied greatly. The Public Health Service Act of 1912 authorized the investigation of pollution in navigable waters, and the Oil Pollution Act of 1924 was enacted to prevent oil discharges into navigable waters. The Water Pollution Control Act (WPCA, PL 80-845), passed in 1948, established some federal authority in abating interstate water pollution.

Because the 1948 Act focused on receiving-water standards instead of effluent standards, and because it was enforced ineffectively, it was not an immediate success. The receiving-water standards only considered the current uses of the water and were difficult to enforce against any single polluter. It usually was not possible to determine who was responsible when a stream standard was exceeded. The early regulations also placed much of the burden of proof on enforcement agencies.

Most of the water pollution control legislation enacted since 1948 has been in the form of amendments to the 1948 Act. Important amendments include:

1956:Made the legislation permanent and funded construction grants for publicly owned treatment works (POTWs)
1961:Increased funding for water quality research and construction grants
1965:Increased construction grants and started research concerning combined sewer overflows
1966:Removed the dollar limit on construction grants
1972:Enacted the most important advances to date, as described more fully below
1977:Extended some of the deadlines established in the 1972 amendments
1988:Required discharge permits for stormwater

The WPCA Amendments of 1972. The Refuse Act of 1899 (33 USC 407) was used in 1970 to establish a discharge permit system. The Refuse Act prohibited the discharge of any material, except sewage and runoff, into navigable waterways without a permit from the Department of the Army. It was written during the "progressive conservation era," when multiple uses of natural resources first became seriously considered. The U.S. Supreme Court upheld the applicability of the law for pollution control in 1966. However, a later decision in 1971 invalidated the program. Because of these difficulties, the WPCA amendments of 1972 (PL 92-500) contained a permit program called the National Pollutant Discharge Elimination System (NPDES). The NPDES is an enforcement scheme through which effluents from point sources are controlled. The permits are required for all point sources and establish discharge limits based on the available control technology. The discharge limits set for each industry were based on a series of studies that characterized each industry's waste effluents and existing controls. Future discharge limits were reduced as all facilities were to obtain the "best available technology economically achievable" (BATEA) by July 1983. New sources were to obtain discharges representative of a "standard of performance" that was much more restrictive. Nonpoint sources were originally exempt from the NPDES permit program, but stormwater discharges are included in the NPDES system as part of CFR 122, 123, 124, and 504, as published in the Federal Register of December 7, 1988.

The NPDES was intended to achieve Congress's goal of no pollutant discharges whatsoever by 1985. Other goals of PL 92-500 included the protection and propagation of fish, shellfish, and wildlife and recreational uses of water by July 1983; the prohibition of the discharge of toxic pollutants; the continuation of funding for publicly-owned treatment works (POTWs); the development of area-wide wastewater treatment management plans; the funding of a major resource and demonstration effort to improve treatment technology; and the protection of the rights of the states to reduce pollution and to plan their water resources uses.

The Clean Water Act of 1977. The Clean Water Act of 1977 (PL 95-217) extended some of the deadlines, but no waivers were allowed for toxic pollutant discharges. The amendments also allowed for ad valorem taxes by municipalities to fund treatment projects, provided incentives for use of innovative technologies, and placed increased emphasis on the area-wide treatment planning studies (Section 208).

Section 208 planning studies recognized the need to control nonpoint pollution in order to meet Congressional goals. This section was an incentive to local governments to develop their own plans, with minimal federal input. These plans were to characterize all point and nonpoint pollutant discharges in designated areas and to develop treatment schemes that would allow the goals to be met. The results of these plans affect the issuance of NPDES permits. Unfortunately, most of these plans were conducted in short time periods with limited technical success. Control measures were recommended with few local demonstrations of their potential success (especially for nonpoint pollution control). Recognizing these technical shortcomings, Congress authorized the Nationwide Urban Runoff Program (NURP) to demonstrate the applicability of various urban runoff control measures in about 30 cities. These studies were completed in 1983, and with these results, the Section 208 plans were to be revised under Section 205g.

EPA Stormwater Control Regulations. In the United States, regulations promulgated under the Clean Water Act are administered by the Environmental Protection Agency (EPA). Many of the laws allow for the EPA to transfer permitting and regulatory authority to the states. The NPDES permit program is an example of this. In general, states must adopt regulations that are at least as stringent as the federal regulations before they are granted regulatory authority.

The EPA regulations to control stormwater runoff were first published in the December 7, 1988 issue of the Federal Register. These regulations initiated a permit process for urban runoff, but the schedules and required reporting information varied depending on land use and the size of the community. The EPA was required by Section 405 of the Water Quality Act of 1987 to establish permit application requirements for municipalities having populations greater than 250,000 and industrial concerns (including construction operations) by February 4, 1989. Permit application requirements for municipalities having populations between 100,000 and 250,000 were to be established by February 4, 1991. The first applications for industrial concerns and large cities were to be submitted by February 4, 1990. The applications for smaller cities were to be filed by February 4, 1992. Permits are now required for cities with a population of less than 100,000 as well, as part of the second phase of the NPDES stormwater permit program.

Phase I Regulations. The Phase I general application requirements stressed descriptive information concerning the drainage area, with minimal runoff monitoring requirements. The permit applications relied mainly on the use of simple models to predict annual discharges of pollutants, and on field analyses for detecting illicit connections and illegal dumping. Permit applications also required a description of any locally required stormwater and construction-site runoff controls. Local municipalities were also to establish an authority for managing stormwater.

"We're lucky. This stream could be next to a paper mill instead of a brewery."

Phase II Regulations. The Clean Water Act 402(p)(6) initial Phase II rule (for small municipalities) was published on August 7, 1995. Its purpose was to designate additional sources of stormwater that needed to be regulated to protect water quality. It requires operators of small municipal separate storm sewer systems (MS4s) to apply for NPDES permits starting in March 2003. Phase II regulations also apply to entities such as universities, departments of transportation, and military bases that operate MS4s. It also affects millions of industrial/commercial facilities and almost all construction activities. Little, if any, monitoring is required by Phase II communities, but they are still responsible for ensuring that water quality regulations are met.

The Stormwater Phase II Rule superseded the August 1995 regulation. The final rule was published in the Federal Register on December 8, 1999. The schedule for implementation of the Phase II regulations calls for general permits to be issued in 2003 with an in-compliance schedule that extends up to five years. Some commercial and industrial groups have also been working with the EPA for special consideration and scheduling of implementation of regulations for their group members.

The Phase II Rule defines a stormwater management program for a small MS4 as a program composed of six elements that, when implemented together, are expected to reduce pollutants discharged into receiving water bodies to the maximum extent possible. These six program elements, or minimum control measures, are

- Public education and outreach on stormwater impacts
- Public involvement/participation
- Illicit discharge detection and elimination
- Construction site runoff control
- Post-construction stormwater management in new development and redevelopment
- Pollution prevention/good housekeeping for municipal operations

For each minimum control measure, permittees will select and implement best management practices and measurable goals that comprehensively address the specific stormwater problems in their respective areas.

Significant Court Cases. Several court cases have considered the question of whether urban runoff is a point source (and therefore required to have a NPDES permit under the original program) or a nonpoint source (and therefore exempt from the permit process, but still subject to applicable discharge standards). Urban runoff usually enters receiving waters through a conduit or ditch. The Natural Resources Defense Council (NRDC) v. Costle et al. (568 F.2D 1369, 1977) case found that uniform NPDES discharge limitations were not a necessary precondition for the inclusion of agricultural, silvicultural, and stormwater runoff point sources into the NPDES program. However, the regulations (40 CFR Section 125.4, 1975) did specifically exempt separate storm sewers containing only storm runoff uncontaminated by any industrial or commercial activity.

The NRDC vs. Costle et al. case validated the NPDES program by setting a clear precedent for considering any pollutant discharges as unlawful. The legal premise of the inherent right to utilize the nation's waterways for the purpose of disposing of wastes was eliminated; however, due in part to technological limitations pollutant discharges are continuing. Also, the stormwater runoff exemption of 40 CFR Section 125.4 was maintained because it was felt that requiring permits on the approximately 100,000 urban runoff point sources would tie up the ability of the EPA to effectively administer the more important industrial and municipal point sources. State authorities were encouraged to implement a plan to use general permits to cover these stormwater discharges. The District of Columbia Court of Appeals concluded that Congress's intent was to require permits for all point sources, but that the EPA was to have flexibility in structuring the permits in the form of general or area permits.

In Pedersen et al. v. Washington State Department of Transportation (611 P.2D 1293, 1980), the court found that "Separate storm sewers, as defined in this section, are point sources subject to the NPDES permit program. Separate storm sewers may be covered either under individual NPDES permits or under the general permit." The State of Washington Court of Appeals found that states are not required to implement general permit programs, and that individual permits can be required if appropriate general permits are not available.

The decision in the United States of America v. Frezzo (642 F.2D 59, 1981) case stated that the intent of the NPDES regulations "is to exclude from the NPDES permit program all natural runoff from agricultural land which results from precipitation events." This case further stated, "When water pollution from irrigation ditches results from precipitation events, that pollution is nonpoint in nature. However, when discharges from irrigation ditches result from the controlled application of water by any person, that pollution is considered point source and subject to the program proposed herein."

These cases established that collected urban runoff is a point source, while agricultural ditches are nonpoint sources during rain events and point sources during irrigation events.

Total Maximum Daily Load (TMDL) Program. Statewide Watershed Assessment Reports mandated by the Clean Water Act and other legislative and regulatory programs identify water quality deficiencies and prioritize watersheds for assessment. The 305(b) report, or State Water Quality Assessment, identifies and prioritizes water bodies with known violations of water quality standards or criteria. The 305(b) reports are prepared by the states on a biennial basis.

When states identify impaired water bodies through 305(b) assessments, they are required to identify the waters and to develop total maximum daily loads (TMDLs) for them. In general, a TMDL is a quantitative assessment of water quality problems, contributing sources, and pollution reductions needed to attain water quality standards. The TMDL specifies the amount that pollution and/or other stressors need to be reduced by in order to meet water quality standards It further allocates pollution control or management responsibilities among sources in a watershed and provides a science-based policy for taking action to restore a water body.

Pollutant loading from stormwater is considered in developing the TMDL. Stormwater discharges that are regulated under the Phase I or Phase II of the NPDES stormwater program are point sources and are included in the Waste Load Allocation portion of the TMDL. Stormwater discharges that do not require an NPDES permit are considered nonpoint sources and are included in the load allocation of the TMDL. When the TMDL is established, it may result in requirements for monitoring and/or treatment. For non-permitted discharges, treatment may involve implementation of one or more Best Management Practices (see Chapter 15).

By mid-1998, all states had EPA-approved Section 303(d)(1) lists, but the content and scope of these lists varied greatly. Development of TMDLs has been initiated at an increasing pace in some states, but most TMDLs are still incomplete. Many of the waters still needing TMDLs are impaired by contributions from both point and nonpoint sources. The EPA has undertaken a variety of steps to strengthen and define the TMDL program. The EPA should be consulted for updates on the list of impaired waters, the status of TMDLs, and policy guidance.

Dredge and Fill Permit Program. Section 404 of the Clean Water Act requires a permit, issued by the U.S. Army Corps of Engineers, for discharge of fill or dredge materials into the waters of the United States. This jurisdiction includes not only navigable waters, but also their tributaries and associated wetlands. The Supreme Court has held that this jurisdiction extends to lands supporting plant growth typical of wetlands, such as willow trees.

The scope of activities covered by Section 404 is much more broad than the traditional dredging and filling of navigable ship channels. Regulated activities include, for example, construction of bridges, dams, buildings, roads, flood control facilities, and shellfish operations.

Section 404 brings a wide range of federal environmental law to bear on any project requiring a permit. In its issuance of a Section 404 permit, the Corps must comply with the Fish and Wildlife Coordination Act, Endangered Species Act, Wild and Scenic Rivers Act, Coastal Zone Management Act, National Environmental Policy Act, and other laws that govern federal activities.

State Laws and Regulations

In the United States, power to regulate the use and control of waters is vested largely in individual states, although there are federal requirements relating to factors such as navigability and reserved water rights. State constitutions or statutes often contain provisions relating to water laws (statutory laws), but most applied water laws are based on precedent. As a result, laws can vary significantly from one state to another, and even within a state from incorporated municipal areas to adjacent unincorporated areas.

Despite the fixed chemical composition of pure water and the interconnections among the various stores in which water exists, the legal system tends to compartmentalize water into various types, with each type being subject to somewhat different sets of water laws. For example, groundwater is generally subject to a different set of laws than surface water in a stream. Diffused water, which results from runoff that has not reached a well-defined channel with a bed and banks, is subject to yet another set of laws. A summary of the main legal doctrines and rules for surface streams and diffused waters is provided in the following subsections. Groundwater laws are not discussed here, as they are not as important in the context of urban stormwater management. The interested reader is referred to Getches (1997) for a thorough introduction to water law.

"So that's where it goes! Well, I'd like to thank you fellows for bringing this to my attention."

Stormwater Regulations in the State of Virginia

Stormwater in Virginia is driven by federal and state laws. National permits, like the EPA's NPDES Phase I and II permits, are implemented through programs under the Virginia Department of Environmental Quality (DEQ). One characteristic of Virginia's stormwater laws is that Virginia is a "Dillon Rule" state. This means that a locality needs to have enabling state legislation to develop ordinances for different topics of concern. If localities want to develop more stringent regulations than the current state minimums, the Dillon Rule allows them to develop those ordinances. In Virginia, the state has enabling legislation for erosion and sediment control ordinances, stormwater management ordinances, and stormwater utility ordinances. This does not force a community to enact these ordinances, but simply provides a framework for developing them. Most Virginia local government have enacted erosion and sediment control ordinances, but not stormwater ordinances.

There are many different Virginia state agencies that play a role in stormwater management. The DEQ plays the primary environmental regulatory role. The Chesapeake Bay Local Assistance Department addresses specific water quality issues and regulations in the portion of Virginia that drains to the Chesapeake Bay. The VA Department of Conservation and Recreation plays an important role by managing state lands and providing guidance in areas such as dam safety and erosion and sediment control. The VA Department of Health plays a role in shellfish sanitation issues and protecting drinking water supplies. The VA Department of Game and Inland Fisheries deals with non-tidal fisheries and boating issues. The VA Marine Resources Commission (VMRC) is the state agency that manages the state portion of the U.S. Army Corps of Engineers' Joint Permit Application for any work including construction, dredging, filling, or excavation in the water or in wetlands. The VA Department of Emergency Management deals with natural disasters and administers FEMA grant programs for areas such as flood mitigation. Finally, the VA Department of Transportation (VDOT) plays an important role in establishing road drainage guidelines. Roads must meet VDOT design guidelines, including stormwater drainage, in order for the road to be maintained by VDOT or for communities to receive state money to maintain their VDOT-approved roads. Therefore, the VDOT design guidelines are often used for most stormwater design standards.

Most of these state agencies have regional offices throughout the state that cover multiple counties and cities. These agencies' regions are not the same, so often designers must get approval from VDOT in one location, DEQ in another location, DCR in a third location, and VMRC at the state capital.

Surface Water Doctrines. There are two main doctrines of water law. These doctrines are known as the riparian and prior appropriation doctrines. Some states have adopted hybrid systems consisting of elements from each of these doctrines, whereas others, such as Louisiana and Hawaii, have almost completely different systems. Most states east of the 100th meridian have adopted the riparian doctrine, whereas most of the western states adhere to the doctrine of prior appropriation. (For further discussion, see Novotny and Olem, 1994.)

Riparian Doctrine. Landowners whose property abuts a natural stream, river, or lake hold riparian rights. By virtue of their physical location, they are entitled to rights not held by nonriparians. When a property is sold or conveyed, the associated riparian rights are transferred to the new owner. Riparian rights entitle a landowner to the full natural flow of a stream, both in quantity and quality. However, riparian rights also require a landowner to use the water reasonably level and without infringing upon the rights of downstream property owners. Riparian rights can be sold as property, but water cannot be transferred from one drainage basin to another. During water shortages, property owners have equal rights to reasonable use of the water, and the available supply must be shared.

Doctrine of Prior Appropriation. During settlement of the American West, it became apparent that the riparian doctrine was not practical in the relatively water-poor west. This gave rise to a first-in-time, first-in-right approach to water resources. Under this doctrine, seniors (those making early appropriations) have priority of water use over juniors (those with later appropriations) during times of shortage; no requirement exists for water sharing. An appropriation is made by giving notice of intent to divert water, diverting water from its natural course, and applying it to a beneficial use. Non-use of a water right can result in loss of the right. Often, water rights are term limited and must be reappropriated. Water rights can be sold as property, and water can be transferred from one drainage basin to another, but restrictions are placed on the amount of water that can be transferred out of a basin so as to protect other appropriators.

Diffused Water Rules. Diffused water is water from precipitation or snowmelt that runs across the land surface without a well-defined course having a channel bed and banks. Two major rules have been applied by the courts to resolve disputes concerning diffused waters: the civil law rule and the common enemy rule. Both of these rules may be viewed as extreme stances, and thus there has been some modification to bring them closer together through a reasonable use rule.

Civil Law Rule. This rule is based on the perpetuation of natural drainage. It places a servitude upon a lower elevation landowner for the drainage of surface water in its natural course and flow from a higher elevation landowner. The higher landowner is not entitled to discharge runoff at a location other than that at which flow occurred historically, nor is the higher landowner entitled to increase the amount of the flow discharged to the lower proprietor.

Common Enemy Rule. Under this rule, diffused waters are viewed as the common enemy of all, and each landowner may take whatever actions are deemed necessary to protect their own interests. Owners whose properties may be damaged have no recourse against their neighbors.

Reasonable Use Rule. Under this rule, each landowner can make reasonable use of his land, even though the flow of diffused water may be altered, thereby causing harm to others. Liability occurs when the interference or alteration of the flow is unreasonable, as judged by a balancing test. Generally, questions concern issues such as whether there was a reasonable necessity to alter the flow to make use of one's land, whether the alteration itself was accomplished in a reasonable manner, and whether the utility of the actor's conduct reasonably outweighs the gravity of harm to others.

Local Regulations

Federal legislation mandates stormwater quality management at most levels of government. These mandates have arisen mainly from the NPDES program and from the Phase I and Phase II regulations that have been promulgated by the EPA. Thus, many of the local laws and regulations concern issues such as erosion and sedimentation, stormwater quality controls, and various codes, regulations, and ordinances relating to detention facilities, land-use zoning, and subdivision development requirements.

Codes, ordinances, and subdivision regulations are usually adopted and published by municipal and/or county governments with the purpose of establishing minimum criteria that must be met by developers of unimproved land parcels. These criteria establish a degree of uniformity from one development to another, especially in terms of the transportation and utility infrastructures. Regulations are often set forth with regard to stormwater management systems, including design-storm frequencies and minimum system requirements.

Zoning regulations are also commonly adopted and published by municipalities and county governments. Zoning maps and master plans are often prepared to guide future development patterns. This planning has advantages not only with respect to ensuring that objectionable development patterns do not occur (for example, a factory next to a prime residential area), but also for the planning of transportation and utility systems. In the context of urban stormwater management, zoning maps may be used to project future land uses in undeveloped areas upstream of a project locale, and hence can help to ensure that the project facilities will meet future needs.

Many states and cities throughout the United States have regulations that require developers to construct runoff control facilities. Many of these regulations were initially used only to regulate runoff flow rates or to reduce erosion due to construction. Some state court cases have led to the requirement for runoff controls as part of the environmental impact statement process (such as Maryland v. U.S. Postal Service, 487 F. 2D 1029, 1973, and Veterans Administration Hospital case in Tennessee 48 FR 11551, 1983). In the opinion for Parsippany v. Costle (503 F. Supp. 314, 1979), it was recommended that municipalities within the drainage basin adopt regulations that stipulate the use of detention and sedimentation basins.

"I do know one thing, gentlemen. If we don't plug into the environment right now, we're going to be missing out on a lot of big envirobucks."

14.3 CHAPTER SUMMARY

Stormwater conveyance systems must be designed to meet minimum requirements specified by local, regional, state, or national review agencies.

In 1977, the UN Mar del Plata Conference placed water on the international agenda. In 1987, the World Commission on Environment and Development agreed on the need to reverse current development trends and popularized the notion of sustainable development. This new awareness resulted in the United Nations Conference on Environment and Development in Rio de Janeiro in 1992, whose major outcome was to create a framework for the work toward sustainable development during the twenty-first century. A broad consensus exists that sustainable development requires the sustainable use and management of water resources in an environmental, economic, and social context.

In developing countries, insufficient urban drainage systems are a major public health concern. Existing channels are often misused as trash dumps during dry weather conditions. When rain occurs, there is a considerable risk of clogging the channels, which results in flooding, spreading of garbage, and pollution of water bodies. A lack of adequate urban land-use planning and enforcement in most developing countries means that technical tools and principles developed for industrialized countries must be adapted to the developing country's present conditions.

The Helsinki Convention on the Protection of the Marine Environment of the Baltic Sea was signed by all of the seven Baltic Sea States in 1974 and has been in force since 1980. It is the first international instrument to deal with all of the different sources of pollution in the Baltic Sea. A major aim is to minimize the load of pollution entering the Baltic Sea from land-based sources.

Until recently, the European Union (EU) has applied a uniform standards approach. This approach has caused concerns about high costs, and questions have been raised about the efficiency of the overall system in meeting water quality goals. Germany, France, Spain, and the United Kingdom have river basin authorities of one kind or another, and those systems are now changing to come into compliance with EU requirements. Nevertheless, they still have some flexibility within their own areas of authority. Such flexibility allows for the setting of priorities and realistic targets consistent with available resources.

In Canada, development proponents are typically subject to legislation at the provincial level. Federal-provincial/territorial agreements exist to permit cursory review and approval of development applications with regard to federal legislation. Canadian stormwater regulations developed, as they did elsewhere, with a focus on flood control.

Historically, stormwater management programs in almost all U.S. communities addressed drainage issues only. Stormwater quality management was brought to the attention of most medium- to large-size cities with the passage of the Clean Water Act and its various amendments, and the associated NPDES permit program.

REFERENCES

Carp, E. ed. 1972. *Proceedings, International Conference on the Conservation of Wetlands and Waterfowl, Ramsar, Iran, 30 January - 3 February 1971*. Slimbridge, UK: International Wildfowl Research Bureau.

Coldewey, W. G., P. Göbel, W. F. Geiger, C. Dierkes, and H. Kories. 2001. "Effects of Stormwater Infiltration on the Water Balance of a City." in *New Approaches Characterizing Groundwater Flow*, Proceedings of the XXXI. International Association of Hydrogeologists Congress, München, 10–14 September 2001. ed. K. P. Seiler and S. Wohnlich. Balkema, Lisse.

Copeland, C. 1999. Clean Water Act: A Summary of the Law, CRS Issue Brief for Congress, RL 30030.

Environment Australia. 2002. *Urban Stormwater Initiative*. Canberra: Environment Australia.

Environment Canada. 1993. *Flooding: Canada's Water Book*. Ottawa, Canada: Ecosystem Sciences and Evaluation Directorate, Economics and Conservation Branch.

European Commission. 1998. "Towards Sustainable Water Resources Management – A Strategic Approach." DG Development and DG External Relations.

European Conference of Sustainable Cites and Towns (ECSCT). 1994. *Charter of European Cites and Towns Towards Sustainability (Aalborg Charter)*. Aalbord, Denmark: ECSCT.

Foundation for Water Research. 1994. *Urban Pollution Management Manual*. Marlow, Great Britain: Foundation for Water Research.

Foundation for Water Research. 1998. *Urban Pollution Management Manual*. 2nd ed. Marlow, Great Britain: Foundation for Water Research.

Getches, D. H. 1997. *Water Law*. 3d ed. St. Paul, Minnesota: West Publishing Co.

Hillary, R. 1994. *The Eco-Management and Audit Scheme: A Practical Guide*. Trans-Atlantic Designs.

International Commission for the Protection of the Rhine (ICPR). 1992. *Ecological master plan for the Rhine*. Coblenz: International Commission for the Protection of the Rhine against Pollution.

International Environmental Technology Center (IETC). 2002. *Environmentally Sound Technology for Wastewater and Stormwater Management: An International Source Book*. London: IWA Publishing.

International Rivers Network (IRN). 2003. IRN Press Release. 27 Feb. 2003.

Martin, P. et al. 2000. *Sustainable Urban Drainage Systems – Design Manual for Scotland and Northern Ireland*. CIRIA Publications,.

Novatny, V. and H. Olem. 1984. *Water Quality, Prevention, Identification and Management of Diffuse Pollution*. New York: Van Nostrand Reinhold.

Research Advisory Panel: The Walkerton Inquiry. 2001. "Wastewater Collection and Treatment – Issue Papar 9." Research Paper Advisory Panel, The Walkerton Inquire. Toronto, ON: Delcan Corporation.

United Nations. 1977. *Report of the United Nations Conference*. Mal del Plata. 14–25 March 1977. New York: United Nations.

United Nations. 1992. *Intergovernmental Negotiating Committee For The Elaboration Of An International Convention To Combat Desertification In Those Countries Experiencing Serious Drought And/Or Desertification, Particularly In Africa.*

United Nations. 1992. *Report of the United Nations Conference on Environment and Development (UNCED)*. Rio de Janeiro, June 1992. New York: United Nations.

United Nations, 1996, *Report of the United Nations Conference on Human Settlements (Habitat II)*, Istanbul, 3–14 June 1996. New York: United Nations.

United Nations. 1997. *Report of the Commission on Sustainable Development on Preparations for the Special Session of the General Assembly for the Purpose of an Overall Review and Appraisal of the Implementation of Agenda 21*. New York: United Nations.

United Nations Development Program (UNDP). 1998. *Capacity Building for Sustainable Management of Water Resources and the Aquatic Environment*. New York: United Nations.

U.S. Environmental Protection Agency. 1991. *Guidance for Water Quality-Base Decisions: The TMDL Process*. EPA 440/4-91-001.

Weber, U. 2000. *The 'Miracle' of the Rhine Courier*. Unesco, Paris.

World Bank. 1993. *Water Resources Management*. Washington, D.C.: International Bank for Reconstruction and Development/The World Bank.

World Bank. 1998. *Pollution Prevention and Abatement Handbook 1998: Toward Cleaner Production*. Washington, D.C.: International Bank for Reconstruction and Development/The World Bank.

World Bank. 2002. *Thailand Environmental Monitor 2001: Water Quality*. Washington, D.C.: The World Bank Group.

World Bank. 2003. *The New Water Resources Strategy*. Washington, D.C.: World Bank Group.

World Commission on Environment and Development (WCED). 1987. *Our Common Future*. Oxford, Great Britain: Oxford University Press.

World Resources Institute. 1997. *The World Resources 1996–97*. Oxford University Press.

Wright, A. M. 1997. *Toward a Strategic Sanitation Approach: Improving the Sustainability of Urban Sanitation in Developing Countries*. Washington, D.C.: International Bank for Reconstruction and Development/The World Bank.

Measures should be taken to protect drainageways in and around construction sites from sediment-laden runoff due to erosion of disturbed soils.

CHAPTER

15

Stormwater Quality Management

Sources of water pollution can be classified as either *point sources* or *nonpoint sources.* With point-source pollution, the discharge of a contaminant can be attributed to a specific location such as a wastewater discharge or chemical spill. In the case of nonpoint-source pollution, the contamination source is distributed over a large area or a large number of point sources (Chin, 2000). Nonpoint sources, such as runoff from urbanized areas, irrigated lands, surface mining operations, and construction sites, are frequently the largest uncontrolled contributors to water pollution. This chapter focuses on the negative effects of stormwater runoff from urbanized areas and presents various techniques, often referred to as *best management practices (BMPs)*, used in mitigating these effects.

The effects of stormwater on receiving waters are a function of the types and concentrations of contaminants in the runoff and the frequency, intensity, and duration of runoff flows. The nature of water-body impairment includes trash accumulation, erosion and sedimentation, and contamination by harmful microorganisms and various chemicals and toxins. Control measures designed to temporarily detain, reroute, or treat stormwater can mitigate these effects, as can measures that eliminate contact between stormwater and pollution sources. Stormwater quality controls have proven to significantly contribute to improving the quality of urban runoff when implemented as part of a comprehensive stormwater management plan.

Section 15.1 of this chapter summarizes the extent to which stormwater affects the quality of receiving waters and describes various environmental effects of stormwater runoff. Section 15.2 discusses hydrology as it relates to stormwater quality, as storm events of different magnitudes tend to impact receiving-water quality in different ways.

Section 15.3 introduces BMPs and presents descriptions of commonly used nonstructural and structural BMPs. The use of BMPs in stormwater management programs for newly developed and existing communities and for construction projects is discussed in Section 15.4.

Although many similarities exist between stormwater controls implemented in different regions, a particular type of stormwater control (such as a filter strip or grassed swale) may be appropriate in one location but not another. It may also be appropriate for some types of sites (for example, residential), but not others. For this reason, and because individual state and local governments develop differing management plans, the discussion of controls in this chapter is rather general. Much of the material presented here is modeled after the plan implemented in the Denver, Colorado (USA) metropolitan area and published by the Urban Drainage and Flood Control District (UDFCD, 1999). These discussions are presented only as examples to illustrate the general process of control implementation. For another example, the reader is referred to the manual published by the Water and Rivers Commission of Western Australia (Water and Rivers Commission, 1998), which is accessible on the Internet.

15.1 ENVIRONMENTAL EFFECTS OF STORMWATER RUNOFF

In the United States, the effect of urban runoff and storm sewers on water quality is described in the 2000 National Water Quality Report (U.S. EPA, 2002). The report summarizes the assessments of approximately 33 percent of U.S. waters, including rivers, streams, lakes, and estuaries, relative to the water quality standards set forth for each body. Waters not meeting water quality standards were classified as impaired. For each impaired body, the leading cause of impairment was identified.

Table 15.1 shows that the urban runoff/storm sewers category was the leading source of impairment for 13 percent of rivers and streams, 18 percent of lakes, and 32 percent of estuaries in the United States. Relative to other sources, urban runoff/storm sewers was the second leading source of impairment for estuaries, third leading source for lakes, and fourth leading source for rivers and streams. Urban stormwater is a significant cause of impaired water in many parts of the world, as well. Environment Canada maintains the website for the United Nations Environment Programme's (UNEP) Global Environment Monitoring System (GEMS) Freshwater Quality Programme. Sixty-nine nations participate in this program, whose activities are aimed at understanding freshwater quality issues. Information on water quality in participating countries is available through the website, including an "Annotated Digital Atlas of Global Water Quality" that lists concentrations for a number of constituents in many major rivers and lakes for the years 1976 through 1990.

Table 15.1 Impaired waters in the United States attributed to urban runoff/storm sewers (U.S. EPA, 2002)

Receiving Body Type	Units	Total Impairment		Urban runoff/storm sewers as leading source		
		Amount	% of Total Assessed	Amount	% of Impaired	Rank[a]
Rivers and streams	miles	269,258	39	34,871	13	4
Lakes	acres	7.7 million	45	1.37 million	18	3
Estuary	mi^2	15.7 million	51	5.05 million	32	2

a. Other sources: municipal point sources, industrial discharges, atmospheric deposition, agriculture, hydrologic modifications, and resource extraction

The adverse effects of stormwater pollutants on receiving waters include acute and chronic toxic effects and secondary effects such as eutrophication. Stormwater also causes changes to the physical properties and flow regime of streams. Table 15.2 lists typical receiving-water problems associated with the long-term accumulation of pollutants and short-term storm events (Pitt, 1995a, b). Many of these problems have been found in urban receiving waters worldwide. Because these problems are so diverse, a variety of individual stormwater controls must be used together to be effective.

Table 15.2 Typical receiving water problems in heavily developed urban drainage basins (Pitt, 1995a, b)

Long-term problems associated with accumulations of contaminants:
Trash accumulation along receiving water corridors
Accumulation of debris and sediment in stormwater conveyance systems
Accumulation of fine-grained and contaminated sediment in receiving waters
Nuisance algal growths from nutrient discharges
Inedible fish, undrinkable water, and shifts to aquatic organisms less sensitive to heavy metals and organic toxicants
Short-term problems associated with high pollutant concentrations or frequent high flows (event related):
Swimming beach closures because of pathogenic microorganisms
Water-quality violations
Property damage from flooding and conveyance system failures
Habitat destruction caused by frequent high flow rates (bed scour, bank erosion, flushing of organisms downstream, etc.)

Pollutants in Urban Stormwater

Although stormwater is often viewed by the public as being as clean as rain, it in fact contains significant quantities of the same types of constituents more commonly associated with municipal and/or industrial wastewater. A summary of some of the common constituents in urban stormwater runoff, their sources, and their effects is provided in Table 15.3.

An example of the effects that stormwater may have on water quality in urban receiving waters is shown in Table 15.4. This table lists the average stormwater constituent concentrations for the major watersheds in the Atlanta, Georgia area. Clearly, the stormwater in this area has elevated concentrations of many important pollutants, with some direct effects. These elevated concentrations, coupled with increased runoff volumes, account for massive quantities of pollutant discharges.

Table 15.5 shows the contaminant percentage contributions of the different flow sources for the South River Basin in Atlanta. Stormwater accounts for most of the pollutant discharges for all but one of the listed constituents. (The exception is fecal coliform bacteria, which are predominantly from CSO storage overflows at one facility.) This example is typical in that stormwater discharges usually account for the majority of pollutant discharges in completely developed urban watersheds.

Table 15.3 Urban runoff pollutants (U.S. EPA, 1993)

Constituents	Sources	Effects
Sediments — TSS, turbidity, dissolved solids	Construction sites, urban/agricultural runoff, landfills, septic fields	Habitat changes, stream turbidity, recreation and aesthetic loss, contaminant transport, bank erosion
Nutrients — nitrate, nitrite, ammonia, organic nitrogen, phosphate, total phosphorus	Lawn/agricultural runoff, landfills, septic fields, atmospheric deposition, erosion	Algal blooms, ammonia toxicity, nitrate toxicity
Pathogens — total and fecal coliforms, fecal streptococci viruses, E. Coli, enterooccus	Urban/agricultural runoff, septic systems, illicit sanitary connections, domestic and wild animals	Ear/intestinal infections, shellfish bed closure, recreational/aesthetic loss
Organic enrichment — BOD, COD, TOC, DO	Urban/agricultural runoff, landfills, septic systems	Dissolved oxygen depletion, odors, fish kills
Toxic pollutants — metals, organics	Urban/agricultural runoff, pesticides/herbicides, underground storage tanks, hazardous waste sites, landfills, illegal disposals, industrial discharges	Toxicity to humans and aquatic life, bioaccumulation in the food chain
Salts — sodium chloride	Urban runoff, snowmelt	Contamination of drinking water, harmful to salt-intolerant plants

Note: TSS = total suspended solids, BOD = biochemical oxygen demand, COD = chemical oxygen demand, TOC = total organic carbon, DO = dissolved oxygen

Table 15.4 Representative average stormwater and combined sewer overflow (CSO) constituent concentrations for major land uses in Atlanta, Georgia (CH2M-Hill, 1998)

Land Use Category	BOD (mg/L)	Fecal Coliform Bacteria (per 100 mL)	Suspended Solids (mg/L)	Ammonia Nitrogen (mg/L)	Phosphorus (mg/L)	Copper (mg/L)	Lead (mg/L)	Zinc (mg/L)
Forest/open space	5	500	70	0.2	0.1	0.00	0.00	0.04
Residential	15	8,700	180	0.2	0.4	0.03	0.03	0.10
Institutional	10	1,400	140	0.5	0.3	0.01	0.02	0.14
Commercial	10	1,400	130	0.5	0.3	0.01	0.02	0.12
Industrial	14	2,300	90	0.4	0.3	0.02	0.02	0.16
Highway	14	1,400	140	0.5	0.3	0.01	0.02	0.14
Construction	13	1,000	1,000	0.2	0.2	0.05	0.02	0.40
CSOs	13	175,000	170	1.3	0.8	0.04	0.04	0.20

Table 15.5 Distribution of pollutant sources for the South River drainage basin, as percentages of total loading (CH2M-Hill, 1998)

Constituent	Stormwater	Base Flow	CSO Storage Overflow	CSO Effluent[a]
Flow	61	21	14	5
BOD	54	4	32	6
Fecal coliform bacteria	10	6	84	<1
Suspended solids	76	1	24	1
Ammonia nitrogen	37	19	28	16
Phosphorus	55	10	31	4
Copper	64	N/A	30	6
Lead	67	N/A	32	1
Zinc	67	N/A	29	4

a. CSO effluent undergoes some treatment prior to discharge.

Clearly, high levels of combined sewer overflow (CSO) control are needed in areas having combined sewers. But urban stream bacterial levels are generally high even in areas affected only by stormwater (fecal coliform levels greater than 10,000 per 100 mL are common), preventing any contact recreation uses. Unfortunately, bacteria levels in stormwater are not amenable to sufficient reduction, although controls such as wet detention ponds and filters may reduce the levels by about 50 to 90 percent (compared to desirable reductions of much greater than 90 percent).

Ponds and lakes, by virtue of their long residence times relative to streams, are particularly susceptible to the adverse effects of contamination in urban runoff. They take longer to recover from pollution episodes and they effectively act as sinks for nutrients, metals, and sediments. Nutrient enrichment, in particular, can result in the undesirable growth of algae and aquatic plants, which, in turn, can cause additional problems related to dissolved oxygen levels and odors.

"Before you do something you may regret, I think you should know that I contain six parts of mercury per million."

Effects on Streams

In addition to the increases in various pollutant concentrations, urbanization results in increases in the frequency and magnitude of runoff events and decreases in stream base flows. These changes result in secondary effects that include alteration of the channel cross-section and stream alignment, increased erosion and stream sediment, and, ultimately, degradation of the aquatic habitat. The subsections that follow describe the effects of urban runoff on streams in more detail.

Flow Regime Alteration. Urbanization decreases the availability of pervious surfaces and surface storage capacity, resulting in an increase in the frequency and severity of flooding. Hydrograph peak flows in urban streams tend to be higher and occur earlier than hydrographs for streams in undeveloped areas that are otherwise similar. Also, the total volume of runoff substantially increases under developed conditions as less rainfall infiltrates into the soil. These flow alterations result in changes in flow regime that can have numerous effects on receiving streams (U.S. EPA, 1999; U.S. EPA, 1997).

Increased peak flows can greatly accelerate the process of streambank erosion, and the increased volume and velocity can overwhelm the natural capacity of the stream network. As an area is urbanized, its streams typically experience an increase in the frequency of bank-full flows. Because bank-full flows are highly erosive, streams may become wider and straighter. Other effects on streams attributable to increased flows include undercut and fallen banks, felled bushes and trees along the banks, and exposed sewer and utility pipes. Sediments from eroding banks and other sources are deposited in areas where the water slows, causing buildup, destruction of *benthic* (bottom) habitat, and decreased capacity for flood conveyance. The decreased capacity ultimately results in greater potential for further erosion.

Urbanization also affects dry-weather flows in streams. Because groundwater recharge is reduced, the water table may be lowered. This change in watershed hydrology reduces the contribution of base flow to the total flow in the stream, which translates into decreased dry-weather flows, pool depths, and ripple intensities.

Habitat Effects. Changes in flow regime, the resulting changes in stream morphology, and the increase in pollutants generally degrade the habitat for aquatic biota. The quality of an aquatic habitat may be measured by the number of organisms, number of species, and types of species present and surviving in the water.

Effects of urban stormwater include a reduction in the total number and diversity of macroinvertebrates and the emergence of more pollutant-tolerant species. A study in the U.S. state of Delaware found that approximately 70 percent of the macroinvertebrate community in streams in undeveloped, forested watersheds consisted of pollution-sensitive mayflies, stoneflies, and caddisflies, as compared to 20 percent in urbanized watersheds (Maxted and Shaver, 1997).

The health of an ecosystem may also be measured by the abundance and variety of fish species present and the presence of native species. A study compared fish populations in urbanized and nonurbanized sections of Coyote Creek, California (Pitt, 1995). The results showed that the native fish were generally replaced by non-native fish in the urbanized section of the stream. Limburg and Schmidt (1990) compared the abundance of fish eggs and larvae with different land uses in New York. The researchers concluded that significant declines in abundance were noted in watersheds having an imperviousness of 10 percent or more.

Receiving-Water Considerations

The goals of an urban stormwater control program may be developed in terms of flow control, pollutant removal, and pollutant source reduction (U.S. EPA, 1999). One of the most difficult issues associated with stormwater management is that of determining reasonable receiving-water goals. It is not practical to believe that stormwater management will enable natural conditions to be attained by receiving waters in highly developed urban watersheds. The radical alteration of the hydrologic cycle and the receiving-water habitat likely prevent any complete restoration to natural conditions. In large watersheds where the urbanized portions comprise a small fraction of the area, nearby streams of excellent quality may exist.

Typically, the greater the extent of urbanization in a watershed, the greater the negative effects on receiving waters. Beyond a certain point, receiving-water uses are severely limited and are not likely to be restored, even with extensive stormwater management. However, in some cases restoration efforts have significantly improved receiving-water conditions. It is practical to expect only limited, but important, uses of urban streams in heavily developed areas. Aesthetically pleasing waterways, some habitat protection for relatively hardy aquatic and riparian life, and safe flow conditions should be the minimum objectives. The basic uses that should be available include:

- Noncontact recreation associated with stream-side parks and greenways. Nuisance odors, noxious sediment accumulations, and trash, for example, should be rare. Hazardous debris should be nonexistent.
- A reasonable assemblage of tolerant aquatic life and riparian animals and plants should be able to exist in the receiving water and surrounding area.
- Dangerous flows and floods should be controlled to minimize property damage and prevent loss of life.

Without extensive knowledge and continuous monitoring, it is foolish to encourage fishing and swimming in waters receiving substantial urban stormwater or at locations near stormwater outfalls in large receiving waters. There is a small but definite health risk associated with pathogens in the runoff. It may be possible for the effects of urbanization to be sufficiently moderated using effective stormwater management tools to allow downstream high-level uses such as fishing and swimming.

The determination of the level of stormwater control required for a receiving body to achieve a specified quality level is a challenging process. In the United States, a Total Maximum Daily Load (TMDL; see Section 14.2, page 598) is a calculation of the maximum amount of a pollutant that a water body can receive and still meet water quality standards. Based on this amount, pollution control responsibilities in a watershed are allocated among pollutant sources. Water quality standards are set by states, territories, and tribes. They identify the uses for each water body [for example, drinking water supply, contact recreation (swimming), and aquatic life support (fishing)], and the scientific criteria to support that use.

15.2 HYDROLOGY FOR STORMWATER QUALITY CONTROL

Small, frequently occurring storm events account for a significant portion of the total annual pollutant load from urban drainage basins. Thus, capture and treatment of the stormwater from such events is the recommended design approach for water-quality enhancement, as opposed to facility designs that focus on less-frequent, larger events. A design that not only controls the large events, but that also acknowledges the importance of small ones, is both possible and desirable.

The basis for developing appropriate stormwater controls is the understanding that specific receiving-water problems are associated with specific rain-depth categories. To identify which rain-depth categories are important for which receiving-water problems, long-term evaluations are needed. To illustrate the importance of relatively small events in stormwater quality management, Pitt et al. (2000) conducted long-term, continuous simulations for many locations in the United States.

An example from these simulations uses Atlanta, Georgia, rainfall data with the SLAMM (Source Loading and Management Model) computer program (Pitt, 1986; Pitt and Voorhees, 1995). The Atlanta simulations were based on eight years of rainfall records for the years 1985 through 1992, containing about 1,000 individual storms. Dry periods of at least six hours were used to separate adjacent storm events, and all rainfall depths were used, with the exception of trace values. (Similar analyses were made using inter-event definitions ranging from 3 to 24 hours, with little difference in the conclusions.) These historical rainfall data were modeled in SLAMM with typical medium-density and strip-commercial developments.

The results of the Atlanta simulations are illustrated in Figure 15.1. The "Accumulative Rain Count" data for a particular rain depth yields the percentage of rainfall events having a total depth less than or equal to that amount. The "Accumulative Commercial Runoff Quantity" and "Accumulative Residential Runoff Quantity" plots provide information on the proportion of runoff from these respective sources for given rainfall depths.

The points that follow summarize the Atlanta evaluation. It should be noted that in most areas of the United States (and in many parts of the world), the specific rain depths shown below would be substantially smaller than those for Atlanta.

- **Less than 0.3 in. (8 mm).** These rains account for most of the events (about 60 percent), but little of the runoff volume (5 to 9 percent), and are therefore the easiest to control. They produce much lower pollutant mass discharges and probably have less effect on receiving waters than do other rains. However, the runoff pollutant concentrations likely exceed regulatory standards for several categories of critical pollutants, especially bacteria and some total recoverable metals. These rains also cause large numbers of overflow events in uncontrolled combined sewers and are very common, occurring once or twice a week. Rain depths less than about 0.1 in. (3 mm) do not produce noticeable runoff.
- **0.3 to 4 in.** These rains account for the majority of the runoff volume (about 85 percent) and produce moderate to high flows. They make up about 40 percent of annual rain events. These rains occur approximately every two weeks, on average, and subject the receiving waters to frequent high-pollutant loads.

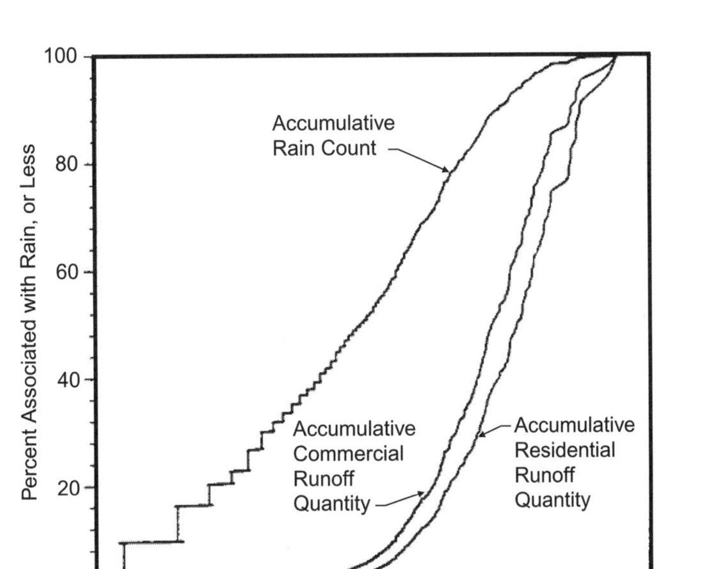

Figure 15.1 Modeled rainfall and runoff accumulative probability distributions for Atlanta, Georgia for the years 1985 to 1992 (Pitt et al., 2000)

- **More than 4 in.** These rains probably produce the most damaging flows from a habitat destruction standpoint and occur, on average, once every several months. These recurring high flows establish a high energy gradient for the stream and cause stream banks to become unstable. Fewer than 2 percent of rainfall events are in this category, but they are responsible for about 5 to 9 percent of the annual runoff and pollutant discharges.
- **Very large rains.** This category is seldom represented in field studies due to the rarity of large events and the typically short duration of most field observations. These rains occur only, on average, once every several decades or less often, and produce extremely large flows. For example, the 8-year monitoring period for Atlanta included a large rain of about 7 in. (an event with less than a 1-percent probability of occurring in any one year, if the rainfall duration is about 1 day). These extreme rains produce only a very small fraction of the annual average discharge. However, when they do occur, great property and receiving-water damage results. The receiving-water damage primarily consists of habitat destruction, sediment scouring, and the flushing of organisms great distances downstream and out of the system. The receiving water can conceivably recover naturally to before-storm conditions within a few years.

These findings and others have resulted in recognition that the design of stormwater-treatment facilities should be based on flows and volumes different from those used in the design of conveyance structures. The volume used to size treatment facilities is called the *water quality capture volume*, which corresponds to the volume of runoff expected to result from a frequently occurring storm event, where the specific frequency is typically defined by local regulations.

In the state of Georgia, for example, the *water quality volume (WQ_V)* sizing criterion specifies a treatment volume equal to the amount of runoff generated by an 85th-percentile storm event, which corresponds to 1.2 in. (30.5 mm) of rainfall for Atlanta. Thus, all runoff from 85 percent of the storms occurring during the course of an average year is treated, as well as a portion of the runoff from all rainfall events greater than 1.2 in. (30.5 mm).

The calculated volume for WQ_v is directly related to the amount of impervious cover on a site (Atlanta Regional Commission, 2001). It is equivalent to a rainfall depth of 1.2 in. multiplied by the volumetric runoff coefficient (R_v) and the site area. R_V is calculated as

$$R_V = 0.05 + 0.009I \tag{15.1}$$

where R_V = volumetric runoff coefficient
I = percentage of area with impervious cover (%)

The water quality volume is then calculated as

$$WQ_V = \frac{1.2R_VA}{12} \tag{15.2}$$

where WQ_V = water quality volume (ac-ft)
A = site area (ac)

For SI units, the equation is

$$WQ_V = 305R_VA \tag{15.3}$$

where WQ_V = water quality volume (m^3)
A = site area (ha)

In the state of New York, the water quality volume WQ_V is sized to capture and treat 90 percent of the average annual stormwater runoff volume (New York State Department of Environmental Conservation, 2001), which corresponds to a precipitation depth P ranging from 0.8 to 1.3 in. (20 to 33 mm) over the state. It is calculated as

$$WQ_V = \frac{PR_VA}{12} \tag{15.4}$$

where WQ_V = water quality volume (ac-ft)
P = depth of 90-percent rainfall event (mm)
A = site area (ac)

For SI units, the equation is

$$WQ_V = 10PR_VA \tag{15.5}$$

where WQ_V = water quality volume (m^3)
P = depth of 90-percent rainfall event (mm)
A = site area (ha)

15.3 STORMWATER QUALITY BEST MANAGEMENT PRACTICES

Best management practice (BMP), as applied to urban runoff management, is a term adopted in the 1970s to represent actions and practices that could be used to reduce urban runoff flow rates and constituent concentrations (WEF, 1998). An effective stormwater management program consists of a series of BMPs that act in concert to reduce the negative impacts of stormwater.

BMPs may be classified as *structural* or *nonstructural*. Structural BMPs are facilities designed to reroute, temporarily detain, or treat stormwater prior to discharge into a receiving body. Examples of structural BMPs are detention ponds, wetlands, porous pavements, filter strips, and grassed swales. Structures designed to increase infiltration, such as porous pavement, are located on-site, but detention and treatment structures may be located away from the point of generation. Structural BMPs also include temporary controls used during construction activities to limit sediment loads in runoff, such as hay bales and silt fences. Stormwater treatment and control structures are typically designed to operate in a passive manner; they have no moving parts and do not require the attention of an operator. However, periodic maintenance is required.

Nonstructural BMPs include a variety of institutional and educational practices designed to reduce the pollutant loads entering the stormwater system (Urbonas, 1999). Some nonstructural BMPs address land development and redevelopment processes, and others focus on educating the public to modify behavior that contributes to pollutant deposition on urban landscapes. Still others search out and disconnect illicit wastewater connections, contain accidental spills, and enforce violations of ordinances designed to prevent the deposition of pollutants on the urban landscape and its uncontrolled transport downstream.

Detailed information on the design of stormwater controls may be found in references such as the *Urban Runoff Quality Management* (WEF, 1998), *Design and Construction of Urban Stormwater Management Systems* (ASCE, 1992), or the *Urban Storm Drainage Criteria Manual* (Urban Drainage and Flood Control District, 1999). The U.S. EPA maintains the "National Menu of Best Management Practices for Storm Water Phase II" on their website as required by the Storm Water Phase II rule (see Section 14.2, page 595). For each BMP, the menu contains a description of the practice, its applicability and effectiveness, and references. Design equations and engineering sketches are also provided for structural BMPs. Though the menu is intended for small municipal, separate storm-sewer systems, the material is applicable to stormwater management in any urban or suburban area.

The National Stormwater BMP Database was developed by a team of stormwater experts associated with the Urban Water Resources Research Council of the American Society of Civil Engineers under a grant from the U.S. Environmental Protection Agency and is available on the Internet. The data fields included in this database have undergone intensive review by many experts and encompass a broad range of parameters, including test-site location; watershed characteristics; climatic data; BMP design and layout; monitoring instrumentation; and monitoring data for precipitation, flow, and water quality. This database is one component of a broader project to identify factors that affect BMP performance, develop measures for assessing BMP performance, and use the findings to implement design improvements (UWRRC, 2001).

Structural BMPs

Structural best management practices may be viewed as reactive approaches to stormwater quality management, in that they seek to reduce peak runoff rates, reroute flows, and/or improve stormwater quality during a runoff event. Descriptions of the more commonly used structural BMPs are presented here. Engineers must ensure that their designs accommodate local conditions such as weather, soils, and stream flow patterns.

Porous Pavement and Porous Pavement Detention. Modular *porous pavement,* which is shown in Figure 15.2, consists of open-void concrete block units placed on gravel bedding. Surface voids in the blocks are filled with sand or a sandy turf. The purpose of porous pavement is to reduce the imperviousness of a site by encouraging rain falling directly on the porous pavement to infiltrate. A variation of this approach is stabilized turf pavement, which consists of plastic rings affixed to a filter fabric and underlain with gravel bedding. Voids between the plastic rings are filled with sand and grass sod or seed. These types of controls are intended for use in low-traffic areas, such as driveways and small parking lots, where traffic loading and frequency do not require more conventional materials such as concrete or asphalt pavement. Porous pavements can significantly reduce runoff rates and volumes.

Figure 15.2
Modular block porous pavement

Credit: Photo courtesy of Unilock

Experience with porous pavement has shown that, when it is properly installed, freeze-thaw cycles in cold climates do not adversely affect its performance. Porous pavement is quite effective at removing sediment and associated constituents such as oil, grease, and metals, but removal rates for dissolved constituents are low to moderate. It has an uneven surface and therefore can be hazardous to those wearing heeled footwear. It can also be rather costly to replace when the gravel and sand layers become clogged.

Porous pavement detention (see Figure 15.3) is similar to porous pavement in that many of the same types of modular block are used, the openings in the blocks are filled with sand, and a granular bedding is used. However, porous pavement designed to function as detention differs from basic porous pavement in a few ways. With porous pavement detention, the porous area is slightly depressed [on the order of 1 to 2 in. (25 to 50 mm)] with respect to the surrounding pavement to create a shallow ponding area for storage of the water quality capture volume. In addition to treating the rain that falls directly on the pavement, the detention area may collect runoff from other parts of the site. The water slowly infiltrates through the void spaces and exits through a perforated pipe underdrain in the bedding material.

Aside from the benefits of additional detention and treatment of larger volumes of stormwater, the advantages and disadvantages of porous pavement detention, as well as its pollutant removal efficiencies, are similar to those for porous pavement.

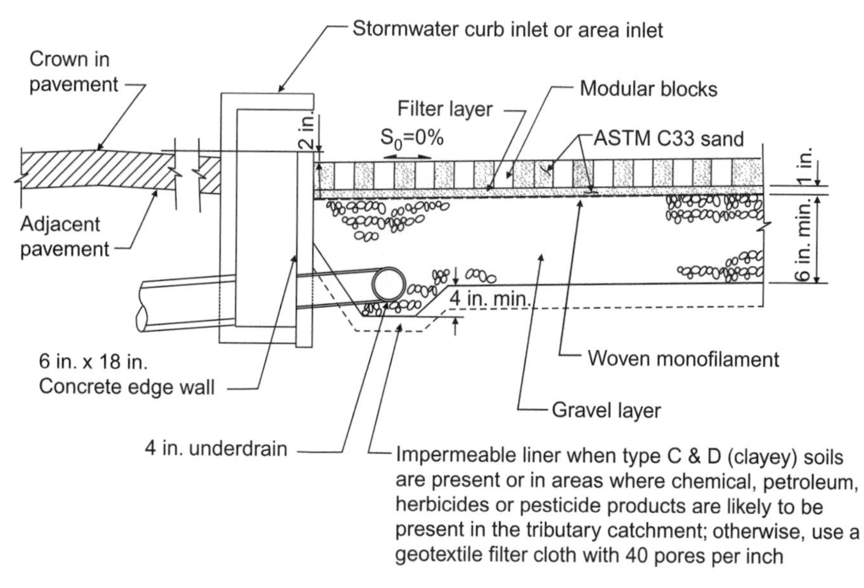

Credit: Urban Drainage and Flood Control District, 2002

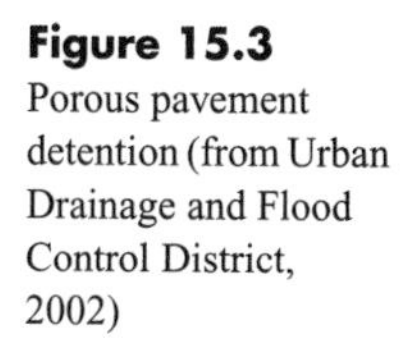

Figure 15.3
Porous pavement detention (from Urban Drainage and Flood Control District, 2002)

Grass Buffer Areas. A *grass buffer*, or *filter strip*, is a uniformly graded and densely vegetated area of turf grass. Sheet flows over the grass tend to infiltrate, and contaminants in the runoff tend to filter out and settle. Nevertheless, these strips provide only marginal pollutant-removal capability and generally should be used in conjunction with other controls. Grass buffers are particularly appropriate between detached sidewalks and curbs and along edges of parking lots where sheet flows occur across the pavement. In addition to stormwater discharge and pollutant control, buffers provide green space and can be aesthetically pleasing. Trees and shrubs can be planted to improve their appearance and provide wildlife habitat.

Over a period of time, the accumulation of sediment in the portion of a grass buffer adjacent to an impervious area (such as a parking lot) can cause runoff to pond on the impervious surface. When this occurs, a portion of the buffer needs to be removed and replaced.

Grass Swales. Grass swales, such as the one shown in Figure 15.4, are small drainageways that convey concentrated flows. They may be used to convey flow along roadway edges in lieu of curb and gutter, through park settings, and away from parking areas. Grass swales should be densely vegetated with turf grass and should have relatively flat side and longitudinal slopes. Swales should force the flow to be slow and shallow, thereby facilitating infiltration and sedimentation. Berms and/or check dams may be placed along swales at regular intervals to further slow flows and promote settling and infiltration.

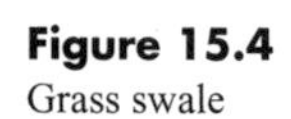

Figure 15.4
Grass swale

Credit: Urban Drainage and Flood Control District, 2002

Some disadvantages of grass swales are that they can become soggy areas, provide breeding areas for mosquitoes, and may require more right-of-way than a curb and gutter or storm sewer. They are also prone to erosion during large storm events if flow velocities are high. Positive attributes are that they are generally less-expensive to construct and maintain than curbs and gutters or sewers and that they provide green space. If wetland species of vegetation establish themselves in a swale, they can contribute to pollution control. Like grass buffers, swales do not provide high pollutant-removal efficiencies, though moderate removal rates of suspended sediments can occur if flow velocities are low.

Stream Buffers and Greenways. Establishing *greenways* along urban stream corridors (such as those shown on the plan in Figure 15.5) is becoming popular in many cities. These greenways can provide important recreational benefits and help protect the stream riparian zone and aquatic habitat. Although greenways are commonly described as areas that provide water quality treatment by filtering stormwater, little filtering occurs in actuality because most stormwater enters the stream through channels and underground pipes. However, buffer areas can provide shade that moderates stream temperatures and provides protection for the stream habitat.

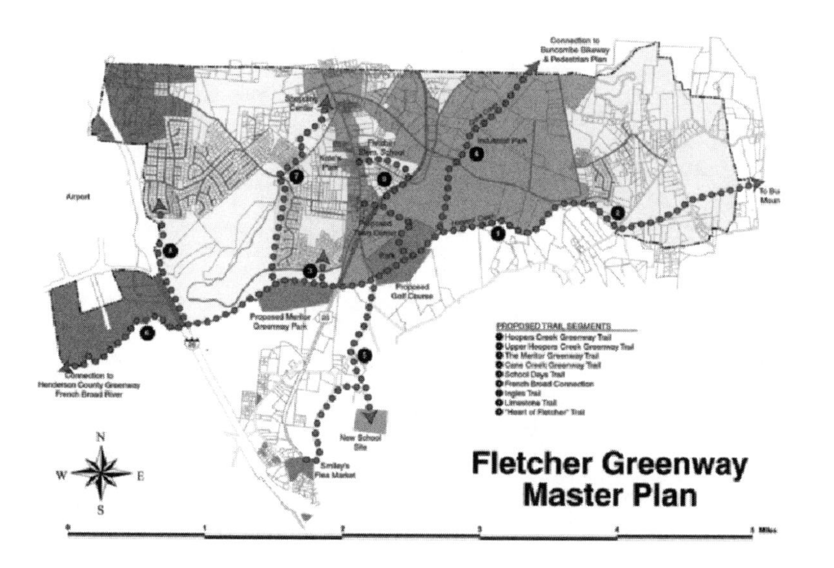

Credit: Used with permission from Town of Fletcher, North Carolina

Figure 15.5 Greenway Master Plan for Fletcher, North Carolina

Porous Landscape Detention. Like porous pavement detention, porous landscape detention consists of slightly depressed areas for temporary detainment of all or a portion of the water quality capture volume. As its name implies, however, porous landscape detention is provided in landscaped rather than paved areas. The low-lying vegetated (or otherwise landscaped) area is underlain by a sand filter bed and perforated-pipe underdrain system. The underdrains discharge the filtered stormwater to a suitable location.

Porous landscape detention facilities (see Figure 15.6) can be easily integrated into islands in parking lots, roadway medians, and roadside swale features. Experience has shown that they work well with common grass species such as bluegrass, whereas extended detention basins (see next section) often become too soggy to maintain bluegrass. Pollutant removal rates can be significant and should be at least equal to those afforded by sand filters. Disadvantages of porous landscape detention facilities are that they can become clogged by excessive amounts of sediment. They must also be kept away from building foundations or other areas where flooding could create problems.

Extended Detention Basins. For development sites or drainage basins of moderate area, with significant water quality capture volume, use of an extended detention basin (see Figure 15.7) may be appropriate. A basin of this type is designed with basically the same approach used to design dry detention ponds (see Chapter 12); however, the pond outlet is usually much smaller than would be required to meet peak outflow design criteria, such as maintaining predevelopment peak flow rates. With an extended detention basin, the outlet is sized to provide the residence time necessary for suspended sediments in the water quality capture volume to settle.

Figure 15.6
Porous landscape detention

Figure 15.7
Plan and profile views of an extended detention basin

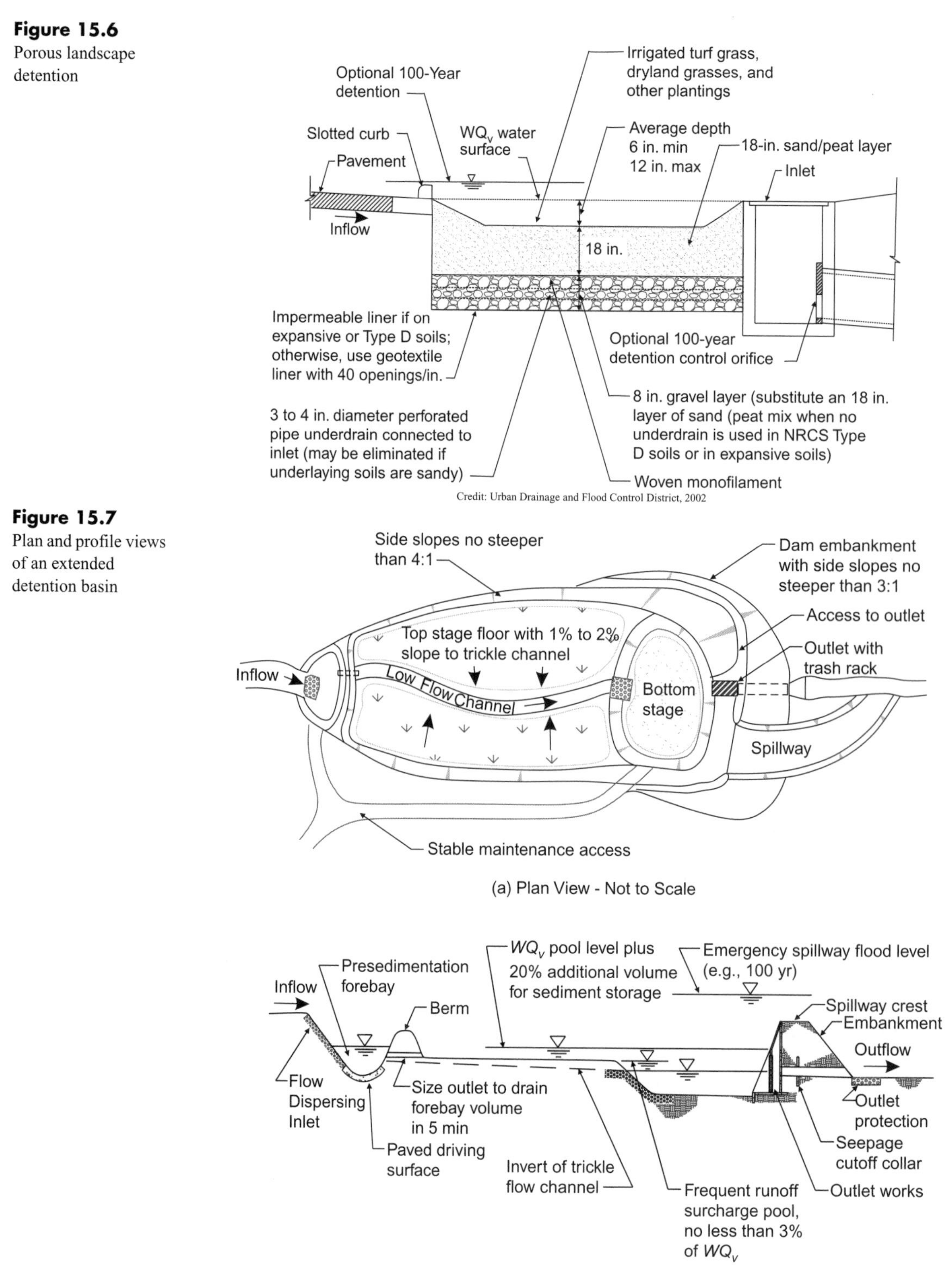

Extended detention basins can also be used to control peak runoff rates for larger storm events if they provide additional storage volume above the pool level associated with the water quality capture volume. The outlet(s) used for this purpose should be located above the pool level corresponding to the water quality capture volume.

Extended detention basins are quite effective in removing sediments, associated heavy metals, and other constituents that have an affinity for sediments, but removal rates for nutrients are low to moderate. Nutrient removal can be enhanced by providing wetland vegetation in the bottom of the basin or through the use of additional controls. Because extended detention basins are designed to permit water to drain very slowly, they may be inundated much of the time during dry-weather periods. This inundation is a potential safety hazard and may create a mosquito-breeding habitat.

Sand Filter Extended Detention Basins. A *sand filter extended detention basin* is an extended detention basin with the addition of a sand filter bed and a gravel layer with an underdrain system beneath. The basin is sized to accept the water quality capture volume and infiltrate it. Pollutant removal is accomplished through the joint mechanisms of settling and filtration. Because the sand bed can clog over time, this type of control is best suited to sites where sediment loads are expected to be low. For instance, a sand filter extended detention basin should not be put into operation during the construction phase of a project. This type of control should also not be used in locations with a base flow.

Wet Detention Ponds. A wet detention pond (see Figure 15.8 and Chapter 12) is a structural control appropriate for large drainage basins. This type of pond has a permanent pool of water below the invert of its outlet structure that can be depleted only through evaporation and infiltration (that is, water below the level of the outlet is retained). Temporary extended detention storage is provided above the permanent pool level to capture runoff and facilitate sedimentation. The water in the permanent pool is partially or completely replaced during a stormwater runoff event.

Maintenance of the permanent pool requires a continual or nearly continual base flow to replenish losses due to evaporation and infiltration. For this reason, a wet detention pond is best suited for use in regional applications where the drainage area is large enough to establish a reliable base flow. At the individual site level, this type of control may not work well unless a constant source of water is provided artificially.

Wet detention ponds can create wildlife habitat opportunities; they offer amenities related to recreation, aesthetics, and open space; and they can be part of a larger flood peak control storage basin. They can also achieve moderate-to-high removal efficiencies for many urban pollutants. Their primary disadvantages include safety concerns, floating litter, scum and algae problems, odors, and mosquito-breeding areas. Removal of accumulated sediment is also more difficult than with alternative types of basins that completely drain between storm events.

Constructed Wetland Basins. A *constructed wetland basin* is a shallow retention pond or wet detention pond. This type of control is appropriate for large drainage areas where a continual or nearly continual base flow is present to sustain the growth of rushes, willows, cattails, and similar wetland species. It is an on-line control that treats runoff through settling and biological uptake.

Figure 15.8
Plan and section of a wet detention pond

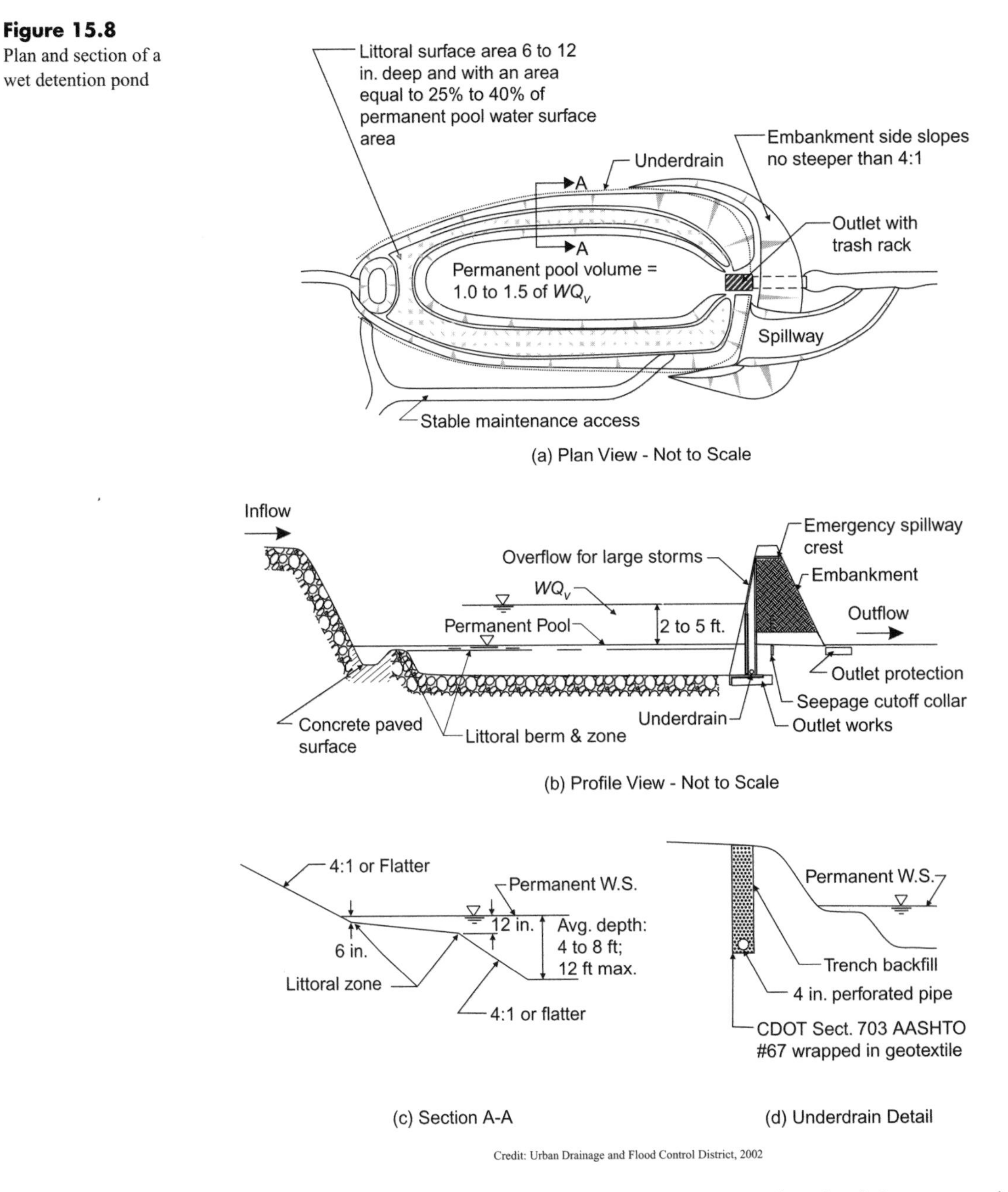

Credit: Urban Drainage and Flood Control District, 2002

In the United States, regulations intended to protect natural wetlands have recognized a distinction between natural and constructed wetlands. Constructed wetlands generally are not permitted on receiving water bodies, nor can they be used to mitigate losses of natural wetlands. Constructed wetlands can be disturbed for maintenance purposes, though the U.S. Army Corps of Engineers has established criteria for the maximum area that may be disturbed. It is recommended that any activity that could disturb a constructed wetland be cleared with the Corps to ensure that it is covered by some type of Section 404 permit (see Section 13.2).

Constructed Wetland Channels. A constructed wetland channel, like a constructed wetland basin, is a control with a continual base flow that sustains reeds, willows, rushes, and similar wetland species. A wetland channel, unlike a wetland basin, does not attempt to temporarily store the water quality capture volume, but rather is designed to promote a slow and shallow flow conducive to settling solids. Because of the slow flow, residence times within the channel reach may be long enough to promote biological uptake by the vegetation. A wetland can be established within an otherwise man-made channel, which can then act as both a conveyance facility and a water-quality-enhancement facility. This design is particularly suited to wide, gently sloping channels.

Like constructed wetland basins, constructed wetland channels are man-made, and, in the United States, they are regulated under a different set of rules than are natural wetlands. However, any planned disturbance of a constructed wetland channel in the United States should first be cleared with the U.S. Army Corps of Engineers.

Covering of Storage and Handling Areas. In locations where toxic or non-toxic potential pollutants can accumulate on exterior surfaces, such as loading docks or outdoor storage and stockpiling facilities, covering the area can eliminate or greatly reduce stormwater wash-off or wind erosion of the accumulated materials. Covers can be temporary or permanent, and can consist of tarpaulins, roofing, completely enclosed structures, or other such features.

Spill Containment and Control. Accidental spills are prone to occur at facilities where bulk liquids are handled and dispensed, such as gasoline filling stations. Spills can have devastating effects on water quality and aquatic ecosystems; therefore, spill containment and control facilities are required on such sites. These facilities can consist of berms or curbs to contain spills on the ground surface where they occur, or devices such as oil and grease traps that take advantage of the different densities of the spilled contaminants and water. In any case, spill-containment facilities should be readily accessible for removal of pollutants. Stormwater collected in the containment structure during a spill event may require treatment.

Critical Source Area Control. Special controls are needed in many *critical source areas* (ultra-urban areas) such as vehicle service areas, heavy-equipment storage and maintenance yards, scrap yards, and areas having frequent, high levels of vehicle activity (convenience stores, for example). These areas usually have substantially greater unit-area stormwater pollutant discharges than other areas in the watershed and should therefore be controlled near the site. Typical critical source area controls include stormwater filters, subsurface multichambered treatment tanks, grit chambers, and oil/water separators. Some of these controls (the filters and multiple treatment units, for example) can provide very high levels of pollutant reduction on the order of 90 percent or greater for many critical pollutants. Poor performance is usually associated with greatly undersized units, bad designs that hinder needed maintenance, and improper understanding of stormwater characteristics.

Source Area Biofiltration Devices. In many situations, significant stormwater control is possible by altering the microscale drainage paths around structures to allow enhanced infiltration. *Rain gardens* are an example in which roof runoff is directed to depressed lawn areas. These gardens are typically only several inches deep and are planted with moisture-tolerant vegetation. They can be designed to easily infiltrate all roof runoff from small to moderate rains. Another biofiltration approach is to amend urban soils with compost or other organic matter to enhance infiltration, capture runoff pollutants, and reduce the adverse effects of soil compaction associated with construction.

Use of Cisterns and Rain Barrels. On-site temporary storage of runoff from relatively clean areas (such as most roofs and residential driveways) in underground cisterns or simple rain barrels allows this water to be used for irrigation during dry periods. In many areas, it is possible to provide several weeks of landscaping irrigation with the stored water while significantly reducing the amount of runoff during rains.

Effectiveness of Structural BMPs

The first concern when investigating stormwater treatment methods is determining the needed level of stormwater control. This issue should be considered carefully as it has a great effect on the cost of the stormwater management program. Problems that need to be addressed range from sewer maintenance issues to protection of many receiving-water uses.

For example, Laplace et al. (1992) recommend that all particles greater than about 1 to 2 mm in diameter be removed from stormwater to prevent deposition in sewers. The specific value is dependent on the energy gradient and hydraulic radius of the flow in the conveyance system. This treatment objective can be achieved easily using a number of cost-effective source-area and inlet treatment practices. In contrast, far greater levels of control are typically required to prevent excessive receiving-water degradation. Specific treatment goals usually specify a reduction in suspended solids concentration of about 80 percent. In most stormwaters, this level of treatment would require the removal of most particulates greater than about 10 μm in diameter.

The selection of a treatment goal must be done with great care. The Engineering Foundation/ASCE Mt. Crested Butte Conference held in 1993 included many presentations describing receiving-water impacts associated with stormwater discharges (Herricks, 1995). Similarly, Pitt (1996) summarized numerous issues concerning potential groundwater impacts associated with subsurface stormwater disposal. These references illustrate the magnitude and variation of typical problems resulting from untreated stormwater. Because of these variations, control programs must be area-specific.

It is important to understand which stormwater controls are suitable for specific site conditions and can achieve the required treatment goals. This knowledge will assist in the realistic evaluation of each practice for technical feasibility, implementation costs, and long-term maintenance requirements and costs. Unfortunately, the reliability and performance characteristics of many of these controls are not well established, with most still in the development stage. Emerging controls can be effective, but there is not a large amount of historical data on which to base designs and be confident that performance criteria will be met under local conditions. The most promising and best understood stormwater control practices are wet detention ponds. Less reliable in terms of predicting performance, but showing promise, are stormwater filters, wetlands, and percolation basins (Roesner et al., 1989). Grass swales have also shown great promise in the EPA's Nationwide Urban Runoff Program (NURP) (U.S. EPA, 1983) and other research projects.

A study of 11 types of stormwater quality- and quantity-control practices used in Prince George's County, Maryland (Metropolitan Washington Council of Governments, 1992) was conducted to examine their performance and longevity. That report concluded that several types of practices had either failed or were not performing as well as intended. Generally, wet ponds, artificial marshes, sand filters, and infiltration trenches achieved moderate to high levels of removal for both particulate and soluble pollutants. Only wet ponds and artificial marshes demonstrated an ability to function for a relatively long time without frequent maintenance. They were also much more robust and functioned adequately under a wider range of marginal conditions. Infiltration basins, porous pavements, grass filters, swales, smaller "pocket" wetlands, dry extended detention ponds, and oil/grit separators were found to perform poorly. Stormwater infiltration techniques had high failure rates, which often could be attributed to poor initial site selection and/or lack of proper maintenance. Other performance problems were likely a function of poor design, improper installation, inadequate maintenance, and/or unsuitable placement of the control. Greater attention to these details would probably reduce the failure rate of these practices.

In addition to performance, important design considerations include

- Safety for maintenance access and operations
- Possible hazards to the general public (for example, drowning)
- Possible public nuisances (such as mosquito breeding)
- Acceptability to the public (For example, does the project enhance area aesthetics and property values?)

The majority of stormwater treatment processes are most effective for the removal of particulates only, especially the settleable-solids fraction. Removal of dissolved or colloidal pollutants is minimal; therefore, pollution prevention or controls at the source (which are frequently nonstructural BMPs) offer a more effective way to reduce dissolved pollutant concentrations. Fortunately, most toxic stormwater pollutants (heavy metals and organic compounds) are also associated with stormwater particulates (Pitt et al., 1995). Therefore, removal of the solids will also remove a large portion of the pollutants of interest. Notable exceptions of potential concern include nitrates, chlorides, zinc, pathogens, 1,3-dichlorobenzene, fluoranthene, and pyrene.

As shown in Table 15.6, structural controls can be very effective for the reduction of stormwater pollutant concentrations if they are correctly designed, constructed, and maintained. Many of these devices also reduce runoff volumes, significantly lowering pollutant mass discharges. Grass buffers, grass swales, porous pavements, and biofiltration devices (porous landscape detention) can all be designed to provide up to a 90-percent reduction in runoff volume. The mass discharge reductions for these devices can therefore be much greater than shown in Table 15.6, which considers concentration reductions alone. Combinations of stormwater controls are even more effective, but removal efficiencies reported in the table should not be viewed as additive.

Table 15.6 Pollutant concentration removal efficiencies, in percent, for common stormwater controls (UDFCD, 2002)

Type of Control	TSS	TP	TN	TZ	TPb	BOD	Bacteria
Grass buffer	10–50	0–30	0–10	0–10	N/A	N/A	N/A
Grass swale	20–60	0–40	0–30	0–40	N/A	N/A	N/A
Modular block porous pavement	80–95	65	75–85	98	80	80	N/A
Porous pavement detention	8–96	5–92	−130–85	10–98	60–80	60–80	N/A
Porous landscape detention	8–96	5–92	−100–85	10–98	60–90	60–80	N/A
Extended detention basin	50–70	10–20	10–20	30–60	75–90	N/A	50–90
Constructed wetland basin	40–94	−4–90	21	−29–82	27–94	18	N/A
Retention pond	70–91	0–79	0–80	0–71	9–95	0–69	N/A
Sand filter extended detention	8–96	5–92	−129–84	10–98	60–80	60–80	N/A
Constructed wetland channel	20–60	0–40	0–30	0–40	N/A	N/A	N/A

Notes: TSS = total suspended solids, TP = total phosphorus, TN = total nitrogen, TZ = total zinc, TPb = total lead, BOD = biochemical oxygen demand. Negative values reflect an increase in pollutant concentration.

Noticeably lacking in Table 15.6 are performance values for nonstructural stormwater controls. Unfortunately, nonstructural controls are difficult to evaluate, even though substantial research has been conducted on some public works practices (street cleaning and catch basins, for example). Although the direct environmental benefits of many nonstructural controls appear to be limited, some practices (such as public education) can play an important role in increasing overall support for a stormwater control program.

Nonstructural BMPs

Nonstructural controls are often called *source controls* or *pollution prevention* because they seek to reduce or eliminate the introduction of contaminants into stormwater. They cannot wholly eliminate pollution—and thus must be implemented along with structural controls—but they can make structural controls more effective by reducing the loadings that structural controls must handle. The principal disadvantage of nonstructural controls is that they require changes in the activities, behaviors, and attitudes of people. Such changes are very difficult to achieve and require sustained efforts on the part of those attempting to implement them. The main benefit, however, is that nonstructural controls are usually less expensive than structural controls.

Nonstructural controls take many forms, but each can be categorized by the purpose it serves:

- Waste minimization
- Good housekeeping
- Preventive maintenance
- Exposure minimization
- Spill prevention
- Public education
- Mitigation

Within these categories, some controls are more appropriate for industrial and commercial sites than for residential sites, and some apply to construction activities. The applicability of various structural and nonstructural BMPs in different situations is discussed further in Section 15.4.

Public works practices are commonly considered to be nonstructural stormwater controls. These practices, which should be considered in all areas, include the use of catch basins with frequent cleaning (with the frequency depending on sump volume and drainage-area characteristics), street cleaning in industrial areas and for leaf control, and field screening procedures to identify and repair inappropriate discharges to the storm drainage system. The subsections that follow discuss these practices in detail.

Catch Basin Cleaning. As described in Section 11.1 (page 416), the typical catch basin configuration consists of an appropriately sized sump with a hood over the outlet pipe. Stormwater bed load and a low to moderate amount of suspended solids (about 30 to 45 percent of the annual load) are trapped in the sump. The larger fraction of the sediment in the flowing stormwater will be trapped in preference to finer material, which has greater amounts of associated pollutants.

The catch basin configuration is robust and should be used in most areas. In almost all full-scale field investigations, this design withstood extreme flows with little scour loss, no significant differences between the quality of the supernatant water (the upper portion of the water column in the structure that is discharged through the outlet) and runoff quality, and minimal insect problems. Catch basins are designed for hydraulic capacity using conventional procedures for the design of grate and outlet dimensions, and the sump is designed based on the desired cleaning frequency (Lager et al., 1997).

Street Cleaning. Basic street cleaning should be conducted about four times a year in industrial and other areas having obvious high dirt loadings in streets. Additional street cleaning should be scheduled in heavily vegetated areas to remove leaves before rains wash them into the drainage system. In areas having long dry periods before the onset of seasonal rains (such as central and southern California), street cleaning may be able to reduce runoff pollutant loadings by a low to moderate amount (possibly up to 30 percent). However, in areas having more consistent rains, very small pollutant reductions would likely result. Recent innovations in street-cleaning technology that allow more efficient removal of fine particles should result in higher levels of stormwater pollution control, especially in areas having large amounts of pavement that can be cleaned.

Inappropriate Discharge Control. Screening procedures (Pitt et al., 1993) for identifying and correcting inappropriate discharges into storm drainage systems are also needed. Many separate storm drainage systems have flowing water during dry weather, with some discharging highly polluted water such as raw sewage, cleaning waters, or industrial wastewaters. These waste discharges are inappropriate and need to be identified and controlled. Communities with successful control programs for inappropriate discharges have seen substantial improvements in receiving-water quality.

Public Education Programs. Public education is an important and common nonstructural component of many stormwater control programs. Education efforts commonly include advertising to inform the general public about nonpoint-source water-pollution problems and help them make proper decisions concerning the protection of runoff quality. Public use of pesticides and fertilizers and disposal of pet wastes and household toxic wastes are examples of areas in which such education can be beneficial.

Unfortunately, monitoring and quantifying the direct water-quality benefits of these programs are difficult. Reported results include decreases in the number of instances of improper disposal of wastes in gutters and inlets and greater care in disposing of pet wastes. However, the greatest benefits of public education programs are support for better enforcement of construction-site erosion and stormwater regulations, and support for increased fees to help pay for these programs.

15.4 APPROACHES TO APPLYING BEST MANAGEMENT PRACTICES

The BMPs described in this chapter may be used to develop an effective stormwater management program for a municipality or regional government. An integrated set of BMPs that includes source controls and on-site and off-site practices must be customized for application to a specific area. In the selection and site-specific design of BMPs, many factors must be considered, including costs, performance efficiency, applicable regulations, land-area requirements, public safety, and aesthetics.

The total design, construction, and maintenance costs should be considered. Generally, nonstructural controls are more cost-effective than structural controls. However, use of nonstructural controls shifts the cost burden from the municipality to property

"So this is the famous environment everyone's so hyped up about?"

owners and residents, and procedures must be implemented to ensure that participation levels are high. Structural BMPs are generally more cost-effective when included in the planning stages of new developments.

The sections that follow describe programs that are suitable for newly developed areas, retrofitting existing areas, and construction projects.

Implementation of Controls in Newly Developing Areas

Stormwater controls established at the time of development can take advantage of simple grading and site layout options that provide significant stormwater benefits at little cost. In addition, the design of the storm drainage system can be easily modified to meet pollution-control objectives as well as drainage objectives. Most important, regional stormwater control facilities can be effectively located and sized without much interference with existing development. The following list describes stormwater management scenarios likely to be effective under various conditions for new developments (Pitt et al., 2000):

- **Low- and very low-density residential developments [greater than 2 ac (0.8 ha) lot sizes]**. Most stormwater should be infiltrated on-site by directing runoff from paved and roof areas to small *bioretention areas* (retention areas consisting of a planted soil bed over an underdrained sand layer). Disturbed soil areas should use compost-amended soils and should otherwise be constructed to minimize soil compaction. Roads should have grass swale drainage to accommodate moderate to large storms.
- **Medium-density developments [¼ to 2 acre (0.1 to 0.8 ha) lot sizes].** Again, most stormwater should be infiltrated on-site by directing runoff from paved and roof areas to small bioretention areas. Paved areas should be minimized, and porous pavements and paver blocks should be used for walkways, driveways, overflow parking areas, and so forth. Disturbed soil areas should have compost-amended soils and be constructed to minimize soil compaction. Grass swales should be used where possible to accommodate runoff from moderate to large storm events. Use of grass swales in commercial and industrial areas must also consider groundwater contamination potential and available space. Runoff from commercial and industrial areas should be controlled using wet detention ponds, and critical source areas that have excessive pollution-generating potential should have special controls.
- **High-density developments.** Combined sanitary- and storm-sewer systems can be effectively used in these areas, provided they are legally acceptable. On-site infiltration of the least-contaminated stormwater (such as that from roofs and landscaped areas) is needed to minimize wet-weather flows. On-site storage of sanitary wastewaters during wet weather (Preul, 1996), extensive use of in-line and off-line storage, and the use of effective high-rate treatment systems can minimize the damage associated with any combined sewer overflows (CSOs). The treatment of the wet-weather flows at the wastewater treatment facility would likely result in less pollutant discharge in these areas than if conventional separate wastewater collection systems were used.

Denver's Urban Drainage and Flood Control District (UDFCD, 2002) recommends a four-step process for the selection of structural controls for newly developed and redeveloped urban areas. Selection of controls for a development site should be made collaboratively between the developer, design engineer, and local jurisdiction. The subsections that follow summarize this process.

Step 1: Employ runoff reduction practices. One of the most effective ways of reducing runoff peaks and volumes from urbanizing areas is to minimize the directly connected impervious areas that contribute to the stormwater conveyance system. In so doing, infiltration of rainfall is promoted, with consequent reductions in pollution and conveyance system costs.

Low-impact development (LID) uses a variety of strategies intended to minimize the amount of stormwater and stormwater pollution from newly developing areas. Reduction of the directly connected impervious areas in a drainage basin can be accomplished in a number of ways. Narrower roadways save not only on the initial capital cost of construction, but also on long-term maintenance costs. For driveways and other low-traffic and parking areas, modular-block porous pavement or reinforced turf can be used in lieu of asphaltic or concrete pavement. Grass buffers, such as narrow landscaped areas between detached sidewalks and curbs, also encourage infiltration and essentially make the walks indirectly connected impervious areas. Use of grass swales instead of curb and gutter (and even storm sewers themselves in some instances) also slows runoff and promotes infiltration. With LID strategies, it is possible in low- and medium-density residential areas to produce runoff rates, volumes, and pollutant mass discharges comparable to those of predevelopment conditions.

Step 2: Provide water quality capture volume. Provision of temporary storage for stormwater (that is, detention storage) is a fundamental requirement for any site where stormwater quality is to be effectively addressed. The purpose of temporary storage is to permit settling of suspended sediments and to provide sufficient residence time for the settling to be effective. As previously discussed, detention, for the purpose of water quality enhancement, should focus on small, frequently occurring storm events. Additional detention volume for the management of storm runoff peaks may be provided either separately or in a surcharge pool area above the water quality pool. The types of structural controls that may be employed to provide the water quality capture volume were described in Section 15.2; they are porous pavement detention, porous landscape detention, extended detention basins, sand filter extended detention basins, constructed wetland basins, and retention ponds.

Step 3: Stabilize drainageways. Erosion of drainageways, whether natural or man-made, is a major source of contaminating sediments and associated pollutants such as phosphorus. Increases in peak rates and volumes of runoff caused by urbanization can lead to uncontrolled bed and bank erosion. Stabilization of channels may be necessary to minimize these potential problems, and can be accomplished in three ways:

- **Construct a grass-, riprap-, or concrete-lined channel.** This method is the traditional engineering approach and can nearly eliminate bed and bank erosion completely. Unfortunately, these channel stabilization practices do not provide water-quality enhancement or wetland habitat. The Urban Drainage and Flood

Control District does not recommend the use of riprap or concrete channels for flood conveyance, but it does recommend the use of rock-lined, low-flow channels for conveyance of small base flows.

- **Avoid bed degradation through the use of grade control structures and by reshaping sharp channel bends and steep banks.** These stabilization efforts should be accomplished in such a way as to minimize disturbances to desirable vegetation and habitat.
- **Employ a constructed wetland channel.** This approach gives special attention to stormwater quality, and is one form of a structural control.

Step 4: Consider the need for additional controls. If a site is being developed or redeveloped for industrial or commercial purposes, specialized types of controls may be needed. Included among these are spill containment and control facilities and protection (that is, covering) of storage and handling areas. Discussions of these controls were presented in Section 15.3.

Stormwater Controls Suitable for Retrofitting in Existing Areas

The ability to construct new stormwater controls in developed areas is severely limited by both the availability of suitable controls and the extent of control that can be accomplished. In addition, retrofitting a control is typically much more costly than if the same control were installed at the time of development. Retrofitting stormwater controls is generally limited to the practices shown in the following list. In general, items near the top of this list are more popular; however, they may not be the least expensive or most effective.

- Public works practices
  - Enhanced street cleaning
  - Increased catch basin cleaning
  - Repairing trash hoods in catch basins
  - Installing catch basin inserts
  - Rebuilding inlets to create catch basins
  - Litter control campaigning, with increases in availability of trash receptacles and trash pickup frequency
  - Public environmental education campaigning
  - Pet waste control enforcement
  - Household toxicant collection
  - Enhanced enforcement of erosion control requirements
  - Modification of public place landscaping maintenance
- Infiltration practices at source areas
  - Routing rooftop drains toward pervious areas
  - Amending soils with compost

- Modification of residential and commercial landscaping (such as constructing small bioretention areas)
- Providing French drains for collection of roof runoff where pervious area is limited
- Replacement of pavement with porous paver blocks

- Outfall wet detention ponds
  - Modification of existing dry detention ponds for enhanced pollution control
  - Enhancement of performance at existing wet ponds (options include modifying the outlet structure, constructing a *forebay* to trap sediment near the inlet, and providing additional treatment using media filtration or wetland treatment)
  - Construction of new wet detention ponds in available areas
- Control runoff at critical source areas (such as vehicle service facilities, scrap yards, etc.)
  - Use of sand perimeter filters
  - Construction of underground sedimentation/filtration units
- Other controls
  - Replacement of exposed galvanized metal with non-polluting material
  - Providing new regional stormwater controls downstream of existing developed areas

Construction Project Best Management Practices

Construction activities that disturb natural soils and protective vegetation make soils more prone to erosion by wind and water. Eroded sediments may be transported onto roadway surfaces and then spread through the site's neighborhood by passing vehicles. The sediment can then clog newly constructed inlets, storm sewers, and natural channels and can cause disruptions to natural aquatic life. Construction BMPs help limit erosion and sedimentation and provide protection to natural and man-made drainageways and stream channels.

Design engineers and construction contractors are often required to develop and obtain approval on erosion and sedimentation control plans for a site before construction begins. Ultimately, however, responsibility for erosion and sedimentation control falls on the owner of the site. Local review agencies should be consulted for guidance and information on specific requirements.

An erosion and sediment control plan consists of three major elements: erosion control measures for disturbed areas, sediment-transport control measures, and drainageway protection. Commonly used practices for addressing each element are described in the subsections that follow. More comprehensive discussions on sediment and erosion measures for construction activities may be found in *Best Management Practices for Erosion and Sediment Control* (Eastern Federal Lands Highway Design, 1995),

Urban Drainage and Flood Control District (2002), and the U.S. EPA's "National Menu of Best Management Practices for Storm Water Phase II" web site.

Erosion Control Measures. Erosion control measures should be used to limit erosion of soil from disturbed areas on a construction site. *Mulching* is the application of plant residues or other materials to the soil surface to provide temporary stabilization. Mulch may be applied to surfaces that are exposed for a short time or during seeding after final grades are established. Mulch protects the soil from rainfall impact, retards overland flow, and promotes the growth of vegetation by protecting the seed and fostering germination. Long-stemmed grass hay or cereal straw are typically used in mulch, but synthetic chemical materials are also available. On steeper slopes, mats, blankets, and nets are sometimes used to anchor the mulch.

Revegetation is the establishment of a viable plant community. Temporary vegetation is typically required on surfaces that will be exposed for one year or longer. Permanent vegetation is required on all areas that will not be covered by buildings, pavement, or other structures. Specifications for topsoil, fertilizer, seed mixtures, and seeding techniques depend on the use of the site and the local climate.

Road cuts, road fills, and parking areas on a construction site are susceptible to erosion after initial grading. The aggregate base course for the pavement should be applied promptly on these areas following grading to prevent erosion. Another result of grading is soil stockpiles; these should be seeded with a temporary or permanent grass cover within 14 days of stockpile completion if the stockpile will remain on the site for more than 60 days. Mulching is recommended to ensure vegetation establishment. If a stockpile is located near a drainageway, additional sediment control measures, such as a temporary diversion dike or silt fence, should be provided (see next subsection).

Sediment Transport Control Measures. Sediment transport control measures are used to limit transport of sediment to off-site properties and downstream receiving waters. One such control measure is a *check dam,* which consists of a small temporary obstruction in a ditch or waterway used to prevent sediment transport by reducing the velocity of flow. Check dams are most commonly constructed of loose rock, logs, or compacted earth.

A temporary *slope drain* is a flexible or rigid conduit used to effectively transport runoff down disturbed slopes. These structures are used during grading operations until the permanent drainage structures are installed and slopes are permanently stabilized. Corrugated plastic pipe or flexible tubing are commonly used. For flatter, shorter slopes, a polyethylene-lined channel may be installed.

Diversions are measures used temporarily or permanently to divert water around an area that is either under construction, being stabilized, or prone to erosion. The diversion of water is accomplished with channels, berms, or temporary culverts.

A *silt fence* is a temporary barrier used to filter sediment from sheet flow. The barrier causes water to pond upstream of the fence and deposit suspended sediment. This is the most common type of perimeter control for removing sediment before it leaves the site. The fence is constructed with a synthetic filter fabric mounted on posts and embedded in the ground.

Temporary sediment traps are small impoundments that detain runoff from small drainage areas [less than 5 ac (2 ha)] so that sediments can settle out. They are usually constructed in drainageways, where a depression is created through excavation and construction of an embankment. The discharge is controlled by a rock spillway or other outlet that slows the release of runoff. Another type of sediment trap involves excavating a depression around a storm sewer inlet structure to detain stormwater and cause sediments to settle before water enters the storm sewer system.

A *sediment basin* is a temporary impoundment used to detain runoff from an area of up to approximately 100 ac (40 ha) to settle sediments before discharging stormwater from the construction site. The design of a sediment basin is similar to that of the extended detention basins and wet detention ponds described in Section 15.3. The outlet is designed to ensure a sufficiently long detention time, and frequently consists of a perforated riser pipe. If permanent detention facilities are planned for the site, the sediment basin can often be constructed so that it can be easily adapted to this purpose by reconfiguring the outlet structure when construction controls are no longer necessary.

Drainageway Protection. *Drainageway protection* measures prevent damage to streams and other drainageways that can result from erosion and sedimentation on a construction site. Crossing streams with construction vehicles should be avoided, if possible, but if a *temporary stream crossing is required, it should be properly constructed to stabilize streambanks and minimize stream damage and sediment loading.* Three options for temporary stream crossings are bridges, culverts, and fords. Bridges cause the least amount of disturbance to a stream's banks and habitat. Culvert crossings should be of sufficient size to pass expected peak flows. Fords should be constructed from a stabilizing material such as large rocks, and should only be used if normal flows are shallow and intermittent across a wide channel.

Temporary channel diversions can be used to keep flowing water away from construction activities, thus significantly reducing sediment movement. Diversion channels may be constructed as lined channels or with pipes.

Storm sewer inlets that are in operation during construction must be protected to prevent sediment in construction site runoff from entering the conveyance system. The most common technique is to construct a *temporary filter* around inlets using straw bales, filter fabric, or rocks. These filters may require cleaning during the construction project. As discussed in the previous subsection, a temporary sediment trap around the inlet is another option.

To protect downstream areas from erosion due to concentrated flows from a construction zone, the outlets of slope drains, culverts, sediment traps, and sediment basins must include some type of erosion protection. The design of outlet protection was discussed in Sections 9.8 and 11.7.

Resources for Stormwater Management

Although the concern about pollution from stormwater originated in Europe, much of the current development is taking place in the United States, Australia, and New Zealand. In these countries, there are several websites giving guidance on the design of environmentally sound drainage systems. Some examples are as follows:

www.rmi.org – This site is maintained by the Rocky Mountain Institute, an entrepreneurial, non-profit organization that fosters the efficient and restorative use of resources to create a more secure, prosperous, and life-sustaining world.

www.metrocouncil.org/environment/Watershed/bmp/manual.htm – The Metropolitan Council of Minneapolis-St. Paul, Minnesota, presents an Urban Small Sites Best Management Practices Manual.

www.wrc.wa.gov.au/public/WSUD_manual – Western Australia's Water and Rivers Commission provides a manual on managing urban stormwater quality.

Very helpful guidance on the selection of stormwater controls (and many other topics related to stormwater management) is available from the Center for Watershed Protection (**www.cwp.org**). They have recently developed a website called *The Stormwater Manager's Resource Center* at **www.stormwatercenter.net**. These websites contain vast amounts of useful information, from past review articles from *Watershed Techniques*, to current stormwater management ordinances. Stormwater characterization, effects, sources, and control information is also presented. These sites are therefore good resources for obtaining a general understanding for stormwater issues that are important when developing a stormwater control program.

Some professional associations work with urban drainage. The International Association for Hydraulic Research (IAHR) and the International Water Association (IWA, formerly IAWQ) have a joint Committee on urban storm drainage, which has been active since 1978. The Committee organizes conferences on Urban Storm Drainage, with the ninth conference held in 2002. Furthermore, it produces a newsletter that can be downloaded at the IAHR website (**www.iahr.org**).

The American Society of Civil Engineers (ASCE) is preparing an extensive database containing data from many research studies that have investigated stormwater controls throughout the United States (**www.bmpdatabase.org**). The ongoing effort of this project involves gathering before-and-after water quality information and presenting it with simple data analysis tools and search engines. As the database expands, it should become an extremely useful tool for selecting the most appropriate control practice for a specific situation. However, a thorough understanding of the limitations of available controls is needed for proper selection; one cannot rely only on test results from other locations, even though those results are indicative of what may be useful.

The proceedings of the 1997 Engineering Foundation Conference in Malmo, Sweden, contains numerous examples of what could be labeled European Best Management Practices. The proceedings from this conference, entitled *Sustaining Urban Water Resources in the 21st Century* (Rowney, Stahre and Roesner, 1999), are available through the American Society of Civil Engineers (ASCE) Bookstore (**www.asce.org/publications**).

Many other websites containing stormwater information are available. The most useful and unbiased information is often obtained from academic and governmental agency sites. The stormwater manuals for the U.S. states of Texas and Maryland are two good examples, and these can be found at **www.txnpsbook.org**, and at **www.mde.state.md.us/environment/wma/stormwatermanual**.

The U.S. EPA provides much information concerning stormwater. The National Menu of Best Management Practices for Storm Water Phase II can be found at **cfpub.epa.gov/npdes/stormwater/menuofbmps/menu.cfm.** Information on recent nonpoint-source water pollution issues can be found at www.epa.gov/owow/nps/urban.html. Links to other sites are included there. The EPA's Office of Research and Development maintains a comprehensive on-line library of historical and recent peer-reviewed research reports at **www.epa.gov/clariton/clhtml/pubord.html,** including many addressing stormwater. In addition, the EPA's Urban Watershed Management Branch sponsors many research projects, and recent peer-reviewed reports are available at **www.epa.gov/ednnrmrl/news-main.htm**.

15.5 CHAPTER SUMMARY

Urban stormwater can negatively affect receiving waters by serving as a source of contaminants and trash and by contributing to erosion. A variety of measures known as "best management practices" or "BMPs" are available to help mitigate the adverse impacts of stormwater. Some controls are more appropriate than others for a specific location and situation, and, because of the wide range of stormwater-related problems, it is often necessary to use multiple controls at a specific location. Contaminants that may be present in stormwater include sediments, nutrients, organic matter, toxic pollutants, and salts.

Small, frequently occurring storms account for a significant portion of the pollutant load. Consequently, designs for water-quality enhancement, unlike typical designs for controlling discharge rate, focus on capturing and treating these smaller events. A sizing criterion known as the water quality capture volume or water quality volume may be used in the design of treatment facilities.

Actions and practices used to reduce runoff flow rates and constituent concentrations are called best management practices, or BMPs. BMPs are classified as either structural or nonstructural. Structural BMPs consist of constructed facilities for detaining, rerouting, or treating stormwater. Examples of structural BMPs include porous pavement, grass buffers and swales, stream buffers and greenways, porous landscape detention, extended detention basins, wet detention ponds, constructed wetlands, spill containment facilities, critical source area controls, biofiltration devices, and rain barrels.

Nonstructural BMPs are institutional and educational practices for producing behavioral changes that result in decreases in pollutant loads. Nonstructural BMPs include catch basin and street cleaning, preventing inappropriate discharges, and educating the public about protecting water quality.

To determine which BMPs to use in a particular case, the level of control required must be established. The more stringent the requirement, the more expensive it will be to achieve, so cost-effectiveness must be considered. Once the treatment goal has been determined for a particular site, the appropriate BMPs can be selected. The selection process is complicated by the fact that the performance characteristics for many BMPs are poorly understood, and because different measures are more applicable to developing areas, existing areas, or construction sites. A number of websites provide up-to-date information on stormwater best management practices.

REFERENCES

American Society of Civil Engineers (ASCE). 1992. *Design and Construction of Urban Storm Water Management Systems*. New York: ASCE.

Atlanta Regional Commission. 2001. *Georgia Stormwater Manual Volume 2: Technical Manual.* Atlanta, Georgia: Atlanta Regional Commission.

Brown, S. A., S. M. Stein, and J. C. Warner. 1996. *Urban Drainage Design Manual.* Hydraulic Engineering Circular No. 22, FHWA-SA-96-078. Washington, D.C.: Federal Highway Administration, U.S. Department of Transportation.

Center for Environmental Research Information. 1995. EPA/625/R-95/003. Cincinnati, Ohio: U.S. Environmental Protection Agency.

CH2MHill (with W. L. Jorden and R&D Environmental). 1998. *East Watershed Impacts Assessment* (also *West Watershed Impacts Assessment*). Atlanta, Georgia: Metro Atlanta Watersheds Initiative, City of Atlanta, Department of Public Works.

Chin, D. A. 2000. *Water-Resources Engineering*. Upper Saddle River, New Jersey: Prentice Hall.

Eastern Federal Lands Highway Design. 1995. *Best Management Practices for Erosion and Sediment Control*. Report No. FHWA-FLP-64-005. Washington, D.C.: U.S. Department of Transportation.

Getches, D. H. 1997. *Water Law*. 3d ed. St. Paul, Minnesota: West Publishing Co.

Herricks, E. E., ed. 1995. *Stormwater Runoff and Receiving Systems: Impact, Monitoring, and Assessment*. Boca Raton, Florida: Engineering Foundation and ASCE, CRC/Lewis.

Lager, J. A., W. G. Smith, W. G. Lynard, R. M. Finn, and E. J. Finnemore. 1997. *Urban Stormwater Management and Technology: Update and Users' Guide*. EPA-600/8-77-014. U.S. Cincinnati, Ohio: Environmental Protection Agency.

Laplace, D., A. Bachoc, Y. Sanchez, and D. Dartus. 1992. "Trunk sewer clogging development—description and solutions." *Water Science and Technology* 25, no. 8: 91–100.

Limburg and R. E. Schmidt. 1990. "Patterns of Fish Spawning in Hudson river Tributaries: Response to an Urban Gradient?" *Ecology* 71, no. 4.

Maxted, J, and E. Shaver. 1997. *The Use of Retention Basins to Mitigate Stormwater Impacts on Aquatic Life, Effects of Water Development and Management on Aquatic Ecosystems*. ed. by L. A. Roesner. New York: American Society of Civil Engineers.

Metropolitan Washington Council of Governments (Anacostia Restoration Team). 1992. *A Current Assessment of Urban Best Management Practices: Techniques for Reducing Nonpoint Source Pollution in the Coastal Zone*. Department of Environmental Programs, Metropolitan Washington Council of Governments, Water Resources Planning Board.

New York State Department of Environmental Conservation. 2001. *Stormwater Management Design Manual*.

Pitt, R. 1986. "The incorporation of urban runoff controls in the Wisconsin Priority Watershed Program." Pp. 290–313 in *Advanced Topics in Urban Runoff Research*, ed. by B. Urbonas and L. A. Roesner. New York: Engineering Foundation and ASCE.

Pitt, R. E., R. I. Field, M. M. Lalor, D. D. Adrian, and D. Barbé. 1993. *Investigation of Inappropriate Pollutant Entries into Storm Drainage Systems: A User's Guide*. Rep. No. EPA/600/R-92/238, NTIS Rep. No. PB93-131472/AS. Cincinnati, Ohio: US EPA, Storm and Combined Sewer Pollution Control Program, Risk Reduction Engineering Lab.

Pitt, R., R. Field, M. Lalor, and M. Brown. 1995. "Urban stormwater toxic pollutants: assessment, sources and treatability." *Water Environment Research* 67, no. 3: 260–275.

Pitt, R., and J. Voorhees. 1993. *Source loading and management model (SLAMM)*. Seminar Publication, National Conference on Urban Runoff Management, Enhancing Urban Watershed Management at the Local, County, and State Levels.

Pitt, R. 1995a. "Biological Effects of Urban Runoff Discharges." Pp. 127–162 in *Stormwater Runoff and Receiving Systems: Impact, Monitoring, and Assessment*, ed. by E. E. Herricks. Boca Raton, Florida: Engineering Foundation and ASCE, CRC/Lewis.

Pitt, R. 1995b. "Effects of Urban Runoff on Aquatic Biota." Pp. 609–630 in *Handbook of Ecotoxicology*, ed. by D. J. Hoffman, B. A. Rattner, G. A. Burton, Jr. and J. Cairns, Jr. Boca Raton, Florida: Lewis Publishers/CRC Press.

Pitt, R. 1996. *Groundwater Contamination from Stormwater*. Chelsea, Michigan: Ann Arbor Press.

Pitt, R., S. Nix, J. Voorhees, S. R. Durrans, and S. Burian. 2000. *Guidance Manual for Integrated Wet Weather Flow (WWF) Collection and Treatment Systems for Newly Urbanized Areas (New WWF Sys-*

tems). Second year project report: model integration and use, Wet Weather Research Program. U.S. Environmental Protection Agency, Cooperative agreement #CX 824933-01-0.

Preul, H. C. 1996. "Combined sewage prevention system (CSCP) for domestic wastewater control." Pp. 193–198 in *Proceedings of the 7th International Conference on Urban Storm Drainage*. Hannover, Germany.

Roesner, L. A., B. R. Urbonas, and M. B. Sonnen, eds. 1989. *Design of Urban Runoff Quality Controls, Proceedings of an Engineering Foundation Conference on Current Practice and Design Criteria for Urban Quality Control*. American Society of Civil Engineers.

Urban Drainage and Flood Control District. 1999. *Urban Storm Drainage Criteria Manual Vol. 3 – Best Management Practices.* Denver, Colorado: Denver Regional Council of Governments.

Urban Water Resources Research Council (UWRRC) of ASCE and Wright Water Engineers. 2001. *National Best Management Practices (BMP) Database*. www.bmpdatabase.org.

U.S. Environmental Protection Agency. 1983. *Final Report for the Nationwide Urban Runoff Program, Water Planning Division*. Washington, D.C.

U.S. Environmental Protection Agency. 1993. *Handbook on Pollution Prevention and Control Planning*. 625/R-93-004.

U.S. Environmental Protection Agency. 1997. *Urbanization and Streams: Studies of Hydrologic Impacts*. EPA-841-R-97-009.

U.S. Environmental Protection Agency 1999. *Preliminary Data Summary of Best Management Practices*. EPA-821-R-99-012.

U.S. Environmental Protection Agency. 2002. *2000 National Water Quality Inventory Report*. EPA-841-R-02-001.

Water and Rivers Commission. 1998. *A Manual for Managing Urban Stormwater Quality in Western Australia.* Perth.

Water Environment Federation. 1998. *Urban Runoff Quality Management.* WEF Manual of Practice No. 23. Alexandria, Virginia: Water Environment Federation.

APPENDIX

A

Physical Properties of Water

The physical properties of water depend on the ambient temperature and pressure of the fluid. In the majority of civil and environmental applications, where ambient pressures of interest do not span a large range, it can be safely assumed that the physical properties of liquid water depend on temperature only. The tables below contain the physical properties of water for both the U.S. customary and the S.I. systems of units.

Table A.1 Physical properties of water at atmospheric pressure (U.S. customary units)

Temperature (°F)	Density (slug/ft^3)	Specific Weight (lb/ft^3)	Kinematic Viscosity (ft^2/s)	Dynamic Viscosity (lb-s/ft^2)	Vapor Pressure (psia)	Bulk Modulus (psi)
40	1.940	62.43	1.664×10^{-5}	3.229×10^{-5}	0.12	2.94×10^5
50	1.940	62.41	1.410×10^{-5}	2.735×10^{-5}	0.18	3.05×10^5
60	1.938	62.37	1.217×10^{-5}	2.359×10^{-5}	0.26	3.11×10^5
70	1.936	62.30	1.059×10^{-5}	2.050×10^{-5}	0.36	3.20×10^5
80	1.934	62.22	0.930×10^{-5}	1.799×10^{-5}	0.51	3.22×10^5
90	1.931	62.11	0.826×10^{-5}	1.595×10^{-5}	0.70	3.23×10^5
100	1.927	62.00	0.739×10^{-5}	1.424×10^{-5}	0.96	3.27×10^5
110	1.923	61.86	0.667×10^{-5}	1.284×10^{-5}	1.28	3.31×10^5
120	1.918	61.71	0.609×10^{-5}	1.168×10^{-5}	1.69	3.33×10^5
130	1.913	61.55	0.558×10^{-5}	1.069×10^{-5}	2.22	3.34×10^5

Table A.2 Physical properties of water at atmospheric pressure (S.I. units)

Temperature (°C)	Density (kg/m^3)	Specific Weight (N/m^3)	Kinematic Viscosity (m^2/s)	Dynamic Viscosity ($N\text{-}s/ft^2$)	Vapor Pressure (kN/m^2)	Bulk Modulus (N/m^2)
5	1000.0	9806	1.519×10^{-6}	1.519×10^{-3}	0.87	2.06×10^{9}
10	999.7	9802	1.308×10^{-6}	1.308×10^{-3}	1.23	2.11×10^{9}
15	999.1	9797	1.141×10^{-6}	1.140×10^{-3}	1.70	2.14×10^{9}
20	998.2	9786	1.007×10^{-6}	1.005×10^{-3}	2.34	2.20×10^{9}
25	997.1	9777	0.897×10^{-6}	0.894×10^{-3}	3.17	2.22×10^{9}
30	995.7	9762	0.804×10^{-6}	0.801×10^{-3}	4.24	2.23×10^{9}
35	994.1	9747	0.727×10^{-6}	0.723×10^{-3}	5.61	2.24×10^{9}
40	992.2	9730	0.661×10^{-6}	0.656×10^{-3}	7.38	2.27×10^{9}
45	990.2	9711	0.605×10^{-6}	0.599×10^{-3}	9.55	2.29×10^{9}
50	988.1	9689	0.556×10^{-6}	0.549×10^{-3}	12.33	2.30×10^{9}

APPENDIX

B

Unit Conversions

The physical units used in quantitative hydrologic and hydraulic analyses and designs have the fundamental dimensions of mass (M), length (L), and time (T). Some quantities have these fundamental dimensions, whereas other quantities have derived units made up of various combinations of the fundamental dimensions. Many of the derived units have their own names, such as miles, acres, Joules, and Watts.

Table B.1, below, contains factors to convert from the U.S. customary system of units to the S.I. system of units. Table B.2 contains factors to convert from one type of unit to another within a given system of units. To convert a quantity having dimensions shown in the left-hand columns of the tables to dimensions shown in the right-hand columns, multiply by the conversion factors in the center columns.

Table B.1 Conversion factors from U.S. customary units to S.I. units

Parameter	U.S. Customary Unit	Conversion	S.I. Unit
Length	ft	0.3048	m
	in.	25.4	mm
	mi (statute)	1609.3	m
	mi (nautical)	1852	m
Area	ft^2	0.0929	m^2
	in^2	645.2	mm^2
Volume	ft^3	0.02832	m^3
	in^3	16,387	mm^3
Velocity	ft/s	0.3048	m/s
	mi/hr	0.44704	m/s
	knot (naut. mi/hr)	0.51444	m/s
Acceleration	ft/s^2	0.3048	m/s^2
Mass	slug	14.594	kg

Table B.1 Conversion factors from U.S. customary units to S.I. units

Parameter	U.S. Customary Unit	Conversion	S.I. Unit
Density	$slug/ft^3$	515.38	kg/m^3
Weight	lb	4.4482	N
Specific Weight	lb/ft^3	157.09	N/m^3
Pressure	lb/ft^2	47.88	N/m^2
	psi	6894.8	N/m^2
Energy, Work	ft-lb	1.3558	J (N-m)
	Btu	1055.1	J (N-m)
Power	ft-lb/s	1.3558	W (N-m/s)
	hp	745.69	W (N-m/s)
Dyn. Viscosity	$lb\text{-}s/ft^2$	47.88	$N\text{-}s/m^2$
Kin. Viscosity	ft^2/s	0.0929	m^2/s

Table B.2 Conversion factors for units within unit systems

Parameter	Unit	Conversion	Unit
Length	ft	12	in
	mi (statute)	5280	ft
Area	ac	43,560	ft^2
	mi^2	640	ac
	ha	10,000	m^2
Volume	ac-ft	43,560	ft^3

APPENDIX

C

Unsteady Stormwater Flow

Although most collection system design is based on steady state hydraulics, flow in sewers is inherently unsteady due to the nature of precipitation and snow melt events. This is further complicated by pump cycling and changing water levels in detention structures. Routing is usually not important in small systems, such as culverts, or in steep systems where the hydrograph is essentially translated downstream.

Changes in flow do not propagate instantaneously through a channel or sewer system, but rather move through the system gradually. Consider what occurs in a pipe or as the flow increases at the upstream end. The increase in flow manifests itself as an increase in water level and velocity, which in turn moves downstream. In addition, the flow regime may change. As a result of a storm event, flow in a storm sewer or open channel begins empty or partially full, changes to a full flow condition, and may even become pressurized as the tailwater rises at the downstream end or a restriction downstream limits the flow.

C.1 BASICS OF UNSTEADY FLOW ANALYSIS

The process of analyzing an unsteady flow event through a sewer system is referred to as *routing the flow.* The *National Engineering Handbook* (Mockus and Styner, 1972) defines routing as "computing the flood at a downstream point from the flood at an upstream point, taking storage into account." As a flow disturbance moves downstream, the peak of the disturbance (the *hydrograph*) tends to flatten. For instance, in the case of an increasing flow, some of the flow goes toward filling the storage volume corresponding to the increase in flow depth in the channel. This process is known as *attenuation*.

For steady, uniform flow, a unique *rating curve* relates depth and flow rate. However, for unsteady flow, the depth deviates from normal depth because the slope of the water surface is greater than that of the channel slope when the flow is increasing, and is flatter than that of the channel slope during decreasing flow. The depth-flow relationship in unsteady flow situations is a *looped rating curve.*

A variety of methods exist for evaluating unsteady flow in conveyance systems, ranging from the most sophisticated numerical solutions of the Saint-Venant equations to simpler approximation methods. Routing methods are grouped into two general categories: hydraulic and hydrologic. *Hydraulic models* analyze both flow and hydraulic grade through the system. *Hydrologic models* analyze flow only and use a simplified approximation to account for momentum effects. Hydrologic models are also referred to as *lumped models* because they calculate flow at one location-the downstream end of a reach. Hydraulic routing models are referred to as *distributed models* because they calculate flow and depth at several cross sections simultaneously (Fread, 1993).

Attenuation of flows is due to temporary storage in a pipe or channel reach. Hydraulic methods explicitly consider the depth of water in determining the volume stored, but the calculations can be quite complex. In many cases, storage in a reach can be approximated using simplified storage equations. Hydrologic methods assume that knowing the inflow into the reach at the current and previous time steps and the outflow at the previous time step is sufficient to determine outflow at the current time step. Once the inflow and outflow are known, flow depth, if needed, can be determined using gradually varied flow evaluation methods, or even normal depth calculations in some cases.

Hydrologic routing methods rely on the storage equation, which is given by

$$I - Q = \frac{dS}{dt} \tag{3.1}$$

where S = storage (ft^3, m^3)
I = inflow (cfs, m^3/s)
Q = outflow (cfs, m^3/s)
t = time (s)

Various methods for calculating the outflow from a channel or pipe reach based on the inflow have been developed. Such methods are generally accurate, provided that downstream conditions do not cause flow to back up in the reach. In other words, tailwater depth and flow restrictions should not significantly reduce the downstream flow rate if a hydrologic method is to be applied. If a downstream restriction does exist, outflow hydrographs predicted by hydrologic routing methods tend to have higher peak flows that arrive at the outlet earlier than those predicted by hydraulic models. This difference is due to the fact that hydrologic models do not adequately account for all factors leading to attenuation in such cases.

The hydraulic routing methods, on the other hand, are based on partial differential unsteady flow equations that allow the flow rate and depth to be computed as functions of both space and time, rather than time only as with the lumped hydrologic methods.

C.2 HYDRODYNAMIC EQUATIONS

The most commonly used equations for modeling unsteady flow in sewers are based on the assumption that the flow is one-dimensional, the pressure distribution is hydrostatic, the length of the channel is much greater than the flow depth, and the water density is constant. Disturbances in a fluid under such conditions are referred to as *shallow-water* or *translatory* waves (Linsley, Kohler, and Paulhus, 1982). The equations used to describe this type of flow are attributed to Jean-Claude Barre de St. Venant. Because of the complexity of these equations, numerous approximations have been developed, some of which are described in Section C.3.

The St. Venant equations consist of two partial differential equations, the first describing continuity and the second describing momentum. These equations can take a variety of forms. For one-dimensional, unsteady flow in a nonprismatic channel or conduit, the equations can be written as (Jin, Coran, and Cook, 2002):

$$\frac{\partial Q}{\partial x} + \frac{\partial(A + A_o)}{\partial t} - q = 0 \tag{3.2}$$

$$\frac{\partial Q}{\partial t} + \frac{\partial(\beta Q^2 / A)}{\partial x} + gA\left(\frac{\partial y}{\partial x} - S_o + S_f + S_i\right) + L = 0 \tag{3.3}$$

where Q = discharge (cfs, m^3/s)
x = distance along the longitudinal axis of the channel (ft, m)
A = active cross-sectional area of flow (ft^2, m^2)
A_o = inactive (off-channel storage) cross-sectional area of flow (ft, m)
t = time (s)
q = lateral inflow or outflow (ft^2/s, m^2/s)
β = coefficient for nonuniform velocity distribution in cross-section
g = gravitational acceleration constant (ft/s^2, m/s^2)
y = flow depth (m, ft)
S_o = slope of channel bed in longitudinal direction
S_f = friction slope
S_i = slope due to severe local expansion effects
L = momentum effect of lateral flow (ft^3/s^2, m^3/s^2)

To solve these equations, initial conditions and boundary conditions are required. The boundary conditions usually consist of the inflow hydrograph, the tailwater condition at the outlet. The gradually varied flow profile or normal depth based on some predetermined flow is used at the initial condition. Because the St. Venant equations cannot be solved analytically, a variety of numerical solutions have been developed over the years. Papers on available numerical solutions include those by Abbott (1979); Chow (1973); Cunge, Holley and Verwey (1980); Fread (1993); French (1985); NOAA (2000); Ponce, Li and Simons (1978); Price (1973); Roesner, Aldrich and Dickinson (1989); and Yen (1996, 2001).

C.3 APPROXIMATIONS OF THE HYDRODYNAMIC EQUATIONS

Because of the difficulty in solving the St. Venant equations, numerous approximations have been developed, which usually involve eliminating terms from the momentum equation. It is possible to rearrange and simplify the St. Venant momentum equation to express discharge as a function of the discharge under normal flow, as shown in Equation C.4 (Weinmann and Laurenson, 1979; French, 1985).

$$Q = Q_N \left(1 - \frac{1}{S_x}\frac{\partial y}{\partial x} - \frac{u}{S_o g}\frac{\partial u}{\partial x} - \frac{1}{S_o g}\frac{\partial u}{\partial t}\right)^{1/2} \tag{3.4}$$

where Q = actual unsteady flow (cfs, m^3/s)
Q_N = flow under normal conditions (cfs, m^3/s)
u = velocity in the longitudinal direction (ft/s, m/s)

When all of the terms in the parentheses are considered, then Equation C.4 corresponds to the St. Venant equations. When the last term in the parentheses is neglected, then the models are referred to as a *diffusion analogy.* When only the first term is considered, the model determines normal depth only and is called *kinematic wave routing.* With each successive level of approximation, the equations become easier to solve but lose some of their ability to represent real phenomena.

Kinematic Wave Routing

If the conditions during the unsteady flow event will not deviate significantly from normal flow, then most terms on the right side of Equation C.4 can be eliminated to yield the kinematic wave solution:

$$\frac{\partial Q}{\partial t} + c\frac{\partial Q}{\partial x} = 0 \tag{3.5}$$

where c = kinematic wave *celerity* (velocity), dQ/dA (ft/s, m/s)

Kinematic wave routing works well when flow changes are slow and the stage-discharge relationship is not significantly looped.

Muskingum Routing

A commonly used form of hydrologic routing is the Muskingum method, which assumes that the storage in the reach can be given by

$$S = K[XI + (1 - X)Q] \tag{3.6}$$

where K = storage constant usually taken as the travel time through the reach (s)
X = relative importance of inflow and outflow in determining storage

This method is presented in more detail in Section 5.8 (page 170).

Muskingum-Cunge Routing

A variation in the Muskingum method proposed by Cunge (1969) allows the routing coefficients to change based on changes in flow and the top width of the water surface, and is referred to as the *Muskingum-Cunge method.* This method combines the desirable features of the kinematic wave method with the benefits of the Muskingum method in that it allows for a limited amount of attenuation. These routing methods are discussed in Ponce, Li, and Simons (1978), U.S. Army Corps of Engineers (1998) and Yen (2001).

Because the Muskingum-Cunge method changes the kinematic-type Muskingum method to a method based on the diffusion analogy. Capable of predicting hydrograph attenuation, the Muskingum-Cunge method can be effectively used as a distributed routing technique. This method uses the following equations for K and X:

$$K = \frac{\Delta x}{c} \quad (3.7)$$

$$X = \frac{1}{2}\left(1 - \frac{Q}{BcS_0\Delta x}\right) \quad (3.8)$$

where B = width of water surface (ft, m)

The depth can be determined by backwater calculations. The Muskingum-Cunge method is not suitable for fast-rising hydrographs and for flows where backwater effects exist.

Convex Routing

Another variation in hydrologic routing is *convex routing,* which uses only flows from the previous time step to determine outflow from a reach (Mockus and Styner, 1972). The flow equation can be written as

$$Q_2 = cI_1 + (1-c)Q_1 \quad (3.9)$$

where c = convex routing coefficient

Note that the convex routing coefficient c is not the same as the wave speed c described previously. Rather, it is the ratio of the time step size to the travel time in the reach. The time step size should be less than the travel time (i.e., $c < 1$). The travel time, however, is not constant, but is a function of velocity. It is usually determined for a velocity corresponding to 50 to 75 percent of maximum flow. Therefore, c is defined by

$$c = \frac{V\Delta t}{L} \quad (3.10)$$

where V = representative velocity (ft/s, m/s)
Δt = time step size (s)
L = length of pipe or channel reach (ft, m)

Weighted Translation Routing

Generally, the travel time should be less than the time step size; however, in some cases, very short pipes would force time step sizes to be unrealistically small. Therefore, it is necessary to have a method that works when $c > 1$. The method presented below (Haestad Methods, 2001) is useful in such situations.

$$Q_2 = \frac{1}{c} I_1 + \left(1 - \frac{1}{c}\right) I_2 \tag{3.11}$$

Translation Routing

In small systems with steep slopes, disturbances essentially move through the pipe with little to no attenuation. To determine the flow at a given time, it is only necessary to look upstream to the flow one time step earlier at position $V\Delta t$. This value will most likely not be at a location between points where the flow is known. Therefore, interpolation will be needed to determine the flow to be passed downstream.

C.4 ROUTING METHOD COMPLICATIONS

Although the preceding equations can account for unsteady flow in channels and pipes, real conveyance systems can include phenomena not accounted for in these flow routing methods. Some of these phenomena include:

- Manholes, inlet boxes, and other junction structures
- Overflows at inlets and manholes
- Regulators and diversions
- Parallel channels and pipes
- Impoundments, especially with multiple outlets
- Flow reversal
- Dry pipes
- Drop structures
- Tidally influenced outlets

Each of these complications requires an adjustment to the flow routing. The adjustments depend on the routing method being used.

REFERENCES

Abbott, M. B. 1979. *Computational Hydraulics: Elements of the Theory of Free Surface Flows*. London: Pitman.

Chow, V. T. 1973. *Open Channel Hydraulics*. New York: McGraw Hill.

Cunge, J. A. 1969. "On the subject of a flood propagation computation method (Muskingum method)." *Journal of Hydraulic Research* 7, no. 2: 205–230.

Cunge, J. A., F. M. Holly, and A. Verwey. 1980. *Practical Aspects of Computational River Hydraulics*. London: Pitman.

Fread, D. L. 1993. "Flow Routing." In Maidment, D. R. ed. *Handbook of Hydrology*. New York: McGraw-Hill.

French, R. H. 1985. *Open-Channel Hydraulics*. New York: McGraw-Hill.

Haestad Methods. 2001. *SewerCAD: Sanitary Sewer Modeling Software*. Waterbury, Connecticut: Haestad Methods.

Jin, M., S. Coran, and J. Cook. 2002. "New One-Dimensional Implicit Numerical Dynamic Sewer and Storm Model." *Proceedings of the 9th International Conference on Urban Drainage*. American Society of Civil Engineers (ASCE).

Linsley, R. K., Jr.; M. A. Kohler; and J. L. H. Paulhus. 1982. *Hydrology for Engineers*. 3rd ed. New York: McGraw-Hill.

Mockus, V. and W. Steiner. 1972. "Flood Routing." *National Engineering Handbook*, Section 4 Hydrology, Chapter 17. Washington, D.C.: Soil Conservation Service, U.S. Department of Agriculture.

National Oceanic and Atmospheric Administration, National Weather Service. 2000. *FLDWAV Computer Program*, Version 2-0-0.

Ponce, V. M., R. M. Li, and D. B. Simons. 1978. "Applicability of Kinematic and Diffusion Models." *Journal of Hydraulic Division, ASCE* 104, HY12: 1663.

Price, R. K. 1973. "A comparison of four numerical methods for flood routing." *Journal of the Hydraulics Division, ASCE* 100, no. 7: 879–899.

Roesner, L. A., J. A. Aldrich, and R. E. Dickinson. 1989. "Storm Water Management Model User's Manual Version 4: Extra Addendum." EPA 600/3-88/001b.

U.S. Army Corps of Engineers (USACE). 1998. *HEC-1 Flood Hydrograph Package User's Manual*. Davis, Ca.: USACE Hydrologic Engineering Center.

Weinmann, D. E. and E. M. Laurenson. 1979. "Approximate Flood Routing methods: A Review." *Journal of Hydraulics Division, ASCE*. 105, HY12: 1521.

Yen, B. C. 1996. "Hydraulics for Excess Water Management." In L. W. Mays, ed. *Water Resources Handbook*. New York: McGraw-Hill.

Yen, B. C. 2001. "Hydraulics of Storm Sewer Systems." In L. Mays, ed. *Stormwater Collection System Design Handbook*. New York: McGraw Hill.

APPENDIX

D

References for Common Stormwater Conveyance System Component Models and Software

Component	Modeling Methods	Typical Input data	Section of Text	Software
Channel with uniform flow	Manning equation, Chézy equation, or Darcy-Weisbach equation	Channel roughness Channel geometry Channel slope Discharge or velocity	7.3	FlowMaster
Channel with gradually varied flow	Direct-step method or standard-step method	Geometry of each cross-section Starting water surface elevation	7.8	StormCAD
Channel with flow routing	Muskingum	Historical flood data or routing coefficients Inflow hydrograph	5.8	PondPack
	Modified Puls	Geometry of outlet Inflow hydrograph	5.8	PondPack
Culvert with inlet control	HDS-5 methodology	Inlet geometry Discharge	9.5	CulvertMaster
Culvert with outlet control	HDS-5 methodology	Pipe geometry Pipe material Inlet geometry Discharge Tailwater elevation	9.5	CulvertMaster
Roadway overtopping	HDS-5 methodology	Roadway profile coordinates Type of road Roadway width Discharge	9.6	CulvertMaster
Gutter flow	Manning equation or Chézy equation	Roadway slopes Curb height Pavement roughness Discharge	10.1	FlowMaster, StormCAD

Component	Modeling Methods	Typical Input data	Section of Text	Software
Storm sewer pipes with open channel flow	Manning equation or Chézy equation; Saint-Venant equations	Pipe roughness Pipe geometry Pipe slope Discharge or velocity	7.3; 7.4; 7.5; 7.8 11.5; Appendix C	StormCAD
Storm sewer pipes with pressure flow	Darcy-Weisbach, Manning, Hazen-Williams, or Chézy equation; Saint-Venant equations	Pipe roughness Pipe geometry Pipe slope Discharge or velocity Tailwater conditions	6.2; 6.3; 11.5; Appendix C	StormCAD
Inlets	Weir and orifice equations	Discharge Inlet type and geometry	10.2	FlowMaster, StormCAD
Manholes	Minor loss equations	Entrance and exit velocities Structure geometry	6.3; 11.4	StormCAD
Diversions/over-flows	Weir and/or orifice equations	Geometry Water surface elevations	7.6; 7.7	StormCAD
System outlets	Weir equations Orifice equations Minor loss equations Energy dissipation model	Boundary conditions such as tail-water data or downstream chan-nel characteristics Geometric characteristics of outlet	6.3; 9.5; 11.7	FlowMaster, StormCAD
Detention ponds	Hydrologic routing and Hydraulic routing	Inflow hydrograph Depth vs. volume for pond Geometry of outlet or stage-discharge curve	12.6; 12.7	PondPack

Bibliography

Abbott, M. B. 1979. *Computational Hydraulics: Elements of the Theory of Free Surface Flows*. London: Pitman.

Akan, A. O. 1989. "Detention Pond Sizing for Multiple Return Periods." *Journal of Hydraulic Engineering* 115 no. 5: 650.

American Association of State Highway and Transportation Officials (AASHTO). 1991. *Model Drainage Manual*. Washington, D.C.: AASHTO.

American Association of State Highway and Transportation Officials (AASHTO). 2000. *Model Drainage Manual*, 2000 Metric Edition. Washington, D.C.: AASHTO.

American Concrete Pipe Association (ACPA). 2000. *Concrete Pipe Design Manual*, 13th Printing (revised). Irving, Texas: ACPA.

American Iron and Steel Institute (AISI). 1980. *Modern Sewer Design*. Washington, D.C.: AISI.

American Iron and Steel Institute (AISI). 1999. *Modern Sewer Design*. 4th ed. Washington, D.C.: AISI.

American Society of Civil Engineers (ASCE). 1977. *Sedimentation Engineering*. ASCE Manuals and Reports on Engineering Practice, No. 54. New York: ASCE.

American Society of Civil Engineers (ASCE). 1992. *Design and Construction of Urban Storm Water Management Systems*. New York: ASCE.

American Society of Civil Engineers (ASCE). 1996. *Hydrology Handbook*. 2d ed.,ASCE Manual of Practice No. 28. New York: ASCE.

Armortec. 2003. Product literature for Armorflex concrete mats. (www.armortec.com).

Anderson E.A. 1973. "National Weather Service River Forecast System – Snow Accumulation and Ablation Model." Report NWS 17. Washington, D.C.: U.S. Department of Commerce.

Anderson E.A. 1976. "A Point Energy and Mass Balance Model of Snow Cover." Report NWS 19. Washington, D.C.: U.S. Department of Commerce.

Ang, A. H.-S., and W.H. Tang. 1975 and 1984. *Probability Concepts in Engineering Planning and Design, Vol. I, Basic Principles,* and *Vol. II, Decision, Risk, and Reliability*. New York: John Wiley.

Aron, G. and D. F. Kibler. 1990. "Pond Sizing for Rational Formula Hydrographs." *Water Resources Bulletin* 26 no. 2: 255.

ASCE Task Committee. 1985. *Detention Outlet Control Structures*. Reston, Virginia: ASCE.

ASCE. 1982. *Stormwater Detention Facilities*. Reston, Virginia: ASCE.

ASCE. 2000. *Joint Conference on Water Resources Engineering and Water Resources Planning and Management*. Minneapolis, Minnesota: ASCE.

Baumgardner, R. H. 1981. "Estimating Required Storage to Reduce Peak Pumping Rates at Storm Water Pumping Stations." Washington, D.C.: Federal Highway Administration Workshop Paper, Office of Engineering.

Bazin, H. 1897. "A New Formula for the Calculation of Discharge in Open Channels." *Mémoire No. 41, Annales des Ponts et Chausées*. Paris 14, Ser. 7, 4th trimester: 20–70.

Bedient, P.B., and W.C. Huber. 1992. *Hydrology and Floodplain Analysis*. 2nd ed. Reading, Massachusetts: Addison Wesley.

Blake, M.E. 1947. *Ancient Roman Construction in Italy from the Prehistoric Period to Augustus*. Washington, D.C.: Carnegie Institution of Washington.

Blevins, R.D. 1984. *Applied Fluid Dynamics Handbook*. New York: Van Nostrand Reinhold.

Bobée, B. 1975. "The Log Pearson Type 3 Distribution and Its Applications in Hydrology," *Water Resources Research* 14, no. 2: 365–369.

Bobée, B., and F. Ashkar. 1991. *The Gamma Distribution and Derived Distributions Applied in Hydrology*. Littleton, Colorado: Water Resources Publications.

Brice, J. C. 1981. *Stability of Relocated Stream Channels*. Report No. FHWA/RD-80/158. U.S. Geological Survey.

Brookes, A. and F. D. Shields. 1996. *River Channel Restoration: Guiding Principles for Sustainable Projects*. New York: John Wiley & Sons.

Brown, S. A., S. M. Stein, and J. C. Warner. 1996. *Urban Drainage Design Manual*. Hydraulic Engineering Circular No. 22, FHWA-SA-96-078. Washington, D.C.: U.S. Department of Transportation.

Burchett, C. 1998. "Open Channel Linings: Choosing the Right Erosion Control Products." Burns & McDonnell Tech Briefs 1998, no. 3. Kansas City, Missouri: Burns & McDonnell (www.burnsmcd.com).

Burian, S. J., S. J. Nix, S. R. Durrans, R. E. Pitt, C. Y. Fan, and R. Field. 1999. "Historical Development of Wet-Weather Flow Management," *Journal of Water Resources Planning and Management* 125, no. 1: 3–13.

Carp, E. ed. 1972. *Proceedings, International Conference on the Conservation of Wetlands and Waterfowl*. Ramsar, Iran, 30 January–3 February 1971. Slimbridge, UK: International Wildfowl Research Bureau.

Center for Environmental Research Information, U.S. Environmental Protection Agency, EPA/625/R-95/003, Cincinnati, Ohio, pp. 225–243, April, 1995.

CH2MHill (with W.L. Jorden and R&D Environmental). September, 1998. *East Watershed Impacts Assessment* (also *West Watershed Impacts Assessment*). Prepared for the Metro Atlanta Watersheds Initiative, City of Atlanta. Atlanta, Georgia: Department of Public Works.

Chang, F. M., and J. M. Normann. September, 1976. *Design Considerations and Calculations for Fishways Through Box Culverts*. Unpublished Report. Washington, D.C.: Federal Highway Administration, U.S. Department of Transportation.

Chang, F. M., R. T. Kilgore, D. C. Woo, and M. P. Mistichelli. 1994. *Energy Losses through Junction Manholes, Vol. 1, Research Report and Design Guide*. McLean, Virginia: Federal Highway Administration.

Chapra, S. C. and R. P. Canale. 1988. *Numerical Methods for Engineers*, 2d ed. New York: McGraw-Hill.

Chapra, S. C. 1997. *Surface Water-Quality Modeling*. New York: McGraw-Hill Book Co.

Chaudhry, M. H. and S. M. Bhallamudi. 1988. "Computation of Critical Depth in Compound Channels." *Journal of Hydraulic Research* 26, no. 4: 377–395.

Chaudhry, M.H. 1993. *Open-Channel Flow*. Englewood Cliffs, New Jersey: Prentice Hall.

Chesbrough, C. S. 1858. "Chicago Sewerage Report of the Results of Examinations Made in Relation to Sewerage in Several European Cities in the Winter of 1856–57." Chicago, Illinois.

Chen, Y. H. and B. A. Cotton. 2000. "Design of Roadside Channels with Flexible Linings.: Hydraulic Engineering Circular no. 15 (HEC-15). McLean, Virginia: Federal Highway Administration (FHWA).

Chin, D.A. 2000. *Water-Resources Engineering*. Upper Saddle River, New Jersey: Prentice Hall.

Chow, V. T. 1959. *Open-Channel Hydraulics*. New York: McGraw Hill.

Chow, V.T., D.L. Maidment, and L.W. Mays. 1988. *Applied Hydrology*, New York: McGraw-Hill.

Coldewey, W. G., Göbel, P., Geiger, W. F., Dierkes, C. & Kories, H. 2001. "Effects of stormwater infiltration on the water balance of a city." In *New Approaches Characterizing Groundwater Flow*, Proceedings of the XXXI. International Association of Hydrogeologists Congress, ed K. P. Seiler and S. Wohnlich. Balkema, Lisse.

Colebrook, C. F. 1939. "Turbulent Flow in Pipes with Particular Reference to the Transition Between Smooth and Rough Pipe Laws." *Journal of the Institution of Civil Engineers of London* 11.

Copeland, C. 1999. "Clean Water Act: A Summary of the Law." CRS Issue Brief for Congress, RL 30030.

Copeland, R. R., D. N. McComas, C. R. Thorne, P. J. Soar, M. M. Jonas, and J. B. Fripp. 2001. *Hydraulic Design of Stream Restoration Projects*. Technical Report ERDC/CHL TR-01-28. Vicksburg, Mississippi: U.S. Army Corps of Engineers, Engineer Research and Development Center.

Corry, M. L, P. L. Thompson, F. J. Watts, J. S. Jones, and D. L. Richards. 1983. *Hydraulic Design of Energy Dissipators for Culverts and Channels*. Hydraulic Engineering Circular No. 14. Washington, D.C.: Federal Highway Administration, U.S. Department of Transportation.

Cunge, J. A. 1969. "On the subject of a flood propagation computation method (Muskingum method)." *Journal of Hydraulic Research* 7, no. 2: 205–230.

Cunge, J. A., F. M. Holly, and A. Verwey. 1980. *Practical Aspects of Computational River Hydraulics*. London: Pitman.

Cunnane, C. 1978. "Unbiased Plotting Positions – A Review." *Journal of Hydrology* 37: 205–222.

David, H. A. 1981. *Order Statistics*. 2d ed. New York: John Wiley.

Doyle, E. 2001. "Wastewater Collection and Treatment – Issue Paper 9." Toronto, Ontario: Research Paper Advisory Panel, The Walkerton Inquire, Delcan Corporation.

Durrans, S. R. and P. A. Brown. 2001. "Estimation and Internet-Based Dissemination of Extreme Rainfall Information" *Transportation Research Record* no. 1743: 41–48.

Durrans, S. R., T. B. M. J. Ouarda, P. F. Rasmussen, and B. Bobée. 1999. "Treatment of Zeroes in Tail Modeling of Low Flows." *Journal of Hydrologic Engineering*, ASCE 4, no. 1: 19–27.

Durrrans, S.R. 1996. "Low-Flow Analysis with a Conditional Weibull Tail Model." *Water Resources Research* 32, no. 6: 1749–1760.

Dyhouse, Gary, J. A. Benn, David Ford Consulting, J. Hatchett, and H. Rhee. 2003. *Floodplain Modeling Using HEC-RAS*. Waterbury, Connecticut: Haestad Methods.

Einstein, H. A. 1934. "The Hydraulic or Cross Section Radius." *Schweizerische Bauzeitung* 103, no. 8: 89–91.

Einstein, H. A. and R. B. Banks. 1951. "Fluid Resistance of Composite Roughness." *Transactions of the American Geophysical Union* 31, no. 4: 603–610.

Engineers Joint Contract Documents Committee (EJCDC). 2002. *Standard General Conditions of the Construction Contract*. Reston, Virginia: ASCE.

Environment Australia. 2003. *Urban Stormwater Initiative.*

Environment Canada. 1993. *Flooding: Canada's Water Book.* Ottawa, Canada: Ecosystem Sciences and Evaluation Directorate, Economics and Conservation Branch.

European Commission. 1998. *Towards Sustainable Water Resources Management – A Strategic Approach.*

European Conference of Sustainable Cites & Towns (ECSCT). 1994. *Charter of European Cites and Towns Towards Sustainablility* (Aalborg Charter). Aalborg, Denmark.

Federal Aviation Agency (FAA). 1970. *Advisory Circular on Airport Drainage*. Report A/C 150-5320-58. Washington, D.C.: U.S. Department of Transportation.

Federal Highway Administration (FHWA). 1977. *Bicycle Safe Grate Inlets Study: Vol. 1 and 2 – Hydraulic and Safety Characteristics of Selected Grate Inlets on Continuous Grade*. FHWA-RD-77-44. Washington, D.C.: U.S. Department of Transportation.

Federal Highway Administration (FHWA). 1982. *Highway Storm Water Pumping Stations*, Vol. 1 and Vol. 2, FHWA-IP-82-17. Washington, D.C.: U.S. Department of Transportation.

Federal Highway Administration (FHWA). 1985. *Hydraulic Design Series No. 5 (HDS-5).* Washington, D.C.: U.S. Department of Transportation.

Federal Interagency Stream Corridor Restoration Working Group. 2001. *Stream Corridor Restoration: Principles, Processes and Practices*. GPO 0120-A. http://www.usda.gov/stream_restoration/.

Field, R., M. Borst, M. Stinson, C-Y. Fan, J. Perdek, D. Sullivan, and T. O'Connor. 1996. *Risk Management Research Plan for Wet Weather Flows*. Edison, New Jersey: U.S. Environmental Protection Agency.

Field, R., M. P. Brown, and W. V. Vilkelis. 1994. *Storm Water Pollution Abatement Technologies*. U.S. Environmental Protection Agency, Office of Research and Development. Cincinnati, Ohio: EPA/600/R-94/129.

Fortier, S., and F. C. Scobey. 1926. "Permissible Canal Velocities." *Transactions, American Society of Civil Engineers* 89: 940–956.

Foundation for Water Research. 1994. *Urban Pollution Management Manual*. Marlow, Great Britain: Foundation for Water Research.

Foundation for Water Research. 1998. *Urban Pollution Management Manual*. 2d ed. Marlow, Great Britain: Foundation for Water Research.

Finnemore, E. J. and J. Franzini. 2000. *Fluid Mechanics with Engineering Applications.* New York: McGraw-Hill.

Franzini, J. and E. J. Finnemore. 1997. *Fluid Mechanics*. 9th ed. New York: McGraw-Hill.

Fread, D. L. 1993. "Flow Routing." In Maidment, D. R. ed. *Handbook of Hydrology.* New York: McGraw-Hill.

Frederick, R. H., V. A. Myers, and E. P. Auciello. 1977. *Five- to 60-Minute Precipitation Frequency for the Eastern and Central United States*. NOAA Technical Memorandum NWS HYDRO-35. Silver Spring, Maryland: National Weather Service, Office of Hydrology.

French, R. H. 1985. *Open-Channel Hydraulics*. New York: McGraw-Hill.

Ganguillet, E., and W. R. Kutter. 1869. "An Investigation to Establish a New General Formula for Uniform Flow of Water in Canals and Rivers" (in German), *Zeitschrift des Oesterreichischen Ingenieur und Architekten Vereines* 21, no. 1: 6–25; no. 2: 46–59. Translated into English as a book by R. Hering and J. C. Trautwine, Jr. 1888 *A General Formula for the Uniform Flow of Water in Rivers and Other Channels*. New York: John Wiley and Sons.

Getches, D. H. 1997. *Water Law.* 3d ed. St. Paul, Minnesota: West Publishing Co.

Glazner, Michael K. 1998. "Pond Routing Techniques: Standard vs. Interconnected." In A. M. Strafaci, ed. *Engineering Hydraulics and Hydrology.* Waterbury, Connecticut: Haestad Methods.

Goodman, A. S. 1984. *Principles of Water Resources Planning*. Englewood Cliffs, NJ: Prentice Hall.

Gray, D. M. and T. D. Prowse. 1993. "Snow and Floating Ice." In R. Maidment, ed. *Handbook of Hydrology.* D New York: McGraw-Hill, Inc.

Green, W. H., and G. Ampt. 1911. "Studies of Soil Physics, Part I: The Flow of Air and Water Through Soils." *Journal of Agricultural Science* 4: 1–24.

Greenwood, J. A., J. M. Landwehr, N. C. Matalas, and J. R. Wallis. 1979. "Probability Weighted Moments: Definition and Relation to Parameters of Several Distributions Expressible in Inverse Form." *Water Resources Research* 15, no. 6: 1049–1054.

Grigg, N. S. 1996. *Water Resources Management: Principles, Regulations, and Cases*. New York: McGraw-Hill.

Gringorten, I. I. 1963. "A Plotting Rule for Extreme Probability Paper." *Journal of Geophysical Research* 68, no. 3: 813–814.

Gumbel, E. J. 1958. *Statistics of Extremes*. New York: Columbia University Press.

Guo, J. C. Y. 1997. *Street Hydraulics and Inlet Sizing Using the Computer Model UDINLET.* Highlands Ranch, Colorado: Water Resources Publications.

Guo, J. C. Y. 2001. "Design of Off-Line Detention Systems." In L. W. Mays, ed. *Stormwater Collection Systems Design Handbook*. New York: McGraw-Hill.

Haan, C. T. 1977. *Statistical Methods in Hydrology*. Ames, Iowa: Iowa State University Press.

Haestad Methods. 2001. *SewerCAD: Sanitary Sewer Modeling Software*. Waterbury, Connecticut: Haestad Methods.

Haestad Methods. 2003a. *PondPack: Detention Pond and Watershed Modeling Software*. Waterbury, Connecticut: Haestad Methods.

Haestad Methods. 2003b. *StormCAD: Storm Sewer Design and Analysis Software*. Waterbury, Connecticut: Haestad Methods.

Harrison, L. J., J. L. Morris, J. M. Normann, and F. L. Johnson. 1972. *Hydraulic Design of Improved Inlets for Culverts*. Hydraulic Engineering Circular No. 13. Washington, D.C.: Federal Highway Administration, U.S. Department of Transportation.

Hathaway, G. A. 1945. "Design of Drainage Facilities." *Transactions, American Society of Civil Engineers* 110: 697-730.

Hayes, D. F., ed. 2001. "Wetlands Engineering and River Restoration." *ASCE Conference Proceeding*. Reno, Nevada: ASCE.

Helvey, J. D and J. H. Patric. 1965. "Canopy and Litter Interception of Rainfall by Hardwoods of the Eastern United States." *Water Resources Research* 1: 193-206.

Helvey, J. D. 1967. "Interception by Eastern White Pine." *Water Resources Research* 3: 723–730.

Henderson, F. M. and R. A. Wooding. 1964. "Overland Flow and Groundwater Flow from a Steady Rain of Finite Duration." *Journal of Geophysical Research* 69, no. 8: 1531–1540.

Herricks, E.E., ed.1995. *Storm Water Runoff and Receiving Systems: Impact, Monitoring, and Assessment*. Boca Raton, Florida: Engineering Foundation and ASCE, CRC/Lewis.

Herschel, C. 1897. "On the Origin of the Chézy Formula." *Journal of the Association of Engineering Societies* 18: 363-368.

Hershfield, D. M. 1961. *Rainfall Frequency Atlas of the United States for Durations from 30 Minutes to 24 Hours and Return Periods from 1 to 100 Years*. Technical Paper No. 40. Washington, D.C.: Weather Bureau, U.S. Department of Commerce.

Hillary, R. 1994. *The Eco-Management and Audit Scheme; A Practical Guide*. Trans-Atlantic Designs.

Hjulstrom, F. 1935. "The Morphological Activity of Rivers as Illustrated by River Fyris." *Bulletin of the Geological Institute* 25, chap. 3.

Holtz, R. D. and W. D. Kovacs. 1981. *An Introduction to Geotechnical Engineering*. Englewood Cliffs, New Jersey: Prentice Hall.

Horton, R. 1919. "Rainfall Interception." *Monthly Weather Review* 47: 602–623.

Horton, R. 1933. "Separate Roughness Coefficients for Channel Bottom and Sides." *Engineering News-Record* 111, no. 22: 652–653.

Horton, R. 1939. "Analysis of Runoff Plot Experiments with Varying Infiltration Capacity." *Transactions, American Geophysical Union* 20: 693–711.

Hosking, J. R. M. 1990. "L-moments: Analysis and Estimation of Distributions Using Linear Combinations of Order Statistics," *Journal of the Royal Statistical Society, Series B* 52, no. 2105–124.

Hosking, J. R. M. and J. R. Wallis. 1997. *Regional Frequency Analysis, An Approach Based on L-moments.* Cambridge University Press, 1997.

Huber, W. C. and Dickinson, R. E. 1988. *Storm Water Management Model User's Manual*. Version 4. EPA-600/3-88-001a (NTIS PB88-236641/AS). Athens, Georgia: U.S. EPA.

Huff, F. A. 1967. "Time Distribution of Rainfall in Heavy Storms." *Water Resources Research* 3, no. 4: 1007–1019.

Huff, F. A. and J. R. Angel. 1992. "Rainfall Frequency Atlas of the Midwest." Illinois State Water Survey Bulletin 71: 141.

Hydraulic Institute. *Engineering Data Book*. Parsippany, New Jersey: Hydraulic Institute.

Hydraulic Institute. *Standards for Vertical Turbine Pumps*. Parsippany, New Jersey: Hydraulic Institute.

Hydraulic Institute. 1998. *Centrifugal/Vertical Pump Intake Design*. Parsippany, New Jersey: Hydraulic Institute.

Hydraulic Institute. 1990. *Hydraulic Institute Engineering Data Book*, 2d ed. Parsippany, New Jersey: Hydraulic Institute.

Interagency Advisory Committee on Water Data (IACWD). 1982. *Guidelines for Determining Flood Flow Frequency*. Bulletin 17B, IACWD.

International Commission for the Protection of the Rhine (ICPR). 1992. *Ecological master plan for the Rhine*. Coblenz: International Commission for the Protection of the Rhine against Pollution.

International Rivers Network (IRN). 2003. IRN Press Release, 27 February 2003.

International Rivers Network, (IRN). 2003. World Bank "Water Strategy Is Dishonest and Reactionary." Berkeley, CA: International Rivers Network.

Izzard, C. F. 1946. "Hydraulics of Runoff from Developed Surfaces." *Proceedings, Highway Research Board* 26: 129–150.

Jain, A. K. 1976. "Accurate Explicit Equation for Friction Factor." *Journal of the Hydraulics Division, ASCE* 102, no. HY5: 674–677.

Jenkinson, A. F. 1955. "The Frequency Distribution of the Annual Maximum (or Minimum) Value of Meteorological Elements." *Quarterly Journal of the Royal Meteorological Society* 81: 158–171.

Jennings, M. E.; W. O. Thomas, Jr.; and H. C. Riggs. 1994. *Nationwide Summary of U.S. Geological Survey Regional Regression Equations for Estimating Magnitude and Frequency of Floods for Ungaged Sites*. Water Resources Investigations Report 94-4002. U.S. Geological Survey.

Jin, M., S. Coran, and J. Cook. 2002. "New One-Dimensional Implicit Numerical Dynamic Sewer and Storm Model." *Proceedings of the 9th International Conference on Urban Drainage*. American Society of Civil Engineers (ASCE).

Johnson, L. E., P. Kucera, C. Lusk, and W. F. Roberts. 1988. "Usability Assessments for Hydrologic Forecasting DSS." *Journal of the American Water Resources Association* 34, no. 1: 43–56.

Johnstone, D. and W. P. Cross. 1949. *Elements of Applied Hydrology*. New York: Ronald Press.

Kamaduski, G. E. and R. H. McCuen. 1979. "Evaluation of Alternative Stormwater Detention Policies." *Journal of Water Resources Planning and Management* 105(WR2): 171.

Kapur, K. C., and L. R. Lamberson, 1977. *Reliability in Engineering Design*. New York: John Wiley.

Karassik, I. J., J. P. Messina, P. Cooper, and C. C. Heald. 2000. *Pump Handbook*, 3d ed. New York: McGraw-Hill.

Keifer, C. J. and H. H. Chu. 1957. "Synthetic Storm Pattern for Drainage Design." *Journal of the Hydraulics Division, ASCE* 84, no. HY4: 1-25.

Kennedy, W. J., Jr., and J. E. Gentle. 1980. *Statistical Computing*. New York: Marcel Dekker.

Kent, K. 1972. "Travel Time, Time of Concentration, and Lag." *National Engineering Handbook*, Section 4: Hydrology, Chapter 15. Washington, D.C.: Soil Conservation Service (SCS), U.S. Department of Agriculture.

Kerby, W. S. 1959. "Time of Concentration of Overland Flow." *Civil Engineering* 60: 174.

Kessler, A. and M. H. Diskin. 1991. "The Efficiency Function of Detention Reservoirs." *Water Resources Research* 27 no. 3: 259.

Kibler, D. F. 1982. "Desk-Top Runoff Methods." pp. 87–135 in D.F. Kibler, ed. *Urban Stormwater Hydrology*. Water Resources Monograph Series. Washington, D.C.: American Geophysical Union.

Kirby, W. 1974. "Algebraic Boundedness of Sample Statistics." *Water Resources Research* 10, no. 2: 220–222.

Kirpich, T. P. 1940. "Time of Concentration of Small Agricultural Watersheds." *Civil Engineering* 10, no. 6: 362.

Knighton, D. 1998. *Fluvial Forms and Processes*. Arnold.

Kottegoda, N. T. and R. Rosso. 1997. *Statistics, Probability, and Reliability for Civil and Environmental Engineers*. New York: McGraw-Hill.

Krishnamurthy, M. and B. A. Christensen. 1972. "Equivalent Roughness for Shallow Channels." *Journal of the Hydraulics Division, ASCE* 98, no. 12: 2257–2263.

Kroll, C. N. and J. R. Stedinger. 1996. "Estimation of Moments and Quantiles Using Censored Data." *Water Resources Research* 32, no. 4: 1005–1012.

Kuczera, G. 1982. "Robust Flood Frequency Models." *Water Resources Research* 18, no. 2: 315–324.

Kuichling, E. 1889. "The Relation Between the Rainfall and the Discharge of Sewers in Populous Districts." *Transactions, American Society of Civil Engineers* 20: 1–60.

Lager, J. A., W. G. Smith, W. G. Lynard, R. M. Finn, and E. J. Finnemore. 1977. *Urban Stormwater Management and Technology: Update and Users' Guide*. U.S. EPA-600/8-77-014. Cincinnati, Ohio: Environmental Protection Agency.

Lane, E. W. 1955. "Design of Stable Channels." *Transactions, American Society of Civil Engineers* 120: 1234–1260.

Laplace, D., A. Bachoc, Y. Sanchez, and D. Dartus. 1992. "Truck Sewer Clogging Development–Description and Solutions." *Water Science and Technology* 25, no. 8: 91–100.

Larson, L. W. and E. L. Peck. 1974. "Accuracy of Precipitation Measurements for Hydrologic Modeling." *Water Resources Research* 10, no. 4: 857–863.

Linsley, R. K. 1943. "Application of the Synthetic Unit Graph in the Western Mountain States." *Transactions American Geophysical Union* 24:581–587.

Linsley, R. K., Jr.; M. A. Kohler; and J. L. H. Paulhus. 1982. *Hydrology for Engineers*. 3rd ed. New York: McGraw-Hill.

Loganathan, G. V., D. F. Kibler, and T. J. Grizzard. 1996. "Urban Stormwater Management." In L. W. Mays, ed. *Water Resources Handbook*. New York: McGraw-Hill.

Lotter, G.K. 1933. "Considerations of Hydraulic Design of Channels with Different Roughness of Walls." *Transactions All Union Scientific Research, Institute of Hydraulic Engineering* 9: 238–241.

Loucks, D. P., J. R. Stedinger, and D. A. Haith. 1981. *Water Resource Systems Planning and Analysis*. Englewood Cliffs, N.J.: Prentice Hall.

Manning, R. 1891. "On the Flow of Water in Open Channels and Pipes." *Transactions, Institution of Civil Engineers of Ireland* 20: 161–207.

Matheussen, B. R. and Thorolfsson, S. T. 2001. "Urban Snow Surveys in Risvollan, Norway, Urban Drainage Modeling.: *Proceedings of the Specialty Symposium of the World and Water Environmental Resources Conference*. Alexandria, VA: ASCE.

Mays, L. W. 2001. *Water Resources Engineering*. New York: John Wiley and Sons.

McCarley, R. W., J. J. Ingram, B. J. Brown, and A. J. Reese. 1990. *Flood Control Channel National Inventory*. MP HL-90-10. Vicksburg, Mississippi: U.S. Army Engineer Waterways Experiment Station.

McCarthy, G. T. 1938. "The Unit Hydrograph and Flood Routing." North Atlantic Division Conference. U.S. Army Corps of Engineers.

McCuen, R. H. 1998. *Hydrologic Analysis and Design*, 2nd ed. Upper Saddle River, NJ: Prentice Hall.

McCuen, R. H., S. L. Wong, and W. J. Rawls. 1984. "Estimating Urban Time of Concentration." *Journal of Hydraulic Engineering* 110, no. 7: 887-904.

McEnroe, B. M. 1992. "Preliminary Sizing of Detention Reservoirs to Reduce Peak Discharge." *Journal of Hydraulic Engineering* 118, no. 11: 1540.

McWhorter, J. C., T. G. Carpenter, and R.N. Clark. 1968. "Erosion control criteria for drainage channels." Study conducted for the Mississippi State Highway Department and the Federal Highway Administration. State College, Mississippi: Department of Agricultural Engineering, Mississippi State University.

Mehrota, S. C. 1983. "Permissible Velocity Correction Factor." *Journal of the Hydraulics Division, ASCE* 109, no. HY2: 305–308.

Merriam, R. 1960. "A Note on the Interception Loss Equation." *Journal of Geophysical Research* 65: 3850–3851.

Merriam, R. A. 1961. "Surface Water Storage on Annual Ryegrass." *Journal of Geophysical Research* 66: 1833-1838.

Merritt, F. S., ed. 1976. *Standard Handbook for Civil Engineers*, 2nd ed. New York: McGraw-Hill.

Metcalf, L. and H. P. Eddy. 1928. *Design of Sewers*. Vol. 1 of *American Sewerage Practice*. New York: McGraw-Hill.

Metropolitan Washington Council of Governments (Anacostia Restoration Team). 1992. *A Current Assessment of Urban Best Management Practices: Techniques for Reducing Nonpoint Source Pollution in the Coastal Zone*. Department of Environmental Programs, Metropolitan Washington Council of Governments, Water Resources Planning Board.

Miller, J. F., R. H. Frederick, and R. J. Tracey. 1973. *Precipitation-Frequency Atlas of the Coterminous Western United States*. NOAA Atlas 2, 11 vols. Silver Spring, Maryland: National Weather Service.

Mockus, V. 1969. "Hydrologic Soil-Cover Complexes." *National Engineering Handbook*, Section 4: Hydrology, Chapter 9. Washington, D.C.: Soil Conservation Service (SCS), U.S. Department of Agriculture.

Mockus, V. 1972. "Estimation of Direct Runoff from Storm Rainfall." *National Engineering Handbook*, Section 4: Hydrology, Chapter 10. Washington, D.C.: Soil Conservation Service (SCS), U.S. Department of Agriculture.

Mockus, V. 1973. "Estimation of Direct Runoff from Snowmelt." *National Engineering Handbook*, Section 4: Hydrology, Chapter 11. Washington, D.C.: Soil Conservation Service (SCS), U.S. Department of Agriculture.

Mockus, V. and W. Steiner. 1972. "Flood Routing." *National Engineering Handbook*, Section 4 Hydrology, Chapter 17. Washington, D.C.: Soil Conservation Service, U.S. Department of Agriculture.

Moody, L. F. 1944. "Friction Factors for Pipe Flow." *Transactions, American Society of Mechanical Engineers* 66.

Morel-Seytoux, H. J., and J. P. Verdin. 1981. *Extension of Soil Conservation Service Rainfall Runoff Methodology for Ungaged Watersheds*. Report FHWA/Rd-81/060. Washington, D.C.: Federal Highway Administration.

Motayed, A.K., and M. Krishnamurthy. 1980. "Composite Roughness of Natural Channels." *Journal of the Hydraulics Division, ASCE* 106, no. 6: 1111-1116.

Muhlhofer, L. 1933. “Rauhigkeitsunteruchungen in einem stollen mit betonierter sohle und unverkleideten wanden.” *Wasserkraft und Wasserwirtschaft* 28(8): 85-88.

National Oceanic and Atmospheric Administration, National Weather Service. 2000. *FLDWAV Computer Program*, Version 2-0-0.

National Research Council. 1991. *Opportunities in the Hydrologic Sciences*. Washington, D.C.: Committee on Opportunities in the Hydrologic Sciences, Water Science and Technology Board, National Academy Press.

National Science and Technology Council. 1997. Committee on Environment and Natural Resources: Fact Sheet, Natural Disaster Reduction Research Initiative.

Natural Resources Conservation Service (NRCS). 1993. “Storm Rainfall Depth.” *National Engineering Handbook*, Part 630: Hydrology, Chapter 4. Washington, D.C.: U.S. Department of Agriculture.

Nix, S. J. and S. R. Durrans. 1996. “Off-Line Stormwater Detention Systems.” *Journal of the American Water Resources Association* 32, no. 6: 1329–1340.

Normann, J. M., R. J. Houghtalen, and W. J. Johnston. 2001. *Hydraulic Design of Highway Culverts*, 2nd ed. Washington, D.C.: Hydraulic Design Series No. 5, Federal Highway Administration (FHWA).

Nunnally, N. R. 1978. “Stream Renovation: An Alternative to Channelization.” *Environmental Management* 2, no. 5: 403.

Oberts, G. 1994. “Influence of Snowmelt Dynamics on Stormwater Runoff Quality.” *Watershed Protection Techniques* 1,g no. 2: 55–61.

Olin, D. A. 1984. *Magnitude and Frequency of Floods in Alabama*. Water Resources Investigations Report 84-4191. U.S. Geological Survey.

Olsen, O. J. and Q. L. Florey. 1952. *Sedimentation Studies in Open Channels: Boundary Shear and Velocity Distribution by Membrane Analogy, Analytical, and Finite-Difference Methods*. U.S. Bureau of Reclamation, Laboratory Report Sp-34, August.

Overton, D. E., and M. E. Meadows. 1976. *Stormwater Modeling*. New York: Academic Press.

Paine, J. N. and A. O. Akan. “Design of Detention Systems.” In L. W. Mays, ed. *Stormwater Collection Systems Design Handbook*. New York: McGraw-Hill.

Patric, J. H. 1966. “Rainfall Interception by Mature Coniferous Forests of Southeast Alaska.” *Journal of Soil and Water Conservation* 21: 229–231.

Pavlovskii, N. N. 1931. “On a Design Formula for Uniform Movement in Channels with Nonhomogeneous Walls.” (in Russian) *Izvestiia Vsesoiuznogo Nauchno-Issledovatel'skogo Insituta Gidrotekhniki* 3: 157–164.

Petratech. 2003. Product literature for Petraflex interlocking blocks. (www.petraflex.com).

Pitt, R. 1986. “The incorporation of urban runoff controls in the Wisconsin Priority Watershed Program.” Pp. 290–313 in B. Urbonas and L. A. Roesner, ed. *Advanced Topics in Urban Runoff Research*. New York: Engineering Foundation and ASCE.

Pitt, R. 1995a. “Biological Effects of Urban Runoff Discharges.” Pp. 127–162 in E. E. Herricks, ed. *Storm water Runoff and Receiving Systems: Impact, Monitoring, and Assessment*. Boca Raton, Florida: Engineering Foundation and ASCE, CRC/Lewis.

Pitt, R. 1995b. “Effects of Urban Runoff on Aquatic Biota” Pp. 609–630 in D. J. Hoffman; B. A. Rattner; G. A. Burton, Jr.; and J. Cairns, Jr., ed. *Handbook of Ecotoxicology*. Boca Raton, Florida: Lewis Publishers/CRC Press.

Pitt, R. 1996. *Groundwater Contamination from Storm water*. Chelsea, Michigan: Ann Arbor Press.

Pitt, R. and J. Voorhees. 1993. “Source loading and management model (SLAMM).” Seminar Publication, National Conference on Urban Runoff Management, Enhancing Urban Watershed Management at the Local, County, and State Levels, March 30–April 2.

Pitt, R. E., R. I. Field, M. M. Lalor, D. D. Adrian, and D. Barbé. 1993. *Investigation of Inappropriate Pollutant Entries into Storm Drainage Systems: A User's Guide*. Rep. No. EPA/600/R-92/238, NTIS Rep. No. PB93-131472/AS, U.S. EPA, Storm and Combined Sewer Pollution Control Program (Edison, New Jersey). Cincinnati, Ohio: Risk Reduction Engineering Lab.

Pitt, R., R. Field, M. Lalor, and M. Brown. 1995. "Urban storm water toxic pollutants: assessment, sources and treatability." *Water Environment Research* 67, no. 3 260–275.

Pitt, R., S. Nix, J. Voorhees, S. R. Durrans, and S. Burian. 2000. *Guidance Manual for Integrated Wet Weather Flow (WWF) Collection and Treatment Systems for Newly Urbanized Areas (New WWF Systems)*. Second year project report: model integration and use, Wet Weather Research Program, U.S. Environmental Protection Agency. Cooperative agreement #CX 824933-01-0.

Ponce, V. M., R. M. Li, and D. B. Simons. 1978. "Applicability of Kinematic and Diffusion Models." *Journal of Hydraulic Division, ASCE* 104, HY12: 1663.

Potter, M. C., and D. C. Wiggert. 1991. *Mechanics of Fluids*. Englewood Cliffs, NJ: Prentice Hall.

Powell, R. W. 1950. "Resistance to Flow in Rough Channels." *Transactions, American Geophysical Union* 31, no. 4: 575–582.

Prandtl, L. 1952. *Essentials of Fluid Dynamics*. New York: Hafner Publ.

Prasuhn, A.L. 1980. *Fundamentals of Fluid Mechanics*. Englewood Cliffs, New Jersey: Prentice-Hall.

Preul, H. C. 1996. "Combined sewage prevention system (CSCP) for domestic wastewater control." Pp. 193–198 in *Proceedings of the 7th International Conference on Urban Storm Drainage*. Hannover, Germany.

Price, R. K. 1973. "A comparison of four numerical methods for flood routing." *Journal of the Hydraulics Division, ASCE* 100, no. 7: 879–899.

Rasmussen, P. F. and D. Rosbjerg. 1991. "Prediction Uncertainty in Seasonal Partial Duration Series." *Water Resources Research* 27, no. 11: 2875–2883.

Rasmussen, P. F., and D. Rosbjerg. 1989. "Risk Estimation in Partial Duration Series." *Water Resources Research* 25, no. 11: 2319–2330.

Rawls, W. J., Ahuja, L. R., Brakensiek, D. L., and Shirmohammadi, A. 1993. "Chapter 5: Infiltration and Soil Water Movement" in D. R. Maidment, ed. *Handbook of Hydrology*. New York: McGraw Hill.

Rawls, W., P. Yates, and L. Asmusse. 1976. *Calibration of Selected Infiltration Equations for the Georgia Coastal Plain*. Technical Report ARS-S-113. Washington, D.C.: U.S. Dept. of Agriculture, Agricultural Research Service.

Ree, W. O. 1949. "Hydraulic Characteristics of Vegetation for Vegetated Waterways." *Agricultural Engineering* 30, no. 4: 184–187 and 189.

Ree, W. O. and V. J. Palmer. 1949. *Flow of Water in Channels Protected by Vegetative Lining*. U.S. Soil Conservation Service, Technical Bulletin No. 967.

Reynolds, O. 1883. "An Experimental Investigation of the Circumstances Which Determine Whether the Motion of Water Shall Be Direct or Sinuous and of the Law of Resistance in Parallel Channels." *Philosophical Transactions of the Royal Society* 174: 935.

Richards, K. 1982. *Rivers: Form and Processes in Alluvial Channels*. New York: Methuen and Co.

Rodriguez-Iturbe, I., and J. B. Valdes. 1979. "The Geomorphologic Structure of Hydrologic Response." *Water Resources Research* 15, no. 6: 1409–1420.

Roesner, L. A., J. A. Aldrich, and R. E. Dickinson. 1989. "Storm Water Management Model User's Manual Version 4: Extra Addendum." EPA 600/3-88/001b.

Roesner, L. A., B. R. Urbonas, and M. B. Sonnen (editors). 1989. "Design of Urban Runoff Quality Controls,. *Proceedings of an Engineering Foundation Conference on Current Practice and Design Criteria for Urban Quality Control*. American Society of Civil Engineers (ASCE).

Rosbjerg, D., H. Madsen, and P. F.Rasmussen. 1992. "Prediction in Partial Duration Series With Generalized Pareto-Distributed Exceedances." *Water Resources Research* 28, no. 11: 3001.

Rouse, H. 1936. "Discharge Characteristics of the Free Overfall." *Civil Engineering* 6, no. 7: 257.

Rouse, H. 1946. *Elementary Fluid Mechanics*. New York: Wiley.

Sanks, R. L. 1998. *Pumping Station Design*, 2nd ed. Boston: Butterworth-Heinemann.

Sauer, V. B.; W. O. Thomas, Jr.; V. A. Stricker; and K. V. Wilson. 1983. *Flood Characteristics of Urban Watersheds in the United States*. Water Supply Paper 2207. U.S. Geological Survey.

Schaake, J. C., Jr.; J. C. Geyer; and J. W. Knapp. 1967. "Experimental Examination of the Rational Method." *Journal of the Hydraulics Division, American Society of Civil Engineers* 93, no. HY6.

Seely, F. B., and J. O. Smith. 1959. *Advanced Mechanics of Materials*. 2nd ed. New York: John Wiley and Sons.

Semadeni-Davies, A. 2000. "Representation of Snow in Urban Drainage Models." *J. Hydrologic Engineering* 5, no. 4: 363-370.

Shames, I. H. 1992. *Mechanics of Fluids*, 3d ed. New York: McGraw-Hill.

Sherman, L. K. 1932. "Stream Flow from Rainfall by the Unit Graph Method." *Engineering News-Record* 108: 501-505.

Shields, F. D., R. R. Copeland, P. C. Klingeman, M. W. Doyle, and A. Simon. 2003. "Design for Stream Restoration." Accepted for *ASCE Journal of Hydraulic Engineering*.

Simon, A. L. 1981. *Practical Hydraulics*. 2d ed. New York: John Wiley & Sons.

Singh, V. P. 1988. *Hydrologic Systems, Rainfall-Runoff Modeling*, vol. I. Englewood Cliffs, New Jersey: Prentice Hall.

Singh, V. P. 1992. *Elementary Hydrology*. Englewood Cliffs, New Jersey: Prentice Hall.

Snider, D. 1972. "Hydrographs." *National Engineering Handbook*, Section 4: Hydrology, Chapter 16. Washington, D.C.: Soil Conservation Service (SCS), U.S. Department of Agriculture.

Snyder, F. F. 1938. "Synthetic Unit Graphs." *Transactions, American Geophysical Union* 19: 447.

Soil Conservation Service (SCS). 1969. Section 4: Hydrology in *National Engineering Handbook*. Washington, D.C.: U.S. Dept. of Agriculture.

Soil Conservation Service (SCS). 1986. *Urban Hydrology for Small Watersheds*. Technical Release 55. Washington, D.C.: U.S. Department of Agriculture.

Soil Conservation Service (SCS). 1992. *TR-20 Computer Program for Project Formulation Hydrology* (revised users' manual draft). Washington, D.C.: U.S. Department of Agriculture.

Stedinger, J. R., and Cohn, T. A. 1986. "Flood Frequency Analysis With Historical and Paleoflood Information." *Water Resources Research* 22, no. 5: 785.

Steel, E. W. and T. J. McGhee. 1979. *Water Supply and Sewerage*. 5th ed. New York: McGraw-Hill.

Stillwater Outdoor Hydraulic Laboratory. 1954. *Handbook of Channel Design for Soil and Water Conservation*. U.S. Soil Conservation Service, SCS-TP-61.

Strickler, A. 1923. "Some Contributions to the Problem of the Velocity Formula and Roughness Factors for Rivers, Canals and Closed Conduits" (in German). *Mitteilungen des Eidgenossischen Amtes fur Wasserwirtschaft* 16.

Swamee, P. K., and A. K. Jain. 1976. "Explicit Equations for Pipe Flow Problems." *Journal of the Hydraulics Division, ASCE* 102, no. HY5: 657-664.

Sweeney, T. L. 1992. *Modernized Areal Flash Flood Guidance*. NOAA Technical Memorandum NWS HYDRO 44. Silver Spring, Maryland: National Weather Service, Office of Hydrology.

Taylor, A.B., and H.E. Schwarz. 1952. "A unit hydrograph lag and peak flow related to basin characteristics." *Trans. Am. Geophys. Union* 33: 235.

U.S. Army Corps of Engineers (USACE). 1959. *Flood Hydrograph Analysis and Computations, Engineering and Design Manuals*. EM 1110-2-1405. Washington, D.C.: USACE.

U.S. Army Corps of Engineers (USACE). 1960. *Routing of Floods Through River Channels*. Engineering and Design Manuals, EM 1110-2-1408. Washington, D.C.: USACE.

U.S. Army Corps of Engineers (USACE). 1989. *Structural and Architectural Design of Pumping Stations*. EM 1110-2-3104. Washington, D.C.: USACE.

U.S. Army Corps of Engineers (USACE). 1998. *HEC-1 Flood Hydrograph Package User's Manual*. Davis, Ca.: USACE Hydrologic Engineering Center.

U.S. Army Corps of Engineers (USACE). 1991. *HEC-2, Water Surface Profiles, User's Manual*. Davis, Ca: USACE Hydrologic Engineering Center.

U.S. Army Corps of Engineers (USACE). 1995. *General Principles of Pumping Station Design and Layout*. EM 1110-2-3102. Washington, D.C.: USACE.

U.S. Army Corps of Engineers (USACE). *HEC-RAS River Analysis System*. Davis, California: USACE Hydrologic Engineering Center.

U.S. Army Corps of Engineers. 1989. *Environmental Engineering for Local Flood Control Channels*. Engineering Manual 1110-21601. Washington, D.C.: USACE.

U.S. Army Corps of Engineers. 1994. *Hydraulic Design of Flood Control Channels*. Engineering Manual 1110-21601. Washington, D.C.: USACE.

U.S. Army Corps of Engineers Hydrologic Engineering Center (HEC). 1998. HEC-1 Flood Hydrograph Package User's Manual. Davis, Ca.: HEC.

U.S. Army Corps of Engineers. 1999. *Mechanical and Electrical Design of Pumping Stations*. EM 1110-2-3105. Washington, D.C.: USACE.

U.S. Bureau of Reclamation (USBR). 1955. *Research Studies on Stilling Basins, Energy Dissipators, and Associated Appurtenances*. Hydraulic Laboratory Report No. Hyd-399. Lakewood, Colorado: USBR.

U.S. Bureau of Reclamation (USBR). 1983. *Design of Small Canal Structures*. Lakewood, Colorado: USBR.

U.S. Bureau of Reclamation (USBR). 1987. *Design of Small Dams*. 3rd ed. Lakewood, Colorado: USBR.

U.S. Bureau of Reclamation (USBR). 1987. *Design of Small Dams*. Washington, D.C.: U.S. Department of the Interior, USBR.

U.S. Environmental Protection Agency. 1991. *Guidance for Water Quality-Base Decisions: The TMDL Process*. EPA 440/4-91-001.

U.S. Environmental Protection Agency. December, 1983. *Final Report for the Nationwide Urban Runoff Program*. Washington, D.C.: Water Planning Division.

U.S. Environmental Protection Agency. September, 1993. *Handbook on Urban Runoff Pollution Prevention and Control Planning*. 625/R-93-004.

U.S. Weather Bureau. 1958. *Rainfall Intensity-Frequency Regime, Part 2 – Southeastern United States*. Technical Paper 29. Washington, D.C.: U.S. Department of Commerce.

United Nations Development Program (UNDP). 1998. *Capacity Building for Sustainable Management of Water Resources and the Aquatic Environment*. New York: United Nations.

United Nations. 1977. Report of the United Nations Conference, Mal del Plata, 14–25 March 1977.

United Nations. 1992. Report of the United Nations Conference on Environment and Development (UNCED), Rio de Janeiro, June 1992.

United Nations. 1997. Report Of The Commission On Sustainable Development On Preparations For The Special Session Of The General Assembly For The Purpose Of An Overall Review And Appraisal Of The Implementation Of Agenda 21.

Urban Drainage and Flood Control District (UDFCD). 1982. *Urban Storm Drainage Criteria Manual*, Vol. 2, Major Drainage. Denver, Colorado: UDFCD. November 1982.

Urban Drainage and Flood Control District (UDFCD). 1999. *Urban Storm Drainage Criteria Manual*, Vol. 3, Best Management Practices. Denver, Colorado: Urban Drainage and Flood Control District. September, 1999.

Urbonas, B. and P. Stahre. 1993. *Stormwater Best Management Practices Including Detention*. Englewood Cliffs, New Jersey: Prentice-Hall.

Urbonas, B. R. and L. A. Roesner. 1993. "Urban Design and Urban Drainage for Flood Control." In D. Maidment, ed. *Handbook of Hydrology*. New York: McGraw-Hill.

van der Vink, G., R.M. Allen, J. Chapin, M. Crooks, W. Fraley, J. Krantz, A.M. Lavigne, A. LeCuyer, E.K. MacColl, W.J. Morgan, B. Ries, E. Robinson, K. Rodriquez, M. Smith, and K. Sponberg. 1998. "Why the United States is Becoming More Vulnerable to Natural Disasters," *EOS, Transactions of the American Geophysical Union* 79, no. 44: 533–537.

Viessman, W., J. Knapp, and G. L. Lewis. 1977. *Introduction to Hydrology*. 2nd ed. New York: Harper & Row.

Viessman, W., Jr., and M. J. Hammer. 1993. *Water Supply and Pollution Control*. 5th ed. New York: Harper Collins.

Virginia Department of Conservation and Recreation (DCR). 1992. *Virginia Erosion and Sediment Control Handbook*. 3rd ed. Richmond, Virginia: DCR.

Vogel, R. M., and N. M. Fennessey, "L-moment Diagrams Should Replace Product Moment."

Wallis, J. R., N. C. Matalas, and J. R. Slack. 1971. "Just a Moment!" *Water Resources Research* 10, no. 2: 211–219.

Walski, T. M. 1984. *Analysis of Water Distribution Systems*. New York: VanNostrand Reinhold.

Wang, Q. J. 1990. "Estimation of the GEV Distribution from Censored Samples by Method of Partial Probability Weighted Moments." *Journal of Hydrology* 120: 103–114.

Wang, Q. J. 1997. "LH Moments for Statistical Analysis of Extreme Events." *Water Resources Research* 33, no. 12: 2841–2848.

Weber, U. 2000. "The 'Miracle' of the Rhine" *Courier*, June 2000, Unesco, Paris.

Weinmann, D. E. and E. M. Laurenson. 1979. "Approximate Flood Routing methods: A Review." *Journal of Hydraulics Division, ASCE*. 105, HY12: 1521.

White, F.M. 1999. *Fluid Mechanics*. 4th ed. New York: McGraw-Hill.

Williams, G. B. 1922. "Flood Discharges and the Dimensions of Spillways in India." *Engineering* (London) 134: 321.

Williams, G. S., and A. H. Hazen. 1933. *Hydraulic Tables*. 3rd ed. New York: John Wiley and Sons.

World Bank. 1993. *Water Resources Management*. Washington, D.C.: International Bank for Reconstruction and Development/The World Bank.

World Bank. 1998. *Pollution Prevention and Abatement Handbook 1998: Toward Cleaner Production*. Washington, D.C.: International Bank for Reconstruction and Development/The World Bank.

World Bank. 2002. *Thailand Environmental Monitor 2001: Water Quality*. Washington, D.C.: The World Bank Group.

World Bank. 2003. *The New Water Resources Strategy*. Washington, D.C.: World Bank Group.

World Commission on Environment and Development (WCED). 1987. *Our Common Future*. Oxford, Great Britain: Oxford University Press.

Yang, C. T. 1976. "Minimum Unit Stream Power and Fluvial Hydraulics." *Journal of Hydraulic Engineering*. 105 (HY7): 769.

Yang, C. T. 1996. *Sediment Transport: Theory and Practice*. New York: McGraw-Hill.

Yen, B. C. 1996. "Hydraulics for Excess Water Management." In L. W. Mays, ed. *Water Resources Handbook*. New York: McGraw-Hill.

Yen, B. C. 2001. "Hydraulics of Storm Sewer Systems." In L. Mays, ed. *Stormwater Collection System Design Handbook*. New York: McGraw Hill.

Young, G. K., S. E. Walker, and F. Chang. 1993. *Design of Basic Bridge Deck Drainage*. Hydraulic Engineering Circular No. 21, GHWA-SA-92-010. Washington, D.C.: U.S. Department of Transportation.

Zinke, P.J. 1967. "Forest Interception Studies in the United States." Pp. 137–160 in W. E. Sopper and H. W. Lull, ed. *Forest Hydrology*. Oxford: Pergamon Press.

Index

A

B

C

D

E

F

G

H

I

N

O

P

R

S

T

U

V

W

Z

CD-ROM Contents

Included with this book is a CD-ROM containing software and documentation. The specific items that can be found on the CD are as follows:

- A 1-pond, 250-acre (100 ha) version of PondPack; a 10-inlet version of StormCAD (Stand-Alone interface); and full versions of CulvertMaster and FlowMaster that you can use to work many of the exercises in the book. Note that the software is licensed for ACADEMIC PURPOSES ONLY. Professional or commercial application is strictly prohibited under the license agreement.
- Installation instructions containing CD-ROM contents and advanced installation details.
- Complete User Manuals in Adobe Acrobat format for each of the software packages—*PondPack.pdf*, *StormCAD.pdf*, *CulvertMaster.pdf*, and *FlowMaster.pdf*. Note that all information in the manuals is also available in the context-sensitive on-line Help systems for each product.
- *Exam_book.pdf*, an Adobe Acrobat file of the examination booklet, which can be filled out and submitted to Haestad Methods for grading and award of CEUs (refer to "Continuing Education Units" on page *xv* for more information).
- Adobe Acrobat software required to view the User Manuals and exam booklet. Installing Acrobat is an option included in the installation routine.

Installation Instructions

Insert the CD into your CD-ROM drive. If you have Autorun enabled on your computer, the installation menu will appear automatically after the CD-ROM is inserted. If Autorun is not enabled, click the Start menu, select Run, enter D:\Autorun, and click OK (if D: is not your CD drive, enter the appropriate letter instead).

A menu will open with installation options. Follow the on-screen instructions.

Haestad Methods' On-Line KnowledgeBase

The KnowledgeBase is an on-line frequently asked questions database on the Haestad Methods website.

Free access to the KnowledgeBase for StormCAD, PondPack, CulvertMaster, and FlowMaster is included with this book and is valid for one year from the date of purchase. To enter the KnowledgeBase, click the Globe button on the toolbar of any of these software packages.

System Requirements

To run the included software, your computer should have at a minimum:

- Pentium III 750 MHz processor
- 128 MB RAM
- 35 MB hard disk space, plus room for data files
- 800 x 600 display resolution with 256 colors

The following are the recommended specifications:

- Fastest available processor
- 256 MB RAM

Technical Support

If the CD-ROM or software provided is not functioning as expected, e-mail Haestad Methods technical support at support@haestad.com, or call +1-203-755-1666 for assistance.

Software License Agreement